MEDICINAL PLANT BIOTECHNOLOGY FOR AMELIORATED SECONDARY METABOLITE PRODUCTION

Medicinal and Industrial Plants and Products

MEDICINAL PLANT BIOTECHNOLOGY FOR AMELIORATED SECONDARY METABOLITE PRODUCTION

Edited by

SAIKAT GANTAIT

Crop Research Unit (Genetics and Plant Breeding), Bidhan Chandra Krishi Viswavidyalaya, Mohanpur, West Bengal, India

JAYOTI MAJUMDER SARKAR

Department of Floriculture and Landscape Architecture, Bidhan Chandra Krishi Viswavidyalaya, Mohanpur, West Bengal, India

SUDHIR SINGH

Nuclear Agriculture & Biotechnology Division, Bhabha Atomic Research Centre (BARC), Mumbai, Maharashtra, India

SANJAY KUMAR SINGH

Division of Horticultural Science, Indian Council of Agricultural Research, Krishi Anusandhan Bhawan – II, New Delhi, India

Series Editor

MANSOUR GHORBANPOUR

Department of Medicinal Plants, Faculty of Agriculture and Natural Resources, Arak University, Arak, Iran

ELSEVIER

ACADEMIC PRESS

An imprint of Elsevier

Academic Press is an imprint of Elsevier
125 London Wall, London EC2Y 5AS, United Kingdom
50 Hampshire Street, 5th Floor, Cambridge, MA 02139, United States

ISBN: 978-0-443-43944-5

For Information on all Academic Press publications
visit our website at https://www.elsevier.com/books-and-journals

Publisher: Andre Wolff
Acquisitions Editor: Chiara Giglio
Editorial Project Manager: Kritika Sharma
Production Project Manager: M. Balakrishnan
Cover Designer: S. Bakyalakshmi

Typeset by MPS Limited, Chennai, India

Fondly dedicated to

Dr. Krishan Lal Chadha, ***Padma Shri Awardee***
(November 15, 1936 to March 23, 2025)

Dr. Krishan Lal Chadha was a distinguished horticulturist of international repute – an able scientist and administrator whose work made outstanding contributions to Indian horticulture. He was a pioneer in bringing new dimensions to horticultural research, development, policy, and management, guided by sound scientific knowledge, a concern for building futuristic research infrastructure, and a vision for science-led horticulture in the service of food and nutritional security and livelihoods. At various times he served as President of the Indian Academy of Horticultural Sciences, New Delhi; Vice President of the National Academy of Agricultural Sciences; and Deputy Director General (Horticulture) and National Professor (Horticulture) of the Indian Council of Agricultural Research. He passed away on March 23, 2025, at the age of 88. His contributions continue to inspire the horticultural community, and many regard him as the Father of the Golden Revolution in India.

Contents

5. Development of synthetic seeds for conservation, storage, and exchange of medicinal plants

Arisha Kausar, Nabia Rehan Mahi, Mariya Anjum, Siddiqa Javed and Anwar Shahzad

6. Biotechnological strategies to enhance phenolic compound biosynthesis in medicinal plants

Amanda Nascimento and Ana Novo Barros

7. Advancing phytochemical biosynthesis through *in vitro* mutagenesis: strategies and applications

Mohamed Addi, Amine Elbouzidi and Mohamed Taibi

8. Elicitation for enhanced production of secondary metabolites and bioactive molecules

Engin Tilkat and Veysel Süzerer

9. Bioreactors: a scalable approach for secondary metabolite production in medicinal plants

Jericó Jabín Bello-Bello

10. Sustainable production of camptothecin using plant cell culture and metabolic engineering

Kishmita Sharma, Saikat Gantait and Sudhir Singh

11. *Rhizobium rhizogenes*-mediated genetic transformation for augmented production of high-valued secondary metabolites in medicinal plants

Sayani Sanyamat, Hiranmoy Bhattacharya, Somroop Chakravarti, Para Ganguly, Saikat Dewanjee and Moumita Gangopadhyay

12. CRISPR/Cas9-mediated gene editing and its application for developing novel genotypes in medicinal plants

Sanjay Kumar Singh, Jayoti Majumder Sarkar and Saikat Gantait

13. Gas chromatography–mass spectrometry-mediated plant metabolomics analysis: recent trends in *in vitro* cultures and potential pharmacological applications

Yashika Bansal, A. Mujib, Mohammad Mohsin and Afeefa Nafees

14. Nanoparticle-mediated approaches for enhancing cryopreservation of medicinal plant germplasm: innovations and future directions

Isaac Baidoo and Paromita Sarbadhikary

List of contributors

Mohamed Addi Laboratoire d'Amélioration des Productions Agricoles, Biotechnologie et Ecophysiologie (LAPABE), Faculté des Sciences, Université Mohammed Premier, Oujda, Morocco

Mahender Aileni Department of Biotechnology, Telangana University, Nizamabad, Telangana, India

Mariya Anjum Plant Biotechnology Section, Department of Botany, Aligarh Muslim University, Aligarh, Uttar Pradesh, India

Sharon Aravind ICAR–Indian Institute of Spices Research, Kozhikode, Kerala, India

Muthukrishnan Arun Department of Biotechnology, Bharathiar University, Coimbatore, Tamil Nadu, India

Honnappa Asangi ICAR–Indian Institute of Spices Research, Regional Station, Appangala, Madikeri, Karnataka, India

Isaac Baidoo Laser Research Centre, Faculty of Health Sciences, University of Johannesburg, Johannesburg, Gauteng, South Africa

Yashika Bansal Cellular Differentiation and Molecular Genetics Section, Department of Botany, Jamia Hamdard, New Delhi, Delhi, India

Ana Novo Barros Centre for the Research and Technology of Agro-environmental and Biological Sciences, CITAB, Inov4Agro, Universidade deTrás-os-Montes e Alto Douro, UTAD, Quinta de Prados, Vila Real, Portugal

Jericó Jabín Bello-Bello SECIHTI-Colegio de Postgraduados Campus Córdoba, Córdoba, México

Hiranmoy Bhattacharya Advanced Pharmacognosy Research Laboratory, Department of Pharmaceutical Technology, Jadavpur University, Kolkata, West Bengal, India

Somroop Chakravarti Department of Biotechnology, School of Life Sciences and Biotechnology, Adamas University, Kolkata, West Bengal, India

Sona Charles ICAR–Indian Institute of Spices Research, Kozhikode, Kerala, India

Balasundaramsaraswathy Chithradevi Department of Botany, PSGR Krishnammal College for Women, Coimbatore, Tamil Nadu, India

Sabina Dahija Faculty of Science, University of Sarajevo, Sarajevo, Bosnia and Herzegovina

Saikat Dewanjee Advanced Pharmacognosy Research Laboratory, Department of Pharmaceutical Technology, Jadavpur University, Kolkata, West Bengal, India

Maruthamuthu Rathinam Elakkiya Department of Botany, PSGR Krishnammal College for Women, Coimbatore, Tamil Nadu, India

Amine Elbouzidi Laboratoire d'Amélioration des Productions Agricoles, Biotechnologie et Ecophysiologie (LAPABE), Faculté des Sciences, Université Mohammed Premier, Oujda, Morocco

Moumita Gangopadhyay Department of Biotechnology, School of Life Sciences and Biotechnology, Adamas University, Kolkata, West Bengal, India; Agrotechnology and Rural Development Centre, Adamas University, Kolkata, West Bengal, India

Para Ganguly Department of Biotechnology, School of Life Sciences and Biotechnology, Adamas University, Kolkata, West Bengal, India

Saikat Gantait Crop Research Unit (Genetics and Plant Breeding), Bidhan Chandra Krishi

Viswavidyalaya, Mohanpur, West Bengal, India

Akshitha H.J. ICAR–Indian Institute of Spices Research, Regional Station, Appangala, Madikeri, Karnataka, India

Saida Ibragic Faculty of Science, University of Sarajevo, Sarajevo, Bosnia and Herzegovina

Siddiqa Javed Plant Biotechnology Section, Department of Botany, Aligarh Muslim University, Aligarh, Uttar Pradesh, India

Phanikanth Jogam Multi-Disciplinary Research Unit, Kakatiya Medical College, Warangal, Telangana, India

Nirmal Joshee Agricultural Research Station, Fort Valley State University, Fort Valley, GA, United States

Erna Karalija Faculty of Science, University of Sarajevo, Sarajevo, Bosnia and Herzegovina

Arisha Kausar Plant Biotechnology Section, Department of Botany, Aligarh Muslim University, Aligarh, Uttar Pradesh, India

Poonam Khatri Agricultural Research Station, Fort Valley State University, Fort Valley, GA, United States

Mohandas Krishnasreya Department of Botany, PSGR Krishnammal College for Women, Coimbatore, Tamil Nadu, India

Rohini M.R. ICAR–Indian Institute of Oil Palm Research, Research Centre, Palode, Trivandrum, Kerala, India

Shivakumar M.S. ICAR–Indian Institute of Spices Research, Regional Station, Appangala, Madikeri, Karnataka, India

Nabia Rehan Mahi Plant Biotechnology Section, Department of Botany, Aligarh Muslim University, Aligarh, Uttar Pradesh, India

Mohammad Mohsin Cellular Differentiation and Molecular Genetics Section, Department of Botany, Jamia Hamdard, New Delhi, Delhi, India

A. Mujib Cellular Differentiation and Molecular Genetics Section, Department of Botany, Jamia Hamdard, New Delhi, Delhi, India

Sankar Mukesh S. ICAR–Indian Institute of Spices Research, Kozhikode, Kerala, India

Afeefa Nafees Cellular Differentiation and Molecular Genetics Section, Department of Botany, Jamia Hamdard, New Delhi, Delhi, India

Amanda Nascimento Academic Unit of Food Engineering, Federal University of Campina Grande (UFCG), Campina Grande, PB, Brazil; SENAI Institute of Technology for Operational Efficiency (IST EO), Federation of Industries of the State of Paraíba (FIEPB), Campina Grande, PB, Brazil

Venkataiah Peddaboina Department of Microbiology, Kakatiya University, Warangal, Telangana, India

Gulab Khan Rohela CSB–Central Sericultural Research and Training Institute, Mysuru, Karnataka, India

Aarthi S. ICAR–Indian Institute of Spices Research, Kozhikode, Kerala, India

Maneesha S.R. ICAR–Indian Institute of Spices Research, Kozhikode, Kerala, India

Dulam Sandhya Department of Biotechnology, Kakatiya University, Warangal, India

Sayani Sanyamat Department of Biotechnology, School of Life Sciences and Biotechnology, Adamas University, Kolkata, West Bengal, India

Birat Sapkota Agricultural Research Station, Fort Valley State University, Fort Valley, GA, United States

Paromita Sarbadhikary Laser Research Centre, Faculty of Health Sciences, University of Johannesburg, Johannesburg, Gauteng, South Africa

Jayoti Majumder Sarkar Department of Floriculture and Landscape Architecture, Bidhan Chandra Krishi Viswavidyalaya, Mohanpur, West Bengal, India

Seethapathy G. Saroja Agricultural Research Station, Fort Valley State University, Fort Valley, GA, United States

Anwar Shahzad Plant Biotechnology Section, Department of Botany, Aligarh Muslim University, Aligarh, Uttar Pradesh, India

Kishmita Sharma Nuclear Agriculture & Biotechnology Division, Bhabha Atomic

Research Centre (BARC), Mumbai, Maharashtra, India; Homi Bhabha National Institute, Anushaktinagar, Mumbai, Maharashtra, India

Sanjay Kumar Singh Division of Horticultural Science, Indian Council of Agricultural Research, Krishi Anusandhan Bhawan – II, New Delhi, India

Sudhir Singh Nuclear Agriculture & Biotechnology Division, Bhabha Atomic Research Centre (BARC), Mumbai, Maharashtra, India; Homi Bhabha National Institute, Anushaktinagar, Mumbai, Maharashtra, India

Nageswara Rao Singisala Department of Botany, Osmania University, Hyderabad, Telangana, India

Veysel Süzerer Department of Pharmacy Services, Vocational School of Health Services, Bingöl University, Bingöl, Türkiye

Divya P. Syamaladevi ICAR–Indian Institute of Spices Research, Kozhikode, Kerala, India

Sheeja T.E. ICAR–Indian Institute of Spices Research, Kozhikode, Kerala, India

Muhammed Azharudheen T.P. ICAR–Indian Institute of Spices Research, Kozhikode, Kerala, India

Mohamed Taibi Laboratoire d'Amélioration des Productions Agricoles, Biotechnologie et Ecophysiologie (LAPABE), Faculté des Sciences, Université Mohammed Premier, Oujda, Morocco

Sureshkumar Tharani GRG BIRAC EYUVA Centre, PSGR Krishnammal College for Women, Coimbatore, Tamil Nadu, India

Engin Tilkat Department of Biology, Faculty of Science and Literature, Batman University, Batman, Türkiye

Vinu V. ICAR–Indian Institute of Spices Research, Kozhikode, Kerala, India

Muhammed Nissar V.A. ICAR–Indian Institute of Spices Research, Kozhikode, Kerala, India

Preface

Plant biotechnology is a modern, efficient approach that applies biological, chemical, and engineering principles to enhance plants and plant-derived products. Medicinal plants, in turn, are rich sources of diverse secondary metabolites—such as phenolics, alkaloids, and pigments—that have been used for human health and disease prevention since ancient Vedic times. This book blends modern biotechnological tools with traditional medicinal practices to overcome the inefficiencies and environmental concerns associated with conventional extraction and cultivation methods. Its comprehensive coverage offers insights into enhanced plant protection, genetic engineering, tissue culture, bioprocessing, and genome editing for improving the production and quality of valuable secondary metabolites. Research on biotechnological innovations—including tissue culture, enzymatic processing, metabolic engineering, and green extraction technologies for medicinal plants—is advancing rapidly. This book presents the latest developments across a broad spectrum of plant biotechnology, focusing on medicinal plant improvement, conservation, restoration, and the prevention of species extinction. These in-depth discussions aim to raise awareness and bridge the gap between human health and the bioavailability of medicinal plants. Emphasis is placed on in vitro applications, with clear elucidation of the biological mechanisms underlying secondary metabolite production and the chromatographic methods used by leading experts worldwide to evaluate it. Designed to optimize and integrate eco-friendly biotechnological methods for large-scale application, this book highlights innovative plant-based approaches that meet the growing demand for natural, health-promoting products.

Saikat Gantait
Jayoti Majumder Sarkar
Sudhir Singh
Sanjay Kumar Singh

About the book

Medicinal plants have served as the backbone of traditional healthcare systems since ancient civilizations and continue to play a crucial role in contemporary medicine. This book focuses on medicinal plant biotechnology, emphasizing the effective utilization of plant-derived bioactive compounds that have significantly contributed to the discovery and development of therapeutic agents for the treatment and prevention of numerous diseases. Comprising 16 chapters, this book explores the diverse properties of secondary metabolites from selected medicinal plants and elucidates their underlying mechanisms of action. It offers comprehensive coverage of advanced biotechnological tools essential for plant micropropagation and improvement, including *in vitro* regeneration, cell suspension culture, *Rhizobium*-mediated genetic transformation, nanotechnological interventions, artificial seed technology, cryopreservation, multiomics approaches, and *in vitro* production and enhancement of secondary metabolites using bioreactor systems. This book places strong emphasis on rapid biotechnology-based breeding strategies, highlighting their importance in plant proliferation, conservation of genetic diversity, and enhanced production of high-value secondary metabolites for sustainable medicinal applications. Drawing on contributions from authors from different countries, it provides a platform for showcasing the role of biotechnology in safeguarding medicinal plant biodiversity. By integrating traditional medicinal wisdom with modern scientific innovation, this volume seeks the attention of researchers, academics, postgraduate students, and industry professionals engaged in medicinal plant research, biotechnology, pharmacognosy, and natural product sciences. This book is presented with the high hope that it will serve as a valuable reference and inspire further innovation in the sustainable utilization of biotechnology across the many facets of medicinal plant resources.

CHAPTER

1

Biotechnological approaches for improved production of secondary metabolites: major bottlenecks and how to address them

Mohandas Krishnasreya[1], *Maruthamuthu Rathinam Elakkiya*[1], *Muthukrishnan Arun*[2], *Sureshkumar Tharani*[3], *and Balasundaramsaraswathy Chithradevi*[1]

[1]Department of Botany, PSGR Krishnammal College for Women, Coimbatore, Tamil Nadu, India
[2]Department of Biotechnology, Bharathiar University, Coimbatore, Tamil Nadu, India
[3]GRG BIRAC EYUVA Centre, PSGR Krishnammal College for Women, Coimbatore, Tamil Nadu, India

1.1 Introduction

Secondary metabolites are organic compounds that lack caloric value and are not involved in the major activities, including growth, development, and reproduction of an organism (Upadhyay et al., 2025). However, they play a crucial role in ecological processes such as pest and disease plant defense mechanisms, pollinator attraction, and influencing allelopathy (Xu & Gaquerel, 2025). This collection of bioactive compounds includes different classes of chemicals like alkaloids, terpenoids, phenolic compounds, flavonoids, and glycosides, most of which are known to have significant importance in medicine, nutrition, and business (Chintada & Golla, 2025). The increasing need for secondary metabolites in various fields is attributed to their therapeutic usefulness against inflammation, cancer, microbes, oxidative damage, and even nerve damage (Javed et al., 2025).

Historically, wild and cultivated plants have provided secondary metabolites. Still, sustainable and large-scale production is hampered by several obstacles including seasonal variations, geographic constraints, slow growth rates, overexploitation of medicinal plants,

Medicinal Plant Biotechnology for Ameliorated Secondary Metabolite Production
DOI: https://doi.org/10.1016/B978-0-443-43944-5.00003-7

and low metabolite content (Singh, Nayak, et al., 2025; Singh, Singh, et al., 2025). Many medicinal plants that produce beneficial chemicals are also endangered, therefore conventional harvesting techniques are environmentally and morally questionable (Alum, 2025). In this case, controlled environments offer a viable, sustainable, and scalable approach for generating secondary metabolites using *in vitro* biotechnological techniques (Anuradha, Umalatha, et al., 2025; Anuradha, Balasubramanya, et al., 2025).

Secondary metabolite production has been demonstrated to be efficiently promoted and enhanced by plant tissue culture techniques, including callus culture, cell suspension culture, organ cultures (including shoots and roots), and hairy root cultures (Nordine, 2025). Numerous biotechnological techniques, such as the use of elicitors, feeding of precursors, genetic modifications, and immobilization techniques, can be used to fine-tune these systems in order to increase yield (Indu et al., 2025). Recently, advancements in metabolic engineering and synthetic biology have allowed scientists to alter biosynthetic pathways, increase the expression of essential enzymes, and even reconstruct these pathways in microbial or foreign systems to enhance metabolite production (Seshadri et al., 2025). Furthermore, the use of omics technologies, including proteomics, metabolomics, and transcriptomics, greatly facilitates the deciphering of the complex regulatory networks that govern metabolite (Varadharajan et al., 2025)

Despite these developments, there are often several challenges in the *in vitro* production of secondary metabolites. According to Anuradha, Umalatha, et al. (2025) and Anuradha, Balasubramanya, et al. (2025), these difficulties include low productivity due to limited flow through secondary metabolic pathways, somaclonal variation, epigenetic changes, a gradual decline in biosynthetic capacity, and challenges in scaling up for commercial production. Moreover, it is still costly and time-consuming to extract metabolites from complex culture matrices (Alves & Pereira, 2025).

Understanding and overcoming these bottlenecks is essential to fully unlock the potential of plant-based *in vitro* systems for producing valuable secondary metabolites. Today, scientists are turning to a blend of advanced tools—like computational biology, metabolic flux analysis, genome editing, and bioprocess optimization—to better understand the challenges and improve efficiency (Elazzazy et al., 2025). In this chapter, we take a deep dive into the range of biotechnological strategies used to boost *in vitro* production of these compounds. We will also shed light on the key roadblocks that often stand in the way and explore innovative ways to navigate around them. Ultimately, this chapter aims to provide a clear view of both the exciting possibilities and the ongoing challenges in using plant biotechnology to meet the world's growing demand for these important natural products.

1.2 Plant tissue culture techniques for secondary metabolite production

1.2.1 Callus culture

Callus culture is a key method in plant tissue culture that entails the *in vitro* generation and upkeep of a disorganized, rapidly growing cell mass known as callus, from plant explants in sterile environments (Sidik et al., 2024). This technique is essential for plant regeneration, genetic modification, studies on somaclonal variation, and the production of secondary metabolites (Majumder et al., 2025).

Callus formation begins with an explant, which can be a segment of leaf, stem, root, cotyledon, hypocotyl, or other plant tissues. Initially, the explant undergoes surface sterilization to remove any microbial contaminants before being placed on a solid culture medium, typically Murashige and Skoog medium. This medium is enriched with a carbon source, usually 3% of sucrose; agar is used as a gelling agent, and particular plant growth regulators (Sidik et al., 2024).

The development of callus is greatly affected by the particular type and concentration of plant growth regulators, particularly auxins such as 2,4-dichlorophenoxyacetic acid (2,4-D) and naphthaleneacetic acid, along with cytokinins like benzylaminopurine (BAP) and kinetin. Generally, a higher ratio of auxins to cytokinins promotes callus formation (Khaskheli et al., 2024).

1.2.1.1 *Maintenance and subculturing*

After initiation, the callus is regularly subcultured, generally every 3–4 weeks, onto new medium with a similar formulation to support its growth and avoid browning or necrosis (Devi et al., 2024). Key factors like temperature (typically between 22°C and 25°C), photoperiod (maintaining dark or low light conditions during initiation), and the composition of the medium play a crucial role in determining the quality and proliferation of the callus (Kohan-Baghkheirati et al., 2025) (Fig. 1.1).

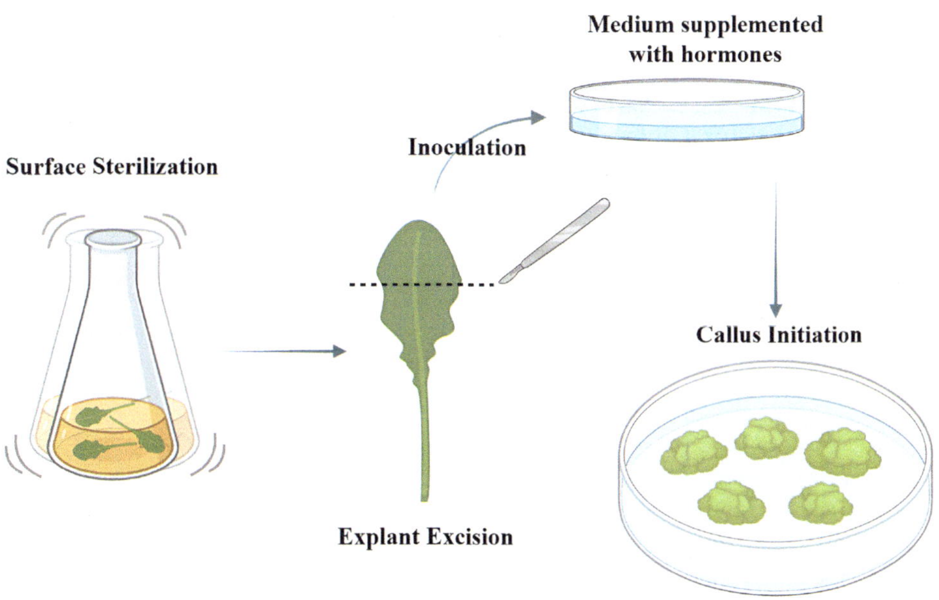

FIGURE 1.1 Callus initiation. Callus induction from leaf explants involving surface sterilization, explant excision, inoculation on hormone-supplemented medium, and subsequent callus formation.

1.3 Suspension culture

Cell suspension culture is an effective and dynamic *in vitro* method that facilitates the growth of plant cells as free-floating aggregates within a liquid nutrient medium, maintained under constant agitation (Najim et al., 2024). This technique originates from friable callus tissue that is initially cultured in a solid medium and later moved to a liquid medium, allowing for rapid cell proliferation in a uniform environment (Nimavat & Parikh, 2024). It is crucial for the large-scale production of valuable secondary metabolites that hold significance in medicinal, cosmetic, and industrial applications (Wadhwa et al., 2024).

1.3.1 Maintenance and subculturing

To start suspension cultures, friable callus derived from explants (leaf, stem, or root) is placed into a sterile liquid medium such as Murashige and Skoog (MS) or Gamborg's B5, frequently enhanced with auxins like 2,4-D to promote cell division. The culture is kept in Erlenmeyer flasks or bioreactors and set on rotary shakers (100–150 rpm) to ensure consistent aeration, nutrient absorption, and to avoid cell settling. Subculturing is conducted regularly (usually every 7–14 days) to maintain robust growth and metabolite production (Sidik et al., 2024) (Fig. 1.2).

1.4 Organ culture

Organ culture involves the *in vitro* growth of plant organs, including shoots, roots, leaves, or floral components, in sterile conditions on an appropriate nutrient medium. In contrast to callus or suspension cultures, which are made up of undifferentiated cells, organ cultures preserve the natural tissue structure and function of the organ (Shukla, 2024). This preservation of structural integrity enhances the organ-specific biosynthetic pathways, rendering organ cultures extremely useful for the focused production of secondary metabolites (Nallasamy et al., 2024) (Fig. 1.3).

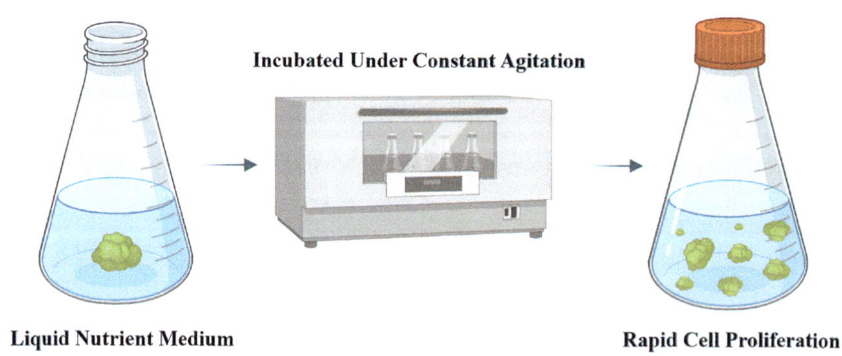

FIGURE 1.2 Cell suspension culture. Establishment of cell suspension culture by transferring callus to liquid nutrient medium and incubating under constant agitation for rapid cell proliferation.

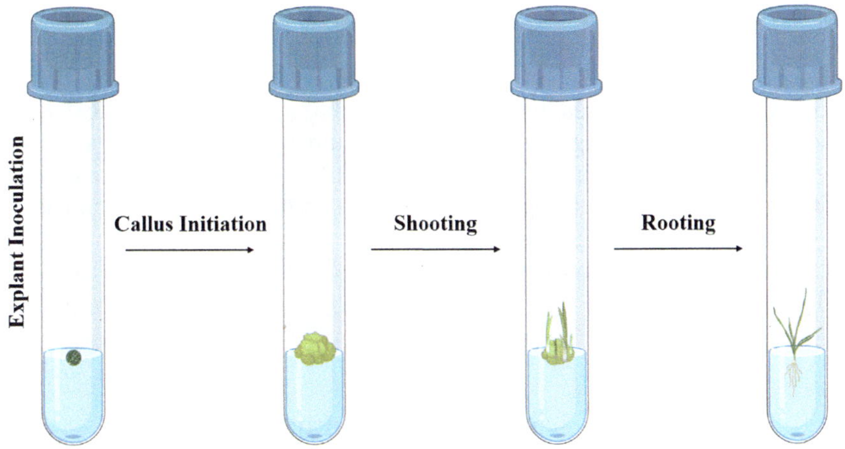

FIGURE 1.3 Organogenesis. Schematic representation of callus-mediated plant regeneration showing explant inoculation, callus initiation, shoot induction, and root formation leading to plantlet development.

1.4.1 Shoot culture

Shoot culture refers to the *in vitro* cultivation and multiplication of shoot tips or axillary buds on a nutrient medium, usually enhanced with cytokinins such as BAP, kinetin, or thidiazuron. These hormones promote the division of meristematic cells and encourage the formation of multiple shoots.

Establishment: Explants, such as nodal segments or shoot tips, are sterilized on the surface and then cultured on solid MS medium that has a high ratio of cytokinins to auxins (Sánchez-Romero, 2024).

Advantages:

- Preserves the natural biosynthetic processes.
- High yield of metabolites resulting from tissue differentiation.
- Reduced somaclonal variation in comparison to callus cultures.

1.4.2 Root culture

Root cultures are established from excised root tips or adventitious roots cultivated *in vitro* , usually under the influence of auxins such as indole-3-acetic acid, indole-3-butyric acid, or naphthaleneacetic acid. These cultures can be sustained in either solid or liquid media and frequently demonstrate consistent growth and metabolic profiles over prolonged durations (Adem et al., 2024).

Advantages:

- Metabolite production specific to organs
- Stability in genetic and biochemical aspects
- Growth in a hairy root system without hormones
- Compatibility for feeding with elicitors and precursors

Organ cultures, especially shoot and root cultures, serve as a significant system in plant biotechnology for the continuous, organ-specific generation of secondary metabolites (Murthy et al., 2024).

The maintenance of organized tissue structure guarantees physiological relevance and biosynthetic accuracy, rendering this method particularly crucial for the conservation and use of rare, endangered, or threatened medicinal plant species (Khezri et al., 2024). By incorporating metabolic engineering and elicitation techniques, organ cultures present a promising opportunity for large-scale metabolite production in the pharmaceutical and nutraceutical industries (Anuradha, Umalatha, et al., 2025; Anuradha, Balasubramanya, et al., 2025).

1.5 Hairy root culture

Hairy root culture is a sophisticated plant tissue culture method that utilizes the genetic transformation capabilities of the soil bacterium *Agrobacterium rhizogenes*. This technique results in the development of rapidly growing, extensively branched root systems that demonstrate genetic stability, a high growth rate, and an improved ability to generate secondary metabolites (Zhu et al., 2024). Hairy root systems have become one of the most effective and dependable *in vitro* platforms for producing root-specific and high-value phytochemicals (Kumar, Kumar, et al., 2024; Kumar, Bhamini, et al., 2024; Kumar, Bharat, et al., 2024).

1.5.1 Mechanism of hairy root culture

The procedure starts with the infection of a sterile plant explant, which is usually a leaf, stem, hypocotyl, or cotyledon, using *A. rhizogenes* that carries a root-inducing plasmid. During the infection process:

- The T-DNA segment of the Ri plasmid is integrated into the plant's genome.
- This T-DNA includes rol (root oncogenic loci) genes such as rolA, rolB, rolC, etc., which play a crucial role in triggering uncontrolled root growth.
- Consequently, hairy roots develop at the site of infection, exhibiting rapid and hormone-independent growth, significant level of lateral branching and plagiotropism, and stable incorporation of transgenes (Zhu et al., 2024).

1.5.2 Establishment of hairy root culture

After induction, the hairy roots are removed and grown in hormone-free MS or B5 liquid or solid medium under sterile conditions. These cultures are generally kept at a temperature of 22°C–25°C in darkness to replicate the conditions found underground for roots. Periodic subculturing is performed to sustain exponential growth and the production of metabolites (Zhu et al., 2024) (Fig. 1.4).

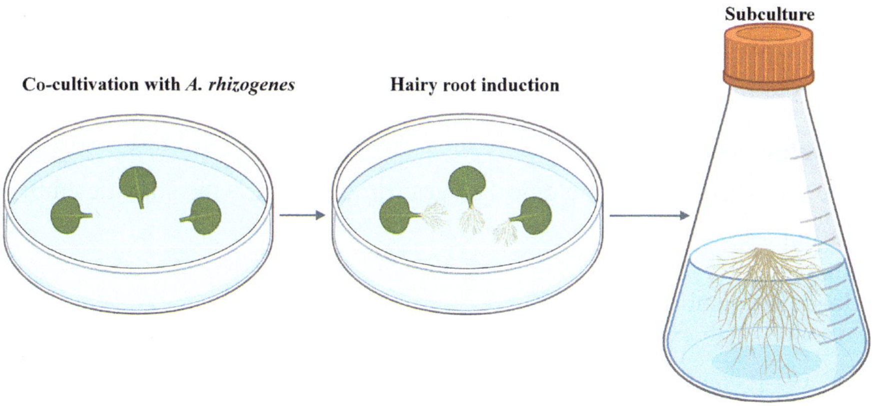

FIGURE 1.4 Hairy root culture. Induction of hairy roots through cocultivation of explants with *Agrobacterium rhizogenes*, followed by root emergence and subculture.

Advantages of hairy root culture

- Genetic and biochemical stability
- High growth rate without hormones
- Enhanced metabolite yield
- Scalability
- Efficient genetic transformation system (Zhu et al., 2024).

1.6 Somatic embryogenesis

Somatic embryogenesis refers to the process where somatic (nonreproductive) plant cells transform into structures resembling embryos that can develop into fully formed plants. These structures, termed somatic embryos, progress through developmental phases akin to zygotic (seed-derived) embryos, but originate from vegetative tissues. This method is crucial in plant tissue culture for purposes such as clonal propagation, genetic transformation, synthetic seed production, and, significantly, the biosynthesis of secondary metabolites (Lelis et al., 2025).

1.6.1 Phases of somatic embryogenesis

The process of somatic embryogenesis (Qin et al., 2025) is divided into the following:

1.6.1.1 Initiation phase

- Explants such as leaf, cotyledon, hypocotyl, or immature embryos are placed on a nutrient medium containing high concentrations of auxins to induce embryogenic callus.
- Embryogenic cells are formed by the dedifferentiation of somatic cells (Qin et al., 2025).

1.6.1.2 *Induction phase*

Proembryogenic masses are formed from the embryogenic cells, which in turn develop to somatic embryos under hormonal influence in controlled condition (Qin et al., 2025).

1.6.1.3 *Developmental phase*

There are four distinct morphological stages through which the somatic embryo passes,

- Globular embryo
- Heart-shaped embryo
- Torpedo-shaped embryo
- Cotyledonary stage embryo

1.6.1.4 *Maturation phase*

- Embyro mature on hormone-free or low-auxin media.
- Addition of abscisic acid or polyethylene glycol promotes proper maturation and desiccation tolerance (Qin et al., 2025).

1.6.1.5 *Germination and conversion phase*

Mature embryos are transferred to germination media and exhibit shoot and root development (Qin et al., 2025).

1.6.2 Biotechnological strategies for metabolite enhancement

Secondary metabolites including alkaloids, flavonoids, phenolics, terpenoids, and glycosides, are compounds with low molecular weights that play essential roles in plant defense and therapeutic uses for humans (Upadhyay et al., 2025). Nevertheless, *in vitro* cultures frequently generate these metabolites at less than optimal levels because they lack the environmental stimuli present in natural settings. To address this issue, various biotechnological approaches have been created to improve the biosynthesis of metabolites in tissue and organ cultures (Alami et al., 2024).

1.6.2.1 *Elicitation*

Elicitation refers to the method of promoting or increasing the production of secondary metabolites through the use of elicitors—substances that trigger plant defense mechanisms. Elicitors simulate either biotic or abiotic stress situations, thereby activating signaling pathways that enhance the expression of genes related to secondary metabolism (Shilpa et al., 2025; Petrova et al., 2024).

1.6.2.1.1 Types of elicitors

1. Biotic elicitors

   Biotic elicitors are derived from living organisms or from the components of their cell wall. They include:
   - Polysaccharides: chitosan (sourced from crustacean shells), pectin, and glucans.
   - Microbial extracts: yeast extract, fragments of fungal cell walls.

- Pathogen-derived molecules: bacterial lipopolysaccharides, fragments of flagellin.
  Mode of action: Biotic elicitors engage with pattern recognition receptors located on plant cell membranes, initiating defense signaling pathways such as MAP kinase cascades and oxidative bursts, which result in the activation of transcription factors that enhance the expression of metabolite biosynthetic genes (Petrova et al., 2024).

2. Abiotic elicitors
   These factors encompass both physical and chemical elements that create stress:
   - Chemical substances: methyl jasmonate (MeJA), salicylic acid (SA), hydrogen peroxide (H_2O_2).
   - Metal ions: Ag^+, Cd^{2+}, Cu^{2+}.
   - Physical stressors: exposure to UV light, thermal shock, osmotic pressure.

Mode of action: Abiotic elicitors frequently imitate environmental stress or function as signaling molecules (for instance, jasmonates serve as stress hormones) that trigger the plant's natural immunity response, enhancing the production of secondary metabolites (Petrova et al., 2024).

1.6.2.2 *Precursor feeding*

Precursor feeding refers to the process of enhancing the culture medium with biochemical intermediates that serve as either direct building blocks or rate-limiting precursors for the desired secondary metabolite (Indu et al., 2025).

1.6.2.2.1 Mechanism

The pathways of secondary metabolites are intricate and frequently constrained by the availability of precursor molecules. By introducing precursors from outside sources:

- The metabolic flow will be redirected toward the target compound,
- Key enzymatic bottlenecks are circumvented, and
- Depletion of intermediates can be prevented.

Care should be given that the precursor must be biocompatible and not be toxic, the timing and dosage must depend on the key biosynthetic activity, and for a synergistic effect, the precursor can be added along with elicitation (Indu et al., 2025).

1.6.2.3 *Genetic transformation*

Genetic transformation refers to the process of modifying the genetic structure of plant cells to enhance the expression of essential biosynthetic enzymes, create new metabolic pathways, or adjust regulatory components to increase the production of secondary metabolites (Wang et al., 2025).

1.6.2.3.1 Overexpression of biosynthetic genes

Rate-limiting enzymes such as:

- Phenylalanine ammonia-lyase in the phenylpropanoid pathway.
- 3-Hydroxy-3-methylglutaryl-CoA reductase in terpenoid pathways.
- Squalene synthase in sterol biosynthesis.

Overexpressing these genes increases flux toward the desired metabolite (Hilal et al., 2024).

1.6.2.3.2 Transcription factor engineering

Certain transcription factors regulate entire clusters of biosynthetic genes:

- Myeloblastosis proteins—one of the largest transcription factor families in plants; they regulate phenylpropanoid pathways, flavonoid biosynthesis, anthocyanin accumulation, and more.
- Basic-helix-loop-helix transcription factors—contain a conserved domain with a basic region and a helix-loop-helix structure; they often interact with MYB proteins to coregulate flavonoid, alkaloid, and terpenoid pathways
- WRKY families—regulate stress responses and are important in activating secondary metabolite pathways like lignin, phenolics, and terpenoids under stress conditions.
- Overexpression can coordinately upregulate multiple pathway enzymes (Li, Chen, et al., 2025; Li, Lin, et al., 2025).

1.6.2.4 Promoter engineering

Strong, constitutive promoters (e.g., CaMV 35S) or inducible promoters enhance gene expression precisely when needed (Al Aboud, 2024).

1.6.2.4.1 Metabolic pathway engineering

Advanced gene editing tools like CRISPR/Cas9 can:

- Knock out competing pathways.
- Silence negative regulators.
- Fine-tune metabolic flux (Al Aboud, 2024).

1.7 Use of endophytes and symbionts

Endophytes are beneficial microorganisms, including fungi and bacteria that live within plant tissues. They play a role in plant metabolism and can enhance the production of secondary metabolites, either directly or indirectly (Muhammad et al., 2024).

1.7.1 Mechanisms

- **Induced systemic resistance:** Endophytes simulate mild biotic stress, activating defense pathways.
- **Production of signaling molecules:** Certain endophytes generate auxins or jasmonates, enhancing secondary metabolism.
- **Symbiotic biosynthesis:** Some endophytes are capable of synthesizing metabolites independently or providing intermediates for pathways (Muhammad et al., 2024).

1.8 Immobilization techniques

Immobilization refers to the process of enclosing plant cells, tissues, or organs within semipermeable matrices. These matrices facilitate the exchange of nutrients and metabolites while limiting the movement of the cells (Najim et al., 2024).

1.8.1 Techniques

- Entrapment in calcium alginate beads.
- Encapsulation in agarose or carrageenan.
- Attachment to solid supports like polyurethane foam.

1.8.2 Benefits

- Repeated use of immobilized cultures for continuous production.
- Higher cell density, leading to increased metabolite yield.
- Protection of shear-sensitive cells in liquid cultures.
- Feasibility of continuous flow bioreactors.

When applied separately or together, elicitation, precursor feeding, genetic transformation, endophyte association, and immobilization act as essential, complementary methods for enhancing the *in vitro* generation of important secondary metabolites (Najim et al., 2024) (Table 1.1).

TABLE 1.1 Summary of biotechnological strategies for the enhanced production of secondary metabolites.

Strategy	Description	Application	References
Elicitation	Use of biotic (e.g., microbial extracts) or abiotic (e.g., UV, heavy metals) factors to stimulate stress responses that activate secondary metabolite pathways.	Methyl jasmonate elicitation to boost paclitaxel in *Taxus* cell cultures	Dehghan et al. (2025)
Precursor feeding	Supplying pathway precursors or intermediates to increase substrate availability for biosynthesis	Feeding tryptophan to enhance indole alkaloids in *Catharanthus roseus*	Miryousefzadeh et al. (2025)
Genetic transformation	Introducing or modifying genes (e.g., overexpression, RNAi) to upregulate key enzymes or silence competing pathways.	Overexpressing HMGR gene for increased terpenoid production.	Zheng et al. (2025)
Use of endophytes and symbionts	Coculturing plant cells with beneficial microorganisms that promote metabolite production via signaling or precursor sharing	Using endophytic fungi to boost vinblastine yield.	Khan and Gond (2024)
Immobilization techniques	Entrapping cells in matrices (e.g., alginate beads) to stabilize production, enable reuse, and maintain high cell densities.	Immobilized *Panax ginseng* cultures for steady ginsenoside output	Zheng and Liang (2024)

1.8.3 Metabolic engineering and synthetic biology

Metabolic engineering and synthetic biology are closely related disciplines that facilitate the accurate redesign and optimization of metabolic pathways in plants, microbes, or cell cultures, aimed at improving the production of valuable secondary metabolites (Rai et al., 2025). Unlike conventional breeding methods or random mutagenesis, these contemporary tools provide a rational, predictable, and finely tuned approach to managing the entire biosynthetic machinery (Inckemann et al., 2024).

1.8.3.1 Gene editing (CRISPR/Cas9 and RNAi)

1.8.3.1.1 CRISPR/Cas9

CRISPR/Cas9 (clustered regularly interspaced short palindromic repeats/CRISPR-associated protein 9) is an advanced, accurate, and adaptable genome editing technology that originates from the adaptive immunity of bacteria.

A single-guide RNA guides the Cas9 nuclease to a designated genomic DNA sequence. Cas9 creates a double-strand break at the specified target location. The cell's inherent repair mechanisms subsequently fix the break through:

- Nonhomologous end joining (NHEJ): a process that is prone to errors and often results in small insertions or deletions (knockouts).
- Homology-directed repair: enables accurate gene insertion or correction when a template is available.

Applications:

Inhibition of competing pathways: This approach redirects metabolic flow toward the desired compounds.

Example: Silencing CYP71AV1 competitors in *Artemisia annua* enhances artemisinin production.

- Activation of dormant genes: CRISPRa (CRISPR activation) technologies can activate hidden biosynthetic gene clusters.
- Multiplex editing: It allows for the simultaneous targeting of multiple genes to rewire metabolic pathways (Bhojiya & Joshi, 2024).

1.8.3.1.2 RNA interference

RNAi is a method of gene silencing that occurs after transcription, effectively reducing the expression of particular genes by breaking down their messenger RNA (mRNA).

Mechanism of action:

Double-stranded RNA (dsRNA) that matches the target gene is introduced into the system. The enzyme Dicer then processes this dsRNA into small interfering RNAs (siRNAs). These siRNAs direct the RNA-induced silencing complex to the corresponding mRNA, resulting in its degradation.

Applications:

Downregulation of negative regulators or competing enzymes.

Example: Suppression of polyphenol oxidase genes to boost polyphenol retention.

- *Catharanthus roseus*: RNAi knockdown of strictosidine β-D-glucosidase to increase vinblastine intermediates (Al Aboud, 2024).

1.8.3.2 *Pathway engineering*

Pathway engineering entails the alteration of complete biosynthetic pathways through the addition, deletion, or modification of several genes to enhance the metabolic flow toward the desired compound.

1. Rate-limiting enzyme overexpression.
2. Side branches blocking-Inhibition of genes rerouting intermediates toward unwanted products.
3. Coexpression of multiple pathway genes.
4. Promoters that are tissue-specific, inducible, or constitutive facilitate accurate expression (Wang et al., 2024).

1.8.3.3 *Heterologous expression systems*

Heterologous expression systems serve as an effective approach in plant biotechnology and synthetic biology for the sustainable and large-scale production of intricate secondary metabolites (Rai et al., 2025).

Rather than depending exclusively on the native plant, which may exhibit a lengthy growth cycle, low metabolite yield, or challenging cultivation conditions, researchers transfer complete or partial biosynthetic pathways into nonnative, rapidly growing microbial hosts such as bacteria (*Escherichia coli*), yeast (*Saccharomyces cerevisiae*), or filamentous fungi (*Aspergillus*, *Streptomyces*) (Chaudhary et al., 2024).

These microbial hosts are genetically manipulable, well-studied, and suitable for fermentation, rendering them excellent "cell factories" for the production of metabolites. In a standard heterologous system, several genes that encode essential pathway enzymes often including rate-limiting cytochrome P450 monooxygenases, methyltransferases, or glycosyltransferases are cloned and coexpressed under robust, inducible promoters to recreate the complete metabolic pathway (Kumar, Kumar, et al., 2024; Kumar, Bhamini, et al., 2024; Kumar, Bharat, et al., 2024).

Advanced synthetic biology tools, including promoter libraries, gene cassettes, modular operons, and genome editing platforms like CRISPR/Cas9, enhance the optimization of gene expression, cofactor regeneration, and precursor supply within microbial hosts (Li, 2025).

A notable example is the microbial synthesis of the anti-malarial compound artemisinin, where genes from *Artemisia annua* were integrated into yeast, allowing for semisynthetic artemisinin production on an industrial scale (Qamar et al., 2024).

Despite facing challenges such as the expression of complex plant enzymes, ensuring appropriate posttranslational modifications, and balancing metabolic flux, heterologous systems present significant potential for the efficient and cost-effective production of valuable phytochemicals, independent of agricultural constraints. These systems effectively connect traditional plant extraction methods with contemporary bio-manufacturing, providing a scalable and environmentally sustainable solution to satisfy the increasing demand for high-value plant-derived bioactives (Singh, 2024).

1.8.4 Role of omics technologies in secondary metabolite production

1.8.4.1 Transcriptomics and proteomics

Omics technologies have transformed our comprehension of the intricate regulatory networks that govern secondary metabolite biosynthesis in plants (Satrio et al., 2024). By offering comprehensive, high-throughput insights across various biological levels, genes, transcripts, proteins, metabolites, and metabolic fluxes, these tools enable researchers to pinpoint critical bottlenecks, uncover new pathway components, and formulate targeted strategies for metabolic engineering (Vitorino, 2024).

Transcriptomics encompasses the thorough examination of gene expression patterns using technologies such as microarrays and next-generation RNA sequencing (RNA-Seq) (Villaseñor-Altamirano et al., 2024). This approach allows for the identification of genes that are either upregulated or downregulated in response to specific culture conditions, developmental phases, or elicitor treatments (Sozoniuk et al., 2025).

In the realm of *in vitro* cultures, transcriptomic investigations can identify transcription factors (such as MYB, bHLH, WRKY) and enzymes specific to pathways that regulate the accumulation of secondary metabolites (Li, Chen, et al., 2025; Li, Lin, et al., 2025). For instance, in *Catharanthus roseus*, transcriptomic profiling has shed light on the fluctuating expression of genes associated with the terpenoid indole alkaloid pathway, enabling targeted overexpression or silencing of these genes (Singh, Nayak, et al., 2025; Singh, Singh, et al., 2025).

In addition, proteomics—the extensive analysis of the complete set of proteins (the proteome) not only identifies the presence of proteins but also examines their posttranslational modifications, abundance, and interactions among key enzymes (Zafar et al., 2024). Given that enzyme activity is the driving force behind metabolite synthesis, proteomic data confirm whether transcripts are translated into functional proteins under particular conditions (Satrio et al., 2024).

Recent advancements in two-dimensional gel electrophoresis, mass spectrometry (MS/MS), and quantitative methods such as iTRAQ or label-free proteomics allow for the correlation of stress treatments, culture stages, or elicitor applications with the activation of essential biosynthetic enzymes, thereby providing practical targets for pathway engineering (Martínez-Esteso et al., 2024).

1.8.5 Metabolomics and fluxomics

While transcriptomics and proteomics indicate potential, metabolomics captures the actual chemical signatures of cellular activities by quantifying a diverse array of primary and secondary metabolites found in cells, tissues, or entire cultures (Alam et al., 2025). By employing high-resolution analytical techniques such as gas chromatography–mass spectrometry, liquid chromatography-mass spectrometry, and nuclear magnetic resonance, metabolomics offers insights into the intracellular levels of essential phytochemicals, their intermediates, and byproducts (Kumar, Kumar, et al., 2024; Kumar, Bhamini, et al., 2024; Kumar, Bharat, et al., 2024). This facilitates the identification of rate-limiting steps, metabolic branch points, and unforeseen bottlenecks within the metabolic pathway (Gharat et al., 2025).

Concurrently, fluxomics advances this analysis by examining the rates at which metabolites are turned over within the cell, illustrating how carbon or nitrogen transitions through primary metabolism into secondary pathways (Mishra & Kumar, 2024). The use of stable isotope labeling (e.g., ^13C, ^15N) in conjunction with computational flux balance analysis enables researchers to measure flux distribution and identify which enzymatic processes require enhancement or redirection to optimize the yields of target compounds. Collectively, metabolomics and fluxomics connect genetic potential with actual biochemical outcomes, forming the foundation of rational metabolic engineering (Fan, 2024).

While each omics layer offers significant insights independently, true advancements arise from integrative omics, which merges transcriptomics, proteomics, metabolomics, and fluxomics into a cohesive systems biology framework. Such integrative studies illuminate how alterations at the gene or protein level translate into metabolic outcomes, facilitating predictive modeling and network analysis of biosynthetic pathways (Joshi, Kumar, Kaushik, 2024; Joshi, Kumar, Kaushik, Kumar, et al., 2024). For example, the integration of multiomics has revealed how environmental factors, culture conditions, and elicitors work together to regulate secondary metabolism (Varadharajan et al., 2025).

By pinpointing master regulators, coexpressed gene modules, or feedback loops, scientists can create more targeted interventions, be it through editing a transcription factor gene, optimizing precursor feeding, or modifying culture conditions (Kumar et al., 2025). This comprehensive viewpoint accelerates the creation of robust, high-yielding cell lines, organ cultures, or engineered microbes, ensuring reliable and scalable production of valuable secondary metabolites (Bhaskar et al., 2024). In the end, integrative omics revolutionizes biotechnological production from a trial-and-error methodology into a systematic, data-informed strategy, advancing the objectives of sustainable and commercially viable phytochemical manufacturing (Liao et al., 2025).

1.8.6 Bottlenecks in *in vitro* production of secondary metabolites

Although plant cell, tissue, and organ cultures hold significant promise as sustainable biofactories for producing high-value phytochemicals, various inherent and practical obstacles hinder their commercial feasibility (Singh, 2024). These issues encompass biological, physiological, and industrial dimensions, necessitating comprehensive strategies to address them. The primary limitations consist of genetic and epigenetic instability, low yields resulting from metabolic flux constraints, intricate downstream processing, and challenges related to reproducibility and scalability (Abdi et al., 2024).

1.8.6.1 *Genetic and epigenetic instability*

One of the primary challenges faced in long-term *in vitro* cultures is genetic instability, which denotes spontaneous alterations in the DNA sequence of cultured cells (Beal et al., 2025). Somaclonal variation that refers to genetic mutations and chromosomal rearrangements that occur during repeated subculturing can result in inconsistent growth, a decline in biosynthetic capability, or even a total inability to produce the desired metabolite (Manchanda et al., 2025).

Alongside genetic alterations, epigenetic modifications (including DNA methylation, histone modification, and RNA-mediated silencing) significantly contribute to the silencing of essential biosynthetic genes over time (Nasir et al., 2024). For instance, in *Taxus* cell cultures that generate paclitaxel, epigenetic silencing of genes for rate-limiting enzymes has been noted after several subcultures, resulting in reduced yields (Yin et al., 2025). The combined impact of these instabilities leads to unpredictable production, necessitating regular screening and rejuvenation methods such as cryopreservation, periodic regeneration from explants, or the application of molecular markers to assess fidelity (Majumder et al., 2025).

1.8.6.2 Low yield and metabolic flux limitations

Even when genetic stability is preserved, *in vitro* cultures frequently generate target secondary metabolites at significantly lower concentrations than those found in the intact parent plant. This discrepancy is primarily attributed to inadequate metabolic flux directed toward the desired biosynthetic pathway (Indu et al., 2025).

In the natural habitat of the plant, stress factors, developmental signals, or intricate tissue interactions often stimulate secondary metabolism signals that are either absent or diminished in controlled culture environments (Alami et al., 2024). Moreover, the presence of essential precursors, cofactors (such as NADPH and ATP), or enzymes involved in the pathway may not be at optimal levels. Additionally, competition from primary metabolic pathways can redirect precursors away from secondary pathways (Di Blasi et al., 2024).

For example, cell suspension cultures of *Catharanthus roseus* generally produce low amounts of vincristine and vinblastine due to the complexity, multistep nature, and stringent regulation of the pathway (Li et al., 2024). Addressing flux limitations necessitates the implementation of strategies like elicitor application, precursor supplementation, pathway engineering, or coexpression of transcription factors to channel more metabolic resources into the synthesis of secondary metabolites (Hu et al., 2024).

1.8.6.3 Difficulty in downstream processing

Even when *in vitro* cultures yield target metabolites, the extraction and purification of these compounds can be both technically demanding and expensive. Many secondary metabolites are found within cells, attached to cell walls, or present in intricate forms like glycosides (Wadhwa et al., 2024b). The process of breaking down cells to liberate these compounds may also result in the release of undesirable byproducts, which can lead to reduced purity and necessitate additional downstream processing steps (Shanu et al., 2024).

Moreover, the yields are frequently so minimal that substantial amounts of biomass or culture medium need to be processed, further increasing extraction costs. For instance, in hairy root cultures, the roots are often situated within complex matrices that hinder recovery (Liu et al., 2025). In contrast to microbial fermentation, where numerous products are secreted outside the cells, plant cultures typically do not possess effective secretion pathways. Implementing strategies such as permeabilization, engineering secretion pathways, or utilizing adsorbents in situ to capture metabolites can assist in alleviating this bottleneck (Kumar & Singh, 2025).

1.8.6.4 *Reproducibility and scalability*

Achieving consistent and reproducible yields while scaling from laboratory flasks to pilot and industrial bioreactors continues to be one of the most significant challenges for commercial utilization (de Mello et al., 2024). *In vitro* cultures exhibit sensitivity to minor alterations in nutrient composition, oxygen levels, shear forces, pH, and various other physical parameters (Turan et al., 2024). What is effective in small shake flasks may not be applicable to large stirred-tank bioreactors, where controlling factors such as mixing and oxygen transfer uniformly pose difficulties (Palladino et al., 2024).

Moreover, heterogeneity within the cell population (e.g., a combination of metabolically active and inactive cells) can result in variations from batch to batch. The risks of contamination also escalate with increased scale (Abdi et al., 2024). Hairy root cultures, for example, frequently face limitations due to oxygen and nutrient diffusion within dense root mats (Barbinta-Patrascu et al., 2024). It is essential to develop scalable bioreactor designs (such as bubble columns and mist reactors for roots) and robust process control systems to guarantee uniform growth, product consistency, and economic feasibility (Anuradha, Umalatha, et al., 2025; Anuradha, Balasubramanya, et al., 2025).

These bottlenecks collectively emphasize the necessity for a comprehensive strategy that integrates advanced culture techniques, metabolic engineering, omics-guided monitoring, and cutting-edge downstream processing (Elazzazy et al., 2025). This integration is essential to fully harness the potential of *in vitro* production systems for secondary metabolites. Tackling these issues will facilitate the development of stable, high-yield, and economically viable production pipelines, providing dependable alternatives to wild harvesting and traditional agriculture (Clagnan et al., 2024).

1.8.7 Elucidation strategies for bottlenecks

To address the ongoing difficulties in the *in vitro* production of secondary metabolites, researchers utilize sophisticated elucidation strategies that integrate contemporary computational tools, experimental evolution, and intelligent process design. These approaches assist in pinpointing the locations of bottlenecks and determining effective solutions to overcome them.

1.8.7.1 *Use of bioinformatics and systems biology*

Bioinformatics tools assist in the analysis of large genomic and transcriptomic datasets to identify significant genes, enzymes, and regulatory elements that play a role in target pathways. Systems biology takes this a step further by combining this data into comprehensive models that replicate the interactions between genes, proteins, and metabolites. Together, these methodologies aid in identifying bottlenecks, such as underexpressed enzymes or inactive gene clusters, and inform targeted genetic or process interventions (Joshi, Kumar, Kaushik 2024; Joshi, Kumar, Kaushik, Kumar, et al., 2024 & Dey et al., 2024).

1.8.7.2 *Metabolic flux analysis*

Metabolic flux analysis is utilized to assess the real movement of metabolites within a cell's pathways (Yoo et al., 2024). By supplying cells with stable isotope-labeled precursors and examining the incorporation of these labels into various compounds, researchers can pinpoint which stages in the pathway are inefficient or sluggish (Bordanaba Florit, 2024). This information aids in developing targeted approaches such as overexpressing specific enzymes or providing limiting precursors to enhance the metabolic flow toward the intended metabolite (Zhang et al., 2024).

1.8.7.3 *Adaptive laboratory evaluation*

Adaptive laboratory evolution is an effective method that allows cultures to organically develop advantageous characteristics through regulated selective pressure. By consistently cultivating cells in environments that promote elevated metabolite production, random mutations that enhance yield can build up over time (Amin et al., 2025). This strategy is particularly beneficial for creating resilient, stable cultures that sustain high productivity over several generations (Chao, 2024).

1.8.7.4 *Integrative bioprocess optimization*

Ultimately, even the well-designed cultures require meticulously optimized growth conditions. Integrative bioprocess optimization merges knowledge from omics and flux analysis with intelligent bioreactor design, along with precise management of nutrients, elicitors, and physical factors such as pH and oxygen levels. This guarantees that enhancements in pathways result in reliable, scalable, and cost-effective production on an industrial scale (Ahlawat et al., 2024).

These strategies together help in converting intricate data into actionable solutions, changing unstable or low-yielding cultures into dependable green biofactories that produce valuable plant-derived compounds.

1.9 Conclusion

The biotechnological methods conducted *in vitro*, such as callus cultures, cell suspensions, organ cultures, hairy root cultures, and somatic embryogenesis, present significant alternatives for the sustainable production of crucial secondary metabolites. Despite the existence of challenges such as genetic instability, low yields, and issues with scalability, contemporary strategies like metabolic engineering, integration of omics, adaptive evolution, and intelligent bioprocess optimization offer effective solutions to address these obstacles. Collectively, these advancements underscore the potential of merging traditional plant tissue culture techniques with innovative tools to create stable, high-yield systems capable of consistently producing valuable phytochemicals for a variety of industrial uses.

References

Abdi, G., Tendulkar, R., Thatte, C., Mishra, S., Desai, B., Surve, S., &, ... Tarighat, M. A. (2024). Scaling up nature's chemistry: A guide to industrial production of valuable metabolites. In *Advances in metabolomics*. Singapore: Springer Nature Singapore, 331–375.

Adem, M., Sharma, L., Shekhawat, G. S., Šafranek, M., & Jásik, J. (2024). Auxin signaling transportation and regulation during adventitious root formation. *Current Plant Biology, 40*, 100385. https://doi.org/10.1016/j.cpb.2024.100385.

Ahlawat, Y. K., Srivastava, V., Samani, M., Harun, S., Rajuloo, V., & Chaudhary, D. (2024). Computational strategies for maximizing biomass and metabolite yields for bioproduction. In L. Kumar, N. Bharadvaja, R. Singh, & R. Anand (Eds.). *Medicinal and Aromatic Plants. Sustainable Landscape Planning and Natural Resources Management* (pp. 143–161). Cham: Springer. https://doi.org/10.1007/978-3-031-60117-0_11.

Al Aboud, N. M. (2024). Unlocking the genetic potential: Strategies for enhancing secondary metabolite biosynthesis in plants. *Journal of the Saudi Society of Agricultural Sciences, 23*(8), 542–554. https://doi.org/10.1016/j.jssas.2024.06.004.

Alam, R., Lee, J. H., Shin, D., Choi, I. K., Kim, S., Lim, H., & Kim, M. S. (2025). Recent advances in single-cell metabolomics using mass spectrometry: Emerging challenges and future perspectives. *Applied Spectroscopy Reviews, 60*(9–10), 852–888. https://doi.org/10.1080/05704928.2025.2483996.

Alami, M. M., Guo, S., Mei, Z., Yang, G., & Wang, X. (2024). Environmental factors on secondary metabolism in medicinal plants: Exploring accelerating factors. *Medicinal Plant Biology, 3*, e016. https://doi.org/10.48130/mpb-0024-0016.

Alum, E. U. (2025). Sustainable harvesting of medicinal plants: Balancing therapeutic benefits with environmental conservation. *Agroecology and Sustainable Food Systems, 49*(3), 380–385.

Alves, M. B., & Pereira, J. F. (2025). Microbial pigment extraction and purification techniques. In L. J. Rather, M. Shahid, & S. Jameel (Eds.). *Microbial colorants: chemistry, biosynthesis and applications* (pp. 177–192). Wiley. https://doi.org/10.1002/9781394287888.ch7.

Amin, M., Tahir, F., Akbar, I., Alsaigh, A. A., Alessa, A. H., Liu, C. G., ... Boopathy, R. (2025). Towards environmental sustainability: Employing adaptive laboratory evolution to develop elite algae strains for industrial and environmental applications. *Current Pollution Reports, 11*, 14.

Anuradha, M., Umalatha, & Balasubramanya, S. (2025). An overview of in vitro secondary metabolite production. In M. Anuradha, & S. Balasubramanya (Eds.). *In vitro production of plant secondary metabolites* (pp. 19–33). Singapore: Springer. https://doi.org/10.1007/978-981-97-8808-8_2.

Anuradha, M., Balasubramanya, S., Subbalakshmi, G., & Shilpa, P. (2025). Commercialization of in vitro secondary metabolite production: Challenges and opportunities. In M. Anuradha, & S. Balasubramanya (Eds.). *In Vitro Production of Plant Secondary Metabolites: Theory and Practice* (pp. 321–346). Singapore: Springer. https://doi.org/10.1007/978-981-97-8808-8_25.

Barbinta-Patrascu, M. E., Bita, B., & Negut, I. (2024). From nature to technology: Exploring the potential of plant-based materials and modified plants in biomimetics, bionics, and green innovations. *Biomimetics, 9*(7), 390.

Beal, M. A., Coughlan, M., Nunnikhoven, A., Corbane, R., Cummings-Lorbetskie, C., Rowan-Carroll, A., ... Meier, M. J. (2025). Impacts of inorganic arsenic exposure on genetic stability of human mesenchymal stromal cells. *Journal of Applied Toxicology, 45*(8), 1555–1571. https://doi.org/10.1002/jat.4785.

Bhaskar, K., Singh, S., Narkhede, S., & Kumar, M. (2024). *Innovative approaches to creating plant-derived pharmaceuticals*. Book Saga Publications.

Bhojiya, A. A., & Joshi, H. (2024). Crispr gene editing for secondary metabolite production: A review. In A. Kumar, S. Arora, S. Ogita, Y. Y. Yau, & K. Mukherjee (Eds.). *Gene Editing in Plants: CRISPR-Cas and Its Applications* (pp. 437–475). Singapore: Springer. https://doi.org/10.1007/978-981-99-8529-6_17.

Bordanaba Florit, G. (2024). Extracellular vesicles as surrogated biomarkers of prostate cancer metabolism: A metabolomics approach to study their role in prostate cancer progression (Doctoral thesis, Universidad del País Vasco/Euskal Herriko Unibertsitatea). ADDI Institutional Repository. http://hdl.handle.net/10810/69184.

Chao, K. (2024). Family farming in climate change: Strategies for resilient and sustainable food systems. *Heliyon, 10*(7), e28599. https://doi.org/10.1016/j.heliyon.2024.e28599.

Chaudhary, R., Nawaz, A., Fouillaud, M., Dufossé, L., Haq, I. U., & Mukhtar, H. (2024). Microbial cell factories: Biodiversity, pathway construction, robustness, and industrial applicability. *Microbiology Research, 15*(1), 247–272.

Chintada, V., & Golla, N. (2025). Exploring the therapeutic potential of bioactive compounds from plant sources. In J. Devi (Ed.). *Biotechnological intervention in production of bioactive compounds: Biosynthesis, characterization and applications* (pp. 229–247). Cham: Springer. https://doi.org/10.1007/978-3-031-76859-0_15.

Clagnan, E., Costanzo, M., Visca, A., Di Gregorio, L., Tabacchioni, S., Colantoni, E., ... Bevivino, A. (2024). Culturomics-and metagenomics-based insights into the soil microbiome preservation and application for sustainable agriculture. *Frontiers in Microbiology, 15* 1473666.

Dehghan, A. K., Zargar, M., Bamneshin, M., Mahmoudieh, M., Safaie, N., Yang, J. L., & Naghavi, M. R. (2025). The effects of Fusarium graminearum cell extracts and culture filtrates on the production of paclitaxel and 10-deacetylbaccatin III in suspension cell cultures of *Taxus baccata* L. *BMC Plant Biology, 25*(1), 229.

Devi, R. S., Dhal, P. B., & Tripathi, S. (2024). Managing browning in plant tissue culture: Strategies and control measures for successful in-vitro propagation and secondary metabolite production. *The Journal of Plant Science Research, 40*(2), 346–356.

Dey, A., Ghosh, S., Majhi, S., Hoque, A. N., & Chakrabarti, A. (2024). Systems biology approaches as a tool in understanding neurodevelopmental disorders: Some case studies. In S. Joshi, R. R. Ray, M. Nag, & D. Lahiri (Eds.). *Systems biology approaches: Prevention, diagnosis, and understanding mechanisms of complex diseases* (pp. 511–536). Singapore: Springer. https://doi.org/10.1007/978-981-99-9462-5_20.

Di Blasi, R., Gabrielli, J., Shabestary, K., Ziarti, I., Ellis, T., Kontoravdi, C., ... Ceroni, F. (2024). Understanding resource competition to achieve predictable synthetic gene expression in eukaryotes. *Nature Reviews Bioengineering, 2*(9), 721–732.

Elazzazy, A. M., Baeshen, M. N., Alasmi, K. M., Alqurashi, S. I., Desouky, S. E., & Khattab, S. M. (2025). Where biology meets engineering: Scaling up microbial nutraceuticals to bridge nutrition, therapeutics, and global impact. *Microorganisms, 13*(3), 566.

Fan, K. (2024). Metabolic engineering in plants: Advancing crop productivity and sustainability through precision pathway manipulation. *ChemRxiv*. https://doi.org/10.26434/chemrxiv-2024-qkkg6.

Gharat, S. A., Tamhane, V. A., Giri, A. P., & Aharoni, A. (2025). Navigating the challenges of engineering composite specialized metabolite pathways in plants. *The Plant Journal, 121*(6), e70100.

Hilal, B., Khan, M. M., & Fariduddin, Q. (2024). Recent advancements in deciphering the therapeutic properties of plant secondary metabolites: Phenolics, terpenes, and alkaloids. *Plant Physiology and Biochemistry, 211*, 108674. https://doi.org/10.1016/j.plaphy.2024.108674.

Hu, R., Teng, X., & Li, Y. (2024). Unleashing plant synthetic capacity: Navigating regulatory mechanisms for enhanced bioproduction and secondary metabolite discovery. *Current Opinion in Biotechnology, 88*, 103148.

Inckemann, R., Chotel, T., Brinkmann, C. K., Burgis, M., Andreas, L., Baumann, J., Sharma, P., Klose, M., Barret, J., Ries, F., Paczia, N., Glatter, T., Willmund, F., Mackinder, L. C. M., & Erb, T. J. (2024). Advancing chloroplast synthetic biology through high-throughput plastome engineering of *Chlamydomonas reinhardtii*. *bioRxiv*. https://doi.org/10.1101/2024.05.08.593163.

Indu, B. K., Hemamalini, Nagamani, J. E., & Anuradha, M. (2025). In vItro strategies for enhanced secondary metabolite synthesis. In M. Anuradha, & S. Balasubramanya (Eds.). *In vitro production of plant secondary metabolites: Theory and practice* (pp. 179–198). Singapore: Springer. https://doi.org/10.1007/978-981-97-8808-8_14.

Javed, A., Hashmi, M. S., Javaid, U., & Amjad, R. (2025). Classification and therapeutic applications of plant secondary metabolites. In U. E. Odoh, S. S. Gurav, & M. O. Chukwuma (Eds.). *Pharmacognosy and phytochemistry: Principles, techniques, and clinical applications* (pp. 189–206). Wiley. https://doi.org/10.1002/9781394203680.ch10.

Joshi, A., Kumar, A., & Kaushik, V. (2024). Functional genomics and network biology. In V. Singh, & A. Kumar (Eds.). *Advances in bioinformatics* (pp. 71–96). Singapore: Springer. https://doi.org/10.1007/978-981-99-8401-5_3.

Joshi, A., Kumar, A., Kaushik, V., Kumar, P., & Dubey, S. (2024). Role of bioinformatics in genome editing. In V. Singh, & A. Kumar (Eds.). *Advances in bioinformatics* (pp. 161–179). Singapore: Springer. https://doi.org/10.1007/978-981-99-8401-5_8.

Khan, U., & Gond, S. K. (2024). Detection of vinblastine and related bioactive compounds from culture extracts of endophytic fungi of *Catharanthus roseus*. *World Journal of Microbiology and Biotechnology, 40*(9), 278.

Khaskheli, A. J., Ali, M., Baloch, S., Mallano, I. A., Memon, Z. F., Imtiaz, F., ... Ali, F. (2024). Regulation of callus induction efficiency in banana using dichlorophenoxyacetic acid in combination with benzylaminopurine. *Journal of Horticulture and Agricultural Sciences, 1*(01), 28–38.

Khezri, M., Asghari-Zakaria, R., & Zare, N. (2024). Plant cell and tissue culture: Propagation, improvement, and conservation of medicinal plants. In N. Kumar (Ed.). *Biosynthesis of natural products in plants: Bioengineering in post-genomics era* (pp. 267–291). Singapore: Springer. https://doi.org/10.1007/978-981-97-2166-5_11.

Kohan-Baghkheirati, E., Vaezi-Kakhki, M. R., & Feizabadi, E. (2025). Callus growth and microcorm formation of *Crocus sativus* L. (Saffron) show medium dependency based on response surface methodology optimization. *Plant Cell, Tissue and Organ Culture (PCTOC), 161*(1), 1–10.

Kumar, A., Bhamini, K., Singh, R. K., & Singh, D. N. (2024). Production of plant-derived natural compounds through hairy root culture. In N. Kumar (Ed.). *Biosynthesis of natural products in plants: Bioengineering in post-genomics era* (pp. 213–239). Singapore: Springer. https://doi.org/10.1007/978-981-97-2166-5_9.

Kumar, K. S., Vanitha, K., Padvala, A., Tiwari, O. N., Ramesh, A., Varma, M. M., & Prakash, P. R. (2025). Bioengineering of oleaginous microorganisms for biosynthesis of fuels and value-added biomolecules. In D. Ghosh, & L. Bhatia (Eds.). *Oleaginous microbes for waste biomass valorization* (pp. 261–307). New York: Apple Academic Press.

Kumar, N., Bharat, J. S., Kumar, A., & Jaitak, V. (2024). Recent advancement in mass-based plant metabolomics: Techniques, tools, and analytical approaches. *Phytochemistry Reviews, 24*, 4013–4050. https://doi.org/10.1007/s11101-024-10028-y.

Kumar, P., & Singh, J. (2025). Perspective and challenges of synergistic removal of toxic contaminants from effluent using different treatment techniques. In S. Kandasamy, M. P. Shah, K. Subbiah, & N. Manickam (Eds.). *Microbial niche nexus sustaining environmental biological wastewater and water-energy-environment nexus* (pp. 419–451). Cham: Springer. https://doi.org/10.1007/978-3-031-62660-9_17.

Kumar, V., Kumar, P., Maity, S. K., Agrawal, D., Narisetty, V., Jacob, S., ... Vivekanand, V. (2024). Recent advances in bio-based production of top platform chemical, succinic acid: An alternative to conventional chemistry. *Biotechnology for Biofuels and Bioproducts, 17*(1), 72.

Lelis, T. M., Santos, I. R., Silva-Cardoso, I. M. A., de Souza, A. L. X., Gomes, A. C. M. M., Mehta, A., & Scherwinski-Pereira, J. E. (2025). Unraveling the occurrence of hyperhydricity in oil palm somatic embryos during somatic embryogenesis process. *Protoplasma, 262*(1), 191–206.

Li, F., Shahsavarani, M., Handy-Hart, C. J., Côté, A., Brasseur-Trottier, X., Montgomery, V., ... Dastmalchi, M. (2024). Characterization of a vacuolar importer of secologanin in *Catharanthus roseus*. *Communications Biology, 7*(1), 939.

Li, H., Chen, N., Zhang, H., & Xu, D. (2025). Multidimensional regulation of transcription factors: Decoding the comprehensive signals of plant secondary metabolism. *Frontiers in Plant Science, 16*, 1522278.

Li, W., Lin, S., Wang, R., Chen, C., Ni, L., Wang, W., & Liang, Z. (2025). Regulation of plant hormones on the secondary metabolism of medicinal plants. *Medicinal Plant Biology, 4*, e020. https://doi.org/10.48130/mpb-0025-0016.

Li, X. (2025). Development of a golden gate assembly-based toolbox and its application for engineering lactiplantibacillus plantarum as monoterpenoid producer. https://nbn-resolving.org/urn:nbn:de:bsz:14-qucosa2-972593

Liao, L., Xie, M., Zheng, X., Zhou, Z., Deng, Z., & Gao, J. (2025). Molecular insights fast-tracked: AI in biosynthetic pathway research. *Natural Product Reports, 42*, 911–936. https://doi.org/10.1039/D4NP00003J.

Liu, C., Ahmad, N., Tao, Y., Hussain, H., Chang, Y., Umar, A. W., & Liu, X. (2025). Reprogramming hairy root cultures: A synthetic biology framework for precision metabolite biosynthesis. *Plants, 14*(13), 1928.

Majumder, S., Igamberdiev, A. U., & Debnath, S. C. (2025). Somaclonal variation and clonal fidelity in commercial micropropagation: Challenges and perspectives. *Agronomy, 15*(6), 1489.

Manchanda, P., Sharma, D., Kaur, G., Kaur, H., & Vanshika (2025). Exploring the significance of somaclonal variations in horticultural crops. *Molecular Biotechnology, 67*, 2185–2203. https://doi.org/10.1007/s12033-024-01214-6.

Martínez-Esteso, M. J., Morante-Carriel, J., Samper-Herrero, A., Martínez-Márquez, A., Sellés-Marchart, S., Nájera, H., & Bru-Martínez, R. (2024). Proteomics: An essential tool to study plant-specialized metabolism. *Biomolecules, 14*(12), 1539.

de Mello, A. F. M., de Souza Vandenberghe, L. P., Herrmann, L. W., Letti, L. A. J., Burgos, W. J. M., Scapini, T., ... Soccol, C. R. (2024). Strategies and engineering aspects on the scale-up of bioreactors for different bioprocesses. *Systems Microbiology and Biomanufacturing, 4*(2), 365–385.

Miryousefzadeh, S., Hosseini, B., & Hemmati, S. (2025). Fungal elicitation and precursor feeding: A novel approach to boost alkaloid yield in *Catharanthus roseus* cell suspension cultures. https://doi.org/10.21203/rs.3.rs-6402841/v1.

Mishra, M. K., & Kumar, A. (2024). Fluxomics and metabolic flux analysis. In I. Mani, & V. Singh (Eds.). *Multi-omics analysis of the human microbiome: From technology to clinical applications* (pp. 171–180). Singapore: Springer. https://doi.org/10.1007/978-981-97-1844-3_8.

Muhammad, M., Basit, A., Ali, K., Ahmad, H., Li, W. J., Khan, A., & Mohamed, H. I. (2024). A review on endophytic fungi: A potent reservoir of bioactive metabolites with special emphasis on blight disease management. *Archives of Microbiology, 206*(3), 129.

Murthy, H. N., Joseph, K. S., Paek, K. Y., & Park, S. Y. (2024). Bioreactor configurations for adventitious root culture: Recent advances toward the commercial production of specialized metabolites. *Critical Reviews in Biotechnology, 44*(5), 837–859.

Najim, A. A., Radeef, A. Y., al-Doori, I., & Jabbar, Z. H. (2024). Immobilization: The promising technique to protect and increase the efficiency of microorganisms to remove contaminants. *Journal of Chemical Technology & Biotechnology, 99*(8), 1707–1733.

Nallasamy, L., Kumar, A., & Swaminathan, A. (2024). Omics and plant tissue culture for secondary metabolite production and synthetic biology. *International Journal of Life Sciences, 13*(2), 61–82. https://doi.org/10.5958/2319-1198.2024.00007.3.

Nasir, L., Yimam, T., & Guadie, D. (2024). Review on epigenetic modification mechanisms and their application for the genetic improvement of plants. *SINET: Ethiopian Journal of Science, 47*(3), 258–286.

Nimavat, N., & Parikh, P. (2024). Innovations in date palm (*Phoenix dactylifera* L.) micropropagation: Detailed review of in vitro culture methods and plant growth regulator applications. *Plant Cell, Tissue and Organ Culture (PCTOC, 159*(1), 6.

Nordine, A. (2025). Trends in plant tissue culture, production, and secondary metabolites enhancement of medicinal plants: A case study of thyme. *Planta, 261*(4), 84.

Palladino, F., Marcelino, P. R. F., Schlogl, A. E., José, Á. H. M., Rodrigues, R. D. C. L. B., Fabrino, D. L., ... Rosa, C. A. (2024). Bioreactors: Applications and innovations for a sustainable and healthy future—A critical review. *Applied Sciences, 14*(20), 9346.

Petrova, M., Miladinova-Georgieva, K., & Geneva, M. (2024). Influence of abiotic and biotic elicitors on organogenesis, biomass accumulation, and production of key secondary metabolites in *Asteraceae* plants. *International Journal of Molecular Sciences, 25*(8), 4197.

Qamar, F., Ashrafi, K., Singh, A., Dash, P. K., & Abdin, M. Z. (2024). Artemisinin production strategies for industrial scale: Current progress and future directions. *Industrial Crops and Products, 218*, 118937.

Qin, Y., Zhang, B., Wang, S., Guo, W., Zhang, Z., Qin, Y., ... Hu, G. (2025). Establishment of somatic embryogenesis regeneration system and transcriptome analysis of early somatic embryogenesis in litchi. *Horticultural Plant Journal, 11*(2), 535–547.

Rai, K. K., Mishra, J., Mishra, A. K., Rai, A. C., & Singh, P. K. (2025). Synthetic and systems biology strategies for enhanced plant secondary metabolite production. *Environmental Sustainability, 8*, 203–222. https://doi.org/10.1007/s42398-025-00357-9.

Sánchez-Romero, C. (2024). Factors affecting tissue culture-induced variations. In C. Sánchez-Romero (Ed.). *Somaclonal variation: Basic and practical aspects* (pp. 37–56). Cham: Springer. https://doi.org/10.1007/978-3-031-51626-9_3.

Satrio, R. D., Fendiyanto, M. H., & Miftahudin, M. (2024). Tools and techniques used at global scale through genomics, transcriptomics, proteomics, and metabolomics to investigate plant stress responses at the molecular level. In M. Shahid, & R. Gaur (Eds.). *Molecular dynamics of plant stress and its management* (pp. 555–607). Singapore: Springer. https://doi.org/10.1007/978-981-97-1699-9_25.

Seshadri, K., Abad, A. N., Nagasawa, K. K., Yost, K. M., Johnson, C. W., Dror, M. J., & Tang, Y. (2025). Synthetic biology in natural product biosynthesis. *Chemical Reviews, 125*(7), 3814–3931. https://doi.org/10.1021/acs.chemrev.4c00567.

Shanu, K., Choudhary, S., Kumari, S., Anu, K., & Devi, S. (2024). Downstream processing for bio-product recovery and purification. In S. Dhagat, S. E. Jujjavarapu, N. Sampath Kumar, & C. Mahapatra (Eds.). *Recent advances in bioprocess engineering and bioreactor design* (pp. 139–169). Singapore: Springer. https://doi.org/10.1007/978-981-97-1451-3_7.

Shilpa, P., Indu, B. K., Subbalakshmi, G., Balasubramanya, S., & Anuradha, M. (2025). Elicitation. In M. Anuradha, & S. Balasubramanya (Eds.). *In vitro production of plant secondary metabolites: Theory and practice* (pp. 257–272). Singapore: Springer. https://doi.org/10.1007/978-981-97-8808-8_20.

Shukla, R. (2024). A review of plant tissue and organ culture. *International Journal of Recent Research in Life Sciences, 11*(1), 1–5.

Sidik, N. J., Agha, H. M., Alkamil, A. A., Alsayadi, M. M. S., & Mohammed, A. A. (2024). A mini review of plant tissue culture: The role of media optimization, growth regulators in modern agriculture, callus induction and the applications. *AUIQ Complementary Biological System, 1*(2), 96–109.

Singh, D., Singh, N., Dwivedi, S., & Trivedi, P. K. (2025). Transcriptional regulation of secondary plant product biosynthesis: Insights into flavonoid, alkaloid, and terpenoid pathways. *Plant Cell, Tissue and Organ Culture (PCTOC, 160*(1), 6.

Singh, H. (2024). Bioengineering for production of biologically active compounds in plants. In N. Kumar (Ed.). *Biosynthesis of natural products in plants: Bioengineering in post-genomics era* (pp. 1–37). Singapore: Springer. https://doi.org/10.1007/978-981-97-2166-5_1.

Singh, P., Nayak, S. L., Sharkar, R., Sarkar, A., Dutta, S., & Paul, D. (2025). Application of modern techniques in cultivation of medicinal plants: A review. *Plant Archives, 25*(1), 1947–1962.

Sozoniuk, M., Petrova, M., Mishev, K., Miladinova-Georgieva, K., & Geneva, M. (2025). Identification and validation of reference genes with stable expression under elicitor treatments of the medicinal plant *Arnica montana* L. *BMC Plant Biology, 25*(1), 546.

Turan, E., Aslantaş, R., Bilgin, J., & Aksu, M. I. (2024). High-pressure homogenization of pomegranate juice: Impact on physicochemical, antioxidant, antimicrobial, and in vitro bioaccessibility properties. *Food Science & Nutrition, 12*(12), 10315–10329.

Upadhyay, R., Saini, R., Shukla, P. K., & Tiwari, K. N. (2025). Role of secondary metabolites in plant defense mechanisms: A molecular and biotechnological insights. *Phytochemistry Reviews, 24*(1), 953–983.

Varadharajan, V., Rajendran, R., Muthuramalingam, P., Runthala, A., Madhesh, V., Swaminathan, G., ... Ramesh, M. (2025). Multi-omics approaches against abiotic and biotic stress—A review. *Plants, 14*(6), 865.

Villaseñor-Altamirano, A. B., Balderas-Martínez, Y. I., & Medina-Rivera, A. (2024). Review of gene expression using microarray and RNA-seq. In D. F. Dluzen, & M. H. M. Schmidt (Eds.). *Rigor and reproducibility in genetics and genomics* (pp. 159–187). Academic Press. https://doi.org/10.1016/B978-0-12-817218-6.00008-5.

Vitorino, R. (2024). Transforming clinical research: The power of high-throughput omics integration. *Proteomes, 12*(3), 25.

Wadhwa, K., Kapoor, N., Kaur, H., Abu-Seer, E. A., Tariq, M., Siddiqui, S., ... Alghamdi, S. (2024). A comprehensive review of the diversity of fungal secondary metabolites and their emerging applications in healthcare and environment. *Mycobiology, 52*(6), 335–387.

Wang, P., Si, H., Li, C., Xu, Z., Guo, H., Jin, S., & Cheng, H. (2025). Plant genetic transformation: Achievements, current status and future prospects. *Plant Biotechnology Journal, 23*(6), 2034–2058.

Wang, S., Zhao, F., Yang, M., Lin, Y., & Han, S. (2024). Metabolic engineering of *Saccharomyces cerevisiae* for the synthesis of valuable chemicals. *Critical Reviews in Biotechnology, 44*(2), 163–190.

Xu, S., & Gaquerel, E. (2025). Evolution of plant specialized metabolites: Beyond ecological drivers. *Trends in Plant Science, 30*(8), 826–836. https://doi.org/10.1016/j.tplants.2025.02.010.

Yin, J. Y., Lai, M., Yu, X. Y., Su, D. D., Xiong, X. Y., & Li, Y. L. (2025). Comprehensive strategies for paclitaxel production: Insights from plant cell culture, endophytic microorganisms, and synthetic biology. *Horticulture Research, 12*(3), uhae346.

Yoo, I., Ahn, I., Lee, J., & Lee, N. (2024). Extracellular flux assay (Seahorse assay): Diverse applications in metabolic research across biological disciplines. *Molecules and Cells, 47*(8), 100095.

Zafar, S., Fatima, S. I., Schmitz, M., & Zerr, I. (2024). Current technologies unraveling the significance of post-translational modifications (PTMs) as crucial players in neurodegeneration. *Biomolecules, 14*(1), 118.

Zhang, X. L., Liu, Z. H., Li, B. Z., & Yuan, Y. J. (2024). Tapping into the natural aromatic potential of microbial lignin valorization towards aromatic fine chemicals. *Green Chemistry, 26*(23), 11378–11405.

Zheng, A., & Liang, H. (2024). Immobilization of bgps-expressing *Escherichia coli* cells coated with ZIF-67 for the production of rare ginsenoside F1. *Biochemical Engineering Journal, 210*, 109432.

Zheng, T., Wei, L., Xiang, J., Wu, J., & Cheng, J. (2025). HMGR modulates strawberry fruit coloration and aroma through regulating terpenoid and anthocyanin pathways. *Foods, 14*(7), 1199.

Zhu, Y., Zhu, X., Wen, Y., Wang, L., Wang, Y., Liao, C., ... Wang, L. (2024). Plant hairy roots: Induction, applications, limitations and prospects. *Industrial Crops and Products, 219*, 119104.

CHAPTER

2

Micropropagation of medicinal plants: a promising approach to produce and conserve phytomedicines

Erna Karalija, Saida Ibragic, and Sabina Dahija
Faculty of Science, University of Sarajevo, Sarajevo, Bosnia and Herzegovina

2.1 Bioreactors and process engineering

Plant micropropagation and *in vitro* production of phytochemicals have increasingly benefited from the temporary immersion system and bioreactor technology. Bioreactors provide a controlled environment for cultivating plant cells, tissues, and organs at scale, enabling improved nutrient delivery, gas exchange, and process control compared to traditional flask or agar culture. However, successful scale-up requires careful consideration of bioreactor design and engineering principles to accommodate the unique shear sensitivity and metabolic needs of plant systems. This chapter reviews the main types of bioreactors used in plant micropropagation and secondary metabolite production, key engineering factors in scaling up these systems, and real-world applications with performance data for medicinal plant cultures.

2.1.1 Types of bioreactors for plant micropropagation and metabolite production

Different types of bioreactors can be found in use for the production of metabolites using micropropagated plants, we will mention some of them.

Stirred-tank bioreactors (STR): The conventional STR (mechanically agitated vessel) has an impeller (e.g., marine blade or Rushton turbine) to mix the liquid medium and disperse gas bubbles (Nienow, 2014). STRs were originally designed for microbes and offer efficient mixing, high oxygen transfer (kLa), and tight control of parameters (pH, temperature, dissolved oxygen). These benefits can be useful for plant cell cultures, and indeed STRs have been employed for certain plant organs (e.g., *Lilium* bulblets, *Stevia* shoot primordia). However, STRs impart significant hydrodynamic shear stress and turbulence near the impeller. Fragile plant cells (which are larger and have thicker cell walls than microbes) often suffer damage in high-shear STR environments. Problems like cell aggregation, foam

Medicinal Plant Biotechnology for Ameliorated Secondary Metabolite Production
DOI: https://doi.org/10.1016/B978-0-443-43944-5.00001-3

generation, and difficulty in harvest are also reported. Strategies to mitigate shear in STRs include using low agitation speeds, specialized impeller designs, or inserts (e.g., looping silicone tubing) to dampen fluid shear. Even so, STRs are less common for micropropagation unless the cultured material is robust. For example, slow-moving paddle impellers have been tested for somatic embryo cultures, but these did not always improve plantlet quality postembryo development. While STRs can provide high mass transfer and precise control, their power input must be limited to protect shear-sensitive plant tissues (often <0.1 W/L specific power input), and alternative designs are often preferred for delicate cultures (Wysokinska & Chmiel, 1997).

Bubble column bioreactors (BCB): Bubble columns are simple pneumatically agitated reactors consisting of a tall cylinder with an air sparger at the bottom. Rising air bubbles alone circulate and oxygenate the liquid, with no mechanical agitator required. This design has low maintenance and operational costs and eliminates moving parts, reducing contamination risk. Bubble columns provide reasonably good heat and mass transfer for their simplicity. They have been used for plant cell suspensions and even organ cultures that can tolerate moderate aeration. The main drawbacks are poor mixing in highly viscous plant cell broths and foam accumulation at high aeration rates. Fluid circulation can be slower than in STRs, leading to longer mixing times and gradients in large columns. Nonetheless, bubble columns are effective for certain applications; for instance, *Taxus* cell cultures for paclitaxel production have been grown in large bubble columns successfully in industrial settings (with antifoam strategies) (Manz et al., 2022; Tomilova et al., 2023). Simplified columns are also sometimes combined with two-phase systems (e.g., adding an in situ organic solvent layer to capture secreted products) to enhance the yield of lipophilic metabolites such as shikonin (Sim & Chang, 1993).

Airlift bioreactors: Airlift bioreactors are a variation of bubble columns that include an internal draft tube or an external loop to direct circulation (Sim & Chang, 1993). Air is sparged at the bottom; the draft tube creates defined up flow (riser) and down flow (downcomer) channels, improving liquid circulation and oxygen distribution compared to a simple bubble column. Airlifts achieve mixing without mechanical impellers, so they maintain low-shear conditions ideal for plant cells. Advantages of airlifts include easy sterilization and scale-up, since only gas flow needs adjustment, and reduced shear damage to cells or somatic embryos (Manickavasagam et al., 2025). Airlifts have been widely applied to suspension cultures of plant cells to produce secondary metabolites. For example, *Panax ginseng* adventitious roots cultured in an airlift demonstrated high growth and saponin accumulation while avoiding mechanical agitation damage. One limitation of airlifts is that at very high aeration rates, the culture can foam, and oxygen transfer may plateau due to limited gas-liquid contact time. Even so, airlift designs (internal or external loop) are among the most popular for large-scale plant cell culture (including the production of *Taxus* cell biomass for paclitaxel) because they combine gentle mixing with adequate oxygenation (Manz et al., 2022). Industrial *Taxus* processes have successfully scaled airlift-type reactors up to ~75,000 L, demonstrating the viability of this design for mass production of plant-derived pharmaceuticals.

Balloon-type bubble bioreactors: Balloon-type bioreactors (also called bulb-type bubble bioreactors) are a modified airlift/bubble column design with a distinctive balloon-shaped vessel and a concentric draft tube (Cui et al., 2014). Air is introduced through a circular sparger at the base, lifting the culture up through the draft tube (riser) and allowing it to circulate down the wider periphery. These reactors have a wide mouth at the top, making it convenient to introduce or harvest large tissue explants or aggregates. Ports for sensors (pH, dissolved oxygen) and gas mixing (to add O_2, CO_2, ethylene, etc., to inlet air) are typically included. Balloon-type reactors have been extensively tailored for organogenic cultures—bulbs, microtubers, corms, rhizomes, and somatic embryos—that require low shear and sometimes partial support. Various cultivation modes can be employed: in complete immersion, explants are fully submerged; in the raft method, explants sit on a mesh raft at the medium surface; and in ebb-and-flood, medium is periodically pumped in and out of the vessel to alternately immerse and aerate the culture. These modes help ensure uniform nutrient exposure while preventing asphyxia or hyperhydricity. Balloon bioreactors have achieved notable successes in micropropagation: For example, using a 5 L balloon-type reactor, *Allium sativum* (garlic) shoot-tip cultures produced ~4,000–5,100 virus-free bulblets (0.02–2 g each) in 12 weeks under different modes (immersion vs. raft). Similarly, *Eleutherococcus* somatic embryos yielded ~128.8 g of mature embryos in a 3 L balloon reactor (2 L medium) within 4 weeks, and orchid protocorm-like bodies of *Oncidium* were efficiently multiplied in balloon reactors using immersion and ebb-and-flood techniques. These examples underscore that balloon-type bioreactors can greatly enhance propagation rates of high-value medicinal plants by combining gentle handling of propagules with improved nutrient delivery.

Temporary immersion bioreactors (TIBs): TIBs (also known as temporary immersion systems, TIS) are specialized bioreactors that periodically immerse the plant tissue in liquid medium and then drain the medium, allowing the culture to reside in a humid, aerated headspace between immersions. A typical TIB design consists of two vessels or compartments: one contains the plant explants, the other serves as a medium reservoir. By applying air pressure or using a pump with solenoid valves, medium is transferred to the explant vessel to submerge the tissues for a programmed duration (e.g., a few minutes) and frequency (e.g., every 3 hours), after which the medium drains back to the reservoir. This intermittent immersion strategy combines the benefits of liquid culture (efficient nutrient uptake) with those of agar/gel culture (exposure to air, preventing continuous flooding stress). TIBs are widely used for shoot proliferation, somatic embryogenesis, and other micropropagation processes—they avoid problems like asphyxiation, hyperhydricity, and accumulation of ethylene that occur in continuously submerged cultures. Studies have shown that plantlets grown in TIBs often have better vigor and acclimatization success than those from static liquid culture. Many TIB configurations exist (Georgiev et al., 2014), including the RITA (a well-known automated temporary immersion vessel), twin-flask systems (e.g., BIT), rocker-based TIS, and commercial platforms like SETIS (Vendrame et al., 2023) and Plantform (Almusawi and Sayegh, 2017). Regardless of design, all provide periodic nutrient contact and ample gas exchange. For example, Etienne and Teisson, 1999 demonstrated improved banana shoot multiplication in one of the first RITA systems. TIBs also minimize hyperhydricity by allowing a gaseous interval, and enriching the headspace with CO_2 or providing light can further promote growth and reduce the need for exogenous sugar (Lyam et al., 2012). A practical advantage is that multiple TIB units can be operated in parallel (each vessel typically 0.5–10 L), which is convenient for commercial labs to scale out production of disease-free plantlets. *Celosia argentea* is one recent example where a TIB regime produced ~63 shoots per explant (6× the yield of semisolid culture) with enhanced leaf biomass and nutrient content. Temporary immersion bioreactors have become a cornerstone of large-scale micropropagation for many medicinal and horticultural species due to their balance of high growth rates and plant quality.

Nutrient mist bioreactors: Nutrient mist or spray bioreactors are a form of gas-phase reactor where the culture is not submerged in liquid, but instead a fine mist of nutrient solution is periodically sprayed or atomized onto the plant tissues or roots (Liu and Guo, 1999; Liu Wang, et al., 1999). The explants are typically supported on a mesh or other support in the growth chamber. A pump or air-blast nozzle generates an aerosol of medium, producing droplets that coat the tissues and then fall into a reservoir to be recycled. In essence, mist bioreactors provide the plants with a humid, aerated environment and frequent feedings of nutrient solution. This design greatly improves gas exchange and oxygen availability to the cultured material, which is beneficial for organs like roots or tubers that are prone to oxygen limitation in liquid culture. Correll et al. (2000) showed that a mist bioreactor could produce well-rooted *Dianthus* plantlets with superior physiological quality. Likewise, carrot somatic embryos were taken from early embryos to fully regenerated plants within a single mist cultivation system. Nutrient mist systems have also been used to grow potato microtubers and *Echinacea* shoots, often yielding faster growth than immersion systems. The challenge with mist bioreactors is scale-up—delivering an even mist across a larger volume and maintaining sterility of aerosolised nutrients can be complex, and thus most mist systems are limited to bench or small pilot scales. Nonetheless, for certain applications (like hairy root cultures requiring high oxygen), mist bioreactors are extremely effective. A modified mist-aeroponic reactor was shown to double the growth of *Panax notoginseng* roots and enhance ginsenoside yield compared to liquid culture. The excellent oxygenation in mist culture can also support higher densities of roots or tissues, albeit at the cost of more sophisticated engineering.

Wave and rocking bioreactors: "Wave" bioreactors are single-use (disposable) systems where cells grow in a flexible plastic bag that is rocked on a platform to create wave-like agitation in the liquid. The wave bioreactor (also called Wave & Undertow, WU) was pioneered for animal cell culture and adapted to plant cells to avoid shear damage from impellers (Terrier et al., 2007). Rocking motions induce mixing and aeration as the waves traveling through the medium continuously circulate the culture and renew the gas-liquid interface. These systems are presterilized plastic bags (often with built-in oxygenation via an overlay or sparger) and thus eliminate cleaning or sterilization steps between batches. For plant cell suspensions, wave bioreactors have successfully been used up to at least 100 L scale at the industrial pilot level. The gentle wave agitation provides a low shear environment and has been shown to support plant cell growth rates comparable to stirred tanks. For example, tobacco and soybean cells producing secondary metabolites were cultured in a wave bioreactor at 100 L working volume, achieving similar biomass doubling times (~2–3 days) as in a 10 L stirred tank. The disposable nature of wave bags is attractive for multiproduct facilities (minimizing cross-contamination) and for processes requiring illumination (the translucent film allows light to enter, useful for photomixotrophic cultures like plant cells that produce carotenoids or anthocyanins). Wave bioreactors are an emerging option in plant cell process engineering when moderate oxygen transfer is sufficient and ease of use is a priority.

Other specialized designs, such as trickle-bed bioreactors (nutrient sprinkling systems) and rotating drum reactors, also exist but are less commonly used in medicinal plant micropropagation.

2.1.2 Engineering principles and scale-up strategies

Expanding plant tissue culture from lab scale (e.g., test tubes or shake flasks) to bioreactors demands careful engineering to maintain a suitable microenvironment. Key considerations include oxygen transfer, mixing efficiency, shear stress, and bioreactor hydrodynamics. Unlike microbes or animal cells, plant cells often grow as dense aggregates and have slower metabolisms, so requirements and tolerances differ. Here we discuss critical engineering parameters and approaches for successful scale-up:

Mass transfer (k_La and oxygen supply): Oxygen is often the limiting nutrient in high-density plant cultures, especially for fast-respiring cells or organized tissues like roots. The volumetric oxygen transfer coefficient (k_La) is a measure of the capacity of a bioreactor to deliver oxygen from the gas phase to the liquid culture. Plant cell suspensions typically require lower k_La values than microbial cultures, but a minimum dissolved O_2 level (often above 20%–30% air saturation) must be maintained to avoid growth inhibition or metabolite changes (Verdú-Navarro et al., 2023). In practice, shake flasks have k_La on the order of ~5 h^{-1}, while well-aerated stirred bioreactors can achieve k_La of 15–30 h^{-1} for plant cell broths under moderate conditions. For instance, in a 10 L stirred tank, increasing aeration from 0.04 to 0.5 vvm (volume air per volume per minute) raised k_La from ~3 h^{-1} to ~29 h^{-1}. Such adjustments ensure the oxygen uptake rate (OUR) of the culture is met. OUR of plant cells can vary with growth phase; in *P. ginseng* root cultures, the specific OUR (qO_2) was high during early exponential growth and then dropped to near zero as sugar became depleted and growth ceased. Continuous monitoring of dissolved O_2 and CO_2 in large bioreactors is therefore important. Scale-up strategies often try to match k_La between scales (scale-up by constant k_La) so that oxygen supply scales with cell population. However, maintaining the same k_La in a larger vessel might require impractically high airflow or agitation if done linearly. Alternatives include enriched oxygen (using O_2-enriched air to boost transfer) or employing microspargers (to create finer bubbles and higher surface area). In one case, a porous sparger with 15 μm pores was used in a 3 L plant cell bioreactor to generate micro-bubbles, achieving high O_2 transfer at low gas flow rates without injuring cells (Manickavasagam et al., 2025). Computational fluid dynamics (CFD) simulations have also been used to predict O_2 concentration fields and optimize sparger design and placement in plant cell bioreactors. Ultimately, sufficient oxygenation must be achieved without overaerating (which can cause foaming or excessive shear from bubble bursting).

Mixing time and homogeneity: Mixing time is the time required to achieve a specified degree of homogeneity after a perturbation (such as nutrient addition) in the bioreactor. Good mixing ensures uniform distribution of nutrients, hormones, and oxygen and prevents local build-up of metabolites or waste (e.g., ethanol from stressed plant cells). In scaling up, mixing time often increases due to larger volumes and lower power densities. Large plant bioreactors might have mixing times of tens of seconds up to a few minutes, which could allow temporary gradients. Keeping mixing times short (e.g., <30 s) is ideal to avoid gradients in pH or sugar that could affect cell differentiation. Engineering solutions include using multiple impellers in tall reactors, increasing impeller diameter relative to tank size, or using circulation pumps in external-loop designs. In pneumatically agitated systems, one must ensure circulation loops (as in airlifts) are effective—for example, increasing the air throughput in an airlift can improve liquid circulation velocity (to a point). Wave bioreactors rely on rocking speed and angle to control mixing intensity; too slow rocking can result in stratification, whereas too vigorous rocking can cause splashing and foaming. Empirical optimization or CFD analysis can determine appropriate mixing conditions. For instance, CFD studies on rocking bioreactors have indicated optimal rocking speeds that balance mixing time (often 5–30 s in small wave bags) against shear level (Wierzchowski & Pilarek, 2024). Ensuring all regions (especially near the vessel bottom or corners) receive flow is important, as plant cell aggregates may settle if mixing is inadequate. If settlement occurs, periodic resuspension (either by momentary high agitation or by flow reversal in airlifts) may be implemented. In summary, scale-up must preserve adequate mixing to keep the culture environment uniform.

Shear sensitivity and shear mitigation: Plant cells and tissues are notoriously sensitive to hydrodynamic shear stress. Mechanical agitation or large bubbles can shear off trichomes, break somatic embryos, or cause cell death accompanied by the release of phenolics and oxidative enzymes (Murthy et al., 2023). In one study, shear from traditional turbine impellers induced oxidative burst in plant cell suspensions, leading to growth inhibition (Huang, 2009). As a rule, bioreactor designs for plant cultures prioritize low shear: absence of rotating impellers (as in airlift, bubble, mist, TIB, wave systems) or use of low-shear impellers (e.g., pitched-blade or paddle impellers at low RPM). CFD models have been very useful in evaluating shear fields in bioreactors with plant cells. For example, a recent CFD study compared impeller designs and found that a custom "setric" impeller created a much lower shear environment than Rushton turbines, while still providing good mixing in a 3 L plant cell reactor. The setric impeller's inclined blades reduced turbulence intensity around the impeller, which was favorable for fragile *Viola* cells (Manickavasagam et al., 2025). This illustrates a key engineering trade-off: often, a slightly lower oxygen transfer or mixing rate is acceptable to markedly reduce shear stress on plant cells. Typical maximum shear rates tolerated by plant cell suspensions are on the order of 10–100 s^{-1} (much lower than microbial cultures). Design heuristics include limiting impeller tip speed (e.g., <1 m/s for plant cells), using gentle sparging (small bubbles or intermittent aeration), and avoiding abrupt contractions in flow paths. Some bioreactors include shear protectants like surfactants or polymers (e.g., Pluronic F-68) to reduce bubble surface tension and cell adhesion, thus minimizing cell damage at the gas-liquid interface. Encapsulation of cells in alginate beads or fibrous bed bioreactors has also been explored to physically shield cells from shear, but these are niche approaches. A practical example of shear mitigation is the Slug bubble bioreactor (SB): it generates large "Taylor bubbles" that rise in a controlled fashion, gently circulating the culture with minimal small-scale turbulence (Terrier et al., 2007). The SB bioreactor was able to scale tobacco cell cultures to 70 L with low shear, by trading off some oxygen efficiency (due to large bubbles) for much gentler fluid motion. Engineering for low shear often defines the feasible operating window (agitation and aeration rates) in plant bioreactors, and modern tools like CFD help identify designs that maximize mixing and oxygen transfer under those constraints.

Power input per volume (P/V): P/V is a scale-up parameter representing how much energy is dissipated in the fluid per unit volume, often related to mixing intensity. In the scale-up of microbial fermentors, maintaining constant P/V is a common strategy to preserve mixing and oxygenation conditions. For plant cell cultures, target P/V values are much lower. Microbial bioreactors may run at 1000–5000 W/m^3 (1–5 W/L) or more, whereas plant cell bioreactors often operate at 10–200 W/m^3 (0.01–0.2 W/L) (Wierzchowski & Pilarek, 2024). Exceeding ~0.1 W/L can lead to shear damage in many plant systems (though exact limits depend on species and culture morphology). During scale-up, keeping the same P/V as a small scale could result in too high shear in a large vessel, so instead, scale-up is sometimes done at constant tip speed or constant shear criteria. Still, monitoring P/V is useful: it can be calculated from agitation power or estimated via CFD. As an example, in a disposable rocking bioreactor, measured P/V ranged from ~10 to 700 W/m^3 depending on rocking speed and fill volume. The upper end of that range (0.7 W/L) would be excessive for most plant cells, whereas the lower end (0.01–0.1 W/L) is in a safer zone. Engineers thus adjust agitation speed, impeller size, or aeration to achieve an effective P/V that supports mixing and oxygenation without overwhelming the cells. In practice, volumetric power input is often kept just high enough to maintain suspension of cells/aggregates. In propagating organized tissues (like shoots or embryos), even lower P/V is used since convective circulation (via bubbles or gentle rocking) is enough to renew the medium around tissues.

Heat transfer and temperature control: Plant cell cultures are typically run at moderate temperatures (20°C–27°C for most tropical species, lower for temperate species), and they are less exothermic than microbial cultures. However, large bioreactors still require cooling systems to remove metabolic and agitation heat. Jacketed vessels or internal coils are standard in STRs and airlifts; disposable bag reactors rely on temperature-controlled enclosures or recirculating air. Because plant cells can be temperature-sensitive (secondary metabolite profiles may shift with temperature), precise control (±0.5°C) is desirable. Some processes intentionally apply temperature shifts to trigger metabolite production (e.g., a shift from 25°C to 18°C to stimulate paclitaxel accumulation in *Taxus* cultures) (Tomilova et al., 2023). Thus bioreactors must accommodate both maintaining a stable temperature and possibly executing programmed changes. Computational models of bioreactors include heat transfer to ensure uniform culture temperature, and scale-up requires ensuring surface area to volume is sufficient for cooling—for instance, large industrial *Taxus* reactors use external heat exchangers because the culture broth is too large for wall cooling alone.

2.1.3 Commercial applications and case studies

Bioreactor technology for medicinal plant cultivation has transitioned from lab demonstrations to commercial practice in several notable cases. Key examples include the production of high-value secondary metabolites (paclitaxel, ginsenosides, alkaloids) and mass propagation of elite plant lines. Below, we highlight some real-world applications, along with performance metrics comparing reactor types or process conditions. Table 2.1 summarizes selected data on yields of important phytochemicals achieved in different bioreactor systems.

TABLE 2.1 Examples of elicitation in medicinal plant cultures, showing the type of elicitor used, the dosage and timing regimen, and the resulting fold-increase in target metabolite yield.

Plant system (culture type)	Elicitor (type)	Elicitor regime (dose and duration)	Metabolite yield increase (vs. control)
Panax ginseng (cell suspension)	Methyl jasmonate (Abiotic)	200 μM MeJA, added at inoculation	Total ginsenosides increased ~28-fold (Xu et al., 2023)
Taxus sp. (cell suspension)	Methyl jasmonate (Abiotic)	100 μM MeJA, after ~2 weeks growth	Paclitaxel titer ~120× higher (0.4→48 mg/L) (Mirjalili and Linden, 1996)
Catharanthus roseus (cell suspension)	*Phytophthora* elicitor (Biotic)	Fungal protein (5% v/v), 48 h exposure	Ajmalicine ~7.9×; Catharanthine ~4× (Srinivasan et al., 2022; Singh et al., 2022)
Withania somnifera (hairy roots)	Methyl jasmonate (Abiotic)	100 μM MeJA, 48 h treatment	Withanolide A ~6.7×; Withaferin A ~4.6× (Mahendran and Vimolmangkang, 2024)
Hypericum perforatum (hairy roots)	Blue light (Physical)	High-intensity blue light, 7 days	Hypericin accumulation ~10× increase (Al-Khayri et al., 2024)
Hypericum perforatum (cell suspension)	ZnO nanoparticles (Abiotic)	100 ppb ZnO NP, 72 h treatment	Hyperforin ~13×; Hypericin ~3× (Pradeep et al., 2024)

2.1.3.1 *Paclitaxel from* Taxus *cell suspension*

The anticancer drug paclitaxel (Taxol) is a flagship example of a plant secondary metabolite produced at an industrial scale via bioreactors. *Taxus* spp. (yew tree) Cell suspensions, when elicited, can accumulate paclitaxel and related taxanes in significant quantities. Phyton Biotech (USA/Germany) developed a process using large stirred tank or airlift bioreactors up to 75 m^3 (75,000 L), successfully producing paclitaxel for the pharmaceutical market (Manz et al., 2022). In research settings, various strategies have been used to boost paclitaxel yields: for instance, a temperature shift from 24°C to 18°C in *Taxus chinensis* cell culture increased paclitaxel concentration from ~50 to 137.5 mg/L (Tomilova et al., 2023). Likewise, combining Methyl jasmonate (MeJA) or coronatine as elicitors with in situ adsorption resins has achieved paclitaxel titers of 20–70 mg/L in lab reactors (Choi et al., 2003; Li et al., 2009). A coculture of *Taxus* cells with *Penicillium* fungus was reported to yield ~25 mg/L paclitaxel in 15 days. In commercial operations, the specific yield per dry cell weight is often low (on the order of 0.01%–0.1% of DW), but large bioreactor volumes compensate. Continuous improvement in bioreactor aeration and feeding strategies (e.g., perfusion to remove inhibitory byproducts) has allowed sustained paclitaxel production over long culture cycles. The success of paclitaxel production in bioreactors underscores the importance of scale-up: the process went from shake flasks to seed reactors to multiple 10,000+ L production reactors, maintaining cell growth and product formation through careful control of shear (low-speed mixing), oxygen (airlift aeration), and feeding (sugar feeding to prolong production phase) (Taura et al., 2014).

Ginsenosides from *Panax ginseng* roots: *P. ginseng* and related species produce ginsenosides (triterpenoid saponins) valued for their medicinal properties. Traditional cultivation is slow (multiyear field growth), so *in vitro* methods are attractive to generate biomass and specific saponins. *P. ginseng* adventitious roots have been cultured in 3–5 L balloon-type and airlift bioreactors with notable success (Saadah et al., 2019). For example, Wang et al. (2016) grew ginseng roots in a 5 L airlift; after 30–40 days, the roots reached peak biomass and total saponin content. The maximum saponin yield occurred around day 30, slightly before peak biomass, suggesting a production phase timed with growth dynamics. Although the exact titer was not stated in the abstract, subsequent analysis showed ginseng adventitious roots can accumulate total ginsenosides up to ~25 mg per g dry weight under optimized conditions (Wang et al., 2013). This is on par with field-grown roots (which have ~3%–5% ginsenosides on dry weight). Bioreactor-grown roots thus can achieve "wild-type" quality in a fraction of the time. Notably, adventitious root cultures contain no microbial genes (unlike hairy roots), making them acceptable for certain markets. Strategies like nutrient feeding and elicitation (e.g., with jasmonates) have roughly doubled ginsenoside content in bioreactors in some studies (Saadah et al., 2019). For instance, Jeong et al. (2009) reported that optimizing inoculum density and aeration in a 5 L bubble-bag bioreactor increased root biomass and ginsenoside production by ~1.5×. *Panax* root cultures are somewhat shear-tolerant (they form intertwined root masses) but still benefit from gentle aeration; balloon or airlift reactors are commonly used to keep roots suspended and oxygenated without mechanical stirring. Some bioreactor platforms (e.g., large-scale mist bioreactors) have been explored to further enhance oxygen delivery to *Panax* roots, yielding faster growth and potentially higher saponin levels. Although not yet at the scale of paclitaxel, commercial production of ginseng metabolites in bioreactors is being pursued, especially in East Asia, to supplement or replace field cultivation.

Ajmalicine and alkaloids from *Catharanthus roseus*: *Catharanthus roseus* (Madagascar periwinkle) produces terpenoid indole alkaloids, including the anticancer drugs vinblastine and vincristine (dimeric alkaloids) and simpler monomeric alkaloids like ajmalicine (antihypertensive). Producing vinblastine/vincristine *in vitro* has been challenging due to low yields and complex biosynthesis requiring tissue differentiation. However, *C. roseus* cell cultures can accumulate substantial amounts of monomeric alkaloids. Elicitation and process innovations have led to very high ajmalicine yields in bioreactor cultures. A striking result was obtained by Almagro et al. (2011), using a 3 L stirred bioreactor; they combined MeJA elicitor with cyclodextrins (to capture and protect secreted alkaloids) and a brief UV exposure, achieving over 1 g/L of ajmalicine in the culture medium (1040 ± 26 mg/L). This is an extraordinarily high titer for a plant secondary metabolite, and it demonstrates the power of coupling chemical elicitors with in situ product removal (cyclodextrin forms inclusion complexes with ajmalicine, preventing feedback inhibition and degradation). By contrast, unelicited *Catharanthus* cultures might yield only a few mg/L of ajmalicine. In the same or similar systems, significant amounts of catharanthine and vindoline (precursor alkaloids to vinblastine) have been produced, though fully assembling vinblastine still requires coupling steps. Bioreactors used for *C. roseus* include STRs (with careful speed control to limit shear) and airlift reactors. A 500 L airlift bioreactor was used in India for *Catharanthus* cell culture, reportedly to produce precursors for semi-synthesis of anticancer drugs (details proprietary). While such a process might be complex (involving multiple elicitors and downstream extraction of cyclodextrin complexes), it offers a template for high-yield production of plant alkaloids in bioreactors.

Shikonin from *Lithospermum* cell/hairy root culture: Shikonin, a naphthoquinone from *Lithospermum erythrorhizon* (gromwell), was one of the first plant metabolites produced commercially via cell culture. In the 1980s, Japanese researchers developed a process with large (20,000 L) bubble column bioreactors for shikonin, which was used as a natural dye and pharmaceutical intermediate. A notable enhancement was the use of a two-phase bioreactor: an organic solvent (like n-hexadecane) was overlaid or circulated to extract shikonin as it was produced, preventing feedback inhibition and toxicity to the cells. In a lab-scale demonstration of this concept, Sim and Chang (1993) grew *Lithospermum* hairy roots in a 2 L bubble column with an in situ n-hexadecane layer, achieving 572.6 mg/L shikonin in 54 days (Sim & Chang, 1993). This corresponds to a production rate of ~10.6 mg/L·day. The final dry biomass was 15.6 g/L, meaning the roots contained a substantial amount of shikonin (some secreted into the solvent, some retained in roots). This data point is an example of a high-value red pigment produced by engineered culture (Table 2.1). Since shikonin is secreted by the roots, using an extractant phase is crucial; without it, yields were about 3× lower due to product feedback. The success of shikonin culture was significant historically—it proved plant cell cultures could be scaled and economically viable. Although *Lithospermum* cell culture for shikonin has waned (the dye market changed), the techniques (e.g., two-phase bioreactors, continuous extraction) are now standard tools in plant cell process engineering.

Artemisinin from *Artemisia annua* hairy roots: Artemisinin is an antimalarial sesquiterpene produced in *A. annua* (sweet wormwood) leaves. Producing it via cell or tissue culture is difficult because the yield in undifferentiated cells is low, and artemisinin is secreted in glandular trichomes in planta. Hairy root culture, however, has shown promise as *A. annua* roots can produce artemisinin in the presence of light (since the pathway is light-dependent). Researchers have applied airlift and mist reactors to *Artemisia* hairy roots. In one study, a modified airlift with internal lighting yielded 22.6 g/L dry root biomass and 374.4 mg/L artemisinin after 20 days (Liu and Guo, 1999; Liu Wang, et al., 1999). This is equivalent to ~16.6 mg artemisinin per g dry weight, which is quite high (commercial field-grown plants average ~5–15 mg/g in leaves). Another report indicated 244.5 mg/L artemisinin with 13.8 g/L DW in a stirred reactor under optimized light and nutrient conditions (Liu et al., 2002). These results (see Table 2.1 for the 374 mg/L case) showcase how bioreactor-grown hairy roots, with proper aeration and illumination, can rival or exceed field production in terms of concentration. The challenge remains scale—to supply global demand, reactors would need to run at thousands of liters. Nonetheless, such bioreactor systems might serve niche needs (off-season production, genetically engineered high-artemisinin lines, etc.) or provide platform knowledge for producing similar terpenoids *in vitro*.

Other cases: Numerous other medicinal plant cultures have been tested in bioreactors. *Echinacea purpurea* cells in a 20 L airlift produced cichoric acid (a phenolic immunostimulant) with yields improved ~30% over flask culture (Saadah et al., 2019). *Hypericum perforatum* (St. John's wort) shoot cultures have been grown in 5 L balloon-type reactors, yielding increased phenolic content (chlorogenic acid, hyperoside) compared to jars. *Bacopa monnieri* shoots in a 3 L balloon reactor showed not only higher biomass but also higher bacoside A (memory-enhancing saponin) accumulation than on solid media (Sharma et al., 2019). These examples underline that bioreactors can influence secondary metabolism—sometimes positively, due to better nutrient access or controlled stress (elicitation), but occasionally negatively if the environment is suboptimal (too much stress can divert metabolism to defense or growth cessation). Thus process optimization (medium, elicitors, environmental factors) is typically done in tandem with scale-up.

The above cases demonstrate the potential of bioreactor-based production. Paclitaxel and shikonin prove that large-scale (multi-thousand-liter) plant cell processes can be economically viable. On the other hand, the ajmalicine example indicates that even in small reactors, intense optimization can lead to gram-per-liter levels, suggesting room for further scale-up or technology transfer to larger units.

In commercial practice, we must consider not just yield but also productivity (rate of production) and consistency. Many plant cell processes are batch or fed-batch, taking 2–6 weeks for a cycle. Semicontinuous strategies (harvesting part of the culture and refeeding) have been tested to improve volumetric productivity (Verdú-Navarro et al., 2023). For instance, *Panax* root cultures have been grown in repeated-batch mode to avoid lag phases between batches. Bioreactor monitoring and control (pH, dissolved oxygen, perhaps even real-time biomass sensors) are increasingly applied to keep the culture in optimal states for growth or production. Additionally, elicitation (with jasmonates, fungal extracts, ozone, etc.) is often done at a specific culture stage to trigger secondary metabolism once sufficient biomass is attained (Saadah et al., 2019). Elicitors can be added to bioreactors in a controlled way (e.g., slowly via pump to avoid shocking the cells osmotically).

The role of bioreactors in the conservation of medicinal plants is also worth mentioning. Bioreactors allow propagation of threatened or slow-growing medicinal species (like some orchids, *Taxus* spp., *Ginseng*) *in vitro*, reducing pressure on wild populations. For example, bioreactor micropropagation of *Taxus* has provided plantlets for reforestation while also serving as a source of paclitaxel-producing cells (Murthy et al., 2023). Similarly, rare Himalayan herbs have been multiplied in TIBs to supply planting stock for cultivation, all under sterile lab conditions. The ability to produce phytochemicals in cells also means fewer field-grown plants need to be harvested for extraction, aiding conservation (Karalija et al., 2025). In summary, bioreactors and process engineering are enabling not only the commercial production of phytomedicines but also the sustainable supply of plant material, aligning with both economic and ecological goals.

2.2 Elicitation for metabolic intensification

2.2.1 Concept and importance of elicitation in phytochemical enhancement

Elicitation is a biotechnological strategy that exploits the plant's innate defense responses to boost the production of secondary metabolites (phytochemicals). In essence, an "elicitor" is any agent (biotic or abiotic) that, when introduced to a plant culture, triggers a stress or defense response, leading the plant to synthesize more of its protective phytochemicals (Karalija et al., 2025; Sangwan et al., 2024). This approach is highly effective and widely used to induce novel metabolite production or to enhance biosynthesis of valuable compounds in plant tissue cultures (Halder et al., 2019). The rationale is that many medicinal secondary metabolites (alkaloids, phenolics, terpenoids, etc.) are naturally produced at low levels under normal conditions, but their biosynthetic pathways can be upregulated by stress signals as part of the plant's defense mechanism (Sangwan et al., 2024). Elicitation is thus a powerful tool to overcome the low yield of phytochemicals in *in vitro* cultures and has been pivotal in making cell and organ cultures commercially viable for producing high-value phytomedicines. In practice, elicitation has been applied to a wide range of medicinal plant systems, from micropropagated shoots to cell suspensions, often achieving multifold increases in target compound content. Not every plant responds equally, and elicitor efficacy must be empirically determined for each species or even each cell line (Makowski et al., 2020). Nonetheless, when optimized, elicitation can dramatically intensify the biosynthesis of therapeutic metabolites without the need for transgenic modifications, making it an indispensable approach in medicinal plant biotechnology.

2.2.2 Types of elicitors: biotic, abiotic, and physical

Elicitors can be classified based on their nature. Broadly, they fall into three categories: biotic (originating from biological organisms or their molecules), abiotic (chemical or environmental factors not of biological origin), and physical (physical stimuli). Each class engages plant defense pathways which can be shown in a generalistic way as in Fig. 2.1.

2.2.2.1 *Biotic elicitors*

Biotic elicitors are molecules derived from living organisms such as pathogens, microbial symbionts, or even plant cell wall fragments. They often mimic an attack by pathogens or

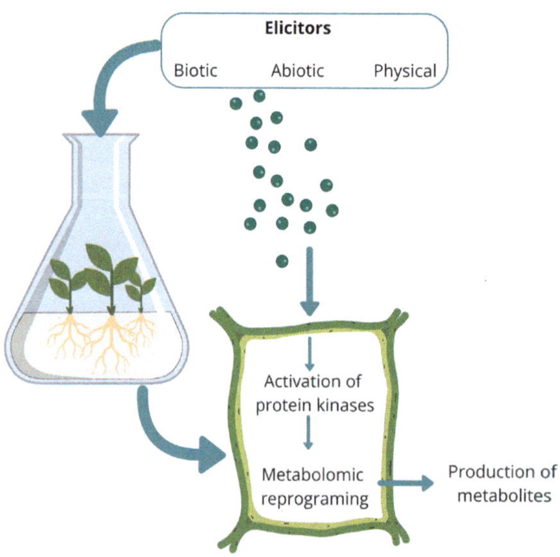

FIGURE 2.1 Generalistic overview of metabolite elicitation.

herbivores. Classic examples include fungal or yeast extracts, polysaccharides like chitosan (from fungal cell walls), microbial lysates, and even glycoproteins or oligosaccharides known as pathogen-associated molecular patterns (PAMPs) (Sangwan et al., 2024). When plant cells perceive these foreign biomolecules, they recognize them as danger signals and activate defense programs. For instance, yeast extract (a common biotic elicitor) added to plant cultures can trigger a cascade of defense responses leading to enhanced secondary metabolism. In one case, yeast extract elicitation of *Hoppea fastigiata* multiple shoot cultures led to elevated xanthone accumulation, a response mediated by a Ca^{2}+-dependent oxidative burst (H_2O_2 generation) that stimulated the downstream pathway (Sangwan et al., 2024). Other biotic elicitors include *Aspergillus* or *Trichoderma* fungal filtrates, bacterial cell suspensions, and polysaccharide fractions like elicitors from *Phytophthora* species. These have been used effectively in many medicinal plants. A notable feature is that biotic elicitors often induce a broad-spectrum defense, simultaneously activating multiple pathways. However, sensitivity varies: some species respond vigorously to a given biotic elicitor, while others show little change (Makowski et al., 2020). Optimal dosing is also critical—too mild a treatment may be ineffective, while too strong can cause cell death. Overall, biotic elicitors are valued for provoking complex defense signals that can lead to significant increases in phytochemical production.

2.2.2.2 *Abiotic elicitors*

Abiotic elicitors are nonbiological chemicals or factors that can induce stress responses in plant cells. These include chemical elicitors such as plant signaling molecules, heavy metals, and various chemical agents, as well as certain growth regulators in stress-inducing roles. Two of the most widely used abiotic elicitors are jasmonic acid and salicylic acid, which are endogenous plant defense hormones. Exogenous application of MeJA (a volatile derivative of jasmonic acid) or salicylic acid (SA) can mimic herbivore or pathogen attack, respectively, and spur the accumulation of secondary metabolites (Sangwan et al., 2024). For example, MeJA is famously effective in *Taxus* cell cultures to induce the anticancer drug paclitaxel; it essentially hijacks the plant's jasmonate signaling pathway (part of the elicitor signal transduction) to turn on paclitaxel biosynthesis genes. Abiotic elicitors also encompass heavy metal ions (e.g., silver, cadmium, cobalt salts) and other chemical stresses (e.g., high salt, osmotic agents). Low doses of heavy metals act as a threat signal, leading plants to bolster their chemical defenses. Studies have shown, for instance, that exposing cell cultures to silver or cadmium can enhance production of certain alkaloids and phenolics as the cells attempt to counter metal-induced oxidative stress (Mirjalili & Linden, 1996). Another category of abiotic elicitors includes various chemical additives that perturb normal growth, such as ethylene-releasing compounds (ethephon) or inhibitors that trigger defense pathways. Even nanoparticle forms of minerals can serve as elicitors: In *Hypericum perforatum* cell suspensions, nanoscale zinc oxide (ZnO) and iron oxide have been used at 100 ppb levels to induce stress, resulting in significantly elevated hypericin and hyperforin concentrations (as detailed in case studies below). In summary, abiotic elicitors are valuable because they are relatively easy to apply in controlled concentrations and often reproducible in their effects. They allow precise timing and dosing to fine-tune the stress response for maximal metabolite yields.

2.2.2.3 *Physical elicitors*

Physical elicitors are forms of physical stress or stimuli that can incite a defensive metabolic response. Plants in nature respond to abiotic stresses like extreme temperatures, UV radiation, wounding, or drought by altering their metabolism; the same principle is exploited *in vitro*. Light stress is a prominent example: exposing cultures to UV-B or specific wavelengths can stimulate the production of UV-screening and antioxidant compounds (often valuable phytochemicals). In *Hypericum perforatum*, which produces the photoreactive pigment hypericin, supplemental UV or blue light has been shown to drastically increase hypericin accumulation (up to 10-fold in some experiments) (Al-Khayri et al., 2024). This is because hypericin biosynthesis in *Hypericum* is closely tied to light cues, and a higher light stress prompts the plant to produce more of this protective pigment. Other physical elicitors include temperature shocks (brief exposure to cold or heat), osmotic stress (adding high levels of sugars or polyethylene glycol to create drought-like conditions), mechanical stress (such as vigorous agitation or sonication of cultures), and hydrodynamic shear in bioreactors. For example, placing cell cultures under osmotic stress with mannitol or sorbitol can lead to increased synthesis of osmoprotectant secondary metabolites (like certain glycosides or polyphenols). A mild cold shock may induce phenylpropanoid pathways as the plant activates its cold-defense mechanisms. Physical elicitation often goes hand-in-hand with other types; for instance, growing cultures on a rotary shaker (imposing shear stress) while also adding a biotic elicitor can have a synergistic effect (Makowski et al., 2020). The use of physical elicitors is particularly attractive because it can sometimes be done without introducing any foreign chemical, simply changing the culture environment. However, these methods require careful control (e.g., duration of UV exposure or severity of temperature change) to avoid damaging the culture. In practice, physical elicitors are frequently combined with chemical ones to maximize the metabolic response.

2.2.3 Mechanistic overview of elicitor action

When an elicitor is introduced to a plant culture, it sets off a chain of physiological and biochemical events inside the plant cells. Though the exact signaling pathways can differ depending on the elicitor and species, the general mechanistic steps are well understood:

- **Perception of the elicitor:** Plant cells must first recognize the elicitor. Biotic elicitors (such as microbial fragments) are often detected by specific cell-surface receptors (e.g., pattern-recognition receptors that bind chitin fragments, ß-glucans), much like an immune system sensor. Abiotic elicitors like MeJA enter cells or bind to carrier proteins (in the case of jasmonates, they are perceived through the JAZ–COI1 receptor complex inside the cell). Physical stresses are "perceived" via stress-sensing mechanisms (e.g., membrane fluidity changes, protein denaturation, or production of reactive oxygen species (ROS) in chloroplasts under high UV). The key point is that the elicitor signal is translated into an intracellular message (Mirjalili & Linden, 1996; Sangwan et al., 2024).
- **Signal transduction and secondary messengers:** Once the elicitor is perceived, a flurry of signaling events is triggered. A very common early event is a transient influx of calcium ions (Ca^{2+}) into the cytoplasm. This Ca^{2+} spike acts as a second messenger that activates

various calcium-dependent protein kinases. Simultaneously, a burst of ROS often occurs—essentially an "oxidative burst" similar to the hypersensitive response in plant immune reactions. For example, in yeast-extract-elicited Hoppea cultures, an elicitor-induced Ca2+ influx was shown to mediate H_2O_2 (ROS) generation, which was crucial for activating downstream biosynthetic enzymes. Nitric oxide (NO) is another possible messenger triggered in some elicitor responses. These molecules (Ca^{2+}, ROS, NO) propagate and amplify the signal within the cell (Sangwan et al., 2024).

- **Activation of protein kinases and signaling cascades:** Elicitor signals frequently converge on kinase cascades, such as the mitogen-activated protein kinase (MAPK) cascade. Activated by the second messengers, a series of phosphorylation events is initiated, ultimately leading to the activation of transcription factors in the nucleus. For instance, jasmonate perception leads to degradation of JAZ repressor proteins, freeing transcription factors like MYC2 to activate gene expression (Mirjalili & Linden, 1996). In general, elicitor-triggered kinases can phosphorylate key regulatory proteins that modulate defense gene transcription.
- **Transcriptional reprogramming:** A hallmark of elicitation is the upregulation of genes involved in secondary metabolite biosynthesis. Specific transcription factors (TFs) are activated by the signaling cascades—examples include AP2/ERF-family TFs (such as ORCA in *Catharanthus*, which regulates terpenoid indole alkaloid biosynthetic genes), basic helix-loop-helix TFs (like MYC2 in the jasmonate pathway), and others depending on the metabolite pathway. These TFs bind to promoter regions of pathway genes (e.g., phenylalanine ammonia-lyase in phenylpropanoid synthesis, or strictosidine synthase in indole alkaloid synthesis) and enhance their transcription (Mirjalili & Linden, 1996). The result is an increased mRNA level for enzymes that synthesize secondary metabolites. In *C. roseus*, for example, fungal elicitation by *Trichoderma* and a nitric oxide donor upregulated several terpenoid indole alkaloid pathway genes (STR, DAT, etc.) by ~two- to fivefold, which correlated with a ~1.9-fold increase in the accumulation of the anticancer alkaloids vinblastine and vincristine (Namdeo et al., 2002). This illustrates how elicitor-triggered TF activation leads directly to higher production of target compounds.
- **Biosynthetic enzyme activation and metabolite accumulation:** As the levels of biosynthetic enzymes rise (and sometimes their activities are also enhanced by posttranslational modification), the metabolic flux through the corresponding pathways increases. Plants often start channeling more primary metabolites into secondary pathways as a defensive measure. In practice, one observes a surge in the accumulation of secondary metabolites after an elicitor treatment—typically detectable in cells or medium within hours to days. A rapid example is the phenylpropanoid pathway: elicitation can cause a rapid PAL enzyme induction, and within 24–48 hours, one might measure significantly higher phenolic content. Many studies document these outcomes quantitatively. As one case, *P. ginseng* cells elicited with MeJA showed a 5.5- to 9.7-fold increase in total ginsenoside saponin content relative to untreated cells, even though overall growth was somewhat suppressed (Xu et al., 2023). In another example, *Withania somnifera* hairy roots treated with MeJA accumulated ~6.7-fold higher withanolide A and ~4.6-fold higher withaferin A within just 48 hours, compared to unelicited controls (Mahendran & Vimolmangkang, 2024). These outcomes are direct consequences of the elicitor-induced activation of the entire enzymatic network leading to those metabolites.

- **Feedback and hormonal cross-talk:** Plant signaling pathways are interconnected. Elicitor responses often involve cross-talk between jasmonate, salicylate, and ethylene signaling. For instance, a strong biotic elicitor might simultaneously trigger a jasmonate burst and a smaller salicylic acid response. These interactions can fine-tune the defense output. In some cases, over-elicitation can lead to feedback inhibition or stress-related growth arrest, so the plant may balance growth and defense.

Overall, the elicitation mechanism is a coordinated defense mimicry—the plant "believes" it is under attack or stress and reallocates resources to produce chemical defenses, which for us are valuable phytomedicines. Modern omics studies have reinforced this understanding by showing global transcriptomic reprogramming upon elicitor treatment (Namdeo et al., 2002). The controlled use of elicitors thus taps into ancient plant defense circuitry for the purpose of metabolic intensification.

2.2.4 Integration of elicitation into *in vitro* cultures and bioreactors

In practical terms, elicitation can be incorporated into plant tissue culture protocols to maximize secondary metabolite output, whether at laboratory scale or in commercial bioreactors. There are several considerations and strategies for effective integration:

Timing and two-stage culture strategies: A common approach is the two-stage culture: an initial growth phase to accumulate biomass, followed by an elicitation phase to induce metabolite production (Xu et al., 2023). During the growth phase (often on standard media with optimal nutrients and hormones), elicitors are withheld to allow cells or plantlets to proliferate. Then, at a defined point (such as when exponential growth slows or a certain biomass is reached), elicitors are introduced to trigger secondary metabolism. For example, in *Taxus* cell suspension cultures used for paclitaxel production, cells might be grown for ~2 weeks, then 100 μM MeJA is added to initiate taxane accumulation (Mirjalili & Linden, 1996). This staggered approach ensures that elicitation (which often slows cell division) does not overly compromise biomass yield. Some studies, however, have found that adding an elicitor at the time of inoculation can also work well for certain species; in *P. ginseng* suspensions, the optimal time to add yeast extract or MeJA was on day 0, which resulted in the highest saponin accumulation (with ginsenoside content reaching 28-fold over control) (Xu et al., 2023). Thus the timing is empirically determined and may differ by system. Whether an immediate or delayed elicitation is better depends on how severely the elicitor stunts growth versus how strongly it induces production.

Dose optimization and delivery: The concentration (or dose) of elicitor and how it's delivered (one-time addition vs. repeated doses) is critical (Karalija et al., 2017; Karalija et al., 2020). In bioreactors or large flasks, liquid elicitors (like MeJA, SA, or yeast extract solutions) are simply added to the culture medium. Solid or particulate elicitors (like fungal mycelium or nanoparticles) are typically suspended or predissolved. Overdosing can lead to toxicity—for instance, *P. ginseng* cultures showed diminished growth and even cell death at very high MeJA levels (above 200–400 μM) (Xu et al., 2023). The goal is to find a dose that maximally triggers the defense response without causing irreversible growth inhibition. In some large-scale systems, a low dose preelicitation (a "priming" dose) is used, followed by a higher dose, to acclimate the culture and amplify the response (Sykłowska-Baranek et al.,

2022). For solid plant tissue cultures (like micropropagated shoots or callus on agar), elicitor solutions can be applied by briefly dipping explants or by adding filter-sterilized elicitor into the agar medium. Physical elicitors *in vitro* require apparatus: for example, UV lamps for radiation or temperature-controlled incubators for heat/cold shock. These can be integrated into the workflow (for example, exposing *Hypericum* cultures to a defined UV-B dose daily for a week).

Bioreactor considerations: Scaling up elicitation to bioreactors (large stirred tanks or airlift bioreactors) is well-established for industrial production of plant secondary metabolites. Many commercial processes (such as for paclitaxel from *Taxus* and shikonin from *Lithospermum*) employ elicitors during the culture. In bioreactors, uniform mixing is important to ensure all cells receive the elicitor. The elicitor can be added to the feed medium or injected at specific time points. One successful example is the large-scale production of paclitaxel: cells of *Taxus* are grown in a two-stage 7,500 L bioreactor, and MeJA is added in the production stage, yielding paclitaxel titers far above what unelicited cultures produce (Demidova et al., 2023). Aeration and agitation are adjusted because elicitation can alter oxygen demand (e.g., a burst of oxidative metabolism during ROS production). Another consideration is foam or shear—some elicitors (like yeast extract) can change medium rheology, so antifoam measures or gentle mixing may be needed. Despite these challenges, numerous studies report that large-scale elicitation is feasible and effective. For instance, *P. ginseng* adventitious root cultures in 3 and 5 L bioreactors responded to both biotic and abiotic elicitors similarly to shake-flask cultures, resulting in increased ginsenoside content with manageable biomass loss (Xu et al., 2023).

Combination of elicitors: In practice, a combination of elicitors is sometimes used to exploit multiple signaling pathways. A biotic elicitor might be paired with an abiotic one for a synergistic effect. For example, combining MeJA with ß-cyclodextrin (a molecule that helps sequester and solubilize secondary metabolites) in *Withania somnifera* hairy root cultures led to a remarkable ~12-fold increase in withaferin A, higher than MeJA alone (Sivanandhan et al., 2013). Similarly, an elicitor could be combined with a precursor feeding strategy (supplying a biosynthetic precursor) to further boost yields (Singh et al., 2022). These integrative strategies are an active area of optimization in bioprocess engineering.

Monitoring and trade-offs: Throughout elicitation, careful monitoring of culture health (growth rate, viability) and metabolite levels is important. There is often a trade-off: Elicitation frequently slows or reduces biomass accumulation even as it increases secondary metabolite concentration (Xu et al., 2023). If the biomass reduction is too severe, the overall yield (metabolite amount = concentration × biomass) could suffer. Therefore elicitation regimens are designed to maximize the product of biomass and metabolite content. In some cases, short elicitation pulses (e.g., 24–48 h exposure) followed by returning to normal medium can allow recovery and repeated induction cycles. Encapsulation of elicitors or slow-release formulations is also being researched to modulate the intensity of the response.

It is also noteworthy that elicited cultures sometimes produce not only more quantity but also a different profile of metabolites. Stress conditions can activate dormant pathways or shunt metabolism to new compounds, occasionally yielding novel or more bioactive analogs (Sangwan et al., 2024). This can be advantageous for obtaining a broader spectrum of phytochemicals. However, one must ensure that the desired medicinal compound is indeed increased and not just its relatives. Thus an integrated elicitation strategy *in vitro* is a fine-tuned process balancing stress and survival, now routinely applied in tissue culture protocols to intensify production of phytomedicines.

2.2.5 Case studies of elicitation in medicinal plants

Below, we highlight real-world examples of elicitation applied to important medicinal plant species. Each case demonstrates quantitative improvements in secondary metabolite yields, illustrating the promise of elicitation in both research and industrial contexts. Key parameters of these examples are summarized in Table 2.1.

2.2.5.1 Panax ginseng (*ginseng*)

Panax ginseng is renowned for its triterpenoid saponins called ginsenosides, which are the major bioactive compounds in ginseng roots. *In vitro* cultures of *P. ginseng* (including cell suspensions and adventitious roots) have been extensively studied for enhanced ginsenoside production. Elicitation has proven exceptionally fruitful in this species. MeJA is one of the most potent elicitors for ginseng cultures. In cell suspension experiments, MeJA treatment can dramatically boost ginsenoside accumulation—one study reported ginsenoside levels reaching 5.5–9.7 times those in control (untreated) cells (Xu et al., 2023). This increase came at the cost of some growth inhibition (a common theme, as MeJA diverts resources to defense), but the net production of ginsenosides was far superior in elicited cultures. Because MeJA is relatively expensive for large-scale use, researchers have explored other elicitors as well. Yeast extract, a cost-effective biotic elicitor, has been used in ginseng cell cultures with great success. Lu et al. found that optimizing the yeast extract concentration and timing led to total saponin content as high as 2.07% DW in cells, which was ~28-fold higher than in the unelicited control. This is a remarkable increase, indicating that practically all ginseng cells ramped up secondary metabolism.

Other biotic elicitors in ginseng include fungal preparations. For example, dried mycelium of *Aspergillus niger* added to ginseng adventitious root cultures significantly enhanced ginsenoside production (with some trade-off in root growth at higher doses). An endophytic bacterium isolated from ginseng was even able to quadruple the ginsenoside content when cocultured with ginseng roots (Xu et al., 2023). Likewise, a homogenate of a pathogenic fungus (*Alternaria panax*) at 200 mg/L elicited a strong response: Within 8 days, treated ginseng roots reached a total ginsenoside content of 29.6 mg/g dry weight, with minimal biomass reduction. To put this in perspective, such elicited *in vitro* roots produced ginsenoside levels comparable to or exceeding those of some field-grown 4–6 year old ginseng roots.

Because of these successes, elicitation is being integrated into ginseng tissue culture protocols aimed at commercial production. Large-scale bioreactor cultures of ginseng have been treated with both biotic and abiotic elicitors; the results show increased ginsenoside yields in bioreactors similar to flask experiments (Xu et al., 2023). One challenge is to maintain root growth while eliciting—a two-stage process (growth then elicitation) is often employed. Additionally, the interaction of elicitors with plant growth regulators needs attention: For instance, MeJA's effect was antagonistic to the synthetic auxin 2,4-D, so removing 2,4-D from the medium improved elicitation outcomes. Overall, *P. ginseng* stands as a prime example of metabolic intensification by elicitation, with multifold gains in medicinal saponin yield that make *in vitro* production much more viable.

2.2.5.2 Taxus *spp. (yew trees)*

Species of *Taxus* (such as *T. chinensis, T. cuspidata, T. brevifolia*) are the source of the famous anticancer compound paclitaxel (Taxol). Taxus cell suspension cultures are a well-known system in which elicitation is essential to achieve useful paclitaxel yields. The groundbreaking discovery was that treating *Taxus* cultures with MeJA induces a massive increase in paclitaxel production (Lenka et al., 2015). For example, in a study by Yukimune et al. (1996), adding 100 μM MeJA to *Taxus* cell suspensions led to paclitaxel accumulation rising from virtually negligible levels (~0.4 mg/L in control) to about 48 mg/L—representing roughly a 120-fold increase in yield. The production of related taxanes like baccatin III also spiked (to ~53 mg/L). This astonishing elicitation effect has been reproduced in various *Taxus* species and cell lines. The mechanism is linked to jasmonate signaling, activating the entire cascade of genes in the paclitaxel biosynthetic pathway; transcriptomic analyses show sharp increases in mRNAs of key pathway enzymes within hours of MeJA treatment (Sohn et al., 2022).

Given MeJA's efficacy, it has become a staple in Taxus bioprocesses. Commercial cell culture methods for paclitaxel (such as those by Phyton Biotech) include jasmonate elicitation in large fermenters to achieve therapeutic yields. Research has also identified other elicitors that can complement or even surpass MeJA's effect. One example is coronatine, a phytotoxin from bacteria that mimics jasmonate's action. Coronatine at only 1 μM can induce taxane production as effectively as 100 μM MeJA, and in some cases with less growth inhibition (Escrich et al., 2021). Salicylic acid, another defense signal, has a more modest effect on Taxus cultures (reports indicate a two- to threefold increase in paclitaxel with salicylate, which is useful but lower than jasmonate's effect). Polysaccharide-rich fungal extracts (e.g., from *Alternaria* or *Fusarium* species) at around 0.1%–0.2% have been shown to increase paclitaxel secretion by a few-fold as well (Goktepe-Atilgan et al., 2023).

Another strategy has been to use precursor feeding (like phenylalanine) alongside elicitation to push the pathway even harder (Perez-Matas et al., 2023). In one study, a combination of MeJA plus a precursor and a NO-donor led to higher paclitaxel than MeJA alone. Taxus cultures also highlight an important practical aspect: elicitation often causes paclitaxel to be released into the medium (perhaps due to MeJA affecting membrane permeability). This is advantageous for extraction. In fact, certain additives (like cyclodextrins or certain surfactants) are used together with elicitors to trap paclitaxel in the medium, simplifying downstream processing (Escrich et al., 2021).

Taxus spp. demonstrate how a single elicitor innovation (MeJA addition) revolutionized the viability of plant cell culture for drug production. Elicited Taxus cultures have yielded paclitaxel at hundreds of milligrams per liter in research settings (Mirjalili & Linden, 1996), and this has translated into industrial application. The Taxus case also underscores careful elicitation management: too high a jasmonate dose can halt cell growth entirely, so optimal intermediate concentrations and timing (often mid-log growth phase) are used to maximize total paclitaxel output (Xu et al., 2023).

2.2.5.3 Catharanthus roseus *(Madagascar periwinkle)*

C. roseus produces the famous anticancer vinca alkaloids (vinblastine and vincristine) as well as other indole alkaloids like ajmalicine and serpentine. These compounds are present in very low levels in the plant, so cell and tissue culture efforts focus on elicitation to boost production. Both biotic and abiotic elicitors have been extensively studied in *Catharanthus* cell suspensions, hairy root cultures, and even microshoot cultures.

On the biotic side, fungal elicitors are particularly effective. For instance, *C. roseus* cell suspensions treated with fungal extracts from *Trichoderma viride* or *Phytophthora* species showed dramatic increases in alkaloid accumulation. In one report, elicitation with a *Phytophthora boehmeriae* cell wall protein (known as PB90) for 48 hours led to a 7.9-fold increase in ajmalicine content and a fourfold increase in catharanthine, compared to unelicited cells (Srinivasan et al., 2022). Multiple fungi have been tested: *Aspergillus niger, Fusarium moniliforme, Trichoderma harzianum*, etc., often applied as a 5% (v/v) culture filtrate or a similar dose (Namdeo et al., 2002). These elicit general defense responses, boosting the indole alkaloid pathway. Interestingly, some fungal elicitors also promote growth in this species—one study found *Aspergillus flavus* extract not only increased vinblastine/vincristine yields but also augmented callus biomass, indirectly benefiting alkaloid totals (Srinivasan et al., 2022).

Abiotic elicitors are equally important in *Catharanthus*. MeJA is known to strongly induce terpenoid indole alkaloids by upregulating TFs like ORCA3. For example, adding 250 μM MeJA to *C. roseus* cell cultures resulted in about a sevenfold increase in ajmalicine content (to 6.34 mg/g DW) and nearly threefold increase in serpentine (to 1.71 mg/g DW) relative to control cells (Almagro et al., 2015). Similarly, salicylic acid at 100–200 μM has been reported to roughly double vincristine and vinblastine yields in some experiments (Stasińska-Jakubas et al., 2024). A particularly potent elicitation strategy combined MeJA with UV-B exposure: the synergy of a chemical (JA-pathway) and physical (light stress) elicitor yielded significantly higher vinblastine and vincristine than either treatment alone (Rady et al., 2021). This underlines how multiple stress signals can converge to amplify the biosynthetic output in *C. roseus*.

Catharanthus cultures also illustrate the need to balance elicitation duration. Namdeo et al. (2002) observed that treating cell suspensions with *T. viride* extract for 48 h gave about a threefold rise in ajmalicine, but extending the treatment longer caused the alkaloid levels to drop, likely due to prolonged stress harming the cells. The optimal elicitation window was therefore around 2 days for that system. Modern studies have dived deep into the molecular response: elicitation (biotic or abiotic) triggers coordinate upregulation of genes like TDC, STR, DAT in the indole alkaloid pathway and activates defense enzymes (peroxidases, catalases) that manage the induced oxidative stress. The net effect is a transient but significant surge in alkaloid synthesis.

In summary, *C. roseus* benefits from a toolkit of elicitors. By choosing the appropriate elicitor type and conditions (e.g., fungal vs JA vs combined treatments) and optimizing exposure time, researchers have achieved multifold increases in valuable alkaloids. This has important implications for producing these anticancer compounds in bioreactor settings as an alternative to field cultivation, which is land- and labor-intensive.

2.2.5.4 Withania somnifera *(ashwagandha)*

Withania somnifera is an important medicinal herb known for withanolides (steroidal lactones like withaferin A and withanolide A) with anti-stress, anticancer, and immunomodulatory activities. Micropropagation and *in vitro* culture of *Withania*, coupled with elicitation, have emerged as strategies to enhance withanolide yield for pharmaceutical use.

Elicitation in *W. somnifera* primarily employs abiotic signals. MeJA has been shown to strongly elevate withanolide production. In hairy root cultures of *Withania*, a recent study demonstrated that 100 μM MeJA treatment for 48 h increased the content of withanolide A to 3.16 mg/g DW, a 6.72-fold rise over control levels, and withaferin A to 4.32 mg/g DW (4.6-fold increase) (Mahendran & Vimolmangkang, 2024). These are substantial improvements achieved in a short time frame. The mechanism involves MeJA upregulating upstream steps of the terpenoid pathway (e.g., the MEP pathway and withanolide biosynthetic genes). Studies have recorded elevated expression of cycloartenol synthase and downstream tailoring enzymes after elicitation, consistent with the higher metabolite output (Thirugnanasambantham & Senthil, 2016).

Biotic elicitors have also been tested. for example, *Penicillium* fungal extract and *Alternaria* have been reported to boost withanolide content in cell cultures, though typically to a lesser extent than jasmonates (often 1.5- to 3-fold increases). One interesting approach combined a biotic elicitor (chitosan, a fungal polysaccharide) with MeJA, finding that dual elicitation gave additive effects on withanolide enhancement (Namdeo et al., 2002). There is also evidence that certain endophytic fungi living in *Withania* can, when added to cultures, modulate secondary metabolism favorably (Pandey et al., 2018).

From an integration perspective, *Withania* shoot cultures have been elicited via foliar spray *in vitro*—e.g., spraying salicylic acid or MeJA onto adventitious shoots. An optimized protocol using foliar SA elicitation yielded significantly more withaferin A in regenerated plantlets than unelicited controls (often on the order of twofold) (Singh et al., 2022). Although a 2× increase is modest relative to some other examples, in a commercial sense, it can be valuable, especially if the treatment is simple. On the higher end, precursor feeding of metabolic intermediates (like sterol precursors) together with elicitation has achieved even larger increases. For instance, one study using squalene (a precursor) plus MeJA saw withanolide A rise by nearly ninefold in cell suspensions (Sivanandhan et al., 2014).

In adventitious root cultures of *Withania*, elicitors have yielded impressive gains: an adventitious root culture treated with a fungal elicitor reportedly achieved withanolide A and B levels 48- and 29-fold higher than baseline, respectively (Singh et al., 2017). Such results, if reproducible, indicate enormous potential for producing these compounds in controlled systems. It should be noted that extreme increases like 48× often involve comparing to very low starting amounts, but nonetheless, the absolute yields can meet or exceed field-grown plant concentrations.

In conclusion, elicitation is a key component of *in vitro Withania* culture protocols when the goal is to produce more withanolides. MeJA is the standout elicitor, but combinations with other signals (biotic or precursor molecules) are an active research area. The ability to multiply withanolide content in a matter of days via elicitation has significant implications for supplying these compounds, given the surging demand for ashwagandha extracts in herbal medicine.

2.2.5.5 Hypericum perforatum *(St. John's wort)*

Hypericum perforatum produces unique naphthodianthrone compounds, chiefly hypericin and pseudohypericin, as well as phloroglucinols like hyperforin. These metabolites are of high interest for their antidepressant and antiviral properties. *In vitro* cultures of *H. perforatum* (shoot cultures, cell suspensions, hairy roots) are often used to study and enhance the production of hypericin and hyperforin. Elicitation has proven to be a valuable strategy in this genus, especially given that hypericin is naturally a stress-inducible compound.

Physical elicitors are particularly relevant for Hypericum. Light stress is a known trigger for hypericin biosynthesis—in planta, hypericin accumulates in dark glands in response to light exposure. In hairy root cultures of *H. perforatum*, experiments have shown that exposing roots to blue light can yield up to a 10-fold increase in hypericin content (Al-Khayri et al., 2024). Likewise, applying UV-B radiation (e.g., 30 minutes of UV-B daily) has been reported to significantly boost hypericin levels (several-fold) compared to dark-grown controls, and often also increases the levels of hyperforin (Tavakoli et al., 2020). These findings reflect the plant's tendency to ramp up its protective pigments under high-light stress.

Among chemical elicitors, salicylic acid (SA) and jasmonic acid have been tested in Hypericum cultures. SA, for instance, at 50–200 μM in cell suspensions induced roughly a twofold increase in total hypericin content (Stasińska-Jakubas et al., 2024). MeJA can also stimulate hyperforin and hypericin production; one study noted that jasmonate elicitation doubled hyperforin in *H. perforatum* shoot cultures relative to control (Shakya et al., 2019). However, the magnitudes from single chemical elicitors are moderate (1.5–3× typically) in Hypericum, perhaps because the basal levels are already significant. Combining elicitors or using novel types can have more impact.

A notable advancement is the use of nanoparticle elicitors in Hypericum. As discussed earlier, extremely low concentrations (parts-per-billion) of metal oxide nanoparticles can induce stress responses. Sharafi et al. (2013) demonstrated that adding ZnO nanoparticles (100 ppb) to *H. perforatum* cell suspensions for 72 h raised hypericin to ~7.9 μg/g (about 3× control) and hyperforin to ~217 μg/g (about 13× control), compared to ~2.1 μg/g and 16.3 μg/g in controls. Iron oxide nanoparticles had a similar effect, with hypericin climbing to ~11.2 μg/g (~5×) and hyperforin to ~196 μg/g (~12×). These are impressive enhancements, particularly for hyperforin, which jumped an order of magnitude. The mechanism is thought to be that nano-Fe and nano-Zn provoke oxidative stress and perhaps release Fe^{2+}/Zn^{2+} ions that act as micronutrient elicitors, thereby activating the phenylpropanoid and mevalonate pathways leading to these compounds.

Biotic elicitors have also been examined: treatments with fungal elicitors (e.g., *Diplococci*, *Phytophthora* spp.) resulted in significant rises in hypericin/hyperforin. One report using three different fungal strains found four- to fivefold increases in hypericin and pseudohypericin levels in treated cell cultures (Al-Khayri et al., 2024). Those increases, however, were often transient—hypericin peaked and then declined, possibly due to its phototoxicity affecting cell viability or its binding to cell debris. Therefore timing the harvest after elicitation is crucial in such cases to capture the maximum yield.

In Hypericum shoot cultures, even hormones like cytokinin have been noted to influence secondary metabolite content. While not a classic "elicitor" in the stress sense, certain cytokinins increased hypericin and hyperforin *in vitro* (likely by developmentally increasing the density of secretory glands in the shoots) (Kwiecień et al., 2023). This highlights that elicitation in a broad sense can include hormonal manipulations that indirectly stress the metabolism.

In summary, *Hypericum perforatum* benefits from elicitation strategies, particularly involving light and novel abiotic elicitors. The ability to multiply hyperforin content via nanoparticle elicitation is a promising avenue for producing this valuable antidepressant compound in bioreactors. Meanwhile, light-based elicitation is practically useful for greenhouse or lab-grown Hypericum microshoots (for instance, brief UV exposure can be applied before harvest to increase product content). The Hypericum case underscores the diversity of elicitor types—from physical to chemical to nanomaterials—that can be exploited to intensify phytochemical production.

2.3 Conclusion

Elicitation has emerged as a cornerstone technique in the micropropagation and *in vitro* production of medicinal plant compounds. Through the intelligent application of biotic, abiotic, and physical elicitors, researchers can mimic nature's stress signals and unlock the metabolic potential of plant cells. The result, as documented in numerous species and systems, is often an order-of-magnitude enhancement in the yields of valuable phytochemicals—transforming otherwise marginal tissue cultures into productive "green factories." Mechanistically, elicitors channel plant defense pathways, activating cascades of signals (Ca^{2+}, ROS, MAPKs) and transcriptional networks that favor secondary metabolite biosynthesis. Integrating elicitation into culture protocols requires balancing act: timing, dosing, and mitigating any negative effects on growth. Yet, when optimized, elicitation strategies have proven their worth, from academic experiments to industrial bioreactors, by providing higher titers of drugs like paclitaxel, ginsenosides, withanolides, and many others in a sustainable, controlled manner. Moving forward, advances in understanding plant signaling and genomics will further refine elicitation approaches—for example, combining elicitors or engineering more responsive cell lines—driving the field of medicinal plant biotechnology toward even greater metabolic intensification for phytomedicine production.

Elicitation exemplifies how leveraging the innate "chemical intuition" of plants—their ability to respond to challenges with a burst of chemistry—an be a powerful means to produce and conserve medicinal compounds. It aligns with the goals of micropropagation not only to propagate plant material but also to enhance the phytochemical arsenal, ensuring a steady and amplified supply of high-value metabolites for pharmaceutical and nutraceutical use. The case studies discussed in this section underscore the practical successes of this approach, while ongoing research continues to expand its applications to new species and novel elicitors, heralding a promising future for the production of phytomedicines through elicited *in vitro* cultures.

References

Al-Khayri, J. M., Narasimha, S. W., Vennapusa, A. R., Nagella, P., Shehata, W. F., & Al-Mssallem, M. Q. (2024). Biotechnological approaches for the production of hypericin and other important metabolites from the genus Hypericum. *Plant Cell, Tissue and Organ Culture (PCTOC), 156*(3), 100.

Almagro, L., Fernández-Pérez, F., & Pedreño, M. A. (2015). Indole alkaloids from *Catharanthus roseus*: Bioproduction and their effect on human health. *Molecules (Basel, Switzerland), 20*(2), 2973–3000.

Almagro, L., López, Perez, A. J., & Pedreño, M. A. (2011). New method to enhance ajmalicine production in *Catharanthus roseus* cell cultures based on the use of cyclodextrins. *Biotechnology Letters, 33*(2), 381–385.

Almusawi, A. H., Sayegh, A. J., Alshanaw, A. M., & Griffis Jr, J. L. (2017). Plantform bioreactor for mass micropropagation of date palm. In Jameel M. Al-Khayri, S. Mohan Jain, & Dennis V. Johnson (Eds.). *Date palm biotechnology protocols volume I: Tissue culture applications* (pp. 251–265). Springer New York.

Choi, H.J., Choi, H.K., Kim, S.I., Yun, J.H., Son, J.S., Chang, M.S., & Lee, H.S. (2003). U.S. Patent No. 6, 589, 765. U.S. Patent and Trademark Office.

Correll, M. J., Wu, Y., & Weathers, P. J. (2000). Controlling hyperhydration of carnations (Dianthus caryophyllus L.) grown in a mist reactor. *Biotechnology and bioengineering, 71*(4), 307–314.

Cui, H. Y., Murthy, H. N., Moh, S. H., Cui, Y., Lee, E. J., & Paek, K. Y. (2014). Protocorm culture of *Dendrobium candidum* in balloon type bubble bioreactors. *Biochemical Engineering Journal, 88*, 26–29.

Demidova, E., Globa, E., Klushin, A., Kochkin, D., & Nosov, A. (2023). Effect of methyl jasmonate on the growth and biosynthesis of C13-and C14-hydroxylated taxoids in the cell culture of yew (*Taxus wallichiana* Zucc.) of different ages. *Biomolecules, 13*(6), 969.

Escrich, A., Almagro, L., Moyano, E., Cusido, R. M., Bonfill, M., Hosseini, B., & Palazon, J. (2021). Improved biotechnological production of paclitaxel in *Taxus media* cell cultures by the combined action of coronatine and calix [8] arenes. *Plant Physiology and Biochemistry, 163*, 68–75.

Etienne, E., Teisson, C., Alvard, D., Lartaud, M., Berthouly, M., Georget, F., & Lorenzo, J. C. (1999). Temporary immersion for plant tissue culture. In Arie Altman, Meira Ziv, & Shamay Izhar (Eds.). *Plant biotechnology and in vitro biology in the 21st century* (pp. 629–632). Springer Netherlands Proceedings of the IXth International Congress of the International Association of Plant Tissue Culture and Biotechnology Jerusalem, Israel, June 14–19, 1998.

Georgiev, V., Schumann, A., Pavlov, A., & Bley, T. (2014). Temporary immersion systems in plant biotechnology. *Engineering in life sciences, 14*(6), 607–621.

Goktepe-Atilgan, I., Dogan, A., & Ari, S. (2023). Enhancement of taxane production in cell suspension culture of *Corylus avellana* L. Kalınkara by elicitation and precursor feeding. *Biotechnology & Biotechnological Equipment, 37*(1), 2255703.

Halder, M., Sarkar, S., & Jha, S. (2019). Elicitation: A biotechnological tool for enhanced production of secondary metabolites in hairy root cultures. *Engineering in Life Sciences, 19*(12), 880–895.

Huang, T. K. (2009). *Development of plant cell suspension cultures as a bioproduction platform for recombinant human therapeutic proteins*. University of California.

Jeong, C. S., Murthy, H. N., Hahn, E. J., Lee, H. L., & Paek, K. Y. (2009). Inoculum size and auxin concentration influence the growth of adventitious roots and accumulation of ginsenosides in suspension cultures of ginseng (Panax ginseng CA Meyer). *Acta Physiologiae Plantarum, 31*(1), 219–222.

Karalija, E., Ćavar Zeljković, S., Tarkowski, P., Muratović, E., & Parić, A. (2017). The effect of cytokinins on growth, phenolics, antioxidant and antimicrobial potential in liquid agitated shoot cultures of *Knautia sarajevensis*. *Plant Cell, Tissue and Organ Culture (PCTOC), 131*(2), 347–357.

Karalija, E., Macanović, A., & Ibragić, S. (2025). Revisiting traditional medicinal plants: Integrating multiomics, in vitro culture, and elicitation to unlock bioactive potential. *Plants, 14*(13) 2029.

Karalija, E., Ćavar, Zeljković, S., & Parić, A. (2020). Harvest time–related changes in biomass, phenolics and antioxidant potential in *Knautia sarajevensis* shoot cultures after elicitation with salicylic acid and yeast. *In Vitro Cellular & Developmental Biology-Plant, 56*(2), 177–183.

Kwiecień, I., Miceli, N., Kędzia, E., Cavò, E., Taviano, M. F., Beerhues, L., & Ekiert, H. (2023). Different types of *Hypericum perforatum*cvs.(Elixir, Helos, Topas) in vitro cultures: A rich source of bioactive metabolites and biological activities of biomass extracts. *Molecules (Basel, Switzerland), 28*(5), 2376.

Lenka, S. K., Nims, N. E., Vongpaseuth, K., Boshar, R. A., Roberts, S. C., & Walker, E. L. (2015). Jasmonate-Responsive expression of paclitaxel biosynthesis genes in Taxus cuspidata cultured cells is negatively regulated by the bHLH transcription factors TcJAMYC1, TcJAMYC2, and TcJAMYC4. *Frontiers in Plant Science, 6*, 115.

Li, Y. C., Tao, W. Y., & Cheng, L. (2009). Paclitaxel production using co-culture of *Taxus suspension* cells and paclitaxel-producing endophytic fungi in a co-bioreactor. *Applied Microbiology and Biotechnology, 83*(2), 233–239.

Liu, C. Z., Guo, C., Wang, Y. C., Ouyang, F., Ye, H. C., & Li, G. F. (1999). Production of artemisinin by Artemisia annua hairy root culture in a IP airlift bioreactor. *Biotechnology Letters, 46*.

Liu, C. Z., Guo, C., Wang, Y. C., & Ouyang, F. (2002). Effect of light irradiation on hairy root growth and artemisinin biosynthesis of *Artemisia annua* L. *Process Biochemistry, 38*(4), 581–585.

Liu, C. Z., Wang, Y. C., Zhao, B., Guo, C., Ouyang, F., Ye, H. C., & Li, G. F. (1999). Development of a nutrient mist bioreactor for growth of hairy roots. *In Vitro Cellular & Developmental Biology-Plant, 35*(3), 271–274.

Lyam, P. T., Musa, M., Jamaleddine, Z., Okere, U. A., & Odofin, W. T. (2012). Temporary immersion bioreactors (TIBs) potential in meeting crop production demand in Nigeria. *Journal of Biology and Life Sciences, 3*(1), 66–86.

Mahendran, G., & Vimolmangkang, S. (2024). Elicitation boosts withaferin A and withanolide A production in *Withania somnifera* (L.) Dunal hairy root cultures. *Plant Cell, Tissue and Organ Culture (PCTOC), 158*(2), 29.

Makowski, W., Tokarz, K. M., Tokarz, B., Banasiuk, R., Witek, K., & Królicka, A. (2020). Elicitation-based method for increasing the production of antioxidant and bactericidal phenolic compounds in *Dionaea muscipula* J. Ellis tissue. *Molecules (Basel, Switzerland), 25*(8), 1794.

Manickavasagam, V. M., Anupindi, K., Bhatt, N., & Srivastava, S. (2025). Characterizing the effect of impeller design in plant cell fermentations using CFD modeling. *Scientific Reports, 15*(1), 9322.

Manz, C., Raorane, M. L., Maisch, J., & Nick, P. (2022). Switching cell fate by the actin–auxin oscillator in *Taxus*: Cellular aspects of plant cell fermentation. *Plant Cell Reports, 41*(12), 2363–2378.

Mirjalili, N., & Linden, J. C. (1996). Methyl jasmonate induced production of taxol in suspension cultures of *Taxus cuspidata*: Ethylene interaction and induction models. *Biotechnology Progress, 12*(1), 110–118.

Murthy, H. N., Joseph, K. S., Paek, K. Y., & Park, S. Y. (2023). Bioreactor systems for micropropagation of plants: Present scenario and future prospects. *Frontiers in Plant Science, 14*, 1159588.

Namdeo, A., Patil, S., & Fulzele, D. P. (2002). Influence of fungal elicitors on production of ajmalicine by cell cultures of *Catharanthus roseus*. *Biotechnology Progress, 18*(1), 159–162.

Nienow, A. W. (2014). Stirring and stirred-tank reactors. *Chemie Ingenieur Technik, 86*(12), 2063–2074.

Pandey, S. S., Singh, S., Pandey, H., Srivastava, M., Ray, T., Soni, S., & Kalra, A. (2018). Endophytes of *Withania somnifera* modulate in planta content and the site of withanolide biosynthesis. *Scientific Reports, 8*(1), 5450.

Perez-Matas, E., Hidalgo-Martinez, D., Escrich, A., Alcalde, M. A., Moyano, E., Bonfill, M., & Palazon, J. (2023). Genetic approaches in improving biotechnological production of taxanes: An update. *Frontiers in Plant Science, 14*, 1100228.

Pradeep, M., Saxena, M., Mondal, D., & Franklin, G. (2024). Do nanoparticles delivered to roots affect plant secondary metabolism? A comprehensive analysis in float seedling cultures of *Hypericum perforatum* L. *Chemosphere, 356*, 141789.

Rady, M. R., Gierczik, K., Ibrahem, M. M., Matter, M. A., & Galiba, G. (2021). Anticancer compounds production in *Catharanthus roseus* by methyl jasmonate and UV-B elicitation. *South African Journal of Botany, 142*, 34–41.

Saadah, I. N., Kristanti, A. N., Hardjo, P. H., & Manuhara, Y. S. W. (2019). Shoots culture of *Gynura procumbens* (Lour.) Merr. in balloon-type bubble-bioreactor influenced by sucrose concentration and inoculum density. *Asian Network for Scientific Information, 18*(2), 85–90.

Sangwan, N. S., Jha, S., & Mitra, A. (2024). Unlocking nature's treasure trove: Biosynthesis and elicitation of secondary metabolites from plants. *Plant Growth Regulation, 104*(1), 1–4.

Shakya, P., Marslin, G., Siram, K., Beerhues, L., & Franklin, G. (2019). Elicitation as a tool to improve the profiles of high-value secondary metabolites and pharmacological properties of *Hypericum perforatum*. *Journal of Pharmacy and Pharmacology, 71*(1), 70–82.

Sharafi, E., Khayam Nekoei, S. M., Fotokian, M. H., Davoodi, D., Mirzaei, H. H., & Hasanloo, T. (2013). Improvement of hypericin and hyperforin production using zinc and iron nano-oxides as elicitors in cell suspension culture of St John's wort (Hypericum perforatum L.). *Journal of Medicinal plants and By-products, 2*(2), 177–184.

Sharma, M., Koul, A., Ahuja, A., & Mallubhotla, S. (2019). Suitability of bench scale bioreactor system for shoot biomass production and bacoside biosynthesis from *Bacopa monnieri* (L.). *Engineering in Life Sciences, 19*(8), 584–590.

Sim, S. J., & Chang, H. N. (1993). Increased shikonin production by hairy roots of *Lithospermum erythrorhizon* in two phase bubble column reactor. *Biotechnology Letters, 15*(2), 145–150.

Singh, M., Agrawal, S., Afzal, O., Altamimi, A. S., Redhwan, A., Alshammari, N., & Khan, S. (2022). Optimization of elicitation conditions to enhance the production of potent metabolite withanolide from *Withania somnifera* (L). *Metabolites, 12*(9), 854.

Singh, P., Guleri, R., Angurala, A., Kaur, K., Kaur, K., Kaul, S. C., & Pati, P. K. (2017). Addressing challenges to enhance the bioactives of *Withania somnifera* through organ, tissue, and cell culture based approaches. *BioMed Research International, 2017*(1), 3278494.

Sivanandhan, G., Kapil Dev, G., Jeyaraj, M., Rajesh, M., Arjunan, A., Muthuselvam, M., & Ganapathi, A. (2013). Increased production of withanolide A, withanone, and withaferin A in hairy root cultures of *Withania somnifera* (L.) Dunal elicited with methyl jasmonate and salicylic acid. *Plant Cell, Tissue and Organ Culture (PCTOC), 114*(1), 121–129.

Sivanandhan, G., Selvaraj, N., Ganapathi, A., & Manickavasagam, M. (2014). Enhanced biosynthesis of withanolides by elicitation and precursor feeding in cell suspension culture of *Withania somnifera* (L.) Dunal in shake-flask culture and bioreactor. *PLoS One, 9*(8), e104005.

Sohn, S. I., Pandian, S., Rakkammal, K., Largia, M. J. V., Thamilarasan, S. K., Balaji, S., & Ramesh, M. (2022). Jasmonates in plant growth and development and elicitation of secondary metabolites: An updated overview. *Frontiers in Plant Science, 13*, 942789 Paclitaxel and baccatin III in Taxus cell suspension cultures. *Nature Biotechnology, 14*(9), 1129-1132.

Srinivasan, A., Sundaram, V., Vidya Muthulakshmi, M., & Srivastava, S. (2022). Multi-fold enhancement in vitamin E (alpha-tocopherol) production via integration of bioprocess optimisation and metabolic engineering in cell suspension of sunflower. *Journal of Plant Biochemistry and Biotechnology, 31*(1), 154–167.

Stasińska-Jakubas, M., Dresler, S., Strzemski, M., Rubinowska, K., & Hawrylak-Nowak, B. (2024). Differentiated response of *Hypericum perforatum* to foliar application of selected metabolic modulators: Elicitation potential of chitosan, selenium, and salicylic acid mediated by redox imbalance. *Phytochemistry, 227*, 114231.

Sykłowska-Baranek, K., Sygitowicz, G., Maciejak-Jastrzębska, A., Pietrosiuk, A., & Szakiel, A. (2022). Application of priming strategy for enhanced paclitaxel biosynthesis in *Taxus× media* hairy root cultures. *Cells, 11*(13), 2062.

Taura, K., Yamamoto, S., Hayashi, S., Furusaki, S., & Shioya, S. (2014). A new bioreactor for paclitaxel production based on foam separation. *Asia-Pacific Journal of Chemical Engineering, 9*(1), 75–80.

Tavakoli, F., Rafieiolhossaini, M., Ravash, R., & Ebrahimi, M. (2020). Subject: UV-B radiation and low temperature promoted hypericin biosynthesis in adventitious root culture of *Hypericum perforatum. Plant Signaling & Behavior, 15*(7), 1764184.

Terrier, B., Courtois, D., Hénault, N., Cuvier, A., Bastin, M., Aknin, A., & Pétiard, V. (2007). Two new disposable bioreactors for plant cell culture: The wave and undertow bioreactor and the slug bubble bioreactor. *Biotechnology and Bioengineering, 96*(5), 914–923.

Thirugnanasambantham, P., & Senthil, K. (2016). In vitro and omics technologies opens a new avenue for deciphering withanolide metabolism in *Withania somnifera. International Journal of Pharmacy and Pharmaceutical Sciences*, 17–26.

Tomilova, S. V., Globa, E. B., Demidova, E. V., & Nosov, A. M. (2023). Secondary metabolism in *Taxus* spp. plant cell culture in vitro. *Russian Journal of Plant Physiology, 70*(3), 23.

Vendrame, W. A., Xu, J., & Beleski, D. G. (2023). Micropropagation of *Brassavola nodosa* (L.) Lindl. using SETIS™ bioreactor. *Plant Cell, Tissue and Organ Culture (PCTOC), 153*(1), 67–76.

Verdú-Navarro, F., Moreno-Cid, J. A., Weiss, J., & Egea-Cortines, M. (2023). The advent of plant cells in bioreactors. *Frontiers in Plant Science, 14*, 1310405.

Wang, J., Man, S., Gao, W., Zhang, L., & Huang, L. (2013). Cluster analysis of ginseng tissue cultures, dynamic change of growth, total saponins, specific oxygen uptake rate in bioreactor and immuno-regulative effect of ginseng adventitious root. *Industrial Crops and Products, 41*, 57–63.

Wang, Y., Guan, Y. M., Lu, B. H., & Gao, J. (2016). First report of ginseng (*Panax ginseng*) root rot caused by *Fusarium acuminatum* in China. *Plant Disease, 100*(2), 525.

Wierzchowski, K., & Pilarek, M. (2024). Disposable rocking bioreactors: Recent applications and progressive perspectives. *Trends in Biotechnology, 42*(3), 261–264.

Wysokinska, H., & Chmiel, A. (1997). Transformed root cultures for biotechnology. *Acta Biotechnologica, 17*(2), 131–159.

Xu, F., Valappil, A. K., Mathiyalagan, R., Tran, T. N. A., Ramadhania, Z. M., Awais, M., & Yang, D. C. (2023). In vitro cultivation and ginsenosides accumulation in Panax ginseng: A review. *Plants, 12*(17), 3165.

Yukimune, Y., Tabata, H., Higashi, Y., & Hara, Y. (1996). Methyl jasmonate-induced overproduction of paclitaxel and baccatin III in Taxus cell suspension cultures. *Nature biotechnology, 14*(9), 1129–1132.

CHAPTER

3

Genetic fidelity assessment of *in vitro*-regenerated medicinal plants: cytological, chromatographic, and molecular marker-based approaches

Phanikanth Jogam[1], *Gulab Khan Rohela*[2], *Mahender Aileni*[3], *Dulam Sandhya*[4], *Nageswara Rao Singisala*[5], *and Venkataiah Peddaboina*[6]

[1]Multi-Disciplinary Research Unit, Kakatiya Medical College, Warangal, Telangana, India [2]CSB-Central Sericultural Research and Training Institute, Mysuru, Karnataka, India [3]Department of Biotechnology, Telangana University, Nizamabad, Telangana, India [4]Department of Biotechnology, Kakatiya University, Warangal, India [5]Department of Botany, Osmania University, Hyderabad, Telangana, India [6]Department of Microbiology, Kakatiya University, Warangal, Telangana, India

3.1 Introduction

Medicinal plants have long served as a cornerstone for plant-based pharmaceutical drug discovery and traditional healthcare systems across the world (Manisha et al., 2025). They continue to play a vital role, particularly in Asian, African, and Latin American regions, where indigenous knowledge systems, herbal formulations, and ethnomedicinal practices remain deeply integrated within the routine health management practices. Nearly 80% of the global population still relies, at least partly, on plant-derived formulations as primary therapeutic sources (Nasim et al., 2022), and a significant proportion of modern drugs also originate directly or indirectly from bioactive phytochemicals isolated from medicinal plants (Halder and Jha, 2023). These biologically active secondary metabolites, including alkaloids, glycosides, phenolics, terpenoids, flavonoids, steroids, coumarins, and essential oils, form the biochemical basis for pharmaceutical applications ranging from anticancer and antidiabetic to neuroprotective, immunomodulatory, and anti-inflammatory therapies.

Medicinal Plant Biotechnology for Ameliorated Secondary Metabolite Production
DOI: https://doi.org/10.1016/B978-0-443-43944-5.00008-6

In vitro regeneration techniques have emerged as powerful biotechnological tools for large-scale multiplication, germplasm conservation, genetic improvement, secondary metabolite enhancement (Kumari, Kumari, et al., 2025), and year-round production of quality planting material, which is independent of seasonal fluctuations (Manohar et al., 2024). However, a major constraint associated with the plant tissue culture-based regeneration is somaclonal variation such as genetic, epigenetic, biochemical, and morphological deviations that may arise during callus induction, organogenesis, somatic embryogenesis, or prolonged subculturing (Majumder et al., 2025). Since minor genomic alterations can influence different biosynthetic pathways and that may alter the phytochemical quantity and quality, maintaining true-to-type fidelity of regenerated plants is extremely crucial to ensure therapeutic consistency and commercial reliability (Kundu et al., 2025).

With the continuing expansion of the global herbal drug market, the demand for genetically uniform, stable, and biochemically consistent medicinal plant material has increased tremendously (Hossain et al., 2022). Conventional methods of propagation are often inadequate due to low multiplication efficiency, seasonal dependency, poor regeneration potential, and unsuitability for large-scale exploitation of threatened or high-value medicinal plants (Ahlawat and Yadav, 2024). Therefore fidelity assessment is a critical requirement for quality assurance, pharmaceutical standardization, industrial scale-up, and regulatory acceptance of plant-derived drugs (Shukla and Bhutani, 2023).

Cytogenetic stability, phytochemical uniformity, and molecular integrity collectively define the therapeutic value and commercial applicability of regenerated medicinal plants (Krishnaswamy, 2024). Thus genetic fidelity assessment must ensure holistic, spanning genome, chromosome, and metabolite levels. Cytology-based tools allow detection of chromosomal aberrations, while chromatographic profiling confirms chemotype authenticity and metabolite stability (Mahanta et al., 2023). Meanwhile, DNA-based molecular markers like random amplified polymorphic DNA (RAPD), inter-simple sequence repeat (ISSR), start codon targeted polymorphism (SCoT), simple sequence repeat (SSR), and amplified fragment length polymorphism (AFLP). provide precise high-resolution confirmation of genetic uniformity (Rohela and Saini, 2023). The integration of these three complementary approaches strengthens micropropagation pipelines and ensures that plant tissue culture-derived materials are reliable for pharmaceutical industries, conservation programs, and research systems.

With the growing commercial demand for standardized herbal products, phytopharmaceuticals, nutraceuticals, plant-based cosmeceuticals, and natural drug lead molecules, the sustainable utilization, rapid multiplication, and conservation of elite, chemically superior genotypes of medicinal plants have become critical priorities for global health, industrial applications, and biodiversity management (Zaman et al., 2025). However, most medicinal plants are slow-growing, seasonal, recalcitrant in seed germination, or threatened due to overharvesting, habitat loss, and climate change (Mykhailenko et al., 2025). As a result, *in vitro* regeneration and plant tissue culture-based propagation techniques have emerged as indispensable biotechnological strategies for mass multiplication, ex situ germplasm conservation, genetic improvement, and favoring continuous supply of high-quality plant biomass for industrial extraction and pharmaceutical production (Wijerathna-Yapa et al., 2025). Ensuring regenerated plants in terms of genetically identical with that of elite parental plants is essential for maintaining their therapeutic efficacy, chemical uniformity, and commercial reliability.

This chapter provides a comprehensive review of the principles, methodologies, advantages, limitations, and application scope of cytological techniques, chromatographic profiling, and molecular marker-based approaches in fidelity assessment of *in vitro* regenerated medicinal plants. The objective is to present a comparative evaluation of these strategies and highlight their significance in ensuring stable genetics, consistent metabolite profiles, and standardized therapeutic outputs for sustainable and globally acceptable phyto-pharmaceutical production.

3.2 *In vitro* regeneration of medicinal plants

In vitro regeneration is a micropropagation technique in which plant explants are cultured in a sterile laboratory setting to develop new plantlets. This process depends on the plant cell's capacity (cellular totipotency) to dedifferentiate and regenerate via *in vitro* morphogenesis called organogenesis or somatic embryogenesis (Desai and Desai, 2022). It is employed for the production of genetically uniform plants, the large-scale propagation of disease-free plants, and the preservation of genetic resources. Organogenesis is the *in vitro* development of plant organs (shoots, roots, or both) from explants (such as leaf, stem, internodes, inflorescence, rhizomes, or root tissues) under controlled sterile conditions, typically on nutrient medium supplemented with plant growth regulators (PGRs), while embryogenesis is the process of creating somatic embryos from somatic cells (without fertilization) that can subsequently develop into complete plantlets. There are two main ways to accomplish the processes of organogenesis and somatic embryogenesis: direct organogenesis, in which the explant generates new organs, like roots/shoots or embryos, without first producing an intermediate mass of unorganized cells (callus). Because it can lower the chance of genetic mutations, commonly referred to as somaclonal variation, this approach is frequently chosen. Conversely, in indirect organogenesis, in this process, the explant forms a callus, which is a disorganized mass of undifferentiated cells. The creation of new organs or embryos is then induced by culturing the callus using various PGRs in the culture media. This technique is very helpful for regenerating a large number of plantlets from a single explant and for genetic modification.

Micropropagation technique is a potential tool of plant biotechnology generally used for mass production of uniform, disease-free, economically important plants of horticulture, agriculture, and medicinal importance year-round with space efficiency (Hasnain et al., 2022). The key steps of organogenesis include selection of explants and sterilization, induction of shoots/roots by direct (without) or indirect (with callus) organogenesis, and acclimatization. While the key steps of somatic embryogenesis include explant selection and sterilization, induction of somatic embryos by direct or indirect embryogenesis, embryo maturation, germination, conversion, and acclimatization of plantlets (Loyola-Vargas and Ochoa-Alejo, 2016). In comparison to the organogenesis route, somatic embryogenesis under *in vitro* conditions is generally considered a more complex developmental pathway and can be a rare/recalcitrant process for many plant species.

Medicinal plants are an indispensable source of pharmaceuticals and nutraceuticals and are used for herbal remedies. Currently, several medicinal plants are facing their overharvesting from the wild, habitat degradation, low seed viability or dormancy, slow growth rates, and poor conventional propagation methods, which threaten many species. Despite this, more than three decades onwards, there are continuous efforts being made for the large scale *in vitro* mass production of several elite medicinal plants due to spiking demand for their phytochemical attributes in the pharmaceutical industry and for medical research (Niazian, 2019; Onipede et al., 2024).

In medicinal plants, organogenesis or somatic embryogenesis is not only widely used for their clonal propagation (mass multiplication) but also facilitates conservation of rare or endangered species, enhancement of secondary metabolites, and creation of plant genotype/ chemotype with high phytoconstituents through genetic methods such as CRISPR/ Cas9, other genome editing tools (ZFNs, TALENs), or metabolic engineering (Mitra et al., 2023).

The achievement of plant organogenesis or somatic embryogenesis in tissue culture relies on a number of factors, including the type and age of the explants, specific plant species genotype, and the composition of the culture medium. By standardizing explant choice, type of culture media, and PGR combinations, it is feasible to achieve high-efficiency *in vitro* regeneration with minimal genetic variation, ensuring quality and consistency of medicinal compounds.

3.3 Genetic fidelity assessment: importance and challenges

Genetic fidelity in the context of *in vitro* regenerated medicinal plants refers to the degree of similarity and true-to-typeness maintained between the regenerated plantlets and their original mother plants at the genetic, biochemical, morphological, and chromosomal levels (Gupta and Chaturvedi, 2022). It ensures that the regenerated plantlets do not deviate from the elite parental stock that was selected for its unique medicinal, agronomic, phytochemical, or pharmaceutical attributes. In short, fidelity assessment confirms that the micropropagated plants remain genetically stable, identical clones, retaining the same therapeutic efficacy, metabolic composition, biosynthetic capacity, and physiological features (Rohela et al., 2022). For medicinal plants that are commercially exploited for pharmaceutically important metabolites, maintaining this genetic purity is crucial, since any deviation in genotype directly influences metabolite profiles, thereby affecting drug potency, safety, toxicity profile, and therapeutic reproducibility (Canter et al., 2005).

One of the major risks associated with plant tissue culture-based mass multiplication is the occurrence of somaclonal variation (Majumder et al., 2025). Somaclonal variation refers to heritable or nonheritable changes that arise spontaneously during *in vitro* culture due to several culture-induced stresses such as high auxin/cytokinin ratios, prolonged callus phases, oxidative stress, nutrient imbalances, repeated subculturing, or chromosomal reorganizations during dedifferentiation and redifferentiation (Ranghoo-Sanmukhiya, 2020). These variations can manifest at various biological levels, ranging from DNA sequence mutations, gene expression shifts, chromosomal rearrangements, epigenetic modifications, and protein profile alterations to significant changes in the accumulation of secondary metabolites (Verma and Shukla, 2015). For medicinal and aromatic plants, even subtle biochemical variation can drastically alter the therapeutic value and commercial applicability of plant-based products, especially plants exploited for alkaloids, glycosides, terpenoids, flavonoids, polyphenols, or resinous compounds (Abiri, 2024). Therefore if genetic fidelity is not assessed rigorously, large-scale micropropagation may run into the risk of propagating variants with inferior chemical compositions, reduced bioactivity, or altered pharmacological attributes.

The genetic stability of *in vitro* cultures is influenced by various biological and technical factors such as the type of explant used (meristems are the safest), duration of culture, type of regeneration pathway (direct organogenesis is generally more stable than callus-mediated pathways), type and concentration of growth hormones, physiological status of donor plant, culture environment, and genomic nature (polyploids, heterozygous, and wild plants often show higher variation) (Long et al., 2022). In addition to these intrinsic factors, methodological inconsistencies, differences in laboratory handling practices, and lack of proper selection strategies further increase the risk of genetic instability. Thus fidelity assessment becomes not just a scientific requirement but a quality assurance mandate in research laboratories, nurseries, commercial tissue culture companies, herbal drug manufacturing industries, and gene banks (Biswas and Kumar, 2023).

A major challenge is that somaclonal variation may not always express morphologically and may remain latent or cryptic. Many variations occur at the molecular or epigenomic levels and are not detectable by phenotypic observations alone (Bairu et al., 2011). Morphology-based screening, though simple, is highly unreliable for early-stage detection. It may identify extreme variants, but minor genomic deviations or subtle metabolite shifts often remain unnoticed until after field transfer or until metabolites are quantified chemically. This highlights the necessity for sensitive, reproducible, quantitative, and high-resolution methods. Detecting subtle variations at the DNA sequence or metabolic pathway level requires advanced molecular marker systems, high-throughput chromatographic platforms, and cytogenetic analytical tools (Manchanda et al., 2025). Another major challenge is ensuring reproducibility of results across different laboratories, different batches of tissue culture material, and different environmental conditions. Variation in protocols often yields inconsistent genetic fidelity results, making standardization a critical requirement.

Ensuring genetic fidelity is further complicated by the fact that different tools provide different layers of information. Chromosomal analysis may confirm ploidy status but may not detect single-nucleotide-level variations. Chromatographic profiling assures phytochemical uniformity, but chemical stability does not always guarantee genetic stability because metabolite production may be influenced by the environment or culture media composition (Mutha et al., 2025). Similarly, molecular marker-based systems provide DNA-level insight but may not directly reflect metabolite yield consistency. Therefore no single approach, when used alone, is sufficient to fully authenticate genetic fidelity. A multifaceted approach combining cytological, biochemical, and molecular profiling is essential for obtaining a complete picture of plant identity and stability (Yu et al., 2022). This integrated assessment approach increases robustness, validation strength, reliability, and scientific acceptance.

The need for robust fidelity assessment has gained further significance with the growing trend of biotechnological interventions in medicinal plants, such as somatic hybridization, haploid breeding, polyploid induction, genetic transformation, gene editing, plant metabolic engineering, endophyte-assisted metabolite boosting, and elicitation-based secondary metabolite enhancement (Mishra and Srivastava, 2025). These interventions increase the probability of unintended gene expression disturbances and unforeseen changes, making systematic fidelity monitoring indispensable. Regulatory authorities and international herbal product standards are increasingly emphasizing genomic traceability and metabolic consistency for medicinal plants (Wang et al., 2024). With global demand rising for plant-based pharmaceuticals, nutraceuticals, herbal extracts, essential oils, plant-derived nanomaterials, and functional foods, genetic fidelity has become a determining factor not only for scientific research reliability but also for regulatory compliance, market acceptability, export certification, and intellectual property protection (Pant et al., 2025).

Another challenge lies in the practical implementation of genetic fidelity assessment in commercial production systems. Many private tissue culture units do not routinely perform the requisite genetic fidelity testing due to cost, lack of expertise, absence of standardized Standard Operating Procedures (SOPs), limited access to molecular laboratories, or time constraints. As a result, somaclonal variants may unknowingly enter the supply chain, leading to poor-quality raw material, compromised drug formulations, and loss of trust in herbal industries (Hossain et al., 2022). There is thus an urgent need for training, awareness-building, the establishment of standardized molecular protocols, commercially viable marker panels for specific high-value medicinal plants, and the development of rapid diagnostic kits for fidelity screening. Governments and quality control bodies must also promote mandatory fidelity certification, similar to seed certification programs in agriculture.

Going forward, the future of fidelity assessment lies in integrating emerging genomic, metabolomic, high-throughput sequencing, and machine learning-based prediction tools. Advanced omics platforms such as whole-genome resequencing, epigenome profiling, transcriptomics, targeted metabolomics, mass spectrometry-based biomarker signatures, and AI-assisted variant prediction models will transform fidelity assessment from conventional screening to precision validation (Ali, 2023). Development of species-specific molecular marker sets, establishment of international reference databases, and harmonization of chromatographic fingerprinting standards are necessary steps toward achieving global quality assurance frameworks.

In conclusion, fidelity assessment is a central component of micropropagation pipelines in medicinal plant biotechnology. It ensures that regenerated plants maintain their genetic and phytochemical identity, which is essential for therapeutic consistency, research authenticity, conservation of elite genotypes, and industrial herbal drug standardization. The challenges associated with the detection of subtle variations, result reproducibility, technical variability, and limitations of individual analytical tools highlight the need for a robust, multilayered, integrated assessment strategy. Combining cytological analysis, chromatographic chemoprofiling, and molecular marker-based DNA fingerprinting provides a comprehensive and reliable platform for fidelity assurance (Savitikadi et al., 2020; Sharma et al., 2018). Strengthening the fidelity assessment approach is therefore vital for the future growth, credibility, and global acceptance of medicinal plant-based biopharmaceutical and phytochemical industries.

3.4 Cytological approaches

Cytological analysis forms one of the most primary, classical, and reliable approaches for evaluating genetic stability in *in vitro* regenerated medicinal plants. Since any genetic instability that develops during tissue culture is often reflected initially at the chromosomal level, cytological methods serve as crucial diagnostic tools to detect large structural anomalies even before they visibly manifest in plant morphology or affect phytochemical composition (Soomro et al., 2025). Cytogenetic fidelity is particularly critical in medicinal plants because chromosomal aberrations may directly influence cellular metabolism, secondary metabolite pathways, enzyme activity, stress response systems, and biosynthetic gene regulation (Kumar, 2021). Thus cytological analysis keeps check on large-scale clonal dissemination of variants with altered pharmaceutical potential.

3.4.1 Chromosome counting and karyotyping

Chromosome counting and karyotyping remain cornerstone cytological techniques in genetic fidelity testing. These techniques involve preparing high-quality metaphase spreads, visualizing chromosomes under the microscope, and comparing number, size, shape, centromere position, and banding pattern between the donor mother plant and *in vitro* raised regenerants. These approaches were extensively applied in medicinal plants such as *Rauwolfia serpentina* (source of reserpine), *Coleus forskohlii* (source of forskolin), *Catharanthus roseus* (anticancer alkaloids), *Withania somnifera* (withanolides), *Tinospora cordifolia* (immunomodulatory lignans), and *Bacopa monnieri* (bacosides), where biochemical traits are highly genotype-dependent (Abdelfatah and Efferth, 2015; Kavitha et al., 2010; Pandey et al., 2017; Pramanik et al., 2021; Prasad et al., 2023). Chromosome analysis can detect chromosomal breakage, deletion, fragmentation, inversion, translocation, diploid to aneuploid shifts, and polyploidization events that are generally noticed during callus-mediated regeneration or prolonged *in vitro* maintenance. Thus karyological uniformity is a direct indicator of genome stability in micropropagated clones (Umesh, 2008).

3.4.2 Flow cytometry for ploidy analysis

Flow cytometry is increasingly preferred in fidelity programs due to its high throughput, speed, precision, and ability to analyze large populations of samples in minimal time. This is especially advantageous in commercial micropropagation industries where thousands of plantlets need to be screened before their field cultivation. Flow cytometry measures nuclear DNA content and can detect shifts in ploidy levels caused by tissue culture-induced somatic polyploidy, chromosome duplication errors, endoreduplication, or genome dosage alterations (Gantait and Mukherjee, 2021). Medicinal plants such as *Artemisia annua*, *Stevia rebaudiana*, *Curcuma longa*, *Mentha arvensis*, *Valeriana jatamansi*, *Costus speciosus*, and *Ocimum sanctum* have been subjected to flow cytometric fidelity screening for genetic uniformity of regenerated clones (Faisal et al., 2014; Liu et al., 2018; Purohit et al., 2015; Ragul et al., 2025; Rewers and Jedrzejczyk, 2016; Zainuddin et al., 2025). Ploidy alteration often results in unpredictable shifts in bioactive metabolite concentrations; therefore ploidy validation prior to field deployment is a critical regulatory and industrial requirement for many high-value medicinal plants.

3.4.3 Mitotic index and cell cycle analysis

Mitotic index determination and cell cycle profiling further strengthen cytogenetic fidelity analysis by measuring mitotic activity and detecting cellular stress response induced by hormonal imbalance, osmotic shock, oxidative stress, and long-term subculture exposure. A reduced mitotic index in regenerated plantlets compared to mother plants may suggest chromosome instability, genotoxicity, and abnormal cell proliferation behavior. This is evident in species like *Phyllanthus amarus*, *Andrographis paniculata*, *Picrorhiza kurroa*, *Datura metel*, and *Salvia officinalis*, where somatic embryos or callus cultures are frequently used and tend to show differential mitotic responses during *in vitro* regeneration (Avani and Rao, 2008; Hassanzadeh et al., 2020; Kuriakose et al., 2014; Mondal et al., 2013; Yusoff et al., 2022). Cell cycle distribution analysis by flow cytometry, useful in reflecting the G1/S/G2/M phase ratio, serves as an early alert method. This analysis is helpful to identify potential variants before chromosomal abnormalities become morphologically evident in *in vitro* regenerants.

Cytological evaluation methods are highly valuable because they are relatively inexpensive, do not require advanced sequencing information, and enable direct visualization of chromosomal architecture. These methods are especially useful in developing countries where tissue culture-based medicinal plant industries are rapidly expanding, wherein resources for high-end molecular screening are limited. However, cytology has its practical limitations; it cannot reliably detect micromutations, point mutations, small insertions/deletions, methylation shifts, transposable element mobility, SNP-level changes, or subtle epigenetic variation that may still significantly alter secondary metabolite pathways. Additionally, some plant species have small-sized chromosomes, polyploid complexity, sticky chromosomes, or high DNA condensation patterns, which make metaphase preparation extremely challenging.

Thus while cytological approaches are indispensable first layer of fidelity screening, when complemented with chromatographic metabolite profiling and DNA marker-based molecular tools would be comprehensive and high-resolution fidelity assurance for quality medicinal plant material. Combined multiplatform analysis ensures that regenerated medicinal plantlets remain genetically pure, chemically consistent, and therapeutically reliable, meeting both industrial quality standards and pharmaceutical validation requirements.

3.5 Molecular marker-assisted approaches to evaluate genetic fidelity

Exposure to synthetic PGRs, long-term callus culture, the use of mutagenic agents or stress conditions, and repeated subculturing can cause genetic changes (mutations, chromosomal rearrangements, altered ploidy levels, and epigenetic alterations) in plants multiplied via *in vitro* culture techniques. The uniformity of active component profiles in micropropagated medicinal plants is decreased by such genetic alterations (i.e., *in vitro* conditions may alter metabolic pathways, resulting in qualitative or quantitative changes in bioactive compounds). The genetic changes in plants grown under *in vitro* culture conditions can be detected by various methods such as molecular markers, RAPD, ISSR, AFLP, SSR, sequence-related amplified polymorphism (SRAP), SCoT, and CAAT box-derived polymorphisms (CBDP) (Table 3.1). Cytological analysis: chromosome counting, karyotyping; flow cytometry: for ploidy changes; and epigenetic profiling: methylation-sensitive markers and metabolic profiling, etc.

In recent times, several reports are available on the assessment of the genetic fidelity status of *in vitro* regenerated medicinal plants, as the critical hallmark of micropropagation is to produce true-to-type plants in relation to mother stock plants (Rohela et al., 2022). Hence, it is necessary to evaluate the genetic uniformity of *in vitro* propagated plants at the molecular level by using various markers. The genetic uniformity in clonally propagated medicinal plants is crucial as they ensure the production and supply of uniform plant clones of commercial significance.

In comparison to biochemical, phenotypical, and physiological markers, molecular marker-assisted genetic uniformity investigations are more robust, accurate, and reliable. Here, in the below section of the chapter, different molecular markers based on the evaluation of *in vitro* regenerated medicinal plants are discussed. Molecular markers such as randomly amplified polymorphic DNA (RAPD), ISSR, AFLP, simple sequence repeats (SSR), SRAP, SCoT, and CBDP are discussed (Table 3.1).

3.5.1 Genetic fidelity evaluation of *in vitro* raised medicinal plants by random amplified polymorphic DNA markers

In plant biotechnology, random amplified polymorphic DNA (RAPD) is a PCR-based molecular marker approach that is frequently used for genetic fingerprinting, diversity analysis, and especially for the identification of somaclonal variation in tissue culture-derived plantlets with the mother plant (Table 3.1). RAPD are random primers (usually 10 base pairs) used to amplify anonymous DNA sequences under low-stringency PCR conditions. The single arbitrary primer amplified fragments are then separated by gel electrophoresis, revealing banding patterns. The presence/absence of bands between samples (scoring of bands) can be interpreted as polymorphisms. The advantages of RAPD are that it is a simple and cost-effective assay, there is no need for prior sequence information, and this assay is rapid and suitable for large sample sizes. While the limitations of RAPD are its lower reproducibility, it cannot distinguish between heterozygous and homozygous loci, and band size alone doesn't confirm identity. However, RAPD markers were employed to assess the genetic uniformity in several medicinal plants, including *B. monnieri* (Tripathi et al., 2012), *Justicia adhatoda* (Kumar et al., 2014), *Rauwolfia tetraphylla* (Rohela et al., 2019), *Solanum trilobatum* (Pendli et al., 2019), *Ranunculus wallichianus* (Srinivasan et al., 2021), and *Dendrobium moschatum* (Tikendra et al., 2019).

3.5.2 Genetic fidelity evaluation of *in vitro* raised medicinal plants by inter-simple sequence repeat markers

It is a PCR-based molecular marker assay used to amplify DNA segments between two microsatellites (simple sequence repeats/SSRs) using primers that are 16–25 bases long. Their primers are designed from di-, tri-, tetra-, or penta-nucleotide repeats [e.g., (GA)n, (AG)n, (AC)n], which are dominant (like RAPD) but more specific, for which they require a DNA

TABLE 3.1 Genetic fidelity analysis using different molecular markers in *in vitro* propagated medicinal plants.

Name of the plant	Type of regeneration	Explants used	Markers used	References
Ajuga parviflora	Direct and Indirect	Leaf, petiole, and root	ISSR, RAPD, and SCoT	Kumari et al. (2025)
Andrographis alata	Direct	Node	ISSR and RAPD	Kadapatti and Murthy (2021)
Andrographis echioides	Direct	Leaf	ISSR	Savitikadi et al. (2020)
Annona reticulate	Direct	Node	ISSR and SCoT	Kudikala et al. (2020)
Anthurium andraeanum	Indirect	Root	ISSR	Zhang et al. (2021)
Artemisia vulgaris	Direct	Node	ISSR and SCoT	Jogam et al. (2020)
Asystasia gangetica	Direct and Indirect	Leaf	SCoT	Dilkalal (2021)
Bacopa monnieri	Direct and Indirect	Shoot tip	ISSR	Pramanik et al. (2021)
Bergenia ligulata	Direct	Leaf and petiole	ISSR	Deeba et al. (2024)
Blepharispermum subsessile	Direct	Cotyledonary node	ISSR	Monalisa et al. (2024)
Boesenbergia rotunda	Direct	Sprouted buds	SSR	Thepthong and Kanjanawattanawong (2025)
Bretschneidera sinensis	Direct	Mature zygotic embryos	ISSR and RAPD	Yan et al. (2024)
Carica papaya	Direct and Indirect	Apical, nodal, petiole, leaf, and root	ISSR and RAPD	Vishal et al. (2024)
Clerodendrum volubile	Direct and Indirect	Leaf, nodal, and petiole	ISSR and SCoT	Olusan et al. (2024)
Chlorophytum borivilianum	Direct	Flower stalk	ISSR and SCoT	Kaushal et al. (2021)
Corallocarpus epigaeus	Direct	Node	ISSR	Vemula et al. (2020)
Corynandra chelidonii	Direct	Node	ISSR	Sirangi et al. (2022)
Crinum malabaricum	Direct	Bulb	ISSR and SCoT	Priyadharshini et al. (2020)
Crocus sativus	Direct	Node	DAMD, IRAP, and SCoT	Gautam and Bhattacharya (2021)
Curcuma longa	Direct	Shoot bud	ISSR and SCoT	Pittampalli et al. (2022)
Curcuma longa	Direct	Unsprouted rhizome buds	ISSR and SCoT	Dhondi et al. (2025)

Curcuma zedoaria	Direct	Rhizome sprouts	ISSR and SCoT	Subramanian et al. (2025)
Dactylorhiza hatagirea	Direct	Shoot meristem	SCoT	Jaryal et al. (2025)
Decalepis salicifolia	Direct	Shoot tip	SCoT	Wasi et al. (2024)
Disanthus cercidifolius	Direct	Sprouting buds	ISSR	Ulvrova et al. (2021)
Dimocarpus longan	Direct	Shoot tip and leaf	AFLP	El-Naby (2022)
Dioscorea pentaphylla	Direct	Node	SCoT	Manokariet al. (2022)
Diplocyclos palmatus	Direct	Node	ISSR and RAPD	Upadhyay et al. (2021)
Drypetes roxburghii	Direct	Root	SCoT	Saini et al. (2025)
Evolvulus alsinoides	Direct	Node	ISSR and RAPD	Awere et al. (2024)
Ficus carica	Direct	Node	DAMD and ISSR	Parab et al. (2021)
Ficus palmate	Direct	Node	SCoT, ISSR, and RAPD	Al-Aizari et al. (2024)
Flemingia macrophylla	Direct	Cotyledon node	ISSR and RAPD	Sirikonda et al. (2020)
Gladiolus grandiflorus	Indirect	Corm sprout	ISSR	Mousavimatin et al. (2025)
Haplophyllum tuberculatum	Direct and Indirect	Immature seed pods	ISSR and RAPD	Alsafran et al. (2022)
Hibiscus sabdariffa	Indirect	Root	DAMD, ISSR, and RAPD	Konar et al. (2019)
Hybanthus enneaspermus	Direct	Shoot tip	RAPD and SCoT	Sathish et al. (2022)
Kaempferia parviflora	Direct	Transverse thin cell layer	CDDP, DAMD, ISSR, and SCoT	Koh et al., 2024
Lagerstroemia speciosa	Direct	Node	ISSR and RAPD	Ahmad et al. (2022)
Lagerstroemia speciosa	Indirect	Leaf	ISSR	Wu et al. (2024)
Lippia javanica	Direct	Node	ISSR and SCoT	Mood et al. (2022)
Malaxis acuminata	Direct	Transverse thin cell layer	SCoT and CBDP	Bhattacharyya et al. (2023)
Moringa concanensis	Direct	Shoot tip	CBDP, ISSR, and SCoT	Gautam et al. (2025)
Muehlenbeckia platyclada	Direct	Node	ISSR and SCoT	Badhepuri et al. (2024)
Nardostachys jatamansi	Direct	Rhizome	DAMD and SCoT	Dhiman et al. (2021)

(*Continued*)

TABLE 3.1 (Continued)

Name of the plant	Type of regeneration	Explants used	Markers used	References
Naregamia alata	Indirect	Laef	ISSR and RAPD	Raja et al. (2022)
Nepeta asterotricha	Direct	Node	ISSR	Zamani et al. (2024)
Nothapodytes foetida	Direct	Cotyledonary node, hypocotyl, and leaf	ISSR and SCoT	Attaluri and Dharavath (2025)
Origanum majorana	Direct	Node	ISSR	Sandhya et al. (2021)
Patchouli	Indirect	Leaf	CBDP and SCoT	Lalthafamkimi et al. (2022)
Pelatantheria insectifera	Direct	Node	ISSR and SCoT	Devi et al. (2025)
Phalaenopsis orchid	Direct	Flower stalk node	RAPD	Sarmah et al. (2024)
Plectranthus amboinicus	Direct	Immature cotyledon	ISSR	Tippani et al., 2025; Faisal 2022
Pterocarpus marsupium	Indirect	immature zygotic embryos	ISSR	Tippani et al. (2019)
Pueraria tuberose	Direct	Node	ISSR and RAPD	Kanthaliya et al. (2025)
Ranunculus wallichianus	Indirect	Leaf	ISSR and RAPD	Srinivasan et al. (2021)
Rauvolfia serpentina	Direct	Shoot tip	ISSR and SCoT	Gantait et al. (2023)
Rauwolfia tetraphylla	Indirect	Leaf, stem	ISSR, RAPD, and SCoT	Rohela et al. (2019)
Rheum ribes	Direct and indirect	Embryo axis, hypocotyl, and cotyledon	SCoT	Aldaif et al. (2025)
Ruta chalepensis	Indirect	Internode	ISSR	Qahtan et al. (2022)
Salvia nemorosa	Direct and Indirect	Leaf	ISSR	Heydari and Chamani (2022)
Solanum erianthum	Direct, Indirect	Leaf	RAPD	Sarkar and Banerjee (2020)
Solanum khasianum	Indirect	Leaf, node, and petiole	SCoT and ISSR	Chirumamilla et al. (2021)
Scaevola taccada	Direct	Node	SCoT	Manokari et al. (2020)
Scaevola taccada	Direct	Node	ISSR	Shekhawat et al. (2021)
Siraitia grosvenorii	Direct	Node	ISSR and RAPD	Patial et al. (2024)
Strychnos potatorum	Direct	Nodal bud	ISSR	Vijaya Kumar et al. (2023)

Solanum viarum	Direct	Leaf, petiole, and internodes	RAPD and SSR	Pandey et al. (2020)
Spathoglottis plicata	Direct	Node	SCoT	
Stevia rebaudiana	Direct	Shoot tip	CDDP, DAMD, ISSR, and SCoT	Subrahmanyeswari et al. (2023)
Tabernaemontana alternifolia	Indirect	Internode	DAMD and ISSR	Shinde et al. (2023)
Tecoma stans	Direct	Shoot tip	RAPD	Hussain et al. (2020)
Tectona grandis	Direct	Auxiliary bud	RAPD, RFLP, and ISSR	Nandy et al. (2025)
Thottea siliquosa	Indirect	Root	SCoT	Vrundha and Thomas (2023)
Tylophora indica	Indirect	Leaves and nodes	SCoT	Mamgain et al. (2022)
Uraria picta	Direct	Cotyledonary node	ISSR and SCoT	Dixit et al. (2024)
Uvaria hamiltonii	Direct	Node	ISSR and RAPD	Mohanty et al. (2025)
Vaccinium vitis	Direct	Leaf	EST-SSR and GSSR	Kundu et al. (2025)
Vanilla planifolia	Direct	Node	SCoT	Manokari et al. (2021)
Vitellaria paradoxa	Direct	Node	ISSR	Attikora et al. (2025)
Withania somnifera	Direct	Shoot tip	ISSR and SCoT	Bandyopadhyay et al. (2025)
Zingiber officinale	Direct	Rhizome bud	ISSR and SSR	George et al. (2022)

sample of 20–100 ng. As the primer sequence designed on microsatellite motifs needs no flanking regions but gives high reproducibility (due to longer, specific primers). The genetic fidelity is carried out by binary scoring, that is, band presence/absence (1/0). ISSR markers were used in several medicinal plants' genetic fidelity (somaclonal variation), such as *Justicia adhatoda* (Kumar et al., 2014), *Dendrobium moschatum* (Tikendra et al., 2019), *Thunbergia coccinea* (Sultana et al., 2022), *Annona reticulata* (Kudikala et al., 2020), *Solena amplexicaulis* (Koppula et al., 2023), *T. cordifolia* (Pandey et al., 2023), *Muehlenbeckia platyclada* (Badhepuri et al., 2024), and *Withania somnifera* (Hiremath et al., 2021; Khabiya et al., 2024) (Table 3.1).

3.5.3 Genetic fidelity evaluation of *in vitro* raised medicinal plants by amplified fragment length polymorphism markers

DNA markers called AFLP (amplified fragment length polymorphism) markers are used to examine genetic variation both within and between species. Without requiring prior knowledge of the DNA sequence, the method entails restricting genomic DNA, ligating adapters, and then selectively amplifying segments using PCR. Because AFLP analysis is highly polymorphic and repeatable, it can be used in a variety of applications, including phylogenetic investigations, population genetics, and genetic mapping. Its advantages are high polymorphism detection power, no need for prior sequence information, genome-wide coverage, reproducibility, and robustness. Its disadvantages are dominant markers (cannot distinguish homozygous dominant vs. heterozygous), labor-intensive and technically demanding, and data analysis can be complex. AFLPs are employed to screen for genetic uniformity in order to guarantee that the propagated plants are true-to-type.

The typical workflow of AFLP involves DNA extraction from the mother (donor) plant and several *in vitro* regenerated plants, following which the restriction digestion, adaptor ligation, preselective and selective amplification, and electrophoresis are carried out. While doing DNA banding pattern comparison, the data analysis is carried out using Jaccard's similarity coefficient, UPGMA clustering, or dendrogram construction. There are several studies evident on AFLP-based genetic uniformity evaluation of *in vitro* regenerated medicinal plants with that of mother stock plants. The reports include genetic relationships among *W. somnifera* genotypes (Negi et al., 2006), genetic diversity in diverse accessions of *B. monnieri* (Srivastava et al., 2016), identification of *Tinospora crispa* and *Tinospora baenzigenri* (Katib et al., 2017), genetic diversity of 60 genotypes of *Aloe vera* (Kumar et al., 2022), distinguishing some medicinally important species of the genus *Phyllanthus* (Bandyopadhyay and Raychaudhuri, 2013), evaluating genetic diversity of *Tribulus terrestris* (Sarwat et al., 2008), and genetic diversity of *Chlorophytum borivilianum* (Tripathi, et al., 2012).

3.5.4 Genetic fidelity evaluation of *in vitro* raised medicinal plants by simple sequence repeat markers

SSR molecular markers are short tandem repeats of DNA sequences (usually 1–6 base pairs long) that are dispersed across the genome that repeatedly repeat a brief DNA motif, such as "AT" or ""GATA," for example, (AT)10 → ATATATATATATATATATAT (Idrees and Irshad, 2014). They have been referred to as simple sequence repeats or microsatellites.

They are among the most popular molecular markers in population studies, plant and animal breeding, and genetics. SSRs are highly polymorphic and helpful for differentiating between genotypes since the number of repetitions can differ between people. PCR primers are designed using the flanking areas of the SSR locus, which are conserved DNA sequences on both sides of the repeat region. DNA extraction from the sample is the first step in the SSR marker analysis process. Next, the SSR area is amplified using certain primers for flanking sequences. Gel electrophoresis or capillary electrophoresis is used to separate the amplified fragments, and genotyping scoring is done according to fragment length, meaning that different lengths correlate to different numbers of repetitions (alleles).

The advantages of SSR markers are that they are highly polymorphic (many alleles per locus), show codominant inheritance (can distinguish homozygotes and heterozygotes), and are reproducible and locus-specific, for analysis requires small amounts of DNA and is used to analyze related species having conserved flanking regions. While the limitations are that development is time-consuming and costly (requires sequence data for primer design) and there is the possibility of null allele formation due to mutations in primer binding sites.

Verifying whether the regenerated plantlets are genetically similar to the mother plant is crucial since uniformity and true-to-type propagation are necessary to preserve the medicinal plants' chemical consistency and therapeutic efficacy. SSR markers can be used in medicinal plant tissue culture studies, as they ensure consistent phytochemical profiles (important for pharmacological activity). Further, it is used to check the maintenance of the genetic identity of elite genotypes used for secondary metabolite production. Moreover, it also aids in the certification of clonally propagated medicinal plants. There are studies available demonstrating the genetic insights into the diversity or genetic evaluations of *in vitro* micropropagated medicinal plants of *Withania somnifera* (Chawla et al., 2025) and *Cannabis sativa* (Ioannidis et al., 2022), respectively (Table 3.1).

3.5.5 Genetic fidelity evaluation of *in vitro* raised medicinal plants by sequence-related amplified polymorphism markers

Open reading frames, or sections of DNA that are expected to be expressed, are amplified by the PCR-based marker system SRAP. SRAP markers are straightforward, effective, and repeatable substitute for other marker systems like RAPD and AFLP (Li and McVetty, 2013). The purpose of SRAP markers is to amplify coding areas (exons) and related regions, such as promoters, introns, and spacers, rather than between exons and noncoding regions. An SRAP primer is usually 17–18 bp long and has two parts: core sequence (13 bp): common across primers; defines binding preference. Selective nucleotides (2–3 bp) at the 3′ end add specificity. Advantages of SARP markers are that they are simple and low cost (no restriction enzymes or radioactive labeling) and reproducible (more stable than RAPD markers). Its applications include genetic diversity and population structure studies, linkage mapping, and marker-assisted selection (MAS), phylogenetic analysis and gene tagging, and identification of quantitative trait loci (QTLs).

Different studies are evident on SRAP-based genetic uniformity evaluation on *in vitro* regenerated medicinal plants with that of mother stock plants. The reports include genetic diversity analysis in medicinal *Chrysanthemum morifolium* (Shao et al., 2010) and *Tetrastigma hemsleyanum* (Peng et al., 2015), and the genetic fidelity of *Atropa acuminata*, a critically endangered medicinal herb regenerant, has been investigated (Rajput and Agrawal, 2020) using SRAP markers.

3.5.6 Genetic fidelity evaluation of *in vitro* raised medicinal plants by start codon targeted polymorphism markers

A molecular marker technique used to examine genetic diversity in plants and fungi is called SCoT markers, or start codon targeted markers. They use a single primer to amplify DNA segments in a PCR assay and are based on a conserved area of DNA that flanks the start (or initiation) codon. Because of their high polymorphism, SCoT markers are useful in a variety of applications, including genetic structure analysis, fingerprinting, and breeding initiatives (Collard and Mackill, 2009). By targeting conserved areas close to gene start codons, the SCoT marker assay provides an economical and effective way to verify the genetic homogeneity of medicinal plants such as *Curcuma longa* (Pittampalli et al., 2022), *Rauwolfia tetraphylla* (Rohela et al., 2019), *Annona reticulata* (Kudikala et al., 2020), *Decalepis salicifolia* (Wasi et al., 2024), *Carthamus tinctorius* (Ejaz et al., 2022), and *Dendrobium nobile* (Bhattacharyya et al., 2013) (Table 3.1).

3.5.7 Genetic fidelity evaluation of *in vitro* raised medicinal plants by CAAT box-derived polymorphism markers

A novel promoter-targeted molecular marker called CBDP uses the nucleotide sequence of the CAAT box, a regulatory element present in eukaryotic gene promoters, to identify genetic variation. Because it employs a single primer that binds to the conserved CAAT box region, it functions similarly to other markers like RAPD in the analysis of genetic diversity and relationships in plants (Singh et al., 2014). The SCoT molecular analysis takes advantage of the fact that the CAAT box is a conserved sequence found upstream of the start codon in many eukaryotic genes. DNA is amplified using a single primer containing the specific nucleotide sequence of the CAAT box. The primer binds to multiple CAAT box sites throughout the genome, and the amplified segments are then separated to create a DNA fingerprint that reveals genetic differences between individuals or populations. The resulting banding pattern, or "polymorphism," can be used to determine genetic diversity, population structure, and relationships. The CBDP molecular marker has been used to evaluate genetic diversity and relationships in medicinal plants, namely among *Salvia* species (Fabriki-Ourang and Karimi, 2019), *Artemisia* species (Jamshidi et al., 2024), *Prosopis cineraria* (Shekhawat et al., 2018), *Bacopa monnieri* (Choudhury et al., 2025), and *Centaurea* species (Atia et al., 2021).

3.6 Chromatographic approaches

Chromatographic approaches play a significant role in the biochemical fidelity assessment of *in vitro* regenerated plants. Molecular marker-based techniques confirm genetic stability, while chromatographic techniques validate the biochemical uniformity of regenerated medicinal plants. The pharmacological efficacy of medicinal plants mostly depends upon their bioactive compound profiles, ensuring that the profiles of secondary metabolites remain stable in concentration with the wild-type plant. The concentration and composition of secondary metabolites can be altered by somaclonal variation or stress conditions during tissue culture (Chetri et al., 2014).

3.6.1 High-performance liquid chromatography for secondary metabolite profiling

High-performance liquid chromatography (HPLC) is one of the most reliable analytical methods for the quantitative and qualitative assessment of secondary metabolites, including terpenoids, alkaloids, phenolics, and flavonoids. This tool is very efficient with high resolution, reproducibility, and sensitivity for compound identification, labeling detailed biochemical comparisons between wild and regenerated plants (Tikendra et al., 2022). HPLC has been widely used for the estimation of biochemical fidelity in species such as *Withania somnifera*, *Catharanthus roseus*, and *Stevia rebaudiana*, confirming stable metabolite patterns among regenerated lines (Chetri et al., 2014). Additionally, the combination of HPLC with photodiode array (PDA) or mass spectrometric detection gives more precise compound quantification, improving its applicability in quality control of micropropagated plants.

3.6.2 Gas chromatography–mass spectrometry for volatile compounds

Gas chromatography–mass spectrometry (GC–MS) is particularly used for measuring volatile and thermally stable compounds like terpenes, essential oils, and aromatic components. The combination of both GC and MS allows compound separation and structural elucidation with high efficiency and accuracy. Previous reports show that the integration of GC–MS-based profiling has confirmed that regenerated plants retain the same volatile composition as mother plants, like in *Mentha arvensis*, *Ocimum sanctum*, and *Zingiber officinale*. This technique is used for validating plant biochemical fidelity for rapid screening of metabolic integrity in aromatic and medicinal plants for future purposes (Hussain and Nathar, 2022).

3.6.3 Thin-layer chromatography for rapid screening

Thin layer chromatography (TLC) is another simple, cost-effective method for preliminary metabolite analysis. It gives fast and effective comparative results of crude extracts, especially where advanced chromatographic tools are unavailable. TLC lacks the quantification accuracy of HPLC or GC–MS; it is useful for only the first-level analytical tool in metabolic fidelity studies of regenerated plants such as *Aloe vera*, *Azadirachta indica*, and *Ginkgo biloba* (Liu et al., 2022).

The most important advantage of chromatographic methods depends on their ability to connect both genotype and phenotype, enabling biochemical confirmation that *in vitro* regenerated plants are equivalent to their parent plants. However, these methods are associated with notable standardization of the extraction of solvents, sample preparation, and detection parameters to ensure accuracy. These methods have some limitations, such as time-consuming procedures, dependence on reference standards, and the significant effect of *in vitro* conditions on secondary metabolite production. However, the integration of HPLC, GC–MS, and TLC gives a robust framework for genetic fidelity assessment by confirming the true-to-type micropropagation in relation to metabolic content. Future studies should focus on the combination of chromatographic data with metabolomics and chemometric modeling, which will further improve the precision and throughput of biochemical genetic fidelity assessment in medicinal industries and plant biotechnology (Xiao et al., 2022).

3.7 Conclusion and future prospects

The fidelity assessment of *in vitro* regenerated plants is entering an era of technological convergence. Future prospects will focus on enhancing accuracy, efficiency, and cost-effectiveness to maintain the genetic, epigenetic, and biochemical stability of regenerated plants. The combination of genomics, proteomics, and metabolomics will give a comprehensive view of the mechanisms of somaclonal variation and allow the prevention of instability during *in vitro* regeneration. The combination of these multiomics techniques will enable researchers to associate molecular changes with functional and biochemical characters, resulting in the reliability of tissue culture-based propagation systems (Xiao et al., 2022).

A critical future direction is associated with the establishment of standardized protocols for genetic fidelity assessment. Currently, some factors affect genetic fidelity studies, including differences in explant source, frequency of subculture, and various marker systems, creating inconsistencies among studies. Development of universally accepted guidelines for genetic stability enables improvement of reproducibility, enhancement of data comparability, and facilitation of regulatory acceptance of medicinally propagated plants. An integrated combination of research institutions, producers, and policy-making rules, including World Health Organization (WHO), Food and Agriculture Organization (FAO), and the International Plant Genetic Resources Institute (IPGRI), will be important to implement internationally recognized quality assurance systems (Majumder et al., 2025). Emerging digital and automation technologies will further transform genetic fidelity assessment into a high-throughput process. The combination of robotic handling, automated imaging, and machine learning algorithms will allow real-time monitoring of morphological and biochemical characters. Future artificial intelligence (AI)-based tools are expected to analyze large datasets from omics platforms, determine genetic fidelity biomarkers, and predict genetic variation among plants (Yan and Wang, 2023). Like this, microfluidic lab systems and sequencing devices will make genetic fidelity assessment more accessible and cost-effective for all future researchers (Ferreira et al., 2023). The regulatory and ethical aspects of plant biotechnology will also gain importance in future genetic fidelity studies. It is important to ensure the biosafety and genetic stability of regenerated medicinal plant materials used for therapeutic and pharmacological industries. Emerging genetic tools like

CRISPR/Cas tools are used in labeling and certification processes for regenerated plant materials (Hahn and Nekrasov, 2019).

The integration of molecular techniques such as methylation-sensitive amplified polymorphism (MASP), whole-genome bisulfite sequencing (WGBS), RNA sequencing, and CRISPR/Cas9 is a reform to minimize somaclonal variation in plant biotechnology. Additionally, the CRISPR/Cas9 tool introduces targeted mutations; it supports the stable maintenance of clonal fidelity associated with somaclonal variation. Various micropropagated companies frequently employ these techniques to validate planting materials, but some limitations remain in making these molecular assays affordable for commercial uses. To counter these limitations, it is necessary to optimize protocols, select accurate genotypes, reduce subculture cycles, and use meristematic explants for stable clonal fidelity (Bairu et al., 2011). The bioreactor system allows a controlled optimum environment that maintains consistent growth conditions for plant tissues, thereby preventing somaclonal variation (Debnath, 2011). Additionally, cryopreservation techniques are also used to maintain genetic stability because this technique is used to store plant tissues at extremely low temperatures (Ballesteros et al., 2024).

Somaclonal variation is significantly influenced by key factors like explant source, culture medium composition, the ratio and type of PGRs, frequency of subcultures, as well as environmental conditions, including temperature and light intensity. To encounter limitations of micropropagated variations, it is important to identify variants at the early stages of their growth. Various techniques, like cytological analysis, morphological analysis, and molecular markers, are used to evaluate epigenetic and genetic variations. However, some somaclonal variants are used as a new agronomic trait for crop improvement; at the same time, it is a significant problem when micropropagated true-to-type plants are expected. The combination of both genetic and epigenetic screening is important to maintain stable clonal fidelity for commercial micropropagation (Majumder et al., 2025).

In conclusion, somaclonal variation is not limited to visual morphological variation but extends to hidden biochemical and genetic modifications. Initially regenerated plants might appear stable, but biochemical and genetic changes manifest under field conditions. Therefore hallmarks can be recommended for better assessment of genetic fidelity, including reducing the callus phase, preferring direct organogenesis, and accurately measuring assays and metabolic profiling (Sarmah et al., 2024). The integrated approaches used regenerate plants that are genetically identical and biochemically reliable for further applications. *In vitro* regenerated plants must ensure the true-to-type nature of their therapeutic value and commercial reliability. The genetic fidelity assessment is emerging from traditional marker-based methods, and that is associated with the integration of biochemical, molecular, and computational methods. These multidisciplinary approaches may help improve overall accuracy and quality assurance of clonal verification in plant tissue culture. However, upcoming genetic fidelity research should focus on developing global standards and AI-based analytics and expanding omics integration to ensure that the true-to-type nature of regenerated plants in terms of metabolically consistent, pharmaceutically authentic, and genetically stable. Additionally, all these approaches will reinforce the credibility of plant biotechnology as a sustainable means for the conservation, propagation, and large-scale production of medicinal plants with assured authenticity, stability, and fitness for the sustainability of phyto-therapeutic agents.

Competing interests

The authors have no relevant financial or nonfinancial interests to disclose.

References

Abdelfatah, S. A., & Efferth, T. (2015). Cytotoxicity of the indole alkaloid reserpine from *Rauwolfia serpentina* against drug-resistant tumor cells. *Phytomedicine International Journal of Phytotherapy and Phytopharmacology, 22*(2), 308–318.

Abiri, R. (2024). Somaclonal variations for improvement in medicinally active plant species. In Azamal Husen, & Manu Pant (Eds.). *Tissue culture techniques and medicinal plants* (pp. 134–161). CRC Press.

Ahlawat, Y. K., Yadav, K., Samani, M., & Chaudhary, D. (2024). Harnessing in-vitro propagation for the sustainable conservation of medicinal plants: Challenges and prospects. *Medicinal and Aromatic Plants: Current Research Status, Value-Addition to Their Waste, and Agro-Industrial Potential, I*, 27–37.

Ahmad, N., Fatima, N., Faisal, M., Alatar, A. A., & Pathirana, R. (2022). Photosynthetic parameters and oxidative stress during acclimation of crepe-myrtle (*Lagerstroemia speciosa* (L.) Pers.) in a meta-topolin-based micropropagation system and genetic fidelity of regenerated plants. *Plants, 11*(9), 1163.

Al-Aizari, A. A., Dewir, Y. H., Ghazy, A. H., Al-Doss, A., & Al-Obeed, R. S. (2024). Micropropagation and genetic fidelity of fegra fig (*Ficus palmata* Forssk.) and grafting compatibility of the regenerated plants with Ficus carica. *Plants, 13*(9), 1278.

Aldaif, M., Ucar, S., Yigider, E., Aydin, M., Gul Kasapoglu, A., & Ilhan, E. (2025). Micropropagation and genetic fidelity of *Rheum ribes* L.: A threatened and endemic medicinal herb from the Turkish Eastern Anatolia region. *Plant Cell, Tissue and Organ Culture (PCTOC), 162*(1), 1.

Ali, H. (2023). Artificial intelligence in multi-omics data integration: Advancing precision medicine, biomarker discovery and genomic-driven disease interventions. *International Journal of Scientific Research Archive, 8*(1), 1012–1030.

Alsafran, M., Wickramanayake, K., Usman, K., & Ahmed, T. (2022). Efficient shoot regeneration of medicinal plant *Haplophyllum tuberculatum* by direct and indirect organogenesis and genetic fidelity assessment using Inter Simple Sequence Repeats markers. *Frontiers in Plant Science, 13*, 995825.

Atia, M. A. M., El-Moneim, D. A., Abdelmoneim, T. K., Reda, E. H., Shakour, Z. T. A., El-Halawany, A. M., & Hegazy, M. E. F. (2021). Evaluation of genetic variability and relatedness among eight Centaurea species through CAAT-box derived polymorphism (CBDP) and start codon targeted polymorphism (SCoT) markers. *Biotechnology & Biotechnological Equipment, 35*(1), 1230–1237.

Attaluri, S., & Dharavath, R. (2025). The synergistic effect of plant growth regulators and polyamines on regeneration of *Nothapodytes foetida*, and assessment of genetic fidelity using ISSR and SCoT markers. *Plant Cell, Tissue and Organ Culture (PCTOC), 160*(2), 42.

Attikora, A. J. P., Silue, S., Kone, M., Silue, N., Kwibuka, Y., Yao, S. D. M., & Lassois, L. (2025). Efficient in vitro regeneration and genetic fidelity analysis of shea tree (*Vitellaria paradoxa* Gaertn) using ISSR markers. *Electronic Journal of Biotechnology, 75*, 28–38.

Avani, G., & Rao, M. V. (2008). In vitrocytogenetic effects of *Andrographis paniculata* (kalmegh) on arsenic. *Phytomedicine International Journal of Phytotherapy and Phytopharmacology, 15*(3), 221–225.

Awere, C. O., Rakkammal, K., Ram, P. J., Kumar, K. P., Ragavan, K., Kumari, R. A., & Ramesh, M. (2024). Micropropagation, genetic fidelity and chromatographic analysis in *Evolvulus alsinoides* (L): A potent multipurpose medicinal plant. *Industrial Crops and Products, 213*, 118444.

Badhepuri, M. K., Beeravelli, P. R., Arolla, R. G., Jogam, P., Rohela, G. K., & Singisala, N. R. (2024). Micropropagation and genetic fidelity analysis using SCoT and ISSR markers in *Muehlenbeckia platyclada* (F. Muell.) meisn. *Plant Cell, Tissue and Organ Culture (PCTOC), 157*(3), 51.

Bairu, M. W., Aremu, A. O., & Van Staden, J. (2011). Somaclonal variation in plants: causes and detection methods. *Plant Growth Regulation, 63*(2), 147–173.

Ballesteros, D., Martínez, M. T., Sanchez-Romero, C., Montalban, I. A., Sales, E., Moncalean, P., & Corredoira, E. (2024). Current status of the cryopreservation of embryogenic material of woody species. *Frontiers in Plant Science, 14*, 1337152.

Bandyopadhyay, S., & Raychaudhuri, S. S. (2013). Development and comparison of RAPD, SCAR and AFLP markers for distinguishing some medicinally important species of the genus Phyllanthus. *Plant Biosystems-An International Journal Dealing with all Aspects of Plant Biology, 147*(1), 12–20.

Bandyopadhyay, S., Subrahmanyeswari, T., Sarkar, J. M., Kamble, S. N., Singh, S., Singh, S. K., & Gantait, S. (2025). In vitro multiplication, flowering, and production of withaferin-A and withanolide-A from meta-topolin-induced genetically stable *Withania somnifera* (L.) Dunal plantlets. *Plant Cell, Tissue and Organ Culture (PCTOC), 161*(3), 62.

Bhattacharyya, P., Kumar, S., Lalthafamkimi, L., Sharma, R., Kumar, D., Singh, D., & Kumar, S. (2023). Molecular and phytomedicinal stability of long term micropropagated *Malaxis acuminata*: An endangered terrestrial orchid of biopharmaceutical importance. *South African Journal of Botany, 155*, 372–382.

Bhattacharyya, P., Kumaria, S., Kumar, S., & Tandon, P. (2013). Start codon targeted (SCoT) marker reveals genetic diversity of *Dendrobium nobile* Lindl., an endangered medicinal orchid species. *Gene, 529*(1), 21–26.

Biswas, P., & Kumar, N. (2023). Application of molecular markers for the assessment of genetic fidelity of *in vitro* raised plants: Current status and future prospects. *Molecular Marker Techniques: A Potential Approach of Crop Improvement*, 233–256.

Canter, P. H., Thomas, H., & Ernst, E. (2005). Bringing medicinal plants into cultivation: Opportunities and challenges for biotechnology. *Trends in Biotechnology, 23*(4), 180–185.

Chawla, R., Sharma, H., Dadheech, A., Jattan, M., Devi, S., Kumar, P., & Singh, T. (2025). Genetic insights into diversity and population structure of ashwagandha (*Withania somnifera* (L) Dunal) using EST-SSR, ISSR and SSR markers: Implications for enhancing agricultural and industrial value. *Industrial Crops and Products, 224*, 120242.

Chetri, S. K., Sardar, P. R., & Agrawal, V. (2014). Micropropagation and validation of genetic and biochemical fidelity amongst regenerants of *Cassia angustifolia* Vahl employing RAPD marker and HPLC. *Physiology and Molecular Biology of Plants, 20*(4), 517–526.

Chirumamilla, P., Gopu, C., Jogam, P., & Taduri, S. (2021). Highly efficient rapid micropropagation and assessment of genetic fidelity of regenerants by ISSR and SCoT markers of *Solanum khasianum* Clarke. *Plant Cell, Tissue and Organ Culture (PCTOC), 144*(2), 397–407.

Choudhury, D. R., Sharma, L., Suma, A., Singh, G. P., & Singh, R. (2025). Exploring the genetic diversity and population structure of *Bacopa monnieri* (L) using random amplified (RAPD and ISSR) and gene-targeted (SCoT and CBDP) markers. *Genetic Resources and Crop Evolution, 72*(1), 1043–1063.

Collard, B. C., & Mackill, D. J. (2009). Start codon targeted (SCoT) polymorphism: A simple, novel DNA marker technique for generating gene-targeted markers in plants. *Plant Molecular Biology Reporter, 27*(1), 86–93.

Debnath, S. (2011). Bioreactors and molecular analysis in berry crop micropropagation–A review. *Canadian Journal of Plant Science, 91*(1), 147–157.

Deeba, F., Singh, P., Warsi, Z. I., Husain, Z., Kumari, M., & Rahman, L. U. (2024). *Bergenia ligulata* (Wall.): Micropropagation, genetic fidelity, and SEM studies. *In Vitro Cellular & Developmental Biology-Plant, 60*(1), 75–84.

Desai, P., Desai, S., Rafaliya, R., & Patil, G. (2022). Plant tissue culture: Somatic embryogenesis and organogenesis. In Avinash Chandra Rai, Ajay Kumar, Arpan Modi, & Major Singh (Eds.). *Advances in plant tissue culture: Current developments and future trends* (pp. 109–130). Academic Press.

Devi, K. S., Devi, H. H., Singh, N. S., & Devi, H. S. (2025). Meta-topolin enhanced in vitro regeneration and genetic fidelity assessment of an endangered orchid, *Pelatantheria insectifera* using scot and ISSR. *Vegetos*, 1–13.

Dhiman, N., Devi, K., & Bhattacharya, A. (2021). Development of low cost micropropagation protocol for *Nardostachys jatamansi*: A critically endangered medicinal herb of Himalayas. *South African Journal of Botany*.

Dhondi, P., Kota, S., Rohela, G. K., & Kasula, K. (2025). Development of efficient micropropagation, assessment of genetic fidelity and biochemical fidelity in *Curcuma longa* L. *Vegetos*, 1–13.

Dilkalal, A., AS, A., & TG, U. (2021). In vitro regeneration, antioxidant potential, and genetic fidelity analysis of *Asystasia gangetica* (L.) T. Anderson. *In Vitro Cellular & Developmental Biology-Plant, 57*(3), 447–459.

Dixit, J., Verma, P., Mishra, P., Tiwari, K. N., Tiwari, S. K., Mishra, S. K., & Singh, J. (2024). Direct shoot regeneration from cotyledonary node of *Uraria picta* (Jacq.) Desv. ex DC., an important plant of dashmula drugs, and assessment of genetic fidelity, metabolic profiling, and anti-diabetic activity. *Industrial Crops and Products, 219*, 119064.

Ejaz, B., Mujib, A., Mamgain, J., Malik, M. Q., Syeed, R., Gulzar, B., & Bansal, Y. (2022). Comprehensive in vitro regeneration study with SCoT marker assisted clonal stability assessment and flow cytometric genome size analysis of *Carthamus tinctorius* L.: An important medicinal plant. *Plant Cell, Tissue and Organ Culture (PCTOC), 148*(2), 403–418.

El-Naby, M. E. S. A. (2022). In vitro propagation for conservation and genetic fidelity of the near threatened *Dimocarpus longan* plant. *Journal of Genetic Engineering and Biotechnology, 20*(1), 130.

Fabriki-Ourang, S., & Karimi, H. (2019). Assessment of genetic diversity and relationships among *Salvia* species using gene targeted CAAT box-derived polymorphism markers. *Journal of Genetics, 98*(3), 75.

Faisal, M., & Alatar, A. A. (2022). Establishment of an efficient in vitro propagation method for a sustainable supply of Plectranthus amboinicus (Lour.) and genetic homogeneity using flow cytometry and SPAR markers. *Horticulturae, 8*(8), 693.

Faisal, M., Alatar, A. A., Hegazy, A. K., Alharbi, S. A., El-Sheikh, M., & Okla, M. K. (2014). Thidiazuron induced *in vitro* multiplication of *Mentha arvensis* and evaluation of genetic stability by flow cytometry and molecular markers. *Industrial Crops and Products, 62*, 100–106.

Ferreira, M. D. S., Rocha, A. D. J., Nascimento, F. D. S., Oliveira, W. D. D. S., Soares, J. M. D. S., Rebouças, T. A., & Amorim, E. P. (2023). The role of somaclonal variation in plant genetic improvement: A systematic review. *Agronomy, 13*(3), 730.

Gantait, S., Das, S., Mahanta, M., & Banerjee, M. (2023). Influence of in vitro culture age on morphology, antioxidant activities, reserpine production, and genetic fidelity in Indian snakeroot (*Rauvolfia serpentina* (L.) Benth. ex Kurz.). *South African Journal of Botany, 162*, 864–872.

Gantait, S., & Mukherjee, E. (2021). Induced autopolyploidy—A promising approach for enhanced biosynthesis of plant secondary metabolites: An insight. *Journal of Genetic Engineering and Biotechnology, 19*(1), 4.

Gautam, N., & Bhattacharya, A. (2021). Molecular marker based assessment of genetic homogeneity within the in vitro regenerated plants of *Crocus sativus* L.–A globally important high value spice crop. *South African Journal of Botany.*

Gautam, N., Faroda, P., Shekhawat, S., & Gupta, A. K. (2025). Effect of gamma radiation on in vitro morphogenesis, anatomy and DNA polymorphism of *Moringa concanensis* Nimmo. *Scientific Reports, 15*(1), 31507.

George, N. M., Raghav, S. B., & Prasath, D. (2022). Direct in vitro regeneration of medicinally important Indian and exotic red-colored ginger (*Zingiber officinale* Rosc.) and genetic fidelity assessment using ISSR and SSR markers. *In Vitro Cellular & Developmental Biology-Plant, 58*(4), 551–558.

Gupta, S., & Chaturvedi, P. (Eds.). (2022). *Commercial scale tissue culture for horticulture and plantation crops* (pp. 1–21). Springer.

Hahn, F., & Nekrasov, V. (2019). CRISPR/Cas precision: Do we need to worry about off-targeting in plants? *Plant Cell Reports, 38*(4), 437–441.

Halder, M., & Jha, S. (2023). Medicinal plants and bioactive phytochemical diversity: A fountainhead of potential drugs against human diseases. In Sumita Jha, & Mihir Halder (Eds.). *Medicinal plants: Biodiversity, biotechnology and conservation* (pp. 39–93). Springer Nature Singapore.

Hasnain, A., Naqvi, S. A. H., Ayesha, S. I., Khalid, F., Ellahi, M., Iqbal, S., & Abdelhamid, M. M. (2022). Plants in vitro propagation with its applications in food, pharmaceuticals and cosmetic industries; current scenario and future approaches. *Frontiers in Plant Science, 13*, 1009395.

Hassanzadeh, F., Zakaria, R. A., & Azad, N. H. (2020). Polyploidy induction in *Salvia officinalis* L. and its effects on some morphological and physiological characteristics. *Cytologia, 85*(2), 157–162.

Heydari, H. R., & Chamani, E. (2022). High-frequency plant regeneration and genetic fidelity assessment of regenerants by molecular and biochemical markers in Woodland Sage (*Salvia nemorosa* L.). *Industrial Crops and Products, 187*, 115404.

Hiremath, C., Philip, R., & Sundaresan, V. (2021). Molecular characterization of India Ginseng *Withania somnifera* (L) using ISSR markers. *Molecular Biology Reports, 48*(5), 3971–3977.

Hossain, C. M., Gera, M. E. E. T. A., & Ali, K. A. (2022). Current status and challenges of herbal drug development and regulatory aspect: A global perspective. *Asian Journal of Pharmaceutical and Clinical Research, 15*(2), 31–41.

Hussain, M. A., Nathar, V. N., & Mushtaq, R. (2022). In vitro plant regeneration and metabolite profiling of an aromatic medicinal plant *Rutagraveolens* L. by using GC-MS. *Current Botany, 13*, 26–33.

Hussain, S. A., Anis, M., & Alatar, A. A. (2020). Efficient in vitro regeneration system for *Tecoma stans* L., using shoot tip and assessment of genetic fidelity among regenerants. *Proceedings of the National Academy of Sciences, India Section B: Biological Sciences, 90*(1), 171–178.

Idrees, M., & Irshad, M. (2014). Molecular markers in plants for analysis of genetic diversity: A review. *European Academic Research, 2*(1), 1513–1540.

Ioannidis, K., Tomprou, I., Mitsis, V., & Koropouli, P. (2022). Genetic evaluation of in vitro micropropagated and regenerated plants of *Cannabis sativa* L. using SSR molecular markers. *Plants, 11*(19), 2569.

Jamshidi, B., Etminan, A., Mehrabi, A. M., Pour-Aboughadareh, A., Shooshtari, L., & Ghorbanpour, M. (2024). Genetic diversity of Artemisia species based on CAAT-box derived polymorphism (CBDP) and start codon targeted (SCoT) markers. *Genetic Resources and Crop Evolution, 71*(7), 3803–3813.

Jaryal, P., Pathak, P., & Warghat, A. R. (2025). An efficient clonal propagation of a medicinally important and endangered Himalayan herb, *Dactylorhiza hatagirea* D. Don Soo using shoot meristem culture and genetic fidelity analysis. *Plant Cell, Tissue and Organ Culture (PCTOC), 160*(1), 20.

Jogam, P., Sandhya, D., Shekhawat, M. S., Alok, A., Manokari, M., Abbagani, S., & Allini, V. R. (2020). Genetic stability analysis using DNA barcoding and molecular markers and foliar micro-morphological analysis of in vitro regenerated and in vivo grown plants of *Artemisia vulgaris* L. *Industrial Crops and Products, 151*, 112476.

Kadapatti, S. S., & Murthy, H. N. (2021). Rapid plant regeneration, analysis of genetic fidelity, and neoandrographolide content of micropropagated plants of *Andrographis alata* (Vahl) Nees. *Journal of Genetic Engineering and Biotechnology, 19*(1), 1–8.

Kanthaliya, B., Joshi, A., Verma, K. K., & Arora, J. (2025). Efficient micropropagation protocol and assessment of genetic fidelity of micropropagated plantlets of *Pueraria tuberosa*. *Plant Biosystems-An International Journal Dealing with all Aspects of Plant Biology, 159*(1), 26–35.

Katib, S., Ruangrungsi, N., Palanuvej, C., Chareonsap, P. P., & Rungsihirunrat, K. (2017). Macroscopic-microscopic characteristics and AFLP marker for identification of *Tinospora crispa* and *Tinospora baenzigeri* endemic to Thailand. *Journal of Health Research, 31*(2).

Kaushal, N., Alok, A., Kajal, M., & Singh, K. (2021). Regeneration and genetic fidelity analysis of *Chlorophytum borivilianum* using flower stalk as explant source. *Advances in Bioscience and Biotechnology, 12*(04), 95.

Kavitha, C., Rajamani, K., & Vadivel, E. (2010). *Coleus forskohlii*: A comprehensive review on morphology, phytochemistry and pharmacological aspects. *Journal of Medicinal Plant Research, 4*(4), 278–285.

Khabiya, R., Choudhary, G. P., Sairkar, P., Silawat, N., Jnanesha, A. C., Kumar, A., & Lal, R. K. (2024). Unraveling genetic diversity analysis of Indian ginseng (*Withania somnifera* (Linn.) Dunal) insight from RAPD and ISSR markers and implications for crop improvement vital for pharmacological and industrial potential. *Industrial Crops and Products, 210*, 118124.

Koh, K. S., Ismail, M. F., Naharudin, N. S., Gantait, S., & Sinniah, U. R. (2024). Harnessing the potential of transverse thin cell layer culture for high-frequency micropropagation of Thai ginseng (*Kaempferia parviflora* Wall. Ex Baker). *Industrial Crops and Products, 213*, 118375.

Konar, S., Adhikari, S., Karmakar, J., Ray, A., & Bandyopadhyay, T. K. (2019). Evaluation of subculture ages on organogenic response from root callus and SPAR based genetic fidelity assessment in the regenerants of *Hibiscus sabdariffa* L. *Industrial Crops and Products, 135*, 321–329.

Koppula, T., Sandhya, D., Rohela, G. K., Kommidi, S., & Mohammed, M. (2023). Micropropagation and assessment of genetic homogeneity of regenerants by ISSR and SCoT markers in *Solena amplexicaulis* (Lam.) Gandhi—a threatened medicinal cucurbit. *In Vitro Cellular & Developmental Biology-Plant, 59*(6), 724–733.

Krishnaswamy, S. (2024). Phytopharmaceutical biotechnology: Integration of botany, pharmacology and plant biotechnology to deliver the best therapeutic potential of herbs. In Sankhadip Bose, Amritesh Chandra Shukla, Mirza R. Baig, & Sabyasachi Banerjee (Eds.). *Concepts in pharmaceutical biotechnology and drug development* (pp. 437–464). Springer Nature Singapore.

Kudikala, H., Jogam, P., Sirikonda, A., Mood, K., & Allini, V. R. (2020). In vitro micropropagation and genetic fidelity studies using SCoT and ISSR primers in *Annona reticulata* L.: An important medicinal plant. *Vegetos, 33*(3), 446–457.

Kumar, A., Mishra, P., Singh, S. C., & Sundaresan, V. (2014). Efficiency of ISSR and RAPD markers in genetic divergence analysis and conservation management of *Justicia adhatoda* L., a medicinal plant. *Plant Systematics and Evolution, 300*(6), 1409–1420.

Kumar, R., Salar, R. K., Naik, P. K., Yadav, M., Kumar, A., Kumar, A., & Chhokar, V. (2022). Elucidation of genetic diversity and population structure of sixty genotypes of *Aloe vera* using AFLP markers. *South African Journal of Botany, 147*, 1146–1155.

Kumar, S. (2021). Impact of ploidy changes on secondary metabolites productions in plants. In Akhileshwar Kumar Srivastava, Vinod Kumar Kannaujiya, Rajesh Kumar Singh, & Divya Singh (Eds.). *Evolutionary diversity as a source for anticancer molecules* (pp. 29–46). Academic Press.

Kumari, A., Kumari, V., Sharma, P., Kumar, D., Sharma, R. K., & Joshi, R. (2025). Development of a robust in vitro propagation protocol, assessment of genetic fidelity, and metabolite analysis of *Ajuga parviflora* Benth. *In Vitro Cellular & Developmental Biology-Plant*, 1–21.

Kumari, A., Sharma, L., Yadav, Y., Karma, A., Singh, J., & Arya, S. C. (2025). Biotechnological strategies for seed germplasm maintenance and plant genetic resource conservation. In Bidhan Roy, Gopal Shukla, Vijay Dunna, Priyanka Sharma, & Prabha Shankar Shukla (Eds.). *Advances in seed quality evaluation and improvement* (pp. 379–403). Springer Nature Singapore.

Kundu, S., Barua, R., Marshall, H. D., Siow, Y. L., Igamberdiev, A. U., & Debnath, S. C. (2025). Somatic embryogenesis in *Vaccinium vitis-idaea* L.: A high-fidelity micropropagation approach producing true-to-type plants with enhanced phytochemical profiles. *Plant Cell, Tissue and Organ Culture (PCTOC), 161*(2), 1–19.

Kuriakose, G. C., Singh, S., Rajvanshi, P. K., Surin, W. R., & Jayabaskaran, C. (2014). In vitro cytotoxicity and apoptosis induction in human cancer cells by culture extract of an endophytic *Fusarium solani* strain isolated from *Datura metel* L. *Pharmaceutica Analytica Acta, 5*(2).

Lalthafamkimi, L., Bhau, B. S., Kumar, S., Mukhia, S., Kumar, R., Banik, D., & Bhattacharyya, P. (2022). Indirect organogenesis-mediated high frequency conversion of non-embryonic synthetic seeds, essential oil profiling and antibacterial activity in genetically stable plants of Patchouli. *3 Biotech, 12*(12), 349.

Li, G., McVetty, P. B., & Quiros, C. F. (2013). SRAP molecular marker technology in plant science. In Sven Bode Andersen (Ed.). *Plant breeding from laboratories to fields*. IntechOpen.

Liu, Z., Guo, S., Xu, J., Zhang, Y., Dong, L., Xiao, S., & Chen, S. (2018). Genome size estimation of Chinese cultured *Artemisia annua* L. *Journal of Plant Biology and Crop Research, 1*(1), 1002.

Liu, Z., Li, G., Zhang, Y., Jin, H., Liu, Y., Dong, J., & Liang, X. (2022). Blending technology based on HPLC fingerprint and nonlinear programming to control the quality of ginkgo leaves. *Molecules (Basel, Switzerland), 27*(15), 4733.

Long, Y., Yang, Y., Pan, G., & Shen, Y. (2022). New insights into tissue culture plant-regeneration mechanisms. *Frontiers in Plant Science, 13*, 926752.

Loyola-Vargas, V. M., & Ochoa-Alejo, N. (2016). Somatic embryogenesis. An overview. In Víctor M. Loyola-Vargas, & Neftalí Ochoa-Alejo (Eds.). *Somatic embryogenesis: Fundamental aspects and applications* (pp. 1–8). .

Mahanta, M., Gantait, S., Mukherjee, E., & Bhattacharyya, S. (2023). *meta*-Topolin-induced mass propagation, acclimatization and cyto-genetic fidelity assessment of gerbera (*Gerbera jamesonii* Bolus ex Hooker f.). *South African Journal of Botany, 153*, 236–245.

Majumder, S., Igamberdiev, A. U., & Debnath, S. C. (2025). Somaclonal variation and clonal fidelity in commercial micropropagation: Challenges and perspectives. *Agronomy, 15*(6), 1489.

Mamgain, J., Mujib, A., Ejaz, B., Gulzar, B., Malik, M. Q., & Syeed, R. (2022). Flow cytometry and start codon targeted (SCoT) genetic fidelity assessment of regenerated plantlets in *Tylophora indica* (Burm. f.) Merrill. *Plant Cell, Tissue and Organ Culture (PCTOC, 150*(1), 129–140.

Manchanda, P., Sharma, D., Kaur, G., Kaur, H., & Vanshika (2025). Exploring the significance of somaclonal variations in horticultural crops. *Molecular Biotechnology, 67*(6), 2185–2203.

Manisha, D. R. B., Begam, A. M., Chahal, K. S., & Ashok, M. A. (2025). Medicinal plants and traditional uses and modern applications. *Journal of Neonatal Surgery, 14*(3).

Manohar, K. A., Shukla, G., Shahina, N. N., Sivasankarreddy, K., Ravuther, S. S., & Chakravarty, S. (2024). Conventional versus non-conventional methods of propagation of forest tree species: Applications and limitations. In Dennis Thomas T., M. K. Razdan, & Ajay Kumar (Eds.). *Biotechnological approaches for sustaining forest trees and their products* (pp. 483–517). Springer Nature Singapore.

Manokari, M., Badhepuri, M. K., Cokulraj, M., Sandhya, D., Dey, A., Kumar, V., & Shekhawat, M. S. (2022). Validation of meta-Topolin in organogenesis, improved morpho-physio-chemical responses, and clonal fidelity analysis in *Dioscorea pentaphylla* L.–An underutilized yam species. *South African Journal of Botany, 145*, 284–292.

Manokari, M., Priyadharshini, S., Cokulraj, M., Jayaprakash, K., Dey, A., Faisal, M., & Shekhawat, M. S. (2022). Polyethylene glycol mediated improved shoot proliferation, foliar morpho-anatomy, and rooting of micropropagated shoots of *Spathoglottis plicata* Blume. *South African Journal of Botany, 146*, 897–904.

Manokari, M., Priyadharshini, S., Jogam, P., & Shekhawat, M. S. (2020). Application of seismo-mechanical stress in attenuation of in vitro induced abnormalities in *Scaevola taccada* (Gaertn.) Roxb. *Biocatalysis and Agricultural Biotechnology, 29*, 101821.

Manokari, M., Priyadharshini, S., Jogam, P., Dey, A., & Shekhawat, M. S. (2021). Meta-topolin and liquid medium mediated enhanced micropropagation via ex vitro rooting in *Vanilla planifolia* Jacks. ex Andrews. *Plant Cell, Tissue and Organ Culture (PCTOC), 146*(1), 69–82.

Mishra, S., Srivastava, V., & Mehrotra, S. (2025). *Elicitation methods for improvement in plant secondary metabolites.* Springer.

Mitra, S., Anand, U., Ghorai, M., Kant, N., Kumar, M., Radha, Jha, N. K., Swamy, M. K., Proćków, J., de la Lastra, J. M. P., & Dey, A. (2023). Genome editing technologies, mechanisms and improved production of therapeutic phytochemicals: Opportunities and prospects. *Biotechnology and Bioengineering, 120*(1), 82–94.

Mohanty, S. P., Das, P. K., Acharya, L., Panda, P. C., & Barik, D. P. (2025). An efficient micropropagation protocol, establishment of genetic and biochemical fidelity of a rare medicinal plant, *Uvaria hamiltonii* Hook. f. & Thomson. *South African Journal of Botany, 179*, 389–399.

Monalisa, K., Behera, S., Pidika, S. P., Madkami, S. K., & Naik, S. K. (2024). Development of micropropagation protocol, assessment of genetic fidelity and phytochemical analysis of *Blepharispermum subsessile* DC.: A conservation concerned medicinal plant of Eastern Ghats, India. *Plant Cell, Tissue and Organ Culture (PCTOC), 157*(1), 14.

Mondal, T. K., Bantawa, P., Sarkar, B., Ghosh, P., & Chand, P. K. (2013). Cellular differentiation, regeneration, and secondary metabolite production in medicinal *Picrorhiza* spp. *Plant Cell, Tissue and Organ Culture (PCTOC), 112*(2), 143–158.

Mood, K., Jogam, P., Sirikonda, A., Shekhawat, M. S., Rohela, G. K., Manokari, M., & Allini, V. R. (2022). Micropropagation, morpho-anatomical characterization, and genetic stability studies in *Lippia javanica* (Burm. f.) Spreng: A multipurpose medicinal plant. *Plant Cell, Tissue and Organ Culture (PCTOC)*, 1–11.

Mousavimatin, S., Mortazavi, S. N., Samiei, L., & Azadi, P. (2025). Indirect regeneration of long-term callus cultures in gladiolus: A protocol for stable genetic fidelity. *Frontiers in Horticulture, 4*, 1571042.

Mutha, R. E., Kalaskar, M., & Khan, Z. G. (2025). Modern analytical techniques for quality control and chemical identification of phytochemicals. *Pharmacognosy and Phytochemistry: Principles, Techniques, and Clinical Applications*, 167–188.

Mykhailenko, O., Jalil, B., McGaw, L. J., Echeverría, J., Takubessi, M., & Heinrich, M. (2025). Climate change and the sustainable use of medicinal plants: A call for "new" research strategies. *Frontiers in Pharmacology, 15*, 1496792.

Nandy, S., Mishra, V., Rajpal, V. R., Singh, A., Devarumath, R., Raturi, D., & Raina, S. N. (2025). Genetic fidelity analysis of tissue culture-derived teak (*Tectona grandis*) elite clones. *The Nucleus*, 1–11.

Nasim, N., Sandeep, I. S., & Mohanty, S. (2022). Plant-derived natural products for drug discovery: Current approaches and prospects. *The Nucleus, 65*(3), 399–411.

Negi, M. S., Sabharwal, V., Wilson, N., & Lakshmikumaran, M. S. (2006). Comparative analysis of the efficiency of SAMPL and AFLP in assessing genetic relationships among *Withania somnifera* genotypes. *Current Science*, 464–471.

Niazian, M. (2019). Application of genetics and biotechnology for improving medicinal plants. *Planta, 249*(4), 953–973.

Olusan, B. M., Ashidi, J. S., Feyisola, R. T., Oseni, O. M., Oyelami, O. K., & Sanusi, A. S. (2024). In vitro direct and indirect organogenesis and genetic fidelity assessment in *Clerodendrum volubile* P. Beauv. *Nigerian Journal of Botany, 37*(2), 275–294.

Onipede, G. O., Fawole, A. O., Odukoya, O. J., & Onuh, J. O. (2024). Production, stability, and industrial commercialization of bioactive phytochemicals. In John Oloche Onuh, & Yashwant V. Pathak (Eds.). *Plant food phytochemicals and bioactive compounds in nutrition and health* (pp. 79–98). CRC Press.

Pandey, S., Maharjan, L., & Pant, B. (2023). In vitro propagation and assessment of genetic homogeneity using RAPD and ISSR markers in *Tinospora cordifolia* (Wild.) Hook. F. & Thoms, an important medicinal plant of Nepal. *Journal of Nepal Biotechnology Association, 4*(1), 27–36.

Pandey, S., Patel, P., Prasad, A., Sawant, S. V., & Misra, P. (2020). Assessment of direct shoot organogenesis and genetic fidelity in *Solanum viarum* Dunal—A commercially important medicinal plant. *In Vitro Cellular & Developmental Biology-Plant, 56*(4), 538–547.

Pandey, V., Ansari, W. A., Misra, P., & Atri, N. (2017). *Withania somnifera*: Advances and implementation of molecular and tissue culture techniques to enhance its application. *Frontiers in Plant Science, 8*, 1390 (2012). Cytogenetic characterization and genome size of the medicinal plant *Catharanthus roseus* (L.) G. Don. *AoB Plants, 2012*, pls002.

Pant, M., Sohrab, S. S., & Husen, A. (Eds.). (2025). *Genetic improvement and conservation practices of medicinal plants.* CRC Press.

Parab, A. R., Lynn, C. B., & Subramaniam, S. (2021). Assessment of genetic stability on in vitro and ex vitro plants of *Ficus carica* var. black jack using ISSR and DAMD markers. *Molecular Biology Reports, 48*(11), 7223–7231.

Patial, M., Devi, K., Sharma, P., Sharma, R. K., Pal, P. K., Kumar, S., & Joshi, R. (2024). Development of robust in vitro propagation protocol and cyto-genetic fidelity assessment of *Siraitia grosvenorii* (monk fruit). *Scientia Horticulturae, 331*, 113142.

Pendli, S., Rohela, G. K., Jogam, P., Bylla, P., Korra, R., & Thammidala, C. (2019). High frequency in vitro plantlet regeneration in *Solanum trilobatum* L., an important ethno-medicinal plant and confirmation of genetic fidelity of R1 plantlets by using ISSR and RAPD markers. *Vegetos, 32*(4), 508–520.

Peng, X., Ji, Q., Fan, S., Zhang, Y., & Zhang, J. (2015). Genetic diversity in populations of the endangered medicinal plant *Tetrastigma hemsleyanum* revealed by ISSR and SRAP markers: Implications for conservation. *Genetic Resources and Crop Evolution, 62*(7), 1069–1078.

Pittampalli, B., Jogam, P., Thampu, R. K., Abbagani, S., & Peddaboina, V. (2022). High-frequency plant regeneration and genetic homogeneity assessment of regenerants by molecular markers in turmeric (*Curcuma longa* L.). *In Vitro Cellular & Developmental Biology-Plant, 58*(1), 169–180.

Pramanik, B., Sarkar, S., Bhattacharyya, S., & Gantait, S. (2021). meta-Topolin-induced enhanced biomass production via direct and indirect regeneration, synthetic seed production, and genetic fidelity assessment of *Bacopa monnieri* (L.) Pennell, a memory-booster plant. *Acta Physiologiae Plantarum, 43*(7), 1–14.

Prasad, A., Patel, P., Kumari, M., Saxena, G., Chakrabarty, D., & Singh, S. S. (2023). A literature update on strategies for harnessing and conserving the bioactive phytochemicals from *Tinospora cordifolia*: Current status, challenges, and future prospects. *Plants for Immunity and Conservation Strategies*, 1–20.

Priyadharshini, S., Kannan, N., Manokari, M., & Shekhawat, M. S. (2020). In vitro regeneration using twin scales for restoration of critically endangered aquatic plant *Crinum malabaricum* Lekhak & Yadav: A promising source of galanthamine. *Plant Cell, Tissue and Organ Culture (PCTOC), 141*(3), 593–604.

Purohit, S., Rawat, V., Jugran, A. K., Singh, R. V., Bhatt, I. D., & Nandi, S. K. (2015). Micropropagation and genetic fidelity analysis in *Valeriana jatamansi* Jones. *Journal of Applied Research on Medicinal and Aromatic Plants, 2*(1), 15–20.

Qahtan, A. A., Faisal, M., Alatar, A. A., & Abdel-Salam, E. M. (2022). Callus-mediated high-frequency plant regeneration, phytochemical profiling, antioxidant activity and genetic stability in *Ruta chalepensis* L. *Plants, 11*(12), 1614.

Ragul, P., Rajagopal, C., Loudursamy, D. K., Thanga Hemavathy, A., Vellaikumar, S., Rajagopal, B., & Karthikeyan, P. (2025). Integrated analysis of genetic diversity in *Costus pictus* through phenotypic characterization, molecular markers and UHPLC-MS-based diosgenin profiling. *Genetic Resources and Crop Evolution*, 1–14.

Raja, H. D., Srinivasan, P., & Tamilvanan, R. (2022). Effect of purified vermicast extract with plant growth regulators on indirect regeneration of *Naregamia alata* Wight & Arn from leaf explants and genetic fidelity analysis of regenerants. *Plant Cell, Tissue and Organ Culture (PCTOC), 151*(1), 23–34.

Rajput, S., & Agrawal, V. (2020). Micropropagation of *Atropa acuminata* Royle ex Lindl.(a critically endangered medicinal herb) through root callus and evaluation of genetic fidelity, enzymatic and non-enzymatic antioxidant activity of regenerants. *Acta Physiologiae Plantarum, 42*(11), 160.

Ranghoo-Sanmukhiya, V. M. (2020). Somaclonal variation and methods used for its detection. In Iram Siddique (Ed.). *Propagation and genetic manipulation of plants* (pp. 1–18). Springer Singapore.

Rewers, M., & Jedrzejczyk, I. (2016). Genetic characterization of *Ocimum* genus using flow cytometry and inter-simple sequence repeat markers. *Industrial Crops and Products, 91*, 142–151.

Rohela, G. K., & Saini, P. (2023). Molecular techniques for the detection of ploidy level and genetic fidelity of regenerated plantlets. In A. Mujib (Ed.). *Genome size and genetic homogeneity of regenerated plants: Methods and applications* (pp. 294–326). Bentham Science Publishers.

Rohela, G. K., Jogam, P., Bylla, P., & Reuben, C. (2019). Indirect regeneration and assessment of genetic fidelity of acclimated plantlets by SCoT, ISSR, and RAPD markers in *Rauwolfia tetraphylla* L.: An endangered medicinal plant. *BioMed Research International*.

Rohela, G. K., Jogam, P., Saini, P., Sandhya, D., Peddaboina, V., & Shekhawat, M. S. (2022). Assessing the genetic stability of *in vitro* raised plants. In Shubhpriya Gupta, & Preeti Chaturvedi (Eds.). *Commercial scale tissue culture for horticulture and plantation crops* (pp. 245–276). Springer Nature Singapore.

Saini, R., Mishra, A. K., Mishra, P., Shukla, P. K., Vishwakarma, J., Tiwari, K. N., & Singh, J. (2025). Micropropagation of *Drypetes roxburghii* (Wall.) Hurusawa—A valuable medicinal plant through root culture and evaluation of its genetic fidelity and metabolomics. *In Vitro Cellular & Developmental Biology-Plant*, 1–21.

Sandhya, D., Jogam, P., Manokari, M., Shekhawat, M. S., Jadaun, J. S., Allini, V. R., & Abbagani, S. (2021). High-frequency in vitro propagation and assessment of genetic uniformity and micro-morphological characterization of *Origanum majorana* L.–A highly traded aromatic herb. *Biocatalysis and Agricultural Biotechnology, 34*, 102024.

Sarkar, J., & Banerjee, N. (2020). Influence of different cytokinins on micropropagation of an important medicinal plant, *Solanum erianthum* D. Don, and assessment of the genetic fidelity of the regenerants. *In Vitro Cellular & Developmental Biology-Plant, 56*(4), 480–490.

Sarmah, D., Mohapatra, P. P., Seleiman, M. F., Mandal, T., Mandal, N., & Wasonga, D. O. (2024). Efficient regeneration of in vitro derived plants and genetic fidelity. *Speed Breeding Systems For Food.*

Sarmah, D., Mohapatra, P. P., Seleiman, M. F., Mandal, T., Mandal, N., Pramanik, K., & Wasonga, D. O. (2024). Efficient regeneration of in vitro derived plants and genetic fidelity assessment of Phalaenopsis orchid. *Frontiers in Sustainable Food Systems, 8*, 1359486.

Sarwat, M., Das, S., & Srivastava, P. S. (2008). Analysis of genetic diversity through AFLP, SAMPL, ISSR and RAPD markers in *Tribulus terrestris*, a medicinal herb. *Plant Cell Reports, 27*(3), 519–528.

Sathish, S., Vasudevan, V., Karthik, S., Ajithan, C., Siva, R., Parthasarathy, S. P., & Manickavasagam, M. (2022). Impact of silver nanoparticles on the micropropagation of *Hybanthus enneaspermus* and assessment of genetic fidelity using RAPD and SCoT markers. *Plant Cell, Tissue and Organ Culture (PCTOC)*, 1–7.

Savitikadi, P., Jogam, P., Rohela, G. K., Ellendula, R., Sandhya, D., Allini, V. R., & Abbagani, S. (2020). Direct regeneration and genetic fidelity analysis of regenerated plants of *Andrographis echioides* (L.)-An important medicinal plant. *Industrial Crops and Products, 155*, 112766.

Shao, Q. S., Guo, Q. S., Deng, Y. M., & Guo, H. P. (2010). A comparative analysis of genetic diversity in medicinal *Chrysanthemum morifolium* based on morphology, ISSR and SRAP markers. *Biochemical Systematics and Ecology, 38*(6), 1160–1169.

Sharma, M. K., Sengar, K., Singh, A., & Sengar, R. S. (2018). Potential applications of molecular markers for genetic diversity and DNA fingerprinting in sugarcane. In Kalpana Sengar (Ed.). *Biotechnology to enhance sugarcane productivity and stress tolerance* (pp. 137–174). CRC Press.

Shekhawat, J. K., Rai, M. K., Shekhawat, N. S., & Kataria, V. (2018). Exploring genetic variability in Prosopis cineraria using two gene targeted CAAT box-derived polymorphism (CBDP) and start codon targeted (SCoT) polymorphism markers. *Molecular Biology Reports, 45*(6), 2359–2367.

Shekhawat, M. S., Priyadharshini, S., Jogam, P., Kumar, V., & Manokari, M. (2021). Meta-topolin and liquid medium enhanced in vitro regeneration in *Scaevola taccada* (Gaertn.) Roxb. *In Vitro Cellular & Developmental Biology-Plant, 57*(2), 296–306.

Shinde, S., Jain, J. R., Kadapatti, S. S., Jang, E. B., Murthy, H. N., & Park, S. Y. (2023). Micropropagation system and their genetic fidelity evaluation from regenerated plants by ISSR and DAMD markers of *Tabernaemontana alternifolia* L., an endangered medicinal woody species. *Horticulture, Environment, and Biotechnology, 64*(5), 849–857.

Shukla, S., Bhutani, R., Das, S., Kapoor, N., & Raman, T. (2023). Secondary metabolite enhancement via *in vitro* techniques and its industrial prospects. In Ravi Singh, & Nitish Kumar (Eds.). *Genetic manipulation of secondary metabolites in medicinal plant* (pp. 307–337). Springer Nature Singapore.

Singh, A. K., Rana, M. K., Singh, S., Kumar, S., Kumar, R., & Singh, R. (2014). CAAT box-derived polymorphism (CBDP): A novel promoter-targeted molecular marker for plants. *Journal of Plant Biochemistry and Biotechnology, 23*(2), 175–183.

Sirangi, S., Jogam, P., Rohela, G. K., Ragan, A., & Raju, V. S. (2022). Micropropagation of endemic *Corynandra chelidonii* var. pallae (Cleomaceae) through nodal explants and validation of their genetic integrity by ISSR markers. *Vegetos, 35*(2), 511–519.

Sirikonda, A., Jogam, P., Ellendula, R., Kudikala, H., Mood, K., & Allini, V. R. (2020). In vitro micropropagation and genetic fidelity assesment in *Flemingia macrophylla* (Willd.) Merr: An ethnomedicinal plant. *Vegetos, 33*(2), 286–295.

Soomro, S. R., Soomro, S. N., Altaf, M. T., Liaqat, W., Nadeem, M. A., Baloch, F. S., & Mohamed, H. I. (2025). Development of tetraploids in tissue culture: Modern techniques and biotechnological innovations. *Plant Cell, Tissue and Organ Culture (PCTOC), 160*(2), 51.

Srinivasan, P., Raja, H. D., & Tamilvanan, R. (2021). Efficient in vitro plant regeneration from leaf-derived callus and genetic fidelity assessment of an endemic medicinal plant *Ranunculus wallichianus* Wight & Arnn by using RAPD and ISSR markers. *Plant Cell, Tissue and Organ Culture (PCTOC)*, 1–8.

Srivastava, A., Garg, G., Sharma, P., Shah, N., Sharma, S., & Shrivastava, N. (2016). Genetic diversity in chemically diverse accessions of *Bacopa monnieri*. *JPC-Journal of Planar Chromatography-Modern TLC, 29*(3), 203–208.

Subrahmanyeswari, T., Gantait, S., Kamble, S. N., Singh, S., & Bhattacharyya, S. (2023). *meta*-Topolin-induced regeneration and ameliorated rebaudioside-A production in genetically uniform candy-leaf plantlets (*Stevia rebaudiana* Bert.). *South African Journal of Botany, 159*, 405–418.

Subramanian, M., Gantait, S., Jaafar, J. N., Ismail, M. F., & Sinniah, U. R. (2025). Micropropagation of white turmeric (*Curcuma zedoaria* (Christm.) Roscoe) and establishment of adventitious root culture for the production of phytochemicals. *Industrial Crops and Products, 223*, 120101.

Sultana, K. W., Das, S., Chandra, I., & Roy, A. (2022). Efficient micropropagation of *Thunbergia coccinea* Wall. and genetic homogeneity assessment through RAPD and ISSR markers. *Scientific Reports, 12*(1), 1683.

Thepthong, K., & Kanjanawattanawong, S. (2025). In vitro production of plantlets and microrhizomes, genetic fidelity assessment, and metabolic profiling of *Boesenbergia rotunda* (L.) Mansf. *Horticulturae, 11*(2), 186.

Tikendra, L., Amom, T., & Nongdam, P. (2019). Molecular genetic homogeneity assessment of micropropagated *Dendrobium moschatum* Sw.-A rare medicinal orchid, using RAPD and ISSR markers. *Plant Gene, 19*, 100196.

Tikendra, L., Sushma, O., Amom, T., Devi, N. A., Paonam, S., Bidyananda, N., & Nongdam, P. (2022). Genetic clonal fidelity assessment of rhizome-derived micropropagated *Acorus calamus* L.–A medicinally important plant by random amplified polymorphic DNA and inter-simple sequence repeat markers. *Pharmacognosy Magazine, 18*(77).

Tippani, R., Nanna, R. S., Mamidala, P., & Thammidala, C. (2019). Assessment of genetic stability in somatic embryo derived plantlets of *Pterocarpus marsupium* Roxb. using inter-simple sequence repeat analysis. *Physiology and Molecular Biology of Plants, 25*(2), 569–579.

Tippani, R., Porika, M., & Thammidala, C. (2025). Clonal fidelity analysis using ISSR markers in *Pterocarpus marsupium* Roxb. plantlets regenerated through direct somatic embryogenesis. *Discover Plants, 2*(1), 1–14.

Tripathi, N., Chouhan, D. S., Saini, N., & Tiwari, S. (2012). Assessment of genetic variations among highly endangered medicinal plant *Bacopa monnieri* (L.) from Central India using RAPD and ISSR analysis. *3 Biotech, 2*(4), 327–336.

Tripathi, N., Saini, N., Nair, P., & Tiwari, S. (2012). Lack of genetic diversity of a critically endangered important medicinal plant *Chlorophytum borivilianum* in Central India revealed by AFLP markers. *Physiology and Molecular Biology of Plants, 18*(2), 161–167.

Ulvrova, T., Vitamvas, J., Cepkova, P. H., Eliasova, K., Janovska, D., Bazant, V., & Viehmannova, I. (2021). Micropropagation of an ornamental shrub *Disanthus cercidifolius* Maxim. and assessment of genetic fidelity of regenerants using ISSR and flow cytometry. *Plant Cell, Tissue and Organ Culture (PCTOC), 144*(3), 555–566.

Umesh, B.T. (2008). Cytogenetical and phytochemical assays on in vitro and in vivo plants of Indian strawberry-*Fragaria indica* L. (*Duchesnea indica* Focke) (Doctoral dissertation, University of Calicut).

Upadhyay, A., Shahzad, A., Ahmad, Z., Alatar, A. A., Guerriero, G., & Faisal, M. (2021). High frequency direct organogenesis, genetic homogeneity, chemical characterization and leaf ultra-structural study of regenerants in *Diplocyclos palmatus* (L.) C. Jeffrey. *Agronomy, 11*(11), 2164.

Vemula, S., Koppula, T., Jogam, P., & Mohammed, M. (2020). In vitro high frequency multiplication and assessment of genetic fidelity of *Corallocarpus epigaeus*: An endangered medicinal plant. *Vegetos, 33*(1), 63–73.

Verma, N., & Shukla, S. (2015). Impact of various factors responsible for fluctuation in plant secondary metabolites. *Journal of Applied Research on Medicinal and Aromatic Plants, 2*(4), 105–113.

Vijaya Kumar, V., Packiaraj, P., & Kannan, P. (2023). An improved in vitro regeneration and assessment of genetic fidelity using ISSR markers in *Strychnos potatorum* L. f. *Vegetos, 36*(2), 661–667.

Vishal, Sidhu, G. S., Gaikwad, P. N., Mann, S. S., Gill, M. S., & Manchanda, P. (2024). Optimized protocol for high-frequency papaya propagation: Morpho-stereomicroscopic analysis and genetic fidelity assessment. *Plant Cell, Tissue and Organ Culture (PCTOC), 156*(3), 81.

Vrundha, C. K., & Thomas, T. D. (2023). Indirect shoot regeneration from root explants, assessment of clonal fidelity of regenerated plants using SCoT primers and antioxidant analysis in *Thottea siliquosa* (Lamk.) Ding Hou. *Plant Cell, Tissue and Organ Culture (PCTOC), 155*(1), 255–266.

Wang, M., Zhang, S., Li, R., & Zhao, Q. (2024). Unraveling the specialized metabolic pathways in medicinal plant genomes: A review. *Frontiers in Plant Science, 15*, 1459533.

Wasi, A., Shahzad, A., & Tahseen, S. (2024). In vitro propagation, SEM analysis, and genetic fidelity assessment using SCoT markers in *Decalepis salicifolia*, an endangered medicinal plant. *South African Journal of Botany, 175*, 304–313.

Wijerathna-Yapa, A., Hiti-Bandaralage, J., & Pathirana, R. (2025). Harnessing metabolites from plant cell tissue and organ culture for sustainable biotechnology. *Plant Cell, Tissue and Organ Culture (PCTOC), 162*(3), 55.

Wu, B., Zhang, N. S., Dixon, B., Sierra, I., Kan, S., Layton, A., & Qin, H. (2024). Reliable callus-induced plantlet regeneration from leaf explants of *Lagerstroemia speciosa* and genetic fidelity assessment through ISSR markers. *Plant Cell, Tissue and Organ Culture (PCTOC), 157*(3), 76.

Xiao, Q., Mu, X., Liu, J., Li, B., Liu, H., Zhang, B., & Xiao, P. (2022). Plant metabolomics: A new strategy and tool for quality evaluation of Chinese medicinal materials. *Chinese Medicine, 17*(1), 45.

Yan, J., & Wang, X. (2023). Machine learning bridges omics sciences and plant breeding. *Trends in Plant Science, 28*(2), 199–210.

Yan, X., Zheng, K., Li, P., Zhong, X., Zhu, Z., Zhou, H., & Zhu, M. (2024). An efficient in vitro organogenesis protocol for the endangered relic tree species *Bretschneidera sinensis* and genetic fidelity assessment using DNA markers. *Frontiers in Plant Science, 15*, 1259925.

Yu, S., Zhao, X., Wang, Y., Jiang, D., Zhang, Y., Hu, L., & Cai, X. (2022). Morphological, cytological, and molecular-based genetic stability analysis of *in vitro*-propagated plants from newly induced aneuploids in caladium. *Agriculture, 12*(10), 1708.

Yusoff, S. D., Kumolosasi, E., Ali, M. M., Qian, Y., & Mohd, N. O. R. S. Y. A. H. I. D. A. (2022). *Camellia sinensis* and *Phyllanthus amarus* ethanol extracts induced apoptosis and cell cycle arrest on human leukemic cell lines. *Sains Malaysiana, 51*(8), 2609–2617.

Zainuddin, Z., Urbi, Z., & Ab Halim, N. M. (2025). *In vitro* regeneration and ISSR-based genetic fidelity evaluation of *Stevia rebaudiana*. *Malaysian Applied Biology, 54*(1), 76–86.

Zaman, W., Ayaz, A., & Park, S. (2025). Climate change and medicinal plant biodiversity: Conservation strategies for sustainable use and genetic resource preservation. *Genetic Resources and Crop Evolution*, 1–34.

Zamani, M., Sonboli, A., Goldansaz, M., & Mirjalili, M. H. (2024). In vitro micropropagation and conservation of endangered medicinal plant *Nepeta asterotricha* Rech. f.(Lamiaceae): genetic fidelity, phytochemical and biological assessment. *Physiology and Molecular Biology of Plants, 30*(1), 67–80.

Zhang, H., Wang, G., Qiao, Y., & Chen, C. (2021). Plant regeneration from root segments of *Anthurium andraeanum* and assessment of genetic fidelity of in vitro regenerates. *In Vitro Cellular & Developmental Biology-Plant, 57*(6), 954–964.

CHAPTER

4

Recent trends in the somatic embryogenesis research of medicinal plants

Birat Sapkota, Poonam Khatri, Seethapathy G. Saroja, and Nirmal Joshee

Agricultural Research Station, Fort Valley State University, Fort Valley, GA, United States

4.1 Introduction

Plant embryogenesis in nature is well established as the culmination of the pollination and fertilization process, leading to the formation of a seed. Embryo and endosperm are developed through a complex, highly organized process of zygotic embryogenesis (Dodeman et al., 1997; Winkelmann, 2016). During this process, a zygote is formed after double fertilization, resulting in the formation of the embryo and the endosperm (Dodeman et al., 1997). On the other hand, the formation of embryos via somatic cells is called somatic embryogenesis. Somatic embryos differ from zygotic embryos as they develop from vegetative or somatic cells instead of zygotes (fusion product of male and female gamete) and thus potentially can be used to produce clones of select genotypes. The ability to produce morphologically well formed and normally developed embryos from somatic cells is a unique characteristic that is found within the plant kingdom (Zimmerman, 1993). Somatic embryogenesis takes place in the presence of plant growth regulators (PGRs), additives, and suitable environmental conditions like photoperiod and temperature. Somatic embryos have a bipolar structure with both apical and basal meristem regions, which are capable of forming shoot and root, respectively. A plant derived from somatic embryos is also called an embling. First record of *in vitro* somatic embryogenesis was reported in carrot (*Daucus carota*) by Reinert (1958). Somatic embryogenesis is a multi step regeneration process starting with the formation of proembryogenic masses, followed by somatic embryo formation, maturation, desiccation, and plant regeneration. Somatic embryos are used for studying the regulation of embryo development, but also as a tool for large scale vegetative propagation and genetic transformation, as well as for long term

Medicinal Plant Biotechnology for Ameliorated Secondary Metabolite Production
DOI: https://doi.org/10.1016/B978-0-443-43944-5.00005-0

storage of germplasm using cryopreservation. In addition, fundamental studies of early embryogenesis are easier to conduct on somatic embryos compared to zygotic embryos, as distinct developmental stages could be observed and studied with ease (Winkelmann, 2016). Although great progress has been made in improving the protocols used, it has been observed that certain treatments can have adverse effects on the embryo quality, impairing germination and further growth of the plant arising from a somatic embryo.

Arabidopsis thaliana has been widely adopted as a model plant for studying embryo development. Several studies have shown that the stages of embryo development consist of zygote, one cell, 2–4 cells, dermatogen, globular, heart, torpedo, bent, and mature (Zhang et al., 2021). Also, a genomic study on *A. thaliana* reported the development of the embryo in three stages based on: (1) zygotic genome activation, (2) pattern formation, and (3) maturation (Armenta-Medina et al., 2021). During somatic embryogenesis, the somatic cells go through a series of morphological and biochemical changes following two developmental pathways: direct and indirect. Direct somatic embryogenesis consists of the formation of embryos directly from somatic cells without undergoing callus development. However, during indirect somatic embryogenesis, the embryogenic callus is formed that develops into somatic embryos (Ramírez-Mosqueda, 2022). The ability of an explant to develop somatic embryos usually depends on the age and type of explant. In *Arabidopsis*, immature and mature zygotic embryos, shoot apices, and flower buds have been used as explants for *in vitro* somatic embryo development (Ikeda-Iwai et al., 2002; Kobayashi et al., 2010; Wójcikowska et al., 2018). Similarly, an investigation on *Sapindus mukorossi* reported the optimized somatic embryo induction when a 6 day old leaf was used as an explant, indicating the role of the age and organ used to initiate somatic embryogenesis (Singh et al., 2015). In addition to the explants, the role of PGRs or phytohormones also influences the induction of somatic embryogenesis. In most of the plant species, the utilization of PGRs such as 2,4 dichlorophenoxyacetic acid (2,4 D) resulted in a higher number of somatic embryos development. A study on *Camellia oleifera* reported that the combination of 2,4 D and thidiazuron (TDZ) showed the best result for direct somatic embryogenesis, while 2,4 D alone was better for callus induction and indirect somatic embryogenesis (Keshvari et al., 2018). Auxins and cytokinins are particularly important, with auxins often required to induce the process and cytokinins helping to promote cell division and shoot formation. The specific ratio of these and other hormones, along with other factors like endogenous level of growth regulators, priming, and genotype, is crucial for successful somatic embryo induction and maturation.

Medicinal plants constitute an important component of vegetation, and roughly 25% of total plants have been estimated to have useful bioactive compounds. This has led to the commercial exploitation of this important wealth in an indiscriminate manner, rendering many medicinal plants rare, threatened, or endangered. There are many reports of somatic embryogenesis in medicinal plants, providing viable multiplication and conservation possibilities for the future (Table 4.1). This chapter provides an insight into the recent trend of research on medicinal plants in general, with special attention to twelve important plant families. Table 4.1 reviews major medicinal plant genera in these 12 families for their micropropagation protocols with emphasis on explant selection and PGR optimization. In addition, a narrative on the plant based and external factors influencing the process of somatic embryogenesis with reference to medicinal plants is also included.

TABLE 4.1 A review of somatic embryogenesis in medicinal plants from dicot and monocot families.

Plant name	Plant habit	Explant	Embryogenic callus/somatic embryo induction in the medium	Somatic embryo germination conditions	Reference
Dicotyledonous families					
Apiaceae					
Centella asiatica (L.) Urb.	Herbaceous creeper	Leaf, stolon	MS + 2.26 and 4.52 μM 2,4 D	–	Joshee et al. (2007)
Coriandrum sativum L.	Herb	Hypocotyl	MS + 1.0 mg L^{-1} 2,4 D	0.25 mg L^{-1} BA + 0.5 mg L^{-1} NAA	Ali et al. (2018)
Cuminum cyminum L.	Herb	Cotyledon	MS + 0.1 mg L^{-1} KN	–	Bahmankar et al. (2017)
Trachyspermum ammi Sprague (Syn. *Carum copticum* (L.) Benth. & Hook.f.)	Herb	Hypocotyl	MS + 1.5 mg L^{-1} 2,4 D and 0.5 mg L^{-1} KN	MS + 1.5 mg L^{-1} 2,4 D and KN	Niazian et al. (2017)
Apocynaceae					
Catharanthus roseus (L.) G. Don	Herb	Stem, leaf, hypocotyl	MS + 4.52 μM 2,4 D	MS + 2.22 μM BAP	Mujib et al. (2014)
Cynanchum annularium (Roxb.) Liede & Khanum (Syn. *Holostemma ada kodien* Schult.)	Vine	Leaf, internode, root	MS + 1 mg L^{-1} 2,4 D	1/10 MS	Martin (2003)
Rauvolfia micrantha Hook.f.	Woody shrub	Root	MS + 3% Sucrose + 100 mg L^{-1} myo inositol, 0.5 mg L^{-1} NAA	MS + 0.05 mg L^{-1} NAA + 0.1 mg L^{-1} BA	Sudha & Seeni (2006)
Hancornia speciosa Gomes	Tree	Leaf and nodal segments	MS salts, putrescine, 45 μM 2,4 D and 22.19 μM BAP	MS salts, PEG (0 and 2%), and BAP (44.39 and 66.58 μM)	Santana et al. (2024)

(*Continued*)

TABLE 4.1 (Continued)

Plant name	Plant habit	Explant	Embryogenic callus/somatic embryo induction in the medium	Somatic embryo germination conditions	Reference
Asteraceae					
Cichorium intybus L.	Herb	Nodal stem, petiole	MS + 1.5 mg L^{-1} KN + 0.5 mg L^{-1} +1 mg L^{-1} IAA + 500 mg L^{-1} CH	MS + 1.5 mg L^{-1} KN + 0.1 mg L^{-1} IAA + 0.5 mg L^{-1} IBA + 500 mg L^{-1} CH	Abdin and Ilah (2007)
Eclipta prostrata (L.) L. (Syn. *Eclipta alba* L. ex B.D.Jacks.)	Herb	Node	MS + 10.75 μM NAA + 9.04 μM 2,4 D	MS	Devendra et al. (2011)
Echinacea purpurea (L.) Moench	Herb	*In vitro* leaf	5 mg L^{-1} BA + 0.5 mg L^{-1} NAA + AC + CM + CH	–	Dehestani-Ardakani et al. (2020)
Sphagneticola calendulacea (L.) Pruski (Syn. *Wedelia calendulacea* Less.)	Herb	Leaf, Internode	MS + 2 mg L^{-1} 2,4 D	MS + 0.1–2 mg L^{-1} NAA	Sharmin et al. (2014)
Fabaceae					
Calliandra tweedii Benth	Shrub	Internode, petiole	MS + 1 μM 2iP	–	Kumar et al. (2002)
Clitoria ternatea L.	Vine	Cotyledon	MS + 2 mg L^{-1} 2,4 D	MS + 2 mg L^{-1} BAP + 0.2 mg L^{-1} NAA+ 3 mg L^{-1} ABA.	Krishna Kumar and Thomas (2012)
Desmodium motorium (Houtt.) Merr.	Shrub	Cotyledon	MS + 2.9 μM IAA + 4.44 μM and 8.88 μM BA	+	Chitra Devi and Narmathabai (2011)
Falcataria falcata (L.) Greuter & R.Rankin (Syn. *Falcataria moluccana* (Miq.) Barneby & J.W.Grimes)	Tree	Leaf	MS + 5 mM TDZ + 1.2 g L^{-1} Proline	–	Sunandar and Supena (2017)

Mucuna pruriens (L.) DC.	Climber	Cotyledon	MS + 6.7 μM 2,4 D	MS + 2.3 μM Kinetin + 5.4 μM α- NAA	Vibha et al. (2009)
Pterocarpus marsupium Roxb.	Tree	Hypocotyl	MS + 5 μM 2,4 D + 1 μM BAP	MS + 0.5 μM BA+ 0.1 μM NAA and 10 μM ABA	Husain et al. (2010)
Cucurbitaceae					
Citrullus lanatus Thunb.	Vine	Leaf	MS + 2.44 μM 2,4 D + 2.27 μM TDZ	–	Vinoth and Ravindhran (2016)
Citrullus colocynthis L.	Vine	Seeds	MS + 1.5 mg L^{-1} BAP + 0.5 mg L^{-1} GA_3	–	Ramakrishna and Shasthree (2016)
Cucumis anguria L.	Vine	Leaf	MS + 4 μM 2,4 D + 1.0 μM KN	MS+ 2% sucrose+ 0.5 μM ABA + 0.2% Gelrite.	Thiruvengadam et al. (2013)
Cucumis sativus L.	Vine	Seeds	MS + 6.8 μM 2,4 D + 2.2 μM BA	MS + 175.2 mM sucrose + 0.5 g L^{-1} activated charcoal.	Vengadesan et al. (2005)
Cucurbita pepo L.	Vine	Cotyledon	MS + 26.85 μM NAA + 4.44 μM BAP	2.89 μM GA_3 +0. 54 μM NAA + 11.42 μM IAA	Urbanek et al. (2004)
Euphorbiaceae					
Euphorbia hirta L.	Herb	Stem	MS +0.4 mg L^{-1} BAP + KN 0.5 mg/L IAA	MS + 0.5 mg/L BAP and 0.1 mg/L KN.	Shekhawat et al. (2018)
Jatropha curcas L.	Shrub	Leaf	MS basal medium 2.3 μM KN and 1.0 μM IBA	MS basal medium + 3% sucrose 2.3 μM KN + 1.0 μM IBA + 13.6 μM adenine sulfate	Jha et al. (2007)

(*Continued*)

TABLE 4.1 (Continued)

Plant name	Plant habit	Explant	Embryogenic callus/somatic embryo induction in the medium	Somatic embryo germination conditions	Reference
Sapium sebiferum Roxb.	Tree	Seeds	MS + 1.0 mg L^{-1} IAA	MS +1.0 mg L^{-1} 6-BA	Hou et al. (2015)
Lamiaceae					
Ocimum basilicum L.	Herb	Leaf	MS + 1 mg L^{-1} 2,4 D	MS basal medium + BAP 1.0 mg L^{-1} +NAA 1 mg L^{-1} +KN 0.5 mg L^{-1}.	Gopi and Ponmurugan (2006)
Salvia officinalis L.	Shrub	Leaf	MS + 1.8–18 mM 2,4 D + KN	–	Kintzios et al. (1999)
Salvia fruticosa Mill.	Shrub	Leaf	MS + 10.5–21 mM NAA + BAP	–	Kintzios et al. (1999)
Thymus hyemalis Lange	Shrub	Node	MS + 4.44 µM BAP + 0.54 µM NAA + 4.65 µM KN	MS	Nordine et al. (2014)
Teucrium polium L.	Herb	Leaf, petioles, stem	MS + 0, 0.5, and 1.0 mg L^{-1} KN + 0, 1.0, and 1.5 mg L^{-1} 2,4 D	–	Saadat et al. (2023)
Lauraceae					
Cinnamomum camphora L.	Tree	Cotyledon	MS + 0.1–3.0 mg L^{-1}ABA	MS + 0.5 mg L^{-1} IBA	Shi et al. (2010)
Rubiaceae					
Cephaelis ipecacuanha A. Richard	Shrub	Leaf	MS+ 2.5 mg L^{-1} mg L^{-1} KN + 1.0 mg L^{-1} 2,4 D + 3% sucrose	MS basal	Rout et al. (2000)
Oldenlandia umbellata L.	Herb	Leaves, shoot apices, stem	MS + 0.5 mg L^{-1} BA + 0.2 mg L^{-1} NAA + 0.1% CM	–	Siva et al. (2009)
Spermacoce hispida L.	Herb	Cotyledonary leaf, shoot tip, hypocotyl, and nodal explants	MS + BAP (0.5, 1.0, 2.0 mg/dm^{-3}) and KN (0.2, 0.4, 0.8 mg/dm^{-3})	–	Deepak et al. (2019)

Rutaceae					
Murraya koenigii (L.) Spreng.	Tree	Zygotic embryos, Cotyledon	MS basal medium + 8.88 μM BA + 2.675 μM NAA	MS basal medium + 8.88 μM BA + 2.675 μM NAA + TDZ	Paul et al. (2011)
Solanaceae					
Capsicum annuum L.	Shrub	Cotyledon, Zygotic embryos, embryo axis	MS + 9 μM 2,4 D	MS + 9 μM 2,4 D + 10% Coconut water	Binzel et al. (1996)
Withania somnifera (L.) Dunal	Shrub	Cotyledonary leaf, internodes, root	MS + 3 mg L^{-1} KN + 2 mg L^{-1} BA+ 2 mg L^{-1} TIBA	MS + 3 mg/L BA + 2 mg L^{-1} IAA + 2 mg L^{-1} IBA	Rani et al. (2004)
Monocotyledonous families					
Orchidaceae					
Dendrobium cv. Chiengmai Pink	Herb	Leaf	½ MS + 18.16 μM TDZ	½ MS basal	Chung et al. (2005)
Epipactis veratrifolia Boiss. & Hohen.	Herb	Protocorm/ Protocorm like bodies	1/8 MS + 0.5 mg L^{-1}ABA + 3 mg L^{-1} BA	Fast medium with organic substances	Moradi et al. (2017)
Malaxis densiflora (A. Rich.) Kuntze	Herb	Protocorm like bodies	½ MS + 3.39 μM 2,4 D + 6.80 μM TDZ	–	Mahendran and Bai (2016)
Rhynchostylis retusa (L.) Blume	Herb	Leaf and root	1.0 mg L^{-1} BAP + 1.5 mg L^{-1} KN	½ MS + 1.5 mg L^{-1} BAP	Islam and Bhattacharjee (2015)
Oncidium spp.	Herb	Leaf; Leaf; Leaf; Leaf tips	½ MS + 10–20 g L^{-1} sucrose + 170 mg L^{-1} NaH_2PO_4 + 0.5 g L^{-1} peptone; ½ MS + 20 g dm^{-3} Sucrose + 85 mg dm^{-3} NaH_2PO_4 + 3 mg dm^{-3} KN; ½ MS + 4.54 μM TDZ + 0.1 μM acetylsalicylic acid; ½ MS + 1 mg dm^{-3} TDZ + 30 60 g dm^{-3} sucrose	½ MS basal; ½ MS basal; ½ MS basal	Chen and Chang (2002), Su et al. (2006), Hong et al. (2008a, 2008b)

(*Continued*)

TABLE 4.1 (Continued)

Plant name	Plant habit	Explant	Embryogenic callus/somatic embryo induction in the medium	Somatic embryo germination conditions	Reference
Zingiberaceae					
Boesenbergia rotunda (L.) Mansf.	Herb	Shoot buds	MS medium, 1 mg L^{-1} 2,4 D and 0.5 mg L^{-1} BAP	1 mg L^{-1} NAA and 3 mg L^{-1} BAP	Wong et al. (2013)
Curcuma amada Roxb.	Herb	Leaf sheath	MS medium, 2.0 mg L^{-1} 2,4 D and 0.5 mg L^{-1} BAP	0.25 mg L^{-1} GA3	Raju et al. (2016)
Curcuma caesia Roxb.	Herb	Sprouted buds	MS medium, 2 mg L^{-1} 2,4 D and 5 mg L^{-1} BAP	5 mg/L BAP and 0.2 mg L^{-1} 2,4 D	Zuraida et al. (2014)
Curcuma longa L.	Herb	Leaf base	MS medium, 4.49 μM 2,4 D	1.44 μM GA_3	Raju et al. (2015)
Elettaria cardamomum (L.) Maton	Herb	Inner core region of rhizome	MS medium, 4.4 μM BAP and 0.5 μM NAA	13.2 μM BAP and 0.5 μM NAA	Manohari et al. (2008)
Hedychium bousigonianum Pierre ex Gagnep	herb	Leaf of *in vitro* germinated seed	MS medium, 9.05 μM 2,4 D and 4.6 μM KN	MS basal salts, 0.6 μM TDZ, and 8.9 μM 6-BAP	Sakhanokho et al. (2009)
Kaempferia galanga L.	Herb	Leaf base	MS medium, 1.5 mg L^{-1} 2,4 D and 1 mg L^{-1} BA	2 mg L^{-1} BA and 0.1 mg L^{-1} NAA	Rahman et al. (2004)
Zingiber officinale Roscoe	Herb	Leaf sheath	MS medium, 9.06 μM 2,4 D, 2.27 μM TDZ	2.22 μM BAP and 2.6 μM NAA	Mehaboob et al. (2019)

ABA, Abscisic acid; *AC*, activated charcoal; *BAP*, 6 benzylaminopurine; *CH*, casein hydrolysate; *CM*, coconut milk; *2,4 D*, 2,4 dichlorophenoxyacetic acid; *GA3*, gibberellic acid; *IAA*, indole 3 acetic acid; *2 iP*, 2 isopentanyl adenine; *KN*, kinetin; *NAA*, 1 naphthaleneacetic acid; *TDZ*, 1 phenyl 3 (1,2,3 thiadiazol 5 yl) urea (thidiazuron); *TIBA*, tri iodobenzoic acid; *PEG*, polyethylene glycol.

4.1.1 History

The observation of asexual reproduction in plants through runners, suckers, tubers, and cuttings, as well as the production of multiple embryos in a single seed (known as polyembryony), has been observed since the 18th century (Tisserat et al., 1979). However, *in vitro* embryogenic tissue cultures in plants were studied since the latter half of the 20th century (McFarland & Kaeppler, 2025). The early observations of the callus, an unorganized mass of cells formed as a reaction to tissue injury in plants that are capable of regenerating into new organs, led to the pioneering studies of *in vitro* plant tissue and cell culture (Carra et al., 2024).

Over one hundred years ago, Austrian botanist Gottlieb Haberlandt provided an idea that a complete and functional plant could be generated from a single cell (Höxtermann, 1997; McFarland & Kaeppler, 2025). However, the hypothesis that it was possible to regenerate plant organs *in vitro* was evident upon researchers discovered the regeneration process in *Oenanthe aquatica* (Waris, 1957) and *D. carota* (Steward et al., 1958). Finally, in the mid 1960s, production of a whole plant from a single cell was demonstrated, which showed the totipotency of plant cells (Vasil & Hildebrandt, 1965a, 1965b). The highest incidence associated with embryos originating in the nucellus has been particularly observed in Rutaceae (Rangan et al., 1968). In addition to that, families such as Cactaceae, Celastraceae, Liliaceae, Myrtaceae, Orchidaceae, Rosaceae, and Solanaceae also possess several species that naturally exhibit asexual embryogenesis, often of nucellar origin (Tisserat et al., 1979).

The dedifferentiation of plant cells is a key process for understanding the plasticity of plant development, enabling insights into regeneration, totipotency, and adaptive responses. Various studies on somatic embryogenesis provide the hypothesis that many mature plant cells retain totipotency and related dedifferentiation to the initial step of the expression of totipotency. It has been evident that although all the diploid cells in an individual have the same genomic DNA, different cell types have distinct cell characteristics, and among them, only few cells (mostly with high number of meristematic cells) are totipotent to become embryos. It is suggested that differences in the ability to generate somatic embryos are driven by different epigenetic changes at different loci of totipotent cells. The totipotent cells have been found to differ from their surrounding somatic cells mainly by having a large nucleus, a large nucleolus, fragmented vacuoles, symplasmic

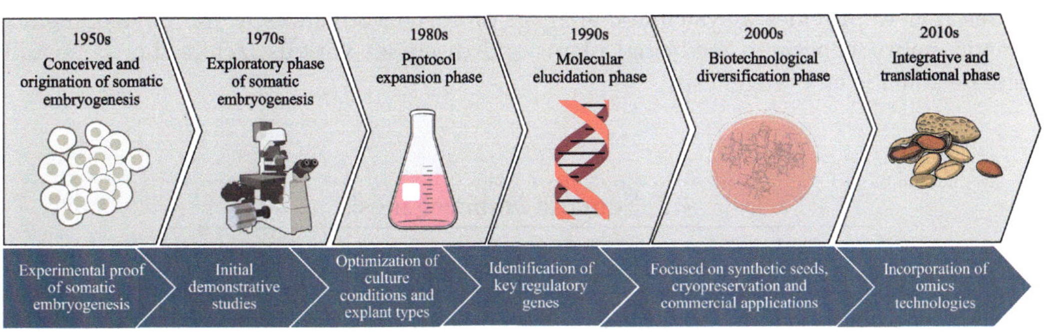

FIGURE 4.1 Distinctive phases in the evolution of somatic embryogenesis research.

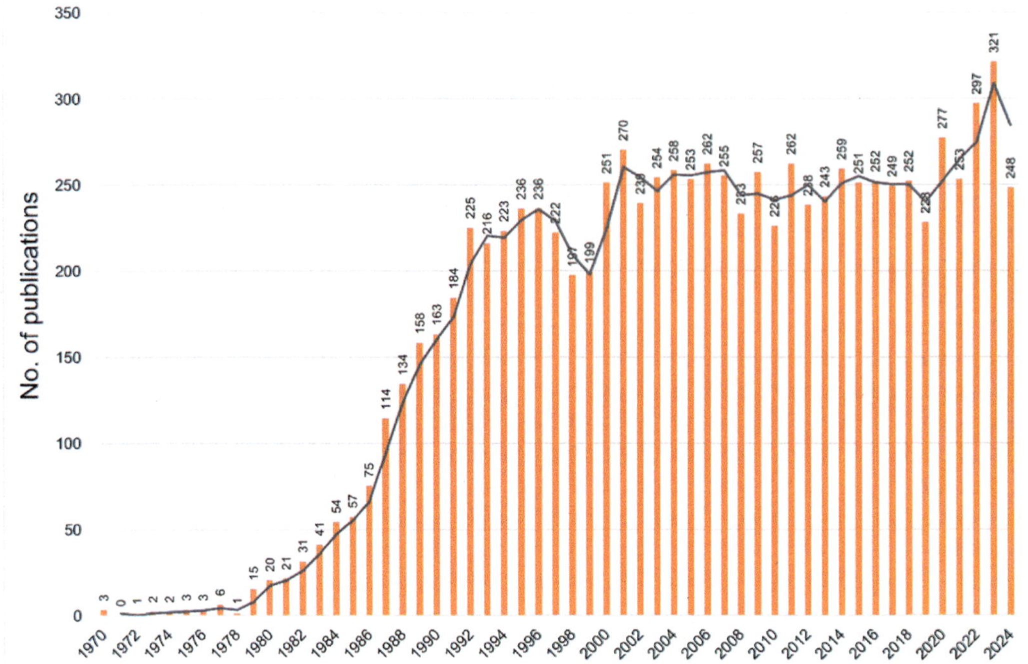

FIGURE 4.2 Number of scientific articles published per year on somatic embryogenesis research from 1970 to 2024. Source: *Adapted from Pérez Molina, J.P., Carvajal Campos, P., & Jiménez, V.M. (2025). Mapping fifty five years of somatic embryogenesis research: A bibliometric analysis of trends, global collaborations, and key advances across more than nine thousand articles.* Plant Cell, Tissue and Organ Culture, *162(2), 1–17.*

isolation, and low levels of heterochromatin (Godel-Jedrychowska et al., 2020; Verdeil et al., 2007; Wang, Luo, et al., 2022; Wang, Shang, et al., 2022). During the past decades, tremendous advances in somatic embryogenesis research have been achieved, which include cryopreservation techniques, the development of bioreactor systems, genetic transformation, and multiomics (Fig. 4.1).

Fig. 4.2 illustrates the steady increase in the volume of publications on the topics of somatic embryogenesis since the year 1970.

This review provides a systematic analysis of current advances in the development of somatic embryogenesis in medicinal plants and discusses various factors that potentially regulate somatic embryogenesis.

4.2 Somatic embryogenesis

4.2.1 Factors regulating somatic embryogenesis

Literature published on somatic embryogenesis thus far indicates that the critically important requirements for successful *in vitro* somatic embryogenesis depend on genotype, medium used, types and concentrations of PGRs, explant used, and culture conditions, especially light.

4.2.1.1 *Genotype*

Genotype plays a crucial role in somatic embryogenesis by determining the potential for, and efficiency of, inducing and developing somatic embryos in dicot and monocots, both. Genetic factors are primary determinants of the success of somatic embryogenesis, with some genotypes exhibiting high embryogenic potential and others showing very poor or no response. Differences in gene expression profiles between genotypes contribute to these varied responses, affecting key processes like callus induction, maturation, and plant regeneration. Importance of genotype and explant selection has been investigated in many crops for somatic embryogenesis and micropropagation in general. A detailed study of 33 soybean lines, showing good regeneration were found to have in their pedigrees one or both highly regenerative 'Manchu' or 'A.K. Harrow' ancestral lines (Parrott et al., 1989). Identifying and selecting genotypes with high somatic embryogenesis potential is a key strategy for improving efficiency in plant propagation and genetic modification programs.

4.2.1.2 *Plant growth regulators*

4.2.1.2.1 Auxins

Auxin, particularly 2,4 D, is a synthetic auxin analog that serves as the primary PGR needed for the somatic embryo induction stage (Zimmerman & Hitchcock, 1942). This is followed by an embryo maturation stage on medium with reduced or no auxins. It is widely reported that providing an external supply of auxin can increase endogenous auxin levels, creating a hormonal gradient that promotes embryogenesis initiation (Tang et al., 2010). For proper development of somatic embryos, removing auxins or lowering their amount in the medium is essential (Liu et al., 1993). Prolonged exposure of explants to high auxin concentrations can disrupt the polar auxin gradient during embryo development, potentially leading to incorrect formation of the embryo's apical basal parts (Liu et al., 1993). While 2,4 D is the most used auxin, others like picloram (Khatri & Joshee, 2024a, 2024b), 3,6 dichloro o anisic acid (dicamba) (Filippov et al., 2006), indole 3 acetic acid (IAA) (Verma et al., 2012), and naphthaleneacetic acid (NAA) (Raju et al., 2014), in combination with cytokinin, are also widely used and effective. However, the choice of auxin type and its concentration for successful somatic embryogenesis varies depending on the plant species.

4.2.1.2.2 Cytokinins

Cytokinins are adenine derived phytohormones that regulate cell proliferation and differentiation. Although cytokinins are commonly used with auxins, it may not be crucial for successful outcomes. It is well established that the type of morphogenesis primarily depends on the ratio of auxins to cytokinins in the medium (Van Yen & Li, 2022). Benzylaminopurine (BAP) in combination with NAA in MS medium induced more somatic embryos than when 6 benzyladenine (BA) was used alone in *Scaevola sericea* (Liang et al., 2020). Combination treatment of TDZ with 2,4 D has induced direct somatic embryogenesis in *Pluchea laneolata* (Kher & Nataraj, 2019).

4.2.1.3 *Other plant growth regulators*

4.2.1.3.1 Ethylene

Ethylene interacts with other PGRs, such as auxins and cytokinins, to regulate somatic embryogenesis. Its effects are species specific and can vary at different developmental stages. For example, in *Medicago sativa* (Alfalfa), ethylene, along with growth regulators, plays an important role in the induction and development of somatic embryos (Kępczyńska & Kępczyński, 2023), whereas in *Hedychium bousigonianum,* inhibiting ethylene enhanced callus growth and increased somatic embryo production (Sakhanokho et al., 2009).

4.2.1.3.2 Explants

The choice of explant is a key factor for successful somatic embryogenesis. Size, age, source, and the orientation in which it is cultured can influence the induction of somatic embryogenesis (George, 1993). The physiological condition of the explant significantly affects the formation of somatic embryos in *Centella asiatica,* and studies have shown notable differences in embryogenic response among mature vegetative tissues and young tissues (Joshee et al., 2007; Khatri & Joshee, 2024a, 2024b). Different types of explants, such as leaf, cotyledon, immature embryo, mature zygotic embryo, stem, shoot tip, internode, and seed, can successfully undergo somatic embryogenesis. However, the type of explant that exhibits more response depends on the plant species.

4.2.1.3.3 Medium composition and pH

Somatic embryos can be obtained on different plant growth media such as Murashige and Skoog (1962) (MS) Medium, Gamborg et al., 1968 (B5) Medium, McCown and Lloyd (1981) Woody Plant Medium (WPM), Schenk and Hildebrandt (1972) (SH) Medium and their altered forms used for plantlet regeneration. The pH of the culture medium and nitrogen type and content play a significant role in regulating somatic embryo frequency and *in vitro* regeneration. The optimal pH range for somatic embryogenesis is 5.4–5.8 (Jianhua et al., 2020).

4.2.1.3.4 Carbon source

Various carbohydrates have been employed in plant tissue culture, with sucrose and glucose recognized as the most effective carbon sources for promoting culture growth (Junaid et al., 2008). However, depending on the plant species, tissue type, or specific requirements, other carbohydrates (sorbitol, galactose, and maltose) can serve as an alternative to sucrose or glucose (Junaid et al., 2008).

4.2.1.3.5 Gelling agents

Gelling agent type and concentration have a significant impact on inducing somatic embryos due to their effect on the water availability, physical properties of the medium used, and embryogenic response of the explant (Triqui et al., 2008). In literature, several gelling agents have been used, such as agar, gelatine, starch derivative gels, silica gel, and gelrite; however, agar remains the most used gelling agent. In *Cucumis sativus,* agar was superior to gelrite in inducing somatic embryos from cotyledon explants. On the other hand, gelrite has been found to be better than agar in maintaining embryogenic lines in *Vitis vinifera* (Triqui et al., 2008). Whereas, in other crops, gelrite had a negative effect as it produced vitrified shoots in *Fraxinus americana* (Huang et al., 1995). Kwack and Fujieda (1988) noted an inhibitory effect of agar (proportional to its concentration) on *Cucurbita moschata* embryogenic response. As a means of reducing abnormal growth in *C. sativus*, Ziv and Gadasi (1986) suggested the use of double layer culture with activated charcoal in the lower agar layer and abscisic acid (ABA) in the top liquid layer. This technique increased the frequency of regenerated plantlets and improved their morphological development. In *Poa pratensis*, the medium solidified with gelrite, the regeneration frequency of the callus was twice as high as on agar solidified medium (Van Ark et al., 1991).

4.2.1.3.6 Photoperiod

Somatic embryogenesis occurs under various light and dark conditions. Each plant species requires specific illumination levels for optimal culture development. According to George (1993), callus formation may occur in darkness, although the requirement for light during initiation varies among species. Once initiated, the callus can typically be subcultured and maintained in dark conditions.

4.2.1.3.7 Amino acids and vitamins

Glutamine, an amino acid, plays a vital role in improved callus induction and stimulates embryogenesis by acting as an organic nitrogen source and a key intermediate in nitrogen assimilation (Abo-El-Nil, 1989).

Casein hydrolysate is a source of phosphate, calcium, vitamins, a mixture of 18 amino acids, and microelements (Gerdakaneh et al., 2011). Zenk et al. (1975) reported that the addition of 4 gm/L casein hydrolysate to the culture medium significantly stimulated callus growth in *Morinda citrifolia*.

Activated charcoal improves somatic embryo induction and development and boosts germination and rooting during somatic embryogenesis (Al-Khayri, 2011; Dipti & Mujib, 2014; Ghazzawy et al., 2017).

4.3 Direct somatic embryogenesis

Direct somatic embryogenesis is the process by which somatic (nonreproductive) cells develop into somatic embryos directly on the explant without going through an intervening callus phase. In this pathway, embryos are produced directly from competent cells within the explant tissue, such as leaf, cotyledon, and stem, under the influence of specific PGRs, mostly involving low concentrations of auxins like 2,4 D or cytokinin rich media. Since this pathway bypasses a callus phase and involves minimal cellular dedifferentiation, it typically yields genetically uniform plantlets and substantially reduces the risk of somaclonal variation (Finer, 1994).

4.4 Indirect somatic embryogenesis

Indirect somatic embryogenesis begins with a dedifferentiation phase during which the explant tissue first produces callus. Callus cells redifferentiate into somatic embryos upon transfer to lower concentrations of auxins or no growth regulators at all (Quiroz-Figueroa and Monforte-González, 2006). This two step process offers greater flexibility regarding the type of explant used and enables the production of many embryos, making it particularly suitable for mass propagation, genetic transformation, and synthetic seed production (George and Hall, 2008).

4.5 Secondary somatic embryogenesis

Secondary somatic embryos, also known as repetitive, cyclic, recurrent, accessory, or adventitious embryos obtained by using primary somatic embryos as an explant, and this developmental process is called secondary somatic embryogenesis. Secondary somatic embryos have been reported to form either directly or indirectly through the hypocotyls, cotyledons, root tips, or epicotyls of the primary somatic embryos (Bogdanović et al., 2021). In plant species, especially in broadleaved trees, embryogenic competence declines over time due to aging and subculturing for several months in *in vitro* cultures. Thus somatic embryogenesis provides a way to restore the embryogenic potential of important productive lines and is routinely used as a method for long term management of such species (Klimaszewska et al., 2009; Lelu-Walter et al., 2018; Pila Quinga et al., 2017). Secondary embryogenesis induction depends on various biotic and abiotic factors. For example, a study on the induction of secondary somatic embryogenesis in *Centaurium erythraea* was successful by application of 2,4-D and N (2 chloro 4 pyridyl) N0 phenylurea and developed secondary somatic embryos which were also readily germinated and converted into plantlets (Bogdanović et al., 2021).

4.6 Developmental aberrations in culture

Although somatic embryogenesis has been successfully achieved in several plant species, a major challenge remains: the frequent occurrence of abnormal embryos that cannot develop into normal plants. These abnormalities may occur due to genetic or epigenetic changes in DNA, which can be influenced by external factors such as the use of PGRs, as well as the mutagenic substances or stress factors applied to the plant tissues (high and low temperatures, drought, salinity, heavy metals). The abnormal embryos generated through genetic changes in DNA are hardly reversible; however, those developed by epigenetic changes may be reversible and be able to produce normal plants in most cases (Garcia et al., 2019).

Similarly, somatic embryos response is also affected by the release of organic molecules by the explants into the culture medium. These molecules that help to induce or modulate the somatic embryos response are polysaccharides, amino acids, growth regulators, vitamins, proteins, and lipophilic molecules (De Jong et al., 1992; De Jong et al., 1993; De Vries et al., 1988; Igasaki et al., 2003; Kobayashi et al., 2001; Kreuger & van Holst, 1993; Matthys-Rochon, 2002). However, some secondary metabolites have been found to inhibit the embryogenic response of the cells. A study on *D. carota* showed that somatic embryogenesis is inhibited by 4 hydroxybenzyl alcohol (4-HBA), a chemical compound of low molecular mass released in response to high-density cell cultures (Higashi et al., 1998; Kobayashi et al., 2000). Its accumulation in embryogenic cell cultures was found to suppress the rapid division of cells, mainly at early globular stages, inhibiting somatic embryogenesis. The processes such as DNA methylation have been found to influence somatic embryogenesis, for example, in *Pinus nigra*, low levels of DNA methylation promoted somatic embryo formation, whereas high levels were associated with the development of nonembryogenic cell lines (Noceda et al., 2009). Similar response was reported in other species, such as *Rosa* x *hybrid* (Xu et al., 2004) and *Eleuterococcus senticosus* (Chakrabarty et al., 2003).

4.7 Molecular determination of somatic embryogenesis

Understanding the mechanisms controlling somatic embryogenesis and its regulation is a key issue. Clonal propagation via somatic embryogenesis is widely used for many important agronomic and medicinal plants and is used for the genetic transformation of economically important plants. The success of transformation is correlated with the embryogenic and regeneration capacity; hence, processes of somatic embryogenesis are of great importance (Leljak-Levanić et al., 2015). However, somatic embryogenesis involves complex and not well characterized genetic inheritance. Despite the similarities between somatic and zygotic embryogenesis in terms of some hormonal, transcriptional, or epigenetic controls, somatic embryogenesis differs from *in vivo* zygotic embryogenesis in several aspects (Nic-Can & Loyola-Vargas, 2016). For instance, gene expression patterns in somatic embryogenesis require reprogramming, which comprises a cascade of genetic signals that switch the expression of various groups of genes on or off (Fehér et al., 2003). To achieve somatic embryogenesis, the totipotency of plant cells is usually induced *in vitro* by stress. Stress conditions that are essential for somatic embryogenesis induction can be carried out by providing a high level of plant hormones in the culture media, wounding, extreme pH, osmotic, heat shock, and treatments with different chemicals (Nic-Can & Loyola-Vargas, 2016). Early somatic embryogenesis is characterized by the induction, which is due to the expression of stress related genes and encoding transcription factors belonging to AP2/ERF, MYB, AUX/IAA, B3 or WUS/WOX families (Wójcikowska et al., 2013).

In general, most of the somatic embryogenesis studies and protocols are based on the large variation in PGR concentration added to the culture media. Specifically, 2,4 D, a synthetic auxin-like PGR and the treatment of plant tissues with 2,4 D involve both stress and the auxin response (Wójcikowska & Gaj, 2017). During the induction, auxin response factor genes display specific expressions that are differently regulated, suggesting that auxin signaling is central in the process. Moreover, YUCCA and AUX/IAA genes that are respectively involved in auxin biosynthesis and response, are transcriptionally regulated during somatic embryogenesis, including by LAFL transcription factors. ABSCISIC ACID INSENSITIVE 3 (ABI3), FUSCA 3 (FUS3), and LEAFY COTYLEDON 2 (LEC2), BABY BOOM (BBM), WUSCHEL (WUS), and WUSCHEL RELATED HOMEOBOX (WOX) Genes, AGAMOUS LIKE 15 (AGL15), WOUND INDUCED DEDIFFERENTIATION 1 (WIND1) are transcription factors involved in somatic embryogenesis programs as well as in the initiation and maintenance of the embryogenic fate of plant cells (Wójcikowska et al., 2013).

Somatic embryogenesis is governed by a multitude of mechanisms, encompassing intricate gene regulatory networks, hormonal signaling pathways, and epigenetic modifications—many of which are still not fully elucidated. The formation of somatic embryos is closely associated with distinct cellular processes. To deepen our understanding of these regulatory systems, it is essential to investigate the gene networks at the single cell level. Comprehensive profiling of the transcriptome, epigenome, and proteome within individual cells is crucial for uncovering the molecular dynamics driving these events. Advancements in single cell analysis and cellular transcriptome technologies are crucial for deepening our understanding of the early gene regulatory networks involved in somatic embryogenesis. These techniques applied to samples enriched in embryonic cells with the help of specific marker genes or specific morphological markers could help to understand the early controls of somatic embryogenesis.

4.8 Somatic embryogenesis and plant recovery

Somatic embryos are transferred from the proliferative medium to a medium that promotes the development of a mature embryo. This often involves adjusting the sucrose concentration and sometimes removing PGRs to induce germination.

4.9 Developmental biology

4.9.1 Stages in somatic embryo development

The process of somatic embryogenesis can be broadly divided into four stages of initiation, proliferation, maturation (including globular, heart shaped, and torpedo shaped stages), and plant regeneration, although each stage can be further subdivided into more sections depending on the genotype and pathway adopted.

4.9.1.1 Establishing embryogenic cultures

Somatic cells have all the genetic information needed to produce a healthy plant. Under stress conditions and plant growth hormones, cells activate an internal signaling pathway that alters gene expression and promotes cell division, which can result in either unorganized callus growth or organized development into somatic embryos (Dudits and Györgyey, 1995). In some plant species, selecting the right explant is essential for establishing embryogenic cell lines (Schiavo, 1995). It has been reported that two key processes, asymmetric cell division and controlled cell growth, are essential for initiating *in vitro* somatic embryogenesis (De Jong et al., 1993; Emons, 1994). Plant hormones can trigger asymmetric cell division by changing the cell's internal balance, such as the pH or electrical fields (Smith & Krikorian, 1990), which leads to an uneven distribution of cellular contents inside the cell following polarity development. Such uneven distribution plays a crucial part in the early stages of somatic embryo formation (De Jong et al., 1993).

4.9.1.2 *Proliferation of embryogenic callus*

After the formation of embryogenic callus, the cells continue to grow to form proembryogenic masses. Initially, this process needs auxin for the multiplication of proembryogenic masses; however, it can inhibit further somatic embryo development if it remains on the same medium (with auxin) for an extended period. It is recommended that either a low concentration or a medium with no auxin is suitable for the somatic embryos initiation (Filonova et al., 2000). Several reports highlighted that the proliferation of embryogenic callus is faster in suspension cultures because explants in liquid medium minimize embryo dependency on the parent tissue, thereby improving embryo yields (Moscatiello et al., 2012).

4.9.1.3 *Maturation of somatic embryos*

Upon the embryogenic callus induction and proliferation, the maturation stage follows, which is a crucial developmental step where early stage embryos undergo morphological and physiological changes to attain desiccation tolerance and capacity to germinate (Kermode, 1990). This process is mainly regulated by ABA, which helps to promote protein storage and lipid accumulation while preventing precocious germination, hence ensuring normal embryo development (Zimmerman & Hitchcock, 1942). However, long exposure to ABA can have a negative effect on plant growth (Von Arnold et al., 2002). In addition, nutritional and osmotic agents such as mannitol, sorbitol, and polyethylene glycol also promote embryo maturation by providing desiccation stress and stimulating storage reserve accumulation (Dodeman et al., 1997; Iraqi & Tremblay, 2001).

4.9.1.4 *Germination and plantlet formation*

Upon germination somatic embryos develop into complete plantlets under *in vitro* conditions, typically on a culture medium without growth regulators. However, in some cases, the presence of auxins and cytokinins can enhance germination. Additionally, supplements such as casein hydrolysate and glutamine have been reported to promote efficient and large scale regeneration of plantlets (Joshi et al., 2009).

4.10 Histology and histochemistry

Somatic embryogenesis is a complex mechanism as it requires the optimization of various factors for its proper development. Histological approaches have been used as a tool to study the anatomical and histochemical changes in the cellular processes (Yeung, 1999). Histological study on embryogenic and nonembryogenic calli of *Pimpinella brachycarpa* after 14 days of culturing showed that the embryogenic calli were firm, rather opaque, and light yellow in color, and the cells were formed as small, compact clusters. However, nonembryogenic calli were observed as friable, semitransparent, and yellow or gray with larger and loosely held clusters (Na et al., 2007). Michaux-Ferrière and Carron (1989) reported embryogenic quality of calli using histological techniques to determine the optimal sequences during embryogenesis that aided in obtaining favorable proembryos. These studies track the origin (unicellular or multicellular), the stages of development (globular, heart, torpedo, and cotyledonary), and the differentiation of tissues like vascular bundles and meristems. By using light and electron microscopy, these investigations provide a detailed understanding of the anatomical and ultrastructural events that occur from the initial somatic cell to a complete embryo. Based on histological investigations, the development of somatic embryos in the oil palm plant was found to originate from perivascular cells of the leaf (Schwendiman et al., 1988). Similarly, a lack of starch and protein reserve in somatic embryos (developed from floral buds) in comparison to zygotic embryos was reported in *Theobroma cacao* (Alemanno et al., 1997).

Histological examination of direct somatic embryogenesis in *Coffea arabica* revealed that the development of the embryo began in the form of a small, isodiametric, densely cytoplasmic cell that underwent a series of organized divisions at approximately one week after the explant was placed in the culture. During indirect somatic embryogenesis, however, the embryogenic cell was observed within the first week of culture. Histological observations further established the unicellular origin of embryos during both direct and indirect somatic embryogenesis pathways (Quiroz-Figueroa et al., 2002).

4.10.1 Embryogenic cell detection by stainings

Cell morphology and embryogenic potential can be identified through double staining. This protocol uses acetocarmine (0.1%) and Evans blue (2%) to distinguish between embryogenic and nonembryogenic cells (Gupta & Durzan, 1987). Embryogenic cells are round, densely cytoplasmic, reactive to acetocarmine, and stain red, while nonembryogenic cells are vacuolated, elongated, permeable to Evans blue, and stain blue (Khatri & Joshee, 2024a, 2024b). It has been reported that acetocarmine detects DNA, chromatin, and macromolecules and stains viable embryogenic cells, whereas Evans blue penetrates ruptured membranes and stains nonviable, nonembryogenic cells (Pádua et al., 2014).

4.10.2 Somatic embryogenesis in medicinal plants

Somatic embryogenesis in medicinal plants is a key method for mass propagation and conservation, with successful systems established for numerous species in more than 100 families (Fig. 4.3).

4.10.2.1 Apiaceae

General description: The Apiaceae (Umbelliferae) is a large and readily identifiable family of flowering plants, consisting of about 430 genera and 3780 species that are divided into

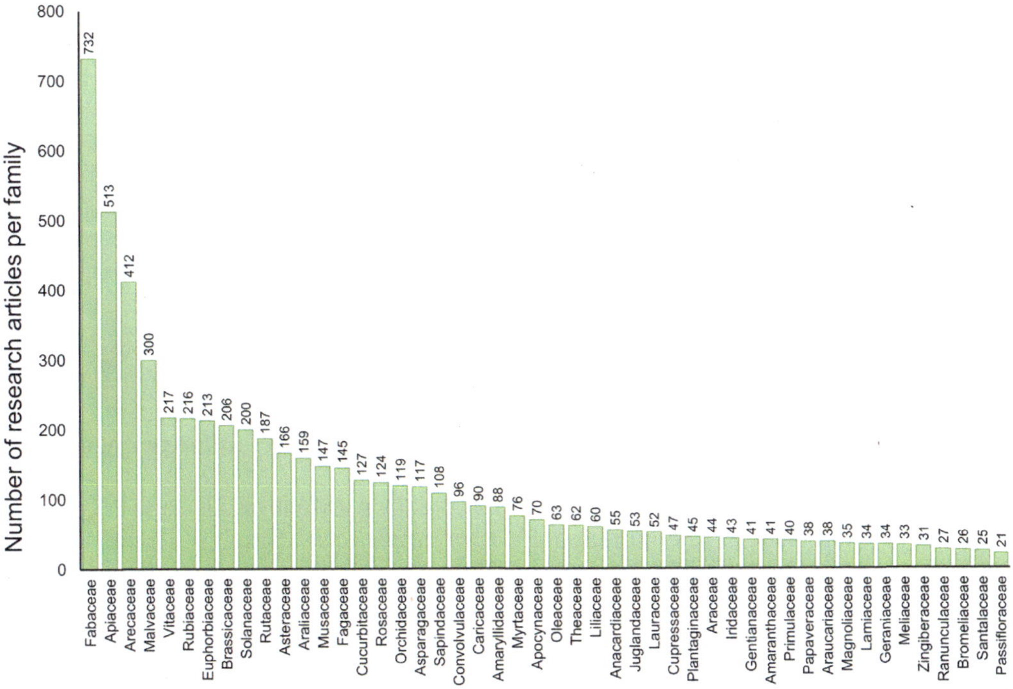

FIGURE 4.3 Number of scientific publications among the top 48 families (out of 180 families) for somatic embryogenesis from 1970 to 2024. Source: *Adapted from Pérez-Molina, J.P., Carvajal-Campos, P., & Jiménez, V.M. (2025). Mapping fifty five years of somatic embryogenesis research: A bibliometric analysis of trends, global collaborations, and key advances across more than nine thousand articles.* Plant Cell, Tissue and Organ Culture, *162(2), 1–17.*

four subfamilies (Calviño et al., 2016) and have a worldwide distribution. The cremocarp consists of two parts that split open in Apiaceae seeds (Wang, Luo, et al., 2022; Wang, Shang, et al., 2022). Many species of Apiaceae are economically important, as foods (e.g., carrot: *D. carota* L., and parsnip: *Pastinaca sativa* L.), herbs (e.g., species of *Angelica* L. and *Bupleurum* L.), and spices (e.g., coriander: *Coriandrum sativum* L., fennel: *Foeniculum vulgare* Mill., and parsley: *Petroselinum crispum* (Mill.) Fuss), as well as some poisonous species (e.g., water hemlock and poison hemlock) (Wen et al., 2021). Essential oils and their components, mainly monoterpenoids and sesquiterpenoids, show antimicrobial, repellent, chemosterilant, antifeeding, and other biocidal activities

Medicinal genera:*C. asiatica* (L.) Urb., *C. sativum* L., *Cuminum cyminum* L., *Trachyspermum ammi* Sprague.

Metabolites reported: phenolic compounds: flavonoids, phenolic acids, coumarin, tannins, apiin, and malonylapiin (Thiviya et al., 2021); polyacetylenes: aliphatic C17 polyacetylenes (Sousa et al., 2021); terpenoids: carotol, trans anethole, and carvone (Wang, Luo, et al., 2022; Wang, Shang, et al., 2022).

Somatic embryogenesis reported in: *C. asiatica* (L.) Urb. (Joshee et al., 2007), *C. sativum* L. (Ali et al., 2018), *C. cyminum* L. (Bahmankar et al., 2017), and *T. ammi* Sprague (Syn. *Carum copticum* (L.) Benth. & Hook.f.) (Niazian et al., 2017).

4.10.2.2 *Apocynaceae*

General description: Apocynaceae is composed of five subfamilies. It encompasses 424 genera with more than 4600 species and is distributed in the tropical, subtropical, and temperate zones. Most Apocynaceae species are shrubs, woody, or herbaceous plants known to produce milky latex. Plants from Apocynaceae family are known globally to possess broad spectrum ethnomedicinal uses.

Medicinal genera: *Alstonia scholaris* (L.) R.Br., *Catharanthus roseus* (L.) G.Don, *Calotropis gigantea* (L.) Dryand., *Holarrhena antidysenterica* Wall. ex A.DC., *Nerium indicum* Mill., *Rauvolfia serpentina* Benth. ex Kurz, and *Thevetia peruviana* (Pers.) K.Schum.
Metabolites reported: The alkaloids isolated and reported from different members are indole, iboga, and vinca.
Somatic embryogenesis reported in: *C. roseus* (L.) G.Don (Mujib et al., 2014), *Cynanchum annularium* (Roxb.) Liede & Khanum (Syn. *Holostemma ada-kodien* Schult.) (Martin, 2003), and *Rauvolfia micrantha* Hook.f. (Sudha & Seeni, 2006).

4.10.2.3 *Asteraceae*

General description: The Asteraceae family is among the largest plant families, classified into three subfamilies, encompassing thousands of species. This family includes approximately 24,000 recognized species and consists of 1600–1700 genera found worldwide, except in Antarctica. Due to its widespread distribution, Asteraceae is considered a cosmopolitan family, with a high concentration of species in temperate, cold temperate, and subtropical regions.

Medicinal genera: *Artemisia annua* Pall., *Achillea* sp., *Blumea laciniata DC. Calendula* sp., *Cichorium intybus* L., *Echinacea purpurea, Eclipta prostrata, Helianthus annuus, Lobelia inflata, Matricaria chamomilla, Sphagneticola calendulacea* (L.) Pruski, *Tussilago farfara*, and *Taraxacum officinale*.
Metabolites reported: Polyphenols, phenolic acids, flavonoids, sesquiterpene lactones, triterpenes, and saponins.
Somatic embryogenesis reported in: *Cichorium intybus* L., (Abdin & Ilah, 2007), *Eclipta prostrata* (L.) L. (Syn. *Eclipta alba* L. ex B.D.Jacks.) (Devendra et al., 2011), *Echinacea purpurea* (L.) Moench (Dehestani-Ardakani et al., 2020), *Sphagneticola calendulacea* (L.) Pruski (Syn. *Wedelia calendulacea* Less.) (Sharmin et al., 2014).

4.10.2.4 *Cucurbitaceae*

General description: Cucurbitaceae is the largest family of vegetable and fruit crops, which includes approximately 125 genera and 960 species. Further, the family is classified into two major subfamilies based on their morphological, cytological, and floral characteristics. Its distribution is essentially tropical and subtropical, with relatively few species reaching the temperate regions of the world. Several therapeutic benefits have been reported of these Cucurbitaceae plants *Benincasa hispida* (wax gourd), *Benincasa fistulosa* (apple gourd), *Coccinia grandis* (ivy gourd), *Lagenaria siceraria* (bottle gourd), *Cucumis melo* (musk melon), *C. sativus* (cucumber), *Cucurbita maxima* (pumpkin), *Cucurbita pepo* (field pumpkin), *Citrullus lanatus* (watermelon), *Luffa acutangula* (ridge gourd), *Luffa cylindrica* (sponge gourd), *Trichosanthes dioica* (pointed gourd), *Trichosanthes cucumerina* (snake gourd), *Momordica dioica* (spine gourd), *Momordica charantia* (bitter gourd), etc.

Medicinal genera: *C. lanatus* Thunb., *Citrullus colocynthis* L., *Cucumis anguria* L., *C. sativus* L., *C. pepo* L.
Metabolites reported: Cucurbitacin, sterols, phenolics, alkaloids, megastigmane, sphingolipid, carotenoids, and flavonoids.
Somatic embryogenesis reported in: *C. lanatus* Thunb. (Vinoth & Ravindhran, 2016), *Citrullus colocynthis* L. (Ramakrishna & Shasthree, 2016), *Cucumis anguria* L. (Thiruvengadam et al., 2013), *C. sativus* L. (Vengadesan et al., 2005), *C. pepo* L. (Urbanek et al., 2004).

4.10.2.5 *Fabaceae*

General description: Fabaceae (Leguminosae) is widely distributed and is one of the largest families in the number of species, with about 765 genera and nearly 20,000 known species. Fabaceae is composed of five subfamilies and can be readily identified by its characteristic fruit (a legume pod) and compound stipulate leaves. Plants represent an important source of vegetable protein. Their consumption has beneficial effects on health by moderating blood sugar levels after meals and improving insulin sensitivity.

Medicinal genera: *Calliandra tweedii* Benth, *Clitoria ternatea* Linn, *Desmodium motorium* (Houtt.) Merr., *Falcataria falcata* (L.) Greuter & R. Rankin, *Mucuna pruriens* (L.) DC., *Pterocarpus marsupium* Roxb.
Metabolites reported: The main secondary metabolites of legumes include alkaloids, NPPA, cyanogens, peptides, phenolics, polyketides, flavonoids, triterpenes, and pinitol.
Somatic embryogenesis reported in: *C. tweedii* Benth (Kumar et al., 2002), *C. ternatea* Linn (Krishna Kumar & Thomas, 2012), *D. motorium* (Houtt.) Merr. (Chitra Devi & Narmathabai, 2011), *F. falcata* (L.) Greuter & R.Rankin (Syn. *Falcataria moluccana* (Miq.) Barneby & J.W.Grimes) (Sunandar & Supena, 2017), *M. pruriens* (L.) DC. (Vibha et al., 2009), and *P. marsupium* Roxb. (Husain et al., 2010).

4.10.2.6 *Euphorbiaceae*

General description: Euphorbiaceae exhibits a global, cosmopolitan distribution with a great diversity of growth forms such as herbs, shrubs, or trees. The Euphorbiaceae comprises three subfamilies spanning ~247 genera with an estimated ~6700 species. The largest genus, Euphorbia, contains over ~2000 species, of which ~650 are succulent.

Medicinal genera: *Euphorbia hirta* L., *Jatropha curcas* L., *Sapium sebiferum* Roxb.
Metabolites reported: Phenolic acids, flavonoid glycosides, alkaloids, terpenes, and cyclic peptides.
Somatic embryogenesis reported in: *E. hirta* L. (Shekhawat et al., 2018), *J. curcas* L. (Jha et al., 2007), and *S. sebiferum* Roxb. (Hou et al., 2015).

4.10.2.7 *Lamiaceae*

General description: The Lamiaceae plant family is generally known as the mint family. They consist of shrubs or herbs that produce and release the aromatic smell, which consists of more than 7200 species belonging to roughly 236 genera (Li et al., 2016). Genera with the highest diversity in species in Lamiaceae are *Salvia*, *Scutellaria*, and *Stachys*.

Medicinal genera: *Hyssopus seravschanicus*, *Leonurus panzerioides*, *Mentha longifolia var. asiatica*, *Nepeta alatavica*, *Ocimum basilicum* L., *Origanum tyttanthum*, *Perovskia scrophulariifolia*, *Phlomis thapsoides*, *Salvia officinalis* L., *Salvia fruticosa* Mill., *Thymus hyemalis* Lange, *Teucrium polium* L., *Scutellaria immaculata*, *S. ramosissima*, *Ziziphora clinopodioides*.
Metabolites reported: Lamiaceae plants contain a wide range of phenolic compounds such as rosmarinic acid, caffeic acid, and flavonoids like luteolin and apigenin. They are also rich in essential oils dominated by terpenoids like monoterpenes (e.g., carvacrol, thymol) and sesquiterpenes (e.g., beta caryophyllene), as well as alkaloids. These compounds are responsible for the plants' strong aromas and diverse medicinal properties.
Somatic embryogenesis reported in: *O. basilicum* L. (Gopi & Ponmurugan, 2006), *Salvia officinalis* L., *S. fruticosa* Mill. (Kintzios et al., 1999), *T. hyemalis* Lange (Nordine et al., 2014), and *T. polium* L. (Saadat et al., 2023).

4.10.2.8 *Lauraceae*

General description: Lauraceae is a large family of woody plants (except for the genus *Cassytha*, which is herbaceous), composed of 50 genera and 2500–3000 species. Lauraceae is classified into two subfamilies and has a distribution in tropical to subtropical areas, including Southeast Asia, the Mediterranean, and Central and South America. Many species yield high quality timber, others spice or aromatic oils, and *Persea americana* Miller is cultivated worldwide for its edible fruits.

Medicinal genera: Camphor (*Cinnamomum camphora* [L.] J. Presl), benzoin (*Lindera benzoin* [L.] Blume), rosewood (*Aniba rosaeodora* Ducke, Lauraceae).
Metabolites reported: Essential oils (terpenoids and phenylpropanoids), alkaloids, flavonoids, and lignans/neolignans.
Somatic embryogenesis reported in: *Cinnamomum camphora* L. (Shi et al., 2010).

4.10.2.9 *Rubiaceae*

General description: The Rubiaceae family has a cosmopolitan distribution, mostly concentrated in the tropics. Being one of the largest in the Magnoliopsida class, it ranks fourth in diversity of species among Angiosperms. It includes approximately 637 genera and 13,000 species. Rubiaceae is further divided into four subfamilies.

Medicinal genera: *Cephaelis ipecacuanha*, *Cinchona* spp., *Coffea* spp. (coffee), *Oldenlandia umbellata*, *Spermacoce hispida*.
Metabolites reported: The secondary metabolites recorded are iridoids, indole alkaloids, anthraquinones, terpenoids (diterpenes and triterpenes), flavonoids, and other phenolic derivatives.
Somatic embryogenesis reported in: *Cephaelis ipecacuanha* A. Richard (Rout et al., 2000), *Oldenlandia umbellata* L. (Siva et al., 2009), and *Spermacoce hispida* L. (Deepak et al., 2019).

4.10.2.10 *Rutaceae*

General description: Rutaceae is a large family (about 162 genera and 2085 species) with great diversity in morphological characters and worldwide distribution. Rutaceae is characterized mainly by the presence of secretory cavities containing aromatic ethereal oils, usually scattered throughout the parenchyma and in the pericarp. The family is best known due to *Citrus* L. (oranges, lemons, tangerines, and grapefruits).

Medicinal genera: *Murraya koenigii* (L.) Spreng., *Dictamnus albus*, *Ptelea trifoliata*, *Fortunella*, *Zanthoxylumalatum*, *Rutagraveolens*, *Chloroxylon swietenia*, and *Amyris* sp.
Metabolites reported: Rutaceae has a vast array of secondary chemical compounds—among them, limonoids, flavonoids, coumarins, volatile oils, and alkaloids. Plants are an important source of aromatic oils (*Boronia*, *Ruta*, and even *Citrus*), drugs (*Pilocarpus*, source of pilocarpine that is used against glaucoma), and bitter beverages used to treat fevers (*Angostura* and *Galipea*).
Somatic embryogenesis reported in: *Murraya koenigii* (L.) Spreng. (Paul et al., 2011).

4.10.2.11 *Solanaceae*

General description: Solanaceae comprises 96 genera and approximately 2400 species distributed among all tropical and temperate regions of the world. Solanaceae stands out among flowering plants for its long history of human domestication and economic importance, comprising many crop species such as potato (*Solanum tuberosum* L.), tomato (*Solanum lycopersicum* L.), eggplant (*Solanum melongena* L.), chili peppers (*Capsicum* spp.), and tobacco (*Nicotiana tabacum* L.), as well as ornamental flowers like petunia (*Petunia x hybrida* (Hook.) Vilm.) and trumpet flower (*Brugmansia* spp.). Solanaceae is commonly known as the "nightshade family" and is composed primarily of perennial herbaceous and woody plants with such marked morphological diversity that it is difficult to find characteristics universally shared by its members.

Medicinal genera: *Capsicum annuum* L., *Solanum nigrum*, *Withania somnifera* (L.) Dunal.
Metabolites reported: The Solanaceae plants produce most notably alkaloids (like tropane alkaloids such as atropine and scopolamine, and steroidal glycoalkaloids like α tomatine), but also include terpenoids, flavonoids, phenols, glycosides, and lactones. Solanaceae contains a wide variety of secondary metabolites, with at least nine types of alkaloids, as well as over 300 types of withanolides. These compounds serve diverse functions, such as plant defense against pests and herbivores.
Somatic embryogenesis reported in: *Capsicum annuum* L. (Binzel et al., 1996), *Withania somnifera* (L.) Dunal (Rani et al., 2004).

4.10.2.12 *Orchidaceae*

General description: The Orchidaceae is perhaps the largest family of flowering plants today and includes perennial herbs, with many forms containing aerial roots. with more than 28,000 accepted species spanning 763 genera (Christenhusz & Byng, 2016).

Medicinal genera: *Acampe papillosa*, *Anoectochilus*, *Coelogyne nitida*, *Calanthe* sp., *Coelogyne* sp., *Cymbidium*, *Dendrobium nobile*, *Dendrobium moschatum*, *Dendrobium densiflorum*, *Ephemerantha*,

Eria, Galeola, Gastrodia, Gymnadenia, Habenaria, Ludisia, Luisia, Nevilia, Thunia, and *Arundina graminifolia*

Metabolites reported: The different bioactive substances identified are alkaloids, phenanthrenes, flavonoids, bibenzyl derivatives, and terpenoids.

Somatic embryogenesis reported in: Plant regeneration via direct somatic embryogenesis in various orchids including *Dendrobium* (Chung et al., 2005), *Epipactis* (Moradi et al., 2017), *Malaxis* (Mahendran & Bai, 2016), *Rhynchostylis* (Islam & Bhattacharjee, 2015), and *Oncidium* (Chen & Chang, 2002; Hong et al., 2008a, 2008b; Su et al., 2006). TDZ was effective in inducing direct somatic embryogenesis from the leaf explants (Chen & Chang, 2002; Hong et al., 2008a, 2008b; Su et al., 2006).

4.10.2.13 *Zingiberaceae*

General description: Zingiberaceae is a family of over 1200 species that span 53 genera. These species are found throughout the tropical and subtropical world, with their primary populations and species diversity centered in Southern and Southeast Asia. The family Zingiberaceae consists of four recognized subfamilies. Four important genera, *Globba, Hedychium, Kaempferia,* and *Zingiber,* belonging to Zingiberoideae. Some of the high economic value species, such as *G. schomburgkii* and *H. coronarium* ornamental and medicinal plants, *K. galanga,* a medicine and flavoring spice, and *Z. officinale,* an important crop and edible food.

Medicinal genera: *Boesenbergia rotunda* (L.) Mansf., *Curcuma amada* Roxb., *Curcuma caesia* Roxb., *Curcuma longa* L., *Elettaria cardamomum* (L.) Maton, *Kaempferia galanga* L., *Zingiber officinale* Roscoe.

Metabolites reported: Gingerols, shogaols, and curcuminoids, as well as a wide variety of terpenes and flavonoids.

Somatic embryogenesis reported in: *Boesenbergia rotunda* (L.) Mansf. (Wong et al., 2013), *Curcuma amada* Roxb. (Raju et al., 2016), *Curcuma caesia* Roxb. (Zuraida et al., 2014), *Curcuma longa* L. (Raju et al., 2015), *Elettaria cardamomum* (L.) Maton (Manohari et al., 2008), *Kaempferia galanga* L. (Rahman et al., 2004), *Zingiber officinale* Roscoe (Mehaboob et al., 2019).

4.11 Application of somatic embryogenesis

Somatic embryogenesis is one of the most efficient and promising approaches for the rapid multiplication of genetically identical plants.

4.11.1 Synthetic seed production

Synthetic seeds are artificially encapsulated somatic embryos obtained from tissue culture, which are usually prepared with sodium alginate or by desiccating the somatic embryos with or without a coating (Kumar, 2000). Artificial seeds, enclosing somatic embryos in a protective coating, have been suggested as a low cost, high volume propagation system (Redenbaugh et al., 1986; Redenbaugh, 1990). To make somatic embryos an effective tool for low cost, large scale plant propagation, an efficient delivery system is essential, and in general, the following four methods have been proposed (Janick, 1993).

(1) Use of a liquid based system to distribute the embryos, (2) encapsulating somatic embryos in a gel made from alginate, (3) drying the somatic embryos to allow easier handling and storage, and (4) drying encapsulated embryos with a water-soluble resin to improve shelf- life and usability.

However, the practical utility of synthetic seeds varies among plant species and depends on the level of advancement in their current propagation methods, as well as the potential for further improvement through biotechnological innovations (Gray and Conger, 1995). The study by Joshee et al. (2007) reported that the somatic embryos of *C. asiatica* were individually encapsulated in a sodium alginate and calcium chloride based matrix to produce synthetic seeds. These synthetic seeds demonstrated viability and germination after storage at 5°C to 8°C for 30 and 60 days, highlighting their potential for conservation and mass propagation.

4.11.2 Metabolite production

Many higher plants are key sources of natural compounds used in agriculture, medicine, food, and fragrance industries (Balandrin & Klocke, 1988). Exploring new plant based chemicals should be essential for sustainable biodiversity use (Phillipson, 1990). Plant tissue culture approaches, especially somatic embryogenesis, offer promising alternatives to traditional farming for producing valuable medicinal compounds (Ramachandra Rao & Ravishankar, 2002). Aslam et al. (2011) showed that low doses of gamma irradiation (20–40 Gy) enhanced the production of alkaloids, specifically vinblastine and vincristine, in *C. roseus* tissues regenerated through somatic embryogenesis. Furthermore, the alkaloid yield was influenced by the developmental stage of somatic embryos, with the germinating stage exhibiting the highest vinblastine accumulation, highlighting the stage specific nature of metabolite synthesis during embryo development. It was also reported in *Panax ginseng* that somatic embryos contained higher levels of ginsenosides Rb1 and Re (triterpene glycosides) than zygotic embryos and uniquely contained ginsenoside Rd. Embryogenic tissues on solid media showed greater Rd and Rg1 content, while cut- up tissues yielded more total ginsenosides than clumps, highlighting their efficiency for metabolite production (Shu et al., 1999).

4.11.3 Virus free plant development

Second, somatic embryogenesis has been the tool to develop the virus free plants from the infected plant materials. Viruses are mostly restricted to vascular tissues, and somatic embryos are developed from nonvascular tissue, which increases the chance of virus elimination. Virus free plants development through somatic embryogenesis has been reported in many plants such as cocoa (*T. cacao*), garlic (*Allium sativum*), grapevine (*V. vinifera*), sugarcane (*Saccharum officinarum*), potato (*Solanum tuberosum*), and black pepper (*Piper nigrum*). Although somatic embryogenesis has been found effective in the elimination of viruses, its combination with chemotherapy (use of chemical agents) has been found more effective than somatic embryogenesis alone (Bhat & Rao, 2020). A study done on garlic reported the overall rate of virus elimination by somatic embryogenesis by 20.7%, 22.9%, and 30.5% for onion yellow dwarf virus, garlic common latent virus, and leek yellow stripe virus, respectively (Kereša et al., 2021).

4.11.4 Germplasm conservation

Medicinal plants serve as both therapeutic source and a significant source of income for many households across the globe (IPBES, 2022), and this has led to overharvesting of several species (Ganie et al., 2019; Larsen & Olsen, 2007). Of the 5411 medicinal plants evaluated by the International Union for the Conservation of Nature (IUCN), 723 species, or 13%, are classified as threatened (RBG Kew, 2020). One of the most appropriate actions for saving exploited plant species is to improve propagation techniques. Therefore somatic embryogenesis offers a potential solution for conserving and multiplying these important medicinal plants. Datta and Jha (2008) also reported the induction of somatic embryos and the formation of plantlets in the endangered *Taxus wallichiana* plant species, which is used for its anticancer properties, specifically for Taxol.

4.12 Industrial adaptation of somatic embryos

Somatic embryogenesis could be a better option for large scale production of clonal plants rather than other clonal propagation methods, such as the rooting of cuttings, as they have a similar growing pattern to zygotic embryos without plagiotropism, as it has been associated with propagation by cuttings. In addition to that, somatic embryos/ somatic embryogenesis cultures (clones) can be cryopreserved for at least 20 years, allowing long-term testing of clones (Egertsdotter et al., 2019). Somatic embryogenesis in conifers has been an area of interest for industries due to their high medicinal and economic value. However, large scale production of somatic embryos has been a problem due to high labor costs (Egertsdotter et al., 2019). Although industries are adopting various techniques such as the use of liquid cultures (in bioreactors), the use of robotics and automation for the harvest of mature embryos, handling of germination and planting (Egertsdotter et al., 2019).

4.13 Conclusions

According to Fehér (2015), the mechanisms underlying auxins and stress induced somatic embryogenesis are not known, but both these conditions induce biosynthesis of endogenous auxins. The synthesis of endogenous auxins is considered a crucial early step in the switch on to totipotent growth. Despite the research developed to understand the cascade of events underlying somatic embryogenesis induction and conversion, these processes are far from being understood. Significant challenges remain, including the high rate of abnormal embryo development, the need for species-specific optimization of culture conditions (media, hormones, and environment), the role of endogenous secondary metabolites on somatic embryogenesis, and a poor understanding of the underlying molecular mechanisms at each stage. Genetic instability, or somaclonal variation, is another concern, although somatic embryos can lead to more genetically stable plants than shoot morphogenesis.

Taking into account that in somatic embryogenesis, the mechanisms responsible for expression of embryogenic competence, in particular the molecular mechanisms of cell fate transition, are of crucial importance for the success of somatic embryogenesis. Through future research, the availability of omics data on the cell fate transition step will help better understand somatic embryogenesis, assisting future application of somatic embryogenesis as a real opportunity for mass *in vitro* propagation, germplasm conservation, genetic improvement, and breeding of plants in general.

Competing interests

The authors have no relevant financial or nonfinancial interests to disclose.

References

Abdin, M. Z., & Ilah, A. (2007). Plant regeneration through somatic embryogenesis from stem and petiole explants of Indian chicory (*Chichorium intybus* L.). *Indian Journal of Biotechnology, 6*, 250–255.

Abo-El-Nil, M. (1989). The effects of amino acid nitrogen on growth of date palm callus. Second symposium on date palm. *K.F.U., Saudi Arabia,* 59–65.

Alemanno, L., Berthouly, M., & Michaux-Ferrière, N. (1997). A comparison between *Theobroma cacao* L. zygotic embryogenesis and somatic embryogenesis from floral explants. *In Vitro Cellular & Developmental Biology-Plant, 33*(3), 163–172.

Ali, M., Mujib, A., Zafar, N., & Tonk, D. (2018). Somatic embryogenesis, biochemical alterations and synthetic seed development in two varieties of coriander (*Coriandrum sativum* L). *Advances in Horticultural Science, 32*(2), 239–248.

Al-Khayri, J. M. (2011). Influence of yeast extract and casein hydrolysate on callus multiplication and somatic embryogenesis of date palm (*Phoenix dactylifera* L). *Scientia Horticulturae, 130*, 531–535.

Armenta-Medina, A., Gillmor, C. S., Gao, P., Mora-Macias, J., Kochian, L. V., Xiang, D., & Datla, R. (2021). Developmental and genomic architecture of plant embryogenesis: From model plant to crops. *Plant Communications, 2*(1).

Aslam, J., Mujib, A., & Sharma, M. (2011). Influence of freezing and non-freezing temperature on somatic embryogenesis and vinblastine production in *Catharanthus roseus* (L.) G. Don. *Acta Physiologiae Plantarum, 33*, 473–480.

Bahmankar, M., Mortazavian, S. M. M., Tohidfar, M., Sadat Noori, S. A., Izadi Darbandi, A., Corrado, G., & Rao, R. (2017). Chemical compositions, somatic embryogenesis, and somaclonal variation in cumin. *BioMed Research International, 2017*(1), 7283806.

Balandrin, M. J., & Klocke, J. A. (1988). Medicinal, aromatic and industrial materials from plants. In Y. P. S. Bajaj (Ed.). *Biotechnology in agriculture and forestry. Medicinal and aromatic plant* (pp. 1–36). Springer-Verlag , vol. 4.

Bhat, A. I., & Rao, G. P. (2020). Virus elimination through somatic embryogenesis. In Alangar Ishwara Bhat, & Govind Pratap Rao (Eds.). *Characterization of plant viruses: Methods and protocols* (pp. 479–489). Springer US.

Binzel, M. L., Sankhla, N., Joshi, S., & Sankhla, D. (1996). Induction of direct somatic embryogenesis and plant regeneration in pepper (*Capsicum annuum* L). *Plant Cell Reports, 15*, 536–540.

Bogdanović, M. D., Ćuković, K. B., Subotić, A. R., Dragićević, M. B., Simonović, A. D., Filipović, B. K., & Todorović, S. I. (2021). Secondary somatic embryogenesis in *Centaurium erythraea* Rafn. *Plants, 10*(2), 199.

Calviño, C. I., Teruel, F. E., & Downie, S. R. (2016). The role of the Southern Hemisphere in the evolutionary history of Apiaceae, a mostly north temperate plant family. *Journal of Biogeography, 43*(2), 398–409.

Carra, A., Wijerathna-Yapa, A., Pathirana, R., & Carimi, F. (2024). Development and applications of somatic embryogenesis in grapevine (*Vitis* spp.). *Plants, 13*(22), 3131.

Chakrabarty, D., Yu, K. W., & Paek, K. Y. (2003). Detection of DNA methylation changes during somatic embryogenesis of Siberian ginseng (*Eleuterococcus senticosus*). *Plant Science, 165*(1), 61–68.

Chen, J. T., & Chang, W. C. (2002). Effects of tissue culture conditions and explant characteristics on direct somatic embryogenesis in *Oncidium* 'Gower Ramsey. *Plant Cell, Tissue and Organ Culture, 69*, 41–44.

Chitra Devi, B., & Narmathabai, V. (2011). Somatic embryogenesis in the medicinal legume *Desmodium motorium* (Houtt.) Merr. *Plant Cell, Tissue and Organ Culture, 106*, 409–418.

Christenhusz, M. J. M., & Byng, J. W. (2016). The number of known plants species in the world and its annual increase. *Phytotaxa, 261*(3), 201–217. https://doi.org/10.11646/phytotaxa.261.3.1.

Chung, H. H., Chen, J. T., & Chang, W. C. (2005). Cytokinins induce direct somatic embryogenesis of *Dendrobium* Chiengmai Pink and subsequent plant regeneration. *In Vitro Cellular & Developmental Biology-Plant, 2005*(41), 765–769.

Datta, M. M., & Jha, S. (2008). Plant regeneration through somatic embryogenesis in *Taxus wallichiana*. *Journal of Plant Biochemistry and Biotechnology, 17*, 37–44.

De Jong, A. J., Cordewener, J., Lo Schiavo, F., Terzi, M., Vandekerckhove, J., Van Kammen, A., & De Vries, S. C. (1992). A carrot somatic embryo mutant is rescued by chitinase. *The Plant Cell, 4*(4), 425–433.

De Jong, A. J., Heidstra, R., Spaink, H. P., Hartog, M. V., Meijer, E. A., Hendriks, T., Schiavo, F. L., Terzi, M., Bisseling, T., Van Kammen, A., & De Vries, S. C. (1993). Rhizobium lipooligosaccharides rescue a carrot somatic embryo mutant. *The Plant Cell, 5*(6), 615–620.

De Vries, S. C., Booij, H., Janssens, R., Vogels, R., Saris, L., LoSchiavo, F., Terzi, M., & Van Kammen, A. (1988). Carrot somatic embryogenesis depends on the phytohormone-controlled presence of correctly glycosylated extracellular proteins. *Genes and Development, 2*, 462–476.

Deepak, K. V., Narayanan, G. S., Prakash, M., Murugan, S., & Anandan, R. (2019). Efficient plant regeneration and histological evaluations of regenerants through organogenesis and somatic embryogenesis in *Spermacoce hispida* L. – An underutilized medicinally important plant. *Industrial Crops and Products, 134*, 292–302.

Dehestani-Ardakani, M., Hejazi, M., & Aliabad, K. K. (2020). Indirect somatic embryogenesis of purple coneflower (*Echinacea purpurea* (L.) Moench): A medicinal-ornamental plant: Evaluation of antioxidant enzymes activity and histological study. *Molecular Biology Reports, 47*(9), 6621–6633.

Devendra, B. N., Srinivas, N., & Reddy, S. (2011). High frequency somatic embryogenesis and plant regeneration in nodal explant cultures of *Eclipta alba* L. Hassk. *Annals of Biological Research, 2*, 143–149.

Dipti, F. S., & Mujib, A. (2014). Morphological anomalies in somatic embryo structure in *Catharanthus roseus*: improving embryo germination by amending plant growth regulators, activated charcoal and sucrose level. *Biotechnology Journal International*, 10–20.

Dodeman, V. L., Ducreux, G., & Kreis, M. (1997). Zygotic embryogenesis versus somatic embryogenesis. *Journal of Experimental Botany, 48*(8), 1493–1509.

Dudits, D., Györgyey, J., Bögre, L., & Bakó, L. (1995). Molecular biology of somatic embryogenesis. In Trevor A. Thorpe (Ed.). *In vitro embryogenesis in plants* (pp. 267–308). Springer Netherlands.

Egertsdotter, U., Ahmad, I., & Clapham, D. (2019). Automation and scale up of somatic embryogenesis for commercial plant production, with emphasis on conifers. *Frontiers in Plant Science, 10*, 109.

Emons, M. C. (1994). Somatic embryogenesis: Cell biological aspects. *Acta Botanica Neerlandica, 43*, 1–14.

Fehér, A. (2015). Somatic embryogenesis–stress-induced remodeling of plant cell fate. *BBA Gene Regulatory Mechanisms, 1849*, 385–402. https://doi.org/10.1016/j.bbagrm.2014.07.005.

Fehér, A., Pasternak, T. P., & Dudits, D. (2003). Transition of somatic plant cells to an embryogenic state. *Plant Cell, Tissue and Organ Culture, 74*, 201–228.

Filippov, M., Miroshnichenko, D., Vernikovskaya, D., & Dolgov, S. (2006). The effect of auxins, time exposure to auxin and genotypes on somatic embryogenesis from mature embryos of wheat. *Plant Cell, Tissue and Organ Culture, 84*, 213–222.

Filonova, L. H., Bozhkov, P. V., & Arnold, S. V. (2000). Developmental pathway of somatic embryogenesis in Picea abies as revealed by time-lapse tracking. *Journal of Experimental Botany, 51*, 249–264.

Finer, J. J. (1994). Plant regeneration via embryogenic suspension cultures. In R. A. Dixon, & R. A. Gonzales (Eds.). *Plant cell culture: A practical approach* (pp. 67–102). Oxford University Press.

Gamborg, O., Miller, R., & Ojimi, K. (1968). Nutrient requirements of suspension cultures of soybean root cells. *Experimental Cell Research, 50*, 151–158.

Ganie, A. H., Tali, B. A., Khuroo, A. A., Reshi, Z. A., & Nawchoo, I. A. (2019). Impact assessment of anthropogenic threats to high-valued medicinal plants of Kashmir Himalaya, Indian. *Journal for Nature Conservation, 50*, 125715.

Garcia, C., Furtado de Almeida, A. A., Costa, M., Britto, D., Valle, R., Royaert, S., & Marelli, J. P. (2019). Abnormalities in somatic embryogenesis caused by 2,4-D: An overview. *Plant Cell, Tissue and Organ Culture, 137*(2), 193–212.

George, E. F. (1993). *Plant propagation by tissue culture part 1: The technology* (2nd ed.). Exergetics Limited, viii–574.

George, E. F., Hall, M. A., & De Klerk, G. J. (2008). *Plant propagation by tissue culture* (3rd ed.). The Background Springer, 65–175.

Gerdakaneh, M., Mozafari, A. A., & Sarabi, B. (2011). Effects of different amino acids on somatic embryogenesis of strawberry (*Fragaria × ananassa* Duch. *Acta Physiologiae Plantarum, 33*, 1847–1852.

Ghazzawy, H. S., Alhajhoj, M. R., & Munir, M. (2017). In vitro somatic embryogenesis response of date palm cv. Sukkary to sucrose and activated charcoal concentrations. *Journal of Applied Horticulture, 19*(12), 91–95.

Godel-Jedrychowska, K., Kulinska-Lukaszek, K., Horstman, A., Soriano, M., Li, M., Malota, K., & Kurczynska, E. U. (2020). Symplasmic isolation marks cell fate changes during somatic embryogenesis. *Journal of Experimental Botany, 71*(9), 2612–2628.

Gopi, C., & Ponmurugan, P. (2006). Somatic embryogenesis and plant regeneration from leaf callus of *Ocimum basilicum* L. *Journal of Biotechnology, 126*(2), 260–264.

Gray, D. J., Conger, B. V., & Compton, M. E. (1995). Somatic embryogenesis and the technology of synthetic seed. In Y. P. S. Bajaj (Ed.). *Biotechnology in agriculture and forestry* (pp. 126–151). Springer Vol. 30.

Gupta, P. K., & Durzan, D. J. (1987). Biotechnology of somatic polyembryogenesis and plantlet regeneration in loblolly pine. *Nature Biotechnology, 5*, 147–151.

Higashi, K., Daita, M., Kobayashi, T., Sasaki, K., Harada, H., & Kamada, H. (1998). Inhibitory conditioning for carrot somatic embryogenesis in high-cell-density cultures. *Plant Cell Reports, 18*(1), 2–6.

Hong, P. I., Chen, J. T., & Chang, W. C. (2008a). Effects of salicylic and acetylsalicylic acid on direct somatic embryogenesis in *Oncidium*. *Journal of Plant Biochemistry and Biotechnology, 17*, 149–153.

Hong, P. I., Chen, J. T., & Chang, W. C. (2008b). Promotion of direct somatic embryogenesis of *Oncidium* by adjusting carbon sources. *Biologia Plantarum, 52*, 597–600.

Hou, J., Wu, Y., Shen, Y., Mao, Y., Liu, W., Zhao, W., Mu, Y., Li, M., Yang, M., & Wu, L. (2015). Plant regeneration through somatic embryogenesis and shoot organogenesis from immature zygotic embryos of *Sapium sebiferum* Roxb. *Scientia Horticulturae, 197*, 218–225.

Huang, L. C., Kohashi, C., Vangundy, R., & Murashige, T. (1995). Effects of common components on hardness of culture media prepared with gelrite. ™. *In Vitro Cellular & Developmental Biology-Plant, 31*(2), 84–89.

Husain, M. K., Anis, M., & Shahzad, A. (2010). Somatic embryogenesis and plant regeneration in *Pterocarpus marsupium* Roxb. *Trees, 24*, 781–787.

Höxtermann, E. (1997). Cellular 'elementary organisms' in vitro: The early vision of Gottlieb Haberlandt and its realization. *Physiologia Plantarum, 100*(3), 716–728.

Igasaki, T., Akashi, N., Ujino-Ihara, T., Matsubayashi, Y., Sakagami, Y., & Shinohara, K. (2003). Phytosulfokine stimulates somatic embryogenesis in *Cryptomeria japonica*. *Plant and Cell Physiology, 44*(12), 1412–1416.

Ikeda-Iwai, M., Satoh, S., & Kamada, H. (2002). Establishment of a reproducible tissue culture system for the induction of *Arabidopsis* somatic embryos. *Journal of Experimental Botany, 53*(374), 1575–1580.

IPBES (2022). Summary for policymakers of the thematic assessment of the sustainable use of wild species of the Intergovernmental Science-Policy Platform on Biodiversity and Ecosystem Services. *Zenodo*. https://doi.org/10.5281/zenodo.6448567.

Iraqi, D., & Tremblay, F. M. (2001). Analysis of carbohydrate metabolism enzymes and cellular contents of sugars and proteins during spruce somatic embryogenesis suggests a regulatory role of exogenous sucrose in embryo development. *Journal of Experimental Botany, 52*, 2301–2311.

Islam, S. S., & Bhattacharjee, B. (2015). Plant regeneration through somatic embryogenesis from leaf and root explants of *Rhynchostylis retusa* (L.) Blume. *Applied Biological Research, 17*, 158–165.
Janick, J. (1993). Agricultural use of somatic embryos. *Acta Horticulturae*, (336), 207–215.
Jha, T. B., Mukherjee, P., & Datta, M. M. (2007). Somatic embryogenesis in *Jatropha curcas* Linn., an important biofuel plant. *Plant Biotechnology Reports, 1*, 135–140.
Jianhua, Y. U., Yan, D. O., Wenyang, L. I., Meng, L. I., & Yan, Z. H. (2020). Effects of pH on physiological characters in somatic embryo induction stage of *Agapanthus praecox*. *Acta Agriculturae Zhejiangensis, 32*, 1405.
Joshee, N., Biswas, B. K., & Yadav, A. K. (2007). Somatic embryogenesis and plant development in *Centella asiatica* L., a highly prized medicinal plant of the tropics. *HortScience, 42*(3), 633–637.
Joshi, R., Shukla, A., & Kumar, P. (2009). In vitro flowering in hill maize: A novel technique for future. *Indian Journal of Plant Physiology, 14*(3), 299–330.
Junaid, A., Mujib, A., Samar, F., & Sharma, M. P. (2008). Culture conditions affect somatic embryogenesis in *Catharanthus roseus* (L.) G. Don. *Plant Biotechnology Reports, 2*(3), 179–189.
Kępczyńska, E., & Kępczyński, J. (2023). Hormonal regulation of somatic embryogenesis in *Medicago* spp. *Plant Cell, Tissue and Organ Culture, 155*, 613–625.
Kereša, S., Kurtović, K., Ban, S. G., Vončina, D., Jerčić, I. H., Bolarić, S., Lazarević, B., Godena, S., Ban, D., & Mihovilović, A. B. (2021). Production of virus-free garlic plants through somatic embryogenesis. *Agronomy, 11*(5), 876.
Kermode, A. R. (1990). Regulatory mechanisms involved in the transition from seed development to germination. *Critical Reviews in Plant Sciences, 9*, 155–195.
Keshvari, T., Najaphy, A., Kahrizi, D., & Zebarjadi, A. (2018). Callus induction and somatic embryogenesis in *Stevia rebaudiana* Bertoni as a medicinal plant. *Cellular and Molecular Biology, 64*(2), 46–49.
Khatri, P., & Joshee, N. (2024a). Effect of picloram and desiccation on the somatic embryogenesis of *Lycium barbarum* L. *Plants, 13*, 151.
Khatri, P., & Joshee, N. (2024b). Somatic embryogenesis and shoot organogenesis in peanut cv. 'Georgia-12Y' and successful transfer to the soil. *PLOS ONE, 19*(12), e0315060.
Kher, M. M., & Nataraj, M. (2019). Direct somatic embryogenesis and shoot regeneration from leaves and internodes of *Pluchea lanceolata* (DC.) C.B. Clarke. *In Vitro Cellular & Developmental Biology – Plant, 55*, 720–724.
Kintzios, S., Nikolaou, A., & Skoula, M. (1999). Somatic embryogenesis and in vitro rosmarinic acid accumulation in *Salvia officinalis* and *S. fruticosa* leaf callus cultures. *Plant Cell Reports, 18*, 462–466.
Klimaszewska, K., Noceda, C., Pelletier, G., Label, P., Rodriguez, R., & Lelu-Walter, M. A. (2009). Biological characterization of young and aged embryogenic cultures of *Pinus pinaster* (Ait.). *In Vitro Cellular & Developmental Biology – Plant, 45*(1), 20–33.
Kobayashi, T., Higashi, K., & Kamada, H. (2001). 4-Hydroxybenzyl alcohol accumulates in suspension-cell cultures and inhibits somatic embryogenesis in carrot. *Physiologia Plantarum, 112*(2), 280–284.
Kobayashi, T., Higashi, K., Sasaki, K., Asami, T., Yoshida, S., & Kamada, H. (2000). Purification from conditioned medium and chemical identification of a factor that inhibits somatic embryogenesis in carrot. *Plant and Cell Physiology, 41*(3), 268–273.
Kobayashi, T., Nagayama, Y., Higashi, K., & Kobayashi, M. (2010). Establishment of a tissue culture system for somatic embryogenesis from germinating embryos of *Arabidopsis thaliana*. *Plant Biotechnology, 27*(4), 359–364.
Kreuger, M., & van Holst, G. J. (1993). Arabinogalactan proteins are essential in somatic embryogenesis of *Daucus carota* L. *Planta, 189*(2), 243–248.
Krishna Kumar, G., & Thomas, T. D. (2012). High-frequency somatic embryogenesis and synthetic seed production in *Clitoria ternatea* Linn. *Plant Cell, Tissue and Organ Culture, 110*, 141–151.
Kumar, S., Agrawal, V., & Gupta, S. C. (2002). Somatic embryogenesis in the woody legume *Calliandra tweedii*. *Plant Cell, Tissue and Organ Culture, 71*, 77–80.
Kumar, U. (2000). *Synthetic seeds for commercial crop production*. Updesh Purohit for Agrobios.
Kwack, S. N., & Fujieda, K. (1988). Somatic embryogenesis in cultured unfertilized ovules of *Cucurbita moschata*. *Journal of the Japanese Society for Horticultural Science, 57*, 34.
Larsen, H. O., & Olsen, C. S. (2007). Unsustainable collection and unfair trade? Uncovering and assessing assumptions regarding central Himalayan medicinal plant conservation. *Biodiversity and Conservation, 16*(6), 1679–1697.

Leljak-Levanić, D., Mihaljević, S., & Bauer, N. (2015). Somatic and zygotic embryos share common developmental features at the onset of plant embryogenesis. *Acta Physiologiae Plantarum, 37*, 1–14.

Lelu-Walter, M. A., Gautier, F., Eliášová, K., Sanchez, L., Teyssier, C., Lomenech, A. M., & Reeves, C. (2018). High gellan gum concentration and secondary somatic embryogenesis: Two key factors to improve somatic embryo development in *Pseudotsuga menziesii* [Mirb]. *Plant Cell, Tissue and Organ Culture, 132*(1), 137–155.

Li, B., Cantino, P. D., Olmstead, R. G., Bramley, G. L., Xiang, C.-L., Ma, Z.-H., Tan, Y.-H., & Zhang, D.-X. (2016). A large-scale chloroplast phylogeny of the Lamiaceae sheds new light on its subfamilial classification. *Scientific Reports-UK, 6*(1), 1–18.

Liang, H., Xiong, Y., Guo, B., et al. (2020). Shoot organogenesis and somatic embryogenesis from leaf and root explants of *Scaevola sericea. Scientific Reports, 10*, 11343. https://doi.org/10.1038/s41598-020-68084-1.

Liu, C. M., Xu, Z. H., & Chua, N. H. (1993). Auxin polar transport is essential for the establishment of bilateral symmetry during early plant embryogenesis. *The Plant Cell, 5*, 621–630.

Mahendran, G., & Bai, V. N. (2016). Direct somatic embryogenesis of *Malaxis densiflora* (A. Rich.) Kuntze. *Journal of Genetic Engineering and Biotechnology, 14*, 77–81.

Manohari, C., Backiyarani, S., Jebasingh, T., Somanath, A., & Usha, R. (2008). Efficient plant regeneration in small cardamom (*Elettaria cardamomum* Maton.) through somatic embryogenesis. *Indian Journal of Biotechnology, 7*, 407–409.

Martin, K. P. (2003). Plant regeneration through somatic embryogenesis on *Holostemma ada-kodien*, a rare medicinal plant. *Plant Cell, Tissue and Organ Culture, 72*, 79–82.

Matthys-Rochon, E. (2002). Fascinating questions raised by the embryonic development in plants. *Biologia, 57*, 1–4.

McCown, B. H., & Lloyd, G. (1981). Woody Plant Medium (WPM)-A mineral nutrient formulation for microculture of woody plant species. *HortScience, 16*, 453.

McFarland, F. L., & Kaeppler, H. F. (2025). History and current status of embryogenic culture-based tissue culture, transformation and gene editing of maize (*Zea mays* L). *The Plant Genome, 18*(1), e20451.

Michaux-Ferrière, N., & Carron, M. P. (1989). Histology of early somatic embryogenesis in *Hevea brasiliensis*: The importance of the timing of subculturing. *Plant Cell, Tissue and Organ Culture, 19*(3), 243–256.

Moradi, S., Dianati Daylami, S., Arab, M., & Vahdati, K. (2017). Direct somatic embryogenesis in *Epipactis veratrifolia*, a temperate terrestrial orchid. *The Journal of Horticultural Science and Biotechnology, 92*, 88–97.

Moscatiello, R., Baldan, B., & Navazio, L. (2013). Plant cell suspension cultures. *Methods in Molecular Biology, 953*, 77–93.

Mujib, A., Ali, M., & Isah, T. (2014). Somatic embryo-mediated mass production of *Catharanthus roseus* in culture vessel (bioreactor) – A comparative study. *Saudi Journal of Biological Sciences, 21*(5), 442–449.

Murashige, T., & Skoog, F. (1962). A revised medium for rapid growth and bioassays with tobacco tissue cultures. *Physiologia Plantarum, 15*, 473–497.

Mehaboob, M. V., Faizal, K., Thilip, C., Raja, P., Thiagu, G., Aslam, A., & Shajahan, A. (2019). Indirect somatic embryogenesis and Agrobacterium-mediated transient transformation of ginger (*Zingiber officinale* Rosc.) using leaf sheath explants. *The Journal of Horticultural Science and Biotechnology, 94*(6), 753–760.

Na, H., Kim, K. W., Kwack, Y., Kim, S. K., & Chun, C. (2007). Comparative anatomy of embryogenic and non-embryogenic calli from *Pimpinella brachycarpa. Journal of Plant Biology, 50*(3), 344–350.

Niazian, M., Noori, S. A. S., Galuszka, P., Tohidfar, M., & Mortazavian, S. M. M. (2017). Genetic stability of regenerated plants via indirect somatic embryogenesis and indirect shoot regeneration of *Carum copticum* L. *Industrial Crops and Products, 97*, 330–337.

Nic-Can, G. I., & Loyola-Vargas, V. M. (2016). The role of the auxins during somatic embryogenesis. In V. M. Loyola-Vargas, & N. Ochoa-Alejo (Eds.). *Somatic embryogenesis: Fundamental aspects and applications* (pp. 171–182). Springer International Publishing ISBN 9783319337050.

Noceda, C., Salaj, T., Pérez, M., Viejo, M., Cañal, M. J., Salaj, J., & Rodriguez, R. (2009). DNA demethylation and decrease on free polyamines is associated with the embryogenic capacity of *Pinus nigra* Arn. cell culture. *Trees, 23*(6), 1285–1293.

Nordine, A., Tlemcani, C. R., & El, M. (2014). Regeneration of plants through somatic embryogenesis in *Thymus hyemalis* Lange, a potential medicinal and aromatic plant. *In Vitro Cellular & Developmental Biology – Plant, 50*, 19–25.

Pádua, M. S., Paiva, L. V., Silva, L. C. D., Livramento, K. D., Alves, E., & Castro, A. H. F. (2014). Morphological characteristics, and cell viability of coffee plant calli. *Rural Science, 44*, 660–665.

Parrott, W. A., Williams, E. G., Hildebrand, D. F., & Collins, G. B. (1989). Effect of genotype on somatic embryogenesis from immature cotyledons of soybean. *Plant Cell, Tissue and Organ Culture, 16*, 15–21.

Paul, S., Dam, A., Bhattacharyya, A., & Bandyopadhyay, T. K. (2011). An efficient regeneration system via direct and indirect somatic embryogenesis for the medicinal tree *Murraya koenigii. Plant Cell, Tissue and Organ Culture, 105*, 271–283.

Phillipson, J. D. (1990). Plants as source of valuable products. In B. V. Charlwood, & M. J. C. Rhodes (Eds.). *Secondary products from plant tissue culture* (pp. 1–21). Clarendon Press.

Pila Quinga, L. A., Pacheco de Freitas Fraga, H., do Nascimento Vieira, L., & Guerra, M. P. (2017). Epigenetics of long-term somatic embryogenesis in *Theobroma cacao* L.: DNA methylation and recovery of embryogenic potential. *Plant Cell, Tissue and Organ Culture, 131*(2), 295–305.

Quiroz-Figueroa, F. R., Fuentes-Cerda, C. F. J., Rojas-Herrera, R., & Loyola-Vargas, V. M. (2002). Histological studies on the developmental stages and differentiation of two different somatic embryogenesis systems of *Coffea arabica. Plant Cell Reports, 20*, 1141–1149. https://doi.org/10.1007/s00299-002-0464-x.

Quiroz-Figueroa, F. R., Monforte-González, M., Galaz-Ávalos, R. M., & Loyola-Vargas, V. M. (2006). Direct somatic embryogenesis in *Coffea canephora*. In Loyola-Vargas Victor M., & Vázquez-Flota Felipe (Eds.). *Plant cell culture protocols* (pp. 111–117). Humana Press.

Rahman, M. M., Amin, M. N., Ahamed, T., Ali, M. R., & Habib, A. (2004). Efficient plant regeneration through somatic embryogenesis from leaf base derived callus of *Kaempferia galanga* L. *Asian Journal of Plant Sciences, 3*(6), 675–678.

Raju, C. S., Aslam, A., Kathiravan, K., Palani, P., & Shajahan, A. (2014). Direct somatic embryogenesis and plant regeneration from leaf sheath explants of mango ginger (*Curcuma amada* Roxb. *In Vitro Cellular & Developmental Biology – Plant, 50*, 752–759. https://doi.org/10.1007/s11627-014-9653-x.

Raju, C. S., Aslam, A., & Shajahan, A. (2016). Germination and storability of calcium-alginate coated somatic embryos of mango ginger (*Curcuma amada* Roxb). *Horticulture, Environment and Biotechnology, 57*, 88–96.

Raju, S. C., Aslam, A., & Shajahan, A. (2015). High-efficiency direct somatic embryogenesis and plant regeneration from leaf base explants of turmeric (*Curcuma longa* L). *Plant Cell, Tissue and Organ Culture, 122*, 79–87.

Ramachandra Rao, S., & Ravishankar, G. A. (2002). Plant cell cultures: Chemical factories of secondary metabolites. *Biotechnology Advances, 20*, 101–153.

Ramakrishna, D., & Shasthree, T. (2016). High efficient somatic embryogenesis development from leaf cultures of *Citrullus colocynthis* (L.) Schrad for generating true type clones. *Physiology and Molecular Biology of Plants, 22*, 279–285.

Ramírez-Mosqueda, M. A. (2022). Overview of somatic embryogenesis. *Somatic Embryogenesis: Methods and Protocols*, 1–8.

Rangan, T. S., Murashige, T., & Bitters, W. P. (1968). *In vitro* initiation of nucellar embryos in monoembryonic *Citrus*. *HortScience, 3*(4), 226–227.

Rani, G., Virk, G. S., & Nagpal, A. (2004). Somatic embryogenesis in *Withania somnifera* (L.) Dunal. *Journal of Plant Biotechnology, 6*(2), 113–118.

RBG Kew (2020). *The state of the world's plants report—2020*. Royal Botanic Gardens.

Redenbaugh, K. (1990). Application of artificial seed to tropical crops. *HortScience, 25*(3), 251–255.

Redenbaugh, K., Paasch, B., Nichol, J., Kossler, M., Viss, P., & Walker, K. (1986). Somatic seeds encapsulation of asexual embryos. *Bio/Technology, 4*, 797–801.

Reinert, J. (1958). Morphogenese und ihre Kontrolle an Gewebekulturen aus Karotten. *Naturwissenschaften, 45*(14), 344–345.

Rout, G. R., Samantaray, S., & Das, P. (2000). In vitro somatic embryogenesis from callus cultures of *Cephaelis ipecacuanha* A. Richard. *Scientia Horticulturae, 86*(1), 71–79.

Sakhanokho, H. F., Rajasekaran, K., & Kelley, R. Y. (2009). Somatic embryogenesis in *Hedychium bousigonianum*. *HortScience, 44*(5), 1487–1490.

Saadat, S., Majd, A., Naseri, L., Iranbakhsh, A., & Jafari, M. (2023). Optimization of somatic embryogenesis, synthetic seed production, and evaluation of genetic fidelity in *Teucrium polium* L. *In Vitro Cellular & Developmental Biology – Plant, 59*(4), 483–496.

Santana, F. V., Freire, G. S., Góis, I. B., Nascimento, L. L. P., Santos, P. A. A., Silva, A. V. C., & Lédo, A. S. (2024). Induction and maturation of somatic embryogenesis in mangaba tree a multipurpose Apocynaceae. *Scientia Plena, 20*(9), 1–9. https://doi.org/10.14808/sci.plena.2024.090202.

Schenk, R. U., & Hildebrandt, A. C. (1972). Medium and techniques for induction and growth of monocotyledonous and dicotyledonous plant cell cultures. *Canadian Journal of Botany, 50*, 199–204.

Schiavo, F. L. (1995). Early events in embryogenesis. In Y. P. S. Bajaj (Ed.). *Somatic embryogenesis and synthetic seed I* (pp. 20–29). Springer Berlin Heidelberg.

Schwendiman, J., Pannetier, C., & Michaux Ferriere, N. (1988). Histology of embryogenic formations during *in vitro* culture of oil palm *Elaeis guineensis*. *Annals of Botany, 62*(1), 43–52.

Sharmin, S. A., Alam, M. J., Sheikh, M. M. I., Sarker, K. K., Khalekuzzaman, M., Haque, M. A., Alam, M. F., & Alam, I. (2014). Somatic embryogenesis and plant regeneration in *Wedelia calendulacea* Less. – An endangered medicinal plant. *Brazilian Archives of Biology and Technology, 57*, 394–401.

Shekhawat, M. S., Manokari, M., & Revathi, J. (2018). Optimization of *in vitro* conditions for induction of somatic embryos and regeneration of plantlets in *Euphorbia hirta* L. *Current Trends in Biotechnology and Pharmacy, 12*(1), 85–95.

Shi, X., Dai, X., Liu, G., Zhang, J., Ning, G., & Bao, M. (2010). Cyclic secondary somatic embryogenesis and efficient plant regeneration in camphor tree (*Cinnamomum camphora* L.). *In Vitro Cellular & Developmental Biology – Plant, 46*, 117–125.

Shu, W., Yoshimatsu, K., Yamaguchi, H., & Shimomura, K. (1999). Somatic embryogenesis and ginsenoside production of *Panax ginseng* in phytohormone-free medium. *Kokuritsu Iyakuhin Shokuhin Eisei Kenkyusho Hokoku=Bulletin of National Institute of Health Sciences,* (117), 140–147.

Singh, R., Rai, M. K., & Kumari, N. (2015). Somatic embryogenesis and plant regeneration in *Sapindus mukorossi* Gaertn. from leaf-derived callus induced with 6-benzylaminopurine. *Applied Biochemistry and Biotechnology, 177*, 498–510.

Siva, R., Rajasekaran, C., & Mudgal, G. (2009). Induction of somatic embryogenesis and organogenesis in *Oldenlandia umbellata* L., a dye-yielding medicinal plant. *Plant Cell, Tissue and Organ Culture, 98*, 205–211.

Smith, D. L., & Krikorian, A. D. (1990). Somatic proembryo production from excised, wounded zygotic carrot embryos on hormone-free medium: Evaluation of the effects of pH, ethylene and activated charcoal. *Plant Cell Reports, 9*, 34–37.

Sousa, R. M. O., Cunha, A. C., & Fernandes-Ferreira, M. (2021). The potential of Apiaceae species as sources of singular phytochemicals and plant-based pesticides. *Phytochemistry, 187*, 112714.

Steward, F. C., Mapes, M. O., & Mears, K. (1958). Growth and organized development of cultured cells. II. Organization in cultures grown from freely suspended cell. *s. American Journal of Botany, 45*(9), 705–708.

Su, Y. J., Chen, J. T., & Chang, W. C. (2006). Efficient and repetitive production of leaf derived somatic embryos of *Oncidium*. *Biologia plantarum, 50*, 107–110.

Sudha, C. G., & Seeni, S. (2006). Spontaneous somatic embryogenesis on *in vitro* root segment cultures of *Rauvolfia micrantha* Hook. f. – A rare medicinal plant. *In Vitro Cellular & Developmental Biology – Plant, 42*, 119–123.

Sunandar, A., & Supena, E. D. J. (2017). Induction of somatic embryogenesis in sengon (*Falcataria moluccana*) with thidiazuron and light treatments. *HAYATI Journal of Biosciences, 24*(2), 105–108.

Tang, Y., Liu, B., Li, X. M., Li, J., & Li, H. X. (2010). Endogenous hormone concentrations in explants and calluses of bitter melon (*Momordica charantia* L.). *Interciencia, 35*, 680–683.

Thiruvengadam, M., Jeyakumar, J. J., Kamaraj, M., Lee, Y. J., & Chung, I. M. (2013). Plant regeneration through somatic embryogenesis from suspension cultures of gherkin (*Cucumis anguria* L.). *Australian Journal of Crop Science, 7*(7), 969–977.

Thiviya, P., Gamage, A., Piumali, D., Merah, O., & Madhujith, T. (2021). Apiaceae as an important source of antioxidants and their applications. *Cosmetics, 8*(4), 111.

Tisserat, B., Esan, E. B., & Murashige, T. (1979). Somatic embryogenesis in angiosperms. *Horticultural Reviews, 1*, 1–78.

Triqui, Z. E. A., Guédira, A., Chlyah, A., Chlyah, H., Souvannavong, V., Haïcour, R., & Sihachakr, D. (2008). Effect of genotype, gelling agent, and auxin on the induction of somatic embryogenesis in sweet potato (*Ipomoea batatas* Lam). *Comptes Rendus Biologies, 331*(3), 198–205.

Urbanek, A., Zechmann, B., & Müller, M. (2004). Plant regeneration via somatic embryogenesis in Styrian pumpkin: Cytological and biochemical investigations. *Plant Cell, Tissue and Organ Culture, 79*, 329–340.

Van Ark, H. F., Zaal, M. A. C. M., Creemers-Molenaar, J., & Van der Valk, P. (1991). Improvement of the tissue culture response of seed-derived callus cultures of *Poa pratensis* L.: Effect of gelling agent and abscisic acid. *Plant Cell, Tissue and Organ Culture, 27*(3), 275–280.

Van Yen, D., & Li, J. (2022). Effect of growth regulators on *in vitro* micropropagation of *Stahlianthus thorelii* Gagnep. *Agriculture, 12*, 1766. https://doi.org/10.3390/agriculture12111766.

Vasil, V., & Hildebrandt, A. C. (1965a). Differentiation of tobacco plants from single, isolated cells in microcultures. *Science, 150*(3698), 889–892.

Vasil, V., & Hildebrandt, A. C. (1965b). Growth and tissue formation from single, isolated tobacco cells in microculture. *Science, 147*(3664), 1454–1455.

Vengadesan, G., Selvaraj, N., Anand, R. P., Gaba, V., & Ganapathi, A. (2005). Ontogeny of somatic embryos in cucumber (*Cucumis sativus* L.). *In Vitro Cellular & Developmental Biology – Plant, 41*, 789–793.

Verdeil, J. L., Alemanno, L., Niemenak, N., & Tranbarger, T. J. (2007). Pluripotent versus totipotent plant stem cells: Dependence versus autonomy? *Trends in Plant Science, 12*(6), 245–252.

Verma, S. K., Sahin, G., Yucesan, B., Eker, I., Sahbaz, N., Gurel, S., & Gurel, E. (2012). Direct somatic embryogenesis from hypocotyl segments of *Digitalis trojana* Ivan and subsequent plant regeneration. *Industrial Crops and Products, 40*, 76–80.

Vibha, J. B., Choudhary, K., Singh, M., Rathore, M. S., & Shekhawat, N. S. (2009). An efficient somatic embryogenesis system for velvet bean [*Mucuna pruriens* (L.) DC.]: A source of anti-Parkinson's drug. *Plant Cell, Tissue and Organ Culture, 99*, 319–325.

Vinoth, A., & Ravindhran, R. (2016). Efficient plant regeneration of watermelon (*Citrullus lanatus* Thunb.) via somatic embryogenesis and assessment of genetic fidelity using ISSR markers. *In Vitro Cellular & Developmental Biology – Plant, 52*(1), 107–115.

Von Arnold, S., Sabala, I., Bozhkov, P., Dyachok, J., & Filonova, L. (2002). Developmental pathways of somatic embryogenesis. *Plant Cell, Tissue and Organ Culture, 69*(3), 233–249.

Wang, F. X., Shang, G. D., & Wang, J. W. (2022). Towards a hierarchical gene regulatory network underlying somatic embryogenesis. *Trends in Plant Science, 27*(12), 1209–1217.

Wang, X. J., Luo, Q., Li, T., Meng, P. H., Pu, Y. T., Liu, J. X., Zhang, J., Liu, H., Tan, G. F., & Xiong, A. S. (2022). Origin, evolution, breeding, and omics of Apiaceae: A family of vegetables and medicinal plants. *Horticulture Research, 9*, uhac076.

Waris, H. (1957). A striking morphogenetic effect of amino acid in seed plant. *Suom Kemistil B, 30*, 121.

Wen, J., Xie, D. F., Price, M., Ren, T., Deng, Y. Q., Gui, L. J., Guo, X. L., & He, X. J. (2021). Backbone phylogeny and evolution of Apioideae (Apiaceae): New insights from phylogenomic analyses of plastome data. *Molecular Phylogenetics and Evolution, 161*, 107183.

Winkelmann, T. (2016). Somatic versus zygotic embryogenesis: Learning from seeds. *In Vitro Embryogenesis in Higher Plants*, 25–46.

Wójcikowska, B., Botor, M., Morończyk, J., Wójcik, A. M., Nodzyński, T., Karcz, J., & Gaj, M. D. (2018). Trichostatin A triggers an embryogenic transition in *Arabidopsis* explants via an auxin-related pathway. *Frontiers in Plant Science, 9*, 1353.

Wójcikowska, B., & Gaj, M. D. (2017). Expression profiling of AUXIN RESPONSE FACTOR genes during somatic embryogenesis induction in *Arabidopsis*. *Plant Cell Reports, 36*, 843–858.

Wójcikowska, B., Jaskóła, K., Gąsiorek, P., Meus, M., Nowak, K., & Gaj, M. D. (2013). LEAFY COTYLEDON2 (LEC2) promotes embryogenic induction in somatic tissues of *Arabidopsis*, via YUCCA-mediated auxin biosynthesis. *Planta, 238*, 425–440.

Wong, S. M., Salim, N., Harikrishna, J. A., & Khalid, N. (2013). Highly efficient plant regeneration via somatic embryogenesis from cell suspension cultures of *Boesenbergia rotunda*. *In Vitro Cellular and Developmental Biology – Plant, 49*(6), 665–673.

Xu, M., Li, X., & Korban, S. S. (2004). DNA methylation alterations and exchanges during in vitro cellular differentiation in rose (*Rosa hybrida* L.). *Theoretical and Applied Genetics, 109*(5), 899–910.

Yeung, E. C. (1999). The use of histology in the study of plant tissue culture systems—Some practical comments. *In Vitro Cellular & Developmental Biology-Plant, 35*(2), 137–143.

Zenk, M. H., El-Shagi, H., & Schulte, U. (1975). Anthraquinone production by cell suspension cultures of *Morinda citrifolia*. *Planta Medica, 28*(S 01), 79–101.

Zhang, M., Wang, A., Qin, M., Qin, X., Yang, S., Su, S., Sun, Y., & Zhang, L. (2021). Direct and indirect somatic embryogenesis induction in *Camellia oleifera* Abel. *Frontiers in plant science, 12*, 644389.

Zimmerman, J. L. (1993). Somatic embryogenesis: A model for early development in higher plants. *Plant Cell, 5*, 1411–1423.

Zimmerman, P. W., & Hitchcock, A. E. (1942). Growth regulation of plants by synthetic compounds. *Contributions from Boyce Thompson Institute, 12*, 321–343.

Ziv, M., & Gadasi, G. (1986). Enhanced embryogenesis and plant regeneration from cucumber (*Cucumis sativus* L.) callus by activated charcoal in solid/liquid double-layer cultures. *Plant Science, 47*(2), 115–122.

Zuraida, A. R., Izzati, K. F. L., Nazreena, O. A., Radziah, C. M. Z. C., Asyikin, S. G. S. N., & Sreeramanan, S. (2014). In vitro regeneration of *Curcuma caesia* plantlets from basal part and via somatic embryogenesis. *Advances in Bioscience and Biotechnology, 5*(4), 363–372.

CHAPTER

5

Development of synthetic seeds for conservation, storage, and exchange of medicinal plants

Arisha Kausar, Nabia Rehan Mahi, Mariya Anjum, Siddiqa Javed, and Anwar Shahzad

Plant Biotechnology Section, Department of Botany, Aligarh Muslim University, Aligarh, Uttar Pradesh, India

5.1 Introduction

Approximately four-fifths of the people residing in developing countries rely on herbal medications as their main source of health. Furthermore, the cumulative demand for herbal medicines and plant-based complexes is accelerating the global consumption of medicinal plants (Cole et al., 2007). Many medicinal plants produce seeds that are sensitive to desiccation (desiccation-sensitive seeds or refractory seeds), whose viability is limited to only a few weeks to months (Gantait et al., 2015). Synthetic seed (synseed) technology is a promising alternative to traditional breeding methods, enabling the production of large quantities of plants and the long-term conservation of important plant genetic resources (Phuong & Phung, 2025). The exact origins of the synthetic seed concept are unclear, but it is believed to have emerged contemporaneously with the early development of somatic embryogenesis (Reinert, 1958; Steward, 1958). Murashige (1977) officially presented the idea at the Horticultural Tissue Culture Symposium in Ghent, Belgium. Kitto and Janick (1982) developed desiccated synseeds by encapsulating carrot somatic embryos (SEs) in water-soluble resin and polyoxyethylene glycol. Since then, research on synthetic seeds has expanded to various plants, including cereals, fruits, vegetables, trees, palms, orchids, ornamentals, and medicinal plants (Abbas et al., 2022; Ghosh et al., 2025; Nicolae et al., 2023; Nongdam, 2016; Rai et al., 2009). This technology not only supports cost-effective plant breeding (Roy & Mandal, 2008) but also ensures genetic uniformity (Saiprasad, 2001), eliminates the need for acclimatization, and enables the use of different plant tissues for seed production (Gantait & Mitra, 2019; Shallal et al., 2020). This method is widely used in

Medicinal Plant Biotechnology for Ameliorated Secondary Metabolite Production
DOI: https://doi.org/10.1016/B978-0-443-43944-5.00013-X

numerous medicinal plants (Ahmad et al., 2025; Fonseka et al., 2019; Moharana et al., 2024; Sharma et al., 2019), crop plants (Hegde et al., 2016; Kumar et al., 2005), endangered plants (Kiran et al., 2025; Kumar & Chandra, 2014; Negahdar et al., 2021), ornamental plants (Ekinci et al., 2019; Maqsood et al., 2021; Pradhan et al., 2014; Ratilal, 2009) and other plant species (Koçak et al., 2025; Prasannan et al., 2025).

Sodium alginate (Na-Alg) is widely used to encapsulate SEs in synthetic seed production due to its cost-effectiveness, minimal toxicity, and excellent gel-forming ability (Ghosh & Haque, 2019). Solidification of the alginate matrix encasing the embryos occurs through an ion exchange mechanism, where sodium ions (Na^+) in the alginate solution are exchanged with calcium ions (Ca^{2+}) in a calcium chloride solution ($CaCl_2$) (Daud et al., 2008). This biodegradable matrix acts as a protective barrier, protecting the tissue fragments from mechanical damage and adverse environmental conditions. It also functions as the endosperm of the SEs and provides important nutrients for their early development (Rihan et al., 2017). In this chapter, we have discussed the different types of propagules utilized for synseed production and their production in some valuable plant families, and we also highlighted the factors that affect the synseed technology.

5.2 Propagules utilized for synseed production

Explant selection is crucial for effective *in vitro* culture and is determined by factors such as developmental stage, health, regenerative potential, and disease presence. Initially, only SE was used for synthetic seed production (Tailor & Joshi, 2024). However, subsequent studies have reported the successful utilization of many plant micropropagules, such as unipolar shoot tips, nodal segments, embryogenic calli and nodules, as well as alternative explants such as protocorms, protocorm-like bodies (PLBs), hairy roots, bulbs, bulblets, and microtubers for synseed production (Gantait & Sinniah, 2013) (Fig. 5.1).

5.2.1 Bipolar propagules

SEs are bipolar in nature, possessing both a radicle and a bulb, making them ideal propagules for synthetic seed technology (Standardardi and Piccioni, 1998). While hydrated embryos must germinate immediately after generation or be kept moist for a short period of time, encapsulated embryos can be used in both hydrated and dry conditions (Talei et al., 2025). Strategies such as osmotic adjustment, embryo weight monitoring, and image-based growth tracking (Pintos et al., 2008) can help improve maturation accuracy, but genotype dependency still limits large-scale cloning of elite lines. SE has been effectively used as an explant in the propagation of many medicinal plants. Ouzhand et al. (2023) prepared *Nigella sativa* synseeds through SE with various combinations of Na-Alg (2%–4%) and $CaCl_2$ (50–100 mM), followed by polymerization and washing to remove excess $CaCl_2$. Encapsulated SE at the cotyledon stage was cultured under controlled germination conditions on Murashige and Skoog (MS) medium augmented with indole butyric acid (IBA) and kinetin (Kn) (Murashige & Skoog, 1962). Abd El-Zaher et al. (2024) successfully produced artificial olive (*Olea europaea*) seeds by encapsulating SE in a 4% Na-Alg solution and solidifying them with 8% calcium nitrate ($Ca(NO_3)_2$) for 45 minutes. This protocol

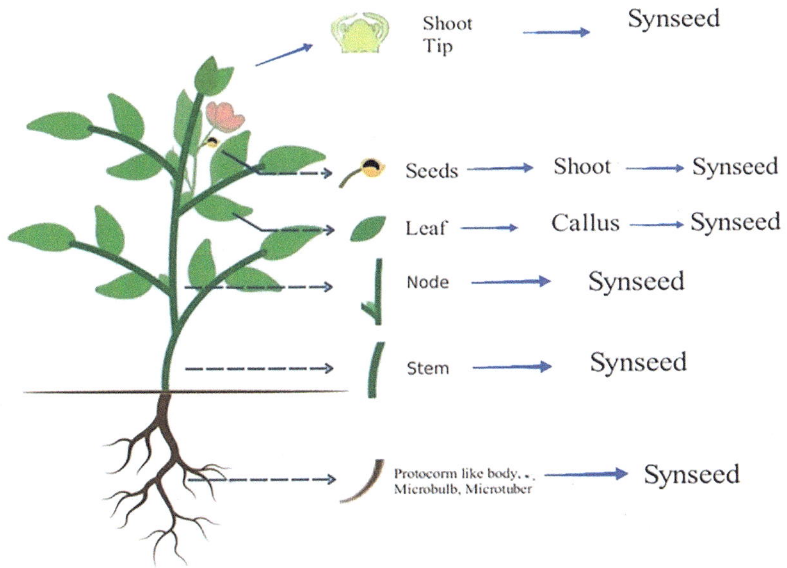

FIGURE 5.1 Different types of propagules/tissues utilized for *in vitro* synseed production.

resulted in an impressive conversion rate of 88.8%, suggesting that encapsulation parameters can be optimized for efficient plant regeneration of woody plants.

Similarly, the production of synthetic seeds of *Phoenix dactylifera* was most effective under conditions of 3% Na-Alg and 2.5% $CaCl_2$, especially when stored at 5°C in a medium supplemented with 20 g/L sucrose. Under these conditions, storage and recovery efficiencies were significantly improved (Bekheet, 2017). To cultivate synthetic seeds of *Bacopa monnieri*, both the synthetic seeds and shoot tips were coated with Na-Alg and $CaCl_2$. These seeds were stored at 4°C and 20°C for 140 days. Interestingly, embryos stored at 25°C (41.7%) showed a significantly higher conversion rate than embryos stored at 4°C (17.3%) (Khilwani et al., 2016). Synseeds of *Citrus sinensis* were formed by coating stigma-derived SE with 3% Na-Alg + 2.5% $CaCl_2$ and storing them at 10°C for 3 months. They were then cultured under controlled conditions on high-sucrose MS medium. Germination and subsequent shoot regeneration were monitored over 1 month. Embryos stored for 1 month yielded the highest number of plantlets, while embryos stored for 2 months showed superior values in growth parameters such as shoot height and leaf number after a 10-month recovery period. These observations suggest that the optimal storage period may vary depending on the desired growth response (Mohamed et al., 2023).

5.2.2 Unipolar propagules

Unipolar plant propagules contain either a shoot or a root pole (apical meristem) and are commonly used in synthetic seed production. These nonembryogenic vegetative parts reduce technical complexity and somaclonal variation. Earlier, Standardi and Piccioni (1998) classified them into three types based on their origin.

5.2.3 Nodes and microshoots

Microcuttings, that is, small nodal explants or shoot tips with apical or axillary buds, are ideal tissue explants for synseed production because they can be easily harvested using a micropropagation system. The use of these explants maintains clonal fidelity and virtually eliminates somaclonal variation (Piccioni, 1997). Encapsulation of such nonembryogenic propagules has been stated in numerous plant types.

Siddique and Bukhari (2018) developed synseeds in *Capparis decidua* by encapsulating nodal explants in 3% Na-Alg and 75 mM $CaCl_2$. This method produced optimal results, with a very high plant regeneration rate (79%) observed on MS medium complemented with 5.0 μM thidiazuron (TDZ) and 0.5 μM indole acetic acid (IAA). They observed that stored synseeds sustained viability for 4 weeks at 4°C, and acclimated seedlings exhibited an 80% survival rate along with enhanced stress tolerance. In *Casuarina equisetifolia*, transparent, uniform synthetic seeds prepared using 4% Na-Alg and 100 mM $CaCl_2$ showed 68.60% seedling conversion. The highest germination rate (77%) occurred on MS medium comprised of 5.0 μM benzyl adenine (BA) and 0.5 μM naphthalene acetic acid (NAA), while the control without plant growth regulators (PGRs) had a germination rate of only 48.6%. Storage for 4 weeks (4°C) resulted in 71% conversion and an average of 3.6 shoots after 8 weeks, although longer storage periods reduced regeneration efficiency (Ahmad et al., 2022). Srivastava et al. (2009) encapsulated *Cineraria maritima* shoot tips and nodal segments in 3% Na-Alg medium supplemented with half-strength MS (½ MS) and 3% $CaCl_2$ and stored them under humid conditions at 25°C ± 2°C for up to 6 months. Regeneration was attained on MS medium without PGRs, and rootless shoots were moved to half-strength MS medium augmented with 1.0 mg/L NAA, resulting in healthy seedlings under greenhouse conditions (Srivastava et al., 2009). A synseed production protocol for *Aerva lanata* using 3% Na-Alg medium and 100 mM $CaCl_2$ showed 100% germination and high regeneration rates after 5 months of storage at 4°C, with the highest response achieved on MS medium containing 3% sucrose. Rooting and acclimation were successful, with a survival rate of 73.3%, but a decline in regeneration rates was observed with prolonged storage. Apparently, exogenous growth regulators were unnecessary (Hassanen, 2021). In the endangered species *Withania coagulans*, microcutting encapsulated in 3% Na-Alg and 100 mM $CaCl_2$ resulted in a 96% regeneration rate, with fully developed plants within 30 days. Rooting was promoted by IBA pulse treatment, and synseeds stored at 4°C for 60 days maintained 72% regeneration potential, making them suitable for brief storage and germplasm exchange (Rathore & Kheni, 2017). In *Spilanthes mauritiana*, a gelling matrix of 4% Na-Alg (CDH) or 3% Na-Alg (Loba Chemie) and 100 mM $CaCl_2$ was optimal for the formation of calcium alginate beads. On MS medium, CDH showed a higher synthetic seed conversion rate (74.4%) than Loba Chemie (62%). Addition of 5.0 μM BA and 0.5 μM IAA increased the conversion rate to 83%, and the synseeds germinated after storage at 4°C for 1–8 weeks. In both *in vitro* and *ex vitro* transformation, the survival rates after acclimation were high at 90% and 80%, respectively (Sharma & Shahzad, 2014). In our lab, we encapsulated nodal segments of *Ajuga bracteosa* in 3% Na-Alg and 100 mM $CaCl_2$ for synseed production, and the synseed was germinated on MS medium containing 1.5 μM BA after 4 weeks of inoculation (Fig. 5.2).

FIGURE 5.2 Germination of synseed of *Ajuga bracteosa*: (A) Synseed inoculated on 1.5 μM BA and (B) synseed germination on 1.5 μM BA after 4 weeks of inoculation.

5.3 Microbulbs, microtubers, rhizomes, and corms

Microbulbs, rhizomes, and PLBs are naturally differentiated storage propagules and are therefore usually converted into plants without complex treatment. Encapsulation of these organs has been reported in several taxa (Saiprasad & Polisetty, 2003). In Malabar daffodil (*Ipsea malabarica*), encapsulated microbulbs were sown either on half-strength hormone-free MS medium or on 6.97 μM Kn, achieving a 100% conversion rate. Kn supplementation nearly doubled shoot yield (7.2 vs. 4.6 shoots per capsule) and improved growth (Martin, 2003). Bekheet (2006) demonstrated the highest recovery rate for encapsulated garlic bulbs (approximately 6 shoots, 6 cm long, 3.8 g fresh weight) on MS + 2 mg/L BA and 2 mg/L NAA.

PLB-based synseeds are particularly valuable because orchid seeds are small and lack endosperm. Protocorms of *Cymbidium giganteum* showed comparable regeneration rates and reduced acclimation costs both *in vitro* (100%) and when directly planted in sterile sand (88%) or a sand-soil mixture (64%) (Corriealri & Tandonalri, 1993). Subsequent studies reported successful PLB encapsulation in several orchid genera, including *Dendrobium*, *Cattleya*, *Oncidium*, *Pias*, and *Spathoglottis* (Saiprasad & Polisetty, 2003). For *Dendrobium*, *Oncidium*, and *Cattleya*, PLBs at the early leaf development stage (13–15 days old) were encapsulated in 3/4-strength MS medium containing 0.44 μM BA and 0.54 μM NAA and then cultured on the same medium or NAA-supplemented medium, achieving 100% plant recovery. High conversion rates have also been reported for other orchids, including 94.9%

for *Vanda coerulea* on IY medium (Sarmah et al., 2010), 95% for *Flickingeria nodosa* on Burgeff medium (Nagananda et al., 2011), and 96.4% for *Aranda* × *Vanda* hybrids grown on half-strength MS medium with 3% Na-Alg and 75 mM $CaCl_2$ and then cultured without PGR (Gantait et al., 2012).

5.4 Embryogenic masses and calli

Embryogenic masses are known for their high regenerative potential and genetic fidelity, making them ideal materials for clonal propagation and genetic transformation. Callus tissue has also been encapsulated despite its disorganized structure and low differentiation potential (Gantait et al., 2015; Wagley et al., 1987). However, *in vitro* maintenance is labor-intensive due to the need for frequent subcultures (Ara et al., 2000). A brief treatment of these tissues with 6-benzylaminopurine (BAP), followed by encapsulation with Na-Alg and storage at 4°C, facilitates handling, and synthetic seeds retain their viability and growth potential for approximately 2 months (Redenbaugh and Fujii, 1991). This method has been effectively used to obtain plantlets of *Anthurium andreanum* and *Arnebia euchroma* (Manjkhola et al., 2005). However, further research is needed to determine whether shelf life can be extended without compromising quality. In particular, synthetic seeds derived from callus have been successfully used to regenerate whole plants of *Allium sativum*, *Rhodiola kirilowii*, and *Carica papaya*, confirming the feasibility of this approach for these species (Kim & Park, 2002; Teixeira da Silva, 2014; Zych et al., 2005) (Table 5.1).

MS = RO=

5.5 Synseed production in some important plant families

Synseed technology has been extensively utilized across numerous plant species, such as crop and ornamental plants, endangered species, and various medicinal plants. However, choosing suitable encapsulation procedures as well as matrix constituents is crucial and varies from plant to plant. Here, we discuss the protocols developed for synseed production in medicinal plants of some families.

5.5.1 Lamiaceae

For synseed production in rosemary (*Rosmarinus officinalis*), torpedo-shaped SEs were coated with 4% Na-Alg and soaked in 100 mM $CaCl_2$ solution for 40 minutes. The highest regeneration rate of 80.55% was attained on MS medium comprised of 1.5 mg/L Kn and 0.5 mg/L BAP (Azarmi et al., 2020). For *Salvia officinalis*, the best results were obtained when apical and axillary buds remained soaked in 1.2% Na-Alg solution for some time, placed in 200 mM $CaCl_2$ solution, and the resulting synseeds were stored for 30 days at 4°C on MS

TABLE 5.1 Development of synseeds from different medicinal plants propagules (mg/L to mg/L throughout).

Type	Propagule(s)	Plant species	Encapsulation matrix	Medium composition	Germination rate (%)	References
Bipolar propagules	Somatic embryos	*Colchicum cilicium*	3% Na-Alg + 4 mg/L 2,4-D	RO medium + 2 mg/l 2,4-D 0.5 mg/L 2iP	27.8	Mendi et al. (2022)
	Somatic embryos	*Iris pallida*	3% Na-Alg + 100 mM $CaCl_2$	½ MS without sucrose + 300 mg/L GSH + 500 mg/l MES	50	Meucci et al. (2024)
	Somatic embryos	*Nigella sativa*	3% Na-Alg + 100 mM $CaCl_2$	MS + 1 mg/L IBA + 2 mg/L Kin	77.7	Ouzhand et al. (2023)
	Somatic embryos	*Citrus sinensis*	3% Na-Alg + 2.5% $CaCl_2$	MS medium without PGRs	80	Mohamed et al. (2023)
	Somatic embryos	*Flickingeria nodosa*	2% Na-Alg + 100 mM $CaCl_2$ + 2% sucrose + Burgeff's N3F basal liquid medium	Burgeff's N3F basal liquid medium + 2 mg/L adenine sulfate + 1 mg/L IAA	82	Nagananda and Satishchandra (2023)
Unipolar propagules	Nodal segments	*Capparis decidua*	3% Na-Alg + 75 mM $CaCl_2$	MS + 5.0 µM TDZ + 0.5 µM IAA	79	Siddique and Bukhari (2018)
	Nodal segments	*Tinospora cordifolia*	2.5% Na-Alg + 100 mM $CaCl_2$	MS + 2.0 mg/l BAP + 0.2 mg/L NAA	94	Singh et al. (2021)
	Nodal segments	*Moringa concanensis*	4% Na-Alg + 100 mM $CaCl_2$	MS + 0.25 mg/L kinetin + 0.1 mg/L IAA	90.4	Gautam et al. (2025)
	Nodal segments	*Casuarina equisetifolia*	4% Na-Alg + 100 mM $CaCl_2$	MS + 5.0 µM BA + 0.5 µM NAA	77	Ahmad et al. (2022)
	Shoot tip	*Cineraria maritima*	½ MS + 3% Na-Alg + 3% $CaCl_2$	MS medium without PGRs	82.3	Srivastava et al. (2009)
	Shoot tip	*Aerva lanata*	3% Na-Alg + 100 mM $CaCl_2$	MS + 1 mg/L BA + 0.4 mg/L NAA	100	Hassanen (2021)
	Shoot tip	*Withania coagulans*	3% Na-Alg + 100 mM $CaCl_2$	MS + 1.11 µM BAP + 0.57 µM IAA	96	Rathore and Kheni (2017)
	Bulblets	*Allium sativum*	3% Na-Alg + 2.5% $CaCl_2$	MS + 2 mg/L BA + 2 mg/L NAA	–	Bekheet (2006)

(*Continued*)

TABLE 5.1 (Continued)

Type	Propagule(s)	Plant species	Encapsulation matrix	Medium composition	Germination rate (%)	References
	Protocorm-like bodies	*Dendrobium* sp.	3% Na-Alg + 75 mM $CaCl_2$	¾ MS medium + 0.44 μM BA + 0.54 μM NAA	100	Saiprasad and Polisetty (2003)
	Protocorm-like bodies	*Oncidium* sp.	3% Na-Alg + 75 mM $CaCl_2$	¾ MS medium + 2.69 μM NAA	100	Saiprasad and Polisetty (2003)
	Protocorm-like bodies	*Cattleya*	3% Na-Alg + 75 mM $CaCl_2$	¾ MS medium + 5.38 μM NAA	100	Saiprasad and Polisetty (2003)
	Protocorm-like bodies	*Dendrobium nobile*	3% Na-Alg + 100 mM $CaCl_2$	½ MS + 7.5% sucrose + 1 mg/L BAP + 0.1 mg/L NAA	78.20	Mohanty et al. (2013)
	Callus	*Rhodiola kirilowii*	5% Na-Alg + 50 mM $CaCl_2$	MS medium without PGRs	100	Zych et al. (2005)

medium comprising 0.3 mg/L BAP (Deja, 2022). In *Salvia splendens*, the highest regeneration rate (63.6%) was accomplished with 5.0 µM BA and 2.5 µM IAA (Sharma et al., 2014). However, Badhepuri et al. (2025) stated that for *Coleus forskohlii*, synseeds cultured in 3% Na-Alg and 75 mM $CaCl_2$ were the most effective encapsulation matrix, with a high regeneration rate (63.3%) in half-strength MS medium augmented with 0.5 mg/L BAP + 0.1 mg/L IAA. In *Satureja kujistanica*, a combination of 2% Na-Alg and 100 mM $CaCl_2$ yielded capsules with a regeneration rate of 88.8% (Asadi et al., 2022). The combination of 100 mM $CaCl_2$ and 3% Na-Alg was also effective against *Teucrium polium* (Saadat et al., 2023), *Pogostemon cablin* (Lalthafamkimi et al., 2022), and *Vitex negundo* (Ahmad & Anis, 2010).

5.5.2 Asteraceae

Synseeds were developed in *Gerbera jamesonii* by encapsulating microshoots and seedlings in 3.0% Na-Alg and 100 mM $CaCl_2$. Encapsulation matrices containing growth regulators (2.0 mg/L BAP + 0.5 mg/L NAA) exhibited germination rates of 74.5% for microshoots and 54.2% for seedlings in SE. High germination rates were observed even after storage at 4°C for 1–3 months (Taha et al., 2008). Salma et al. (2019) showed that torpedo-shaped seedlings were effectively encapsulated in *Eclipta alba* in 2.5% Na-Alg and 75 mM $CaCl_2$, achieving a germination rate of 93.3%. The best results (86.6%) were obtained when encapsulated embryos were stored at a low temperature of 8°C for 60 days. Seeds artificially produced from shoot tips of *Stevia rebaudiana* on MS medium comprising 0.5 mg/l BA, 3% Na-Alg, and 75 mM $CaCl_2$ resulted in 9.47 shoots with a length of 10.94 mm and 64.33 leaves (Subrahmanyeswari et al., 2023). Ali et al. (2012) also found that a gel matrix comprised of 3% Na-Alg and 100 mM $CaCl_2$ (for 15 minutes) remained effective in the production of round, transparent beads of uniform diameter in *Stevia rebaudiana* (at 4°C). They also found that seeds pretreated with potassium nitrate (KNO_3) germinate faster than untreated seeds. Sharma et al. (2009) reported that the best results were attained with nodal segments of *Spilanthes acmella* using 4% Na-Alg and 100 mM $CaCl_2$, and that the highest germination rate (87.8%) was accomplished by the addition of 1.0 µM BA and 0.5 µM NAA to the MS medium after 6 weeks of culture. Wang FuYuan et al. (2011) also observed very high germination rates (100%) and seedling conversion rates (92%) for synseeds produced from *Atractylodes macrocephala* axillary buds coated with 2.5% Na-Alg and 2% $CaCl_2$. Furthermore, supplementing the matrix with 0.5 mg/L IAA and 30 g/L sucrose significantly enhanced growth. However, Sujatha and Kumari (2008) reported that a combination of 2% Na-Alg and 50 mM $CaCl_2$ effectively produced high-quality beads suitable for encapsulating nodal segments of *Artemisia vulgaris*.

5.5.3 Orchidaceae

Encapsulation conditions vary depending on the orchid species. In *Dendrobium sonia*, encapsulation with 3% Na-Alg and treatment with 75 mM $CaCl_2$ for 30 minutes resulted in 100% conversion of PLBs in plants (Saiprasad & Polisetty, 2003). In contrast, in *Geodorum densiflorum*, a PLB conversion of up to 88% was achieved by encapsulation with 4% Na-Alg and stabilization with 50 mM $CaCl_2$ for 15–20 minutes (Datta et al., 1999). Mahdavi et al.

(2023) generated a procedure for growing uniform, round, and durable synthetic seeds in *Phalaenopsis* orchids using Na-Alg and $CaCl_2$. They used Na-Alg solution in half-strength MS medium containing 2 g/L peptone and then cultivated the synthetic seeds in half-strength MS medium containing 3% sucrose and 0.7% agar. They found that the highest germination rate (100%) was achieved when 4% Na-Alg was combined with a 100 mM $CaCl_2$ solution. Furthermore, their study found that storing synthetic seeds at 25°C under LED light, especially green light, was the most effective condition for maintaining viability for 180 days. These results emphasize the importance of optimizing both the Na-Alg and $CaCl_2$ solutions for each orchid species. Furthermore, Khor et al. (1998) showed that incorporating chitosan or gelatin into Na-Alg as part of a bilayer encapsulation system improved the quality of synthetic seeds of *Spathoglottis plicata*. These results underscore the importance of optimizing both the Na-Alg and $CaCl_2$ solutions for each orchid species.

5.5.4 Solanaceae

Oceania et al. (2015) found that a gel matrix composed of 2% Na-Alg and 100 mM $CaCl_2$ was highly effective in encapsulation of tomato shoot tips. Tarek et al. (2022) also reported 100% viability for *Brunfelsia pauciflora* using 1% Na-Alg and 2.5% $CaCl_2$ for a long period (7 weeks). However, the highest viability (100%) was achieved using 3% high Na-Alg and 2.5% low $CaCl_2$ at the shortest storage time (3 weeks). When the storage time was extended to 7 weeks, viability gradually decreased from 100% to 22.11% (Tarek et al., 2022). However, Lakshmana Rao and Singh (1991) observed that 3% Na-Alg and 75 mM $CaCl_2$ showed the best response for encapsulation of SE of *Solanum melongena*. Singh et al. (2006a) developed a method to produce *Withania somnifera* plants from encapsulated shoot tips. 3% Na-Alg and 75 mM $CaCl_2$ were used for encapsulation, and the best regeneration results were obtained after 5 weeks of culture on MS medium comprised of 0.5 mg/L IBA. Huda et al. (2007) also used *Solanum melongena* SE and found that the highest germination rate was attained on MS medium containing 1.0 mg/L BAP and 0.1 mg/L gibberellic acid 3 (GA_3). Moreover, Nyende et al. (2003) also demonstrated that hollow synseed beads are an effective alternative for the propagation of potato. In their study, precultured shoot tips on MS media were encapsulated in calcium alginate to form hollow synseed capsules. Surprisingly, all encapsulated shoot tips showed a 100% regeneration rate after 180 days.

5.5.5 Euphorbiaceae

Upadhyay et al. (2014) produced synseeds from nodal segments of *Phyllanthus fraternus* and found that a polymerization time of up to 15 minutes in a matrix of 3% Na-Alg and 100 mM $CaCl_2$ was ideal for synthetic seed formation. They achieved the highest seedling conversion rate (92.5%) in half-strength MS solid medium devoid of plant growth regulators (PGRs), and optimal multiple shoot growth was attained in half-strength MS medium with 0.5 mg/L BAP. Singh et al. (2006b) reported that the most effective gel complex formation in *Phyllanthus amarus* was achieved by adding 3% Na-Alg and 75 mM $CaCl_2$. They observed the maximum conversion rate of 90% in encapsulated shoot tips after 5 weeks of culture in MS medium, even without PGRs.

5.5.6 Apiaceae

Bisht et al. (2015) reported that encapsulating *Angelica glauca* SEs by immersing them in a solution containing 3% Na-Alg and 100 µM $Ca(NO_3)_2$ for 30 minutes protected the SEs and promoted the formation of synthetic seeds. In *Serinum tenuifolium*, the addition of MS medium containing 3% sucrose, 2.0 µM BA, and 2.0 µM IBA to an alginate matrix (3% Na-Alg and 100 mM $CaCl_2$) significantly improved the germination rate (77.8%) of synthetic seeds (Joshi et al., 2006). In *Anetum graveolens*, a mixture of 100 mM $CaCl_2$ and 3% Na-Alg also produced the best results for synthetic seeds (Dhir et al., 2014) (Table 5.2).

5.6 Other plant species

When using embryos for synseed production, Na-Alg is typically applied at a concentration of 2%–3%. However, even concentrations as low as 1.5% Na-Alg have been shown to effectively encapsulate spherical SEs of *Cyclamen persicum* (Winkelmann et al., 2004). Bead hardening is typically performed with 50–100 mM $CaCl_2$ dihydrate, with exposure times ranging from 20 minutes for *Cyclamen persicum* (Winkelmann et al., 2004) to 60 minutes for *Pelargonium holtorum* (Gill et al., 1994). High regeneration efficiency (58.5%) was observed in *Tecomella undulata* when a gelling matrix containing 3% Na-Alg and 100 mM $CaCl_2$ was added to the MS medium augmented with 2.5 µM BA and 0.5 µM IAA (Shaheen & Shahzad, 2015). The encapsulated segments were moved to a medium containing MS and 10.0 µM BA, resulting in an average of 6.5 shoot formation (Shaheen & Shahzad, 2015). Several studies have shown that not only the Na-Alg concentration but also the concentration and duration of exposure to the chelating agent significantly influence seedling germination rates. For example, in *Paulownia elongata*, reducing the Na-Alg concentration from 3% to 1% resulted in a significant reduction in SE survival and seedling transformation rates (Ipekci & Gozukirmizi, 2003). Optimal germination was achieved when the beads were polymerized in 50 mM $CaCl_2$ for 30 minutes. The highest SE induction rates (leaf: 50.1%, internode: 29.8%) were accomplished in MS medium containing 3% sucrose, 0.6% phytagel, 500 mg/L casein, and 10 mg/L TDZ. However, these values were lower than the SE induction rates achieved by 10-minute exposure to 60 mM or 80 mM $CaCl_2$ (Ipekci & Gozukirmizi, 2003).

Studies have consistently shown that a combination of 100 mM CaCl2 and 3% Na-Alg provides the most favorable results for the formation of spherical synthetic seed beads. This formulation has been successfully used in numerous medicinal plants, such as *Simmondsia chinensis* (Kumar et al., 2010). *Rauvolfia tetraphylla* (Faisal et al., 2013), *Balanites aegyptiaca* (Varshney & Anis, 2014), *Simmondsia chinensis* (Bala et al., 2020), *Pterocarpus marsupium* (Ahmad et al., 2023), and *Ficus palmata* (Al-Aizari et al., 2025), and synthetic seeds are very effective with high regeneration rates.

TABLE 5.2 *In vitro* synseed production in some important medicinal plants.

Family	Plant species	Encapsulation matrix	Medium composition	Germination rate (%)	References
Lamiaceae	*Rosmarinus officinalis*	4% Na-Alg + 100 mM $CaCl_2$	MS medium + 1.5 mg/L kinetin + 0.5 mg/L BAP	80.5	Azarmi et al. (2020)
	Salvia officinalis	1.2% Na-Alg + 200 mM $CaCl_2$	MS medium + 0.3 mg/L BAP	85	Deja (2022)
	Salvia splendens	4% Na-Alg + 100 mM $CaCl_2$	MS medium + 5.0 μM BA + 2.5 μM IAA	63.6	Sharma et al. (2014)
	Coleus forskohlii	3% Na-Alg + 75 mM $CaCl_2$	½ MS medium + 0.5 mg/L BAP + 0.1 mg/L IAA	63.3	Badhepuri et al. (2025)
	Satureja khuzistanica	2% Na-Alg + 100 mM $CaCl_2$	2.0 μM TDZ in MS liquid medium	88.8	Asadi et al. (2022)
Asteraceae	*Gerbera jamesonii*	3% Na-Alg + 100 mM $CaCl_2$	MS medium + 2.0 mg/L BAP + 0.5 mg/l NAA	74.5	Taha et al. (2008)
	Stevia rebaudiana	3% Na-Alg + 75 mM $CaCl_2$	MS medium + 0.5 mg/L BA	–	Subrahmanyeswari et al. (2023)
	Spilanthes acmella	4% Na-Alg + 100 mM $CaCl_2$	MS medium + 1.0 μM BA + 0.5 μM NAA	87.8	Sharma et al. (2009)
	Atractylodes macrocephala	2.5% Na-Alg + 2% $CaCl_2$	MS medium + 0.5 mg/L IAA	100	Wang FuYuan et al. (2011)
Orchidaceae	*Dendrobium* sp.	3% Na-Alg + 75 mM $CaCl_2$	¾ MS medium + 0.44 μM BA + 0.54 μM NAA	100	Saiprasad and Polisetty (2003)
	Geodorum densiflorum	4% Na-Alg + 50 mM $CaCl_2$	Knudson medium + 15% (v/v) coco milk + 2 g/L peptone + 2 mg/L BAP + 1 mg/l NAA	88	Datta et al. (1999)
Solanaceae	*Withania somnifera*	3% Na-Alg + 75 mM $CaCl_2$	MS medium + 0.5 mg/L IBA	87	Singh et al. (2006a)
	Solanum melongena	4% Na-Alg + 1.4% $CaCl_2$	MS medium without PGRs	70	Huda et al. (2007)

Euphorbiaceae	*Phyllanthus fraternus*	3% Na-Alg + 100 mM $CaCl_2$	Half-strength MS medium without PGRs	92.5	Upadhyay et al. (2014)
	Phyllanthus amarus	3% Na-Alg + 75 mM $CaCl_2$	MS medium without PGRs	90	Singh et al. (2006b)
Apiaceae	*Selinum tenuifolium*	3% Na-Alg + 100 mM $CaCl_2$ + MS medium + 2.0 μM BA + 2.0 μM IBA	MS medium + 2.0 μM BA + 2.0 μM IBA	77.8	Joshi et al. (2006)
	Anethum graveolens	3% Na-Alg + 100 mM $CaCl_2$	MS medium without PGRs	83	Dhir et al. (2014)

5.7 Factors affecting synseed production

The effective germination of synthetic seeds is dependent on a variety of parameters, each of which needs to be precisely optimized to ensure high conversion rates into plantlets that are susceptible to survival. However, the successful germination of synthetic seeds is influenced by a multitude of factors that must be precisely optimized to ensure high conversion rates into viable plantlets (Fig. 5.3).

5.7.1 Storage temperature

The stabilization and prolonged viability of propagules remain central goals in synthetic seed (synseed) technology. Low-temperature storage, particularly at 4°C, is widely recognized for its effectiveness in minimizing metabolic activity within encapsulated embryos or microshoots, thereby conserving internal energy reserves and delaying premature (Sharma et al., 2013; Yücesan, 2019). This temperature regime has been extensively reported as optimal for short- to mid-term synseed preservation across a variety of species, including *Rubus fruticosus* (Regni et al., 2024), *Taraxacum pieninicum* (Kamińska et al., 2018), and *Rauvolfia serpentina* (Gantait & Kundu, 2017). Supporting these findings, Al-Qurainy et al. (2014) investigated the response of nodal segments of *Ochradenus baccatus*, which, after encapsulation, were stored at 4°C, 8°C, and 12°C for 16 weeks. The highest conversion rate to viable plantlets was attained at 4°C on MS medium comprised of 1.0 μM BA, underscoring the advantages of cold storage in maintaining regenerative capacity. However, optimal storage temperature can vary based on species and encapsulation type. Saadat et al. (2023) reported that *Teucrium polium* reached its maximum germination rate

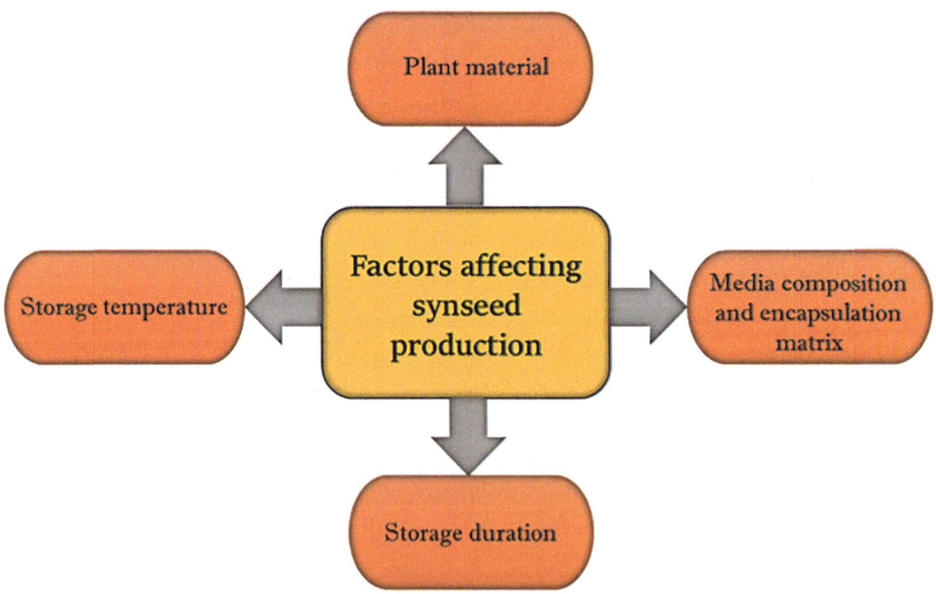

FIGURE 5.3 Factors affecting *in vitro* synseed production in plants.

(96.6%) after 4 weeks of storage at 25°C and 3 weeks at 4°C. However, a temperature of –18°C negatively affected seed germination. Kader et al. (2022) examined the impact of storage temperature on the growth of synthetic seeds and found that seeds stored at 24°C exhibited higher germination rates and regrowth potential throughout the storage period than seeds stored at 4°C. Furthermore, evidence supports the notion that temperature requirements are species-specific (Sota et al., 2023). In *Juglans regia*, propagation at 25°C was optimal for shoot quality and recovery, while a moderate storage period of 4 months at 8°C was possible, though with only minimal signs of necrosis.

5.7.2 Storage duration

Storage duration also affects the regrowth potential and viability of synseeds. Longer storage periods often lead to reduced germination. This is mainly due to physiological and structural changes such as dehydration, depletion of internal energy stores, and changes in the encapsulation matrix (Ahmad et al., 2025). A steady decrease in viability has been observed with increasing storage duration. For example, Kalsoom et al. (2023) investigated the effects of storage duration on alginate-encapsulated nodal buds of *Solanum tuberosum*. Their results showed a significant decrease in viability over time, with the highest germination rate (83.3%) recorded for synthetic seeds sown immediately after encapsulation. This rate decreased significantly to 63.33% after 45 days of storage and then further decreased after 60 days (59.3%) and 75 days (43%). Similar results have been reported in several experiments where long-term storage significantly reduced the viability and regenerative capacity of synseeds (Kader et al., 2022; Wasi et al., 2024).

5.8 Medium composition and encapsulation matrix

Encapsulation efficiency and the quality of synseeds are highly affected by the composition of the gelling matrix, particularly the concentrations of Na-Alg and $CaCl_2$. Hamza (2013) demonstrated that both the morphology and texture of synthetic banana seeds are significantly influenced by these two components. This study found that dropping a 4% Na-Alg solution into a 100 mM $CaCl_2$ solution produced uniformly round, hard synthetic seeds. Furthermore, the exposure time of the $CaCl_2$ solution was crucial; a 10-minute treatment resulted in 100% germination after 16 weeks of storage. Similar results were reported for *Spilanthes mauritiana*, confirming that a combination of 4% Na-Alg and 100 mM $CaCl_2$ is the most effective combination for producing ideal calcium alginate beads (Sharma & Shahzad, 2014). Lower concentrations of Na-Alg resulted in misshapen, overly soft synthetic seeds, which compromised their structural integrity. Kader and Ghosh (2022) also optimized a slightly different formulation for *Acacia auriculiformis* and recommended 3% Na-Alg and 75 mM $CaCl_2$ for effective encapsulation. Another important element, sucrose, also plays an important role in seed recovery and preservation. Tsvetkov et al. (2006) emphasized that the absence of sucrose led to significantly reduced regeneration. This was supported by Shilpha et al. (2021), who found that adding 1.5%–2% sucrose helped achieve a high rate of plant recovery (over 86%) in *Solanum trilobatum*. In contrast, synseeds stored with only 0.5% sucrose remained green and viable but had a much lower success rate (around 55%) in plant regeneration. Synthetic seed storage capability was enhanced in *Dalbergia sissoo* and *Eclipta alba* by low sucrose concentrations (less than 3%) (Ray & Bhattacharya, 2010; Singh & Chand, 2010).

5.8.1 Plant materials

SEs are regarded as ideal candidates for synthetic seed production due to their inherent bipolar nature, distinguished by the simultaneous occurrence of both shoot and root ends (Standardi & Piccioni, 1998). This structural advantage, along with their ability to be clonally propagated in large quantities, makes SEs highly desirable for encapsulation in artificial seed systems. The concept of synthetic seeds, "encapsulated SEs enclosed in a protective matrix," was first proposed as a commercially successful as well as scalable substitute to conventional propagation techniques (Redenbaugh, 1990). Despite their theoretical advantages, the practical application of encapsulated SEs has often shown limited regeneration success compared to other explant types. For instance, Prewein and Wilhelm (2003) reported a relatively low regeneration rate of 26% in *Quercus robur* synseeds cultivated on P24 medium augmented with 0.9 µM BA and 0.1 µM IBA. However, more favorable outcomes have also been documented. In *Hemidesmus indicus*, encapsulated SE showed complete (100%) regeneration on MS medium augmented with 2 µM BA and 0.5 µM IBA. This suggests that optimized hormonal conditions significantly improved regenerative capacity (Cheruvathur et al., 2013). The developmental stage of the SE also plays an important role in germination and conversion efficiency. Mature embryos typically show superior performance due to their advanced physiological and morphological preparation. Overall, SE offers significant advantages in synthetic seed production due to its bipolarity and scalability. However, its successful application depends heavily on factors such as embryo maturity, media composition, and hormonal balance.

Adventitious roots can arise spontaneously or be induced by environmental or physiological stimuli and originate from various tissues, including stems, leaves, and even root regions that predate the endothelium. The regeneration potential of hairy roots has also been investigated in the generation of synseeds of therapeutic plants. In pioneering investigations, Piątczak and Wysokińska (2013) and Rawat et al. (2013) demonstrated the efficacy of hairy root culture as a novel platform for synseed production in *Centaurium erythraea* and *Picrorhiza kurrooa*, achieving regeneration rates of 86% and 73%, respectively. Similarly, synseeds produced at 4 weeks of age from *in vitro*-induced adventitious roots of *Hemidesmus indicus* obtained from leaf discs showed high germination efficiency (89%) following culture on MS medium soaked with 2.0% Na-Alg, solidified with 100 mM $CaCl_2$, and augmented with 3.0 mg/L IBA (Manokari & Shekhawat, 2024). Among nonembryogenic explants, shoot tips have shown considerable promise due to high mitotic activity in the meristematic region (Ballester et al., 1997). However, in *Dendrobium* orchids, synseeds derived from PLBs achieved 80% regeneration, whereas shoot-tip-derived synseeds yielded lower success rates, with ≤50% plantlet formation (Teixeira da Silva et al., 2015). Besides, nodal segments are also commonly used as explants in micropropagation and are widely applied in synseed technology, primarily due to their ease of production once the culture system is established (Piccioni & Standardi, 1995). Nodal explants are better in synseed production compared to other types of explants. In *Enicostema axillare*, Loganathan et al. (2022) reported successful synseed production from both nodal and root explants. Nodal explants exhibited high germination rates and maintained viability for over 80 days, while root explants were comparatively less responsive.

5.9 Conclusion

The *in vitro* synseed production technique offers numerous advantages that make it a valuable tool in plant biotechnology. It is one of the most affordable ways to deliver plantlets and aids in reducing the production costs. It holds great promise for large-scale micropropagation and represents a valuable method for the direct application of artificial seedlings *in vivo*, particularly for germplasm conservation through cryopreservation. However, a major limitation is that synthetic seeds can't be sown directly into the soil or in nonsterile conditions, which restricts their broader practical use.

Acknowledgments

The authors, Arisha Kausar (E-Certificate No.: 24 J/03/01630), Nabia Rehan Mahi (24D/01/01583), and Mariya Anjum (22D/23J00472), are thankful to the University Grants Commission, India, for providing financial support in the form of Junior Research Fellowship.

References

Abbas, M. K., Mahood, H. E., & Alhasan, A. S. (2022). Production of synthetic seeds in vegetable crops: A review. *IOP Conference Series: Earth and Environmental Science, 1060*, 012099. https://doi.org/10.1088/1755-1315/1060/1/012099.

Abd El-Zaher, M. H., Abd El-Wahab, S. M., Saleh, S. S., & Al-Maghrabi, N. N. (2024). Synthetic seeds production by encapsulated somatic embryo of white olive cultivar. *Journal of Global Innovations in Agricultural Sciences, 12*, 1109–1114. https://doi.org/10.22194/JGIAS/24.1428.

Ahmad, A., Ansari, F. A., Anis, M., & Khan, A. S. (2023). Micropropagation of *Pterocarpus marsupium* Roxb. through synthetic seeds and its novel antibiofilm activities against ESKAPE pathogens. *Industrial Crops and Products, 198*, 116681. https://doi.org/10.1016/j.indcrop.2023.116681.

Ahmad, N., & Anis, M. (2010). Direct plant regeneration from encapsulated nodal segments of *Vitex negundo*. *Biologia Plantarum, 54*, 748–752. https://doi.org/10.1007/s10535-010-0134-8.

Ahmad, Z., Babich, O., Sukhikh, S., Yadav, V., Ramakrishnan, M., Firdaus, F., & Shahzad, A. (2025). Unlocking the biotechnological potential of *Decalepis arayalpathra*: Exploring synthetic seed production, metabolic profiling, genetic stability, and the impact of photosynthetic photon flux density on acclimatization. *BMC Plant Biology, 25*, 189. https://doi.org/10.1186/s12870-025-06174-w.

Ahmad, Z., Yadav, V., Shahzad, A., Emamverdian, A., Ramakrishnan, M., & Ding, Y. (2022). Micropropagation, encapsulation, physiological, and genetic homogeneity assessment in *Casuarina equisetifolia*. *Frontiers in Plant Science, 13*, 905444. https://doi.org/10.3389/fpls.2022.905444.

Al-Aizari, A. A., Dewir, Y. H., Al-Obeed, R. S., Sallam, M., Almutairi, K. F., & Al-Doss, A. (2025). *In vitro* conservation through cold storage and genetic fidelity of the recovered Fegra fig (*Ficus palmata* Forssk.) micropropagated plantlets. *South African Journal of Botany, 184*, 319–328. https://doi.org/10.1016/j.sajb.2025.06.010.

Ali, A., Gull, I., Majid, A., Saleem, A., Naz, S., & Naveed, N. H. (2012). *In vitro* conservation and production of vigorous and desiccate tolerant synthetic seeds in *Stevia rebaudiana*. *Journal of Medicinal Plants Research, 6*, 1327–1333. https://doi.org/10.5897/JMPR11.1443.

Al-Qurainy, F., Nadeem, M., Khan, S., Alansi, S., Tarroum, M., & Al-Ameri, A. (2014). Synseed production for storage and conservation of *Ochradenus baccatus* delile. *Pakistan Journal of Botany, 46*, 897–902.

Ara, H., Jaiswal, U., & Jaiswal, V. S. (2000). Synthetic seed: Prospects and limitations. *Current Science, 25*, 1438–1444. https://www.jstor.org/stable/24104316.

Asadi, R., Abdollahi, M. R., Moosavi, S. S., & Mirzaie-Asl, A. (2022). Alginate encapsulation of micro-cuttings in endangered *Satureja khuzistanica* species: A promising method for obtaining genetically stable plants with high rosmarinic acid content. *Plant Cell, Tissue and Organ Culture, 151*, 307–320. https://doi.org/10.1007/s11240-022-02353-x.

Azarmi, R., Tahami, S. K., Farjaminezhad, R., & Pourbeyrami Hir, Y. (2020). A protocol for synthetic seed production in *Rosmarinus officinalis*. *Iran. Journal of Genetics and Plant Breeding, 9*, 126–133. https://doi.org/10.30479/IJGPB.2022.16249.1304.

Badhepuri, M. K., Aileni, M., Beeravelli, P. R., Durgam, S. K., Sandhya, D., Shekhawat, M. S., & Singisala, N. R. (2025). Production of non-embryogenic synthetic seeds for short-term cold storage and ISSR marker-assisted genetic fidelity analysis of *Coleus forskohlii* (Willd) Briq. plants. *Plant Cell, Tissue and Organ Culture, 160*, 1. https://doi.org/10.1007/s11240-025-02977-9 -0.

Bala, R. A., Laura, J. S., & Beniwal, V. S. (2020). Large scale production of female plants of *Simmondsia chinensis* (link) Schneider through synthetic seed technology. *Plant Cell Biotechnology and Molecular Biology, 21*, 47–53.

Ballester, A., Janeiro, L. V., & Vieitez, A. M. (1997). Cold storage of shoot cultures and alginate encapsulation of shoot tips of *Camellia japonica* L. and *Camellia reticulata* Lindley. *Scientia Horticulturae, 71*, 67–78. https://doi.org/10.1016/S0304-4238(97)00074-5.

Bekheet, S. A. (2006). A synthetic seed method through encapsulation of *in vitro* proliferated bulblets of garlic (*Allium sativum* L.). *Arabian Journal of Biotechnology, 9*, 415–426.

Bekheet, S. A. (2017). Encapsulation of date palm somatic embryos: Synthetic seeds. In Jameel M. Al-Khayri, S. Mohan Jain, & Dennis V. Johnson (Eds.). *Date palm biotechnology protocols volume II: Germplasm conservation and molecular breeding* (pp. 71–78). Humana Press. https://doi.org/10.1007/978-1-4939-7159-6_7 (Methods in Molecular Biology, Vol. 1638).

Bisht, A. K., Bhatt, A., Dhar, U., Bhatt, A., & Dhar, U. (2015). Note on somatic embryogenesis and synthetic seed production in *Angelica glauca*: A valuable medicinal plant of Himalaya. *Journal of Medicinal Plant Research, 9*, 419–425. https://doi.org/10.5897/JMPR2014.5326.

Cheruvathur, M. K., Najeeb, N., & Thomas, T. D. (2013). *In vitro* propagation and conservation of Indian sarsaparilla, *Hemidesmus indicus* LR Br. through somatic embryogenesis and synthetic seed production. *Acta Physiologiae Plantarum/Polish Academy of Sciences, Committee of Plant Physiology Genetics and Breeding, 35*, 771–779. https://doi.org/10.1007/s11738-012-1117-5.

Cole, I. B., Saxena, P. K., & Murch, S. J. (2007). Medicinal biotechnology in the genus *Scutellaria*. *In Vitro Cellular & Developmental Biology. Plant: Journal of the Tissue Culture Association, 43*, 318–327. https://doi.org/10.1007/s11627-007-9055-4.

Corriealri, S., & Tandonalri, P. (1993). Propagation of *Cymbidium giganteum* Wall. through high frequency conversion of encapsulated protocorms under *in vivo* and *in vitro* conditions. *Indian Journal of Experimental Biology, 31*, 61–64.

Datta, K. B., Kanjilal, B., & Sarker, D. D. (1999). Artificial seed technology: Development of a protocol in *Geodorum densiflorum* (Lam) Schltr.–an endangered orchid. *Current Science, 25*, 1142–1145.

Daud, N., Taha, R. M., & Hasbullah, N. A. (2008). Artificial seed production from encapsulated micro shoots of *Saintpaulia ionantha* Wendl. (African violet). *Journal of Applied Sciences, 8*, 4662–4667.

Deja, A. K. (2022). Synthetic seeds technology of *Salvia officinalis* as a method for short-term storage and in vitro propagation of valuable genotypes. *Herba Polonica, 68*, 25–29. https://doi.org/10.2478/hepo-2022-0024.

Dhir, R., Shekhawat, G. S., & Alam, A. (2014). Improved protocol for somatic embryogenesis and calcium alginate encapsulation in *Anethum graveolens* L.: A medicinal herb. *Applied Biochemistry and Biotechnology, 173*, 2267–2278. https://doi.org/10.1007/s12010-014-1032-x.

Ekinci, H., Çiftçi, Y.Ö., & Nadarajan, J. (2019). Medium-and long-term conservation of ornamental plants using synthetic seed technology. *Synthetic Seeds: Germplasm Regeneration, Preservation and Prospects*, 259–281. https://doi.org/10.1007/978-3-030-24631-0_11.

Faisal, M., Alatar, A. A., & Hegazy, A. K. (2013). Molecular and biochemical characterization in *Rauvolfia tetraphylla* plantlets grown from synthetic seeds following *in vitro* cold storage. *Applied Biochemistry and Biotechnology, 169*, 408–417. https://doi.org/10.1007/s12010-012-9977-0.

Fonseka, D. L., Wickramaarachchi, W. W., & Madushani, R. P. (2019). Synthetic seed production as a tool for the conservation and domestication of *Celastrus paniculatus*: A rare medicinal plant. *Annual Research & Review in Biology, 32*, 1–8. https://doi.org/10.9734/ARRB/2019/v32i430091.

Gantait, S., Bustam, S., & Sinniah, U. R. (2012). Alginate-encapsulation, short-term storage and plant regeneration from protocorm-like bodies of Aranda Wan Chark Kuan 'Blue'× *Vanda coerulea* Grifft. ex. Lindl. (Orchidaceae). *Plant Growth Regulation, 68*, 303–311. https://doi.org/10.1007/s10725-012-9699-x.

Gantait, S., & Kundu, S. (2017). Does synthetic seed storage at higher temperature reduce reserpine content of *Rauvolfia serpentina* (L.) Benth. Ex Kurz.? *Rendiconti Lincei, 28*, 679–686. https://doi.org/10.1007/s12210-017-0637-8.

Gantait, S., Kundu, S., Ali, N., & Sahu, N. C. (2015). Synthetic seed production of medicinal plants: A review on influence of explants, encapsulation agent and matrix. *Acta Physiologiae Plantarum/Polish Academy of Sciences, Committee of Plant Physiology Genetics and Breeding, 37*, 1–2. https://doi.org/10.1007/s11738-015-1847-2.

Gantait, S., & Mitra, M. (2019). Applications of synthetic seed technology for propagation, storage, and conservation of orchid germplasms. In M. Faisal, & A. Alatar (Eds.). *Synthetic seeds* (pp. 301–321). Springer. https://doi.org/10.1007/978-3-030-24631-0_13.

Gantait, S., & Sinniah, U. R. (2013). Storability, post-storage conversion and genetic stability assessment of alginate-encapsulated shoot tips of monopodial orchid hybrid Aranda Wan Chark Kuan 'Blue'× *Vanda coerulea* Grifft. ex. Lindl. *Plant Biotechnology Reports, 7*, 257–266. https://doi.org/10.1007/s11816-012-0257-9.

Gautam, N., Faroda, P., & Gupta, A. K. (2025). Encapsulation, short-term storage and SCoT based genetic homogeneity assessment of *Moringa Concanensis* Nimmo.: A multipurpose wild drumstick. *South African Journal of Botany, 184*, 995–1000. https://doi.org/10.1016/j.sajb.2025.07.011.

Ghosh, B., & Haque, S. M. (2019). Synthetic seeds: An alternative approach for clonal propagation to avoiding the heterozygosity problem of natural botanical seeds. *Synthetic Seeds: Germplasm Regeneration, Preservation and Prospects*, 77–112. https://doi.org/10.1007/978-3-030-24631-0_4.

Ghosh, P. K., Ghorai, S., Bhunia, R., Samanta, A., Bose, S., Mondal, M. S., Roy, S., & Hazra, S. (2025). Synthetic seed technology and *in vitro* conservation of vegetable germplasm: A review. *Vegetos, 24*, 1–20. https://doi.org/10.1007/s42535-025-01321-y.

Gill, R., Senaratna, T., & Saxena, P. K. (1994). Thidiazuron-induced somatic embryogenesis enhances viability of hydrogel-encapsulated somatic embryos of geranium. *Journal of Plant Physiology, 143*, 726–729. https://doi.org/10.1016/S0176-1617(11)81165-X.

Hamza, E. M. (2013). Factors affecting synseeds formation and germination of banana cultivar Grande Naine. *World Applied Sciences Journal, 25*, 1390–1399. https://doi.org/10.5829/idosi.wasj.2013.25.10.13411.

Hassanen, S. A. (2021). *In vitro* preservation by encapsulation of shoot tips of *Aerva lanata* (L.) Juss. ex Schult. as a rare medicinal plant. *International Journal of Research in Agricultural Sciences, 8*, 2348–3997.

Hegde, V., Makeshkumar, T., Sheela, M. N., Chandra, C. V., Koundinya, A. V., Anil, S. R., Muthuraj, R., & Darshan, S. (2016). Production of synthetic seed in cassava (*Manihot esculenta* Crantz). *Journal of Root Crops, 42*, 5–9.

Huda, A. K., Rahman, M., & Bari, M. A. (2007). Effect of carbon source in alginate bead on synthetic seed germination in eggplant (*Solanum melongena* L.). *Journal of Plant Science, 2*, 538–544.

Ipekci, Z., & Gozukirmizi, N. (2003). Direct somatic embryogenesis and synthetic seed production from *Paulownia elongata*. *Plant Cell Reports, 22*, 16–24. https://doi.org/10.1007/s00299-003-0650-5.

Joshi, M., Dhar, U., & Manjkhola, S. (2006). Somatic embryogenesis and plantlet regeneration from encapsulated embryos of *Selinum tenuifolium*. *Journal of Horticultural Science & Biotechnology, 81*, 570–574. https://doi.org/10.1080/14620316.2006.11512107.

Kader, A., Sinha, S. N., & Ghosh, P. (2022). A strategy for development of genetically stable synseeds of *Azadirachta indica* A. Juss. (Neem) suitable for *in vitro* storage. *Plant Cell, Tissue and Organ Culture, 151*, 47–58. https://doi.org/10.1007/s11240-022-02329-x.

Kalsoom, T., Ahmed, T., Khan, M. A., Hasanuzzaman, M., Ahmed, M., & Werbrouck, S. P. (2023). *In vitro* synthetic seed production of potato under different fungicide levels and storage intervals. *Phyton-International Journal of Experimental Botany, 92*. https://doi.org/10.32604/phyton.2023.030846.

Kamińska, M., Gołębiewski, M., Tretyn, A., & Trejgell, A. (2018). Efficient long-term conservation of *Taraxacum pieninicum* synthetic seeds in slow growth conditions. *Plant Cell, Tissue and Organ Culture, 132*, 469–478. https://doi.org/10.1007/s11240-017-1343-z.

Khilwani, B., Kaur, A., Ranjan, R., & Kumar, A. (2016). Direct somatic embryogenesis and encapsulation of somatic embryos for *in vitro* conservation of *Bacopa monnieri* (L.) Wettst. *Plant Cell, Tissue and Organ Culture, 127*, 433–442. https://doi.org/10.1007/s11240-016-1067-5.

Khor, E., Ng, W. F., & Loh, C. S. (1998). Two-coat systems for encapsulation of *Spathoglottis plicata* (Orchidaceae) seeds and protocorms. *Biotechnology and Bioengineering, 59*, 635–639.

Kim, M. A., & Park, J. K. (2002). High frequency plant regeneration of garlic (*Allium sativum* L.) calli immobilized in calcium alginate gel. *Biotechnology and Bioprocess Engineering, 7*, 206–211. https://doi.org/10.1007/BF02932971.

Kiran, V. E., Aswathi, N. V., & Thomas, T. D. (2025). Micropropagation and conservation of *Rotala malabarica* Pradeep Joseph and Shivarajan, a critically endangered aquatic herb, endemic to Western Ghats. *Plant Cell, Tissue and Organ Culture, 160*, 26. https://doi.org/10.1007/s11240-025-02971-1.

Kitto, S. L., & Janick, J. (1982). Polyox as an artificial seed coat for asexual embryos. *HortScience, 17*, 488.

Koçak, M., Yılmaz, M. C., Kuzğun, C., Sirke, S. T., & Yildiz, M. (2025). Synthetic seed production in *Crataegus monogyna* L. and prediction of regeneration of synthetic seeds with machine learning algorithms. *Plant Cell, Tissue and Organ Culture, 161*, 1–3. https://doi.org/10.1007/s11240-025-03049-8.

Kumar, M. A., Vakeswaran, V., & Krishnasamy, V. (2005). Enhancement of synthetic seed conversion to seedlings in hybrid rice. *Plant Cell, Tissue and Organ Culture, 81*, 97–100. https://doi.org/10.1007/s11240-004-2448-8.

Kumar, S., Rai, M. K., Singh, N., & Mangal, M. (2010). Alginate-encapsulation of shoot tips of jojoba [*Simmondsia chinensis* (Link) Schneider] for germplasm exchange and distribution. *Physiology and Molecular Biology of Plants, 16*, 379–382. https://doi.org/10.1007/s12298-010-0043-5.

Kumar, V., & Chandra, S. (2014). High frequency somatic embryogenesis and synthetic seed production of the endangered species *Swertia chirayita*. *Biologia (Lahore, Pakistan), 69*, 186–192. https://doi.org/10.2478/s11756-013-0305-0.

Lakshmana Rao, P. V., & Singh, B. (1991). Plantlet regeneration from encapsulated somatic embryos of hybrid *Solanum melongena* L. *Plant Cell Reports, 10*, 7–11. https://doi.org/10.1007/BF00233023.

Lalthafamkimi, L., Bhau, B. S., Kumar, S., Mukhia, S., Kumar, R., Banik, D., & Bhattacharyya, P. (2022). Indirect organogenesis-mediated high frequency conversion of non-embryonic synthetic seeds, essential oil profiling and antibacterial activity in genetically stable plants of Patchouli. *3 Biotech, 12*, 349. https://doi.org/10.1007/s13205-022-03302-3.

Loganathan, K., Sankaran, P., Subbiah, S., & Cholairaj, R. (2022). Synthetic seed production and germplasm conservation of *Enicostema axillare* (Lam.) *raynal* ssp. *littoralis* (Blume) *raynal* using nodal and root explants. *Current Applied Science and Technology, 22*, 10–55003. https://doi.org/10.55003/cast.2022.06.22.001.

Mahdavi, Z., Daylami, S. D., Fadavi, A., & Vahdati, K. (2023). Artificial seed production of *Phalaenopsis* orchid: Effect of encapsulation materials, temperature, light spectra, and storage period. *Plant Cell, Tissue and Organ Culture, 155*, 797–808. https://doi.org/10.1007/s11240-023-02600-9.

Manjkhola, S., Dhar, U., & Joshi, M. (2005). Organogenesis, embryogenesis, and synthetic seed production in *Arnebia euchroma*—A critically endangered medicinal plant of the Himalaya. *In Vitro Cellular & Developmental Biology. Plant: Journal of the Tissue Culture Association, 41*, 244–248. https://doi.org/10.1079/IVP2004612.

Manokari, M., & Shekhawat, M. S. (2024). Synseeds of *in vitro* roots as inoculum for storage and mass production of adventitious roots in *Hemidesmus indicus* (L.) R. Br. *Plant Cell, Tissue and Organ Culture, 159*, 51. https://doi.org/10.1007/s11240-024-02889-0.

Maqsood, M., Khusrau, M., Mujib, A., & Kaloo, Z. A. (2021). Synthetic seed technology in some ornamental and medicinal plants: An overview. *Propagation and Genetic Manipulation of Plants*, 19–31. https://doi.org/10.1007/978-981-15-7736-9_2.

Martin, K. P. (2003). Clonal propagation, encapsulation and reintroduction of *Ipsea malabarica* (Reichb. f.) JD Hook., an endangered orchid. *In Vitro Cellular & Developmental Biology. Plant: Journal of the Tissue Culture Association, 39*, 322–326. https://doi.org/10.1079/IVP2002399.

Mendi, N. Y., Özdemir, Ş., Ugur, S., Sevindik, B., İzgu, T., Kaya, E., Tutuncu, M., Çuruk, P., Şimsek, Ö., & Koyuncu, O. (2022). *In vitro* regeneration and synthetic seed production of *Colchicum cilicicum* grown naturally in Turkey. *Alatarım, 21*, 18–26.

Meucci, A., Ghelardi, C., Chietera, G., & Mensuali, A. (2024). Synthetic seed production and slow growth storage of *in vitro* cultured plants of *Iris pallida* Lam. *Horticulturae, 10*, 272. https://doi.org/10.3390/horticulturae10030272.

Mohamed, A. E., Gomaa, A. H., Abd-El-Zaher, M. H., Reda, A. E., & Girgis, N. D. (2023). Protocol for plant regeneration from encapsulated somatic embryos of *Citrus sinensis* L. *Bulletin of the National Research Centre, 47*, 101. https://doi.org/10.1186/s42269-023-01081-2.

Mohanty, P., Nongkling, P., Das, M. C., Kumaria, S., & Tandon, P. (2013). Short-term storage of alginate-encapsulated protocorm-like bodies of *Dendrobium nobile* Lindl.: An endangered medicinal orchid from North-east India. *3 Biotech, 3*, 235–239. https://doi.org/10.1007/s13205-012-0090-4.

Moharana, A., Dwebedi, A. S., Rout, K. K., Naik, S. K., & Barik, D. P. (2024). Non-embryogenic synthetic seed production of *Lawsonia inermis* L. and HPTLC analysis of lawsone. *Vegetos, 14*, 1–2. https://doi.org/10.1007/s42535-024-01023-x.

Murashige, T. (1977). Plant cell and organ cultures as horticultural practices. In *Proceedings of the symposium on tissue culture for horticultural purposes* (Ghent, Belgium, 6–9 September 1977). https://doi.org/10.17660/ActaHortic.1977.78.1

Murashige, T., & Skoog, F. (1962). A revised medium for rapid growth and bio assays with tobacco tissue cultures. *Physiologia Plantarum*, 473–497. http://doi.org/10.1111/j.1399-3054.1962.tb08052.x.

Nagananda, G. S., Satishchandra, N., & Rajath, S. (2011). Regeneration of encapsulated protocorm like bodies of medicinally important vulnerable orchid *Flickingeria nodosa* (Dalz.) Seidenf. *International Journal of Botany, 7*, 310–313. https://doi.org/10.3923/ijb.2011.310.313.

Nagananda, G. S., & Satishchandra, N. (2023). *In vitro* direct plant regeneration from somatic embryos developed from the leaf explants of *Flickingeria nodosa* (Dalz.) Seidenf. through synthetic seed technology. *South African Journal of Botany, 153*, 77–82. https://doi.org/10.1016/j.sajb.2022.12.012.

Negahdar, N., Hashemabadi, D., & Kaviani, B. (2021). *In vitro* conservation and cryopreservation of *Buxus sempervirens* L., a critically endangered ornamental shrub. Russian Journal of Plant Physiology, *68*, 661–668. https://doi.org/10.1134/S1021443721040117.

Nicolae, I. C., Venat, O., Peticilă, A. G., & Hoza, D. (2023). Conservation of plant germplasm using synthetic seed technology-review. *Scientific Papers. Series B. Horticulture, 67*.

Nongdam, P. (2016). Development of synthetic seed technology in plants and its applications: A review. *International Journal of Current Science, 19*, 86–101.

Nyende, A. B., Schittenhelm, S., Mix-Wagner, G., & Greef, J. M. (2003). Production, storability, and regeneration of shoot tips of potato (*Solanum tuberosum* L.) encapsulated in calcium alginate hollow beads. *In Vitro Cellular & Developmental Biology. Plant: Journal of the Tissue Culture Association, 39*, 540–544. https://doi.org/10.1079/IVP2003442.

Oceania, C., Doni, T., Tikendra, L., & Nongdam, P. (2015). Establishment of efficient *in vitro* culture and plantlet generation of tomato (*Lycopersicon esculentum* Mill.) and development of synthetic seeds. *Journal of Plant Sciences, 10*, 15–24. https://doi.org/10.3923/jps.2015.15.24.

Ouzhand, B., Mohayeji, M., Pourseyedi, S., & Abdolshahi, R. (2023). Dormancy breaking, indirect somatic embryogenesis, and encapsulation of somatic embryos in Black Zira. *Plant Cell, Tissue and Organ Culture, 155*, 197–208. https://doi.org/10.1007/s11240-023-02571-x.

Phuong, H. T., & Phung, T. Q. (2025). Encapsulated fronds technology modified for long-term preservation of *Lemna aequinoctialis*. *Plant Cell, Tissue and Organ Culture, 161*, 18. https://doi.org/10.1007/s11240-025-03035-0.

Piątczak, E., & Wysokińska, H. (2013). Encapsulation of *Centaurium erythraea* Rafn–An efficient method for regeneration of transgenic plants. *Acta Biologica Cracoviensia Series Botanica, 55*, 37–44. https://doi.org/10.2478/abcsb-2013-0022.

Piccioni, E. (1997). Plantlets from encapsulated micropropagated buds of M. 26 apple rootstock. *Plant Cell, Tissue and Organ Culture, 47*, 255–260. https://doi.org/10.1007/BF02318980.

Piccioni, E., & Standardi, A. (1995). Encapsulation of micropropagated buds of six woody species. *Plant Cell, Tissue and Organ Culture, 42*, 221–226. https://doi.org/10.1007/BF00029990.

Pintos, B., Bueno, M. A., Cuenca, B., & Manzanera, J. A. (2008). Synthetic seed production from encapsulated somatic embryos of cork oak (*Quercus suber* L.) and automated growth monitoring. *Plant Cell, Tissue and Organ Culture, 95*, 217–225. https://doi.org/10.1007/s11240-008-9435-4.

Pradhan, S., Tiruwa, B., Subedee, B. R., & Pant, B. (2014). *In vitro* germination and propagation of a threatened medicinal orchid, *Cymbidium aloifolium* (L.) Sw. through artificial seed. *Asian Pacific Journal of Tropical Biomedicine, 4*, 971–976. https://doi.org/10.12980/APJTB.4.2014APJTB-2014-0369.

Prasannan, P., Varghese, S., Palanivel, V., Surendran, A., & Ramasubbu, R. (2025). *In vitro* regeneration and conservation of *Elaeocarpus blascoi* Weibel, an endemic and critically endangered tree, using synthetic seeds. *South African Journal of Botany, 184*, 734–747. https://doi.org/10.1016/j.sajb.2025.06.017.

Prewein, C., & Wilhelm, E. (2003). Plant regeneration from encapsulated somatic embryos of pedunculate oak (*Quercus robur* L.). *In Vitro Cellular & Developmental Biology. Plant: Journal of the Tissue Culture Association, 39*, 613–617. https://doi.org/10.1079/IVP2003453.

Rai, M. K., Asthana, P., Singh, S. K., Jaiswal, V. S., & Jaiswal, U. (2009). The encapsulation technology in fruit plants—A review. *Biotechnology Advances, 27*, 671–679. https://doi.org/10.1016/j.biotechadv.2009.04.025.

Rathore, M. S., & Kheni, J. (2017). Alginate encapsulation and *in vitro* plantlet regeneration in critically endangered medicinal plant, *Withania coagulans* (Stocks) Dunal. In: *Proceedings of the National Academy of Sciences, India Section B: Biological Sciences, 87*, 129–134. https://doi.org/10.1007/s40011-015-0577-y.

Ratilal, P.H. (2009). *Tissue culture plant and synthetic seed production of gladiolus variety American Beauty* (Doctoral dissertation, Navsari Agricultural University, ASPEE College of Horticulture and Forestry). http://krishikosh.egranth.ac.in/handle/1/5810044002.

Rawat, J. M., Rawat, B., & Mehrotra, S. (2013). Plant regeneration, genetic fidelity, and active ingredient content of encapsulated hairy roots of *Picrorhiza kurrooa* Royle ex Benth. *Biotechnology Letters, 35*, 961–968. http://doi.org/10.1007/s10529-013-1152-3.

Ray, A., & Bhattacharya, S. (2010). Storage and conversion of *Eclipta alba* synseeds and RAPD analysis of the converted plantlets. *Biologia Plantarum, 54*, 547–550. https://doi.org/10.1007/s10535-010-0097-9.

Redenbaugh, K. (1990). Application of artificial seed to tropical crops. *HortScience, 25*, 251–255.

Redenbaugh, K., Fujii, J., Slade, D., Viss, P., & Kossler, M. (1991). Artificial seeds—Encapsulated somatic embryos. In Y. P. S. Bajaj (Ed.). *High-tech and micropropagation I (Biotechnology in Agriculture and Forestry, Vol. 17)* (pp. 395–416). Springer. https://doi.org/10.1007/978-3-642-76415-8_22.

Regni, L., Micheli, M., Pino, A. M., Facchin, S. L., Rabica, E., Camilloni, L., Cesarini, A., & Proietti, P. (2024). Blackberry synthetic seeds storage: Effects of temperature, time, and sowing substrate. *Plant Cell, Tissue and Organ Culture, 158*, 17. https://doi.org/10.1007/s11240-024-02816-3.

Reinert, J. (1958). Morphogenese und ihre Kontrolle an Gewebekulturen aus Carotten. *The Science of Nature, 45*, 344–345. https://doi.org/10.1007/BF00640240.

Rihan, H. Z., Kareem, F., El-Mahrouk, M. E., & Fuller, M. P. (2017). Artificial seeds (principle, aspects and applications). *Agronomy, 7*, 71.

Roy, B., & Mandal, A. B. (2008). Development of synthetic seeds involving androgenic and pro-embryos in elite indica rice. *Indian Journal of Biotechnology, 7*, 515–519.

Saadat, S., Majd, A., Naseri, L., Iranbakhsh, A., & Jafari, M. (2023). Optimization of somatic embryogenesis, synthetic seed production, and evaluation of genetic fidelity in *Teucrium polium* L. *In Vitro Cellular & Developmental Biology. Plant: Journal of the Tissue Culture Association, 59*, 483–496. https://doi.org/10.1007/s11627-023-10360-6.

Saiprasad, G. V. (2001). Artificial seeds and their applications. *Resonance, 6*, 39–47. https://doi.org/10.1007/BF02839082.

Saiprasad, G. V., & Polisetty, R. (2003). Propagation of three orchid genera using encapsulated protocorm-like bodies. *In Vitro Cellular & Developmental Biology. Plant: Journal of the Tissue Culture Association, 39*, 42–48. https://doi.org/10.1079/IVP2002360.

Salma, U., Kundu, S., Ali, M. N., & Mandal, N. (2019). Somatic embryogenesis-mediated plant regeneration of *Eclipta alba* (L.) Hassk. and its conservation through synthetic seed technology. *Acta Physiologiae Plantarum / Polish Academy of Sciences, Committee of Plant Physiology Genetics and Breeding, 41*, 1. https://doi.org/10.1007/s11738-019-2898-6 -0.

Sarmah, D. K., Borthakur, M., & Borua, P. K. (2010). Artificial seed production from encapsulated PLBs regenerated from leaf base of *Vanda coerulea* Grifft. ex. Lindl.–An endangered orchid. *Current Science, 10*, 686–690. https://www.jstor.org/stable/24111822.

Shaheen, A., & Shahzad, A. (2015). Nutrient encapsulation of nodal segments of an endangered white cedar for studies of regrowth, short term conservation and ethylene inhibitors influenced *ex vitro* rooting. *Industrial Crops and Products, 69*, 204–211. https://doi.org/10.1016/j.indcrop.2015.02.032.

Shallal, H. H., Stănică, F., Peticilă, A. G., Butcaru, A. C., & Nicolae, C. I. (2020). The use of artificial seed technology in the production of horticultural plants. *Scientific Papers. Series B. Horticulture, 64.*

Sharma, N., Gowthami, R., & Pandey, R. (2019). Synthetic seeds: A valuable adjunct for conservation of medicinal plants. *Synthetic Seeds: Germplasm Regeneration, Preservation and Prospects,* 181–216. https://doi.org/10.1007/978-3-030-24631-0_7.

Sharma, S., & Shahzad, A. (2014). Synseed production in *Spilanthes mauritiana* DC. for short-term storage and germplasm exchange. *British Biotechnology Journal, 4*, 696.

Sharma, S., Shahzad, A., & da Silva, J. A. (2013). Synseed technology—A complete synthesis. *Biotechnology Advances, 31*, 186–207. https://doi.org/10.1016/j.biotechadv.2012.09.007.

Sharma, S., Shahzad, A., Kumar, J., & Anis, M. (2014). *In vitro* propagation and synseed production of scarlet salvia (*Salvia splendens*). *Rendiconti Lincei, 25*, 359–368. https://doi.org/10.1007/s12210-014-0308-y.

Sharma, S., Shahzad, A., & Sahai, A. (2009). Artificial seeds for propagation and preservation of *Spilanthes acmella* (L.) Murr., a threatened pesticidal plant species. *International Journal of Plant Developmental Biology, 3*, 62–65.

Shilpha, J., Pandian, S., Largia, M. J., Sohn, S. I., & Ramesh, M. (2021). Short-term storage of *Solanum trilobatum* L. synthetic seeds and evaluation of genetic homogeneity using SCoT markers. *Plant Biotechnology Reports, 15*, 651–661. https://doi.org/10.1007/s11816-021-00709-x.

Siddique, I., & Bukhari, N. A. (2018). Synthetic seed production by encapsulating nodal segment of *Capparis decidua* (Forsk.), *in vitro* regrowth of plantlets and their physio biochemical studies. *Agroforestry Systems, 92*, 1711–1719. https://doi.org/10.1007/s10457-017-0120-7.

Singh, A. K., & Chand, S. (2010). Plant regeneration from alginate encapsulated somatic embryos of *Dalbergia sissoo* Roxb. *Indian Journal of Biotechnology, 9*, 319–324.

Singh, A. K., Sharma, M., Varshney, R., Agarwal, S. S., & Bansal, K. C. (2006b). Plant regeneration from alginate-encapsulated shoot tips of *Phylianthus amarus* Schum and Thonn, a medicinally important plant species. *In Vitro Cellular & Developmental Biology. Plant: Journal of the Tissue Culture Association, 42*, 109–113. https://doi.org/10.1079/IVP2005735.

Singh, A. K., Varshney, R., Sharma, M., Agarwal, S. S., & Bansal, K. C. (2006a). Regeneration of plants from alginate-encapsulated shoot tips of *Withania somnifera* (L.) Dunal, a medicinally important plant species. *Journal of Plant Physiology, 163*, 220–223. https://doi.org/10.1016/j.jplph.2005.06.013.

Singh, S., Tripathi, M. K., Tiwari, S., Tripathi, N., Tejovathi, G., & Ahuja, A. (2021). Encapsulation of nodal segments for propagation and short-term storage of giloe (*Tinospora cordifolia* Willd.): A medicinally important plant species. *Current Journal of Applied Science and Technology, 40*, 15–24. https://doi.org/10.9734/CJAST/2021/v40i3031542.

Sota, V., Benelli, C., Myrselaj, M., Kongjika, E., & Gruda, N. S. (2023). Short-term conservation of *Juglans regia* L. via synthetic seed technology. *Horticulturae, 9*, 559. https://doi.org/10.3390/horticulturae9050559.

Srivastava, V., Khan, S. A., & Banerjee, S. (2009). An evaluation of genetic fidelity of encapsulated microshoots of the medicinal plant: *Cineraria maritima* following six months of storage. *Plant Cell, Tissue and Organ Culture, 99*, 193–198. https://doi.org/10.1007/s11240-009-9593-z.

Standardi, A., & Piccioni, E. (1998). Recent perspectives on synthetic seed technology using nonembryogenic *in vitro*-derived explants. *International Journal of Plant Sciences, 159*, 968–978.

Steward, F. C. (1958). Growth and organized development of cultured cells. II. Organization in cultures grown from freely suspended cells. *American Journal of Botany, 45*, 705–708.

Subrahmanyeswari, T., Laha, S., Kamble, S. N., Singh, S., Bhattacharyya, S., & Gantait, S. (2023). Alginate encapsulation of shoot tips and their regeneration for enhanced mass propagation and germplasm exchange of genetically stable *Stevia rebaudiana* Bert. *Sugar Tech, 25*, 542–551. https://doi.org/10.1007/s12355-022-01194-4.

Sujatha, G., & Kumari, B. R. (2008). Micropropagation, encapsulation and growth of *Artemisia vulgaris* node explants for germplasm preservation. *South African Journal of Botany, 74*, 93–100. https://doi.org/10.1016/j.sajb.2007.09.002.

Taha, R. M., Hasbullah, N. A., & Awal, A. (2008). Production of synthetic seeds from micro shoots and somatic embryos of *Gerbera jamesonii* bolus ex. Hook f. *VI International Symposium on In Vitro Culture and Horticultural Breeding. Acta Hortic,* (829), 91–98. https://doi.org/10.17660/ActaHortic.2009.829.12.

Tailor, A., & Joshi, A. (2024). The role of synthetic seeds in germplasm conservation of forest tree species. In Dennis Thomas T., M. K. Razdan, & Ajay Kumar (Eds.). *Biotechnological approaches for sustaining forest trees and their products* (pp. 103–150). Springer Nature Singapore. https://doi.org/10.1007/978-981-97-4363-6_5.

Talei, D., Shams, A., & Khayam Nekouei, M. (2025). A comprehensive review of *Cannabis* as a crucial pharmaceutical plant and its efficient propagation methods. *ACS Agricultural Science & Technology, 5*, 1191–1214. https://doi.org/10.1021/acsagscitech.4c00483.

Tarek, A. A., Shaaban, S. A., Taha, L. S., Hashish, K. I., Gabr, A. M., & Ali, A. I. (2022). Encapsulation of *Brunfelsia pauciflora* micropropagated shoots as a suitable conservation protocol. *Journal of Pharmaceutical Negative Results, 13*, 1864–1878. https://doi.org/10.47750/pnr.2022.13.S03.283.

Teixeira da Silva, J. A., Cardoso, J. C., Dobránszki, J., & Zeng, S. (2015). *Dendrobium* micropropagation: A review. *Plant Cell Reports, 34*, 671–704. https://doi.org/10.1007/s00299-015-1754-4.

Teixeira da Silva, J. A. (2014). Callus induction in papaya (*Carica papaya* L.) and synseed production for low temperature storage and cryopreservation. *Folia Horticulturae, 26*, 155. https://doi.org/10.1515/fhort-2015-0007.

Tsvetkov, I., Jouve, L., & Hausman, J. F. (2006). Effect of alginate matrix composition on regrowth of *in vitro*-derived encapsulated apical microcuttings of hybrid aspen. *Biologia Plantarum, 50*, 722–724. https://doi.org/10.1007/s10535-006-0115-0.

Upadhyay, R., Kashyap, S. P., Singh, C. S., Tiwari, K. N., Singh, K., & Singh, M. (2014). Ex situ conservation of *Phyllanthus fraternus* Webster and evaluation of genetic fidelity in regenerates using DNA-based molecular marker. *Applied Biochemistry and Biotechnology, 174*, 2195–2208. https://doi.org/10.1007/s12010-014-1175-9.

Varshney, A., & Anis, M. (2014). Synseed conception for short-term storage, germplasm exchange and potentialities of regeneration genetically stable plantlets of desert date tree (*Balanites aegyptiaca* Del.). *Agroforestry Systems, 88*, 321–329. https://doi.org/10.1007/s10457-014-9685-6.

Wagley, L. M., Gladfelter, H. J., & Phillips, G. C. (1987). *De novo* shoot organogenesis of *Pinus eldarica* Medw. *in vitro*: II. Macro-and micro-photographic evidence of *de novo* regeneration. *Plant Cell Reports, 6*, 167–171. https://doi.org/10.1007/BF00268469.

Wang, F. Y., Tang, N., & Ben, A. L. (2011). Study on the preparation technique on synthetic seeds of *Atractylodes macrocephala* Koidz, 2, 48–50.

Wasi, A., Shahzad, A., & Tahseen, S. (2024). Synthesis of nonembryonic synseed, short term conservation, phytochemical evaluation and assessment of genetic stability through SCoT markers in *Decalepis salicifolia*. *Plant Cell, Tissue and Organ Culture, 156*, 101. http://doi.org/10.1007/s11240-024-02719-3.

Winkelmann, T., Meyer, L., & Serek, M. (2004). Germination of encapsulated somatic embryos of *Cyclamen persicum*. *HortScience, 39*, 1093–1097.

Yücesan, B. (2019). Synseed: A new trend in seed technology. In Mohammad Faisal, & Abdulrahman A. Alatar (Eds.). *Synthetic seeds: Germplasm regeneration, preservation and prospects* (pp. 61–75). Springer International Publishing. https://doi.org/10.1007/978-3-030-24631-0_3.

Zych, M., Furmanowa, M., Krajewska-Patan, A., Łowicka, A., Dreger, M., & Mendlewska, S. (2005). Micropropagation of *Rhodiola kirilowii* plants using encapsulated axillary buds and callus. *Acta Biologica Cracoviensia Series Botanica, 47*, 83–87.

CHAPTER

6

Biotechnological strategies to enhance phenolic compound biosynthesis in medicinal plants

Amanda Nascimento[1,2] *and Ana Novo Barros*[3]

[1]Academic Unit of Food Engineering, Federal University of Campina Grande (UFCG), Aprígio Veloso Avenue, 882, Campina Grande, PB, Brazil [2]SENAI Institute of Technology for Operational Efficiency (IST EO), Federation of Industries of the State of Paraíba (FIEPB), Campina Grande, PB, Brazil [3]Centre for the Research and Technology of Agro-environmental and Biological Sciences, CITAB, Inov4Agro, Universidade deTrás-os-Montes e Alto Douro, UTAD, Quinta de Prados, Vila Real, Portugal

6.1 Introduction

Medicinal plants are widely recognized as prolific sources of bioactive compounds with remarkable pharmacological and nutraceutical potential. Among these, phenolic compounds stand out as one of the most structurally diverse and biologically significant classes of secondary metabolites (Kisiriko et al., 2021; Kumar et al., 2022; Riaz et al., 2023).

Phenolics play essential physiological roles in plants, including defense against pathogens, ultraviolet protection, and regulation of oxidative stress. In humans, they are associated with antioxidant, antiinflammatory, antimicrobial, and cardioprotective properties (Merecz-Sadowska et al., 2025; Rajesh et al., 2024; Sun & Shahrajabian, 2023; Zheng et al., 2025). Because of their dual ecological and therapeutic importance, phenolic compounds have attracted growing interest from the pharmaceutical, cosmetic, and food industries (Sadeghi et al., 2024; Sun & Shahrajabian, 2023).

However, their biosynthesis and accumulation in plants are strongly influenced by pedigree and genetics, developmental, and environmental factors, which limit consistent production under field conditions (Gallego-Narbón et al., 2024; Yu et al., 2025; Zagoskina et al., 2023). Seasonal variation, plant maturity, nutrient availability, and abiotic stressors can cause significant fluctuations in both the quantity and qualitative profile of phenolics. These limitations have intensified interest in biotechnological strategies that provide controlled,

Medicinal Plant Biotechnology for Ameliorated Secondary Metabolite Production
DOI: https://doi.org/10.1016/B978-0-443-43944-5.00006-2

sustainable, and high-yield production systems (Martínez-Chávez et al., 2024; Singh et al., 2023; Vojnović et al., 2024; Yusoff et al., 2022).

Plant biotechnology offers a versatile platform to overcome the constraints of conventional cultivation by enabling targeted manipulation of the biosynthetic pathways responsible for secondary metabolite formation (Kowalczyk et al., 2020; Reshi et al., 2023). *In vitro* culture systems—including callus, cell suspension, and hairy root cultures—allow continuous production of phenolics, independent of climatic or geographic variability (Bai et al., 2025; Boobalan et al., 2024; Clapa et al., 2025; Lee et al., 2019). Moreover, these systems enable fine control over growth regulators, nutrient composition, and elicitation treatments, enhancing both metabolic yield and reproducibility (Humbal & Pathak, 2023; Nguyen et al., 2025; Pasternak & Steinmacher, 2024).

An overview of these integrated strategies is presented in Fig. 6.1, which illustrates the transition from environmental variability to controlled biotechnological systems that enable the enhanced production of phenolic compounds.

Several medicinal and aromatic plants (MAPs) have been extensively studied for their phenolic profiles under both conventional and biotechnological production systems, including *Rosmarinus officinalis*, *Thymus vulgaris*, *Origanum vulgare*, *Salvia officinalis*, *Matricaria chamomilla*, and *Mentha* spp. (de Elguea-Culebras et al., 2023; Giordano et al., 2025; Ivanova et al., 2025; Mokhtari et al., 2023). These species are known for high levels of rosmarinic acid, caffeic acid derivatives, flavonoids, and phenolic terpenes, but their yields can vary upto two- to fourfold depending on environmental pressure or postharvest handling. *In vitro* platforms often show increased recovery of specific phenolic(s); for instance, elicited hairy root cultures can serve as highly efficient sources of phenolic antioxidants (Malarz et al., 2023), while callus cultures of *Curcuma longa* exhibit enhanced curcuminoid accumulation under optimized conditions (Iqbal et al., 2024). Similarly, optimized shoot and callus systems of *Hypericum perforatum* show increased production of phenolic acids and related metabolites (Kwiecień et al., 2015). Such examples highlight how biotechnological approaches not only stabilize production but can also increase overall recovery and purity of target compound(s), providing a reliable alternative to traditional *in vivo* cultivation.

Recent developments in molecular biology and omics technologies have further expanded the potential for phenolic enhancement (Choudhary et al., 2025; Gómez-Rocal et al., 2025; Machitescu et al., 2024). Metabolic engineering, coupled with transcriptomic and proteomic analyses, enables the identification and regulation of key genes and enzymes involved in the shikimate and phenylpropanoid pathways (Dong & Lin, 2020; Khan & Boehr, 2025; Kukil & Lindberg, 2024; Utomo et al., 2024). Furthermore, CRISPR/Cas-based genome editing tools have opened new perspectives for the targeted modification of biosynthetic networks, offering unprecedented precision in enhancing metabolite fluxes (Devi et al., 2023; Dhir et al., 2024; Utomo et al., 2021).

Within this context, the integration of biotechnological, molecular, and process-based strategies represents a promising route toward the sustainable exploitation of medicinal plants as renewable biofactories for phenolic compounds (Fierăscu et al., 2020; Guo et al., 2023; Kisiriko et al., 2021; Mohamed et al., 2022). This chapter provides an overview of the biosynthetic mechanisms underlying phenolic production and examines recent advances in plant biotechnology aimed at enhancing their accumulation, with emphasis on *in vitro*

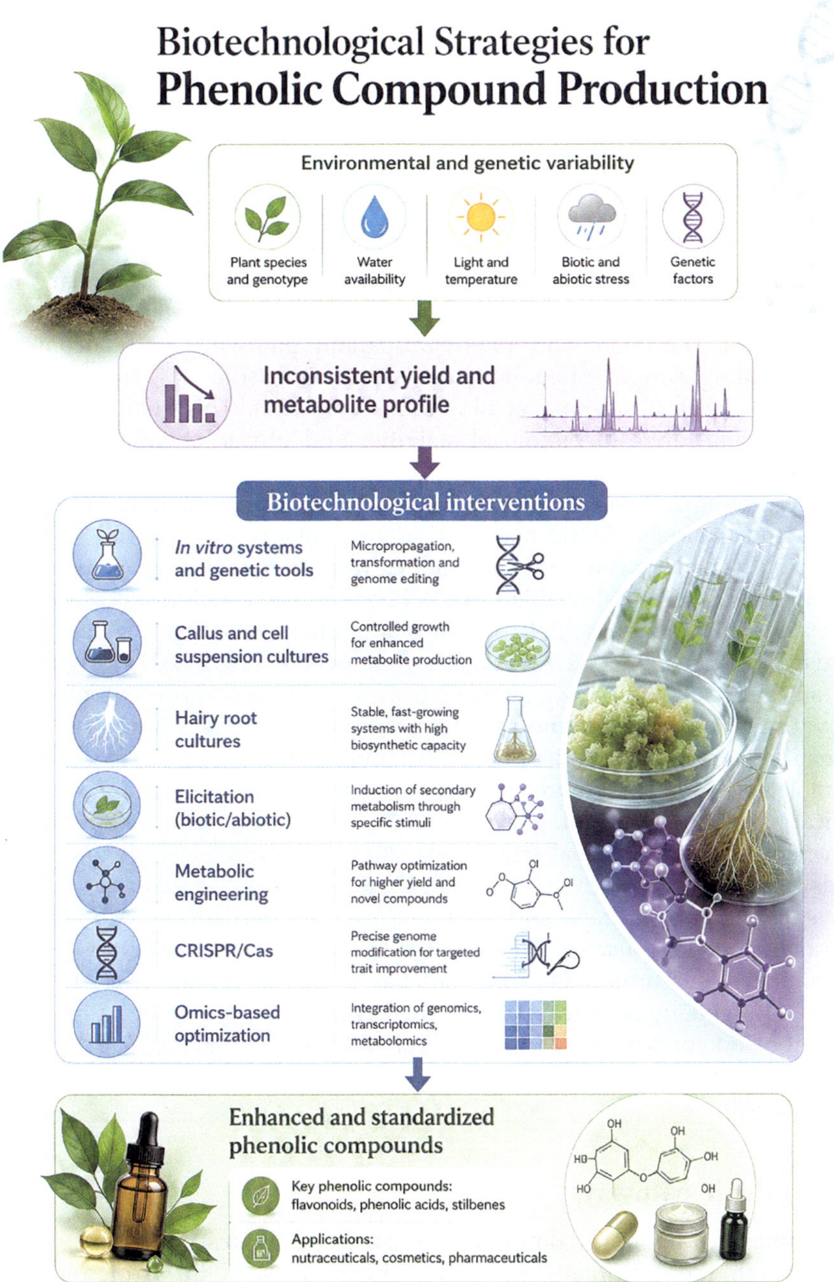

FIGURE 6.1 Conceptual overview of strategies for enhancing phenolic compound production in medicinal plants.

culture systems, elicitation strategies, and metabolic engineering approaches, particularly in relation to mechanistic insights, scalability, and industrial applicability.

6.2 Biosynthesis of phenolic compounds

6.2.1 General classification and chemical diversity

Phenolic compounds comprise a structurally diverse family of secondary metabolites characterized by one or more hydroxyl groups attached to aromatic rings. They originate primarily from the shikimate and phenylpropanoid pathways and encompass several subclasses, including simple phenolic acids, flavonoids, stilbenes, lignans, and tannins (Marchiosi et al., 2020; Zagoskina et al., 2023; Zeb, 2021). This structural heterogeneity underlies their wide range of biological activities and physicochemical properties (Chen et al., 2024; Sadeghi et al., 2024; Tang et al., 2024; Zhang et al., 2024).

Among these groups, flavonoids, such as flavonols, flavones, flavanones, and anthocyanins, are the most widely distributed in medicinal plants and are key determinants of pigmentation, ultraviolet protection, and antioxidant defense (Jiang & Grotewold, 2024; Jomová et al., 2025; Sun & Shahrajabian, 2023). Phenolic acids, including caffeic, chlorogenic, and ferulic acids, play fundamental roles in plant defense mechanisms and exhibit strong radical-scavenging capacity (Khan et al., 2024; Manivel & Chen, 2021; Mughal et al., 2024; Zhong et al., 2025). Stilbenes (e.g., resveratrol) and lignans represent more specialized phenolics with potent health-promoting effects in both plants and humans.

The chemical versatility of phenolic compounds enables them to act as antioxidants, signaling molecules, and modulators of plant–microbe interactions, thereby contributing to physiological regulation and ecological adaptation (Al-Khayri et al., 2023; Mendonça et al., 2024; Runeberg et al., 2022; Valletta et al., 2021). The chemical diversity of phenolic compounds can be illustrated by the heterogeneous profiles observed across several MAPs. For instance, *Rosmarinus officinalis* and *Salvia officinalis* are particularly rich in phenolic acids such as rosmarinic acid; *Camellia sinensis* contains high levels of catechins, while *Curcuma longa* accumulates curcuminoids with notable antioxidant capacity. These examples demonstrate how genetic background and predominant biosynthetic pathways shape the phenolic composition of different MAPs (de Elguea-Culebras et al., 2023; Jiang & Grotewold, 2024; Manivel & Chen, 2021; Mokhtari et al., 2023; Mughal et al., 2024; Sun & Shahrajabian, 2023).

6.2.2 Biosynthetic pathways

The biosynthesis of phenolic compounds in plants originates from primary metabolism, primarily through the shikimate pathway, which provides the aromatic amino acids phenylalanine, tyrosine, and tryptophan. Among these, phenylalanine serves as the main precursor of the phenylpropanoid pathway, a central route in phenolic biosynthesis (Rehan, 2021; Santos-Sánchez and Salas-Coronado, 2019; Shende et al., 2024; Zagoskina et al., 2023).

The first committed step in this pathway is catalyzed by phenylalanine ammonia-lyase (PAL), which deaminates phenylalanine to form cinnamic acid. This intermediate subsequently undergoes hydroxylation and activation by cinnamate-4-hydroxylase (C4H) and 4-coumarate-CoA ligase (4CL), generating p-coumaroyl-CoA, a key branch-point molecule that connects multiple downstream biosynthetic routes (Huang et al., 2024; Santiago et al., 2024; Schröpfer et al., 2025; Schwarze & Petersen, 2024).

From this metabolic node, several specialized branches diverge, leading to the formation of distinct phenolic subclasses: the flavonoid pathway (via chalcone synthase, CHS), the stilbene pathway (via stilbene synthase), and the lignan pathway (mediated by dirigent proteins and peroxidases) (Lewis et al., 2024; Tong et al., 2024). This intricate network is finely regulated by enzyme specificity, substrate availability, and subcellular compartmentalization, ensuring the precise synthesis of phenolic derivatives in response to developmental and environmental cues (Catoni et al., 2025; Liu et al., 2023; Sharma et al., 2019).

Differences among MAPs are also reflected in the relative predominance of the phenylpropanoid, flavonoid, or complex polyphenol pathways. In *Ginkgo biloba*, the flavonoid pathway is especially active, leading to the accumulation of characteristic ginkgolides and bilobalide, whereas in *Echinacea purpurea*, the biosynthesis of caffeic acid derivatives, particularly chicoric acid, is more pronounced. These metabolic distinctions help explain the diverse therapeutic applications attributed to different species (de Elguea-Culebras et al., 2023; Jiang & Grotewold, 2024; Manivel & Chen, 2021; Mokhtari et al., 2023; Mughal et al., 2024; Sun & Shahrajabian, 2023).

6.2.3 Key enzymes and regulatory genes

The regulation of phenolic metabolism results from a complex interaction between structural genes that encode biosynthetic enzymes and transcription factors that control their expression. Structural genes such as PAL, C4H, 4CL, CHS, chalcone isomerase (CHI), flavanone-3-hydroxylase (F3H), flavonol synthase (FLS), and dihydroflavonol-4-reductase (DFR) constitute the core enzymatic machinery involved in the synthesis of phenolic compounds (Kaur et al., 2025; Salami et al., 2024; Wang et al., 2024).

The transcriptional regulation of these pathways is primarily mediated by the MYB, bHLH, and WD40 families of transcription factors. These regulators often form MYB–bHLH–WD40 (MBW) complexes that act as central control units, activating or repressing key genes in the phenylpropanoid pathway (Bulanov et al., 2025). Hormonal signals, such as jasmonic acid, salicylic acid (SA), and ethylene, interact with environmental stimuli including light, temperature, and abiotic or biotic stress to fine-tune gene expression. This integration allows plants to adjust phenolic accumulation in response to changing environmental conditions (Ghosh & Roychoudhury, 2024; Song et al., 2023).

Recent transcriptomic and proteomic analyses have revealed multiple isoforms and paralogs of the enzymes involved, highlighting the metabolic flexibility and evolutionary adaptation of phenolic pathways across plant species.

The activity of key enzymes such as PAL, CHS, C4H, or F3H can vary markedly among MAPs, reflecting species-specific metabolic regulation (Ninkuu et al., 2025). For example, *Hypericum perforatum* exhibits high CHS expression associated with the production of flavonols and benzophenone derivatives, while *Ocimum basilicum* shows strong PAL activity that promotes the accumulation of rosmarinic acid. Such differences underscore how gene regulation ultimately shapes phenolic profiles.

6.2.4 Regulation of phenolic metabolism under stress conditions

Phenolic metabolism responds dynamically to both biotic and abiotic stresses. Exposure to pathogens, herbivory, ultraviolet radiation, drought, or nutrient limitation often stimulates the accumulation of specific phenolic subclasses as part of the plant's defense response. Reactive oxygen species (ROS) generated under these conditions act as secondary messengers, activating defense-related transcription factors and inducing the expression of genes within the phenylpropanoid pathway (Jan et al., 2021; Wawrosch & Zotchev, 2021; Zhao et al., 2023).

Ultraviolet-B (UV-B) radiation enhances flavonoid biosynthesis, functioning as a photoprotective mechanism, whereas pathogen attack promotes phytoalexin formation through the activation of PAL and related enzymes (Hao et al., 2022; Singh et al., 2023). Likewise, elicitation by jasmonates or SA increases the synthesis of caffeic and ferulic acid derivatives, further reinforcing plant defense responses (Jeyasri et al., 2023).

Understanding these molecular regulatory mechanisms provides a foundation for the targeted manipulation of phenolic biosynthesis through biotechnological tools, particularly via elicitation and metabolic engineering approaches discussed in subsequent sections.

The influence of environmental stress on phenolic metabolism is well documented in several MAPs. In Thymus vulgaris, UV-B exposure enhances the accumulation of phenolic compounds and flavonoids (Ramírez-Alvarado et al., 2020). In Melissa officinalis, moderate water deficit has been associated with increased rosmarinic acid content (Radácsi et al., 2016). In *Artemisia annua*, salinity combined with SA treatment can enhance the production of the sesquiterpene lactone artemisinin, illustrating how stress signals interact with specialized metabolism (Aftab et al., 2011; Li et al., 2014). These examples highlight how stress-driven metabolic adjustments are species-dependent.

6.3 *In vitro* culture systems for phenolic production

In vitro culture systems provide a controlled platform for enhancing the production, stability, and recovery of phenolic compounds from MAPs (Murthy et al., 2014). Depending on the plant species and culture conditions, substantial differences may occur in metabolic productivity, extraction efficiency, and phenolic profiles. Several studies have demonstrated that *in vitro* cultures frequently outperform field-grown plants in both phenolic yield and consistency, particularly when elicitation or targeted optimization is applied (Ramirez-Estrada et al., 2016).

6.3.1 Callus and cell suspension cultures

Callus and cell suspension cultures are widely used for phenolic production due to their rapid growth and ease of manipulation (Jomová et al., 2025; Sun & Shahrajabian, 2023).

Phenolic yields can vary considerably among species. In *Salvia miltiorrhiza*, callus cultures enriched in tanshinones and phenolic acids have shown 2–3-fold increases in total phenolic content when exposed to appropriate plant growth regulators (Manivel & Chen, 2021). Cell cultures of *Vitis vinifera* have demonstrated higher recovery of anthocyanins compared with field-grown tissues under optimized nutrient conditions (Jiang & Grotewold, 2024; Mughal et al., 2024). In *Hypericum perforatum*, suspension cultures produce significantly higher flavonoid levels than callus tissue (Kwiecień et al., 2015). These examples highlight that callus and suspension cultures often achieve higher extractability and metabolic productivity, especially when combined with optimized hormone regimes.

An overview of the main *in vitro* systems employed for phenolic production is presented in Fig. 6.2 illustrating the technological progression from natural plant systems to scalable bioreactors.

Plant cell culture technology represents a robust and versatile platform for the production of phenolic compounds under controlled and reproducible conditions. Callus cultures, derived from dedifferentiated plant tissues, are frequently employed as the initial material to establish cell suspension cultures capable of continuous metabolite synthesis (Mohamed et al., 2022; Smetanska, 2018).

The biosynthetic performance of these cultures depends on several critical parameters, including the choice of explant, the composition of the nutrient medium, the type and concentration of plant growth regulators, and physical conditions such as light, temperature, and agitation (Valletta et al., 2021; Razzaq et al., 2022). Optimized ratios of auxins (2,4-D, NAA) and cytokinins (BA, kinetin) can substantially influence phenolic accumulation by modulating the cellular differentiation state and the activity of key enzymes within the phenylpropanoid pathway (Zhang et al., 2024).

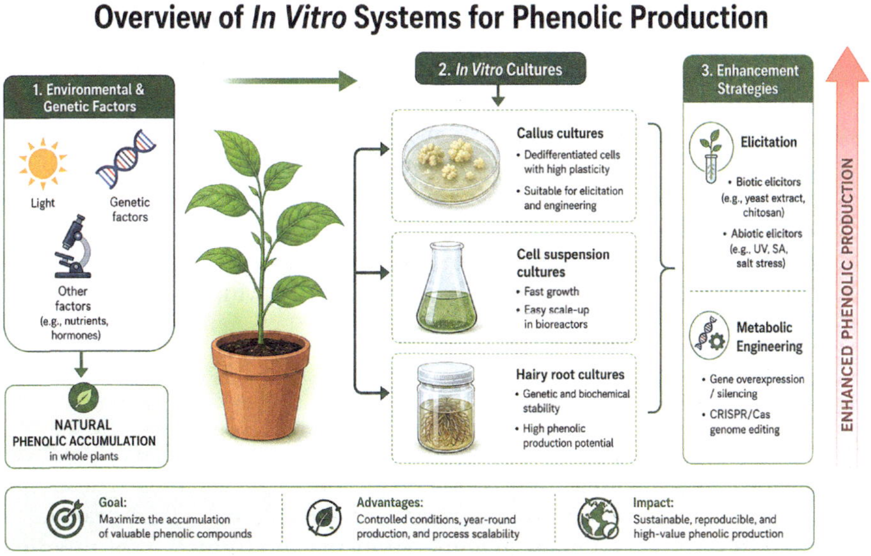

FIGURE 6.2 Overview of *in vitro* systems for phenolic production.

Suspension cultures, in particular, offer significant advantages in terms of scalability and uniform exposure to elicitors and nutrients, making them suitable for both mechanistic studies and industrial applications. However, phenolic productivity frequently decreases during repeated subculturing, necessitating the periodic re-selection of high-yielding cell lines or the use of elicitors to sustain metabolic activity and secondary metabolite biosynthesis (Al-Khayri et al., 2023; Arya et al., 2022).

6.3.2 Organ and tissue cultures

Organ-specific cultures, including shoot, root, and leaf systems, retain partial tissue differentiation, enabling the production of organ-dependent phenolic profiles that differ from those of dedifferentiated cells. Root cultures, in particular, tend to accumulate higher levels of phenolic acids and lignans due to the natural localization of phenylpropanoid metabolism in root tissues. Similarly, shoot cultures have been successfully employed to enhance flavonoid biosynthesis by exploiting their photosynthetic activity and responsiveness to light quality (Umar et al., 2022; Malarz et al., 2023).

The preservation of morphogenic potential in organ cultures often results in more consistent metabolite yields over time compared with callus or suspension systems. By optimizing growth regulators, carbon sources, and light regimes, it is possible to redirect metabolic fluxes toward desired phenolic subclasses. Therefore organ cultures represent an important intermediate platform between conventional field cultivation and fully dedifferentiated cell systems (Jimenez; Khanam et al., 2022; Babich et al., 2020; Mohamed et al., 2022).

Organ cultures (including shoots, roots, and leaves) maintain partial or full tissue differentiation, which often results in phenolic profiles more similar to those found in intact plants. Several studies report that differentiated tissues frequently accumulate higher or more stable levels of phenolic compounds compared with undifferentiated callus, especially when culture conditions, growth regulators, or elicitors are optimized. For instance, organ cultures of aromatic species such as Rosmarinus and Ocimum commonly exhibit enhanced rosmarinic acid accumulation under controlled *in vitro* environments, while root- and shoot-derived cultures of medicinal plants like Echinacea can produce substantial amounts of caffeic acid derivatives. Similarly, differentiated cultures often respond strongly to environmental cues (e.g., light, hormones, precursors), leading to improved metabolic stability and higher phenolic recovery. Altogether, organ cultures offer chemically stable, species-specific profiles and constitute a reliable platform for moderate to high metabolite productivity (Espinosa-Leal et al., 2018; Murthy et al., 2014; Ramirez-Estrada et al., 2016).

6.3.3 Hairy root cultures

Among all *in vitro* systems, hairy root cultures are recognized as one of the most efficient and genetically stable platforms for phenolic compound production. These cultures are established through transformation with *Agrobacterium rhizogenes*, which transfers a portion of its Ri plasmid DNA into the plant genome, inducing the formation of rapidly growing, highly branched roots with autonomous growth capacity (Sharma et al., 2019).

Hairy roots maintain stable secondary metabolism over extended culture periods without the need for exogenous growth regulators and often exhibit enhanced phenolic accumulation compared with nontransformed roots. For example, *Rosmarinus officinalis* hairy roots show elevated levels of rosmarinic acid, whereas *Hypericum perforatum* and *Ocimum basilicum* produce increased quantities of flavonoids and phenolic acids under optimized conditions.

Their genetic and biochemical stability makes hairy roots ideal models for studying regulatory mechanisms of phenolic biosynthesis, testing elicitors, and scaling up metabolite production in bioreactors. Furthermore, they provide a valuable system for integrating metabolic engineering and omics-based approaches to enhance phenolic pathway fluxes (Malarz et al., 2023; Zhang et al., 2024).

Hairy root cultures generated through *Agrobacterium rhizogenes* transformation are widely recognized as highly efficient platforms for the production of phenolic compounds, mainly due to their rapid growth, genetic stability, and remarkable biosynthetic capacity. Numerous studies have demonstrated that transformed roots frequently outperform nontransformed roots in the accumulation of phenolic metabolites. This has been reported in species such as *Withania somnifera* and *Taraxacum officinale*, where enhanced levels of total phenolics and hydroxycinnamic acids were associated with optimized cultivation conditions. In *Origanum vulgare* and other members of the Lamiaceae family, significant increases in phenolic metabolites, particularly rosmarinic acid, have also been observed following elicitation treatments, especially with jasmonates. Moreover, the sustained productivity of hairy root cultures over prolonged cultivation cycles further highlights their potential as reliable and scalable systems for the continuous production and recovery of high-value phenolic compounds (Malarz et al., 2023; Ramirez-Estrada et al., 2016).

6.3.4 Bioreactor systems for large-scale production

The successful transition from laboratory-scale cultures to industrial production requires bioreactor systems capable of sustaining high biomass productivity and metabolic activity. Various configurations, such as stirred-tank, airlift, and bubble column bioreactors, have been adapted for the cultivation of plant cells and hairy roots. Critical operational parameters, including aeration, pH, dissolved oxygen, nutrient availability, and shear stress, must be carefully optimized to prevent cellular damage while promoting metabolite accumulation (Bai et al., 2025; Bruat et al., 2025).

Recent advances in process control, integrating sensor feedback and real-time monitoring of phenolic content, have significantly improved process reproducibility. Immobilization of cells or roots in polymeric matrices, along with continuous feeding of precursors such as phenylalanine or cinnamic acid, can further enhance production yields. However, large-scale phenolic biosynthesis remains challenging due to the complexity of secondary metabolism and the delicate balance between growth and differentiation (Rattan and Kumar, 2022). Despite these constraints, bioreactor-based culture systems remain a cornerstone of modern plant biotechnology, offering scalable and sustainable solutions for the production of high-value phenolics with pharmaceutical and nutraceutical applications (Mohamed et al., 2022; Motolinía-Alcántara et al., 2021; Verdú-Navarro et al., 2023).

A comparative summary of the principal *in vitro* and bioreactor systems used for phenolic compound production is presented in Table 6.1, highlighting their operational principles, advantages, and limitations.

Scaling up *in vitro* cultures to bioreactor systems is fundamental for achieving production levels compatible with industrial applications, particularly when the target compounds are phenolics. In *Panax ginseng*, cell cultures grown in controlled bioreactor conditions have shown marked increases in total phenolic yield, largely due to precise regulation of aeration and agitation that enhances both biomass accumulation and metabolic activity (Thanh et al., 2014). Large-scale systems have also been successfully applied to other medicinal species, where optimized bioreactor configurations contribute to higher metabolite productivity and more efficient extraction processes. Reviews focusing on adventitious root cultures emphasize that bioreactors provide stable, reproducible platforms capable of sustained phenolic production over extended cycles, making them highly suitable for industrial standardization (Khanam et al., 2022). Overall, bioreactor-based approaches offer a robust strategy to increase yield, improve extraction efficiency, and support scalable downstream processing.

6.4 Elicitation strategies

Elicitation is one of the most effective and widely used biotechnological approaches to enhance phenolic biosynthesis in MAPs. By simulating biotic or abiotic stress, elicitors activate defense-related metabolic pathways, inducing measurable increases in both the concentration and diversity of phenolic compounds. The magnitude of the response depends on the species, the type of elicitor, and the physiological state of the culture (Humbal & Pathak, 2023; Jeyasri et al., 2023; Kandoudi et al., 2022; Ramirez-Estrada et al., 2016).

6.4.1 Concept and mechanisms of elicitation

Elicitation refers to the process by which specific stimuli, known as elicitors, trigger or enhance the biosynthesis of secondary metabolites in plant cells and tissues. This strategy exploits the natural defense mechanisms of plants, in which the perception of stress signals activates a cascade of biochemical responses leading to the accumulation of protective compounds such as phenolics, alkaloids, and terpenoids (Selwal et al., 2025; Jeyasri et al., 2023). Upon recognition of elicitors by cell-surface or intracellular receptors, a series of early signaling events occurs, including ion fluxes, generation of ROS, and activation of mitogen-activated protein kinase (MAPK) cascades. These signaling pathways ultimately regulate the transcription of key biosynthetic genes, particularly those involved in the phenylpropanoid and flavonoid pathways. The temporal and spatial dynamics of these responses determine the type and concentration of phenolic compounds synthesized (Pratyusha & Sarada, 2022; Zhang et al., 2024).

Elicitation therefore represents a nontransgenic, cost-effective, and environmentally sustainable approach to enhance metabolite yields, applicable across various *in vitro*

TABLE 6.1 Comparative overview of *in vitro* culture and bioreactor systems for phenolic compound production.

System type	Main characteristics	Advantages	Limitations	Representative references
Callus culture	Dedifferentiated tissue grown on solid media; often used as source for cell suspensions.	Easy to initiate; good for screening elicitors and hormones.	Low metabolite yield; genetic instability with subcultures.	Smetanska (2018), Mohamed et al. (2022)
Cell suspension culture	Cells grown in liquid medium under agitation.	Scalable; uniform nutrient and elicitor exposure.	Loss of productivity over time; shear sensitivity.	Pasternak and Steinmacher (2024), Rattan and Kumar (2022)
Organ culture	Retain partial differentiation (root, shoot, leaf).	Stable metabolite profile; mimic natural synthesis.	Slower growth; less scalable.	Hussain et al. (2022), Mohamed et al. (2022)
Hairy root culture	Agrobacterium rhizogenes-induced roots with autonomous growth.	High stability; enhanced phenolic production.	Genetic variability; difficult to scale.	Boobalan et al. (2024), Bruat et al. (2025)
Bioreactor system	Controlled cultivation (stirred-tank, airlift, bubble column).	High biomass productivity; process automation.	Complex scale-up; cost-intensive optimization.	Bai et al. (2025), Verdú-Navarro et al. (2023), Motolinía-Alcántara et al. (2021)

TABLE 6.2 Main classes of elicitors used to stimulate phenolic compound biosynthesis in plant *in vitro* systems.

Elicitor type	Examples	Mode of action/Effect on phenolic pathway	Representative references
Biotic	Chitosan, yeast extract, pectin, fungal extracts	Induce oxidative burst, activate PAL, enhance flavonoid biosynthesis	Bruat et al. (2025)
Abiotic (chemical)	Methyl jasmonate (MeJA), salicylic acid (SA), hydrogen peroxide	Activate MAPK signaling, increase ROS, upregulate CHS and C4H	Jeyasri et al. (2023)
Abiotic (physical)	UV-B radiation, temperature stress, osmotic stress	Stimulate photoprotective phenylpropanoids, induce flavonoid accumulation	Singh et al. (2023)
Combined elicitation	MeJA + chitosan, SA + UV-B	Synergistic activation of MYB/bHLH regulators and ROS pathways	Kandoudi et al. (2022), Bulanov et al. (2025)

PAL, Phenylalanine ammonia-lyase; *MAPK*, mitogen-activated protein kinase; *ROS*, reactive oxygen species.

systems, including callus, suspension, and hairy root cultures (Boobalan et al., 2024; Bruat et al., 2025; Humbal & Pathak, 2023).

A summary of the main classes of elicitors, their mechanisms of action, and representative examples is presented in Table 6.2.

6.5 Language change in the highlighted portion

Hairy root cultures generated through *Agrobacterium rhizogenes* transformation are widely recognized as highly efficient platforms for the production of phenolic compounds, mainly due to their rapid growth, genetic stability, and remarkable biosynthetic capacity. Numerous studies have demonstrated that transformed roots frequently outperform non-transformed roots in the accumulation of phenolic metabolites. This has been reported in species such as *Withania somnifera* and *Taraxacum officinale*, where enhanced levels of total phenolics and hydroxycinnamic acids were associated with optimized cultivation conditions. In *Origanum vulgare* and other members of the Lamiaceae family, significant increases in phenolic metabolites, particularly rosmarinic acid, have also been observed following elicitation treatments, especially with jasmonates. Moreover, the sustained productivity of hairy root cultures over prolonged cultivation cycles further highlights their potential as reliable and scalable systems for the continuous production and recovery of high-value phenolic compounds. Elicitors stimulate plant defense responses and activate key enzymes involved in the phenylpropanoid and flavonoid biosynthetic pathways, including phenylalanine ammonia-lyase (PAL), chalcone synthase (CHS), and polyphenol oxidase (PPO), thereby promoting the synthesis and accumulation of phenolic compounds. In medicinal species such as *Ocimum sanctum*, *Leonurus cardiaca*, and *Scutellaria lateriflora*, different elicitation strategies have been associated with enhanced production of flavonoids and phenolic acids, accompanied by measurable improvements in antioxidant capacity. The magnitude of these responses depends on factors such as plant species, elicitor type, concentration, exposure time, and physiological status of the culture. Nevertheless, the available evidence consistently demonstrates that elicitation represents an effective and robust strategy for intensifying phenolic metabolism in *in vitro* systems (Humbal & Pathak, 2023; Jeyasri et al., 2023; Kandoudi et al., 2022; Ramirez-Estrada et al., 2016).

6.5.1 Biotic elicitors

Biotic elicitors originate from living organisms or their metabolic products, including microbial cell wall fragments, polysaccharides, and signaling molecules. They mimic pathogen attacks, activating plant defense mechanisms and stimulating secondary metabolism (Boobalan et al., 2024; Jeyasri et al., 2023).

Among the most widely used biotic elicitors are chitosan, yeast extract, pectin fragments, and bacterial lipopolysaccharides. Chitosan, a deacetylated derivative of chitin, is particularly effective in enhancing the synthesis of flavonoids and phenolic acids through the induction of oxidative burst and activation of PAL. Similarly, yeast extract and fungal elicitors derived from *Aspergillus niger* or *Rhizopus oryzae* have been shown to upregulate phenylpropanoid-related genes and increase total phenolic content in cell cultures of *Ocimum basilicum*, *Daucus carota*, and *Vitis vinifera* (Bai et al., 2025; Bruat et al., 2025; Hussain et al., 2022).

The elicitor concentration, exposure time, and developmental stage of the culture are key parameters determining the magnitude and specificity of the metabolic response. In several systems, synergistic combinations of biotic elicitors have been employed to maximize phenolic accumulation by activating multiple defense-related signaling pathways (Kandoudi et al., 2022; Razzaq et al., 2025).

Biotic elicitors, including yeast extract, chitosan, fungal polysaccharides, and bacterial cell wall fragments, are widely employed to activate defense-related pathways and stimulate phenolic biosynthesis in *in vitro* plant systems. In cell cultures of *Echinacea purpurea*, yeast extract treatments have been associated with increased accumulation of caffeic acid derivatives, accompanied by improved extraction efficiency. In callus cultures of *Salvia officinalis*, chitosan applications have promoted the accumulation of rosmarinic acid, whereas fungal-derived elicitors applied to *Glycyrrhiza glabra* cultures have been shown to enhance total phenolic and flavonoid contents, frequently in parallel with increased antioxidant activity. Overall, these elicitors trigger rapid and intense metabolic responses, making biotic elicitation a particularly suitable strategy for short-term enhancement of phenolic compound production in plant culture systems (Humbal & Pathak, 2023; Jeyasri et al., 2023; Ramirez-Estrada et al., 2016).

6.5.2 Abiotic elicitors

Abiotic elicitors include physical or chemical factors that induce controlled stress in plant cells, thereby enhancing secondary metabolism. Representative examples comprise UV radiation, temperature shifts, osmotic stress, heavy metals, and plant hormones such as SA and jasmonic acid (Hao et al., 2022; Jeyasri et al., 2023).

Among these, methyl jasmonate (MeJA) and SA are widely recognized as potent inducers of phenolic biosynthesis. These molecules act as endogenous signaling mediators that activate transcription factors associated with defense-related genes, leading to increased expression of PAL, C4H, and CHS. Exposure to UV-B radiation similarly promotes flavonoid and anthocyanin accumulation through photoreceptor-mediated signaling, while osmotic stress induced by mannitol or NaCl redirects carbon flux toward secondary metabolism as part of adaptive stress responses (Chen et al., 2024; Singh et al., 2023).

Although abiotic elicitation can markedly enhance phenolic content, the process must be carefully optimized to avoid excessive stress that could compromise cell viability or reduce biomass productivity (Arya et al., 2022; Kandoudi et al., 2022; Yusoff et al., 2022).

Abiotic elicitors, including UV-B radiation, salicylic acid, jasmonates, salt stress, and other osmotic agents, frequently promote substantial increases in phenolic compound biosynthesis by modulating oxidative and hormonal signaling pathways. In *Rosmarinus officinalis*, for instance, cultures exposed to UV-B radiation exhibit significant increases in total phenolic content and rosmarinic acid levels, reflecting the activation of stress-responsive metabolic pathways associated with light exposure (Zhao et al., 2024). In *Artemisia annua*, salicylic acid treatment under stress conditions has been shown to enhance the production of secondary metabolites, including phenolic compounds related to the phenylpropanoid pathway. Similarly, in *Melissa officinalis*, controlled NaCl application stimulates the accumulation of total phenolics, flavonoids, and phenolic acids, accompanied by enhanced antioxidant activity, highlighting the ability of osmotic stress to modulate plant metabolic responses (Hawrylak-Nowak et al., 2021). Collectively, these studies demonstrate that abiotic elicitors induce robust and species-specific metabolic responses, resulting in consistent improvements in phenolic compound production in *in vitro* culture systems.

6.5.3 Combined elicitation and synergistic effects

The simultaneous or sequential application of multiple elicitors, known as combined elicitation, has emerged as a promising strategy to achieve synergistic enhancement of phenolic production. Combinations such as MeJA with chitosan, or SA with UV-B irradiation, have demonstrated additive or even synergistic effects on the accumulation of total phenolics and specific subclasses such as flavonols or stilbenes (Bulanov et al., 2025; Kandoudi et al., 2022).

Synergy arises from the convergence of distinct signaling pathways, resulting in stronger transcriptional activation of biosynthetic and regulatory genes. For example, the combined action of MeJA and chitosan can amplify MAPK signaling, increase ROS production, and upregulate MYB and bHLH transcription factors, thereby enhancing the metabolic flux through the phenylpropanoid pathway (Bulanov et al., 2025; Ghosh & Roychoudhury, 2024; Pratyusha & Sarada, 2022; Zhang et al., 2024).

However, successful combined elicitation requires careful calibration of elicitor doses, exposure duration, and timing relative to the growth phase of the culture. Excessive elicitation can trigger programmed cell death or metabolic imbalances, underscoring the need for a mechanistic understanding of signal integration and metabolic feedback regulation (Arya et al., 2022; Kandoudi et al., 2022).

Combined elicitation strategies, integrating biotic and abiotic stimuli, often lead to metabolic responses that exceed those obtained with single elicitors. Studies in medicinal plants show that simultaneous application of compounds such as MeJA and yeast extract can intensify phenolic accumulation in transformed root systems, including species such as *Withania somnifera*. In Lamiaceae cell cultures, combining UV-B radiation with chitosan enhances the biosynthesis and recovery of rosmarinic acid, reflecting the activation of both oxidative and hormonal pathways. Suspension cultures of *Lavandula angustifolia* likewise respond strongly when osmotic stress is applied together with SA, resulting in higher levels of hydroxycinnamic acids. Collectively, these findings demonstrate that targeted combination treatments can substantially boost metabolite yields while maintaining reproducibility, an essential requirement for larger-scale applications (Chemat et al., 2012; Jeyasri et al., 2023; Krol et al., 2025; Ramirez-Estrada et al., 2016).

6.6 Metabolic engineering and molecular approaches

Metabolic engineering and molecular biology provide powerful tools to enhance the biosynthesis of phenolic compounds in MAPs. By directly modifying genes, enzymes, and regulatory networks, these strategies can increase phenolic accumulation, improve extraction efficiency, and generate stable, reproducible yields under controlled conditions (Devi et al., 2023).

6.6.1 Gene overexpression and silencing

Metabolic engineering aims to redirect or enhance the flow of metabolites through biosynthetic pathways by manipulating key genes and enzymes. In the case of phenolic compounds, overexpression of structural genes such as PAL, C4H, 4CL, CHS, CHI, and DFR has proven effective in increasing the production of phenolic acids and flavonoids in various plant systems (Kaur et al., 2025; Salami et al., 2024; Wang et al., 2024).

For instance, overexpression of PAL in *Nicotiana tabacum* and *Arabidopsis thaliana* led to elevated levels of cinnamic acid derivatives, whereas increased expression of CHS enhanced flavonoid accumulation in *Petunia* sp. and *Glycine max* (Lewis et al., 2024; Tong et al., 2024).

Conversely, gene silencing approaches, through RNA interference (RNAi) or antisense constructs, can suppress competitive pathways, thereby increasing precursor availability for phenolic biosynthesis. Silencing of C3H or F5H, for example, redirects metabolic flux toward the accumulation of specific phenolic intermediates (Rehan, 2021; Zagoskina et al., 2023). The combined use of overexpression and silencing strategies allows precise control of pathway branching and metabolite specificity, offering a rational framework for engineering phenolic profiles in medicinal plants (Devi et al., 2023; Khan & Boehr, 2025; Utomo et al., 2024).

Overexpression or silencing of key biosynthetic genes enables precise modulation of phenolic pathways and can substantially alter both the quantity and profile of accumulated metabolites. In *Salvia miltiorrhiza*, the overexpression of genes encoding early enzymes of the phenylpropanoid pathway, such as phenylalanine ammonia-lyase (PAL) and cinnamate 4-hydroxylase (C4H), has been associated with enhanced production of high-value phenolic derivatives, including rosmarinic and salvianolic acids. In *Camellia sinensis*, gene silencing strategies targeting enzymes such as flavanone 3-hydroxylase (F3H) have been employed to redirect metabolic flux toward catechin biosynthesis, promoting the accumulation of bioactive flavan-3-ols. Similarly, in *Hypericum perforatum*, targeted upregulation of chalcone synthase (CHS) expression results in elevated flavonoid levels, accompanied by improved antioxidant activity and extraction efficiency. These examples demonstrate how gene-level interventions can intensify phenolic compound accumulation and enhance the recovery of bioactive metabolites in plant culture systems (Devi et al., 2023; Ninkuu et al., 2025; Pratyusha & Sarada, 2022).

6.6.2 CRISPR/Cas-based genome editing

The emergence of CRISPR/Cas (Clustered Regularly Interspaced Short Palindromic Repeats/CRISPR-associated proteins) technology has revolutionized metabolic engineering, providing a highly efficient and specific method for targeted genome modification (Dhir et al., 2024; Khan & Boehr, 2025; Utomo et al., 2021).

This system enables the introduction of gene knockouts, knock-ins, or point mutations to modulate enzyme activity or regulatory function within phenolic biosynthetic pathways.

Applications of CRISPR/Cas9 have been reported in several model and medicinal plants, including *Solanum lycopersicum*, *Capsicum annuum*, and *Oryza sativa*, resulting in altered phenolic content and composition (Devi et al., 2023; Khan & Boehr, 2025; Utomo et al., 2021). Targeted disruption of transcriptional repressors or competing enzymes, such as FLS or COMT, has enhanced the accumulation of flavonols and monolignols (Kaur et al., 2025; Salami et al., 2024; Wang et al., 2024).

Beyond gene editing, CRISPR-based transcriptional regulation systems (CRISPRa/CRISPRi) enable dynamic modulation of multiple genes simultaneously, allowing coordinated activation of biosynthetic clusters (Dhir et al., 2024; Utomo et al., 2024). The precision, efficiency, and adaptability of CRISPR/Cas systems make them indispensable tools for next-generation metabolic engineering, although challenges related to off-target effects, transformation efficiency, and regulatory acceptance remain under discussion (Devi et al., 2023; Khan & Boehr, 2025).

CRISPR/Cas-based genome editing has emerged as a powerful tool for the precise manipulation of metabolic genes involved in phenolic biosynthesis. This technology enables the targeted modulation of structural enzymes and transcriptional regulators associated with the phenylpropanoid pathway, thereby enhancing specific metabolic fluxes. In different medicinal and aromatic plants (MAPs), genome-editing strategies have been applied to regulate the activity of genes that compete with or limit phenolic compound biosynthesis, promoting the accumulation of flavonoids, phenolic acids, and other high-value bioactive metabolites. Recent studies have also demonstrated that the editing of key enzymes, including phenylalanine ammonia-lyase (PAL), cinnamate 4-hydroxylase (C4H), 4-coumarate-CoA ligase (4CL), and chalcone synthase (CHS), can significantly modify phenolic profiles, resulting in cultures with enhanced antioxidant capacity and improved extraction efficiency. The high precision, reproducibility, and ability to redirect metabolic pathways make CRISPR/Cas technology a highly promising approach for intensifying phenolic compound production in medicinal and aromatic plants (Dhir et al., 2024; Kaur et al., 2025; Utomo et al., 2021).

6.6.3 Omics-assisted metabolic engineering

The integration of omics technologies, including genomics, transcriptomics, proteomics, and metabolomics, has transformed the understanding of phenolic metabolism and facilitated the identification of regulatory bottlenecks (Choudhary et al., 2025; Gómez-Rocal et al., 2025; Machitescu et al., 2024). High-throughput transcriptome analyses allow the discovery of coexpressed genes and transcription factors associated with phenylpropanoid biosynthesis, while metabolomic profiling provides insights into pathway fluxes and metabolite accumulation patterns (Kaur et al., 2025; Salami et al., 2024; Wang et al., 2024).

Proteomic studies reveal posttranslational modifications and enzyme–protein interactions that fine-tune phenolic biosynthesis, and systems biology approaches help model metabolic fluxes under different conditions (Liu et al., 2023; Zagoskina et al., 2023).

These integrated datasets support the rational design of metabolic interventions, enabling the identification of target genes for overexpression, silencing, or editing (Devi et al., 2023; Khan & Boehr, 2025; Utomo et al., 2024).

By combining omics data with computational network modeling, researchers can predict the systemic effects of genetic modifications and avoid unintended trade-offs between growth and secondary metabolism, ultimately advancing the design of high-yielding phenolic production systems (Choudhary et al., 2025; Gómez-Rocal et al., 2025).

Omics technologies, including genomics, transcriptomics, proteomics, and metabolomics, have expanded the capacity to identify regulatory points and metabolic limitations associated with phenolic biosynthesis in MAPs. In *Salvia officinalis*, metabolomic profiling has been used to pinpoint steps that limit rosmarinic acid formation, supporting targeted metabolic adjustments that increase phenolic accumulation in cultured tissues. In *Withania somnifera*, transcriptomic information has guided the modulation of key enzymes involved in secondary metabolism, resulting in higher production of phenolic derivatives linked to withanolide pathways and improved extraction efficiency. Studies in *Camellia sinensis* that combine metabolomic and transcriptomic datasets have also contributed to optimizing catechin biosynthesis, improving reproducibility between batches, and strengthening the recovery of specific bioactive compounds. Altogether, omics-based strategies provide detailed mechanistic insight and support rational interventions aimed at achieving more efficient and predictable phenolic production (Choudhary et al., 2025; Gómez-Rocal et al., 2025; Jiang & Grotewold, 2024; Lewis et al., 2024).

6.6.4 Synthetic biology approaches for pathway reconstruction

Synthetic biology offers new frontiers for phenolic production through the reconstruction of biosynthetic pathways in heterologous hosts such as *Escherichia coli*, *Saccharomyces cerevisiae*, and *Nicotiana benthamiana* (Khan & Boehr, 2025; Utomo et al., 2024). This approach allows the assembly of multiple biosynthetic genes into synthetic operons or expression cassettes, bypassing the limitations of plant-based systems (Dhir et al., 2024; Kukil & Lindberg, 2024).

Engineered microbial or plant chassis can produce targeted phenolic compounds under optimized fermentation conditions, facilitating large-scale production of molecules such as resveratrol, caffeic acid, and naringenin (Devi et al., 2023; Utomo et al., 2024). Advanced tools such as modular cloning systems (Golden Gate, Gibson Assembly) and promoter libraries enable precise control of gene expression, while dynamic regulation circuits maintain metabolic balance and prevent precursor depletion (Khan & Boehr, 2025; Nguyen et al., 2025).

The convergence of synthetic biology with metabolic engineering thus provides a powerful framework for designing customized biosynthetic platforms, bridging the gap between natural diversity and industrial demand for high-value phenolic compounds (Devi et al., 2023; Utomo et al., 2024).

Synthetic biology has expanded the possibilities for reconstructing entire biosynthetic pathways in heterologous hosts and engineered plant cultures. In *Nicotiana benthamiana*, the reconstitution of phenolic pathways has resulted in production levels comparable to those of native medicinal plants, with high efficiency during extraction. Microbial platforms engineered with genes from *Salvia miltiorrhiza* and *Ginkgo biloba* have also achieved phenolic yields approaching the theoretical maximum, demonstrating the feasibility of industrial-scale production in controlled biological systems. In plant cell cultures of *Ocimum sanctum*, the implementation of synthetic regulatory circuits has improved metabolic flux and contributed to greater stability of phenolic compounds during cultivation. Collectively, these advances show that synthetic biology can minimize environmental variability and support consistent, high-yield production suitable for applications in the pharmaceutical, nutraceutical, and cosmetic sectors (Lewis et al., 2024; Nguyen et al., 2025; Utomo et al., 2021; Utomo et al., 2024).

6.7 Biotechnological optimization and scale-up

Scaling up phenolic production from *in vitro* cultures of MAPs requires careful optimization of culture conditions, bioreactor design, and downstream processing. These factors directly influence biomass yield, metabolite accumulation, and recovery efficiency, ensuring that laboratory-scale improvements translate into industrially viable outputs (Motolinía-Alcántara et al., 2021; Smetanska, 2018; Verdú-Navarro et al., 2023).

6.7.1 Optimization of culture conditions

The optimization of culture conditions is a decisive factor for enhancing phenolic compound biosynthesis in *in vitro* systems (Valletta et al., 2021). Variables such as nutrient composition, carbon source, pH, temperature, and light regime profoundly influence both biomass accumulation and secondary metabolism. Nitrogen limitation or modification of the carbon-to-nitrogen ratio often triggers phenylpropanoid pathway activation, increasing total phenolic and flavonoid content (Zhang et al., 2024; Moreno-Garcia et al., 2019).

Elicitation strategies, including the addition of MeJA, SA, chitosan, or yeast extract, stimulate stress-related metabolic responses, promoting the accumulation of phenolics as defense metabolites (Al-Khayri et al., 2023; Arya et al., 2022). Similarly, precursor feeding, using phenylalanine, cinnamic acid, or coumaric acid, enhances substrate availability for biosynthetic enzymes (Rattan and Kumar, 2022). Systematic factorial design and response surface methodologies have been successfully employed to determine optimal culture parameters, leading to reproducible and scalable improvements in phenolic yields across diverse medicinal plant species (Mohamed et al., 2022; Smetanska, 2018).

Culture parameters such as nutrient balance, pH, temperature, light regime, and growth regulators play a decisive role in directing phenolic biosynthesis in MAPs. In *Salvia miltiorrhiza* cell suspensions, the combined adjustment of nitrogen and carbon sources with optimal auxin–cytokinin ratios resulted in a 2.5-fold increase in rosmarinic acid content, accompanied by higher extraction efficiency. In *Ocimum basilicum*, callus cultures exposed to optimized light intensity and timely elicitor application showed approximately 50% increases in total flavonoids, demonstrating the impact of abiotic modulation on metabolite yield. Similarly, fine-tuning temperature and pH conditions in *Hypericum perforatum* cultures improved secondary metabolite stability, enhancing phenolic recovery by up to 30% compared with standard conditions. These findings reinforce that even small adjustments in culture conditions can translate into substantial gains in phenolic accumulation and extraction yield. A synthesis of these improvements across different MAPs is presented in Table 6.3, which summarizes phenolic production, percentage gains relative to controls, and recovery efficiency (Clapa et al., 2025; Hussain et al., 2022; Kwiecień et al., 2015).

6.7.2 Bioreactor design and scale-up strategies

Transitioning from laboratory to industrial production requires the careful design of bioreactor systems that support high-density cultures while maintaining metabolic activity (Motolinía-Alcántara et al., 2021; Verdú-Navarro et al., 2023). Bioreactors for plant cell, tissue, or hairy root cultures, such as stirred-tank, airlift, and bubble column reactors, provide controlled environments for oxygen transfer, shear stress management, and nutrient delivery (Bai et al., 2025).

Hairy root cultures, in particular, exhibit high genetic and biochemical stability and are often preferred for phenolic compound production (Sharma et al., 2019; Grąbkowska et al., 2025). Scale-up strategies involve optimization of aeration, agitation, and medium exchange rates to sustain metabolic efficiency (Bruat et al., 2025). The implementation of fed-batch or perfusion systems allows dynamic regulation of nutrient input, maintaining elicitation phases without compromising cell viability (Mohamed et al., 2022; Rattan and Kumar, 2022).

Recent advances in sensor technology and process automation enable real-time monitoring of metabolite accumulation, improving reproducibility and reducing production costs in large-scale phenolic biosynthesis (Yu et al., 2025).

Transitioning from flask-based cultures to bioreactor systems enables larger-scale production while maintaining, and in many cases improving, phenolic yields. Studies

TABLE 6.3 Effects of culture parameters on phenolic production, relative gain, and recovery efficiency in *in vitro* MAP cultures.

MAP species	Culture type	Treatment/Factor	Total phenolics (mg/g DW)	Gain versus control (%)	Recovery (%)	References
Salvia miltiorrhiza	Cell suspension	Auxin + cytokinin	45.2 ± 1.3	+150%	82	Xing et al. (2015)
Ocimum basilicum	Callus	High-intensity light + elicitor	28.5 ± 0.9	+50%	75	Nadeem et al. (2019)
Hypericum perforatum	Callus	pH 5.8 + 25 °C	36.7 ± 1.1	+80%	78	Kwiecień et al. (2015)
Lavandula angustifolia	Hairy root	Temporary immersion bioreactor	52.3 ± 2.0	+60%	80	Yahya (2024)

with *Echinacea purpurea* suspension cultures demonstrate that 5–10 L stirred-tank bioreactors can achieve phenolic levels equal to or higher than those obtained in small laboratory flasks, with improved consistency in recovery efficiency due to better control of aeration and nutrient dynamics (Motolinía-Alcántara et al., 2021). Similarly, *Salvia officinalis* hairy root cultures cultivated in airlift bioreactors exhibit substantial increases in rosmarinic acid accumulation compared to static systems, reflecting the benefits of enhanced oxygen transfer and circulation (Verdú-Navarro et al., 2023). In *Lavandula angustifolia*, the use of temporary immersion bioreactors promotes a 45%–60% increase in total phenolics, resulting in higher extraction yields and more reproducible metabolite profiles (Yahya, 2024). Effective scale-up therefore depends on optimizing aeration, agitation, nutrient delivery, and elicitor timing to maximize both productivity and downstream recovery.

6.7.3 Downstream processing and product recovery

Efficient recovery and purification of phenolic compounds represent a major technological challenge due to their structural diversity and coextraction with other metabolites (Fierăscu et al., 2020; Mohamed et al., 2022; Smetanska, 2018). Conventional solvent extraction remains widely used but may compromise product stability or purity. Modern alternatives such as supercritical fluid extraction, pressurized liquid extraction, and membrane-based separation have demonstrated higher selectivity and environmental compatibility (Gómez-Rocal et al., 2025; Machitescu et al., 2024; Yusoff et al., 2022).

Chromatographic methods, including preparative HPLC, macroporous resin adsorption, and countercurrent chromatography, enable fractionation of phenolic subclasses such as flavonols, stilbenes, and phenolic acids (Guo et al., 2023; Tang et al., 2024; Zhang et al., 2024). Integration of extraction and purification into continuous bioprocessing pipelines minimizes product degradation and facilitates standardization (Dong & Lin, 2020; Martínez-Chávez et al., 2024).

The adoption of green extraction techniques not only improves sustainability but also aligns with the regulatory demand for environmentally responsible production of bioactive compounds (Merecz-Sadowska et al., 2025; Zheng et al., 2025).

Efficient extraction, purification, and stabilization strategies are essential to ensure that the high phenolic yields obtained from *in vitro* cultures are effectively translated into usable bioactive products. In *Withania somnifera* cultures, optimized solvent extraction followed by solid-phase purification has been shown to recover more than 80% of targeted phenolic constituents, while preserving antioxidant capacity, demonstrating the importance of selecting appropriate purification steps (Chemat et al., 2012). Ultrasound-assisted extraction applied to *Ginkgo biloba* cell cultures markedly increased flavonoid recovery by approximately 70% compared with conventional maceration, highlighting the impact of energy-assisted methods on extraction efficiency (Yusoff et al., 2022). Similarly, freeze-drying of *Ocimum sanctum* extracts contributes to the preservation of phenolic integrity, ensuring that gains achieved during culture optimization are retained during formulation and storage (Sun & Shahrajabian, 2023). Effective downstream processing is therefore essential for converting enhanced *in vitro* phenolic production into consistent, high-quality outputs for cosmetic, nutraceutical, and pharmaceutical applications.

6.7.4 Techno-economic and industrial considerations

Scaling up phenolic production from biotechnological systems requires a comprehensive evaluation of economic feasibility, regulatory compliance, and market competitiveness (Motolinía-Alcántara et al., 2021; Verdú-Navarro et al., 2023). Cost analysis must account for raw material sourcing, culture maintenance, energy inputs, and purification efficiency (Fierăscu et al., 2020; Mohamed et al., 2022; Smetanska, 2018).

Biotechnological production is particularly advantageous for rare or unstable phenolics that are difficult to extract from natural sources (Guo et al., 2023; Martínez-Chávez et al., 2024). Integration with agro-industrial side streams, such as grape pomace or olive mill by-products, can significantly reduce production costs and enhance the circular economy (Merecz-Sadowska et al., 2025; Yusoff et al., 2022). Moreover, adherence to Good Manufacturing Practices (GMP) and quality control protocols ensures batch consistency and bioactivity retention (Gómez-Rocal et al., 2025; Zheng et al., 2025).

The convergence of bioprocess optimization, green chemistry, and industrial biotechnology thus defines a sustainable route for the large-scale production of phenolic compounds, bridging research innovation with commercial application (Dong & Lin, 2020; Machitescu et al., 2024; Martínez-Chávez et al., 2024).

Economic feasibility, process reproducibility, and sustainability are fundamental requirements for the industrial adoption of *in vitro* systems for phenolic production. Analyses of *Salvia miltiorrhiza* bioreactor cultures indicate that optimized *in vitro* platforms can reduce production costs by 30%–50% per milligram of phenolics when compared with field-grown plants, mainly due to higher productivity and greater control of culture conditions. The combined use of elicitation, metabolic engineering, and bioreactor optimization further increases volumetric yield and reduces batch-to-batch variability, supporting consistent extraction performance. Sustainable practices, including the recycling of culture media and the use of low-toxicity elicitors, also improve environmental indicators and operational costs, reinforcing the potential of *in vitro* strategies for scalable and environmentally responsible production of MAP-derived phenolics (Motolinía-Alcántara et al., 2021; Smetanska, 2018; Verdú-Navarro et al., 2023).

6.8 Analytical characterization and bioactivity assessment

Comprehensive analytical profiling is essential to validate the effectiveness of biotechnological strategies, quantify improvements in phenolic production, and establish clear relationships between compound composition and biological activity. Reliable characterization ensures that increases obtained *in vitro* correspond to genuine enhancements in functional bioactivity. Advances in chromatographic and spectrometric techniques have strengthened this connection by allowing precise identification and quantification of phenolics, supporting the integration of biochemical, metabolic, and functional data in MAP research (Machitescu et al., 2024).

6.8.1 Analytical profiling of phenolic compounds

Accurate identification and quantification of phenolic compounds are essential for understanding biosynthetic outcomes and ensuring product standardization (Fierăscu et al., 2020; Guo et al., 2023; Tang et al., 2024; Zhang et al., 2024). High-performance liquid chromatography (HPLC) coupled with diode-array detection (DAD) and mass spectrometry (MS or MS/MS) remains the most widely applied technique for profiling phenolic acids, flavonoids, stilbenes, and lignans. Liquid chromatography–mass spectrometry (LC–MS/MS) provides structural elucidation based on fragmentation patterns, while ultraperformance liquid chromatography (UPLC) enhances resolution and throughput. Additionally, complementary techniques such as nuclear magnetic resonance (NMR) and Fourier transform infrared spectroscopy (FTIR) assist in confirming molecular structures and detecting functional group modifications induced by metabolic engineering (Chen et al., 2024; Zheng et al., 2025).

The diversity of analytical techniques used for phenolic profiling is summarized in Table 6.4, which compares their detection principles, applications, and main advantages for qualitative and quantitative analysis.

Quantitative fingerprinting using chemometric tools (PCA, PLS-DA) allows discrimination between biotechnological and naturally derived phenolic profiles, providing robust analytical validation for process optimization and quality control (Gómez-Rocal et al., 2025; Martínez-Chávez et al., 2024).

Advanced chromatographic and spectroscopic methods provide the precision required for accurate identification and quantification of phenolic metabolites. High-performance liquid chromatography has been effectively applied to quantify rosmarinic acid in *Salvia miltiorrhiza* cultures, revealing increases of more than twofold under optimized conditions. LC–MS/MS analysis of *Ocimum basilicum* cell suspensions has demonstrated targeted accumulation of flavonoids, with extraction recoveries surpassing 75% after elicitor application. In *Ginkgo biloba* hairy roots, quantitative HPLC profiling showed that bioreactor-grown cultures produced 40%–50% more ginkgolide-related phenolics than flask cultures, indicating that scale-up can enhance rather than limit metabolite accumulation. These analytical approaches confirm that both quantitative gains and recovery efficiencies improve when phenolic production is supported by controlled *in vitro* conditions and biotechnological intervention (Gómez-Rocal et al., 2025; Jiang & Grotewold, 2024; Machitescu et al., 2024).

6.8.2 Determination of antioxidant activity

The antioxidant potential of phenolic compounds is a key indicator of their biological efficacy (Chen et al., 2024; Merecz-Sadowska et al., 2025). *In vitro* assays such as DPPH$^{\bullet}$ and ABTS$^{\bullet+}$ radical scavenging, ferric reducing antioxidant power (FRAP), and oxygen radical absorbance capacity (ORAC) are widely employed to quantify the capacity of extracts or purified molecules to neutralize ROS. Kinetic-based methods, which assess radical scavenging dynamics, provide deeper insight into the reaction mechanisms and stability of antioxidant activity (Yusoff et al., 2022; Zhang et al., 2024). The integration of electrochemical techniques, including cyclic voltammetry, offers an additional layer of mechanistic understanding by evaluating redox potential and electron-donating capacity.

TABLE 6.4 Main analytical techniques applied for phenolic compound identification and quantification.

Analytical method	Principle/Detection	Application	Advantages/Limitations	References
HPLC–DAD	Separation by polarity and UV detection	Routine quantification of phenolic acids and flavonoids	Accurate but time-consuming	Fierăscu et al. (2020), Gómez-Rocal et al. (2025)
LC–MS/MS	Ion fragmentation and mass detection	Structural elucidation of complex phenolics	High sensitivity, expensive	Zhong et al. (2025)
UPLC	High-pressure separation	Rapid fingerprinting	High throughput, costly	Zhang et al. (2024)
NMR	Magnetic resonance	Structural confirmation	Comprehensive, requires large sample	Valletta et al. (2021)
FTIR	Vibration spectrum	Functional group analysis	Fast, nondestructive	Guo et al. (2023)

HPLC–DAD, High-performance liquid chromatography coupled with diode-array detection; LC–MS/MS, liquid chromatography–mass spectrometry; UPLC, ultraperformance liquid chromatography; NMR, nuclear magnetic resonance; FTIR, Fourier transform infrared spectroscopy.

These assays, combined with analytical profiling, establish a direct relationship between phenolic composition, structure–activity correlations, and antioxidant efficacy, guiding the selection of high-value bioactive targets for therapeutic or nutraceutical applications (Fierăscu et al., 2020; Tang et al., 2024).

Phenolic content is closely associated with antioxidant capacity, and this relationship can be demonstrated through several *in vitro* assays. In *Hypericum perforatum* cell cultures, DPPH and ABTS radical scavenging tests showed proportional increases in antioxidant activity that reflected the rise in total phenolic concentration. FRAP assays performed with *Withania somnifera* extracts indicated a 30%–45% increase in reducing capacity in elicited cultures compared with untreated controls. Correlation analyses consistently reveal strong positive associations between total phenolics and antioxidant activity, often with R^2 values above 0.9, confirming that improved biosynthesis leads to enhanced functional potential. The functional performance of the accumulated phenolics was further evaluated through antioxidant, antiinflammatory, and antimicrobial assays. A summary of these results is presented in Table 6.5, which highlights the strong relationship between phenolic enrichment and biological activity (Mokhtari et al., 2023; Riaz et al., 2023; Sun & Shahrajabian, 2023).

6.8.3 Evaluation of antiinflammatory, antimicrobial, and anticancer activities

Beyond their antioxidant properties, phenolic compounds exhibit a broad spectrum of biological functions relevant to pharmacology and preventive medicine (Bulanov et al., 2025; Sharma et al., 2019). *In vitro* studies using macrophage or epithelial cell models assess antiinflammatory effects through inhibition of pro-inflammatory mediators such as nitric oxide, TNF-α, and interleukins.

A concise summary of the main biological activities and experimental models used to evaluate phenolic compounds is presented in Table 6.6.

Antimicrobial assays, such as disk diffusion, broth microdilution, and biofilm inhibition tests, have demonstrated the efficacy of flavonoids and phenolic acids against pathogenic bacteria and fungi (Fierăscu et al., 2020; Zheng et al., 2025). Similarly, cytotoxicity and apoptosis induction in cancer cell lines (e.g., HeLa, MCF-7, and HT-29) are routinely evaluated using MTT or resazurin assays to elucidate anticancer potential (Chen et al., 2024; Guo et al., 2023).

The integration of molecular docking and in silico modeling further supports mechanistic insights into enzyme inhibition and receptor interaction, complementing experimental findings with predictive biological validation (Martínez-Chávez et al., 2024; Tang et al., 2024).

Bioactivity assays are essential to determine the therapeutic relevance of the phenolics recovered from optimized *in vitro* cultures. Antiinflammatory activity of *Ocimum sanctum* extracts has been demonstrated through the inhibition of nitric oxide production in macrophage models, with elicited cultures showing nearly twofold greater inhibition than untreated controls. Antimicrobial tests against *Staphylococcus aureus* and *Escherichia coli* indicated that increased flavonoid accumulation in extracts of *Ginkgo biloba* and *Salvia officinalis* was associated with stronger bacterial growth inhibition, confirming the functional gains associated with enhanced secondary metabolism. Cytotoxicity assays using human cancer cell lines also revealed improved selective anticancer activity in extracts enriched in

TABLE 6.5 Biological activity of phenolic-rich extracts from *in vitro* MAP cultures.

MAP species	Total phenolics (mg/g DW)	DPPH (IC_{50}, μg/mL)	ABTS (IC_{50}, μg/mL)	Antiinflammatory (% NO inhibition)	Antimicrobial (mm inhibition)	References
Salvia officinalis	45.2	25.6	30.2	60	12 (*S. aureus*)	Kolac et al. (2017)
Ocimum basilicum	28.5	40.0	48.0	50	14 (*E. coli*)	Zhakipbekov et al. (2024)
Hypericum perforatum	36.7	32.1	35.5	55	10 (*S. aureus*)	Kwiecień et al. (2015)
Lavandula angustifolia	52.3	22.0	24.5	58	11 (*E. coli*)	Choi et al. (2023)

MAP, Medicinal and aromatic plants.

TABLE 6.6 Main biological activities of phenolic compounds and representative assays used for their evaluation.

Biological activity	Mechanism/Biological target	Common assays	Representative compounds	References
Antioxidant	Radical scavenging, redox balance	DPPH, ABTS, FRAP, ORAC	Catechin, gallic acid	Zhong et al. (2025)
Antiinflammatory	Inhibition of NF-κB, cytokines	NO inhibition, ELISA	Quercetin, luteolin	Rajesh et al. (2024)
Antimicrobial	Cell wall and membrane disruption	Disk diffusion, MIC	Rosmarinic acid, caffeic acid	Sadeghi et al. (2024)
Anticancer	Apoptosis induction, ROS modulation	MTT, resazurin, flow cytometry	Resveratrol, naringenin	Zheng et al. (2025)

polyphenols, particularly those obtained from *Hypericum perforatum* and *Withania somnifera* cultures. Together, these results show that increases in phenolic content correlate strongly with enhanced biological activity, reinforcing the therapeutic relevance of biotechnological optimization (Hussain et al., 2022; Kolac et al., 2017; Mokhtari et al., 2023; Riaz et al., 2023; Zhakipbekov et al., 2024).

6.8.4 Correlation between phytochemical composition and bioactivity

Establishing correlations between phytochemical profiles and biological effects is crucial for rationalizing bioactivity-driven production strategies (Merecz-Sadowska et al., 2025; Zhang et al., 2024). Multivariate statistical analyses and network pharmacology approaches allow the identification of key compounds responsible for observed biological outcomes (Dong & Lin, 2020; Gómez-Rocal et al., 2025).

These models enable the differentiation between synergistic and additive effects, providing evidence for the holistic efficacy of complex phenolic mixtures often produced by biotechnological systems. Moreover, structure–activity relationship (SAR) analyses contribute to the identification of structural motifs that enhance specific bioactivities, supporting future metabolic engineering efforts (Chen et al., 2024; Fierăscu et al., 2020). Ultimately, the integration of chemical, biological, and computational analyses provides a comprehensive framework for evaluating the functionality of biotechnologically derived phenolic compounds, aligning analytical rigor with therapeutic relevance (Martínez-Chávez et al., 2024; Zheng et al., 2025).

Linking analytical data to biological outcomes is essential for selecting MAP species and treatments capable of producing high-value phenolics. Multivariate analyses such as PCA and PLS have shown that specific phenolic subclasses, including rosmarinic acid and catechins, are the primary contributors to antioxidant and antimicrobial activities across different *in vitro* cultures. Improvements in recovery and yield obtained through metabolic engineering, elicitation, or scale-up were consistently reflected in bioactivity assays, demonstrating that process optimization produces measurable functional gains. These findings highlight the importance of integrating quantitative analytical profiling with biological evaluation and process refinement to achieve reproducible and high-value phenolic production suitable for pharmaceutical, nutraceutical, and cosmetic applications (Gómez-Rocal et al., 2025; Choi et al., 2023; Ninkuu et al., 2025).

6.9 Challenges, limitations, and future perspectives

Despite remarkable progress in the biotechnological production of phenolic compounds, numerous challenges still constrain their translation from laboratory research to industrial application (Motolinía-Alcántara et al., 2021; Verdú-Navarro et al., 2023). Technical limitations include low productivity, genetic instability of engineered lines, and difficulties in maintaining metabolite yields under scale-up conditions (Mohamed et al., 2022; Smetanska, 2018).

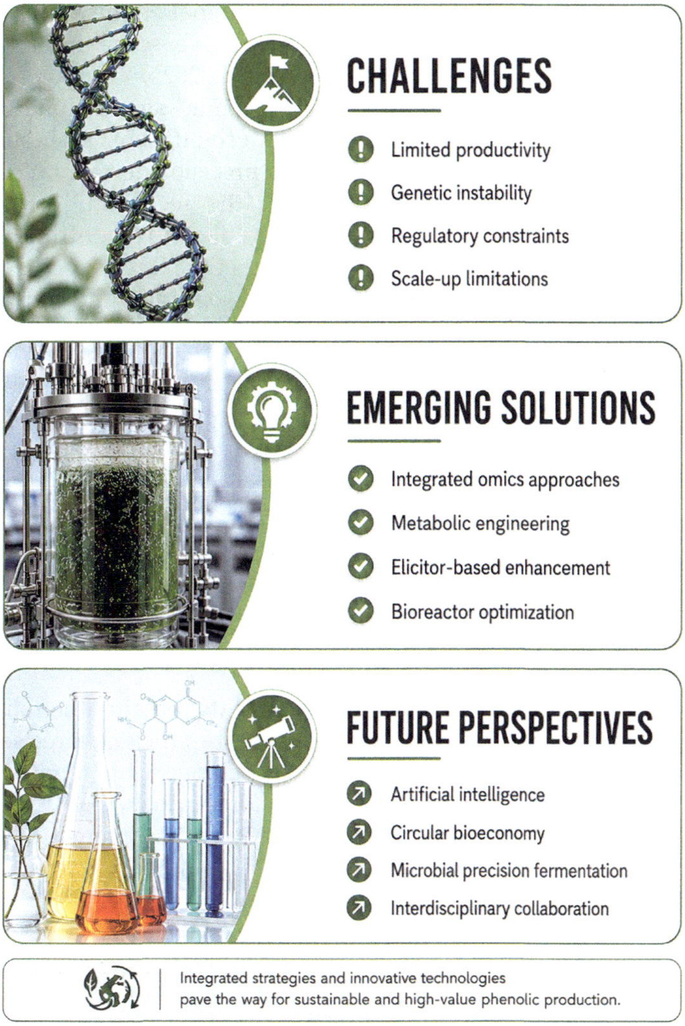

FIGURE 6.3 Roadmap of challenges, emerging solutions, and future perspectives for sustainable phenolic production.

The complex regulation of phenylpropanoid pathways, often involving feedback inhibition, metabolite compartmentalization, and cross-talk with primary metabolism, further complicates pathway optimization (Zhang et al., 2024). Moreover, regulatory and ethical constraints surrounding genetically modified organisms (GMOs) remain significant barriers to commercial adoption, particularly for food and nutraceutical applications (Zheng et al., 2025).

The integration of omics data, genomic, transcriptomic, proteomic, and metabolomic, into computational modeling frameworks represents a promising direction for overcoming these bottlenecks (Chen et al., 2024; Gómez-Rocal et al., 2025; Martínez-Chávez et al., 2024). Systems biology and metabolic flux analysis can identify regulatory nodes and predict the effects of genetic or environmental perturbations, supporting more rational engineering of phenolic biosynthesis (Zhang et al., 2024). From a sustainability perspective, scalability and environmental footprint remain critical concerns (Merecz-Sadowska et al., 2025; Yusoff et al., 2022). The transition toward circular bioeconomy models, leveraging agro-industrial residues and renewable carbon sources, offers a path toward eco-efficient phenolic production (Fierăscu et al., 2020; Guo et al., 2023). Additionally, precision fermentation and synthetic biology hold transformative potential by enabling the reconstitution of complete biosynthetic pathways in microbial or cell-free systems, allowing controlled, high-yield production of targeted phenolics (Tang et al., 2024; Zheng et al., 2025).

A schematic overview summarizing the major challenges, emerging solutions, and future opportunities for sustainable phenolic production is presented in Fig. 6.3.

Future progress will depend on interdisciplinary collaboration between plant biotechnologists, computational biologists, and process engineers (Verdú-Navarro et al., 2023). Integrating these fields will accelerate the shift from discovery to application, promoting the development of next-generation phenolic bioproducts with verified efficacy, stability, and sustainability (Chen et al., 2024; Martínez-Chávez et al., 2024; Verdú-Navarro et al., 2023).

6.10 Conclusions

Biotechnological innovation has transformed the strategies used to obtain phenolic compounds, providing production routes that are simultaneously more efficient, precise, and environmentally responsible. Advances in metabolic and genetic engineering, together with omics-based insights, have expanded our capacity to modulate biosynthetic pathways and substantially increase phenolic yields in both plant and microbial systems. The

TABLE 6.7 Key biological activities and representative phenolic compounds from medicinal plants.

Activity	Example compounds	Notable effects	Citations
Antioxidant	Flavonoids, phenolic acids	Free radical scavenging, cell protection	Riaz et al. (2023), Sun and Shahrajabian (2023), Tang et al. (2024)
Antiinflammatory	Flavonoids, xanthones	COX-2 inhibition, reduced inflammation	Sun and Shahrajabian (2023), Merecz-Sadowska et al. (2025), Zheng et al. (2025)
Anticancer	Polyphenols, tannins	Inhibition of tumor growth	Riaz et al. (2023), Sun and Shahrajabian (2023), Rajesh et al. (2024)
Cardioprotective	Flavonoids, stilbenes	Vascular health, reduced risk factors	Sun and Shahrajabian (2023), Rajesh et al. (2024)
Antimicrobial	Phenolic acids, flavonoids	Defense against pathogens	Kisiriko et al. (2021), Sun and Shahrajabian (2023), Rajesh et al. (2024)

approaches discussed throughout this chapter, including *in vitro* culture optimization, elicitation, CRISPR-mediated editing, and synthetic biology, demonstrate the feasibility of directing secondary metabolism toward the production of high-value phenolics.

These technological advances are particularly relevant to the pharmaceutical, nutraceutical, and cosmetic sectors, where phenolic compounds serve as potent antioxidants, antiinflammatory agents, antimicrobials, and multifunctional bioactives. The diversity of biological properties associated with phenolic subclasses is summarized in Table 6.7, which highlights representative compounds and their reported activities across medicinal plants.

Looking ahead, the implementation of scalable and sustainable production systems will depend on an integrated framework that combines computational modeling, green extraction processes, and harmonized regulatory standards. As biotechnology continues to evolve, phenolic compounds remain at the forefront of a new era of sustainable bioactive molecule production, bridging plant science, engineering, and industrial innovation to address emerging health and environmental needs.

Competing interests

The authors have no relevant financial or nonfinancial interests to disclose.

References

Aftab, T., Khan, M. M. A., Teixeira da Silva, J. A., Idrees, M., Naeem, M., & Moinuddin (2011). Role of salicylic acid in promoting salt stress tolerance and enhanced artemisinin production in *Artemisia annua L. Journal of Plant Growth Regulation, 30*, 425–435. https://doi.org/10.1007/s00344-011-9205-0.

Al-Khayri, J., Mascarenhas, R., Harish, H., Gowda, Y., Lakshmaiah, V., Nagella, P., Al-Mssallem, M., Alessa, F., Almaghasla, M., & Rezk, A. (2023). Stilbenes, a versatile class of natural metabolites for inflammation—An overview. *Molecules, 28*. https://doi.org/10.3390/molecules28093786.

Arya, S., Patel, A., Ghosh, A., & Rani, R. (2022). Green approach for the recovery of secondary metabolites from in vitro plant cultures: A review. *Frontiers in Plant Science, 13*, 987986. https://doi.org/10.3389/fpls.2022.987986.

Bai, X., Lee, H.-S., Han, J.-E., Murthy, H. N., & Park, S.-Y. (2025). Enhancement of phenolic and polyacetylene accumulation in *Lobelia chinensis* (*Chinese lobelia*) plantlet cultures through yeast extract and salicylic acid elicitation. *Horticulturae, 11*(6), 612. https://doi.org/10.3390/horticulturae11060612.

Babich, O., Sukhikh, S., Pungin, A., Ivanova, S., Asyakina, L., & Prosekov, A. (2020). Modern trends in the in vitro production and use of callus, suspension cells and root cultures of medicinal plants. *Molecules, 25*(24), 5805. https://doi.org/10.3390/molecules25245805.

Boobalan, S., Srinivasan, R., Keerthanaa, T., et al. (2024). Enhancing the aervine and methylaervine production in vitro cultures of *Aerva javanica* (Burm. F.) Schult via elicitors and *Agrobacterium rhizogenes*-mediated hairy root cultures. *Journal of Plant Biochemistry and Biotechnology, 33*, 353–366. https://doi.org/10.1007/s13562-024-00897-x.

Bulanov, A., Andreeva, E., Tsvetkova, N., & Zykin, P. (2025). Regulation of flavonoid biosynthesis by the MYB-bHLH-WDR (MBW) complex in plants and its specific features in cereals. *International Journal of Molecular Sciences, 26*. https://doi.org/10.3390/ijms26020734.

Bruat, M., Heng, S., Colas, C., Sena-Belez, M., Chefdor, F., Héricourt, F., Depierreux, C., Morabito, D., Malik, S., Destandau, E., Carpin, S., & Lamblin, F. (2025). Otimização dos parâmetros de crescimento da cultura de raízes pilosas do álamo híbrido (*Populus tremula* × *Populus alba*) para a produção de metabólitos antioxidantes especializados. *Plant Cell, Tissue and Organ Culture, 161*, 31. https://doi.org/10.1007/s11240-025-03064-9.

Catoni, M., Gomez, T., & Probst, A. (2025). Plant epigenetics: Controlling genome expression to integrate developmental and environmental cues. *Journal of Experimental Botany, 76*, 2389–2393. https://doi.org/10.1093/jxb/eraf134.

Chen, Z., Świsłocka, R., Choińska, R., Marszałek, K., Dąbrowska, A., Lewandowski, W., & Lewandowska, H. (2024). Exploring the correlation between the molecular structure and biological activities of metal–phenolic compound complexes: Research and description of the role of metal ions in improving the antioxidant activities of phenolic compounds. *International Journal of Molecular Sciences, 25*. https://doi.org/10.3390/ijms252111775.

Chemat, F., Vian, M. A., & Cravotto, G. (2012). Green extraction of natural products: Concept and principles. *International Journal of Molecular Sciences, 13*(7), 8615–8627. https://doi.org/10.3390/ijms13078615.

Choi, M., Yoon, J., Yang, S. H., Kim, J. K., & Park, S. U. (2023). Phenolic compound production and antioxidant activity in hairy root cultures of *Salvia plebeia*. *Plants, 12*, 3840. https://doi.org/10.3390/plants12223840.

Choudhary, R., Ahmad, F., Kaya, C., Upadhyay, S., Muneer, S., Kumar, V., Meena, M., Liu, H., Upadhyay, H., & Seth, C. (2025). Decrypting proteomics, transcriptomics, genomics, and integrated omics for augmenting the abiotic, biotic, and climate change stress resilience in plants. *Journal of Plant Physiology, 305*, 154430. https://doi.org/10.1016/j.jplph.2025.154430.

Clapa, D., Hârța, M., Radomir, A. M., Peticilă, A. G., Leopold, L., Ranga, F., & Sumedrea, D. I. (2025). Effects of culture period and plant growth regulators on in vitro biomass production and phenolic compounds in seven species of Hypericum. *Plants, 14*(15), 2437.

de Elguea-Culebras, G. O., Panamá-Tapia, L. A., Melero-Bravo, E., Cerro-Ibáñez, N., Calvo-Martínez, A., & Sánchez-Vioque, R. (2023). Comparison of the phenolic composition and biological capacities of wastewater from *Origanum vulgare* L., *Rosmarinus officinalis* L., *Salvia lavandulifolia* Vahl. and *Thymus mastichina* L. resulting from two hydrodistillation systems: Clevenger and MAE. *Journal of Applied Research on Medicinal and Aromatic Plants, 34*, 100480. https://doi.org/10.1016/j.jarmap.2023.100480.

Devi, A., Devi, K., Devi, P., Devi, M., & Das, S. (2023). Metabolic engineering of plant secondary metabolites: Prospects and its technological challenges. *Frontiers in Plant Science, 14*. https://doi.org/10.3389/fpls.2023.1171154.

Dhir, N., Sharma, P., Asija, R., Singh, M., & Kumawat, B. (2024). CRISPR technology in chalcone derivatives: Innovations in biosynthetic pathways and therapeutic potential. *International Journal of Pharmaceutical Quality Assurance, 15*(4), 54. https://doi.org/10.25258/ijpqa.15.4.54.

Dong, N., & Lin, H. (2020). Contribution of phenylpropanoid metabolism to plant development and plant–Environment interactions. *Journal of Integrative Plant Biology, 63*(1), 180–209. https://doi.org/10.1111/jipb.12930.

Espinosa-Leal, C. A., Puente-Garza, C. A., & García-Lara, S. (2018). In vitro plant tissue culture: Means for production of biologically active compounds. *Planta, 248*(1), 1–18. https://doi.org/10.1007/s00425-018-2910-1.

Fierăscu, R., Fierăscu, I., Orțan, A., Georgiev, M., & Sieniawska, E. (2020). Innovative approaches for recovery of phytoconstituents from medicinal/aromatic plants and biotechnological production. *Molecules, 25*. https://doi.org/10.3390/molecules25020309.

Gallego-Narbón, A., Narbona, E., Coca-de-la-Iglesia, M., & Valcárcel, V. (2024). Geography and associated bioclimatic factors differentially affect leaf phenolics in three ivy species (*Hedera* L.) across the Iberian Peninsula. *Perspectives in Plant Ecology, Evolution and Systematics, 65*, 125822. https://doi.org/10.1016/j.ppees.2024.125822.

Ghosh, P., & Roychoudhury, A. (2024). Molecular basis of salicylic acid–phytohormone crosstalk in regulating stress tolerance in plants. *Brazilian Journal of Botany*. https://doi.org/10.1007/s40415-024-00983-3.

Giordano, T., Cerasa, G., Marotta, I., Conte, M., Orlando, S., Salamone, A., Mammano, M., Greco, C., & Tsolakis, H. (2025). Toxicity of essential oils of *Origanum vulgare, Salvia rosmarinus*, and *Salvia officinalis* against *Aculops lycopersici*. *Plants, 14*. https://doi.org/10.3390/plants14101462.

Gómez-Rocal, S., Cruz-Carrión, Á., Morales, D., García-Ruiz, A., Suárez, M., & Arola-Arnal, A. (2025). Foodomics approaches: New insights in phenolic compounds analysis. *Food Research International, 208*, 116168. https://doi.org/10.1016/j.foodres.2025.116168.

Guo, M., Lv, H., Chen, H., Dong, S., Zhang, J., Liu, W., He, L., Yu, H., Chen, S., & Luo, H. (2023). Strategies on biosynthesis and production of bioactive compounds in medicinal plants. *Chinese Herbal Medicines, 16*, 13–26. https://doi.org/10.1016/j.chmed.2023.01.007.

Hao, J., Lou, P., Han, Y., Zheng, L., Lu, J., Chen, Z., Ni, J., Yang, Y., & Xu, M. (2022). Ultraviolet-B irradiation increases antioxidant capacity of pakchoi (*Brassica rapa* L.) by inducing flavonoid biosynthesis. *Plants, 11*(6), 766. https://doi.org/10.3390/plants11060766.

Hawrylak-Nowak, B., Dresler, S., Stasińska-Jakubas, M., Wójciak, M., Sowa, I., & Matraszek-Gawron, R. (2021). NaCl-induced elicitation alters physiology and increases accumulation of phenolic compounds in *Melissa officinalis* L. *International Journal of Molecular Sciences, 22*, 6844. https://doi.org/10.3390/ijms22136844.

Huang, X., Tang, H., Wei, X., He, Y., Hu, S., Wu, J., Xu, D., Qiao, F., Xue, J., & Zhao, Y. (2024). The gradual establishment of complex coumarin biosynthetic pathway in *Apiaceae*. *Nature Communications, 15*. https://doi.org/10.1038/s41467-024-51285-x.

Humbal, A., & Pathak, B. (2023). Harnessing nanoparticle-mediated elicitation in plant tissue culture: A promising approach for secondary metabolite production. *Plant Cell, Tissue and Organ Culture, 155*, 385–402. https://doi.org/10.1007/s11240-022-02460-4.

Hussain, M. S., Fareed, S., Ansari, S., Rahman, M., Ahmad, I., & Saeed, M. (2022). Root cultures: A boon for the production of valuable secondary metabolites of medicinal plants. *Plants, 11*(3), 439. https://doi.org/10.3390/plants11030439.

Ivanova, S., Nalbantova, V., Benbassat, N., Dzhoglova, V., Dzhakova, Z., Koleva, N., Vasilev, V., Grekova-Kafalova, D., & Ivanov, K. (2025). Comparison between the chemical composition of essential oils isolated from biocultivated *Salvia rosmarinus* Spenn. (*Rosmarinus officinalis* L.) and some commercial products. *Pharmacia*. https://doi.org/10.3897/pharmacia.72.e140155.

Iqbal, M., Aftab, Z.-e-H., Anjum, T., Rizwana, H., Akram, W., Aftab, A., Sajid, Z. A., & Li, G. (2024). Nano-Integrated plant tissue culture to increase the rate of callus induction, growth, and curcuminoid production in *Curcuma longa*. *Plants, 13*, 1819. https://doi.org/10.3390/plants13131819.

Jan, R., Asaf, S., Numan, M., Lubna, & Kim, K.-M. (2021). Plant secondary metabolite biosynthesis and transcriptional regulation in response to biotic and abiotic stress conditions. *Agronomy, 11*(5), 968. https://doi.org/10.3390/agronomy11050968.

Jeyasri, R., Muthuramalingam, P., Karthick, K., Shin, H., Choi, S. H., & Ramesh, M. (2023). Methyl jasmonate and salicylic acid as powerful elicitors for enhancing the production of secondary metabolites in medicinal plants: An updated review. *Plant Cell, Tissue and Organ Culture (PCTOC, 153*(3), 447–458. https://doi.org/10.1007/s11240-023-02485-8.

Jiang, N., & Grotewold, E. (2024). Flavonoids and derived anthocyanin pigments in plants—Structure, distribution, function, and methods for quantification and characterization. *Cold Spring Harbor Protocols*. https://doi.org/10.1101/pdb.top108516.

Jimenez, E. Comparison of secondary plant metabolite production in cell suspension, callus culture and temporary immersion system. https://doi.org/10.1007/1-4020-3200-5_39.

Jomová, K., Alomar, S., Valko, R., Líška, J., Nepovimova, E., Kuča, K., & Valko, M. (2025). Flavonoids and their role in oxidative stress, inflammation, and human diseases. *Chemico-Biological Interactions*, 111489. https://doi.org/10.1016/j.cbi.2025.111489.

Kandoudi, W., Radácsi, P., & Németh-Zámboriné, É. (2022). Elicitation of medicinal plants in vivo—Is it a realistic tool? The effect of methyl jasmonate and salicylic acid on Lamiaceae species. *Horticulturae, 8*(1), 5. https://doi.org/10.3390/horticulturae8010005.

Kaur, A., Sharma, K., Pawar, S., & Sembi, J. (2025). Genome-wide characterization of PAL, C4H, and 4CL genes regulating the phenylpropanoid pathway in *Vanilla planifolia*. *Scientific Reports, 15*. https://doi.org/10.1038/s41598-024-81968-w.

Khanam, M. N., Anis, M., Javed, S., Mottaghipisheh, J., & Csupor, D. (2022). Adventitious root culture—an alternative strategy for secondary metabolite production: A review. *Agronomy, 12*(5), 1178. https://doi.org/10.3390/agronomy12051178.

Kolac, U. K., Ustuner, M. C., Tekin, N., Ustuner, D., Colak, E., & Entok, E. (2017). The anti-inflammatory and antioxidant effects of *Salvia officinalis* on lipopolysaccharide-induced inflammation in rats. *Journal of Medicinal Food, 20*(12), 1193–1200. https://doi.org/10.1089/jmf.2017.0035.

Krol, A., Kokotkiewicz, A., Krolicka, A., Hinc, K., Badura, A., Lorenc, A., Marzec-Wroblewska, U., Bucinski, A., Kuzma, L., & Luczkiewicz, M. (2025). Adventitious and hairy root cultures of *Salvia apiana* as a source of rosmarinic acid. *International Journal of Molecular Sciences, 26*, 3138. https://doi.org/10.3390/ijms26073138.

Khan, K., Saleem, M., Afzal, S., Hussain, I., Ameen, F., & Fahad, S. (2024). Ferulic acid: Therapeutic potential due to its antioxidant properties, role in plant growth, and stress tolerance. *Plant Growth Regulation*. https://doi.org/10.1007/s10725-024-01243-w.

Khan, S., & Boehr, D. (2025). Multi-enzyme synergy and allosteric regulation in the shikimate pathway: Biocatalytic platforms for industrial applications. *Catalysts, 15*(8), 718. https://doi.org/10.3390/catal15080718.

Kisiriko, M., Anastasiadi, M., Terry, L., Yasri, A., Beale, M., & Ward, J. (2021). Phenolics from medicinal and aromatic plants: Characterisation and potential as biostimulants and bioprotectants. *Molecules, 26*. https://doi.org/10.3390/molecules26216343.

Kowalczyk, T., Wieczfinska, J., Skała, E., Śliwiński, T., & Sitarek, P. (2020). Transgenesis as a tool for the efficient production of selected secondary metabolites from plant in vitro cultures. *Plants, 9*(2), 132. https://doi.org/10.3390/plants9020132.

Kukil, K., & Lindberg, P. (2024). Metabolic engineering of *Synechocystis* sp. PCC 6803 for the improved production of phenylpropanoids. *Microbial Cell Factories, 23*, 57. https://doi.org/10.1186/s12934-024-02330-3.

Kumar, B., Misra, A., & Srivastava, S. (2022). Bioactive phenolic compounds from Indian medicinal plants for pharmaceutical and medical aspects. *IntechOpen*. https://doi.org/10.5772/intechopen.108879.

Kwiecień, I., Szydłowska, A., Kawka, B., Sędzicka-Kubica, A., & Królikowska, K. (2015). Accumulation of biologically active phenolic acids in agitated shoot cultures of *Hypericum perforatum*. *Plant Cell, Tissue and Organ Culture, 123*, 273–281. https://doi.org/10.1007/s11240-015-0830-3.

Lee, J., Oh, M., Kang, H., Kim, J., Park, S., Jeong, H., Lee, S., & Kwak, S. (2019). UV regulates the expression of phenylpropanoid biosynthesis genes in cucumber (*Cucumis sativus* L.) in an organ- and spectrum-dependent manner. *Photochemical & Photobiological Sciences, 18*(2), 407–418. https://doi.org/10.1039/C8PP00480K.

Lewis, J., Jacobo, E., Palmer, N., Vermerris, W., Sattler, S., Brozik, J., Sarath, G., & Kang, C. (2024). Structural and interactional analysis of the flavonoid pathway proteins: Chalcone synthase, chalcone isomerase and chalcone isomerase-like protein. *International Journal of Molecular Sciences, 25*. https://doi.org/10.3390/ijms25115651.

Li, L., Wei, S., Shen, J., Liu, Y., Shi, X., Chen, J., & Wang, Y. (2014). The physiological response of *Artemisia annua* L. to salt stress and salicylic acid treatment. *Physiology and Molecular Biology of Plants, 20*(2), 161–169. https://doi.org/10.1007/s12298-014-0228-4.

Liu, Y., Singh, S., Pattanaik, S., Wang, H., & Yuan, L. (2023). Light regulation of the biosynthesis of phenolics, terpenoids, and alkaloids in plants. *Communications Biology, 6*. https://doi.org/10.1038/s42003-023-05435-4.

Machitescu, A., Stanga, C., Stănilă, A., & Diaconeasa, Z. (2024). Emerging technologies in polyphenol analysis: From traditional methods to -omics approaches in hop and medicinal plants. *Hop and Medicinal Plants, 32*(1–2), 56–101. https://doi.org/10.15835/hpm.v32i1-2.15019.

Malarz, J., Yudina, Y. V., & Stojakowska, A. (2023). Hairy root cultures as a source of phenolic antioxidants: Simple phenolics, phenolic acids, phenylethanoids, and hydroxycinnamates. *International Journal of Molecular Sciences, 24*, 6920. https://doi.org/10.3390/ijms24086920.

Manivel, P., & Chen, X. (2021). Chlorogenic, caffeic, and ferulic acids and their derivatives in foods. In Jianbo Xiao, Satyajit D. Sarker, & Yoshinori Asakawa (Eds.). *Handbook of dietary phytochemicals*Springerhttps://doi.org/10.1007/978-981-13-1745-3_22-1.

Marchiosi, R., dos Santos, W. D., Constantin, R. P., de Lima, R., Soares, A., Finger-Teixeira, A., Mota, T., Oliveira, D., de Paiva Foletto-Felipe, M., Abrahão, J., & Ferrarese-Filho, O. (2020). Biosynthesis and metabolic actions of simple phenolic acids in plants. *Phytochemistry Reviews, 19*, 1–42. https://doi.org/10.1007/s11101-020-09689-2.

Martínez-Chávez, L., Hernández-Ramírez, M., Feregrino-Pérez, A., & Escalante, K. (2024). Cutting-Edge strategies to enhance bioactive compound production in plants: Potential value of integration of elicitation, metabolic engineering, and green nanotechnology. *Agronomy, 14*(12), 2822.

Mendonça, E., Xavier, J., Fragoso, M., Silva, M., Escodro, P., Oliveira, A., Tucci, P., Saso, L., & Goulart, M. (2024). E-Stilbenes: General chemical and biological aspects, potential pharmacological activity based on the Nrf2 pathway. *Pharmaceuticals, 17*. https://doi.org/10.3390/ph17020232.

Merecz-Sadowska, A., Sadowski, A., Zielińska-Bliźniewska, H., Zajdel, K., & Zajdel, R. (2025). Network pharmacology as a tool to investigate the antioxidant and anti-inflammatory potential of plant secondary metabolites—A review and perspectives. *International Journal of Molecular Sciences, 26*. https://doi.org/10.3390/ijms26146678.

Mokhtari, R., Fard, M., Rezaei, M., Moftakharzadeh, S., & Mohseni, A. (2023). Antioxidant, antimicrobial activities, and characterization of phenolic compounds of thyme (*Thymus vulgaris* L.), Sage (*Salvia officinalis* L.), and thyme–Sage mixture extracts. *Journal of Food Quality*. https://doi.org/10.1155/2023/2602454.

Mohamed, T., Ali, S., Elshamy, A., Saleh, I., Ibrahim, M., Atia, M., Alshammari, S., Mohamed, A., Hussien, T., Hamed, A., Saedi, H., Abdelazim, N., Shams, K., Efferth, T., Saker, M., Paré, P., & Hegazy, M. (2022). Plant cell cultures: An enzymatic tool for polyphenolic and flavonoid transformations. *Phytomedicine: International Journal of Phytotherapy and Phytopharmacology, 100*, 154019. https://doi.org/10.1016/j.phymed.2022.154019.

Moreno-Garcia, L., Gariépy, Y., Barnabé, S., & Raghavan, G. S. V. (2019). Effect of environmental factors on the biomass and lipid production of microalgae grown in wastewaters. *Algal Research, 41*, 101521. https://doi.org/10.1016/j.algal.2019.101521.

Motolinía-Alcántara, E. A., Barba-de la Rosa, A. P., Hernández-Sánchez, H., Bautista-Rosales, P. U., & Muñoz-Ochoa, M. (2021). Engineering considerations to produce bioactive compounds from plant cell suspension culture in bioreactors. *Plants, 10*(12), 2762. https://doi.org/10.3390/plants10122762.

Mughal, A., Jabeen, N., Ashraf, K., Sultan, K., Farhan, M., Hussain, M., Deng, G., Alsudays, I., Saleh, M., Tariq, S., & Zaman, Q. (2024). Exploring the role of caffeic acid in mitigating abiotic stresses in plants: A review. *Plant Stress, 10*, 100487. https://doi.org/10.1016/j.stress.2024.100487.

Murthy, H. N., Lee, E. J., & Paek, K. Y. (2014). Produção de metabólitos secundários a partir de culturas de células e órgãos: Estratégias e abordagens para o aumento da biomassa e acúmulo de metabólitos. *Plant Cell Tiss Organ Cult, 118*, 1–16. https://doi.org/10.1007/s11240-014-0467-7.

Nadeem, M., Abbasi, B. H., Younas, M., Ahmad, W., Zahir, A., & Hano, C. (2019). LED-enhanced biosynthesis of biologically active ingredients in callus cultures of *Ocimum basilicum*. *J Photochem Photobiol B, 190*, 172–178. https://doi.org/10.1016/j.jphotobiol.2018.09.011 Epub 2018 Sep 15. PMID: 30268421.

Ninkuu, V., Aluko, O. O., Yan, J., Zeng, H., Liu, G., Zhao, J., Li, H., Chen, S., & Dakora, F. D. (2025). Phenylpropanoids metabolism: Recent insight into stress tolerance and plant development cues. *Frontiers in Plant Science, 16*, 1571825. https://doi.org/10.3389/fpls.2025.1571825.

Nguyen, N., Kennedy, V., Lee, J., Chan, N., & Chan, C. (2025). Programming nutrient detection with modular regulators for dynamic control of microbial biosynthesis. *ACS Synthetic Biology, 14*(3), 781–793. https://doi.org/10.1021/acssynbio.3c00570.

Pasternak, T. P., & Steinmacher, D. (2024). Plant growth regulation in cell and tissue culture in vitro. *Plants, 13*(2), 327. https://doi.org/10.3390/plants13020327.

Pratyusha, D. S., & Sarada, D. V. L. (2022). MYB transcription factors – Master regulators of phenylpropanoid biosynthesis and diverse developmental and stress responses. *Plant Cell Reports, 41*(12), 2245–2260. https://doi.org/10.1007/s00299-022-02927-1.

Radácsi, P., Szabó, K., Szabó, D., Trócsányi, E., & Németh-Zámbori, É. (2016). Effect of water deficit on yield and quality of lemon balm (*Melissa officinalis* L. *Zemdirbyste-Agriculture, 103*(4), 385–390. https://doi.org/10.13080/z-a.2016.103.049.

Rajesh, R. U., Praveen, S., Shukla, K., Nath, K., Tiwari, R., Upadhyay, R., Saini, R., Shukla, P., & Tiwari, K. N. (2024). Role of secondary metabolites in plant defense mechanisms: A molecular and biotechnological insight. *Phytochemistry Reviews, 24*, 953–983. https://doi.org/10.1007/s11101-024-09976-2.

Ramírez-Alvarado, R. A., Herrera-Arévalo, A. O., Ortiz-Rojas, Y., & Pérez-Rodríguez, C. P. (2020). Impact of ultraviolet B radiation applications on some secondary metabolites in thyme (*Thymus vulgaris* L). *Revista Colombiana de Ciencias Hortícolas, 14*(3), 416–423. https://doi.org/10.17584/rcch.2020v14i3.12082.

Ramirez-Estrada, K., Vidal-Limon, H., Hidalgo, D., Moyano, E., Golenioswki, M., Cusidó, R. M., & Palazon, J. (2016). Elicitation, an effective strategy for the biotechnological production of bioactive high-added value compounds in plant cell factories. *Molecules, 21*(2), 182. https://doi.org/10.3390/molecules21020182 PMID: 26848649; PMCID: PMC6273650.

Rattan, S., Kumar, A., Kumar, D., & Warghat, A. (2022). Enhanced production of phenylethanoids mediated through synergistic approach of precursor feeding and light regime in cell suspension culture of *Rhodiola imbricata* (Edgew). *Applied Biochemistry and Biotechnology, 194*, 3242–3260. https://doi.org/10.1007/s12010-022-03914-8.

Razzaq, A., Zafar, M. M., Ali, A., Ihsan, L., Qadir, F., Khan, M. N., Zhang, Y., Gao, L., Cong, H., Iqbal, R., Jiang, X., & Qiao, F. (2025). Elicitor-mediated enhancement of secondary metabolites in plant species: a review. *Frontiers in Plant Science, 16*. https://doi.org/10.3389/fpls.2025.1706600.

Riaz, M., Khalid, R., Afzal, M., Anjum, F., Fatima, H., Zia, S., Rasool, G., Egbuna, C., Mtewa, A., Uche, C., & Aslam, M. (2023). Phytobioactive compounds as therapeutic agents for human diseases: A review. *Food Science & Nutrition, 11*, 2500–2529. https://doi.org/10.1002/fsn3.3360.

Reshi, Z. A., Ahmad, W., Lukatkin, A. S., & Javed, S. B. (2023). From nature to lab: A review of secondary metabolite biosynthetic pathways, environmental influences, and in vitro approaches. *Metabolites, 13*(8), 895. https://doi.org/10.3390/metabo13080895.

Rehan, M. (2021). *Biosynthesis of diverse class flavonoids via shikimate and phenylpropanoid pathway*. IntechOpen. https://doi.org/10.5772/intechopen.96512.

Runeberg, P., Ryabukhin, D., Lagerquist, L., Rahkila, J., & Eklund, P. (2022). Transformations and antioxidative activities of lignans and stilbenes at high temperatures. *Food Chemistry, 404*, 134641. https://doi.org/10.1016/j.foodchem.2022.134641.

Sadeghi, A., Rajabiyan, A., Nabizade, N., Meygolinezhad, N., & Ahmady, A. (2024). Seaweed-derived phenolic compounds as diverse bioactive molecules: A review on identification, application, extraction and purification strategies. *International Journal of Biological Macromolecules,* 131147. https://doi.org/10.1016/j.ijbiomac.2024.131147.

Salami, M., Heidari, B., Batley, J., Wang, J., Tan, X., Richards, C., & Tan, H. (2024). Integration of genome-wide association studies, metabolomics, and transcriptomics reveals phenolic acid- and flavonoid-associated genes and their regulatory elements under drought stress in rapeseed flowers. *Frontiers in Plant Science, 14*. https://doi.org/10.3389/fpls.2023.1249142.

Santiago, A., Romero, P., Martínez, A., Martínez, M., Echeverria, J., Selles, S., Alvarez-Urdiola, R., Zhang, C., Navarro-Payá, D., Pizzio, G., Micó, E., Samper, A., Morante, J., Cigliano, R., Manzano, D., Bru, R., & Matus, J. (2024). Biosynthesis of oxyresveratrol in mulberry (*Morus alba*L.) is mediated by a group of p-coumaroyl-CoA 2′-hydroxylases acting upstream of stilbene synthases. *bioRxiv*. https://doi.org/10.1101/2024.04.04.588114.

Santos-Sánchez, N., Salas-Coronado, R., Hernández-Carlos, B., & Villanueva-Cañongo, C. (2019). Shikimic acid pathway in biosynthesis of phenolic compounds. In Marcos Soto-Hernández, Rosario García-Mateos, & Mariana Palma-Tenango (Eds.). *Plant physiological aspects of phenolic compounds*IntechOpenhttps://doi.org/10.5772/intechopen.83815.

Selwal, N., Supriadi, K., Rahayu, F., Sukmadjaja, D., Khamidah, A., Budiaarto, K., Antarlina, S., Tripatmasari, M., & Wani, A. K. (2025). Elicitation strategies for enhanced secondary metabolite synthesis in plant cell cultures and its role in plant defense mechanism. *Plant Gene, 41*, 100485. https://doi.org/10.1016/j.plgene.2024.100485.

Sharma, A., Shahzad, B., Rehman, A., Bhardwaj, R., Landi, M., & Zheng, B. (2019). Response of phenylpropanoid pathway and the role of polyphenols in plants under abiotic stress. *Molecules, 24*. https://doi.org/10.3390/molecules24132452.

Schwarze, C., & Petersen, M. (2024). Phenylalanine ammonia-lyases and 4-coumaric acid coenzyme A ligases in *Chara braunii, Marchantia polymorpha,* and *Physcomitrium patens* as extant model organisms for plant terrestrialization. *The Plant Journal, 117*(2), 356–370. https://doi.org/10.1111/tpj.16950.

Schröpfer, S., Haselmair-Gosch, C., Böttcher, C., Molitor, C., Keilwagen, J., Eidenberger, L., Marinovic, S., Hackl, M., Spornberger, A., Walliser, B., Schlosser, C., Stich, K., Rompel, A., Flachowsky, H., & Halbwirth, H. (2025). Identification of a hydroxycinnamoyl-CoA double bond reductase (HDR) affirms multiple pathways for dihydrochalcone formation in apple. *Plant Biotechnology Journal*. https://doi.org/10.1111/pbi.70225.

Shende, V., Bauman, K., & Moore, B. (2024). The shikimate pathway: Gateway to metabolic diversity. *Natural Product Reports, 41*(4), 664–684. https://doi.org/10.1039/d3np00037k.

Smetanska, I. (2018). Sustainable production of polyphenols and antioxidants by plant in vitro cultures. *In: Methods in Molecular Biology,* 225–269. https://doi.org/10.1007/978-3-319-32004-5_2-1.

Singh, H., Desimone, M., Pandya, S., Jasani, S., George, N., Adnan, M., Aldarhami, A., Bazaid, A., & Alderhami, S. (2023). Revisiting the green synthesis of nanoparticles: Uncovering influences of plant extracts as reducing agents for enhanced synthesis efficiency and its biomedical applications. *International Journal of Nanomedicine, 18*, 4727–4750. https://doi.org/10.2147/IJN.S464000.

Song, W., Shao, H., Zheng, A., Zhao, L., & Xu, Y. (2023). Advances in roles of salicylic acid in plant tolerance responses to biotic and abiotic stresses. *Plants, 12*. https://doi.org/10.3390/plants12193475.

Sun, W., & Shahrajabian, M. (2023). Therapeutic potential of phenolic compounds in medicinal plants—Natural health products for human health. *Molecules, 28*. https://doi.org/10.3390/molecules28041845.

Tang, S., Wang, B., Liu, X., Xi, W., Yue, Y., Tan, X., Bai, J., & Huang, L. (2024). Structural insights and biological activities of flavonoids: Implications for novel applications. *Food Frontiers, 5*(3), 744–758. https://doi.org/10.1002/fft2.494.

Thanh, N. T., Murthy, H. N., Pandey, D. M., Yu, K. W., Hahn, E. J., & Paek, K. Y. (2014). Optimization of *Panax ginseng* adventitious root culture in air-lift bioreactors for enhanced biomass and metabolite production. *Biochemical Engineering Journal, 90*, 125–131. https://doi.org/10.1016/j.indcrop.2014.06.036.

Tong, Y., Li, N., Zhou, S., Zhang, L., Xu, S., & Zhou, J. (2024). Improvement of chalcone synthase activity and high-efficiency fermentative production of (2S)-naringenin via in vivo biosensor-guided directed evolution. *ACS Synthetic Biology*. https://doi.org/10.1021/acssynbio.3c00570.

Utomo, J., Barrell, H., Kumar, R., Smith, J., Brant, M., De La Hoz Siegler, H., & Ro, D. (2024). Reconstructing curcumin biosynthesis in yeast reveals the implication of caffeoyl-shikimate esterase in phenylpropanoid metabolic flux. *Metabolic Engineering, 81*, 151–162. https://doi.org/10.1016/j.ymben.2024.02.011.

Utomo, J., Hodgins, C., & Ro, D. (2021). Multiplex genome editing in yeast by CRISPR/Cas9—A potent and agile tool to reconstruct complex metabolic pathways. *Frontiers in Plant Science, 12*, 719148. https://doi.org/10.3389/fpls.2021.719148.

Umar, A. H., Rafi, M., Ratnadewi, D., Sulistyaningsih, Y. C., & Hamim, H. (2022). henolics profile and antioxidant activities of in vitro propagules and field-raised plant organs of Curculigo latifolia. *Journal of Applied Pharmaceutical Science*. https://doi.org/10.7324/japs.2023.55995.

Valletta, A., Iozia, L., & Leonelli, F. (2021). Impact of environmental factors on stilbene biosynthesis. *Plants, 10*, 100. https://doi.org/10.3390/plants10010090.

Verdú-Navarro, F., Corral-Martínez, P., Seguí-Simarro, J. M., & Gómez-Gómez, L. (2023). The advent of plant cells in bioreactors. *Frontiers in Plant Science, 14*, 10749943. https://doi.org/10.3389/fpls.2023.10749943.

Vojnović, Đ., Maksimović, I., Tepić Horecki, A., Milić, A., Šumić, Z., Žunić, D., Adamović, B., & Ilin, Ž. (2024). Biostimulants improve bulb yield, concomitantly affecting the total phenolics, flavonoids, and antioxidant capacity of onion (*Allium cepa*). *Horticulturae, 10*(4), 391. https://doi.org/10.3390/horticulturae10040391.

Wang, S., He, G., Liu, Y., Wang, Y., Fu, C., Xu, H., Hu, R., & Li, S. (2024). A P1-like MYB transcription factor boosts biosynthesis and transport of C-glycosylated flavones in duckweed. *International Journal of Biological Macromolecules, 235*, 134138. https://doi.org/10.1016/j.ijbiomac.2024.134138.

Wawrosch, C., & Zotchev, S. B. (2021). Produção de metabólitos secundários bioativos de plantas por meio de tecnologias in vitro—Estado atual e perspectivas. *Applied Microbiology and Biotechnology, 105*, 6649–6668. https://doi.org/10.1007/s00253-021-11539-w.

Xing, B., Yang, D., Guo, W., Liang, Z., Yan, X., Zhu, Y., & Liu, Y. (2015). Ag^+ as a more effective elicitor for production of tanshinones than phenolic acids in *Salvia miltiorrhiza* hairy roots. *Molecules, 20*(1), 309–324. https://doi.org/10.3390/molecules20010309.

Yu, H., Liao, J., Jiang, Y., Zhong, M., Tao, S., Chai, S., Wang, L., Lin, L., Yang, R., Deng, X., Zhang, Y., Pu, X., Liu, M., & Zhang, L. (2025). Ecotype-specific phenolic acid accumulation and root softness in *Salvia miltiorrhiza* are driven by environmental and genetic factors. *Plant Biotechnol J, 23*(6), 2224–2241.

Yahya, M. (2024). Trends and insights in *Lavandula angustifolia* research: A critical literature review (CLR) and bibliometric analysis of scopus and pubmed databases. *International Journal of Life Sciences and Biotechnology, 7*(3), 161–179. https://doi.org/10.38001/ijlsb.1514950.

Yusoff, I., Taher, Z., Rahmat, Z., & Chua, L. (2022). A review of ultrasound-assisted extraction for plant bioactive compounds: Phenolics, flavonoids, thymols, saponins and proteins. *Food Research International, 157*, 111268. https://doi.org/10.1016/j.foodres.2022.111268.

Zagoskina, N., Zubova, M., Nechaeva, T., Kazantseva, V., Goncharuk, E., Katanskaya, V., Baranova, E., & Aksenova, M. (2023). Polyphenols in plants: Structure, biosynthesis, abiotic stress regulation, and practical applications (Review). *International Journal of Molecular Sciences, 24*.

Zhakipbekov, K., Turgumbayeva, A., Akhelova, S., et al. (2024). Antimicrobial and other pharmacological properties of *Ocimum basilicum* (Lamiaceae). *Molecules, 29*(2), 388. https://doi.org/10.3390/molecules29020388.

Zhang, B., Lan, W., Wang, Z., Shao, Z., & Xie, J. (2024). Modified chitosan with different phenolic acids: Characterization, physicochemical properties, and biological activity. *Food Chemistry, 441*, 138337. https://doi.org/10.1016/j.foodchem.2023.138337.

Zhao, Y., Liu, G., Yang, F., et al. (2023). Multilayered regulation of secondary metabolism in medicinal plants. *Molecular Horticulture, 3*, 11. https://doi.org/10.1186/s43897-023-00059-y.

Zhao, Z., Liu, S., Yun, C., Liu, J., Yao, L., & Wang, H. (2024). Melatonin alleviates UV-B stress and enhances phenolic biosynthesis in rosemary (*Rosmarinus officinalis*) callus. *Physiologia Plantarum, 176*(4), e14453. https://doi.org/10.1111/ppl.14453.

Zheng, X., Zhang, X., & Zeng, F. (2025). Biological functions and health benefits of flavonoids in fruits and vegetables: A contemporary review. *Foods, 14*. https://doi.org/10.3390/foods14010123.

Zeb, A. (2021). Biosynthesis of phenolic antioxidants. In Alam Zeb (Ed.). *Phenolic Antioxidants in Foods: Chemistry, Biochemistry and Analysis*Springerhttps://doi.org/10.1007/978-3-030-74768-8_11.

Zhong, J., Ran, Q., Han, Y., Gan, L., & Dong, C. (2025). Biosynthetic mechanisms of plant chlorogenic acid from a microbiological perspective. *Microorganisms, 13*. https://doi.org/10.3390/microorganisms13051114.

CHAPTER

7

Advancing phytochemical biosynthesis through *in vitro* mutagenesis: strategies and applications

Mohamed Addi, Amine Elbouzidi, and Mohamed Taibi

Laboratoire d'Amélioration des Productions Agricoles, Biotechnologie et Ecophysiologie (LAPABE), Faculté des Sciences, Université Mohammed Premier, Oujda, Morocco

7.1 Introduction

7.1.1 Importance of secondary metabolites in medicinal plants

Secondary metabolites represent a structurally and functionally diverse class of bioactive compounds synthesized by plants as part of their adaptive responses to biotic and abiotic stresses. Unlike primary metabolites, which are essential for growth and development, secondary metabolites are not directly required for survival but play crucial roles in ecological interactions, including defense against herbivores and pathogens, attraction of pollinators, and competition with neighboring plants (Hartmann, 2007; Wink, 2010). These compounds encompass major classes such as alkaloids, flavonoids, terpenoids, phenolics, saponins, and glycosides, many of which are endowed with therapeutic activities (Elbouzidi, Haddou, et al., 2025). Alkaloids like vincristine and vinblastine from *Catharanthus roseus* are widely used in cancer chemotherapy (van der Heijden et al., 2004). Flavonoids such as quercetin and kaempferol demonstrate antioxidant, cardioprotective, and antiinflammatory effects (Panche et al., 2016). Terpenoids, including artemisinin from *Artemisia annua*, exhibit potent antimalarial activity, while phenolics like resveratrol from grapes (*Vitis vinifera*) are renowned for their anticancer and antiaging properties (Bourgaud et al., 2001; Lattanzio, 2013).

Beyond pharmaceutical relevance, secondary metabolites also have industrial and agricultural applications. In agriculture, compounds such as pyrethrins (derived from *Chrysanthemum cinerariifolium*) function as natural insecticides, while phenolic acids act as allelopathic agents that suppress weed growth (Duke et al., 2000). In the food industry,

Medicinal Plant Biotechnology for Ameliorated Secondary Metabolite Production
DOI: https://doi.org/10.1016/B978-0-443-43944-5.00011-6

metabolites like essential oils are exploited as natural flavorings, preservatives, and antioxidants to extend shelf life and enhance sensory properties of food products (Bakkali et al., 2008). The global market for phytopharmaceuticals and nutraceuticals continues to expand, driven by consumer preference for natural products and evidence-based validation of plant-derived compounds (Cragg & Newman, 2013). This demand highlights the importance of developing biotechnological strategies—such as metabolic engineering, elicitor treatments, and cell culture systems—to boost the yield, stability, and diversity of bioactive secondary metabolites (Ochoa-Villarreal et al., 2016; Ramawat et al., 2009; Verpoorte & Alfermann, 2000).

7.1.2 Limitations of conventional breeding for phytochemical enhancement

Conventional plant breeding has historically contributed to crop improvement, including the selection of varieties with elevated levels of secondary metabolites. For example, selective breeding in *Capsicum annuum* has increased capsaicinoid content, while breeding in *Mentha* species has enhanced essential oil profiles (Nirmal Babu & Divakaran, 2015). However, the approach faces significant limitations when applied to the targeted enhancement of complex phytochemical traits.

First, long generation times and the requirement for successive cycles of crossing and selection slow the development of improved varieties, especially in perennial medicinal plants (Pilbeam et al., 2020). Second, the natural genetic diversity available for metabolite traits is often limited, restricting the scope for selection (Kumar, 2018). Third, secondary metabolite accumulation is typically a polygenic trait with low heritability and is strongly influenced by environmental conditions such as light, temperature, and nutrient availability (Smith et al., 2023). For instance, alkaloid content in *Papaver somniferum* and essential oil yield in *Ocimum basilicum* vary significantly under different agro-climatic conditions, complicating the breeding process (Facchini, 2001; Simon et al., 1999). Moreover, conventional breeding lacks the precision required for manipulating specific biosynthetic pathways. Many pathways involve multiple enzymes and regulatory networks, and improving one metabolite often leads to trade-offs in others due to pathway competition (Verpoorte & Memelink, 2002). For example, attempts to enhance anthocyanin content in certain crops often impact flavonol levels due to shared precursors in the flavonoid biosynthetic pathway. Additionally, linkage drag—the coinheritance of undesirable traits with the targeted trait—further reduces the efficiency of conventional breeding in phytochemical enhancement (Kole, 2020).

Given these challenges, conventional breeding alone is inadequate for the rapid and targeted enhancement of secondary metabolites. Modern approaches such as mutagenesis, genetic transformation, CRISPR-based genome editing, and metabolic engineering are increasingly being integrated with traditional methods to overcome these limitations and to develop elite medicinal plant varieties with optimized phytochemical profiles (Alseekh et al., 2021; Ramawat et al., 2009).

7.1.3 Emergence of *in vitro* mutagenesis as a tool for phytochemical improvement

In vitro mutagenesis has emerged as an efficient alternative to conventional breeding for the improvement of secondary metabolite production in medicinal plants. Unlike traditional field-based mutagenesis, which is constrained by environmental variability and long generation times, *in vitro* approaches allow for the controlled application of mutagenic agents to isolated plant tissues such as callus, suspension cultures, and meristems under sterile conditions (Jain, 2005; Shu & Forster, 2012). Both physical mutagens (e.g., gamma rays, X-rays, and ion beams) and chemical mutagens (e.g., ethyl methanesulfonate [EMS], sodium azide [NaN_3], and colchicine) have been employed to generate heritable changes in cultured tissues that can subsequently be regenerated into whole plants (Mba, 2013). The major advantage of *in vitro* mutagenesis lies in its capacity to induce targeted variability in secondary metabolite pathways by creating mutations that activate silent genes, enhance flux through biosynthetic routes, or deregulate transcriptional and enzymatic control points (Alves et al., 2014). For example, gamma-irradiated callus cultures of *Stevia rebaudiana* produced mutant lines with significantly higher stevioside and rebaudioside-A contents compared to the wild type (Sharma et al., 2023). Similarly, EMS-induced hairy root cultures of *Catharanthus roseus* resulted in mutants with elevated yields of vinblastine and vincristine, two alkaloids of critical pharmaceutical importance (Das et al., 2020).

Furthermore, the use of *in vitro* systems permits high-throughput mutant screening under reproducible culture conditions, reducing the time and space required for cultivar development compared to field-based mutagenesis (Riaz & Gul, 2015). By combining mutagenesis with analytical tools such as high-performance liquid chromatography (HPLC), LC–MS, and NMR, researchers can rapidly identify mutants with altered phytochemical profiles. This approach not only accelerates the discovery of novel variants with enhanced metabolite content but also provides access to unique chemical diversity that may not be available through natural variation or conventional breeding.

Consequently, *in vitro* mutagenesis is increasingly recognized as a strategic tool in plant metabolic engineering. Its integration with molecular biology and omics technologies—such as transcriptomics, metabolomics, and genome-wide association studies—further enhances its potential to dissect and optimize complex biosynthetic networks for sustainable production of high-value phytochemicals (Razzaq & Wishart, 2022; Wijerathna-Yapa et al., 2022).

7.1.4 Chapter objectives and scope

This chapter seeks to provide a comprehensive understanding of *in vitro* mutagenesis as a versatile approach for enhancing secondary metabolite biosynthesis in medicinal plants. The specific objectives are fourfold:

1. To describe the types and mechanisms of mutagenic agents—including physical mutagens such as gamma rays and ion beams and chemical agents such as EMS and NaN_3—highlighting their modes of action on plant genomes and their applications in tissue culture systems.
2. To examine the impact of induced mutations on metabolic pathways and gene regulation, focusing on how mutations can alter the expression of structural genes, regulatory transcription factors, and key enzymes involved in secondary metabolism.
3. To explore the integration of mutagenesis with modern omics platforms and genome editing technologies, such as metabolomics-driven mutant screening, CRISPR/Cas-mediated precision editing, and systems biology approaches for pathway optimization.

4. To present case studies of successful applications in which *in vitro* mutagenesis has led to enhanced production of alkaloids, flavonoids, terpenoids, and phenolics, emphasizing its role in pharmaceutical, nutraceutical, and industrial sectors.

This chapter places particular emphasis on the synehanrgistic potential of combining tissue culture, mutagenesis, and metabolic engineering. By leveraging high-throughput screening, molecular markers, and functional genomics, *in vitro* mutagenesis can be harnessed not only to improve yields of established compounds but also to uncover novel bioactive metabolites. Ultimately, this integrative strategy offers a sustainable and innovative route for expanding the phytochemical repertoire of medicinal plants in response to growing global demands for natural products.

7.2 Fundamentals of *in vitro* mutagenesis

7.2.1 Concept and mechanism

In vitro mutagenesis refers to the deliberate induction of genetic alterations in plant cells or tissues under controlled laboratory conditions. Unlike natural mutation, which is random and often rare, induced mutagenesis uses external agents—known as mutagens—to increase the frequency of mutation events. These mutations may result in base substitutions, insertions, deletions, or chromosomal rearrangements that can lead to observable phenotypic changes and potentially beneficial traits, such as elevated levels of secondary metabolites (Jain, 2010).

The process typically begins with the selection of explants (e.g., callus, leaf discs, or meristems), which are then exposed to mutagenic treatments. Following exposure, the tissues are cultured under suitable conditions to regenerate whole plants (Misra & Saema, 2016). The regenerated plants, or somaclones, are screened for desirable traits, often linked to enhanced phytochemical production. The underlying mechanisms by which mutagenesis influences phytochemical pathways include gene activation or inactivation, alteration in promoter regions affecting gene expression, and modulation of regulatory elements involved in secondary metabolism (Singh, 2023).

7.2.2 Classification of mutagens

Mutagens are agents capable of inducing genetic mutations by altering the structure or sequence of DNA. In the context of plant biotechnology, particularly *in vitro* mutagenesis, these agents are utilized to create genetic variation that is crucial for crop improvement, functional genomics, and phytochemical enhancement (Kolakar et al., 2018; Xu et al., 2012). Mutagens are generally categorized into physical, chemical, and biological classes based on their origin and mechanism of action.

7.2.2.1 Physical mutagens

Physical mutagens include ionizing and nonionizing radiation sources that induce mutations by directly interacting with the atomic structure of DNA, often leading to chromosomal rearrangements, breaks, or base modifications (Table 7.1. A diverse range of

TABLE 7.1 Physical mutagens and their application in plant secondary metabolites enhancement.

Mutagen type	Source/characteristics	Mechanism of action	Applications in plants	Notable examples	References
Gamma rays	Emitted from radioisotopes (Co-60, Cs-137); highly penetrative; suitable for whole plant systems and dense tissues	Induces double-strand DNA breaks, chromosomal deletions, inversions, translocations; causes knockout or novel alleles	Used in mutation breeding, enhancement of secondary metabolites, whole plant irradiation	Increased withanolide content in *Withania somnifera*; higher alkaloid yield in *Catharanthus roseus*; enhanced capsaicinoids in *Capsicum annuum*	Kodym and Afza (2003), Mangaiyarkarasi and Girija, 2014, Asif (2023)
X-rays	Generated via electron bombardment; lower energy and penetration than gamma rays	Causes DNA strand scission, chromosome fragmentation, and micronucleus formation	More suited for superficial tissues, meristematic regions, and *in vitro* cultures	Induction of leaf variegation and novel flower morphology in *Chrysanthemum morifolium*; resistance traits in *Nicotiana tabacum*	Broertjes and van Harten (2013), Kodym and Afza (2003)
UV-C (200–280 nm)	Nonionizing radiation; minimal penetration depth	Causes formation of pyrimidine dimers (especially thymine dimers), DNA helix distortion, point mutations	Suitable for surface-level mutagenesis in callus, suspension cultures, and microbial strains	Mutagenesis in *Mentha arvensis* for altered essential oil composition; enhanced pigment production in *Amaranthus spp.*	Oladosu et al. (2016), Rai et al. (2011)
UV-B/UV-A (280–400 nm)	Less energetic than UV-C; causes indirect damage via ROS	Triggers oxidative stress pathways; induces ROS leading to single-strand breaks, lipid peroxidation	Used to mimic abiotic stress, induce secondary metabolite pathways	Flavonoid biosynthesis enhancement in *Arabidopsis thaliana*; lignin accumulation in *Populus spp.*	Hideg et al. (2013), Jenkins (2009)
Heavy-ion beams	Ion beams from particle accelerators (e.g., carbon, argon)	Cause dense ionization tracks, complex DNA damage (clustered lesions)	High mutation frequency with low chimerism; used for precision breeding	Mutation of flowering genes in *Lotus japonicus*; anthocyanin increase in *Gerbera jamesonii*	Guo et al. (2024)

(*Continued*)

TABLE 7.1 (Continued)

Mutagen type	Source/characteristics	Mechanism of action	Applications in plants	Notable examples	References
Microwaves	Nonionizing, low-energy electromagnetic radiation	Generates thermal and nonthermal effects; causes membrane disruption, metabolic alterations	Limited use in plant mutagenesis; more often for seed germination or stress pre-treatment	Induced early flowering and improved antioxidant activity in *Trigonella foenum-graecum*	Ma et al. (2021)
Neutron irradiation	Emitted from nuclear reactors; high LET (linear energy transfer)	High mutagenic efficiency; breaks DNA and causes point mutations, deletions	Rare due to infrastructure needs; effective in seed mutagenesis	Mutagenesis in *Oryza sativa* and *Arachis hypogaea* for salinity tolerance	Domingo et al. (2016)

physical mutagens has been employed in plant biotechnology, each differing in source, mechanism of action, and suitability for particular tissues or experimental systems. Understanding these parameters is essential for optimizing mutagenic efficiency and tailoring treatments to specific research goals. Gamma rays are among the most extensively used mutagens in plant improvement programs. Emitted from radioisotopes such as cobalt-60 and cesium-137, gamma rays possess high penetration power, making them suitable for whole plants, seeds, and dense tissue systems. Their primary mutagenic effect involves the induction of double-strand DNA breaks, which may lead to chromosomal rearrangements such as deletions, inversions, and translocations, ultimately generating knockout alleles or novel variants (Kodym & Afza, 2003). In medicinal plants, gamma irradiation has been successfully applied to enhance secondary metabolites, such as increased withanolide content in *Withania somnifera*, higher alkaloid yields in *Catharanthus roseus*, and improved capsaicinoid accumulation in *Capsicum annuum* (Asif, 2023; Mangaiyarkarasi & Girija, 2014).

X-rays, produced by electron bombardment, operate at lower penetration compared to gamma rays, restricting their effects primarily to superficial tissues and meristematic regions. X-ray irradiation induces DNA strand scission and chromosome fragmentation, which often results in micronucleus formation (Broertjes & van Harten, 2013). Due to their limited penetration, X-rays are widely used in *in vitro* cultures for ornamental plant improvement. Examples include the induction of variegated leaves and novel flower morphologies in *Chrysanthemum morifolium*, as well as resistance traits in *Nicotiana tabacum* (Kodym & Afza, 2003). Ultraviolet (UV) radiation represents another important category of mutagenic agents. UV-C (200–280 nm) is nonionizing but highly effective in inducing pyrimidine dimers, particularly thymine dimers, which distort the DNA helix and cause point mutations. Its limited penetration confines its utility to surface-level mutagenesis, making it particularly suited for callus and suspension cultures. Applications include the alteration of essential oil composition in *Mentha arvensis* and pigment enhancement in *Amaranthus* spp. (Oladosu et al., 2016; Rai et al., 2011). By contrast, UV-A and UV-B (280–400 nm) have less energy but act indirectly through the generation of reactive oxygen species (ROS). These species trigger oxidative stress, leading to single-strand breaks and activation of stress-responsive pathways (Jenkins, 2009). Such treatments have been shown to upregulate flavonoid biosynthesis in *Arabidopsis thaliana* and lignin accumulation in *Populus* spp. (Hideg et al., 2013).

Heavy-ion beams generated from particle accelerators (e.g., carbon or argon ions) have gained prominence due to their unique ability to create dense ionization tracks that produce complex DNA damage, including clustered double-strand breaks. Unlike conventional mutagens, heavy-ion beams offer high mutation frequencies with relatively low chimerism, making them particularly effective in precision breeding (Guo et al., 2024). Applications include the mutation of flowering-related genes in *Lotus japonicus* and increased anthocyanin accumulation in *Gerbera jamesonii* (Sun & Xin, 2025). Microwave radiation represents a nonionizing mutagen with relatively limited application in plant mutagenesis (Ma et al., 2021). Its effects are mediated through both thermal and nonthermal mechanisms, resulting in membrane disruption, metabolic alterations, and stress induction. While its use is more common in seed germination enhancement and abiotic stress priming, mutagenesis studies have reported induced early flowering and improved antioxidant activity in *Trigonella foenum-graecum* (Bashir et al., 2012).

Finally, neutron irradiation, characterized by high linear energy transfer (LET), is capable of inducing efficient DNA breaks, point mutations, and deletions. Although its application is rare due to the requirement of specialized infrastructure such as nuclear reactors, neutron mutagenesis remains effective for seed treatment. Reported cases include the improvement of salinity tolerance in *Oryza sativa* and *Arachis hypogaea* (Joint, 2018).

7.2.2.2 Chemical mutagens

Chemical mutagens represent one of the most versatile classes of agents used in plant biotechnology due to their ability to induce high-frequency point mutations at relatively low concentrations. Compared to physical mutagens, they often result in fewer chromosomal aberrations and more targeted base substitutions, making them especially useful for creating mutant libraries and conducting functional genomics (Table 7.2. Their efficacy and specificity, however, are influenced by chemical properties, tissue type, treatment duration, and metabolic activation within the plant system.

EMS is the most widely used alkylating agent in plant mutagenesis. It is highly effective at low concentrations (0.1%–2%) and stable in aqueous solutions, allowing controlled exposure. EMS predominantly alkylates guanine at the O^6 position, leading to G/C → A/T transitions through base mispairing during DNA replication (Till et al., 2003). Its high mutagenic efficiency, coupled with relatively low lethality, makes it the preferred mutagen for forward and reverse genetics. In *Arabidopsis thaliana*, EMS-mutagenized populations have formed the basis for large-scale TILLING (Targeting Induced Local Lesions IN Genomes) platforms (Kurowska et al., 2011). In crop systems, EMS has been applied to increase lycopene content in *Solanum lycopersicum*, providing mutants valuable for nutritional improvement. NaN_3 is another frequently used mutagen, notable for its relatively simple treatment protocols and low frequency of chromosomal aberrations compared to alkylating agents. In plant tissues, NaN_3 undergoes metabolic activation to produce azidoalanine, which incorporates into DNA and induces base substitutions during replication (Khan et al., 2009). It is particularly effective in cereals and legumes, where it has been applied for inducing agronomically important traits such as salt tolerance in *Triticum aestivum*. In medicinal plants like *Hyoscyamus muticus*, NaN_3 treatment has led to enhanced alkaloid production, demonstrating its potential for phytochemical improvement (Wani and Dar, 2018).

N-methyl-*N*′-nitro-*N*-nitrosoguanidine (MNNG) is a potent nitrosated guanidine derivative that functions as a strong alkylating agent. It methylates DNA bases, notably guanine, producing O^6-methylguanine, which mispairs with thymine, causing transition mutations (Muyibi & Jimat, 2014). Despite its mutagenic strength, its high cytotoxicity has limited its direct application in higher plants. Instead, MNNG is more commonly employed in algal and microbial systems (Gichner & Veleminský, 1982). Diethyl sulfate (DES), chemically related to EMS, is another alkylating mutagen but less stable and more volatile (Pavan et al., 2011). It ethylates purine bases and induces point mutations with occasional frameshifts. Although less commonly used today, DES has been applied in seed mutagenesis and callus culture systems.

Hydroxylamine represents a more specific mutagen, targeting cytosine residues to produce hydroxylated cytosine that pairs with adenine, resulting in G/C → A/T

TABLE 7.2 Chemical mutagens and their application in plant secondary metabolites enhancement.

Mutagen type	Chemical nature/ properties	Mechanism of action	Applications in plants	Notable examples	References
Ethyl methanesulfonate (EMS)	Alkylating agent; stable in aqueous solution; active at low concentrations (0.1%–2%)	Ethylates guanine at O^6 position → G/C to A/T transitions; base mispairing during replication	High-efficiency mutagenesis; widely used in TILLING platforms and mutant library creation	EMS-mutagenized *Arabidopsis thaliana* population used for functional genomics; enhanced lycopene content in *Solanum lycopersicum*	Kurowska et al. (2011), Till et al. (2003)
Sodium azide (NaN_3)	Azide compound; metabolically converted into mutagenic azidoalanine	Causes point mutations via base substitution and mispairing during DNA replication	Suitable for cereals, legumes, and model plants; induces minimal chromosomal aberrations	Increased alkaloid content in *Hyoscyamus muticus*; mutagenesis in *Triticum aestivum* for salt tolerance	Wani and Dar (2018)
***N*-methyl-*N′*-nitro-*N*-nitrosoguanidine (MNNG)**	Nitrosated guanidine derivative; potent alkylating agent	Methylates DNA bases → O^6-methylguanine pairing with thymine → transition mutations; high cytotoxicity	Limited use in plants due to toxicity; applied in algae and microbial systems	Mutagenesis in *Chlorella vulgaris* for lipid overproduction; UV-MNNG combined treatment in *Lemna minor*	Gichner and Velemínský (1982)
Diethyl sulfate (DES)	Volatile alkylating agent; similar to EMS	Ethylates purine bases; causes point mutations and occasional frameshifts	Used in seed treatment and callus cultures; less stable than EMS	Used in *Pisum sativum* to induce early flowering; mutagenesis in *Brassica napus*	Pavan et al. (2011)
Hydroxylamine	Hydroxylating agent specific to cytosine	Hydroxylates cytosine → mispairing with adenine → G/C to A/T transitions	Less commonly used; primarily for microbial or molecular studies	Site-directed mutagenesis in chloroplast DNA of *Chlamydomonas reinhardtii*	Przibilla et al. (1991), Humphreys et al. (1976)
Azide-alkyl combinations (e.g., NaN_3 + EMS)	Synergistic mixture; increases mutation frequency	Combines alkylation and base mispairing with azide-induced miscoding	Enhanced mutagenic efficiency in cereals and legumes	Applied in *Lens culinaris* and *Cicer arietinum* to improve protein content and stress tolerance	Atacan et al. (2018)

transitions. Its use in higher plant systems is limited due to relatively low efficiency, but it has played an important role in microbial and molecular studies. Notably, it has been applied in *Chlamydomonas reinhardtii* to induce targeted mutations in chloroplast DNA, providing insights into photosynthetic gene function (Reuter & Berg, 1974). Finally, combinatorial treatments, such as NaN_3 combined with EMS, have been explored to enhance mutagenic efficiency. These mixtures synergize the miscoding effects of azide with the alkylation activity of EMS, leading to higher mutation frequencies and broader genetic variability. Applications in legumes such as *Lens culinaris* and *Cicer arietinum* have produced mutants with improved protein content and enhanced stress tolerance (Böhmer et al., 2023).

7.2.2.3 *Biological mutagens and comparisons with physical and chemical mutagens*

Biological mutagens represent a less frequently discussed but equally significant category of mutagenic agents in the context of *in vitro* mutagenesis and plant breeding. Unlike physical and chemical mutagens, biological mutagens exert their mutagenic effect primarily through infection or insertional mutagenesis, often involving pathogenic or symbiotic organisms such as viruses, bacteria, transposons, and even certain fungi (Table 7.3. Their activity leads to genomic instability, DNA rearrangements, or gene disruptions that can result in heritable mutations, some of which may manifest as beneficial agronomic traits.

One of the most studied biological mutagens in plants is the class of transposable elements (TEs), also referred to as "jumping genes." These sequences have the unique ability to change their positions within the genome and often carry with them genetic cargo that can activate, silence, or modify gene expression (McClintock, 1948). The movement of transposons such as *Ac/Ds* (activator/dissociation) and *Mutator* elements in maize, or *Tnt1* retrotransposons in *Medicago truncatula*, has facilitated the discovery of gene function and has been exploited as an insertional mutagenesis tool in functional genomics (D'Erfurth et al., 2003).

Viruses, particularly DNA and RNA viruses, can also induce mutations by integrating viral genetic material into the plant genome or disrupting regulatory pathways through viral protein interactions. For example, geminiviruses have been engineered to deliver genome editing tools (like CRISPR-Cas9) to induce targeted mutations (Baltes et al., 2014). Similarly, *Agrobacterium tumefaciens* is a well-known bacterial mutagenic agent through its Ti plasmid, which causes crown gall disease by integrating T-DNA into the plant genome (Gelvin, 2003). This property has been exploited in genetic engineering to introduce desired traits. Moreover, some bacterial toxins and fungal metabolites can act indirectly as mutagens by interfering with DNA replication or repair mechanisms, leading to mutagenesis. The biological origin of these agents also raises concerns about host-pathogen interactions, making their use more context-specific.

Biological mutagens, although less controlled than physical or chemical agents, offer unique advantages, particularly for gene discovery, targeted insertional mutagenesis, and the induction of novel phenotypes that may not be accessible through other means. Their role is increasingly being recognized in modern plant mutagenesis protocols, especially in combination with tissue culture systems and genome editing platforms.

TABLE 7.3 Comparative overview of physical, chemical, and biological mutagens in *in vitro* mutagenesis.

Type	Examples	Mode of action	Advantages	Limitations
Physical	Gamma rays, X-rays, UV light, neutron radiation	Induce double-strand breaks, point mutations, chromosomal aberrations	High penetrability (γ-rays), reproducibility, minimal contamination risk	Requires irradiation facility, may cause large-scale damage
Chemical	EMS, MMS, sodium azide, nitrosoureas, base analogs	Alkylation, deamination, intercalation, base substitution, Indels	High mutation frequency, cost-effective, easy to apply	Toxicity, hazardous handling, off-target effects
Biological	Transposons (*Ac/Ds*, *Mutator*), *Agrobacterium tumefaciens*, geminiviruses	Gene disruption, T-DNA integration, insertional mutagenesis, virus-induced gene silencing, or activation	Precise insertional mutagenesis, functional genomics, compatibility with *in vitro* tissue culture	Unpredictable integration, host defense response, potential pathogenicity

7.2.3 Factors influencing mutagenic effectiveness

Mutagenesis in *in vitro* systems is a powerful tool for inducing genetic variability and developing novel traits in plants. However, its success depends on a range of biological and technical parameters that determine both mutation frequency and tissue regeneration capacity. Optimizing these factors is crucial to balance effective mutagenesis with the survival of viable explants.

The dosage and exposure duration of mutagens are among the most critical determinants of mutagenic effectiveness. An optimal concentration must be identified that maximizes mutation frequency without causing excessive cytotoxicity. The LD_{50} assay is commonly used to determine this threshold, ensuring approximately 50% survival of treated tissues (Yadav & Redden, 2018). For example, in *Chrysanthemum morifolium*, treatment with gamma irradiation at doses above 30 Gy severely reduced shoot regeneration, while moderate doses (10–20 Gy) induced useful phenotypic variability without complete lethality Datta 2023. Similarly, in rice (*Oryza sativa*), prolonged exposure to EMS significantly increased mutation frequency but also led to reduced seed germination and survival Shamshad et al., 2023. These findings emphasize that both dose and exposure time interact synergistically to influence mutation outcomes, requiring fine adjustments depending on the species and mutagen employed.

The developmental stage and physiological state of the explant also determine mutagenic responsiveness. Meristematic and embryogenic tissues, due to their high metabolic activity and DNA replication rates, are particularly susceptible to mutagenic changes (Shu & Forster, 2012). For instance, in bananas (*Musa* spp.), irradiated shoot tips produced a wide range of somaclonal variants, whereas mature leaf explants exhibited poor regeneration after mutagen exposure (Kodym & Afza, 2003). Callus and suspension cultures are also widely used, as seen in barley (*Hordeum vulgare*), where mutagenic treatment of embryogenic callus yielded high mutation frequencies that were stably inherited in regenerated plants (Jain, 2005). Thus the careful selection of responsive tissues enhances the efficiency of mutation induction and recovery of viable mutants.

Different cultivars or accessions of the same species often display variable sensitivity to mutagens. This genotype-dependent response reflects differences in DNA repair capacity, antioxidant defense systems, and tolerance to cellular stress. In wheat (*Triticum aestivum*), gamma irradiation at identical doses produced varying survival and mutation rates across different cultivars, highlighting the need for genotype-specific optimization (Han et al., 2016). Similarly, in cassava (*Manihot esculenta*), EMS treatments revealed significant variation in mutation frequency between landraces, necessitating prior screening to identify responsive genotypes (Chavarriaga-Aguirre et al., 2016). Such variability underscores the importance of preliminary trials before large-scale mutagenesis, ensuring that chosen genotypes can withstand mutagen exposure while expressing heritable mutations.

The regeneration of mutagenized tissues is profoundly influenced by the culture environment. Nutrient composition, hormonal balance, and physical conditions collectively determine whether treated explants can recover and express stable mutations. Auxin-to-cytokinin ratios are particularly critical: in rice, a balanced combination of 2,4-D and kinetin supported the regeneration of EMS-treated callus (Serrat et al., 2014). Physical factors such as light quality also play a role; for example, exposure to red light improved the survival and regeneration of gamma-irradiated sugarcane (*Saccharum officinarum*) callus (Nikam et al., 2015). These findings demonstrate that posttreatment culture conditions must be optimized to support recovery and ensure the fixation of induced mutations.

Mutagenic effectiveness in *in vitro* systems is shaped by interrelated factors: dosage and exposure duration, tissue developmental stage, genotype sensitivity, and optimized culture conditions. Successful mutagenesis requires precise calibration of these parameters to generate high mutation frequencies while sustaining tissue viability. Lessons from various crop systems—ranging from rice and wheat to banana and chrysanthemum—illustrate that mutagenesis cannot be generalized across species but must be tailored to the biological and physiological context of each system.

7.2.4 Screening and selection of mutants

Following mutagenesis, the regeneration of plant tissues provides an opportunity to identify and select mutants with desirable traits. This process involves a combination of visual phenotypic assessment, biochemical and phytochemical analyses, and molecular marker techniques to ensure the selection of mutants with enhanced characteristics.

7.2.4.1 *Visual phenotypic markers*

Initial selection of plant mutants following mutagenesis is frequently based on easily observable phenotypic traits that may indicate underlying beneficial genetic changes. Among the most commonly screened traits are alterations in leaf morphology and pigmentation, which can serve as indicators of modifications in photosynthetic capacity, chlorophyll biosynthesis, or secondary metabolite production, including phenolics and flavonoids (Asif, 2023). Changes in leaf shape, size, or coloration can also reveal disruptions in developmental regulatory genes or pigment biosynthetic pathways, providing early markers for potentially valuable mutants. Variations in plant stature, such as dwarfism or gigantism, often result from mutations in genes regulating cell elongation, division, or hormone signaling pathways, including gibberellins and auxins, and may have implications for biomass accumulation and agricultural productivity (Ramawat et al., 2009). Similarly, deviations in flowering time—either early or delayed—can reflect alterations in photoperiod sensitivity or vernalization requirements, offering opportunities to adapt crops to specific environmental conditions or growing seasons (Kumar, 2018). Moreover, the selection of mutants exhibiting tolerance to abiotic stresses, such as salinity, drought, or heavy metal exposure, is particularly crucial for developing cultivars capable of thriving in marginal or challenging environments. Such stress-tolerant mutants often exhibit physiological and biochemical modifications, including enhanced antioxidant capacity, osmolyte accumulation, or improved ion homeostasis (Roychowdhury & Tah, 2013). Collectively, these phenotypic markers provide an initial, practical framework for identifying promising mutants before proceeding to more detailed biochemical, molecular, or genomic analyses.

7.2.4.2 *Biochemical and phytochemical assays*

Following phenotypic screening, biochemical and phytochemical analyses are integral for characterizing the metabolic profiles of plant mutants. Techniques such as HPLC, thin-layer

TABLE 7.4 Analytical techniques for phytochemical profiling of plant mutants.

Technique	Principle	Application in plant mutants	Example/reference
High-performance liquid chromatography (HPLC)	Separates, identifies, and quantifies components in a mixture based on differential interactions with stationary and mobile phases	Analysis of secondary metabolites such as withanolides; quantitative profiling of bioactive compounds	Developed HPLC method for withaferin A and withanolide D in *Withania somnifera* extracts (Ganzera et al., 2003)
Thin-layer chromatography (TLC / HPTLC)	Separates compounds on a thin layer of stationary phase; qualitative and semiquantitative assessment	Detecting presence and purity of specific phytochemicals; separation of flavonoids, phenolics, and withanolides	Validated HPTLC method for simultaneous separation of withanolides and phenolic acids in *Withania somnifera* (Tomar et al., 2019)
Gas chromatography–mass spectrometry (GC–MS)	Combines gas-liquid chromatography and mass spectrometry for the separation and identification of volatile/semivolatile compounds	Comprehensive profiling of bioactive metabolites including withanolides and withanosides	GC–MS analysis of *Withania somnifera* extracts (Zellner et al., 2025)
Spectrophotometry	Measures absorbance or transmission of light by a sample; quantifies compounds based on their optical properties	Assessment of antioxidant activity, total phenolic content, and other nutraceutical compounds	General use in plant mutant analysis; widely applied for rapid biochemical screening

chromatography (TLC), gas chromatography–mass spectrometry (GC–MS), and spectrophotometry enable both qualitative and quantitative analysis of bioactive compounds, providing insights into the biochemical alterations resulting from mutagenesis (Table 7.4).

7.2.4.2.1 Targeted phytochemicals in mutant profiling

Key targets in the biochemical analysis of plant mutants include the following:

- **Flavonoids and phenolics:** These compounds are known for their antioxidant properties and contribute to the plant's defense mechanisms. Their levels can be influenced by genetic modifications.
- **Alkaloids and saponins:** Alkaloids possess various pharmacological activities, while saponins have been shown to have antimicrobial and anticancer properties. Mutations can lead to altered concentrations of these compounds.
- **Antioxidant activity:** The capacity of a plant extract to scavenge free radicals is a crucial indicator of its potential health benefits. Mutants with enhanced antioxidant activity may offer improved therapeutic properties.

- **Nutraceutical compounds:** Compounds such as withanolides in *Withania somnifera* are of significant interest due to their various health-promoting effects. Mutations may lead to increased or decreased production of these compounds.

7.2.4.2.2 Case study: *Withania somnifera* mutants

In a study by Ganzera et al. (2003), TLC and HPLC were utilized to identify *Withania somnifera* mutants with altered withanolide profiles. The analysis revealed significant variations in the withanolide content among different mutants, highlighting the potential of mutagenesis in enhancing the phytochemical profile of this medicinal plant (Ganzera et al., 2003). Similarly, a study by Tomar et al. (2019) employed HPTLC to analyze withanolides and phenolic acids in *Withania somnifera* extracts, providing a comprehensive understanding of the phytochemical variability in this species (Tomar et al., 2019).

Biochemical and phytochemical analyses are essential for evaluating the medicinal and nutritional potential of plant mutants. Techniques such as HPLC, TLC, GC–MS, and spectrophotometry offer detailed insights into the metabolic alterations resulting from mutagenesis. By targeting specific phytochemicals, researchers can identify mutants with enhanced bioactive compound profiles, paving the way for the development of improved plant varieties with superior therapeutic properties.

7.2.4.3 *Molecular marker analysis*

Following phenotypic and biochemical analyses, molecular markers offer a more precise and reliable method for identifying and characterizing mutations in plant breeding programs (Table 7.5. These markers detect genetic polymorphisms at the DNA level, facilitating the association of specific traits with genotypic variations.

Common molecular marker techniques include the following:

- **RAPD (random amplified polymorphic DNA):** A PCR-based method that amplifies random segments of DNA using short primers, revealing polymorphisms based on the presence or absence of amplification products.
- **ISSR (inter simple sequence repeats):** Utilizes primers anchored in microsatellite regions to amplify regions between simple sequence repeats, providing high levels of polymorphism and reproducibility.
- **SSR (simple sequence repeats):** Also known as microsatellites, these markers target regions with repeating units of 1–6 base pairs, offering high polymorphism and codominant inheritance, making them ideal for genetic mapping and marker-assisted selection.

These molecular markers have been instrumental in various studies assessing genetic diversity and stability in plant species. For instance, a study by Saha et al. (2016) assessed genetic stability in tissue-cultured plants using RAPD and ISSR markers, confirming the reliability of these markers in detecting genetic variations (Saha et al., 2016). Research by Hussein et al. (2023) compared SSR and RAPD markers, highlighting the higher genetic diversity parameters obtained with SSR markers, underscoring their efficacy in genetic studies (Hussein et al., 2023).

TABLE 7.5 Characteristics, applications, and advantages of RAPD, ISSR, and SSR markers in plant mutation studies.

Marker type	Full name	Inheritance	Polymorphism level	Primer requirement	Applications	Advantages
RAPD	Random amplified polymorphic DNA	Dominant	Moderate	Short arbitrary primers	Genetic diversity studies, mutation detection	Simple, cost-effective, no prior sequence information required
ISSR	Inter simple sequence repeats	Dominant	High	Microsatellite-based primers	Genetic diversity, fingerprinting, mutation analysis	High reproducibility, high polymorphism, no sequence information needed
SSR	Simple sequence repeats	Codominant	Very high	Sequence-specific primers	Genetic mapping, marker-assisted selection, mutation analysis	High polymorphism, codominant inheritance, ideal for genetic studies

7.2.5 Advantages and limitations of *in vitro* mutagenesis

In vitro mutagenesis has become a cornerstone in plant biotechnology, offering a controlled environment to induce genetic variability for crop improvement. This technique involves exposing plant tissues—such as calli, embryos, or meristems—to physical or chemical mutagens, leading to mutations that can result in desirable traits. The advantages and limitations of *in vitro* mutagenesis are critical to understanding its application in plant breeding (Table 7.6).

In vitro mutagenesis has emerged as a vital tool in plant biotechnology, enabling the rapid generation of genetic diversity within a single generation. This approach significantly accelerates the development of novel traits compared to conventional breeding, which is particularly valuable in species with long generation times or those that are recalcitrant to hybridization. For example, treatment with EMS has been effectively employed to induce point mutations in various plant species, facilitating the identification of mutants with improved agronomic and qualitative traits (Mmbando, 2024). A key advantage of *in vitro* mutagenesis is its capacity for tissue-specific application. Mutagens can be targeted to particular plant tissues, such as embryos, meristems, or calli, allowing localized induction of mutations without affecting the entire genome. This precision has been demonstrated in *Capsicum annuum*, where gamma radiation applied to callus cultures resulted in mutations that enhanced fruit color and size, illustrating the potential for generating desirable phenotypes while minimizing unintended genomic changes.

Moreover, *in vitro* mutagenesis can be integrated with modern genomic and metabolic tools, enhancing its utility as a discovery platform. Coupling mutagenesis with genomic selection, metabolomics, and genome editing allows for the identification of mutants with specific genetic profiles and altered metabolic pathways, enabling the development of plants with improved nutritional content, stress tolerance, and secondary metabolite production (Scholthof et al., 1996). This integrative approach transforms mutagenesis from a random technique into a more targeted strategy for functional genomics and plant improvement.

TABLE 7.6 Advantages and limitations of *in vitro* mutagenesis.

Aspect	Advantages	Limitations
Generation of diversity	Rapid induction of novel traits within a single generation	Mutations are random, leading to both beneficial and deleterious changes
Tissue specificity	Targeted application to specific tissues like embryos or calli	Potential for somaclonal variation affecting trait stability
Integration with modern tools	Facilitates integration with genomic selection and genome editing tools	Requires sophisticated infrastructure and expertise
Cost and scalability	Cost-effective, especially in species with long generation times	Labor-intensive screening processes for trait evaluation
Trait stability	Potential for stable trait expression with proper selection	Some mutations may not be stably inherited or may exhibit epigenetic silencing

From a practical perspective, *in vitro* mutagenesis is cost-effective and scalable. It reduces the reliance on extensive field trials, shortens the breeding cycle, and facilitates the mass production of mutant lines with desirable traits. This is particularly advantageous for large-scale breeding programs and for crops with extended generation times, offering an efficient alternative to traditional hybridization (Kodym & Afza, 2003). Despite these advantages, *in vitro* mutagenesis also presents several limitations. The random nature of most physical and chemical mutagens can produce both beneficial and deleterious mutations, necessitating labor-intensive and often resource-heavy screening to identify favorable variants. NaN_3, for instance, is a potent mutagen capable of inducing a wide range of genetic changes, some of which may be deleterious (Raina et al., 2021). Additionally, somaclonal variation—genetic changes arising from tissue culture itself—can complicate the attribution of observed traits to induced mutagenesis, potentially confounding selection.

Screening for desirable traits, especially those related to secondary metabolites, often requires sophisticated analytical techniques such as HPLC and gas chromatography–mass spectrometry (GC–MS). These methods, while highly informative, are labor-intensive and demand specialized expertise, limiting accessibility in some research and breeding contexts (Ramawat et al., 2009). Furthermore, the stability of induced mutations can be problematic. Some mutations may not be heritable or may be subject to epigenetic silencing, resulting in the loss of desired traits over successive generations.

7.3 Impact of mutagenesis on phytochemical pathways

Mutagenesis serves as a pivotal tool in plant biotechnology, facilitating the generation of genetic diversity that can lead to enhanced production of secondary metabolites. These compounds, including alkaloids, flavonoids, terpenoids, and phenolic acids, play crucial roles in plant defense, reproduction, and adaptation to environmental stresses (Akula & Ravishankar, 2014; Verpoorte & Memelink, 2002). The impact of mutagenesis on phytochemical pathways can be categorized into metabolic pathway reprogramming, transcriptional regulation of biosynthetic genes, and mutagenesis-induced enzyme variability (O'Connor, 2015; Singh, Yadav, et al., 2025).

7.3.1 Metabolic pathway reprogramming

Mutagenesis can reconfigure metabolic fluxes of secondary metabolite pathways by upregulating biosynthetic genes or disrupting negative regulators. For instance, mutations in genes of the shikimate and phenylpropanoid pathways have been shown to enhance phenolic production in *Arabidopsis* and medicinal plants (Goossens et al., 2003; Rajyaguru and Maheshala, 2021). Similarly, mutagenesis targeting chalcone synthase (CHS), a key enzyme in flavonoid biosynthesis, can increase flavonol and anthocyanin accumulation (Addi et al., 2022). EMS-induced mutants in *Capsicum annuum* showed elevated capsaicinoid and flavonoid contents due to pathway deregulation (Hasan et al., 2025). Additionally, gamma irradiation has been reported to enhance terpenoid biosynthesis in *Mentha piperita*, leading to essential oil profiles with higher menthol content (Prasad et al., 2024).

Thus mutagenesis provides opportunities to redirect plant metabolic fluxes toward the accumulation of targeted bioactive compounds with nutritional and therapeutic applications (Zhang et al., 2014).

7.3.2 Transcriptional regulation of biosynthesis genes

Secondary metabolite biosynthesis is tightly controlled by transcription factors (TFs), including *MYB, bHLH,* and *WRKY* families. Mutagenesis can alter TF binding sites or the regulatory regions of biosynthetic genes, leading to changes in metabolite accumulation. For example, mutations in *AtMYB75/PAP1* enhanced anthocyanin accumulation in *Arabidopsis thaliana* (Borevitz et al., 2000). EMS-induced mutants of *Catharanthus roseus* showed transcriptional upregulation of terpenoid indole alkaloid biosynthesis genes, increasing vinblastine and ajmalicine yields (van der Fits & Memelink, 2000). Similarly, mutagenesis impacting *MYC2* has been linked to modulation of jasmonate-responsive genes, altering flavonoid and phenolic profiles (Zhang et al., 2014). Moreover, mutations in promoters regulating phenylalanine ammonia-lyase, the first committed enzyme in the phenylpropanoid pathway, have been shown to redirect carbon flux toward phenolic compounds (Rajyaguru & Maheshala, 2021). Understanding the interplay between TFs and biosynthetic genes under mutagenic influence offers valuable strategies for metabolic engineering.

7.3.3 Mutagenesis-induced enzyme variability

Enzyme activity and specificity are central to the diversity of secondary metabolites. Mutagenesis introduces amino acid substitutions in enzyme coding sequences, altering catalytic efficiency or substrate selectivity. For example, site-directed mutagenesis of terpene synthases has yielded novel terpenoids with unique structural and bioactive properties (O'Connor, 2015). In alkaloid biosynthesis, mutations in strictosidine synthase variants in *C. roseus* altered substrate acceptance, diversifying alkaloid products (Kastner et al., 2008). Similarly, altered chalcone isomerase mutants led to variations in flavonoid subclasses with enhanced antioxidant activity (Morita et al., 2014). Enzyme variability induced through random or targeted mutagenesis provides opportunities to expand the chemical repertoire of plants, with implications in pharmaceuticals, nutraceuticals, and industrial biotechnology.

7.4 Applications and case studies

7.4.1 Enhanced yield of pharmacologically active compounds

In vitro mutagenesis has emerged as a promising strategy to improve the biosynthesis of pharmacologically active compounds in medicinal plants. Secondary metabolites such as alkaloids, terpenoids, and phenolics often accumulate in low amounts under natural conditions, limiting their direct exploitation. Induced mutagenesis, particularly when combined with tissue culture approaches, can generate genetic variability that enhances metabolite yield while maintaining clonal stability.

One of the most studied systems is *C. roseus*, a major source of indole alkaloids like vincristine and vinblastine, which are used in cancer chemotherapy. Callus cultures subjected to chemical mutagens such as EMS and physical mutagens like gamma irradiation have shown enhanced production of ajmalicine, serpentine, and vindoline (van der Heijden et al., 2004; Zárate & Verpoorte, 2007). Chemical and physical mutagens have demonstrated efficacy in enhancing the production and composition of secondary metabolites in medicinal plants. For instance, gamma radiation applied to *Catharanthus roseus* callus cultures significantly increased the levels of vincristine and vinblastine—valuable anticancer alkaloids—especially at moderate doses, indicating that tissue-specific irradiation can be used to boost alkaloid yields *in vitro* (Mujib et al., 2022). Moreover, seeds of *C. roseus* exposed to varying gamma-ray doses displayed notable increases in catharanthine and vindoline concentrations, as well as morphological variations, illustrating that irradiation can simultaneously modulate both phytochemical profiles and plant development (Hassan et al., 2017). A similar approach has been evaluated in *Hypericum perforatum* (St. John's Wort), a well-known producer of hypericin with pharmacological relevance. *In vitro* culture systems have been optimized for this species to facilitate hypericin accumulation in specific cellular compartments, such as vacuoles and organelles, suggesting that tissue culture methods—not necessarily mutagenesis—offer effective routes to enhance compound production (Bais et al., 2002). These examples highlight the effectiveness of induced mutagenesis in redirecting metabolic fluxes toward improved yields of high-value secondary metabolites. Similarly, irradiation-induced mutants showed not only higher yields of bioactive metabolites but also improved tolerance to oxidative stress, an important factor in sustained metabolite accumulation under *in vitro* conditions (Tharani & Mutharaian, 2025). Beyond these, mutagenesis has also been successfully applied in *Artemisia annua* to improve the production of artemisinin, an antimalarial sesquiterpene lactone (Ferreira & Janick, 1995; Ikram & Simonsen, 2017). Mutant cell lines obtained through EMS and UV treatments displayed higher artemisinin content compared to wild-type cultures, demonstrating the role of induced mutagenesis in enhancing the yield of scarce but pharmacologically important metabolites.

These studies collectively highlight the utility of *in vitro* mutagenesis in generating high-yielding variants of medicinal plants. Such advancements not only strengthen the pharmaceutical supply chain but also reduce pressure on natural plant populations, making it a valuable tool for both healthcare and conservation.

7.4.2 Conservation of rare medicinal species

The conservation of medicinal plants has become an urgent global priority due to the overexploitation of natural populations, habitat loss, and climate change. Traditional harvesting practices, coupled with rising demand from the pharmaceutical and nutraceutical industries, have placed many species at risk of extinction. *In vitro* mutagenesis, when integrated with tissue culture technologies, provides a powerful approach for the conservation and sustainable utilization of these rare genetic resources.

By applying mutagenic agents such as gamma rays, UV-B radiation, or chemical mutagens to explants, seeds, or callus cultures, genetic variability can be introduced into otherwise clonally propagated material. This artificial variability creates opportunities to select variants with enhanced resistance to biotic and abiotic stresses, thereby reducing reliance on wild populations (Shah et al., 2020). For example, studies on *Podophyllum hexandrum*, an endangered Himalayan medicinal plant producing podophyllotoxin (a precursor of anticancer drugs), have demonstrated the usefulness of gamma irradiation in generating variants with better regeneration efficiency and stress tolerance (Dhiman et al., 2016). These advances not only safeguard natural stands of the species but also provide reliable *in vitro* sources of bioactive compounds. Similarly, *Rauwolfia serpentina*, a critical source of reserpine and other antihypertensive alkaloids, faces severe depletion in its natural habitat. Tissue culture combined with mutagenesis has been applied to establish elite lines with improved biomass production and alkaloid content (Roychowdhury & Tah, 2013). These mutant lines can serve as ex situ conservation reservoirs and reduce destructive root harvesting from the wild.

In the case of *Withania somnifera* (ashwagandha), a medicinal plant of global economic importance, induced mutagenesis *in vitro* has led to the development of variants with enhanced withanolide content and improved tolerance to salinity and drought (Das et al., 2010). Such improvements provide resilience against environmental fluctuations and enhance the feasibility of large-scale cultivation, thereby alleviating pressure on natural populations. Beyond enhancing phytochemical yields, *in vitro* mutagenesis contributes to the conservation of genetic diversity. By generating a spectrum of mutants with altered morphophysiological traits, researchers can expand the available gene pool, which is essential for long-term conservation strategies (Shah et al., 2020). The ex situ maintenance of these variants in gene banks or *in vitro* repositories ensures the preservation of both genetic and biochemical diversity.

Therefore *in vitro* mutagenesis serves a dual role in conservation: (1) protecting endangered species from overexploitation by providing alternative cultivated sources, and (2) expanding the genetic base for breeding and adaptation. This aligns with the goals of the Convention on Biological Diversity (CBD) to ensure sustainable use and equitable sharing of plant genetic resources.

7.4.3 Commercial implications

In vitro mutagenesis has emerged as a pivotal tool in enhancing the commercial viability of medicinal plants. By inducing genetic variability through chemical or physical mutagens, this technique enables the development of plant variants with improved traits, such as increased secondary metabolite production, enhanced stress tolerance, and uniform growth characteristics. These advancements are particularly significant in the context of the burgeoning global market for medicinal and aromatic plants, which, as of 2023, was estimated at US$68 billion annually and is growing by 10%–20% each year (Zamani et al., 2025). One notable example is the application of gamma irradiation in *C. roseus* to enhance alkaloid production. Studies have demonstrated that irradiation can lead to increased yields of vincristine and vinblastine, alkaloids with significant anticancer properties (Sherpa and Devadas, 2022). Similarly, in *Hypericum perforatum* (St. John's Wort), mutagenesis has been employed to enhance the production of hypericin, a compound known for its antidepressant properties (Hasan et al., 2025).

The commercial benefits of *in vitro* mutagenesis extend beyond the enhancement of phytochemical profiles. The ability to produce uniform and high-quality plant material through tissue culture techniques facilitates large-scale cultivation, ensuring a consistent supply of medicinal plants. This is particularly crucial for meeting the demands of the pharmaceutical and cosmetic industries, where consistency and quality are paramount. Moreover, the integration of *in vitro* mutagenesis with biotechnological advancements, such as genetic transformation and metabolic engineering, holds promise for the development of plant varieties with tailored profiles of bioactive compounds. This approach not only enhances the commercial value of medicinal plants but also contributes to the sustainability of their production by reducing dependence on wild harvesting and mitigating the risk of overexploitation.

In conclusion, *in vitro* mutagenesis serves as a cornerstone in the commercial cultivation of medicinal plants, offering avenues for the enhancement of desirable traits, ensuring a steady supply of high-quality plant material, and contributing to the economic growth of the medicinal plant industry.

7.4.4 Regulatory and ethical considerations

In vitro mutagenesis, as a tool for enhancing the production of secondary metabolites in medicinal plants, intersects with various regulatory and ethical frameworks that ensure the safety, efficacy, and sustainability of biotechnological applications.

7.4.4.1 *Regulatory oversight*

The regulatory landscape for *in vitro* mutagenesis varies globally, influenced by national policies and international agreements. In the European Union, the regulation of genetically modified organisms (GMOs) is stringent, with Directive 2001/18/EC requiring comprehensive risk assessments before the release of GMOs into the environment. This directive emphasizes the precautionary principle, ensuring that any potential risks to human health and the environment are thoroughly evaluated before approval (Kritikos, 2017).

In contrast, countries like the United States and Canada have adopted more lenient approaches, focusing on the end product rather than the process used to create it. This results in a regulatory environment where gene-edited plants may not be classified as GMOs, provided they do not contain foreign DNA, even if gene editing techniques were employed (Shaffer & Pollack, 2009).

7.4.4.2 *Ethical considerations*

Ethical concerns surrounding *in vitro* mutagenesis in medicinal plants encompass various aspects, including the following:

- **Safety and risk assessment**: Ensuring that mutagenic treatments do not inadvertently produce harmful compounds or lead to genetic instability in plants is paramount. Regulatory bodies require extensive testing to confirm the safety of mutagenized plant products for human consumption (Jităreanu et al., 2023).
- **Environmental impact**: The potential for mutagenized plants to crossbreed with wild relatives or nonmutagenized crops raises concerns about unintended ecological

consequences. Regulatory frameworks often mandate environmental risk assessments to address these issues (McHughen & Smyth, 2008).

- **Consumer rights and transparency**: Ethical considerations also involve the right of consumers to be informed about the nature of the products they consume. Labeling requirements, such as those in the European Union, ensure that consumers are aware of the presence of GMOs or mutagenized plants in the market (Costa-Font et al., 2008).
- **Cultural and societal values**: Different societies may have varying perceptions of biotechnology and its applications. Ethical frameworks must consider these cultural differences to ensure that biotechnological advancements are aligned with societal values and norms (Louwaars & Jochemsen, 2021).

7.5 Future prospects and challenges

7.5.1 Combining mutagenesis with AI and predictive modeling

The integration of *in vitro* mutagenesis with artificial intelligence (AI) and predictive modeling heralds a new era in phytochemical biosynthesis. Mechanistic metabolic models combined with machine learning–guided strategies significantly enhance genotype-to-phenotype prediction accuracy. For instance, a study on yeast metabolism showed AI-driven designs improved tryptophan titer by 74% and productivity by 43% relative to training designs (Zhang et al., 2020). On the frontier of enzyme engineering, platforms integrating machine learning and large language models with biofoundry automation achieved dramatic activity improvements—up to 90-fold substrate preference enhancement in Arabidopsis enzymes—within four iterative rounds using fewer than 500 variants (Singh & Lane, 2025). AI also plays an expanding role in natural-product discovery and pathway elucidation. The utilization of AI/ML can mine "genomic dark matter" to uncover biosynthetic gene clusters in nonmodel species, enabling novel pathway discovery and predictive retrosynthesis (McConnachie & Nguyen, 2025). Machine learning approaches, such as supervised and unsupervised learning, have been applied to predict genes associated with flavonoid or terpenoid biosynthesis in medicinal plants, offering powerful predictors to guide mutagenesis (Zhang & Yang, 2025).

Together, AI enables both high-confidence selection of mutagenesis targets and analysis of mutants, mitigating the historic bottlenecks in mutant screening. In the future, we anticipate autonomous platforms that iteratively propose, build, and evaluate mutant libraries—melding predictive modeling with automated strain construction and characterization—to accelerate discovery, reduce experimental load, and unlock novel phytochemical biosynthetic capabilities.

7.5.2 High-throughput phenotyping for mutant screening

High-throughput phenotyping platforms are transforming mutant evaluation in phytochemical biosynthesis. An example is ChronoRoot 2.0—an AI-powered, open-source platform offering multi-organ tracking, real-time quality control, and automated root-system architecture analysis. Such systems enable phenotyping of temporal developmental traits across many genotypes in high throughput (Gaggion et al., 2025). In parallel, biosensor-enabled screening approaches integrated with mechanistic and machine-learned models allow rapid phenotype characterization. One study employed high-throughput biosensor screening tied to ML algorithms for thousands of metabolic pathway designs, resulting in significantly improved metabolite yields (Zhang et al., 2020).

Future platforms could incorporate imaging, metabolite capture, and automated data pipelines to assess mutant libraries at scale. For phytochemicals, this might involve using fluorescent reporters, Raman spectroscopy, or mass-spectrometry-based sensors to detect compound accumulation in thousands of mutants. Real-time AI-assisted analysis would refine prioritization, enabling iterative rounds of mutagenesis guided by phenotype intelligence. Such integration paves the way for routine screening of extensive mutant panels, accelerating selection of high-performing lines.

7.5.3 Climate-resilient phytochemical production

Achieving stable phytochemical production under environmental stresses is increasingly important. *In vitro* mutagenesis offers a route to engineer tolerant, high-metabolite-producing strains. While direct studies on climate-resilient *in vitro* systems remain limited, frameworks exist within AI-driven genotype–environment–phenotype modeling. For example, multi-omics integrated AI frameworks predict phenotypes across environments, guiding selection of tolerant designs (Wu & Xie, 2025). Hairy root cultures themselves often exhibit genetic and biochemical stability and robustness to varied growth conditions—making them inherently suitable for resilient metabolite production systems (Biswas et al., 2023). Engineering these cultures via mutagenesis or genome editing (e.g., CRISPR) could further enhance tolerance to stresses like temperature shifts, osmotic changes, or elicitor treatments while maintaining metabolite yield (Liu et al., 2025; Elbouzidi et al., 2024; Elbouzidi, Taibi, et al., 2025).

Looking ahead, coupling mutagenesis with AI-informed stress modeling and robust culture platforms (e.g., engineered hairy roots) could deliver phytochemical production systems resilient to climate-induced variability. This may involve editing key stress-response regulators or transporter genes that stabilize metabolite synthesis pathways under stress, ensuring consistent output despite environmental fluctuations.

7.5.4 Bioreactor-based scalable cultures

Scaling from laboratory flasks to industrial-scale bioreactors remains challenging, particularly with delicate systems like hairy-root cultures. Hairy roots—root cultures transformed via *Agrobacterium rhizogenes*—feature rapid growth, high stability, and strong metabolite productivity but require tailored bioreactor designs for scale-up (Biswas et al., 2023). The literature describes successful implementations in mist, airlift, bubble-column, and stirred-tank bioreactors tailored to reduce shear stress and optimize mass transfer (Biswas et al., 2023). Specific case studies highlight improved alkaloid yields: *Atropa belladonna* hairy roots in stirred-tank reactors produced significantly more curcumin than shake-flask cultures (2.3-fold increase) (Biswas et al., 2023); in *Brugmansia candida*, a modified

1.5 L stirred tank outperformed Erlenmeyer flasks for scopolamine and anisodamine production (Biswas et al., 2023). Yet, optimization remains necessary to balance oxygenation, nutrient delivery, shear sensitivity, and biomass handling.

To enhance scalability, combining mutagenesis (e.g., selecting root lines with favorable growth and resilience traits) with bioreactor engineering, process optimization, and AI-assisted real-time monitoring (e.g., for metabolite titer, biomass, oxygen utilization) could streamline scale-up. Future developments may include modular bioreactors with automated feedback systems that adapt culture conditions to maintain high productivity.

7.5.5 Integration into global pharmaceutical supply chains

Transitioning mutagenesis-derived phytochemical production platforms into global pharmaceutical supply chains entails multiple challenges: ensuring consistent quality, regulatory approval, and efficient distribution. Quality control must guarantee batch-to-batch consistency in phytochemical composition, purity, and safety—particularly important for pharmaceutically relevant compounds. Standardization protocols, validated analytical methods, and Good Manufacturing Practice (GMP) compliance will be essential.

Regulatory landscapes require clear documentation of production strains (including mutagenesis and genetic modifications), culture systems, and downstream processing. Mutagenesis approaches (e.g., CRISPR or random mutagenesis) must meet guidelines around GMOs, depending on jurisdiction, with transparency about modifications, absence of off-target effects, and environmental containment. Integration into supply chains also demands scalability and cost-efficiency. Bioreactor-based production must be linked to purification workflows, packaging, stability testing, and distribution logistics. Partnerships with contract development and manufacturing organizations or pharmaceutical firms will be pivotal. Moreover, global distribution requires compliance with international regulatory frameworks (e.g., FDA, EMA), plus addressing transportation, cold-chain requirements (if applicable), intellectual property, and equitable access considerations.

In the future, securing supply chain integration could involve developing standardized production modules (e.g., plug-and-play bioreactor setups), validated QA/QC pipelines, and regulatory dossiers. Embedding traceability—via digital twins, blockchain-enabled provenance, or metadata-rich batch records—could enhance transparency and trust across stakeholders, enabling smooth adoption of mutagenesis-derived phytochemicals in pharmaceutical markets.

7.6 Conclusion

In vitro mutagenesis stands at the intersection of tradition and innovation, offering a powerful platform for expanding the phytochemical potential of medicinal plants. By introducing targeted or random genetic variability, researchers have developed elite variants with enhanced biosynthetic capacities, yielding valuable improvements in alkaloids, flavonoids, terpenoids, and phenolics. The methodologies reviewed in this chapter—from physical agents such as gamma rays and ion beams to chemical mutagens like EMS and NaN_3—demonstrate how deliberate mutagenesis can remodel structural genes, transcriptional regulators, and biosynthetic enzymes, ultimately reshaping secondary metabolism. Coupled with advances in molecular biotechnology, including omics-based mutant screening, systems biology frameworks, and CRISPR/Cas genome editing, mutagenesis has matured into a dynamic strategy for rationally optimizing metabolite production. At the same time, the translation of these laboratory successes into industrial and pharmaceutical applications requires careful consideration of scalability, regulation, and quality assurance. While tissue culture–based systems such as hairy roots and cell suspensions offer promising production platforms, their deployment at commercial scale necessitates robust bioreactor designs, validated purification pipelines, and compliance with GMP. Global regulatory landscapes further demand transparent documentation of mutagenesis techniques, genetic stability, and safety assessments, particularly where GMOs are involved.

Looking to the future, integration into pharmaceutical supply chains will hinge on technological innovation and collaborative frameworks. High-throughput phenotyping, AI-guided predictive modeling, and digital solutions such as blockchain-enabled traceability or digital twins may provide the necessary infrastructure for ensuring consistency, transparency, and trust. Equally important is the imperative of equitable access, ensuring that phytochemicals derived from mutagenesis benefit not only industrialized markets but also underserved populations worldwide.

Ultimately, the synergy between *in vitro* mutagenesis, plant tissue culture, and modern biotechnological tools represents a sustainable and forward-looking path for phytochemical biosynthesis. By addressing technical, regulatory, and ethical challenges in parallel, this field is poised to transform medicinal plants into reliable biofactories for natural products—bridging the gap between experimental innovation and global therapeutic impact.

References

Addi, M., Elbouzidi, A., Abid, M., Tungmunnithum, D., Elamrani, A., & Hano, C. (2022). An overview of bioactive flavonoids from citrus fruits. *Applied Sciences, 12*(1), 29. https://doi.org/10.3390/app12010029.

Akula, R., & Ravishankar, G. A. (2014). Influence of abiotic stress signals on secondary metabolites in plants. *Plant Signaling & Behavior, 6*(11), 1720–1731. https://doi.org/10.4161/psb.6.11.17613.

Alseekh, S., Aharoni, A., Brotman, Y., Contrepois, K., D'Auria, J., Ewald, J., C. Ewald, J., Fraser, P. D., Giavalisco, P., Hall, R. D., Heinemann, M., Link, H., Luo, J., Neumann, S., Nielsen, J., Perez de Souza, L., Saito, K., Sauer, U., Schroeder, F. C., ... Fernie, A. R. (2021). Mass spectrometry-based metabolomics: A guide for annotation, quantification and best reporting practices. *Nature Methods, 18*(7), 747–756. https://doi.org/10.1038/s41592-021-01197-1, http://www.nature.com/nmeth/.

Alves, C. Q., David, J. P., David, J. M., & Bahia, M. V. (2014). Mutagenesis and its application to plant secondary metabolism. *Plant Cell, Tissue and Organ Culture, 116*(1), 1–17.

Asif, A. (2023). *Mutagenesis in medicinal plants.* Informa UK Limited, 211–241. http://doi.org/10.1201/9781003179139-11.

Atacan, K., Özacar, M., & Özacar, M. (2018). Investigation of antibacterial properties of novel papain immobilized on tannic acid modified Ag/CuFe2O4 magnetic nanoparticles. *International Journal of Biological Macromolecules, 109*, 720–731. https://doi.org/10.1016/j.ijbiomac.2017.12.066.

Bais, H. P., Walker, T. S., McGrew, J. J., & Vivanco, J. M. (2002). Factors affecting growth of cell suspension cultures of *Hypericum perforatum* L. (St. John's wort) and production of hypericin. *In Vitro Cellular and Developmental Biology - Plant, 38*(1), 58–65. https://doi.org/10.1079/IVP2001253.

Bakkali, F., Averbeck, S., Averbeck, D., & Idaomar, M. (2008). Biological effects of essential oils – A review. *Food and Chemical Toxicology, 46*(2), 446–475. https://doi.org/10.1016/j.fct.2007.09.106.

Baltes, N. J., Gil-Humanes, J., Cermak, T., Atkins, P. A., & Voytas, D. F. (2014). DNA replicons for plant genome engineering. *The Plant Cell, 26*(1), 151–163. https://doi.org/10.1105/tpc.113.119792, http://www.plantcell.org/content/26/1/151.full.pdf+html.

Bashir, S., Wani, A. A., & Nawchoo, I. A. (2012). Studies on the induction of mutations in fenugreek (*Trigonella foenum-graecum* L.).

Biswas, D., Chakraborty, A., Mukherjee, S., & Ghosh, B. (2023). Hairy root culture: A potent method for improved secondary metabolite production of Solanaceous plants. *Frontiers in Plant Science, 14*. https://doi.org/10.3389/fpls.2023.1197555.

Böhmer, Stefanie, Reimund Goss, C. M., Severin Sasso, M. G., & Anja Hemschemeier, T. H. (2023). Chlamydomonas reinhardtii mutants deficient for old yellow enzyme 3 exhibit increased photooxidative stress. *Plant Direct, 7*(1). https://doi.org/10.1002/pld3.480.

Borevitz, J. O., Xia, Y., Blount, J., Dixon, R. A., & Lamb, C. (2000). Activation tagging identifies a conserved MYB regulator of phenylpropanoid biosynthesis. *The Plant Cell, 12*(12), 2383–2393. https://doi.org/10.1105/tpc.12.12.2383, http://www.plantcell.org/.

Bourgaud, F., Gravot, A., Milesi, S., & Gontier, E. (2001). Production of plant secondary metabolites: A historical perspective. *Plant Science, 161*(5), 839–851. https://doi.org/10.1016/s0168–9452(01)00490-3.

Broertjes, C., & van Harten, A. M. (2013). *Applied mutation breeding for vegetatively propagated crops (Vol. 12)*. Elsevier Science.

Chavarriaga-Aguirre, P., Brand, A., Medina, A., Prías, M., Escobar, R., Martinez, J., Díaz, P., López, C., Roca, W. M., & Tohme, J. (2016). The potential of using biotechnology to improve cassava: a review. *In Vitro Cellular & Developmental Biology - Plant, 52*(5), 461–478. https://doi.org/10.1007/s11627-016-9776-3.

Costa-Font, M., Gil, J. M., & Traill, W. B. (2008). Consumer acceptance, valuation of and attitudes towards genetically modified food: Review and implications for food policy. *Food Policy, 33*(2), 99–111. https://doi.org/10.1016/j.foodpol.2007.07.002.

Cragg, G. M., & Newman, D. J. (2013). Natural products: A continuing source of novel drug leads. *Biochimica et Biophysica Acta - General Subjects, 1830*(6), 3670–33695. https://doi.org/10.1016/j.bbagen.2013.02.008.

Das, A., Datta, A. K., Bhattacharya, A., Bhattacharyya, A., & Ghose, S. (2010). EMS Induced Mutagenesis in Poshita and Jawahar 22 of Withania somnifera (L.) Dunal (Solanaceae). *CYTOLOGIA, 75*(3), 305–311. https://doi.org/10.1508/cytologia.75.305.

Das, A., Sarkar, S., Bhattacharyya, S., & Gantait, S. (2020). Biotechnological advancements in *Catharanthus roseus* (L.) G. Don. *Applied Microbiology and Biotechnology, 104*(11), 4811–4835. https://doi.org/10.1007/s00253-020-10592-1.

Datta, S. K. (2023). *Mutation Breeding for Sustainable Food Production and Climate Resilience*. Springer Nature Singapore, 565–598. https://doi.org/10.1007/978-981-16-9720-3_19.

Dhiman, N., Patial, V., Bhattacharya, A. (2016). In vitro approaches for conservation and sustainable utilization of podophyllum hexandrum and picrorhiza kurroa: endangered medicinal herbs of western himalaya. In: Anis, M., Ahmad, N. (eds) *Plant Tissue Culture: Propagation, Conservation and Crop Improvement*. Springer, Singapore. https://doi.org/10.1007/978-981-10-1917-3_3.

Domingo, C., Lalanne, E., Catalá, M. M., Pla, E., Reig-Valiente, J. L., & Talón, M. (2016). Physiological Basis and Transcriptional Profiling of Three Salt-Tolerant Mutant Lines of Rice. *Frontiers in Plant Science, 7*. https://doi.org/10.3389/fpls.2016.01462.

Duke, Dayan, Romagni, & Rimando (2000). Natural products as sources of herbicides: Current status and future trends. *Weed Research, 40*(1), 99–111. https://doi.org/10.1046/j.1365-3180.2000.00161.x.

D'Erfurth, I., Cosson, V., Eschstruth, A., Rippa, S., Messinese, E., Durand, P., Trinh, H., Kondorosi, A., & Ratet, P. (2003). Rapid inactivation of the maize transposable element En/Spm in *Medicago truncatula. Molecular Genetics and Genomics, 269*(6), 732–745. https://doi.org/10.1007/s00438-003-0889-0.

Elbouzidi, A., Taibi, M., Baraich, A., Haddou, M., Mothana, R. A., Alsufyani, S. A., Darwish, H. W., Molinié, R., Fontaine, J. X., Fliniaux, O., Mesnard, F., & Addi, M. (2025). Elicitor-driven enhancement of phenolic compounds in geranium callus cultures: Phytochemical profiling via LC-MS/MS and biological activities. *Frontiers in Chemistry, 13*. https://doi.org/10.3389/fchem.2025.1537877, http://journal.frontiersin.org/journal/chemistry.

Elbouzidi, A., Haddou, M., Baraich, A., Taibi, M., Hachlafi, N. E., Pareek, A., Mesnard, F., & Addi, M. (2025). Biochemical insights into specialized plant metabolites: Advancing cosmeceutical applications for skin benefits. *Journal of Agriculture and Food Research, 19*, 101651. https://doi.org/10.1016/j.jafr.2025.101651.

Elbouzidi, A., Taibi, M., Baraich, A., Haddou, M., Loukili, E. H., Asehraou, A., Mesnard, F., & Addi, M. (2024). Enhancing secondary metabolite production in *Pelargonium graveolens* hort. Cell cultures: Eliciting effects of chitosan and jasmonic acid on bioactive compound production. *Horticulturae, 10*(5), 521. https://doi.org/10.3390/horticulturae10050521.

Facchini, P. J. (2001). Alkaloid biosynthesis in plants: Biochemistry, cell biology, molecular regulation, and metabolic engineering applications. *Annual Review of Plant Biology, 52*, 29–66. https://doi.org/10.1146/annurev.arplant.52.1.29, http://www.annualreviews.org/journal/arplant.

Ferreira, J. F. S., & Janick, J. (1995). Production and detection of artemisinin from *Artemisia Annua*. *Acta Horticulturae*, (390), 41–50. https://doi.org/10.17660/ActaHortic.1995.390.5.

Gaggion, N., Bonazzola, R., Legascue, M. F., Mammarella, M. F., Rodriguez, F. S., Aballay, F. E., Catulo, F. B., Barrios, A., Accavallo, F., Villarreal, S. N., Crespi, M., Ricardi, M. M., Petrillo, E., Blein, T., Ariel, F., & Ferrante, E. (2025). ChronoRoot 2.0: An open AI-powered platform for 2D temporal plant phenotyping. *arXiv*. https://doi.org/10.48550/arXiv.2504.14736, https://arxiv.org.

Ganzera, M., Choudhary, M. I., & Khan, I. A. (2003). Quantitative HPLC analysis of withanolides in *Withania somnifera*. *Fitoterapia, 74*(1-2), 68–76. https://doi.org/10.1016/s0367-326x(02)00325-8.

Gelvin, S. B. (2003). Agrobacterium-mediated plant transformation: The biology of T-DNA delivery. *Plant Physiology, 133*(4), 1000–1008.

Gichner, T., & Veleminský, J.i.ř. (1982). Genetic effects of N-methyl-N′-nitro-N-nitrosoguanidine and its homologs. *Mutation Research/Reviews in Genetic Toxicology, 99*(2), 129–242. https://doi.org/10.1016/0165-1110(82)90057-4.

Goossens, A., Häkkinen, S. T., Laakso, I., Seppänen-Laakso, T., Biondi, S., De Sutter, V., Lammertyn, F., Nuutila, A. M., Söderlund, H., Zabeau, M., Inzé, D., & Oksman-Caldentey, K.-M. (2003). A functional genomics approach toward the understanding of secondary metabolism in plant cells. *Proceedings of the National Academy of Sciences, 100*(14), 8595–8600. https://doi.org/10.1073/pnas.1032967100.

Guo, X., Ren, J., Zhou, X., Zhang, M., Lei, C., Chai, R., Zhang, L., & Lu, D. (2024). Strategies to improve the efficiency and quality of mutant breeding using heavy-ion beam irradiation. *Critical Reviews in Biotechnology, 44*(5), 735–752. https://doi.org/10.1080/07388551.2023.2226339.

Han, B., Gu, J., Zhao, L., Guo, H., Xie, Y., Zhao, S., Song, X., Han, L., & iu L, L. (2016). Factors Affecting the Radiosensitivity of Hexaploid Wheat to γ-Irradiation: Radiosensitivity of Hexaploid Wheat (Triticum aestivum L.). *PLoS One, 11*(8), e0161700. https://doi.org/10.1371/journal.pone.0161700.

Hartmann, T. (2007). From waste products to ecochemicals: Fifty years research of plant secondary metabolism. *Phytochemistry, 68*(22-24), 2831–2846. https://doi.org/10.1016/j.phytochem.2007.09.017.

Hasan, N., Choudhary, S., Joshi, D. C., Naaz, N., Sharma, N., Bhinda, M. S., & Laskar, R. A. (2025). A comparative study of mutagenic efficacy of EMS and MMS in inducing quantitative and cytological variations in *Capsicum annuum* L. var. NS1101. *Discover Plants, 2*(1). https://doi.org/10.1007/s44372-025-00271-8.

Hassan, M., Badr, M., Abd El- Maksoud, B., & El – Shennawy, O. (2017). Effect of gamma radiation on genetic improvement against salinity in *Catharanthus roseus* plants. *Journal of the Advances in Agricultural Researches, 22*(3), 424–455. https://doi.org/10.21608/jalexu.2017.243819.

Hideg, E., Jansen, M. A. K., & Strid, A. (2013). UV-B exposure, ROS, and stress: Inseparable companions or loosely linked associates? *Trends in Plant Science, 18*(2), 107–115. https://doi.org/10.1016/j.tplants.2012.09.003.

Humphreys, G. O., Willshaw, G. A., Smith, H. R., & Anderson, E. S. (1976). Mutagenesis of plasmid DNA with hydroxylamine: Isolation of mutants of multi-copy plasmids. *Molecular and General Genetics MGG, 145*(1), 101–108. https://doi.org/10.1007/bf00331564.

Hussein, M. A. A., Eid, M., Rahimi, M., Filimban, F. Z., & El-Moneim, D. A. (2023). Comparative assessment of SSR and RAPD markers for genetic diversity in some mango cultivars. *PeerJ. 11*. https://doi.org/10.7717/peerj.15722, http://peerj.com//articles/15722/.

Ikram, N. K. B. K., & Simonsen, H. T. (2017). A review of biotechnological artemisinin production in plants. *Frontiers in Plant Science, 8*. https://doi.org/10.3389/fpls.2017.01966, http://journal.frontiersin.org/article/10.3389/fpls.2017.01966/full.

Jain, S. M. (2005). Major mutation-assisted plant breeding programs supported by FAO/IAEA. *Plant Cell, Tissue and Organ Culture, 82*(1), 113–123. https://doi.org/10.1007/s11240-004-7095-6.

Jain, S. M. (2010). Mutagenesis in crop improvement under the climate change. *Romanian Biotechnological Letters, 15*(2), 88–106. http://ebooks.unibuc.ro/biologie/RBL/rbl1vol15Supplement/12%20S.%20MOHAN%20JAIN.pdf.

Jenkins, G. I. (2009). Signal transduction in responses to UV-B radiation. *Annual Review of Plant Biology, 60*, 407–431. https://doi.org/10.1146/annurev.arplant.59.032607.092953, http://arjournals.annualreviews.org/doi/pdf/10.1146/annurev.arplant.59.032607.092953.

Jităreanu, A., Trifan, A., Vieriu, M., Caba, I. C., Mârțu, I., & Agoroaei, L. (2023). Current trends in toxicity assessment of herbal medicines: A narrative review. *Processes, 11*(1). https://doi.org/10.3390/pr11010083, http://www.mdpi.com/journal/processes.

Joint, F. A. O. (2018). FAO/IAEA International Symposium on Plant Mutation Breeding and Biotechnology. Book of Abstracts (No. IAEA-CN-263). Joint FAO/IAEA Division of Nuclear Techniques in Food and Agriculture, Vienna (Austria).

Kastner, U., Hallard, D., & Zenk, M. H. (2008). Mutagenesis of strictosidine synthase from *Catharanthus roseus*: Enzyme variability and alkaloid diversity. *Phytochemistry, 69*(2), 235–242.

Khan, S., Al-Qurainy, F., & Anwar, F. (2009). Sodium azide: A chemical mutagen for enhancement of agronomic traits of crop plants. *Environmental Science and Pollution Research, 4*, 1–21.

Kodym, A., & Afza, R. (2003). Physical and chemical mutagenesis. *Plant Functional Genomics*, 189–203.

Kolakar, S. S., Nadukeri, S., Jakkeral, S. A., Lakshmana, D., Hanumanthappa, M., & Gangaprasad, S. (2018). Role of mutation breeding in improvement of medicinal and aromatic crops. *Journal of Pharmacognosy and Phytochemistry, 7*(3S), 425–429.

Kole, C. (2020). *Genomic designing of climate-smart fruit crops genomic designing of climate-smart fruit crops*. Springer International Publishing, 1–404. http://doi.org/10.1007/978-3-319-97946-5, https://link.springer.com/book/10.1007/978-3-319-97946-5.

Kritikos, M. (2017). *The 1990/220 deliberate release directive: Early implementation and revision*. Springer Nature, 113–160. http://doi.org/10.1057/978-1-137-31446-8_4.

Kumar, N. (2018). *Biotechnological approaches for medicinal and aromatic plants: Conservation, genetic improvement and utilization biotechnological approaches for medicinal and aromatic plants: Conservation, genetic improvement and utilization*. Springer Singapore, 1–665. http://doi.org/10.1007/978-981-13-0535-1.

Kurowska, M., Daszkowska-Golec, A., Gruszka, D., Marzec, M., Szurman, M., Szarejko, I., & Maluszynski, M. (2011). TILLING–A shortcut in functional genomics. *Journal of Applied Genetics, 52*(4), 371–390. https://doi.org/10.1007/s13353-011-0061-1.

Lattanzio, V. (2013). Phenolic compounds: Introduction. In K. G. Ramawat, & J. M. Merillon (Eds.). *Natural products: Phytochemistry, botany and metabolism of alkaloids, phenolics and terpenes* (pp. 1543–1580). Springer Berlin Heidelberg. http://doi.org/10.1007/978-3-642-22144-6 http://doi.org/10.1007/978-3-642-22144-6_57.

Liu, C., Ahmad, N., Tao, Y., Hussain, H., Chang, Y., Umar, A. W., & Liu, X. (2025). Reprogramming hairy root cultures: A synthetic biology framework for precision metabolite biosynthesis. *Plants, 14*(13), 1928. https://doi.org/10.3390/plants14131928.

Louwaars, N., & Jochemsen, H. (2021). An ethical and societal analysis for biotechnological methods in plant breeding. *Agronomy, 11*(6), 1183. https://doi.org/10.3390/agronomy11061183.

Ma, L., Kong, F., Sun, K., Wang, T., & Guo, T. (2021). From classical radiation to modern radiation: Past, present, and future of radiation mutation breeding. *Frontiers in Public Health, 9*. https://doi.org/10.3389/fpubh.2021.768071.

Mangaiyarkarasi, R., Girija, M., Gnanamurthy, S., & Don, C. L. G. (2014). Mutagenic effectiveness and efficiency of gamma rays and ethyl methane sulphonate in *Catharanthus roseus*. *International Journal of Current Microbiology and Applied Sciences, 3*(5), 881–889.

Mba, C. (2013). Induced mutations unleash the potentials of plant genetic resources for food and agriculture. *Agronomy, 3*(1), 200–231. https://doi.org/10.3390/agronomy3010200.

McClintock, B. (1948). Mutable loci in maize. *Carnegie Inst Wash Year Book, 47*, 155–169.

McConnachie, M., Nguyen, T. A. M., Kim, T., Nguyen, T. D., & Dang, T. T. T. (2025). From data to discovery: Leveraging big data in plant natural products biosynthesis research. *Plant Journal, 122*, 6. https://doi.org/10.1111/tpj.70288, http://onlinelibrary.wiley.com/journal/10.1111/(ISSN)1365-313X.

McHughen, A., & Smyth, S. (2008). US regulatory system for genetically modified [genetically modified organism (GMO), rDNA or transgenic] crop cultivars. *Plant Biotechnology Journal, 6*(1), 2–12. https://doi.org/10.1111/j.1467-7652.2007.00300.x.

Misra, P., & Saema, S. (2016). Plant tissue culture for in vitro mutagenesis, large-scale propagation, and genetic transformation. In Mohammad Anis, & Naseem Ahmad (Eds.). *Plant tissue culture: Propagation, conservation and crop improvement* (pp. 309–342). Springer Singapore. http://doi.org/10.1007/978-981-10-1917-3 http://doi.org/10.1007/978-981-10-1917-3_14.

Mmbando, G. S. (2024). The recent possible strategies for breeding ultraviolet-B-resistant crops. *Heliyon, 10*, 6. https://doi.org/10.1016/j.heliyon.2024.e27806, https://www.sciencedirect.com/science/journal/24058440.

Morita, Y., Takagi, K., Fukuchi-Mizutani, M., Ishiguro, K., Tanaka, Y., Nitasaka, E., Nakayama, M., Saito, N., Kagami, T., Hoshino, A., & Iida, S. (2014). A chalcone isomerase-like protein enhances flavonoid production and flower pigmentation. *The Plant Journal, 78*(2), 294–304. https://doi.org/10.1111/tpj.12469.

Mujib, A., Fatima, S., & Malik, M. Q. (2022). Gamma ray–Induced tissue responses and improved secondary metabolites accumulation in *Catharanthus roseus*. *Applied Microbiology and Biotechnology, 106*(18), 6109–6123. https://doi.org/10.1007/s00253-022-12122-7, https://www.springer.com/journal/253.

Muyibi, S. A., & Jimat, D. N. (2014). Mutagenesis on cyanobacteria for high CO, uptake: A review. *Journal of Pure and Applied Microbiology, 8*, 761–768.

Nikam, A. A., Devarumath, R. M., Ahuja, A., Babu, H., Shitole, M. G., & Suprasanna, P. (2015). Radiation-induced in vitro mutagenesis system for salt tolerance and other agronomic characters in sugarcane (Saccharum officinarum L.). *The Crop Journal, 3*(1), 46–56. https://doi.org/10.1016/j.cj.2014.09.002.

Nirmal Babu, K., Divakaran, M., Raj, R. P., Anupama, K., Peter, K. V., & Sarma, Y. R. (2015). Biotechnological approaches in improvement of spices: A review. In Bir Bahadur, Manchikatla Venkat Rajam, Leela Sahijram, & K. V. Krishnamurthy (Eds.). *Plant biology and biotechnology: Volume II: Plant genomics and biotechnology* (pp. 487–516). Springer India. http://doi.org/10.1007/978-81-322-2283-5 http://doi.org/10.1007/978-81-322-2283-5_25.

Ochoa-Villarreal, M., Howat, S., Hong, S. M., Jang, M. O., Jin, Y. W., Lee, E. K., & Loake, G. J. (2016). Plant cell culture strategies for the production of natural products. *BMB Reports, 49*(3), 149–158. https://doi.org/10.5483/BMBRep.2016.49.3.264, http://bmbreports.org/jbmb/jbmb_files/%5B49-3%5D1603311453_(149-158)04_BMB-15-264.pdf.

O'Connor, S. E. (2015). Engineering of secondary metabolism. *Annual Review of Genetics, 49*, 71–94. https://doi.org/10.1146/annurev-genet-120213-092053, http://arjournals.annualreviews.org/loi/genet.

Oladosu, Y., Rafii, M. Y., Abdullah, N., Hussin, G., Ramli, A., Rahim, H. A., Miah, G., & Usman, M. (2016). Principle and application of plant mutagenesis in crop improvement: A review. *Biotechnology and Biotechnological Equipment, 30*(1), 1–16. https://doi.org/10.1080/13102818.2015.1087333, http://www.tandfonline.com/doi/pdf/10.1080/13102818.2015.1087333.

Panche, A. N., Diwan, A. D., & Chandra, S. R. (2016). Flavonoids: An overview. *Journal of Nutritional Science, 5*. https://doi.org/10.1017/jns.2016.41.

Pavan, S., Schiavulli, A., Appiano, M., Marcotrigiano, A. R., Cillo, F., Visser, R. G. F., Bai, Y., Lotti, C., & Ricciardi, L. (2011). Pea powdery mildew er1 resistance is associated to loss-of-function mutations at a MLO homologous locus. *Theoretical and Applied Genetics, 123*(8), 1425–1431. https://doi.org/10.1007/s00122-011-1677-6.

Pilbeam, D., Nehe, A. S., Foulkes, M. J., Angus, J. F., George, T. S., White, P. J., & Erro, J. (2020). *Achieving Sustainable Crop Nutrition, 76*.

Prasad, P., Gupta, A., Singh, V., & Kumar, B. (2024). Impact of induced mutation-derived genetic variability, genotype and varieties for quantitative and qualitative traits in *Mentha species*. *International Journal of Radiation Biology, 100*(2), 151–160. https://doi.org/10.1080/09553002.2023.2263595.

Przibilla, E., Heiss, S., Johanningmeier, U., & Trebst, A. (1991). Site-specific mutagenesis of the D1 subunit of photosystem II in wild-type Chlamydomonas. *The Plant Cell, 3*(2), 169–174. https://doi.org/10.1105/tpc.3.2.169.

Rai, M. K., Kalia, R. K., Singh, R., Gangola, M. P., & Dhawan, A. K. (2011). Developing stress tolerant plants through in vitro selection – An overview of the recent progress. *Environmental and Experimental Botany, 71*(1), 89–98. https://doi.org/10.1016/j.envexpbot.2010.10.021.

Raina, A., Ansari, S. B., Khursheed, S., Wani, M. R., Khan, Samiullah, & Bhat, T. A. (2021). Mutagens, their types and mechanism of action with an emphasis on sodium azide and gamma radiations. In Tariq Ahmad Bhat (Ed.). *Mutagenesis, cytotoxicity and crop improvement: Revolutionizing food science* (pp. 1–11). Cambridge Scholars Publishing.

Rajyaguru, R., Maheshala, N., Mootapally, C., Nathani, N., Tomar, R., Bhalani, H., & Sharma, P. (2021). CRISPR-Cas9 approaches to enhance contents of plant secondary metabolites. In Mohd. Shahnawaz (Ed.). *Biotechnological approaches to enhance plant secondary metabolites: Recent trends and future prospects* (pp. 133–151). CRC Press. http://doi.org/10.1201/9781003034957-9, https://www.taylorfrancis.com/books/10.1201/9781003034957.

Ramawat, K. G., Dass, S., & Mathur, M. (2009). *The chemical diversity of bioactive molecules and therapeutic potential of medicinal plants*. Springer Science and Business Media LLC, 7–32. http://doi.org/10.1007/978-3-540-79116-4_2.

Razzaq, A., Wishart, D. S., Wani, S. H., Hameed, M. K., Mubin, M., & Saleem, F. (2022). Advances in metabolomics-driven diagnostic breeding and crop improvement. *Metabolites, 12*, 6. https://doi.org/10.3390/metabo12060511, https://www.mdpi.com/2218-1989/12/6/511/pdf?version=1654164954.

Reuter, & Berg (1974). Book reviews. *Theoretical and Applied Genetics, 44*(2), 95–96. https://doi.org/10.1007/bf00277960.

Riaz, A., & Gul, A. (2015). Plant mutagenesis and crop improvement. In Khalid Rehman Hakeem (Ed.). *Crop production and global environmental issues* (pp. 181–209). Springer International Publishing. http://doi.org/10.1007/978-3-319-23162-4 http://doi.org/10.1007/978-3-319-23162-4_8.

Roychowdhury, R., & Tah, J. (2013). Mutagenesis-a potential approach for crop improvement. In Khalid Rehman Hakeem, Parvaiz Ahmad, & Munir Ozturk (Eds.). *Crop improvement: New approaches and modern techniques* (pp. 149–187). Springer US. http://doi.org/10.1007/978-1-4614-7028-1 http://doi.org/10.1007/978-1-4614-7028-1_4.

Saha, S., Adhikari, S., Dey, T., & Ghosh, P. (2016). RAPD and ISSR based evaluation of genetic stability of micropropagated plantlets of *Morus alba* L. variety S-1. *Meta Gene, 7*, 7–15. https://doi.org/10.1016/j.mgene.2015.10.004.

Scholthof, H. B., Scholthof, K. B. G., & Jackson, A. O. (1996). Plant virus gene vectors for transient expression of foreign proteins in plants. *Annual Review of Phytopathology, 34*, 299–323. https://doi.org/10.1146/annurev.phyto.34.1.299.

Serrat, X., Esteban, R., Guibourt, N., Moysset, L., Nogués, S., & Lalanne, E. (2014). EMS mutagenesis in mature seed-derived rice calli as a new method for rapidly obtaining TILLING mutant populations. *Plant Methods, 10*(1), 5. https://doi.org/10.1186/1746-4811-10-5.

Shaffer, G. C., & Pollack, M. A. (2009). *The EU regulatory system for GMOs uncertain risks regulated*. Taylor and Francis, 269–294. http://doi.org/10.4324/9780203884850-24, http://www.tandfebooks.com/doi/book/10.4324/9780203884850.

Shah, D., Kamili, A. N., Wani, A. A., Majeed, U., Wani, Z. A., Sajjad, N., & Ahmad, P. (2020). Promoting the accumulation of scopolamine and hyoscyamine in Hyoscyamus niger L. through EMS based mutagenesis. *PLOS ONE, 15*(5), e0231355. https://doi.org/10.1371/journal.pone.0231355.

Shamshad, A., Rashid, M., Jankuloski, L., Ashraf, K., Sultan, K., Alamri, S., Siddiqui, M. H., Munir, T., & Zaman, Q. (2023). Effect of ethyl methanesulfonate mediated mutation for enhancing morpho-physio-biochemical and yield contributing traits of fragrant rice. *PeerJ, 11*, e15821. https://doi.org/10.7717/peerj.15821.

Sharma, S., Gupta, S., Kumari, D., Kothari, S. L., Jain, R., & Kachhwaha, S. (2023). Exploring plant tissue culture and steviol glycosides production in *Stevia rebaudiana* (Bert.) Bertoni: A review. *Agriculture, 13*(2), 475. https://doi.org/10.3390/agriculture13020475.

Sherpa, R., Devadas, R., Bolbhat, S. N., Nikam, T. D., & Penna, S. (2022). Gamma radiation induced in-vitro mutagenesis and isolation of mutants for early flowering and phytomorphological variations in *Dendrobium* 'emma white' *Plants, 11*, 22. https://doi.org/10.3390/plants11223168, https://www.mdpi.com/journal/plants.

Shu, Q. Y., Forster, B. P., & Nakagawa, H. (2012). *Plant mutation breeding and biotechnology*. CABI. http://doi.org/10.1079/9781780640853.0000.

Simon, J. E., Morales, M. R., Phippen, W. B., Vieira, R. F., & Hao, Z. (1999). Basil: A source of aroma compounds and a popular culinary and ornamental herb. *Perspectives on New Crops and New Uses, 16*, 499–505.

Singh, H. (2023). *Enhancement of plant secondary metabolites by genetic manipulation*. Springer Science and Business Media LLC, 59–90. http://doi.org/10.1007/978-981-99-4939-7_4.

Singh, N., Lane, S., Yu, T., Lu, J., Ramos, A., Cui, H., & Zhao, H. (2025). A generalized platform for artificial intelligence-powered autonomous enzyme engineering. *Nature Communications, 16*, 1. https://doi.org/10.1038/s41467-025-61209-y.

Singh, R., Yadav, R., & Fatima, S. K. (2025). *The engineering of medicinal plants: Future prospects and limitations*. Elsevier BV, 79–102. http://doi.org/10.1016/b978-0-443-22264-1.00005-0.

Smith, E. N., van Aalst, M., Tosens, T., Niinemets, Ü., Stich, B., Morosinotto, T., Alboresi, A., Erb, T. J., Gómez-Coronado, P. A., Tolleter, D., Finazzi, G., Curien, G., Heinemann, M., Ebenhöh, O., Hibberd, J. M., Schlüter, U., Sun, T., & Weber, A. P. M. (2023). Improving photosynthetic efficiency toward food security: Strategies, advances, and perspectives. *Molecular Plant, 16*(10), 1547–1563. https://doi.org/10.1016/j.molp.2023.08.017, https://www.cell.com/molecular-plant/home.

Sun, H., Xin, J., Song, H., Chen, L., Yang, D., Yang, H., Deng, X., Liu, J., Cui, R., Su, Y., Dong, G., & Yang, M. (2025). Harnessing genomic and molecular biology resources for genetic improvement of lotus: Current achievements and future directions. *Horticulture Advances, 3*, 1. https://doi.org/10.1007/s44281-024-00055-2.

Tharani, S., Mutharaian, V. N., Thirugnanasampandan, R., Moin, S., & Devi, B. S. C. (2025). Mutagenesis for enhanced secondary metabolites production in medicinal plants. In Jen-Tsung Chen (Ed.). *Biotechnology, multiple omics, and precision breeding in medicinal plants* (pp. 101–118). CRC Press. http://doi.org/10.1201/9781003475491-9, https://www.taylorfrancis.com/books/edit/10.1201/9781003475491.

Till, B. J., Colbert, T., Tompa, R., Enns, L. C., Codomo, C. A., Johnson, J. E., Reynolds, S. H., Henikoff, J. G., Greene, E. A., Steine, M. N., Comai, L., & Henikoff, S. (2003). High-throughput TILLING for functional genomics. *Methods in Molecular Biology (Clifton, N.J.), 236*, 205–220. https://doi.org/10.1385/1-59259-413-1:205.

Tomar, V., Beuerle, T., & Sircar, D. (2019). A validated HPTLC method for the simultaneous quantifications of three phenolic acids and three withanolides from *Withania somnifera* plants and its herbal products. *Journal of Chromatography B, 1124*, 154–160. https://doi.org/10.1016/j.jchromb.2019.06.009.

van der Fits, L., & Memelink, J. (2000). ORCA3, a jasmonate-responsive transcriptional regulator of plant primary and secondary metabolism. *Science (New York, N.Y.), 289*(5477), 295–297. https://doi.org/10.1126/science.289.5477.295.

van der Heijden, R., Jacobs, D. I., Snoeijer, W., Hallard, D., & Verpoorte, R. (2004). The *Catharanthus* alkaloids: Pharmacognosy and biotechnology. *Current Medicinal Chemistry, 11*(5), 607–628. https://doi.org/10.2174/0929867043455846.

Verpoorte, R., & Alfermann, A. W. (2000). *Metabolic engineering of plant secondary metabolism*. Kluwer Academic Publishers.

Verpoorte, R., & Memelink, J. (2002). Engineering secondary metabolite production in plants. *Current Opinion in Biotechnology, 13*(2), 181–187. https://doi.org/10.1016/s0958-1669(02)00308-7.

Wani, M. R., Dar, A. R., Tak, A., Amin, I., Shah, N. H., Rehman, R., Baba, M. Y., Raina, A., Laskar, R., Kozgar, M. I., & Khan, S. (2018). Chemo-induced pod and seed mutants in mungbean (*Vigna Radiata* L. Wilczek). *SAARC Journal of Agriculture, 15*(2), 57–67. https://doi.org/10.3329/sja.v15i2.35161.

Wijerathna-Yapa, A., Ramtekey, V., Ranawaka, B., & Basnet, B. R. (2022). Applications of in vitro tissue culture technologies in breeding and genetic improvement of wheat. *Plants, 11*(17), 2273. https://doi.org/10.3390/plants11172273.

Wink, M. (2010). *Annual plant reviews, functions and biotechnology of plant secondary metabolites*. Wiley-Blackwell.

Wu, Y., & Xie, L. (2025). AI-driven multi-omics integration for multi-scale predictive modeling of genotype-environment-phenotype relationships. *Computational and Structural Biotechnology Journal, 27*, 265–277. https://doi.org/10.1016/j.csbj.2024.12.030, https://www.sciencedirect.com/science/article/pii/S2001037024004513.

Xu, L., Najeeb, U., Naeem, M. S., Wan, G. L., Jin, Z. L., Khan, F., & Zhou, W. J. (2012). In vitro mutagenesis and genetic improvement. *Technological Innovations in Major World Oil Crops, 2*, 151–173. https://doi.org/10.1007/978-1-4614-0827-7_6, http://doi.org/10.1007/978-1-4614-0827-7.

Yadav, S. S., Redden, R. J., Hatfield, J. L., Ebert, A. W., Hunter, D., & Yadav, S. S. (2018). Breeding new generation genotypes for conservation agriculture in maize-wheat cropping systems under climate change. In Shyam S. Yadav, Robert J. Redden, Jerry L. Hatfield, Andreas W. Ebert, & Danny Hunter (Eds.). *Food security and climate change* (pp. 189–228). Wiley. http://doi.org/10.1002/9781119180661.ch10.

Zamani, S., Fathi, M., Ebadi, M.-T., & Máthé, Á. (2025). Global trade of medicinal and aromatic plants. A review. *Journal of Agriculture and Food Research, 21*, 101910. https://doi.org/10.1016/j.jafr.2025.101910, https://www.sciencedirect.com/science/article/pii/S2666154325002819.

Zárate, R., & Verpoorte, R. (2007). Strategies for the genetic modification of the medicinal plant *Catharanthus roseus* (L.) G. Don. *Phytochemistry Reviews, 6*(2-3), 475–491. https://doi.org/10.1007/s11101-006-9020-6.

Zellner, L., Wierer, M., Himmelsbach, M., Schatzmann, B., Thalhamer, B., & Klampfl, C. W. (2025). Determination of withanolides and withanosides in ashwagandha based products using HPLC-drift-tube-ion-mobility quadrupole time-of-flight mass spectrometry. *Electrophoresis, 46*(5-6), 340–346. https://doi.org/10.1002/elps.202400188, http://onlinelibrary.wiley.com/journal/10.1002/(ISSN)1522-2683.

Zhang, J., Petersen, S. D., Radivojevic, T., Ramirez, A., Pérez-Manríquez, A., Abeliuk, E., Sánchez, B. J., Costello, Z., Chen, Y., Fero, M. J., Martin, H. G., Nielsen, J., Keasling, J. D., & Jensen, M. K. (2020). Combining mechanistic and machine learning models for predictive engineering and optimization of tryptophan metabolism. *Nature Communications, 11*(1). https://doi.org/10.1038/s41467-020-17910-1, http://www.nature.com/ncomms/index.html.

Zhang, J., Yang, Y., Si, J., Chen, D., & Chuan Dong, Z. H. (2025). Artificial intelligence in the discovery and modification of biological elements in medicinal plants. *Medicinal Plant Biology, 4*, 1. https://doi.org/10.48130/mpb-0025-0010 0–0.

Zhang, Y., Butelli, E., & Martin, C. (2014). Engineering anthocyanin biosynthesis in plants. *Current Opinion in Plant Biology, 19*, 81–90. https://doi.org/10.1016/j.pbi.2014.05.011.

CHAPTER

8

Elicitation for enhanced production of secondary metabolites and bioactive molecules

Engin Tilkat[1] *and Veysel Süzerer*[2]

[1]Department of Biology, Faculty of Science and Literature, Batman University, Batman, Türkiye
[2]Department of Pharmacy Services, Vocational School of Health Services, Bingöl University, Bingöl, Türkiye

8.1 Introduction

Plants synthesize a wide variety of secondary metabolites (SMs) that, although not essential for primary metabolism, play critical roles in plant defense, adaptation to environmental stress, and ecological interactions. These compounds also hold immense value in the pharmaceutical, nutraceutical, and cosmetic industries. However, conventional extraction of SMs from field-grown mature plants presents several challenges, including low yields, seasonal variability, environmental dependency, and overharvesting of slow-growing or endangered species such as *Panax ginseng* and *Taxus* spp. (Ramirez-Estrada et al., 2016). For instance, *P. ginseng* requires several years to accumulate sufficient levels of ginsenosides, while *Taxus* trees take decades to produce measurable amounts of taxol, thereby making large-scale production inefficient and unsustainable.

In recent decades, the need for alternative, sustainable sources of SMs has led to the development of *in vitro* plant culture technologies. Plant tissue culture systems—including callus, cell suspensions, organ cultures, and hairy roots—have emerged as versatile platforms for metabolite production. These systems offer numerous advantages: environmental control, independence from climate and seasonality, genetic and biochemical stability, and the possibility of continuous, contaminant-free production. They also enable the use of metabolic engineering and external interventions such as elicitors to redirect and amplify specific biosynthetic pathways.

Despite these advantages, one major limitation of *in vitro* cultures is the relatively low baseline production of SMs under normal, nonstressed culture conditions. Plants naturally produce these compounds as a response to biotic or abiotic stress, and such signals are

Medicinal Plant Biotechnology for Ameliorated Secondary Metabolite Production
DOI: https://doi.org/10.1016/B978-0-443-43944-5.00010-4

largely absent in artificial, sterile environments. To overcome this, elicitation has been developed as a strategic intervention.

Elicitors are substances that mimic natural stress cues—ranging from microbial components and polysaccharides to chemical signals and physical stressors—and trigger the plant's innate defense responses. This, in turn, leads to the activation of SM biosynthesis through complex signaling cascades involving reactive oxygen species (ROS), calcium influx, mitogen-activated protein kinases (MAPKs), and phytohormonal cross-talk (Fazili et al., 2022; Guru et al., 2022). These mechanisms not only activate the expression of biosynthetic genes but also modulate precursor flux and metabolic channeling.

The concept of elicitation in plant biotechnology is not new. Early studies in the 1980s and 1990s demonstrated that fungal elicitors could enhance phytoalexin production in cell cultures. Since then, the field has expanded considerably, encompassing a wide variety of biotic and abiotic elicitors, more sophisticated culture platforms, and an improved molecular understanding of the regulatory networks involved. As previously delineated, SMs are key components in plants' adaptation to extreme conditions (pathogens, herbivores, diverse environmental stress factors, etc.). Moreover, SMs are produced in response to biotic and abiotic stress (Fig. 8.1, Table 8.1) (Khan et al., 2025).

This chapter aims to provide a comprehensive yet focused synthesis of current knowledge on elicitor-based strategies to enhance SM production in plant biotechnology. It covers the types and sources of elicitors, their underlying molecular mechanisms, application strategies in different culture systems, and documented case studies with practical outcomes. Moreover, the chapter discusses critical challenges such as scalability and reproducibility, while also outlining future directions involving omics integration, bioprocess optimization, and digital control systems. By bridging molecular insights with applied strategies, this chapter seeks to support the development of high-yield, sustainable production platforms for valuable bioactive phytochemicals.

8.2 Elicitors and their classification

Elicitors are external agents that stimulate plant defense responses and activate SM biosynthesis by mimicking environmental or biological stress signals. Based on their origin, elicitors are broadly categorized into two groups: biotic and abiotic. Biotic elicitors are derived from living organisms, including microbial cell wall components such as chitin, chitosan, β-glucans, and lipopolysaccharides, as well as microbial proteins like flagellin, harpins, or elicitins. Some biotic elicitors also include plant-derived molecules released during cell damage, known as damage-associated molecular patterns, such as oligogalacturonides. These agents are recognized by the plant immune system as nonself or danger signals, triggering immune-like responses that overlap with secondary metabolism activation (Ramirez-Estrada et al., 2016; Jha & Yadav, 2023; Petrova et al., 2024). The classification of elicitors is summarized in Fig. 8.2, which clearly distinguishes abiotic elicitors (further divided into physical and chemical categories) from biotic elicitors (exogenous microbial-derived and endogenous plant-derived sources). This visual summary complements the detailed explanation given above.

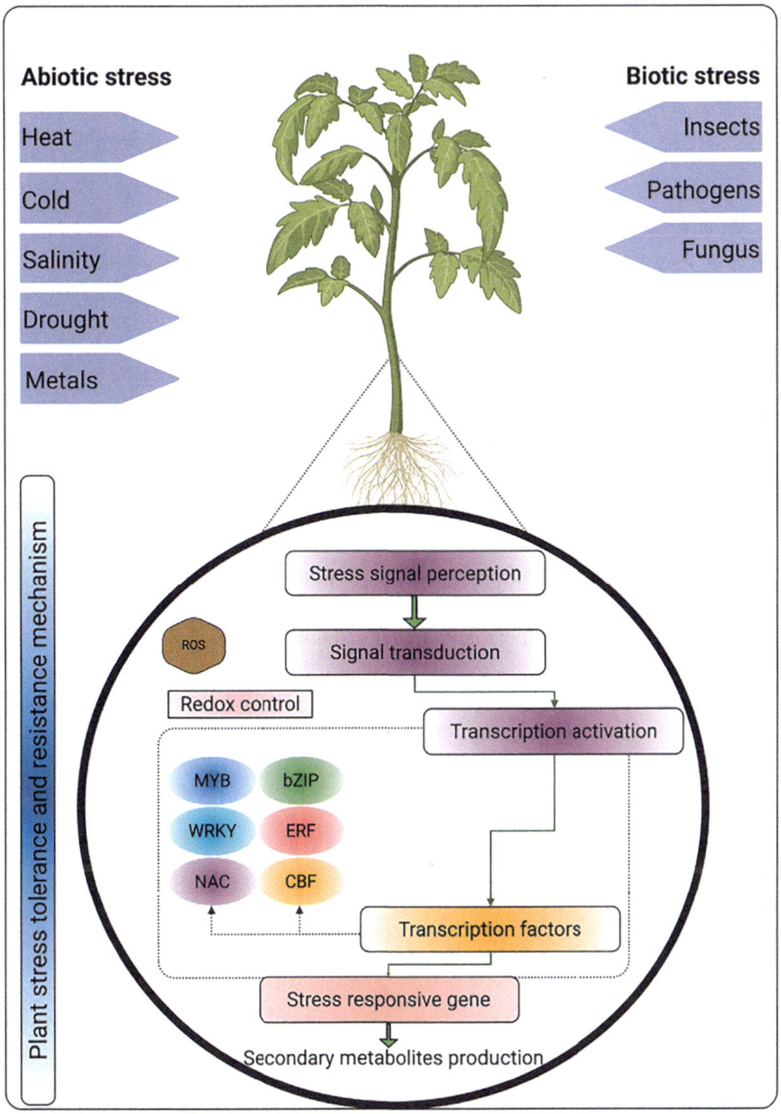

FIGURE 8.1 Diagram illustrating the tolerance and resistance mechanisms in plants under biotic and abiotic stress. Source: *From Khan, A., Kanwal, F., Ullah, S., Fahad, M., Tariq, L., Altaf, M. T., Riaz, A., & Zhang, G. (2025). Plant secondary metabolites—Central regulators against abiotic and biotic stresses.* Metabolites, 15*(4), 276; https://doi.org/10.3390/metabo15040276; open access MPDI; CC-BY.*

The figure clearly illustrates how abiotic elicitors comprise physical [e.g., Ultraviolet (UV), temperature, drought) and chemical (e.g., heavy metals [HMs], nanoparticles [NPs], pH) subgroups, while biotic elicitors include exogenous microbial components and endogenous plant-derived signals. This visual framework supports the subsequent discussion on elicitor-specific signaling and application strategies. Abiotic elicitors, on the other

TABLE 8.1 SMs and their functions.

SMs group	SMs name	Effects	Plant source
Terpenes	Monoterpenes	Antimicrobial, hypotensive, antiinflammatory, antipruritic, antigerminative, antiplasmodial, antiesophageal cancer, and anticandidal.	Chrysanthemum, cumin, pepper, mint, eucalyptus
	Diterpenes	Anticancer, antiinflammatory, antifungal, and antiviral	Codiaeum, *Hura phyllanthus*
	Triterpenes	Anticancer, antioxidant, antiinflammatory, antiatherosclerotic, antiviral, hepatoprotective, and immunomodulatory activity	Rosemary, ginseng, berries, ginkgo
	Polyterpenes	Antimicrobial, cytotoxic, antiinflammatory, and antioxidant	*Bruce banner*
Phenolics	Phenolics acids	Antimicrobial, anticancer, antiinflammatory, and antimutagenic	Tea, berries, coffee, apple
	Flavonoids	Antiinflammatory, antioxidant, and antiproliferative	Onions, broccoli, soybeans, apples, tea, grapes
	Stilbenes	Anticancer, antiinflammatory, antiaging, antioxidant, and antiatherogenic	Peanut, grapes, berries
	Coumarins	Antimicrobial, anticancer, antimitotic, antioxidant, antiinflammatory, and anticoagulant	*Daphne oleoides*, *Ferula sinkiangensis*, *Sarcandra glabra*, *Morinda citrifolia*
	Lignans	Antiestrogenic, antioxidant, anticancer, and antimicrobial	Sunflower seeds, garlic, broccoli, asparagus, apricot, strawberry, rice
	Tannins	Antidiabetic, antioxidant, antimicrobial, and anticancer	Tea, coffee, berries, pomegranate, grapes
Nitrogen-containing SMs	Alkaloids	Anesthetics, cardioprotective, antimicrobial, antitussive, antitumour, antihypertensive, and antiinflammatory	*Berberis vulgaris*, *Capsicum* species, *Erythrina* species
	Cyanogenic glycosides	Antitussive, anticancer, antineoplastic, antiinflammatory, antioxidant, antiatherosclerotic, and antiulcer	Almond, cassava, sorghum, spinach, flaxseed
	Nonprotein amino acid	Cytotoxicity, antimicrobial, antitumor, and antioxidant	Mucuna seeds, *Leucaena leucocephala*, *Blighia sapida*,
Sulfur-containing SMs	Glutathione	Antioxidant and anticancer	Spinach, avocados, okara
	Glucosinolate	Antiinflammatory, antimicrobial, antioxidant, chemoprotective, and anticancer	Cabbage, horseradish, broccoli

Adapted from Khan, A., Kanwal, F., Ullah, S., Fahad, M., Tariq, L., Altaf, M. T., Riaz, A., & Zhang, G. (2025). Plant secondary metabolites—Central regulators against abiotic and biotic stresses. *Metabolites, 15*(4), 276; https://doi.org/10.3390/metabo15040276.

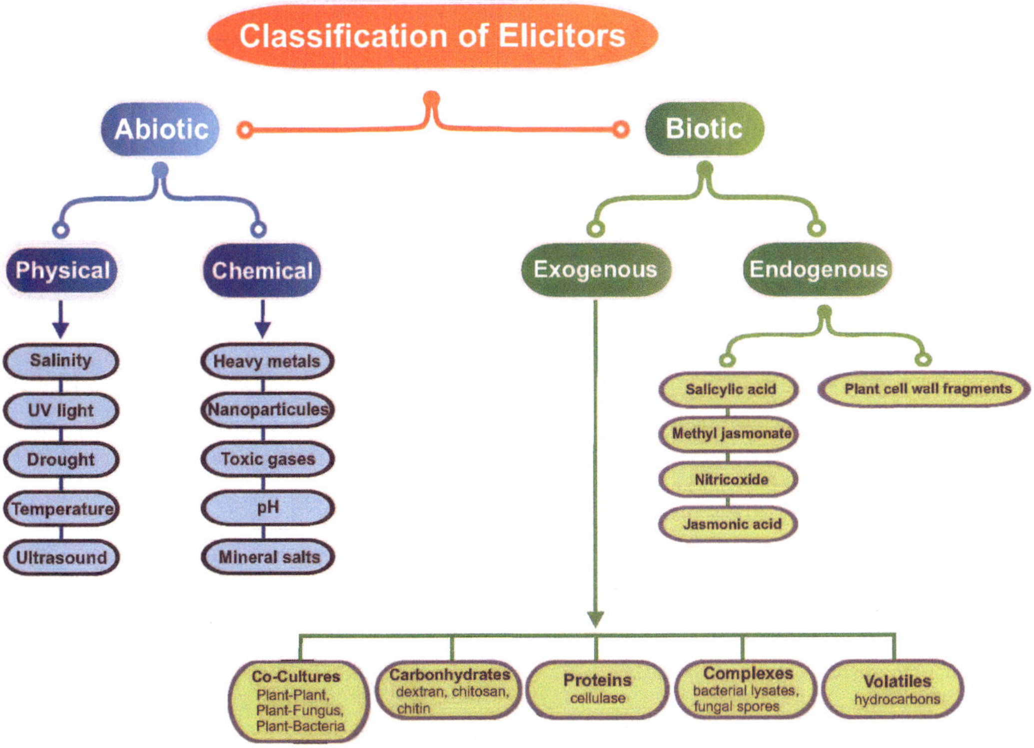

FIGURE 8.2 Classification of elicitors by origin and type. Source: *Adapted from Ayaz-Tilkat, E., Hoşer, A., Süzerer, V., & Tilkat, E. (2023). Influence of salinity on* in vitro *production of terpenes: A review.* IntechOpen; *http://doi.org/10.5772/intechopen. 111813; open access IntechOpen; CC-BY.*

hand, are nonbiological substances or physical factors that induce similar responses. They include chemical compounds such as salicylic acid (SA), methyl jasmonate (MeJA), silver (Ag^+), cobalt (Co^{2+}), and copper (Cu^{2+}) ions, as well as physical stimuli like -UV-B radiation, temperature shifts, osmotic stress, and light spectrum alterations (Halder et al., 2019; Guru et al., 2022). NPs such as Ag^+ or silicon dioxide (SiO_2) are also emerging as abiotic elicitors due to their capacity to induce ROS and modulate cellular pathways.

Elicitors are broadly categorized into biotic and abiotic types, each with distinct origins and functional roles. Detailed examples and comparative features of major elicitor classes are summarized in Table 8.2.

Regardless of their origin, elicitors initiate signal perception through pattern recognition receptors (PRRs) located on the plasma membrane or within plant cells. Upon recognition, early signaling events are triggered, including calcium influx, membrane depolarization, and rapid generation of reactive oxygen and nitrogen species. These are followed by the activation of MAPKs accumulation of defense-related phytohormones such as jasmonic acid (JA) and SA, and upregulation of transcription factors like WRKY, MYB, ERF, and bHLH. These transcription factors drive the expression of secondary metabolism genes encoding

TABLE 8.2 Effects of elicitor applications on SMs production in various plant cell and tissue culture systems.

Plant species (culture type)	Elicitor type and application	Elicitation effects (best treatment)	References
Calendula officinalis (suspension culture)	Chitosan (50 mg/L) + Pectin (2 mg/L), 48 h	Oleanolic acid 0.07 mg/g	Wiktorowska et al. (2010)
Morinda citrifolia (adventitious root culture)	Chitosan (0.2 mg/L), 48 h	Anthraquinones ↑ 45%, phenolics ↑ 8%, flavonoids ↑ 12%	Baque et al. (2012)
Gymnema sylvestre (suspension culture)	*Aspergillus niger* filtrate (200 mg/L), 72 h	Gymnemic acid 8.8 mg/g	Chodisetti et al. (2013)
Calendula officinalis L. (suspension culture)	AgNPs (0.4 mM), 24 h	Saponin ↑ 177%	Ghanati and Bakhtiarian (2014)
Ophiorrhiza mungos (suspension culture)	YE (50 mg/L), 10 d	Camptothecin ↑13X	Deepthi and Satheeshkumar (2016)
Panax ginseng (adventitious root culture)	*Aspergillus niger* extract (200 mg/L), 72h	Total ginsenosides ↑ 3.5X	Li et al. (2016)
Gymnema sylvestre (suspension culture)	Endophytic *Polyancora globosa* (DPFM) (75 mg/100 mL), 72 h	Gymnemic acid ↑ 10.5X	Netala et al. (2016)
Salvia miltiorrhiza (hairy root culture)	YE (%2.5), 12h	Total tanshinones ↑ 4.47X	Zhou et al. (2017)
Panax sikkimensis (suspension culture)	Trichoderma harzianum (2.5%), 7 d	Total ginsenoside ↑ 3.2X	Biswas et al. (2018)
Dracocephalum kotschyi (hairy root culture)	SiO_2 NP (100 mg/L), 48 h SiO_2 NP (200 mg/L), 24 h SiO_2 NP (50 mg/L), 24 h SiO_2 NP (200 mg/L), 48 h SiO_2 NP (100 mg/L), 24 h SiO_2 NP (50 mg/L), 24 h SiO_2 NP (100 mg/L), 24 h SiO_2 NP (200 mg/L), 48 h	Rosmarinic acid 1016.6 μg/g Gallic acid 214.13 μg/g Cinnamic acid 0.54 μg/g Rutin 3.71 μg/g Cirsimaritin 0.94 μg/g Isokaempferid 1.2 μg/g Penduletin 0.43 μg/g Xanthomicrol 1.04 μg/g	Nourozi et al. (2019)
Withania somnifera (L.) Dunal (hairy root culture)	Chitosan (100 mg/L), 7 d	Withaferin-A ↑ 4.03X	Thilip et al. (2019)
Dionaea muscipula J. Ellis (shoot culture)	*Cronobacter sakazakii* lysate (2.5%), 6 d or 7 d	Plumbagin 69.82 mg/g (2.5%, 6d), caffeic acid ↑ 2.43X (2.5%, 7d)	Makowski et al. (2020)
Rubia tinctorum L. (adventitious root culture)	SA (20 μM), 7 d	Alizarin 4.42 mg/g, purpurin 0.73 mg/g	Demirci et al. (2021)
Atropa acuminata (hairy root culture)	Coronatine (0.5 μm), 7 d	Scopolamine 10.95 mg/g	Fattahi et al. (2021)
Chelidonium majus (suspension culture)	MeJA (100 μM), 24h and 72 h	Chelidonine 9.86 μg/mg 72 h Sanguinarine 3.4 μg/mg 24 h	Hashemi and Naghavi (2021)

(Continued)

TABLE 8.2 (Continued)

Plant species (culture type)	Elicitor type and application	Elicitation effects (best treatment)	References
Papaver armeniacum L. (hairy root culture)	MeJA and SA (100 and 200 μM), 24 h and 72 h	Papaverine 60 μg/g (100 μM SA, 48 h) Noscapine 60.33 μg/g (100 μM SA, 48 h) Thebaine 41 μg/g (100 μM MeJA, 48 h) Codeine 120 μg/g (100 μM MeJA, 48 h) Morphine 150.67 μg/g (100 μM MeJA, 48 h)	Sharifzadeh Naeini et al. (2021)
Carum carvi (callus culture)	AgNPs (50 mg/L), 48 h AgNPs (100 mg/L), 24 h AgNPs (50 mg/L), 24 h	Thymol 63.96% Carvacrol 19.31% p-Cymene 2.08%	Dehghani-Aghchekohal et al. (2022)
Nothapodytes nimmoniana (suspension culture)	Chitin (10%), 10 d	Camptothecin ↑ 11.48X	Keshavan et al. (2022)
Manihot esculenta Crantz (callus culture)	Glutathione (2.4 mM/L), 30 d	Total phenols ↑ 64X (4mM/L) Total flavonoids ↑ 11X (2mM/L)	Arafa and Aly (2023)
Solanum surratense (suspension culture)	NaCl, 150 mM (7 d)	Solasodine 0.181 mg/g	Deshmukh et al. (2023)
Primula veris subsp. *veris* (adventitious root culture)	L-proline (100 mg/L), 60 d MeJA (25 μM), 30 d JA (100 μm), 60 d JA (25 μm) JA + L-proline (100 mg/L), 30 d	Primeverin 35.7 mg/g Primulaverin 38.1 mg/g Primulic acid I 27 mg/g Primulic acid II 5.9 mg/g	Sarropoulou et al. (2023)
Pistacia lentiscus L. (callus culture)	$AgNO_3$ (4 mg/L), 28 d leaf callus $AgNO_3$ (4 mg/L), 28 d leaf callus MeJA (1 mg/L), 28 d root callus	Ursonic cid 0.187 μg/g Oleanonic acid 0.038 μg/g Masticadienolic acid 0.008 μg/g	Tilkat et al. (2023)
Withania somnifera L. (callus culture)	Chitosan (100 and 150 mg/L), 30 d Cont. L., 30 d UV (60 and 90 min)	27-OH-Withanolide-A ↑ 160% (chitosan 100 mg/L) Withanolide-B ↑ 255% (chitosan 100 mg/L) Withaferin-A ↑ 82% (Cont. L.) Withanone ↑ 215% (Cont. L.) Withanolide-A ↑ 99% (UV 60 min) 27-OH-Withanolide-B ↑ 40% (UV 90 min)	Abed and Taha (2024)
Hypericum perforatum L. (shoot culture)	ZnO NPs (50 ppm), 48 h	Hyperisin ↑ 2.13X	Pradeep et al. (2024)

(*Continued*)

TABLE 8.2 (Continued)

Plant species (culture type)	Elicitor type and application	Elicitation effects (best treatment)	References
Abeliophyllum distichum (callus culture)	MeJA (50 μm), 21 d	Verbascoside 135.03 mg/g	Choi et al. (2025)
Prunus laurocerasus L. (shoot culture)	nZVI (50 mg/L), 28 d	Quercetin ↑ 10.5× (50 mg/L nZVI) Narcissin ↑ ~2× (50 mg/L nZVI) Nicotiflorin ↑ 75× (50 mg/L nZVI)	Kılıç et al. (2025)

AgNPs, Silver NPs; *Cont. L.*, continuous light; *d*, day; *h*, hour; *JA*, jasmonic acid; *min*, minute; *NaCl*, sodium chloride; *nZVI*, nanosized zero-valent iron; *SA*, salicylic acid; *SiO_2 NP*, silicon dioxide NP; UV, Ultraviolet; *YE*, yeast extract; *ZnO NPs*, zinc NPs.

enzymes such as phenylalanine ammonia-lyase (PAL), chalcone synthase (CHS), terpene synthases, and cytochrome P450 monooxygenases (Khan et al., 2021).

The effectiveness of elicitors varies depending on the plant species, tissue type, culture conditions, and target metabolite. Biotic elicitors often elicit broader immune responses, while abiotic agents like MeJA may induce more targeted metabolic reprogramming. Furthermore, elicitor combinations have demonstrated synergistic effects, where dual or sequential applications—for instance, yeast extract (YE) with Ag^+ ions or MeJA with UV-B radiation—result in significantly higher metabolite accumulation than single treatments alone (Halder et al., 2019; Zhou et al., 2021).

As summarized in Table 8.2, biotic elicitors such as yeast extract, chitosan, *Pseudomonas* spp., *Rhizobium* spp., and *Aspergillus niger* are distinguished by their source, mode of application, and pronounced effects on SM accumulation.

Elicitor selection must therefore consider not only the type and dosage of the elicitor but also its compatibility with the culture system and the desired biosynthetic pathway. In recent years, efforts have been made to map elicitor-responsive pathways through transcriptomics and metabolomics, revealing that elicitor specificity is tightly linked to the plant's innate signaling architecture. While the full mechanistic complexity remains to be elucidated in many species, the classification and strategic application of elicitors provide a powerful means to enhance the biosynthesis of pharmacologically relevant SMs in controlled environments.

8.3 Biosynthesis and roles of SMs

SMs are structurally diverse compounds produced by plants through specialized metabolic pathways that diverge from primary metabolism. While these metabolites are not essential for basic cellular functions such as growth and reproduction, they play crucial roles in plant survival under both biotic and abiotic stress conditions. Additionally, many SMs possess significant pharmaceutical, nutraceutical, and industrial value.

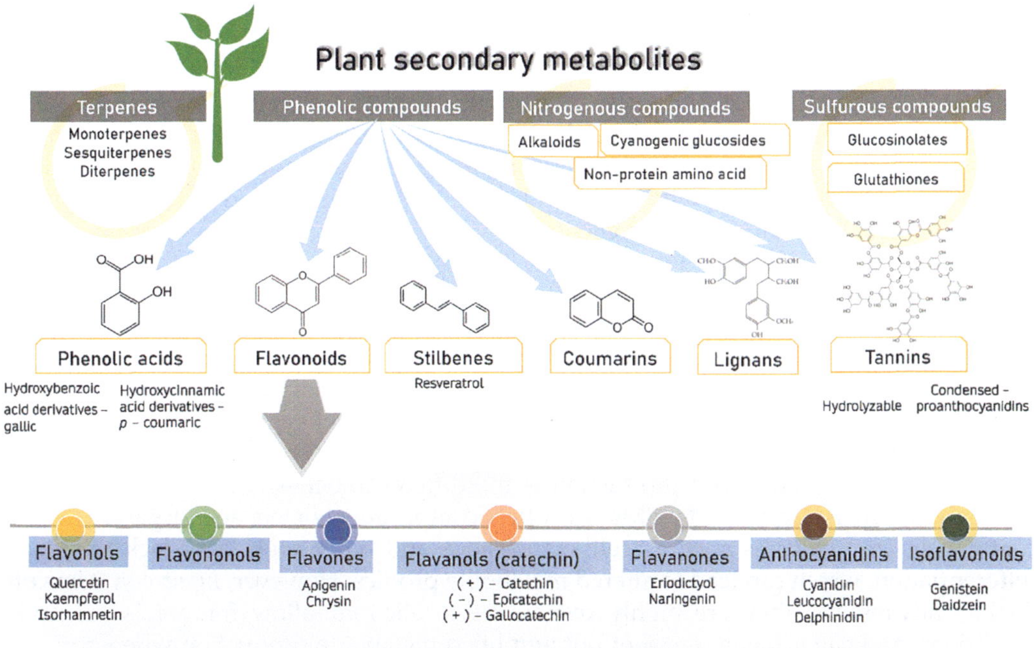

FIGURE 8.3 Main biosynthetic pathways leading to plant SMs (phenolics, terpenoids, alkaloids/polyketides). Source: *Adapted from Razak, A. M., Tan, J. K., Mohd Said, M., & Makpol, S. (2023). Modulating effects of Zingiberaceae phenolic compounds on neurotrophic factors and their potential as neuroprotectants in brain disorders and age-associated neurodegenerative disorders: a review.* Nutrients, 15*(11), 2564.https://doi.org/10.3390/nu15112564.*

Fig. 8.3 illustrates the primary biosynthetic pathways leading to major classes of plant SMs, including phenolic compounds, terpenoids, and nitrogenous compounds.

The biosynthesis of SMs originates from a limited number of central metabolic pathways, including the shikimate pathway, the isoprenoid pathways (mevalonate [MVA] and methylerythritol phosphate [MEP]), and the polyketide pathway. Each of these pathways gives rise to major classes of SMs. Phenylpropanoids and flavonoids are synthesized from phenylalanine via the shikimate pathway, while alkaloids such as camptothecin and camalexin also derive from aromatic amino acids like tryptophan and tyrosine. Terpenoids—including ginsenosides, artemisinin, and taxol—are formed from isopentenyl diphosphate through either the cytosolic MVA or plastidial MEP pathway (Reshi et al., 2023; Han & Miao, 2024). Polyketides such as anthraquinones and stilbenes are synthesized via polyketide synthases that utilize acetyl coenzyme A building blocks (Bisht et al., 2021).

SMs fulfill a variety of biological functions in plants. Many act as phytoalexins or deterrents against herbivores and pathogens, such as alkaloids in *Hyoscyamus* species or camalexin in *Arabidopsis*. Others serve as antioxidants that mitigate oxidative damage caused by UV radiation, salinity, or HM exposure. Compounds like flavonoids and anthocyanins also participate in signaling, pollinator attraction, and allelopathic interactions with neighboring plants (Al Aboud, 2024).

The regulation of SM biosynthesis is complex and responsive to both developmental and environmental cues. Transcriptional control is largely orchestrated by families of transcription factors such as MYB, WRKY, and bHLH, which are activated through signaling pathways involving calcium influx, ROS generation, and phytohormonal cascades. These transcription factors regulate structural genes including PAL, CHS, DXS, and various cytochrome P450s (Reshi et al., 2023). Furthermore, epigenetic modifications such as histone acetylation and DNA methylation also influence the responsiveness of SM biosynthetic genes to external stimuli, including elicitors.

Elicitors can significantly influence SM biosynthesis by activating these regulatory networks. Through interaction with PRR or internal sensors, elicitors induce signal transduction cascades that modulate gene expression and enzyme activity, thereby redirecting metabolic flux toward specific pathways. For instance, the elicitation of flavonoids often involves increased CHS activity, while terpenoid biosynthesis is enhanced through upregulation of squalene synthase or dammarenediol synthase genes (Khan et al., 2021; Petrova et al., 2024).

A key consideration in SM production is the difference between *in vitro* and in planta biosynthesis. In natural plants, SMs are often compartmentalized and organ-specific. In contrast, *in vitro* cultures such as cell suspensions and hairy roots may lack this spatial differentiation, which can lead to altered metabolite profiles. However, *in vitro* systems offer distinct advantages: they are highly responsive to elicitors, allow for precise control of conditions, and can achieve transient but amplified metabolite production when optimized appropriately (Narayani & Srivastava, 2017).

In summary, understanding the biosynthetic origin and physiological roles of SMs is essential for designing effective elicitor-based strategies. By integrating knowledge of core metabolic pathways with regulatory mechanisms and elicitor responsiveness, researchers can develop targeted interventions for the efficient and sustainable production of high-value phytochemicals in plant biotechnology systems.

8.4 Molecular mechanisms of elicitor action

Elicitors initiate their effect by being perceived at the plant cell surface or inside the cell, often through PRRs that recognize conserved microbial or damage-associated molecules—a process akin to pattern-triggered immunity (PTI) (Yoshioka et al., 2023; Goura et al., 2025; Pršić & Ongena, 2020; Selwal et al., 2025; Wang et al., 2025). Binding of elicitors such as chitosan, flagellin, or oligosaccharides triggers early events including ion fluxes, especially a rapid Ca^{2+} influx, plasma membrane depolarization, and medium alkalinization. These are typically followed by a quick burst of ROS, catalyzed by NADPH oxidases, which act as both defense and signaling molecules (Zhao et al., 2005).

ROS accumulation rapidly activates MAPK cascades, which amplify the signal through phosphorylation events and converge on downstream transcription factors and enzymes (Liu & He, 2017; Kolupaev et al., 2023; Manna et al., 2023; Yoshioka et al., 2023). Concurrently, small signaling molecules such as cyclic nucleotides, inositol phosphates, nitric oxide, and lipid-derived signals such as phosphatidic acid and lysophosphatidylcholine modulate this signaling web.

As signaling progresses, phytohormones including JA, SA, ethylene, and abscisic acid become synthesized and activated (Adedayo et al., 2025; Rabeh et al., 2025; Thilakarathne et al., 2025; Wang et al., 2025). JA, in particular, interacts with gasotransmitters and regulates defense genes in a complex network (Ding et al., 2022; Kolupaev et al., 2023; Sangwan et al., 2024). This hormone signaling cross-talk influences transcription factors such as WRKY, MYB, ERF, and bHLH, which bind to promoter regions of biosynthetic genes, including PAL, CHS, and terpene synthases (Zhao et al., 2005).

Specifically, WRKY transcription factors play a central role in relaying elicitor signals into metabolic responses. They contain a conserved WRKY motif and zinc-finger domain, and often interact with other proteins (e.g., VQ motif proteins, histone modifiers) to modulate transcriptional reprogramming of defense and SM biosynthetic pathways.

At the cellular level, these combined signaling events result in the activation of genes responsible for SM biosynthesis. Enzymes such as PAL, CHS, terpene synthases, cytochrome P450 monooxygenases, and polyketide synthases are transcriptionally upregulated. Metabolic flux is redirected toward biosynthetic "bottlenecks," enhancing the accumulation of defense-related compounds like phytoalexins, flavonoids, terpenoids, and alkaloids.

Finally, epigenetic modifications are emerging as another layer of regulation in elicitor responsiveness. Changes in histone acetylation and DNA methylation profiles can prime plant cells for more rapid and robust transcriptional responses upon elicitor challenge, improving both temporal and quantitative aspects of SM accumulation.

8.5 Case studies: elicitors

One of the most compelling ways to understand elicitor impact is through worldwide applications in plant culture systems. The following examples illustrate how specific elicitors have been successfully used to enhance SM production, highlighting experimental conditions, outcomes, and mechanistic insights.

8.5.1 Chitosan in cell culture of *Psoralea corylifolia*

In suspension cultures of *Psoralea corylifolia*, chitosan applied at 25–300 mg/L on day 16 of cultivation resulted in a significant increase in psoralen accumulation. The optimal elicitor concentration and treatment duration (16–24 hours) were determined based on the growth stage, with cell dry weight and metabolite levels measured posttreatment using standard assays (Ahmed & Baig, 2014).

8.5.2 Chitosan elicitation in *Lippia alba*

Chitosan treatment in *Lippia alba* cultures not only induced a rise in essential oils and phenolic compounds but was also shown to upregulate antioxidant enzyme activity, suggesting activation of the ROS–MAPK–TF signaling cascade (Silva et al., 2022).

8.5.3 Chitosan and pectin in *Morinda citrifolia* adventitious roots

Combined elicitation using 0.2 mg/mL chitosan and pectin significantly boosted the biosynthesis of anthraquinones, flavonoids, and phenolics by up to 103 mg/g dry weight, despite a minor reduction in root growth. A two-stage elicitation protocol (adding elicitor on day 28 and harvesting after 2 days) was optimized to balance metabolite yield and biomass (Baque et al., 2012).

8.5.4 MeJA and SA in hairy root cultures

In *Phyllanthus acuminatus* hairy roots, treatments with 50 μM SA or MeJA doubled total phenols and tripled mucic acid derivatives. The metabolomic profiling method revealed targeted changes in phenylpropanoid and flavonoid biosynthesis, clearly demonstrating the metabolic shift induced by elicitors (Benavides et al., 2024).

8.5.5 MeJA in peppermint (*Mentha* × piperita)

Foliar application of MeJA significantly increased essential oil content and total phenolic compounds in peppermint, under both controlled environment and field conditions. This demonstrates the potential of elicitors *in vivo* cultivation systems as well as *in vitro* (Kandoudi et al., 2023).

8.5.6 Combined MeJA and $ZnSO_4$ in barley seedlings

Seedling treatments with 100 μM MeJA or 4 mM $ZnSO_4$ enhanced phenolic acid content by approximately 35%, significantly increasing PAL and C4H enzyme activities. These enhancements highlight how combining phytohormonal elicitors with mineral salts can stimulate enzyme-mediated metabolic flux (Tian et al., 2024).

8.5.7 Insights from proteomic analysis: MeJA in tea plants

Preharvest MeJA application in tea plants (*Camellia sinensis*) induced significant changes in aroma-related metabolites and enzymes. Proteomic data revealed upregulation of terpene synthases and glycosylation pathways, showcasing how elicitation reprograms both primary and secondary metabolism in leaves (Shi et al., 2021).

8.5.8 Nanoparticles application of SMs production

Recently, there has been a growing tendency to employ NPs for enhancing SMs biosynthesis. Among the most commonly used in this context are AgNPs, SiO_2 NPs, titanium dioxide NPs (TiO_2 NPs), and ZnO NPs (Rani et al., 2025). For example, in medicinal and fragrance plant *Dracocephalum kotschyi*, adventitious root cultures were exposed to SiO_2 NPs at varying concentrations and durations. These treatments resulted in significant increases in the levels of various SMs—namely, rosmarinic acid, gallic acid, and cinnamic acid—relative to untreated controls (Nourozi et al., 2019).

A comprehensive summary of major elicitor types, their mechanisms of action, and representative examples is provided in Table 8.2.

8.6 Optimization in plant culture systems

Maximizing SM production in plant *in vitro* systems—such as hairy-root, cell suspension, and organ cultures—demands precise optimization of both biological parameters and engineering conditions, particularly when aiming for scalable, reproducible production in bioreactors. These models provide controlled environments suitable for elicitor-mediated enhancement (Zhao et al., 2005; Dubey et al., 2022).

Hairy-root cultures, induced via *Agrobacterium rhizogenes*, are especially valued for genetic stability, robust growth, and hormone-free development, making them ideal platforms for producing diverse SMs. However, successful scale-up to bioreactors requires careful management of oxygen transfer, nutrient gradients, and shear stress resilience (Stiles & Liu, 2013; Khan et al., 2018).

Optimizing culture conditions is key to maximizing SM production in plant systems, as it involves tuning multiple interacting factors—such as elicitor concentration, exposure duration, pH, temperature, and medium composition. Instead of laborious one-factor-at-a-time approaches, researchers increasingly rely on response surface methodology (RSM) with experimental designs like central composite design (CCD) or Box–Behnken to model these interactions efficiently (Kaur et al., 2020).

Since culture parameters interact in complex ways, researchers increasingly employ statistical modeling techniques such as RSM alongside experimental designs like CCD and Box–Behnken. These tools allow for simultaneous optimization of multiple variables—elicitor concentration, exposure time, medium pH, carbon source, temperature—thereby enhancing both biomass accumulation and metabolite yields (Kaur et al., 2020).

8.6.1 Plant-based response surface methodology case study: *Swertia paniculata* shoot cultures

Kaur et al. (2020) utilized RSM (CCD) and artificial neural networks to optimize SA and chitosan treatments in *Swertia paniculata* shoot cultures—aiming to boost amarogentin, swertiamarin, and mangiferin levels. Treatment ranges were 1–20 mM SA and 1–20 mg/L Chitosan over 14 days: Amarogentin increased from $0.17 \rightarrow 0.435\%$ DW; Swertiamarin increased from $\sim 1 \rightarrow 4.99\%$ DW; Mangiferin rose from $\sim 2.55 \rightarrow 4.36\%$ DW

Quadratic regression models showed excellent fit ($R^2 \geq 0.998$), confirming RSM's power in elicitor optimization (Kaur et al., 2020)

Fig. 8.4 highlights the optimal elicitor combinations—visible as dark-green contours—that maximize swertiamarin production in *Swertia* shoot cultures.

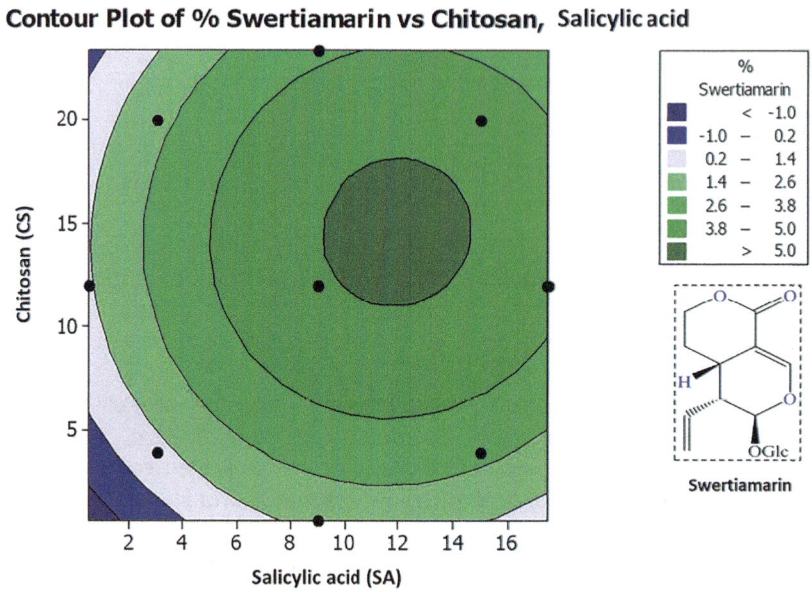

FIGURE 8.4 Contour plot illustrating the interactive effects of salicylic acid (A; 1–20 mM) and chitosan (B; 1–20 mg/L) on swertiamarin production (% DW) in *S. paniculata* shoot cultures. The optimal region (~9 mM SA + 12 mg/L chitosan) is shown in *dark green. Adapted under CC-BY from Kaur, P., Gupta, R. C., Dey, A., Malik, T., & Pandey, K. D. (2020). Optimization of salicylic acid and chitosan treatment for bitter secoiridoid and xanthone glycosides production in shoot cultures of* Swertia paniculata *using response surface methodology and artificial neural network.* BMC Plant Biology, 20, *225; https://doi.org/10.1186/s12870-020-02410-7.*

8.6.2 Broader applications in plant suspension cultures

RSM has also been applied in *Azadirachta indica* (neem) cell suspension systems, optimizing YE concentration and timing. This led to improved yields of azadirachtin (16.08 mg/g DW on day 2), mevalonic acid, and squalene (4.53 mg/L on day 4) (Farjaminezhad & Garoosi, 2020, 2021).

8.6.3 Engineering considerations and scale-up

Bioreactor configuration plays a crucial role: stirred-tank, airlift, and sprinkle systems offer distinct advantages in aeration and shear stress control. For instance, stirred-tank bioreactors supported successful tropane alkaloid production in hairy-root cultures of *Brugmansia candida* and *Datura stramonium.*

Biological variables—explant age, *A. rhizogenes* strain, coculture duration, and medium pH—are also pivotal. In *Verbascum* species, optimized induction protocols achieved transformation efficiencies exceeding 80% (Amini et al., 2024).

Elicitor regimes (type, concentration, timing, combinations) must be carefully moderated: overdose may be cytotoxic, underdose ineffective. Dose–response and time-course trials aligned with RSM help define optimal protocol windows. Elicitors like MeJA and Ag^{+} ions require system-specific calibration to enhance production without harming tissue (Narayani & Srivastava, 2017).

8.7 Future directions and integration

Future developments in elicitor-mediated SM production are increasingly focused on integrating advanced biotechnological and engineering strategies to overcome current limitations and enhance system scalability and efficiency. One promising avenue is the combination of elicitation with in situ product recovery systems. Techniques such as resin adsorption or aqueous two-phase systems allow for semicontinuous harvesting of bioactive compounds. These methods help maintain low intracellular toxicity, minimize feedback inhibition, and ultimately improve product yields. A notable example is the use of resin-containing media in hairy root cultures of *Ophiorrhiza pumila*, which enabled enhanced camptothecin production by facilitating the exudation of metabolites into the surrounding medium (Saito et al., 2001). Furthermore, recent reviews have underscored the value of exudation technologies and absorbent systems in enabling continuous metabolite capture in both cell and hairy root cultures (Cai et al., 2012).

Another critical development area involves the application of omics and computational tools to optimize elicitation strategies. The integration of transcriptomics, metabolomics, and flux modeling facilitates a more rational and targeted approach, enabling precise control over the timing and dosage of elicitors to maximize the flux toward desired SM pathways. Although multiomics approaches are more established in animal systems, their utility in plant culture systems is increasingly recognized (Jamil et al., 2020). For instance, studies on hairy root cultures have demonstrated that correlations between transcriptomic and metabolomic data can effectively guide the design of elicitor treatments (Gardinassi et al., 2018; Halder et al., 2019).

Advancements in synthetic biology also offer exciting opportunities for enhancing elicitor responsiveness and metabolite yield. By engineering elicitor signaling pathways—such as inserting stronger PRRs or knocking out negative regulators—it is possible to significantly improve both the robustness and efficiency of SM production systems.

The integration of ML and digital bioprocessing technologies provides further innovation potential. ML models, when paired with real-time sensors and predictive control systems, can greatly enhance the reproducibility, scalability, and efficiency of elicitation processes (Ibrahim et al., 2023). These tools support the development of smart, automated platforms capable of maintaining consistent process conditions even at industrial scales.

Despite the high value of plant-based systems, the complexity and cost associated with maintaining such cultures have driven interest in alternative production platforms. In this context, microbial chassis—such as engineered strains of *Escherichia coli* or *Saccharomyces cerevisiae*—offer promising routes for heterologous production of specific metabolites. However, the complete biosynthetic reconstruction of complex plant metabolites remains a significant challenge, as many of these compounds require compartmentalization and specific enzymatic modifications unique to plant systems.

8.8 Concluding remarks

Elicitor-based strategies present a precise and environmentally sustainable approach to enhancing SM production in plant systems. However, the transition from laboratory-scale experiments to industrial-scale production platforms requires multidisciplinary innovations. The convergence of elicitation techniques with modern bioreactor designs, omics-based analytics, computational modeling, and synthetic biology—operated within digitally monitored environments—holds great promise for establishing robust, scalable biofactories capable of producing pharmaceuticals, nutraceuticals, and other high-value natural products.

Substantial evidence demonstrates that both biotic and abiotic elicitors, including MeJA, microbial extracts, and metal ions, can consistently trigger multifold increases in valuable compounds such as ginsenosides, taxanes, flavonoids, and alkaloids. Hairy-root and cell suspension cultures have proven especially effective for such enhancements (Halder et al., 2019). Mechanistically, elicitors activate complex plant defense pathways—including calcium influx, ROS bursts, MAPK cascades, and jasmonic acid/SA signaling—that ultimately upregulate key genes involved in SM biosynthesis (Jalmi & Sinha, 2015).

Nevertheless, industrial-scale implementation faces several hurdles, such as reproducibility issues, risk of contamination, uneven elicitor distribution, and the need for cost-effective processes. The application of Quality by Design frameworks to plant systems is still in its early stages (Humbal & Pathak, 2023), highlighting the need for further research and development.

Looking forward, several strategies are emerging to address these challenges. These include continuous metabolite harvesting through in situ adsorption systems (e.g., resins, cyclodextrins), which have proven effective in camptothecin production systems; integration of omics data and computational models to inform the timing and intensity of elicitor applications (Selwal et al., 2025); synthetic biology approaches aimed at enhancing receptor sensitivity and metabolite transport; real-time digital monitoring for process validation; and exploration of microbial platforms for metabolites amenable to heterologous production. However, for complex metabolites requiring specific plant-based enzymatic steps, plant systems remain indispensable.

In conclusion, elicitor-driven enhancement of plant SMs represents a compelling, eco-friendly strategy for sustainable production. Achieving industrial viability will depend on the seamless integration of plant science, engineering, and digital technologies—ultimately enabling the development of next-generation biofactories for natural product synthesis.

References

Abed, A. S., & Taha, A. J. (2024). Biotic and abiotic elicitors induce biosynthesis and accumulation of some withanolide compounds in the callus culture of *Withania somnifera* L. *Egyptian Journal of Botany, 64*(1), 245–256. https://doi.org/10.21608/ejbo.2023.153598.2083.

Adedayo, A. A., Musser, R., Aanaenson, M., & Babalola, O. O. (2025). The biochemical and molecular mechanisms of plants: A review on insect herbivory. *Plant Signaling & Behavior, 20*(1), 2439248. https://doi.org/10.1080/15592324.2024.2439248.

Ahmed, S. A., & Baig, M. M. (2014). Biotic elicitor enhanced production of psoralen in suspension cultures of *Psoralea corylifolia* L. *Saudi Journal of Biological Sciences, 21*(5), 499–504. https://doi.org/10.1016/j.sjbs.2013.12.008.

Al Aboud, N. M. (2024). Unlocking the genetic potential: Strategies for enhancing secondary metabolite biosynthesis in plants. *Journal of the Saudi Society of Agricultural Sciences*. https://doi.org/10.1016/j.jssas.2024.06.004.

Amini, S., Fattahi, M., & Nazemiyeh, H. (2024). Optimization of induction and hairy root culture establishment in two mullein species, *Verbascum erianthum* and *Verbascum stachydiforme*. *Scientific Reports, 14*, 5636. https://doi.org/10.1038/s41598-024-56331-8.

Arafa, N. M., & Aly, U. I. (2023). Increasing growth and accumulation of metabolites for radical scavenging potential in *Manihot esculenta* Crantz calli cultures by abiotic elicitors. *Egyptian Journal of Chemistry, 66*(13), 1093–1102. https://doi.org/10.21608/ejchem.2023.203330.7806.

Ayaz Tilkat, E., Hoşer, A., Süzerer, V., & Tilkat, E. (2023). *Making Plant Life Easier and Productive Under Salinity - Updates and Prospects*. IntechOpen. 10.5772/intechopen.111813.

Baque, M. A., Shiragi, M. H. K., Lee, E. J., & Paek, K. Y. (2012). Elicitor effect of chitosan and pectin on the biosynthesis of anthraquinones, phenolics and flavonoids in adventitious root suspension cultures of *Morinda citrifolia* L. *Australian Journal of Crop Science, 6*(9), 1349–1355.

Benavides, K., Sánchez-Kopper, A., Jiménez-Quesada, K., Perez, R., & Garro-Monge, G. (2024). Evaluation of salicylic acid and methyl jasmonate as elicitors in *Phyllanthus acuminatus* hairy roots by non-targeted analysis using high-resolution mass spectrometry. *Molecules (Basel, Switzerland), 29*(1), 80. https://doi.org/10.3390/molecules29010080.

Bisht, R., Bhattacharyya, A., Shrivastava, A., & Saxena, P. (2021). An overview of the medicinally important plant type III PKS derived polyketides. *Frontiers in Plant Science, 12*. https://doi.org/10.3389/fpls.2021.746908.

Biswas, T., Pandey, S. S., & Maji, D. (2018). Enhanced expression of ginsenoside biosynthetic genes and in vitro ginsenoside production in elicited *Panax sikkimensis* (Ban) cell suspensions. *Protoplasma, 255*, 1147–1160. https://doi.org/10.1007/s00709-018-1219-z.

Cai, Z., Kastell, A., Knorr, D., & Smetanska, I. (2012). Exudation: An expanding technique for continuous production and release of secondary metabolites from plant cell suspension and hairy root cultures. *Plant Cell Reports, 31*(3), 461–477. https://doi.org/10.1007/s00299-011-1165-0.

Chodisetti, B., Rao, K., Gandi, S., & Giri, A. (2013). Improved gymnemic acid production in the suspension cultures of *Gymnema sylvestre* through biotic elicitation. *Plant Biotechnology Reports, 7*(4), 519–525. https://doi.org/10.1007/s11816-013-0290-3.

Choi, D., Park, Y. W., Kang, J., Jung, E. S., & Lee, H. (2025). Effects of elicitation on *Abeliophyllum distichum* leaf callus and changes in verbascoside content. *Plants, 14*(9), 1386. https://doi.org/10.3390/plants14091386.

Deepthi, S., & Satheeshkumar, K. (2016). Enhanced camptothecin production induced by elicitors in the cell suspension cultures of *Ophiorrhiza mungos* Linn. *Plant Cell, Tissue and Organ Culture, 124*(3), 483–493. https://doi.org/10.1007/s11240-015-0908-y.

Dehghani-Aghchekohal, Z., Omidi, M., Azizinezhad, R., & Etminan, A. (2022). Stimulation of secondary metabolites and γ-terpinene synthase by silver nanoparticles in callus cultures of *Carum carvi*. *Applied Biochemistry and Biotechnology, 194*, 3228–3241. https://doi.org/10.1007/s12010-022-03879-8.

Demirci, T., Aras Ascı, Ö., & Göktürk Baydar, N. (2021). Influence of salicylic acid and L-phenylalanine on the accumulation of anthraquinone and phenolic compounds in adventitious root cultures of madder (*Rubia tinctorum* L.). *Plant Cell Tiss Organ Cult, 144*, 313–324. https://doi.org/10.1007/s11240-020-01952-w.

Deshmukh, V., Deshpande, J., & Wani, M. (2023). Elicitation based enhancement of solasodine production in in-vitro cultures of different *Solanum species*. *Journal of Environmental Biology, 44*(2), 167–174. https://doi.org/10.22438/jeb/44/2/MRN-4011.

Ding, L. N., Li, Y. T., Wu, Y. Z., Li, T., Geng, R., Cao, J., Zhang, W., & Tan, X. L. (2022). Plant disease resistance-related signaling pathways: Recent progress and future prospects. *International Journal of Molecular Sciences, 23*(24), 16200. https://doi.org/10.3390/ijms232416200.

Dubey, N., Singh, K., Pankaj, U., & Pandey, V. C. (2022). Role of plant-microbial secondary metabolites in stress mitigation: Current knowledge and future directions. *Microbial Based Land Restoration Handbook, 1*, 23–42.

Farjaminezhad, R., & Garoosi, G. (2021). Improvement and prediction of secondary metabolites production under yeast extract elicitation of *Azadirachta indica* cell suspension culture using response surface methodology. *AMB Express, 11*(1), 43. https://doi.org/10.1186/s13568-021-01203-x.

Farjaminezhad, R., & Garoosi, G.-A. (2020). Establishment of green analytical method for ultrasound-assisted extraction of azadirachtin, mevalonic acid and squalene from cell suspension culture of *Azadirachta indica* using response surface methodology. *Industrial Crops and Products, 144*, 111946. https://doi.org/10.1016/j.indcrop.2019.111946.

Fattahi, F., Shojaeiyan, A., Palazon, J., Moyano, E., & Torras-Claveria, L. (2021). Methyl-β-cyclodextrin and coronatine as new elicitors of tropane alkaloid biosynthesis in *Atropa acuminata* and *Atropa belladonna* hairy root cultures. *Physiologia Plantarum, 172*(4), 2098–2111. https://doi.org/10.1111/ppl.13444.

Fazili, M. A., Bashir, I., Ahmad, M., et al. (2022). *In vitro* strategies for the enhancement of secondary metabolite production in plants: A review. *Bulletin of the National Research Centre, 46*, 35. https://doi.org/10.1186/s42269-022-00717-z.

Gardinassi, L. G., Arévalo-Herrera, M., Herrera, S., Cordy, R. J., Tran, V., Smith, M. R., et al. (2018). Integrative metabolomics and transcriptomics signatures of clinical tolerance to *Plasmodium vivax* reveal activation of innate cell immunity and T cell signaling. *Redox Biology, 17*, 158–170. https://doi.org/10.1016/j.redox.2018.04.011.

Ghanati, F., & Bakhtiarian, S. (2014). Effect of methyl jasmonate and silver nanoparticles on production of secondary metabolites by *Calendula officinalis* L (Asteraceae). *Tropical Journal of Pharmaceutical Research, 13*, 1783–1789. https://doi.org/10.4314/tjpr.v13i11.2.

Goura, K., Legrifi, I., Kallali, N., et al. (2025). Beyond survival: The role of secondary metabolites in plant defense mechanisms. *Journal of Crop Health, 77*, 121. https://doi.org/10.1007/s10343-025-01183-z.

Guru, A., Dwivedi, P., Kaur, P., & Pandey, D. K. (2022). Exploring the role of elicitors in enhancing medicinal values of plants under in vitro condition. *South African Journal of Botany, 149*, 1029–1043 10.1016/j.sajb.2021.10.014.

Halder, M., Sarkar, S., & Jha, S. (2019). Elicitation: A biotechnological tool for enhanced production of secondary metabolites in hairy root cultures. *Engineering in Life Sciences, 19*(12), 880–895. https://doi.org/10.1002/elsc.201900058.

Han, T., & Miao, G. (2024). Strategies, achievements, and potential challenges of plant and microbial chassis in the biosynthesis of plant secondary metabolites. *Molecules (Basel, Switzerland), 29*(9), 2106. https://doi.org/10.3390/molecules29092106.

Hashemi, S. M., Naghavi, M. R., Bakhshandeh, E., Ghorbani, M., Priyanatha, C., & Zandi, P. (2021). Effects of abiotic elicitors on expression and accumulation of three candidate benzophenanthridine alkaloids in cultured greater celandine cells. *Molecules, 26*, 1395. https://doi.org/10.3390/MOLECULES26051395.

Humbal, A., & Pathak, B. (2023). Influence of exogenous elicitors on the production of secondary metabolite in plants: A review ("VSI: secondary metabolites"). *Plant Stress, 8*, 100166. https://doi.org/10.1016/j.stress.2023.100166.

Ibrahim, B., Konduah, J. O., & Ahenkorah, I. (2023). Predicting reservoir temperature of geothermal systems in Western Anatolia, Turkey: A focus on predictive performance and explainability of machine learning models. *Geothermics, 112*, 102727. https://doi.org/10.1016/j.geothermics.2023.102727.

Jalmi, S. K., & Sinha, A. K. (2015). ROS mediated MAPK signaling in abiotic and biotic stress: Striking similarities and differences. *Frontiers in Plant Science, 6*, 769. https://doi.org/10.3389/fpls.2015.00769.

Jamil, I. N., Remali, J., Azizan, K. A., Nor Muhammad, N. A., Arita, M., Goh, H. H., & Aizat, W. M. (2020). Systematic multi-omics integration (MOI) approach in plant systems biology. *Frontiers in Plant Science, 11*, 944. https://doi.org/10.3389/fpls.2020.00944.

Jha, S., & Yadav, A. (2023). Assessment of carbon and fullerene nanomaterials for sustainable crop plants growth and production. In A. Husen (Ed.). *Plant biology, sustainability and climate change: Engineered nanomaterials for sustainable agricultural production, soil improvement and stress management* (pp. 145–160). Academic Press. doi.org/10.1016/B978-0-323-91933-3.00015-5.

Kandoudi, W., Tavaszi-Sárosi, S., & Németh-Zámboriné, E. (2023). Inducing the production of secondary metabolites by foliar application of methyl jasmonate in peppermint. *Plants, 12*(12), 2339. https://doi.org/10.3390/plants12122339.

Kaur, P., Gupta, R. C., Dey, A., Malik, T., & Pandey, K. D. (2020). Optimization of salicylic acid and chitosan treatment for bitter secoiridoid and xanthone glycosides production in shoot cultures of *Swertia paniculata* using response surface methodology and artificial neural network. *BMC Plant Biology, 20*, 225. https://doi.org/10.1186/s12870-020-02410-7.

Keshavan, B., Santosh Srinivas, N., Muthu Tamizh, M., Vairamani, M., & Pachaiappan, R. (2022). *In vitro* elicitation of *Camptothecin* by challenging with biotic elicitors in *Nothapodytes nimmoniana* (J.Graham) Mabb. *South African Journal of Botany, 144*, 325–331. https://doi.org/10.1016/J.SAJB.2021.08.039.

Khan, A., Kanwal, F., Ullah, S., Fahad, M., Tariq, L., Altaf, M. T., Riaz, A., & Zhang, G. (2025). Plant secondary metabolites—Central regulators against abiotic and biotic stresses. *Metabolites, 15*(4), 276. https://doi.org/10.3390/metabo15040276.

Khan, S. A., Siddiqui, M. H., & Osama, K. (2018). Bioreactors for hairy roots culture: A review. *Current Biotechnology, 7*(6), 417–427. https://doi.org/10.2174/2211550108666190114143824.

Khan, T., Khan, M. A., Karam, K., Ullah, N., Mashwani, Z. U. R., & Nadhman, A. (2021). Plant in vitro culture technologies: A promise into factories of secondary metabolites against COVID-19. *Frontiers in Plant Science, 12*. https://doi.org/10.3389/fpls.2021.610194 610194.

Kılıç, G.Ç., Şeker, M. G., Gutul, T., Süzerer, V., Dursun, İ.Ö., & Çiftçi, Y. (2025). The influence of nanosized zero-valent iron (nZVI) on the micropropagation, antioxidant activity, and phenolic compound content of cherry laurel (*Prunus laurocerasus* L.). *Plant Cell Tiss Organ Cult, 160*, 68. https://doi.org/10.1007/s11240-025-02968-w.

Kolupaev, Y. E., Yastreb, T. O., & Dmitriev, A. P. (2023). Signal mediators in the implementation of jasmonic acid's protective effect on plants under abiotic stresses. *Plants, 12*(14), 2631. https://doi.org/10.3390/plants12142631.

Li, J., Liu, S., Wang, J., Li, J., Liu, D., Li, J., & Gao, W. (2016). Fungal elicitors enhance ginsenosides biosynthesis, expression of functional genes as well as signal molecules accumulation in adventitious roots of *Panax ginseng* C. A. Mey. *Journal of Biotechnology, 239*, 106–114. https://doi.org/10.1016/j.jbiotec.2016.10.011.

Liu, Y., & He, C. (2017). A review of redox signaling and the control of MAP kinase pathway in plants. *Redox Biology, 11*, 192–204. https://doi.org/10.1016/j.redox.2016.12.009.

Makowski, W., Tokarz, K. M., Tokarz, B., Banasiuk, R., Witek, K., & Królicka, A. (2020). Elicitation-based method for increasing the production of antioxidant and bactericidal phenolic compounds in *Dionaea muscipula* J. Ellis tissue. *Molecules (Basel, Switzerland), 25*(8), 1794. https://doi.org/10.3390/molecules25081794 2020.

Manna, M., Rengasamy, B., & Sinha, A. K. (2023). Revisiting the role of MAPK signalling pathway in plants and its manipulation for crop improvement. *Plant, Cell & Environment, 46*(8), 2277–2295. https://doi.org/10.1111/pce.14606.

Narayani, M., & Srivastava, S. (2017). Elicitation: A stimulation of stress in in vitro plant cell/tissue cultures for enhancement of secondary metabolite production. *Phytochemistry Reviews, 16*, 1227–1252. https://doi.org/10.1007/s11101-017-9534-0.

Netala, V. R., Kotakadi, V. S., Gaddam, S. A., Ghosh, S. B., & Tartte, V. (2016). Elicitation of gymnemic acid production in cell suspension cultures of *Gymnema sylvestre* R.Br. through endophytic fungi. *3 Biotech, 6*(2), 232. https://doi.org/10.1007/s13205-016-0555-y.

Nourozi, E., Hosseini, B., Maleki, R., & Mandoulakani, B. A. (2019). Pharmaceutical important phenolic compounds overproduction and gene expression analysis in *Dracocephalum kotschyi* hairy roots elicited by SiO_2 nanoparticles. *Industrial Crops and Products, 133*, 435–446. https://doi.org/10.1016/j.indcrop.2019.03.053.

Petrova, M., Miladinova Georgieva, K., & Geneva, M. (2024). Influence of abiotic and biotic elicitors on organogenesis, biomass accumulation, and production of key secondary metabolites in Asteraceae plants. *International Journal of Molecular Sciences, 25*(8), 4197. https://doi.org/10.3390/ijms25084197.

Pradeep, M., Saxena, M., Mondal, D., & Franklin, G. (2024). Do nanoparticles delivered to roots affect plant secondary metabolism? A comprehensive analysis in float seedling cultures of *Hypericum perforatum* L. *Chemosphere, 356*, 141789. https://doi.org/10.1016/j.chemosphere.2024.141789.

Pršić, J., & Ongena, M. (2020). Elicitors of plant immunity triggered by beneficial bacteria. *Frontiers in Plant Science, 11*, 594530. https://doi.org/10.3389/fpls.2020.594530.

Rabeh, K., Hnini, M., & Oubohssaine, M. (2025). A comprehensive review of transcription factor-mediated regulation of secondary metabolites in plants under environmental stress. *Stress Biology, 5*, 15. https://doi.org/10.1007/s44154-024-00201-w.

Ramirez-Estrada, K., Vidal-Limon, H., Hidalgo, D., Moyano, E., Golenioswksi, M., Cusidó, R. M., & Palazón, J. (2016). Elicitation, an effective strategy for the biotechnological production of bioactive high-added value compounds in plant cell factories. *Molecules (Basel, Switzerland), 21*(2), 182. https://doi.org/10.3390/molecules21020182.

Rani, N., Boora, R., Singh, Y., Choudhary, Y., Devi, S., Mohan, N., Kumari, S., & Grewal, S. (2025). Nano-elicitation: an emerging and potential technique for enhanced production of pharmaceutically important secondary metabolites in plants. *Nanotechnology for Environmental Engineering, 10*(3), 42. https://doi.org/10.1007/s41204-025-00444-6.

Razak, A. M., Tan, J. K., Mohd Said, M. M., & Makpol, S. (2023). Modulating effects of zingiberaceae phenolic compounds on neurotrophic factors and their potential as neuroprotectants in brain disorders and age-associated neurodegenerative disorders: A review. *Nutrients, 15*(11), 2564. 10.3390/nu15112564.

Reshi, Z. A., Ahmad, W., Lukatkin, A. S., & Javed, S. B. (2023). From nature to lab: A review of secondary metabolite biosynthetic pathways, environmental influences, and in vitro approaches. *Metabolites, 13*(8), 895. https://doi.org/10.3390/metabo13080895.

Saito, K., Sudo, H., Yamazaki, M., et al. (2001). Feasible production of camptothecin by hairy root culture of *Ophiorrhiza pumila*. *Plant Cell Reports, 20*, 267–271. https://doi.org/10.1007/s002990100320.

Sangwan, N. S., Jha, S., & Mitra, A. (2024). Unlocking nature's treasure trove: Biosynthesis and elicitation of secondary metabolites from plants. *Plant Growth Regulation, 104*, 1–4. https://doi.org/10.1007/s10725-024-01184-4.

Sarropoulou, V., Sarrou, E., Angeli, A., Martens, S., Maloupa, E., & Grigoriadou, K. (2023). Species-specific secondary metabolites from *Primula veris* subsp. *veris* obtained in vitro adventitious root cultures: An alternative for sustainable production. *Sustainability, 15*(3), 2452. https://doi.org/10.3390/su15032452.

Selwal, N., Supriadi, K., Rahayu, F., Sukmadjaja, D., Khamidah, A., Budiaarto, K., et al. (2025). Elicitation strategies for enhanced secondary metabolite synthesis in plant cell cultures and its role in plant defense mechanism. *Plant Gene, 41*, 100485. https://doi.org/10.1016/j.plgene.2024.100485.

Sharifzadeh Naeini, M., Naghavi, M. R., Bihamta, M. R., et al. (2021). Production of some benzylisoquinoline alkaloids in *Papaver armeniacum* L. hairy root cultures elicited with salicylic acid and methyl jasmonate. *In Vitro Cellular & Developmental Biology – Plant, 57*, 261–271. https://doi.org/10.1007/s11627-020-10123-7.

Shi, J., Wang, J., Lv, H., Peng, Q., Schreiner, M., Baldermann, S., & Lin, Z. (2021). Integrated proteomic and metabolomic analyses reveal the importance of aroma precursor accumulation and storage in methyl jasmonate-primed tea leaves. *Horticulture Research, 8*(1), 95. https://doi.org/10.1038/s41438-021-00528-9.

Silva, P. T. S., Souza, L. M., Morais, M. B., Moraes, M. M., da Câmara, C. A. G., & Ulisses, C. (2022). Effect of biotic elicitors on the physiology, redox system, and secondary metabolite composition of *Lippia alba* cultivated in vitro. *South African Journal of Botany, 147*, 415–424. https://doi.org/10.1016/j.sajb.2022.01.042.

Stiles, A. R., & Liu, C. Z. (2013). Hairy root culture: Bioreactor design and process intensification. In Pauline M. Doran (Ed.). *Biotechnology of hairy root systems (Vol. 134)* (pp. 91–114). Springer https://doi.org/10.1007/10_2013_18.

Thilakarathne, A. S., Liu, F., & Zou, Z. (2025). Plant signaling hormones and transcription factors: Key regulators of plant responses to growth, development, and stress. *Plants, 14*(7), 1070. https://doi.org/10.3390/plants14071070.

Thilip, C., Mehaboob, V. M., & Varutharaju, K. (2019). Elicitation of withaferin-A in hairy root culture of *Withania somnifera* (L.) Dunal using natural polysaccharides. *Biologia (Lahore, Pakistan), 74*, 961–968. https://doi.org/10.2478/s11756-019-00236-9.

Tian, X., Zhang, R., Yang, Z., & Fang, W. (2024). Methyl jasmonate and zinc sulfate induce secondary metabolism and phenolic acid biosynthesis in barley seedlings. *Plants, 13*(11), 1512. https://doi.org/10.3390/plants13111512.

Tilkat, E., Süzerer, V., Asan, H. S., Ertaş, A., Yilmaz, M. A., Demir, E., Hoşer, A., & Onay, A. (2023). Enhanced production of anticancer triterpenoids in optimized *Pistacia lentiscus* L. callus cultures via methyl jasmonate and silver nitrate elicitation. *South African Journal of Botany, 163*, 65–74. https://doi.org/10.1016/j.sajb.2023.10.025.

Wang, G., Sun, X., Li, Y., Wang, Y., & Jin, C. (2025). The role of UV B radiation in modulating secondary metabolite biosynthesis and regulatory mechanisms in medicinal plants. *BioResources, 20*(2), 4776–4797. https://doi.org/10.15376/biores.20.2.Wang.

Wiktorowska, E., Długosz, M., & Janiszowska, W. (2010). Significant enhancement of oleanolic acid accumulation by biotic elicitors in cell suspension cultures of *Calendula officinalis* L. *Enzyme and Microbial Technology, 46*(1), 14–20. https://doi.org/10.1016/j.enzmictec.2009.09.002.

Yoshioka, H., Hino, Y., Iwata, K., Ogawa, T., Yoshioka, M., Ishihama, N., & Adachi, H. (2023). Dynamics of plant immune MAPK activity and ROS signaling in response to invaders. *Physiological and Molecular Plant Pathology, 125*, 102000. https://doi.org/10.1016/j.pmpp.2023.102000.

Zhao, J., Davis, L. C., & Verpoorte, R. (2005). Elicitor signal transduction leading to production of plant secondary metabolites. *Biotechnology Advances, 23*(4), 283–333. https://doi.org/10.1016/j.biotechadv.2005.01.003.

Zhou, W., Huang, Q., Wu, X., Zhou, Z., Ding, M., Shi, M., Huang, F., Li, S., Wang, Y., & Kai, G. (2017). Comprehensive transcriptome profiling of *Salvia miltiorrhiza* for discovery of genes associated with the biosynthesis of tanshinones and phenolic acids. *Scientific Reports, 7*(1), 10554. https://doi.org/10.1038/s41598-017-10215-2.

Zhou, W., Shi, M., Deng, C., Lu, S., Huang, F., Wang, Y., & Kai, G. (2021). The methyl jasmonate-responsive transcription factor SmMYB1 promotes phenolic acid biosynthesis in *Salvia miltiorrhiza*. *Horticulture Research, 8*(10), 10. https://doi.org/10.1038/s41438-020-00443-5.

CHAPTER

9

Bioreactors: a scalable approach for secondary metabolite production in medicinal plants

Jericó Jabín Bello-Bello

SECIHTI-Colegio de Postgraduados Campus Córdoba, Córdoba, México

9.1 Introduction

Biotechnology refers to any technological application that uses biological systems, living organisms, or their derivatives to create or modify products or processes for specific purposes (CDB, 1992). Plant biotechnology is one of the scientific disciplines that has shown the most development in recent years. It offers alternative solutions to a range of challenges related to the sustainable and efficient use of plants by humans. This science is a tool for increasing both the quantity and quality of plant-based foods, as well as for obtaining new plant-based products with diverse applications. One key application of plant biotechnology is the production of secondary metabolites. These metabolites are not essential for the plant's basic functions but play important ecological roles, such as providing defense against pathogens, herbivores, and abiotic stress (Dhiman et al., 2025). Plant biotechnology is based on plant cell tissue and organ culture (PCTOC), an alternative approach for the *in vitro* production of high-value natural compounds using plant cell and organ culture.

PCTOC involves the manipulation of plant material under controlled, artificial conditions within an aseptic environment (Bello-Bello et al., 2025). PCTOC applications can be classified into five categories: (1) plant health, which consists of obtaining pathogen-free plants by meristem culture, thermotherapy, and cryotherapy; (2) *in vitro* conservation of germplasm using minimum growth and cryopreservation techniques; (3) genetic breeding via genetic transformation, somatic hybridization, mutagenesis, and *in vitro* selection; (4) *in vitro* propagation of plants; and (5) production of secondary metabolites with high industrial value (Abdalla et al., 2022; Khezri et al., 2024). Fig. 9.1 shows the different applications of PCTOC.

Medicinal Plant Biotechnology for Ameliorated Secondary Metabolite Production
DOI: https://doi.org/10.1016/B978-0-443-43944-5.00009-8

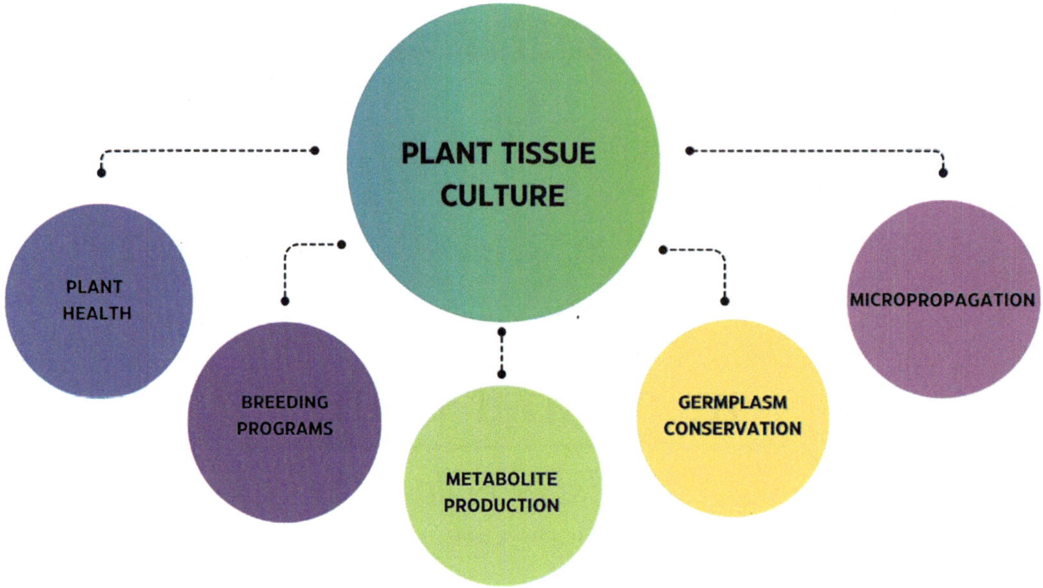

FIGURE 9.1 Plant tissue culture applications in biotechnology.

Currently, several types of plant tissue culture are practiced, which vary to a greater or lesser degree from one another. The most important ones are mentioned and defined below:

1 Whole plant culture
2 Organ culture
3 Tissue or explant culture
4 Cell culture
5 Coculture

Whole plant culture (1) involves the use of complete specimens. Examples of this type of culture include the final stages of micropropagation or when plants are maintained *in vitro* as a source of axenic tissue to establish other types of culture or for genetic transformation. Organ culture (2) refers to the culture of isolated plant organs. For example, the culture of normal or *Agrobacterium rhizogenes*-transformed roots is used for the *in vitro* production of metabolites of interest. On the other hand, tissue culture (3) includes callus tissue culture. This refers to the *in vitro* maintenance of undifferentiated cell masses known as calluses or callus tissue. This type of tissue has countless applications in physiological, biochemical, and genetic studies, as well as in the production of secondary compounds. Cell culture (4) involves the culture of single cells or small, undifferentiated cell clusters. This type of culture is performed in a liquid medium under constant agitation. This culture is ideal for physiological studies at a cellular level, for the *in vitro* production of metabolites of interest, and for the *in vitro* selection of desirable variants. Coculture (5) refers to the culture of any plant tissue with another organism, usually a microorganism. This type of culture is used for the genetic transformation of plant tissues using biological systems. Without a doubt, it is an alternative for producing secondary metabolites using PCTOC.

9.2 *In vitro* production of secondary metabolites

Metabolites are classified as primary or secondary based on their roles in plant physiology. Primary metabolites (amino acids, lipids, and carbohydrates) are directly necessary for plant growth, participating in fundamental processes such as photosynthesis, translocation, and respiration. On the other hand, secondary metabolites (terpenes, alkaloids, and phenolic compounds) are biosynthetically derived from primary metabolites through modifications like methylation, glycosylation, and hydroxylation. Secondary metabolites mediate the plant's interaction with its environment, particularly in response to different stresses, contributing to the plant's survival (Kiran & Tiwari, 2024; Twaij & Hasan, 2022). Secondary metabolites produce various compounds that are important for multiple biochemical and biophysical cell processes, and they are an important source of industrial and medicinal products (Salam et al., 2023).

Due to the significant economic and scientific interest in secondary metabolites derived from medicinal plants, the use of alternative production methods, such as *in vitro* plant cell, tissue, and organ cultures under aseptic conditions, has gained widespread acceptance. Research on the *in vitro* production of secondary compounds aims to develop efficient and controllable models to study their biosynthesis, including associated metabolic pathways, key enzymes, and the regulatory mechanisms of secondary metabolism.

The use of plant cell and tissue culture to obtain secondary metabolites offers several advantages over conventional extraction from whole plants. For example, external factors like soil composition or climate do not affect production, and cells are not threatened by microbial pathogens or insects. Moreover, any plant, even those classified as threatened, can be maintained for secondary metabolite production, and the process can be semiautomated to reduce production costs (Efferth, 2019).

The use of liquid media to culture cell suspensions has several advantages compared to semisolid media. The use of semisolid media prevents uniform exposure of all cells within the callus tissue to the culture medium, resulting in nutrient and gas availability gradients. These gradients can disrupt both cellular growth and metabolic activity. Furthermore, the gelling agent may release toxic impurities, and the plant tissue itself can produce toxic metabolites that accumulate within the gelling matrix, negatively impacting the plant tissue. One strategy to enhance the production of secondary metabolites is the use of liquid media in bioreactors.

9.3 Design and functional characteristics of different types of bioreactors

Bioreactors are containers designed for the large-scale production of cells, tissues, organs, or plantlets grown in liquid media. Two types of culture systems use bioreactors: (1) temporary immersion systems, where explants are only temporarily submerged in culture media, and (2) permanent immersion systems, where explants are continuously submerged in culture media. Both culture systems can be used for the fabrication of secondary metabolites in different medicinal plant species. However, permanent immersion systems are mainly used to grow cells and roots for secondary metabolite production. These systems operate using two

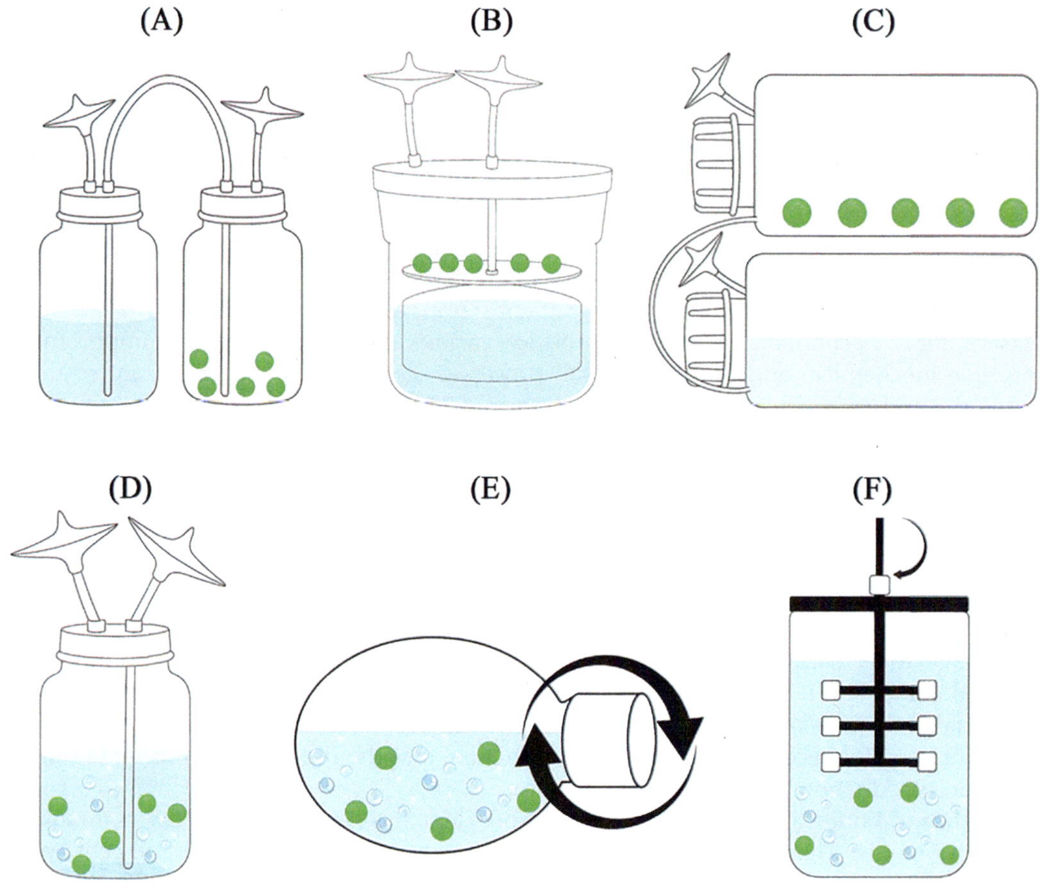

FIGURE 9.2 Examples of different types of bioreactors for secondary metabolites fabrication in medicinal plants (A–C) temporary immersion systems, and (D–F) permanent immersion systems. (A) Temporary immersion bioreactor, (B) recipient for automated temporary immersion, (C) SETIS bioreactor, (D) bubble column bioreactors, (E) mechanical stirring, (F) Propella bioreactors. Green circle: explants.

mechanisms: mechanical agitation (Propella bioreactors, rotating drum system) and airflow agitation (bubble column bioreactors, stirred-tank bioreactor, airlift bioreactor). Temporary immersion systems use semiautomated bioreactors designed to cultivate explants exposed in liquid culture media for a specific time and frequency (Fig. 9.2).

9.4 Bioreactor cultures for secondary metabolites

The biotechnology industry aims to develop and implement a wide variety of efficient bioreactor systems for the large-scale production of secondary metabolites (Table 9.1). Bioreactors offer several advantages, including rapid biomass accumulation, high levels of

TABLE 9.1 Examples of increased production of secondary metabolites in different plant species cultivated in bioreactors.

Plant species	Cultivating type	Type of bioreactor	Secondary metabolites	References
Cyclopia genistoides	Leaves	TIB	Mangiferin (xanthone C-glucoside)	Joubert et al. (2003)
Genista tinctoria	Herb	SAB	Genistin (isoflavone glucoside)	Łuczkiewicz et al. (2004)
Securinega suffruticosa	Herb	SAB	Securinine (indolizidine alkaloid)	Raj and Łuczkiewicz (2008)
Stevia rebaudiana	Leaves	RB	Stevioside (diterpene glycoside)	Goyal et al. (2010)
Centaurium erythraea	Herb	GPB	Gentiopicroside (secoiridoidglucoside)	Aberham et al. (2011)
Hypericum perforatum	Herb	BB	Hypericin (naphthodianthrone)	Murthy et al. (2014)
Leucojum aestivum	Leaves	TIB	Galanthamine (amaryllidaceae alkaloid)	Berkov et al. (2014)
Leonurus sibiricus	Roots	SB	Phenolic acids	Sitarek et al. (2018)
Pueraria tuberosa	Tubers	SAB	Puerarin (isoflavone c-glucoside)	Kanthaliya et al. (2019)
Dracocephalum forrestii	Shoots	RITA and PB	Chlorogenic acid, acacetin rhamnosyl-trihexoside, rosmarinic acid, acacetin acetylrhamnosyl-trihexoside	Weremczuk-Jezyna et al. (2020)
Nasturtium officinale	Microshoots	PB	Total polyphenols, total flavonoids, total glucosionolates	Klimek-Szczykutowicz et al. (2021)
Senna obtusifolia	Roots	SB	Pentacyclic triterpene (betulinic acid)	Kowalczyk et al. (2021)
Salvia viridis	Shoots	PB	Polyphenol	Grzegorczyk-Karolak et al. (2022)
Ruta montana	Seeds	PB	Coumarins (xanthotoxin and bergapten)	Szewczyk et al. (2023)
Sageretia thea	Callus	BB	Phenolics and flavonoids	Kim et al. (2023)
Agastache rugosa	Root	RITA and PB	Rosmarinic acid-rich	Kozłowska et al. (2024)
Catharanthus roseus	Shoots	NSB	Vindoline, vincristine, vinblastine, catharanthine, ajmalicine, tabersonine	Szymańska et al. (2025)
Humulus lupulus	Uninodal microcuttings	PB	Polyphenol	Gianguzzi et al. (2025)

GPB, Gas phase bioreactor; TIB, temporary immersion bioreactor; SAB, simple aeration bioreactor; BB, balloon bioreactor; RB, roller bioreactor; SB, sprinkle bioreactor; RITA, recipient for automated temporary immersion; PB, Plantform bioreactors; NSB, nutrient sprinkle bioreactor.

bioactive compounds, and the ability to semiautomate the production process under controlled conditions. The efficiency of biosynthesis and the quality of metabolites greatly depend on the control and optimization of production process parameters.

The use of semiautomated bioreactors allows for the control of pH, temperature, oxygen, and carbon dioxide levels (Szymańska et al., 2025), as well as real-time monitoring of critical variables, which supports consistent production (Pechlivani et al., 2023). Several plant species can increase metabolite synthesis under bioreactor-controlled conditions. In *Stevia rebaudiana*, a significant increase in the accumulation of stevioside, rebaudioside A, phenolic acids, and flavonoids has been demonstrated through cultivation in a roller bioreactor (Goyal et al., 2010). Similarly, the production of gentiopicroside (a glycosylated secoiridoid) in *Centaurium erythraea* has been optimized using culture systems in a gas-phase bioreactor (Aberham et al., 2011). In *Morinda citrifolia*, a significant increase in the accumulation of phenolics and flavonoids has been demonstrated through cultivation in a balloon-type bubble bioreactor (Baque et al., 2013). In *Echinacea anguistifolia*, the accumulation of phenolics, flavonoids, and caffeic acid derivatives has been demonstrated through cultivation in a balloon-type bubble bioreactor (Cui et al., 2013). In *Lobelia chinensis*, a significant increase in the accumulation of phenolics, including catechin, phloretic acid, salicylic acid, myricetin, naringin, coumaric acid, linarin, and chlorogenic acid, has been demonstrated through cultivation in a continuous immersion bioreactor (Bai et al., 2025). Among the secondary metabolites successfully produced using cell cultures at an industrial scale are mangiferin (a xanthone C-glucoside) (Joubert et al., 2003), genistin (a glycosylated isoflavone (Łuczkiewicz et al., 2004), securinine (an indolizidine alkaloid) (Raj & Łuczkiewicz, 2008), and isoflavones (Ozyigit et al., 2023).

With recent technological advances, bioreactors have evolved from simple containers to complex integrated systems with semiautomation capabilities, digital monitoring, and real-time feedback. This transformation has enabled the precise regulation of biochemical reactions under standardized conditions, driving the efficient production of pharmaceuticals, enzymes, food additives, and biofuels (Fatima et al., 2021).

9.5 Industrial applications

Genetic transformation protocols for plant tissues are used for a wide range of plant species that produce metabolites of industrial importance. Transformed cultures exhibit specific characteristics that make them suitable for large-scale cultivation in high-capacity bioreactors for the production of various secondary metabolites in enclosed spaces. Several types of bioreactors have been successfully used for the cultivation of transformed roots (hairy roots), which offer substantial advantages such as rapid and autonomous growth, genetic and biochemical stability, and a remarkable biosynthetic capacity for bioactive compounds (Biswas et al., 2023). Transformed shoots of *Dracocephalum forrestii* have shown a significant increase in the accumulation of phenolic acids and flavonoids have been demonstrated through cultivation in a temporary immersion system (RITA and Plantform) (Weremczuk-Jezyna et al., 2020). Transformed shoots of *Agastache rugosa* have shown a significant increase in the accumulation of caffeoylquinic acid (neochlorogenic acid), p-coumaric acid, dicaffeoyl-hexose isomer I, apigenin glucoside, and rosmarinic acid as demonstrated through cultivation in a temporary immersion system (RITA and Plantform) (Kozłowska et al., 2024).

Selecting the appropriate type of bioreactor is a critical factor for the successful industrial production of transformed cells and tissues. However, the effective scaling of these processes depends not only on bioreactor design, but also on a series of physiological and technical factors that enable commercial scale production. These factors include tissue morphology and their sensitivity to mechanical stress (Häkkinen & Oksman-Caldentey, 2018). For industrial purposes, metabolite production is commonly carried out in bubble column bioreactors, stirred-tank bioreactors, airlift bioreactors, and nutrient mist systems (Dhiman et al., 2018). In gas-phase systems, such as mist bioreactors, the tissues are kept exposed to air, which facilitates adequate oxygen supply, while nutrients are delivered through aerosol and mist systems (Georgiev et al., 2013; Srikantan & Srivastava, 2018). On the other hand, in liquid-phase bioreactors, like stirred tank, airlift, and bubble column bioreactors, tissues remain submerged in a liquid nutrient medium (Wang & Wu, 2013). Additionally, disposable or single-use bioreactors, such as wave bioreactors, have emerged as promising alternatives due to their ease of use, elimination of cleaning and sterilization processes, and low operational costs (Gutierrez-Valdes et al., 2020; Lehmann et al., 2014; Maschke et al., 2024).

9.6 Conclusions

Bioreactors can be used to fabricate secondary metabolites in medicinal plants. However, a deep understanding of the mechanisms controlling their synthesis and accumulation is necessary to achieve greater control over production. The development and optimization of bioreactor systems specific to transformed tissues are key strategies for the efficient and controlled industrial-scale production of secondary metabolites.

References

Abdalla, N., El-Ramady, H., Seliem, M. K., El-Mahrouk, M. E., Taha, N., Bayoumi, Y., Shalaby, T. A., & Dobránszki, J. (2022). An academic and technical overview on plant micropropagation challenges. *Horticulturae, 8*(8). https://doi.org/10.3390/horticulturae8080677, https://doi.org/10.3390/horticulturae8080677.

Aberham, A., Pieri, V., Croom, E. M., Ellmerer, E., & Stuppner, H. (2011). Analysis of iridoids, secoiridoids and xanthones in *Centaurium erythraea, Frasera caroliniensis* and *Gentiana lutea* using LC-MS and RP-HPLC. *Journal of Pharmaceutical and Biomedical Analysis, 54*(3), 517–525. https://doi.org/10.1016/j.jpba.2010.09.030.

Bai, X., Lee, H.-S., Han, J.-E., Murthy, H. N., Kwon, H.-J., Yeon, S.-H., & Park, S.-Y. (2025). Stimulating the synthesis of phenolics and polyacetylenes in *Lobelia chinensis* Lour. plantlets using various bioreactor culture systems: A comparison of parameter effects. *Plant Cell, Tissue and Organ Culture, 160*(2). https://doi.org/10.1007/s11240-025-02989-5.

Baque, M. A., Shiragi, M. H. K., Moh, S.-H., Lee, E.-J., & Paek, K.-Y. (2013). Production of biomass and bioactive compounds by adventitious root suspension cultures of *Morinda citrifolia* (L.) in a liquid-phase airlift balloon-type bioreactor. *In Vitro Cellular and Developmental Biology - Plant, 49*(6), 737–749. https://doi.org/10.1007/s11627-013-9555-3.

Bello-Bello, J. J., Mancilla-Álvarez, E., & Spinoso-Castillo, J. L. (2025). Scaling-up procedures and factors for mass micropropagation using temporary immersion systems. *In Vitro Cellular and Developmental Biology - Plant, 61*(2), 321–332. https://doi.org/10.1007/s11627-024-10480-7.

Berkov, S., Ivanov, I., Georgiev, V., Codina, C., & Pavlov, A. (2014). Galanthamine biosynthesis in plant in vitro systems. *Engineering in Life Sciences, 14*(6), 643–650. https://doi.org/10.1002/elsc.201300159.

Biswas, D., Chakraborty, A., Mukherjee, S., & Ghosh, B. (2023). Hairy root culture: A potent method for improved secondary metabolite production of Solanaceous plants. *Frontiers in Plant Science, 14.* https://doi.org/10.3389/fpls.2023.1197555.

CBD. (1992). Seventh negotiating session / fifth meeting of the Intergovernmental Negotiating Committee for a Convention on Biological Diversity.

Cui, H.-Y., Abdullahil Baque, M., Lee, E.-J., & Paek, K.-Y. (2013). Scale-up of adventitious root cultures of *Echinacea angustifolia* in a pilot-scale bioreactor for the production of biomass and caffeic acid derivatives. *Plant Biotechnology Reports, 7*(3), 297–308. https://doi.org/10.1007/s11816-012-0263-y.

Dhiman, N., Patial, V., & Bhattacharya, A. (2018). Applications of hairy root cultures. biotechnological approaches for medicinal and aromatic plants: Conservation. *Genetic Improvement and Utilization, Genetic Improvement and Utilization.*

Dhiman, V. K., Singh, D., Dhiman, V. K., Verma, S. K., & Pandey, H. (2025). *Reactive oxygen species (ROS) and biotic stress: Evolving roles under climate change climate change and biotic factors: A molecular approach.* Apple Academic Press, 201–218. https://www.appleacademicpress.com/climate-change-and-biotic-factors-a-molecular-approach/9781774918487.

Efferth, T. (2019). Biotechnology applications of plant callus cultures. *Engineering, 5*(1), 50–59. https://doi.org/10.1016/j.eng.2018.11.006.

Fatima, S., Faryad, A., Ataa, A., Joyia, F. A., & Parvaiz, A. (2021). Microbial lipase production: A deep insight into the recent advances of lipase production and purification techniques. *Biotechnology and Applied Biochemistry, 68*(3), 445–458. https://doi.org/10.1002/bab.2019.

Georgiev, M. I., Eibl, R., & Zhong, J.-J. (2013). Hosting the plant cells in vitro: Recent trends in bioreactors. *Applied Microbiology and Biotechnology, 97*(9), 3787–3800. https://doi.org/10.1007/s00253-013-4817-x.

Gianguzzi, V., Leto, L., Agosti, A., Di Fazio, A., Marra, F. P., Cirlini, M., & Chiancone, B. (2025). Influence of sucrose and immersion time on *Humulus lupulus* L., cv columbus, plantlet in vitro proliferation and potentially bioactive compound content. *Plants, 14*(4). https://doi.org/10.3390/plants14040537.

Goyal, Samsher, & Goyal (2010). Stevia (*Stevia rebaudiana*) a bio-sweetener: A review. *International Journal of Food Sciences and Nutrition, 61*(1), 1–10. https://doi.org/10.3109/09637480903193049.

Grzegorczyk-Karolak, I., Staniewska, P., Lebelt, L., & Piotrowska, D. G. (2022). Optimization of cultivation conditions of *Salvia viridis* L. shoots in the plantform bioreactor to increase polyphenol production. *Plant Cell, Tissue and Organ Culture, 149*(1–2), 269–280. https://doi.org/10.1007/s11240-021-02168-2.

Gutierrez-Valdes, N., Häkkinen, S. T., Lemasson, C., Guillet, M., Oksman-Caldentey, K.-M., Ritala, A., & Cardon, F. (2020). Hairy root cultures—A versatile tool with multiple applications. *Frontiers in Plant Science, 11.* https://doi.org/10.3389/fpls.2020.00033.

Häkkinen, S. T., & Oksman-Caldentey, K.-M. (2018). Progress and prospects of hairy root research. *Hairy Roots: An Effective Tool of Plant Biotechnology,* 3–19. https://doi.org/10.1007/978–981-13–2562-5_1.

Joubert, E., Otto, F., Grüner, S., & Weinreich, B. (2003). Reversed-phase HPLC determination of mangiferin, isomangiferin and hesperidin in Cyclopia and the effect of harvesting date on the phenolic composition of C. genistoides. *European Food Research and Technology, 216*(3), 270–273. https://doi.org/10.1007/s00217-002-0644-5.

Kanthaliya, B., Joshi, A., & Arora, J. (2019). Evaluation of isoflavonoid content in context to tuber size and seed biology study of *Pueraria tuberosa* (Roxb.ex.Willd.) DC: A vulnerable medicinal plant. *Vegetos, 32*(3), 247–253. https://doi.org/10.1007/s42535-019-00042-3.

Khezri, M., Asghari-Zakaria, R., & Zare, N. (2024). Plant cell and tissue culture: Propagation, improvement, and conservation of medicinal plants. *Biosynthesis of Natural Products in Plants: Bioengineering in Post-genomics Era,* 267–291. https://doi.org/10.1007/978–981-97–2166-5_11.

Kim, J.-H., Han, J.-E., Murthy, H. N., Kim, J.-Y., Kim, M.-J., Jeong, T.-K., & Park, S.-Y. (2023). Production of secondary metabolites from cell cultures of *Sageretia thea* (Osbeck) M.C. Johnst. Using balloon-type bubble bioreactors. *Plants, 12*(6). https://doi.org/10.3390/plants12061390.

Kiran, K., & Tiwari, I. M. (2024). *Progress in biotechnological applications via in vitro propagation techniques: As a source of secondary metabolites production.* Springer Science and Business Media LLC, 11–25. http://doi.org/10.1007/978-3-031–64601-0_2.

Klimek-Szczykutowicz, M., Dziurka, M., Blažević, I., Đulović, A., Miazga-Karska, M., Klimek, K., Ekiert, H., & Szopa, A. (2021). Precursor-boosted production of metabolites in nasturtium officinale microshoots grown in plantform bioreactors, and antioxidant and antimicrobial activities of biomass extracts. *Molecules (Basel, Switzerland), 26*(15). https://doi.org/10.3390/molecules26154660.

Kowalczyk, T., Sitarek, P., Toma, M., Rijo, P., Domínguez-Martín, E., Falcó, I., Sánchez, G., & Śliwiński, T. (2021). Enhanced accumulation of betulinic acid in transgenic hairy roots of *Senna obtusifolia* growing in the sprinkle bioreactor and evaluation of their biological properties in various biological models. *Chemistry and Biodiversity, 18*(8). https://doi.org/10.1002/cbdv.202100455.

Kozłowska, W., Piątczak, E., Kolniak-Ostek, J., Kochan, E., Pencakowski, B., Stafiniak, M., Bielecka, M., Płachno, B. J., Strzemski, M., Matkowski, A., & Zielińska, S. (2024). Upscaling biomass production of rosmarinic acid-rich hairy root cultures of *Agastache rugosa* (Fisch. & C.A.Mey.) Kuntze. *Plant Cell, Tissue and Organ Culture, 156*(2). https://doi.org/10.1007/s11240-023-02626-z.

Lehmann, N., Dittler, I., Lämsä, M., Ritala, A., Rischer, H., Eibl, D., Oksman-Caldentey, K. M., & Eibl, R. (2014). Disposable bioreactors for cultivation of plant cell cultures. *Production of Biomass and Bioactive Compounds Using Bioreactor Technology,* 17–46. https://doi.org/10.1007/978-94-017-9223-3_2, http://doi.org/10.1007/978-94-017–9223-3.

Łuczkiewicz, M., Głód, D., Baczek, T., & Buciński, A. (2004). LC-DAD UV and LC-MS for the analysis of isoflavones and flavones from in vitro and in vivo biomass of *Genista tinctoria* L. *Chromatographia, 60*(3–4), 179–185.

Maschke, R. W., Seidel, S., Rossi, L., Eibl, D., & Eibl, R. (2024). Disposable bioreactors used in process development and production processes with plant cell and tissue cultures. *Advances in Biochemical Engineering/Biotechnology, 188,* 119–144. https://doi.org/10.1007/10_2024_249.

Murthy, H. N., Kim, Y.-S., Park, S.-Y., & Paek, K.-Y. (2014). Hypericins: Biotechnological production from cell and organ cultures. *Applied Microbiology and Biotechnology, 98*(22), 9187–9198. https://doi.org/10.1007/s00253-014-6119-3.

Ozyigit, I. I., Dogan, I., Hocaoglu-Ozyigit, A., Yalcin, B., Erdogan, A., Yalcin, I. E., Cabi, E., & Kaya, Y. (2023). Production of secondary metabolites using tissue culture-based biotechnological applications. *Frontiers in Plant Science, 14.* https://doi.org/10.3389/fpls.2023.1132555.

Pechlivani, E. M., Pemas, S., Kanlis, A., Pechlivani, P., Petrakis, S., Papadimitriou, A., Tzovaras, D., & Hatzistergos, K. E. (2023). Enhanced growth of bacterial cells in a smart 3D printed bioreactor. *Micromachines, 14*(10). https://doi.org/10.3390/mi14101829.

Raj, & Łuczkiewicz (2008). Securinega suffruticosa. *Fitoterapia, 79*(6), 419–427. https://doi.org/10.1016/j.fitote.2008.02.011.

Salam, U., Ullah, S., Tang, Z.-H., Elateeq, A. A., Khan, Y., Khan, J., Khan, A., & Ali, S. (2023). Plant metabolomics: An overview of the role of primary and secondary metabolites against different environmental stress factors. *Life (Chicago, Ill. 1978), 13*(3). https://doi.org/10.3390/life13030706.

Sitarek, P., Kowalczyk, T., Picot, L., Michalska-Hejduk, D., Bijak, M., Białas, A. J., Wielanek, M., Śliwiński, T., & Skała, E. (2018). Growth of *Leonurus sibiricus* L. roots with over-expression of AtPAP1 transcriptional factor in closed bioreactor, production of bioactive phenolic compounds and evaluation of their biological activity. *Industrial Crops and Products, 122,* 732–739. https://doi.org/10.1016/j.indcrop.2018.06.059.

Srikantan, C., & Srivastava, S. (2018). Bioreactor design and analysis for large- scale plant cell and hairy root cultivation. *Hairy Roots: An Effective Tool of Plant Biotechnology,* 147–182. https://doi.org/10.1007/978–981-13–2562-5_7.

Szewczyk, A., Marino, A., Taviano, M. F., Cambria, L., Davì, F., Trepa, M., Grabowski, M., & Miceli, N. (2023). Studies on the accumulation of secondary metabolites and evaluation of biological activity of in vitro cultures of *Ruta montana* L. in temporary immersion bioreactors. *International Journal of Molecular Sciences, 24*(8). https://doi.org/10.3390/ijms24087045.

Szymańska, G., Urbaniak, K., Rosicka-Kaczmarek, J., Miśkiewicz, K., Lipert, A., & Kochan, E. (2025). Dynamics of growth and production of indole anti-cancer alkaloids in *Catharanthus roseus* shoot cultures cultivated in flasks and nutrient sprinkle bioreactor. *Industrial Crops and Products, 230.* https://doi.org/10.1016/j.indcrop.2025.121048.

Twaij, B. M., & Hasan, M. N. (2022). Bioactive secondary metabolites from plant sources: Types, synthesis, and their therapeutic uses. *International Journal of Plant Biology, 13*(1), 4–14. https://doi.org/10.3390/ijpb13010003.

Wang, J. W., & Wu, J. Y. (2013). Effective elicitors and process strategies for enhancement of secondary metabolite production in hairy root cultures. *Advances in Biochemical Engineering/Biotechnology, 134*, 55–89. https://doi.org/10.1007/10_2013_183.

Weremczuk-Jezyna, I., Lisiecki, P., Gonciarz, W., Kuzma, lukasz, Szemraj, M., Chmiela, M., & Grzegorczyk-Karolak, I. (2020). Transformed shoots of *Dracocephalum forrestii* W.W. Smith from different bioreactor systems as a rich source of natural phenolic compounds. *Molecules (Basel, Switzerland), 25*(19). https://doi.org/10.3390/molecules25194533.

CHAPTER

10

Sustainable production of camptothecin using plant cell culture and metabolic engineering

Kishmita Sharma[1,2], *Saikat Gantait*[3], *and Sudhir Singh*[1,2]

[1]Nuclear Agriculture & Biotechnology Division, Bhabha Atomic Research Centre (BARC), Mumbai, Maharashtra, India [2]Homi Bhabha National Institute, Anushaktinagar, Mumbai, Maharashtra, India [3]Crop Research Unit (Genetics and Plant Breeding), Bidhan Chandra Krishi Viswavidyalaya, Mohanpur, West Bengal, India

10.1 Introduction

Plants create a wide range of organic substances called secondary metabolites, sometimes referred to as specialized metabolites, natural products, or phytochemicals, which are not directly necessary for their main metabolic activities of growth, development, or reproduction (Erb & Kliebenstein, 2020; Ozyigit et al., 2023). These compounds fulfill key ecological and defensive functions in mediating interactions with the environment, providing selective advantages that enhance survivability and reproductive success (Deepthi & Satheeshkumar, 2017; Erb & Kliebenstein, 2020). These secondary metabolites primarily serve as active components that form the foundation of a plant's medicinal properties. These compounds may offer protection against environmental challenges and provide defense against herbivores and pathogens. In higher plants, a broad spectrum of secondary metabolites is derived from primary metabolites like carbohydrates, lipids, and amino acids (Akula & Ravishankar, 2014). Unlike primary metabolites such as carbohydrates, proteins, and nucleic acids that support core cellular activities, secondary metabolites have species-specific distributions and are essential to plants' evolutionary success because they enable both mutualistic and antagonistic relationships with other organisms (Biswas et al., 2023). Plant secondary metabolites are broadly classified into three major groups based on their biosynthetic pathways: phenolic compounds, terpenoids, polysaccharides, phytoalexins, and nitrogenous compounds such as alkaloids (Reshi et al., 2023). These compounds are made through intricate metabolic pathways that start from primary metabolic intermediates.

Medicinal Plant Biotechnology for Ameliorated Secondary Metabolite Production
DOI: https://doi.org/10.1016/B978-0-443-43944-5.00012-8

These include acetyl-CoA, shikimic acid, mevalonic acid, and 1-deoxylulose 5-phosphate (Teoh, 2016). Genetic, environmental, and developmental factors closely control the production and buildup of secondary metabolites. This production is often triggered by various biotic and abiotic stresses, including pathogen attack, herbivory, UV radiation, drought, and salinity (Biswas et al., 2023).

Alkaloids are naturally occurring organic compounds containing nitrogen, having a low molecular weight (Bhushan et al., 2023). Among the diverse array of alkaloids that constitute this important class of secondary metabolites, camptothecin stands out as a particularly significant monoterpene indole alkaloid with remarkable pharmacological properties (Banadka et al., 2024). The following sections will provide a comprehensive examination of camptothecin, exploring its biosynthesis, structural characteristics, biological activities, and biotechnological applications in detail.

10.1.1 Camptothecin and its uses

Camptothecin (CPT) is considered one of the most promising anticancer medications. This water-insoluble compound is a quinoline alkaloid, featuring a distinctive pentacyclic structure. It is produced through the monoterpenoid indole alkaloid (TIA) pathway, utilizing tryptophan and terpenoid precursors (Sirikantaramas et al., 2013). The critical components of CPT necessary for its activity include the alpha-hydroxy lactone ring, the pyridine section of the D-ring, the lactone part of the E-ring, the conformation at C-20, and the planarity of the five-membered ring system (Sadre et al., 2016).

Research on CPT and clinical trials conducted in China for stomach cancer, certain leukemia types, and bladder cancer revealed that the water-soluble carboxylate form of CPT is significantly less effective in antitumor activity compared to the water-insoluble lactone form. However, by the 1970s, clinical trials involving CPTs were largely halted due to challenges in addressing the water insolubility of the lactone form, low response rates, high toxicity levels (such as myelosuppression, gastrointestinal toxicities, and hemorrhagic cystitis), and an unclear mechanism of action for CPT. Given CPT's high toxicity and insolubility, its derivatives have been developed (Li et al., 2017).

10.1.2 Analogs of camptothecin

The limited clinical application of camptothecin is due to its water insolubility, rapid transformation of its active lactone form into an inactive carboxylate form under physiological conditions, emergence of drug resistance, and off-target side effects associated with prolonged use. Consequently, there is an urgent need to develop new treatment strategies (Padhi & Behera, 2020). Camptothecin analogs designed for targeted enzymatic activation in tumor cells possess enhanced water solubility, reduced cytotoxicity, stability in blood at physiological pH, and susceptibility to specific enzymatic cleavage (Li et al., 2006).

Topotecan (Hycamtin), irinotecan (Camptosar or CPT-11), and belotecan (Liu et al., 2015) are FDA-approved and clinically used semisynthetic water-soluble derivatives of camptothecin (Fig. 10.1). Topotecan is approved for the treatment of ovarian cancer, small-cell lung cancer, and cervical cancer. Irinotecan is widely used as a first-line agent for metastatic colorectal cancer, particularly in combination regimens like FOLFIRI (Schöffski et al., 2024).

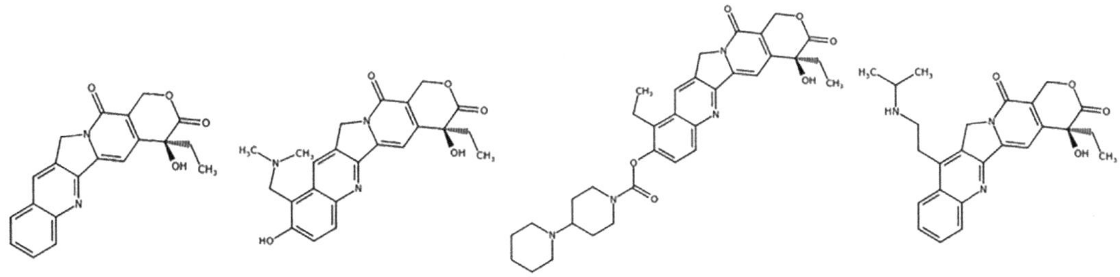

FIGURE 10.1 Structure of approved analogs of camptothecin.

Belotecan is approved and has demonstrated clinical benefit in patients with small-cell lung cancer and ovarian cancer, especially in Asian populations.

Besides these three, further analogs are undergoing clinical trials but are yet to be approved. Exatecan (DX-8951f) has demonstrated superior efficacy compared to SN-38, topotecan, and irinotecan in preclinical cancer models, and it can overcome multidrug resistance; however, it is not widely approved clinically (Ishii et al., 2000). Deruxtecan (DXd), an exatecan derivative with enhanced topoisomerase I inhibition potency, has achieved clinical success as the cytotoxic payload in trastuzumab deruxtecan, which demonstrated superior efficacy over trastuzumab emtansine in the randomized phase 3 DESTINY-Breast03 trial (Cortés et al., 2022)and received FDA approval for multiple HER2-positive indications. Gimatecan (7-t-butoxyiminomethylcamptothecin), an orally active analog, has exhibited significant activity in recurrent ovarian and gastric cancer in Phase II trials (Schöffski et al., 2024). Numerous studies have highlighted the significance of CPT derivatives with modifications at the quinoline ring. Most structural model studies have focused on modifications at positions C-7, C-9, and C-10, which can enhance potency, as CPT does not generally tolerate modifications at position C-12 (Li et al., 2006).

10.1.3 Mode of action

Camptothecin (CPT) is a well-known anticancer compound whose cytotoxic activity arises from its interaction with DNA topoisomerase I (Topo I), an enzyme essential for relieving torsional strain in DNA during replication and transcription. Normally, Topo I transiently cleaves one strand of the DNA helix to relieve supercoiling and then rapidly religates it, keeping the cleavage intermediate at very low steady-state levels (Li et al., 2006). CPT interferes with this process by stabilizing the covalent Topo I–DNA intermediate formed during cleavage, thereby preventing religation (Hsiang et al., 1985). This stabilization leads to persistent Topo I–DNA cleavage complexes that collide with advancing replication forks, generating lethal DNA lesions. Importantly, CPT exerts its inhibitory effect only when Topo I is bound to DNA, functioning as an uncompetitive inhibitor that does not interact with the free enzyme or DNA alone (Hertzberg et al., 1989). This replication fork collision and DNA damage are especially lethal during the S-phase. CPT has also been shown to arrest cells in both S and G2 phases, enhancing its anticancer activity (Tsao et al., 1992). The resulting obstruction of DNA metabolism processes—including replication, transcription, repair, and chromatin remodeling—triggers apoptosis, and hence CPT is classified as a topoisomerase "poison" (Svejstrup et al., 1991). Moreover, CPT sensitivity correlates with Topo I expression levels, and resistance often arises from mutations in the Topo I gene that disrupt drug binding (Li et al., 2006). Besides its anticancer properties, CPT also shows antiviral capabilities by targeting DNA viruses such as adenoviruses and herpesviruses, as well as single-stranded parvoviruses that depend on host Topo I (Pantazis et al., 1999). Additionally, it suppresses hypoxia-inducible factor-1 (HIF-1), a crucial survival element in tumor hypoxia, with three out of four known HIF-1 inhibitors being derivatives of CPT (Rapisarda et al., 2002). Beyond cancer treatment, CPT derivatives exhibit a wide range of biological activities, including antiparasitic, antipsoriasis, antimicrobial, pesticide, and antifungal effects. Further research highlights their involvement in nitric oxide signaling within endothelial cells (Huang et al., 2017), as well as their antiangiogenic and neurotoxic effects (Clements et al., 1999; Uday Bhanu & Kondapi, 2010), and their role in inducing apoptosis through mitochondrial dysfunction, ROS production, neddylation inhibition, and activation of the caspase cascade (Chou et al., 2017).

10.1.4 Drug delivery systems

The last two decades have already witnessed the explosive growth of nanotechnology, resulting in a plethora of innovations in therapeutic domains, with the primary focus being the ability to target tumor tissues either passively or actively. For a potential anticancer agent such as CPT, high chemotherapeutic efficacy and accurate tumor targeting are equally crucial.

Recent advancements in targeted delivery of CPT to cancer cells have focused on improving its solubility, stability, and therapeutic index while minimizing systemic toxicity. To overcome these challenges, a range of nanotechnology-based delivery platforms has been developed (Table 10.1). Passive and active targeting strategies using nanoparticles have been explored to enhance CPT's efficacy against various solid tumors (Padhi & Behera, 2020).

Porous silicon nanoparticles functionalized with EGFR-targeting antibodies demonstrated efficacy in reducing breast cancer metastasis in a humanized mouse model(Landgraf et al., 2020). Lipid-based, polymeric, and inorganic nanoformulations have also been investigated for enhancing drug loading, stability, and tumor-specific accumulation (Alshammari et al., 2023). Among these, polymeric micelles offer high loading efficiency and protection from reticuloendothelial clearance, making them attractive for clinical translation (Farhoudi et al., 2024). Superparamagnetic iron oxide nanoparticles (SPIONs), stabilized with dextran and conjugated with folate (FA-DEX-SPION), achieved selective uptake in AT3B-1 cancer cells, showing high CPT loading capacity and effective therapeutic activity without off-target cytotoxicity (Al-Musawi et al., 2020). In parallel, antibody–drug conjugates (ADCs) such as trastuzumab deruxtecan and sacituzumab govitecan exploit receptor-mediated uptake to deliver CPT derivatives directly to tumor cells, significantly reducing gastrointestinal toxicity. Liposomal irinotecan (Onivyde) has further illustrated how encapsulation can stabilize release, prolong circulation, and leverage the enhanced permeability and retention (EPR) effect for tumor targeting (Gong et al., 2025). Collectively, these molecular targeting and nanocarrier-based strategies represent a transformative paradigm in CPT delivery, offering improved precision, safety, and clinical applicability in cancer chemotherapy.

TABLE 10.1 Camptothecin (CPT) nanoformulations.

S. no.	Nanoformulation type	Animal model/cell lines	Cancer type	Effect on tumor	Mechanism of action	References
1.	CPC@IR780 micelles and CPD@IR780 micelles	Mice bearing intracranial Luc-U87 glioma	Brain	Reduced tumor growth, less side effects	Ability to cross the BBB and target glioma cells via αv β integrin and neuropilin-1-mediated ligand transportation *in vitro* and *in vivo*	Lu et al. (2020)
2.	CPT-loaded Bom/PEG	C6 tumor-bearing rats	Brain	Slow tumor progression, higher therapeutic efficacy	GRPR selectivity, cellular uptake, and tumor-selective delivery to C6 glioma cells	Kanazawa et al. (2020)
3.	NP/CPT	4T1 tumor-bearing mice	Breast	Reduced tumor growth	Biocompatible, tumor-specific, overexpressing sialic acids (SA) in tumor cells interacts strongly with boronopicolinic acid coating NPs	Hao et al. (2021c)
4.	CPT-SS-APBA, BBSA/CPT-SS-APBA NPs	4 T1 tumor-bearing mice	Breast	Reduced tumor growth and progression	Drug release at physiological pH and GSH-rich cytosol	Hao et al. (2021a)
5.	ACI and PCI	4T1 breast tumor-bearing mice	Breast	Reduced tumor volume	Uptake via an endo-/lysosome-mediated mechanism by tumor cells, and prolonged retention	Cheng et al. (2020)
6.	HSD NGs	4T1 tumor-bearing mice	Breast	Reduced tumor growth	Drug release in response to the reductive tumor microenvironment	Y. Lu et al. (2021)
7.	CCO micelles	MCF-7 tumor-bearing nude mice	Breast	Enhanced tumor permeability	GSH-responsive and OEGMA act as a hydrophilic block	Bai et al. (2021)
8.	CSP-CPT	MCF-7 tumor-bearing mice	Breast	Slow tumor progression, higher therapeutic efficacy	Release free CPT in response to GSH	Sun et al. (2021)
9.	CNF	BALB/c 4T1 breast tumor mice	Breast	Reduction in metastasis	Elevates ROS in mitochondria, inducing pyroptopsis	Gao et al. (2025)
10.	CPT_3@$ATRA_3$@DL-HMCON@HA	BALB/c mice bearing 4T1 SP	Breast	Inhibition of tumor growth, recurrence, and metastasis	Cu^+ from the nanoparticle core triggers Fenton-like reactions and GSH depletion, leading to ferroptosis in nonCSCs	Li et al. (2024)

(Continued)

TABLE 10.1 (Continued)

S. no.	Nanoformulation type	Animal model/cell lines	Cancer type	Effect on tumor	Mechanism of action	References
11.	CPT-CS-NPs	Female Wistar rats (DMBA-induced model)	Breast	Reduced tumor volume, restored histological structure of mammary glands,	Encapsulation in chitosan nanoparticles protects CPT from hydrolysis and degradation, allowing slow release	Paul et al. (2024)
12.	HA-coated CPT nanocrystals	Kun Ming mice	Breast and Liver	Inhibited cell migration	Intrinsic mitochondria-mediated apoptosis by an increase of Bax to Bcl-2 ratio and upregulation of P53	Wang et al. (2020)
13.	CPT-CS@GO/Fe_3O_4	HepG2 and SMMC-7721 cancer cells	Liver	Increased cancer cell death	Controlled release profile that is still effective under normal conditions but potentially enhanced in acidic tumor environments.	Sukumar et al. (2025)
14.	GalNAc-3-CPT	BALB/c nude mice and C57BL/6 mice	Liver	Inhibited tumor growth, volume, and weight	Targets the ASGR overexpressed on hepatocytes	Y. Li et al. (2025)
15.	BL_Oc-SS-CPT	BALB/c-nu male mice	Colorectal and liver	Inhibited tumor growth and migration	Gradually released CPT in the presence of excessive MMP-2/9 and GSH in cancer cells	Mai et al. (2022)
16.	CPT@CNPs	Sprague Dawley rats	Colorectal	Reduced cancer cell viability	Biocompatible with controlled and sustained release behavior	Hameed et al. (2025)
17.	TPP-PCL-CPT-PEG-BTN	HaCaT and HeLa cells	Cervical	High apoptotic effect	Mitochondrial dysfunction via ROS generation, ATP depletion	Nandi et al. (2025)
18.	HBSPs and MGLs-CPT	*In vitro* (HeLa cells)	Cervical	~100% cell death	HIFU-induced CPT release	J. Fan et al. (2024)
19.	ICFES	C57BL/6 mice with murine melanoma	Skin	Slow increase in tumor size and volume	Induces hyperthermia and produces singlet oxygen, arresting tumor growth	Lee and Huang (2025)
20.	HA-ZE-CPT NPs	Melanoma-induced BALB/c mice	Skin	Reduced tumor volume, high apoptotic effect	pH-responsive *in vitro* release, upregulation of pro- and antiapoptotic proteins	Batool et al. (2025)

10.1.5 Demand for camptothecin

Because of their broad spectrum of targets, the demand for camptothecin derivatives is rising in therapeutics. The global market for CPT analogs has grown quickly as a result of the derivatives' enhanced pharmacological efficacy. Irinotecan and topotecan, two CPT analogs, had combined sales of 2.2 billion US dollars in 2008, and this number is predicted to rise further (Lorence & Nessler, 2004; Lu et al., 2009). Currently, only 600 kg of CPT is produced annually worldwide, whereas the global market requires approximately 3000 kg of CPT. The biotechnological intensification (tissue culture, cell lines, endophytes, metabolic engineering) is a major research focus to close the supply gap; however, none have yet replaced or significantly supplemented the global CPT supply derived from trees (Banadka et al., 2024).

10.2 Plant sources and supply limitations

10.2.1 Primary plant sources

It was first isolated from the stem wood of *Camptotheca acuminata* by botanists working in the USDA's Plant Introduction Division (Wall & Wani, 1996). C. *acuminata* is a tree native to China, and its bark is a recognized Chinese traditional medicine. The process of CPT discovery was well reviewed by Dr. Monroe Wall and Mansukh Wani (Wall & Wani, 1996), the codiscoverers of CPT and Taxol. Camptothecin was mainly isolated from woody plants, including *C. acuminata* and *Nothapodytes nimmoniana* (Kai et al., 2015; Wall et al., 1966). But the above species grow slowly and fail to meet the demand for resources, while the total chemical synthesis of CPT hasn't been put into commercial production (Kai et al., 2015). Because of the increasing demand for CPT, the search for other CPT-producing sources has been of great interest. There are some promising candidates identified and being used for the extraction, including *Camptotheca lowreyana, Camptotheca yunnanensis, Ervatamia heyneana, Merrilliodendron megacarpum, Mostuea brunonis,* and many *Ophiorrhiza* species (Sharma et al., 2024; Sirikantaramas et al., 2013). Camptothecin and its related compounds are at present obtained by extraction from intact plants, but fungal endophytes may be an alternative source of production, like *Entrophospora infrequens, Meyerozyma* sp., *Diaporthe* sp., *Alternaria alternata, Aspergillus* sp., *Trichoderma atriviride,* and *Fusarium solani* (Doan et al., 2024). Unfortunately, commercial CPT products derived from endophytes have not achieved success, as endophytic fungi tend to be unreliable producers, often losing or reducing theirs CPT production ability over generations. The exception to this is the *Alternaria burnii* NCIM1409 strain, which was isolated from *N. nimmoniana* (Mohinudeen et al., 2021).

10.2.2 Supply chain challenges

Currently, commercial exploitation of camptothecin largely depends on extraction from natural woody plant resources such as *C. acuminata* and *N. nimmoniana* (Kai et al., 2015).

However, these plants typically show low camptothecin concentrations and have long growth cycles, and are thus not suitable for modern biotechnological applications. Several attempts have been made to establish robust plant cell culture systems to provide a platform for high CPT production. Among them are hairy-root cultures of *O. pumila*, an herbal plant of the *Ophiorrhiza* genus in the Rubiaceae family. Plants of *O. pumila* produce camptothecin naturally in various organs, such as roots, stems, and leaves (Hao et al., 2021a).

The species of *Ophiorrhiza* vary in habit from creeping herbs to small shrubs, reaching heights of up to 3 m, and growing in humid regions at different altitudes. Most of them are erect, suffrutescent, some are prostrate, and a few are creeping (Deb & Mondal, 2015).

Only a limited number of naturally occurring alkaloids have undergone thorough laboratory testing. Additionally, the natural production of alkaloids is quite susceptible to changes in weather conditions, which restricts their production. Nonetheless, the recent interest from commercial entities in this area has spurred the investigation of methods to modify alkaloid production (Bhushan et al., 2023).

10.3 Camptothecin distribution in different parts of the plant

10.3.1 Organ-specific accumulation patterns

C. acuminata exhibits distinct organ-specific distribution patterns, with branches, roots, and stems accumulating significantly higher camptothecin concentrations than leaves (Lorence & Nessler, 2004). Bark consistently shows higher concentrations than wood, with dried roots containing 0.51 mg/g camptothecin compared to 0.42 mg/g in shoots (Liu & Adams, 1996). Young leaves contain the highest CPT levels (4–5 mg/g), approximately 50% higher than seeds and 250% higher than bark (Lorence & Nessler, 2004). CPT content in young leaves was much higher than in old leaves and was logarithmically related to leaf area and leaf biomass (Liu & Adams, 1996).

Nothapodytes nimmoniana exhibits a different accumulation pattern, with roots showing the highest CPT content at 1.87 mg/g, followed by shoots at 1.07 mg/g (Mingzhang et al., 2011). CPT accumulates predominantly in the roots (1.91 ± 0.02 mg/g), with lower yields in stembark (1.02–1.23 mg/g), twigs (0.09–0.18 mg/g), and minimal levels in leaves (0.04–0.08 mg/g), irrespective of extraction method (Degambada et al., 2023).

Ophiorrhiza pumila demonstrates unique organ-specific accumulation with stems, roots, and seed pods containing higher CPT contents than leaves, flowers, and ovaries (Lee et al., 2022). Roots maintain the highest CPT concentrations throughout all developmental stages, being 35% higher at the seed-ripening stage compared to other developmental phases.

Miquelia dentata shows exceptional CPT accumulation in seeds, containing 10–14 mg/g dry weight camptothecin, representing the highest reported content among all CPT-producing plants (Malik and Laura, 2014). Other species show varied patterns: *E. heyneana* produces CPT primarily in roots (Gunasekera et al., 1979), *Ixora coccinea* shows higher concentrations in mature leaves (5 mg/g) compared to young leaves (4 mg/g) (Malik and Laura, 2014).

10.3.2 Developmental and seasonal variations

Seasonal variations significantly impact CPT accumulation across species. In *N. nimmoniana*, the monsoon season (August) demonstrates higher CPT levels in bark compared to summer (May), with some locations showing over 50% difference between the highest and lowest seasonal content (Pai et al., 2013). Amboli region consistently shows the highest CPT accumulation across all seasons, yielding 13.37 mg/g during monsoon (Pai et al., 2013). In *C. acuminata*, the impact of developmental stages is notably significant, with younger leaves exhibiting higher concentrations of CPT compared to older ones, showing a linear decrease from the branch tip to the seventh leaf. Nonetheless, within the same growing season, there are no notable differences in CPT levels between younger and older tissues (Liu & Adams, 1996).

For *O. pumila*, the ideal time for harvesting is 63 days posttransplantation, as CPT content peaks during the seed-ripening phase. At this stage, the CPT concentration in roots rises by 35% compared to the vegetative and flowering stages (Lee et al., 2022). Across various seasons, younger plant parts consistently show higher CPT levels than mature ones (Malik and Laura, 2014). In *O. rugosa* var. decumbens, differentiated tissues accumulate significantly more camptothecin than undifferentiated callus, with shoots at 2.2 mg/g and roots at 2.4 mg/g, while callus only shows trace amounts. CPT levels further increase during development, reaching a peak of 3.4 mg/g in the inflorescence at flowering (Sharma et al., 2024). Temporally, strictosidine and key intermediates reach their highest levels in stems, whereas camptothecin primarily concentrates in roots, indicating possible transport mechanisms or varying biosynthetic activities between organs (Yamazaki et al., 2003). These developmental and seasonal variations indicate that optimal harvesting strategies must consider both plant age and environmental conditions to maximize CPT yield.

10.4 Biotechnological production approaches

10.4.1 Plant tissue culture systems

Utilizing plant tissue culture of medicinal plants serves both for mass manufacturing and the preservation of rare and endangered species. A wide range of medicinal plants rich in these metabolites is involved in the formation of secondary metabolites, including quinones, tannins, steroids, alkaloids, terpenoids, and phenylpropanoids. The need for these secondary metabolites produced from plants through *in vitro* transgenic and stable plant culture in industry for commercial purposes has been greatly facilitated by the adoption of modern biotechnology processes.

Since the CPT-producing plants are being cut down by pharmaceutical companies, important plants like *N. nimmoniana* and *C. acuminata* are endangered. Since our current CPT requirement is higher than what is available, it is essential to look for alternate sources for CPT production. While a search for new plant resources for CPT production is being pursued, plant cell and organ cultures have been recognized as an alternate source for continuous production of CPT. The suspension cultures produce comparatively low levels of CPT, although growth regulators can slightly alter levels (Roja & Heble, 1994; Thengane et al., 2003).

10.4.1.1 *Callus culture and clonal selection*

High-yielding camptothecin-producing cell lines are developed through systematic selection strategies that exploit the inherent biochemical heterogeneity in plant cell cultures. The cell aggregate cloning method has proven particularly effective, as demonstrated with *O. mungos*, where the selected OMC3 cell line achieved significantly higher CPT production (0.095 ± 0.002 mg/g) compared to original cultures. Medium optimization plays a crucial role, with strategic manipulation of carbon sources (5% sucrose for CPT production vs 3% for biomass), nitrogen ratios (40/20 mM ammonium/nitrate for optimal CPT yield), and phosphate concentrations enabling targeted enhancement of either growth or metabolite production. For *N. nimmoniana*, optimal callus induction using 8.29 μM picloram with 3% sucrose yielded 9.5 μg g^{-1} CPT, while the combination of 4.52 μM 2,4-D with 2.22 μM BAP achieved 13 mg/g CPT content in fruit-derived cultures (Deepthi & Satheeshkumar, 2017).

10.4.1.2 *Cell suspension cultures*

Cell suspension cultures are advantageous for camptothecin production due to their scalability and process control, with protocols requiring friable callus as the initial material and careful optimization of culture conditions. Scaling up to commercial levels involves crucial factors such as choosing the right bioreactor (with bubble-type reactors being preferred), considering geometric scaling effects, and optimizing process parameters like dissolved oxygen levels, shear stress management, and volumetric oxygen mass transfer coefficients (Titova et al., 2024). Optimizing growth kinetics through various cultivation regimes, especially semicontinuous systems, maximizes productivity, while mathematical modeling combined with reactor dynamics aids in predicting scale-up effects. High-yielding cell line selection integrated with optimized cultivation conditions, as shown with OMC3 cultures using a 14 g/L inoculum size, can lead to significant productivity enhancements, achieving up to 1.12 ± 0.08 mg/g CPT production when paired with elicitation strategies (Deepthi & Satheeshkumar, 2017). However, despite their scalability and rapid proliferation, cell suspension cultures often accumulate camptothecin at very low levels, making it necessary to employ strategies such as elicitation and precursor feeding to boost secondary metabolite production.

10.4.2 Media optimization strategies and elicitation techniques

The cell suspensions and other culture systems can be made to produce higher levels of secondary metabolites through elicitation and precursor feeding, which are effective strategies to enhance the biosynthesis of camptothecin (CPT) and other valuable compounds *in vitro*. While undifferentiated cell cultures often show limited metabolite accumulation, the application of elicitors has been shown to overcome this limitation by inducing plant defense responses and redirecting metabolic fluxes toward secondary metabolite pathways. According to Namdeo (2007), elicitors are biological, pharmacological, or physical substances that imitate stress signals in order to activate the plant's natural defense mechanisms and increase the synthesis of metabolites. Alkaloids, terpenoids, and phenolics are examples of secondary metabolites that plants naturally produce to defend against infections and herbivores. By deliberately inducing these stress responses, pharmaceuticals can be produced (Pandey et al., 2022). Jasmonic acid (JA) and methyl jasmonate (MeJa) are two chemical elicitors that have been extensively researched due to their broad-spectrum capacity to cause the buildup of secondary metabolites. Both substances function as signaling molecules in both biotic and abiotic stress responses, and it has been demonstrated that applying them exogenously greatly increases the synthesis of coumarins, taxanes, phenolic phytoalexins, and alkaloids in a variety of plant species (Liu et al., 2015). Details of different precursors and elicitors used to increase the yield of CPT in CPT-producing plants and endophytes are presented in Table 10.2 and 10.3, respectively.

Overall, these studies highlight that elicitation—through jasmonates, sugars, resins, and metal ions—offers a powerful means to boost CPT production *in vitro*. However, optimization is often species- and tissue-specific, necessitating multivariate approaches to maximize metabolite yield across different plant systems (Sirikantaramas et al., 2013). When combined, culture media optimization and elicitor strategy have been shown to increase camptothecin yields from *in vitro* systems. However, these approaches alone often face limitations in terms of long-term stability, scalability, and consistent metabolite productivity. In order to get over these limitations and create more sustainable platforms for the production of secondary metabolites, a lot of focus has been placed on hairy root culture systems, which combine high biosynthetic capacity with genetic stability.

10.4.3 Hairy root culture systems

10.4.3.1 *Agrobacterium rhizogenes* transformation

Agrobacterium rhizogenes, a soil bacterium, induces hairy root cultures, which are one of the most dependable and effective methods for sustainably producing secondary metabolites. Hairy roots, in contrast to traditional plant cultivation, have unique characteristics like fast growth, hormone-independent proliferation, and widespread branching, all of which work together to provide a very stable system for metabolite production (Cui et al., 2015; X. Fan et al., 2022). The production of camptothecin (CPT) has been extensively studied in hairy root cultures of *Ophiorrhiza* species. Over the past two decades, several *A. rhizogenes* strains, including C58C19 (Shi et al., 2020), TISTR 1450 (Ya-ut et al., 2011), 15834 (Saito et al., 2001), and LBA 9402 (Kamble et al., 2011), have been successfully employed to transform *Ophiorrhiza* and establish high-yielding hairy root lines. Notably, *O. pumila* hairy roots demonstrated significant improvement in CPT accumulation, with (Saito et al., 2001) reporting a 16-fold increase in biomass within 5weeks, accompanied by CPT production as the major alkaloid. Further studies confirmed that CPT yield continued to rise during cultivation, reaching its maximum at 50 days (). Moreover, transgenic hairy root systems allow metabolic engineering strategies to be applied effectively. For instance, overexpression of *G10H* and *STR* genes in *O. pumila* hairy roots significantly enhanced CPT accumulation, with coexpression yielding up to 1.77 mg/g dry weight, surpassing both wild-type and single-gene lines (Cui et al., 2015). Since CPT mainly exists in the roots of resource plants, which are further rarely distributed, harvesting their roots as a drug source may result in the extinction. Thus, *in vitro* hairy root culture is regarded as a promising alternative to wild plant harvesting, offering a sustainable, scalable, and resource-conserving method for camptothecin production.

TABLE 10.2 Camptothecin (CPT) production in various plant species under different *in vitro* culture systems.

Plant species	Culture	Elicitation/Media modification	Effect on CPT yield	References
Nothapodytes nimmoniana	Callus	NAA (10.74 µM) and BA (2.22 µM)	31.3 mg/g	Fulzele et al. (2001)
		2,4-D	Reduction in CPT	
		20 Gy radiation	0.98 mg/g	Improvement of Anticancer Drug Camptothecin Production by Gamma Irradiation on Callus Cultures of Nothapodytes Foetida (2015)
		$CaCl_2$ (25 mM)	3.176 ± 0.32 mg/g	Isah et al. (2022)
Camptotheca acuminata	Callus	JA (50 µM)	0.004 mg/g	Hengel et al. (1994)
	Cell suspension	JA (50 µM)	7.1×10^{-5} mg/g	Song and Byun (1998)
		Sucrose 6% (w/v)	2.9×10^{-6} mg/g	Kim et al. (1999)
		40 mM nitrogen with 5:1 NH_4^+/NO_3^-	6.3 mg/L	Pan et al. (2004)
	Plantlets	MeJA (10 µM)	0.25 mg/g	Pu et al. (2022)
		PEG (5 g/L)	0.21 mg/g	
		AgNO3 (50 µM)	0.26 mg/g	
Ophiorrhiza mungos	*In vitro* shoots	MeJA (150 µM)	0.47 mg/g	Jisha (2006)
	Multiple shoots	Sucrose (3%)	0.02 mg/g	
		Coconut water (20%)	0.4 mg/g	
		Ammonium nitrate (25 µM)	0.4 mg/g	
		2 Gy radiation	0.87 mg/g	
	Cell suspension	Yeast Extract	0.80 ± 0.005 mg/g	Deepthi and Satheeshkumar (2016)
		Silver nitrate	0.52 ± 0.003 mg/g	
	In vitro plants	MeJA (150 µM)	2.3 mg/g	Nagesha (2018)
		SA (50 µM)	1.5 mg/g	

Ophiorrhiza rugosa	Hairy roots	infected with strain LBA9402	0.09 mg/g	Kamble et al. (2011)
Ophiorrhiza eriantha		50 μM MeJA, 1 Gy and 10 mg/L chitosan	0.251, 0.422, and 0.29 mg/g	Ni et al. (2011)
Ophiorrhiza kuroiwa	Hairy roots	100 μM MeJA	0.20–0.25 mg/g	Asano et al. (2004)
Pyrenacantha volubilis	Root culture	Yeast extract (50 ppm)	5.13 ± 0.21 mg/g	Sasidharan et al. (2023)
		Ammonium nitrate (1240 ppm)	2.19 ± 0.2 mg/g	
	Green root culture	Salicylic acid (100 ppm)	4.93 ± 0.21 mg/g	

TABLE 10.3 Camptothecin (CPT) production reported from endophytic strains under varying culture conditions.

Strain	Container	Elicitation	CPT content	References
Entrophospora infrequens	Shake flask	Control (SB)	5.03×10^{-3} mg/g	Amna et al. (2012)
		Tryptophan 0.01 M	5.0×10^{-3} mg/g	
		Tryptamine 0.01 M	5.0×10^{-3} mg/g	
		Geraniol 0.01 M	5.0×10^{-3} mg/g	
		Citral 0.01 M	5.0×10^{-3} mg/g	
		Mevalonic acid 0.01 M	5.0×10^{-3} mg/g	
		L-Leucine 0.025 M	5.0×10^{-3} mg/g	
		Tryptophan + Geraniol (0.01 M each)	5.0×10^{-3} mg/g	
		Tryptophan + Citral (0.01 M each)	5.0×10^{-3} mg/g	
		Tryptophan + Mevalonic acid (0.01 M each)	5.0×10^{-3} mg/g	
		Tryptophan + L-Leucine (0.01 M + 0.025 M)	1.258×10^{-2} mg/g	
Fusarium solani	Shake flask	ethanol	4.02×10^{-2} mg/L	Venugopalan and Srivastava (2015)
		ethanolic extract of *C. roseus*	2.968×10^{-2} mg/L	
F. solani	Shake flask	PDA (tryptamine (0.5 mM) BSA (0.075 mM))	4.19×10^{-5} mg/g	Venugopalan et al. (2016)
F. oxysporum	Shake flask	Response surface methodology	0.283 mg/g	Bhalkar et al. (2015)
Colletotrichum fructicola and Corynespora cassiicola	Shake flask	Coculture	0.146 mg/g	Bhalkar et al. (2016)
Trichoderma atroviride	Shake flask	Methyl jasmonate, optimized fermentation conditions, XAD16, and adsorbent resin	9.85×10^{-5} -1.42×10^{-4} mg/g	Pu et al. (2013)
Nodulisporium sp.	Bioreactor	Scaling up	4.5×10^{-2} mg/g	Rehman et al. (2009)

10.4.3.2 *Advantages of hairy roots over cell cultures*

Hairy root cultures of *O. pumila* significantly outperform cell suspension cultures in camptothecin production, which can further be enhanced by genetic engineering approaches, elicitation, and bioreactor cultivation (Banadka et al., 2024). Standard hairy root cultures in *O. pumila* produce 1–2 mg/g dry weight camptothecin, and peak camptothecin accumulation occurs at 40 days of cultivation, reaching 1.93 mg/flask in tissue. In total, 40% of total camptothecin is secreted into the culture medium by 60 days (Hao et al., 2021a). The hairy roots of *O. pumila* were grown in a 3 L bioreactor equipped with a stainless mesh (Sudo Hiroshi, 2002). From the tissue cultivated for over 8 weeks in the reactor, 22 mg CPT was obtained, with approximately 3.6 mg secreted in the media. The biosynthesis of secondary metabolites is described as "very low" in cell suspension culture compared to hairy roots, and cell suspension cultures derived from hairy roots show dramatically reduced camptothecin production (Asano et al., 2013). In comparative terms, suspension cell cultures of *C. acuminata* were found to produce much lower CPT levels, below 0.05 mg/g, highlighting the superior metabolic potential of hairy roots. Along with much increased expression of important camptothecin biosynthesis genes (*OpTDC, OpSTR, OpSLS*, and *OpG10H*), the metabolomic study verified that the majority of camptothecin pathway intermediates concentrate preferentially in hairy roots. Hairy roots function better than undifferentiated cell suspension cultures because of their differentiated cellular architecture, specialized metabolic machinery, and active expression of camptothecin biosynthesis pathways (Yamazaki et al., 2013).

10.4.4 Bioreactor technology: a potential alternative

Bioreactor-based production of camptothecin (CPT) is crucial for meeting pharmaceutical demand while mitigating the overexploitation of natural sources, such as *C. acuminata, N. nimmoniana*, and *O. pumila*. Once explants with superior camptothecin production are identified, the next crucial step is to establish stable, high-yielding cell lines for efficient bioprocess development. This process typically begins with creating cell suspension cultures by fine-tuning hormone combinations in the growth medium. The resulting cultures are then carefully screened to identify elite cell lines that exhibit enhanced secondary metabolite production. Advancements in bioreactor technology have addressed the increasing pharmaceutical demand for CPT by enabling controlled, sustainable production from plant cell, tissue, and hairy root cultures (Hussain et al., 2012). Bioreactor systems allow for precise manipulation of environmental variables and elicitor addition, fostering the biosynthesis and accumulation of CPT *in vitro*.

10.4.4.1 *Scale-up considerations*

The initiation of cell lines requires careful optimization of growth regulators, with studies demonstrating that specific hormone combinations significantly influence both biomass accumulation and CPT biosynthesis. Cell suspension cultures are typically established from high-producing callus tissues using modified Murashige and Skoog (MS) medium supplemented with optimized concentrations of auxins and cytokinins. The establishment phase involves rigorous screening protocols where individual cell lines are evaluated through cell aggregate cloning methods to isolate clones with superior metabolite production profiles (Deepthi & Satheeshkumar, 2017; Murali et al., 2023). The screening includes systematic evaluation of parameters, including growth kinetics, biomass yield, and metabolite accumulation patterns. The screening methodology incorporates both quantitative analysis of CPT content and assessment of cell line stability over multiple subcultures to ensure consistent production characteristics (Deepthi & Satheeshkumar, 2017). After identifying the cell line, the next step is to refine its growth conditions through flask-scale optimization studies. These experiments help determine the culture environment that supports the highest CPT accumulation.

10.4.4.2 Process optimization

Elicitation is an important way to increase secondary metabolite production in selected cell lines. It works by adding elicitors at the right stage of growth so that metabolite production is improved without harming the cells. This is why both the concentration of the elicitor and the timing of its addition need to be carefully optimized. Moving from flask cultures to larger bioreactors brings challenges like plant cells' sensitivity to shear stress, slow growth rates, and low oxygen demands. While airlift bioreactors work well for less sensitive cells, specialized bioreactors such as wave and undertow, and slug bubble reactors reduce shear for delicate tissues like hairy roots (Gonçalves & Romano, 2018). Critical parameters for bioreactor scale-up include maintenance of appropriate dissolved oxygen levels, pH control, temperature regulation, and agitation rates that promote adequate mixing without causing cellular damage (Banadka et al., 2024).

To maximize CPT yields, process optimization employs media composition adjustment, elicitor and adsorbent (e.g., Diaion HP-20) addition, as well as manipulation of plant hormone regimes and harvesting times (Song & Byun, 1998). Elicitors like methyl jasmonate and yeast extract have been demonstrated to markedly enhance CPT accumulation, especially when introduced during particular growth phases. To maintain constant CPT titers as culture volume grows, standard procedure entails progressively increasing working volumes, refining agitation and aeration techniques, and accounting for plant cell aggregation tendencies (Terrier et al., 2007). Crucially, selection and preservation of high-yielding lines via tissue culture and genetic manipulation remain foundational during scale-up (Gonçalves & Romano, 2018).

10.5 Camptothecin biosynthetic pathway elucidation

Understanding the biosynthetic pathways and their complex regulatory networks in medicinal plants is essential for the successful development of biotechnological approaches to enhance secondary metabolite production (Gonçalves & Romano, 2018). To further enhance camptothecin production by biotechnological engineering, it is crucial to fully dissect its biosynthetic pathway and molecular regulatory mechanisms (Singh et al., 2020). However, the camptothecin biosynthetic pathway is complex and only partly resolved. The elucidation of camptothecin biosynthetic pathways has been historically challenging due to their complex multi-step enzymatic transformations and tissue-specific expression patterns. This field has undergone a revolution since the onset of the omics era, specifically integrative genomics, transcriptomics, and metabolomics approaches, which have made it possible to identify biosynthetic gene clusters systematically, analyze coexpression networks, and thoroughly characterize pathway intermediates and regulatory mechanisms (Pu et al., 2022; Zhang et al., 2023). Camptothecin biosynthesis uses intermediates of the iridoid and shikimate pathways, which converge to form the central precursor molecule strictosidine and then, further, via majorly unknown chemical reactions, form camptothecin (Fan et al., 2022).

10.5.1 Formation of tryptamine via the shikimate pathway

The shikimate pathway provides the indole precursor tryptamine through a series of well-orchestrated enzymatic reactions beginning from chorismate. The first committed step is catalyzed by anthranilate synthase (*AS*), which consists of an α-subunit (*ASA*) and a β-subunit (*ASB*). AS converts chorismate to anthranilate, initiating tryptophan biosynthesis. In *C. acuminata*, two *ASA* isoforms have been identified: *CaASA1*, whose transcription correlates with CPT accumulation, and *CaASA2*, which shows constitutively low expression (Lu et al., 2005). Downstream intermediates include 5-phosphoribosylanthranilate and indole-glycerol phosphate (Sirikantaramas et al., 2013). Tryptophan synthase (*TSB*) completes the pathway by producing tryptophan: its α-subunit generates indole, while the β-subunit condenses this with serine. In *C. acuminata*, *CaTSB* expression parallels camptothecin levels during seedling development, suggesting a tightly coordinated regulation of tryptophan and CPT biosynthesis (Lu & McKnight, 1999). Tryptophan is then converted to tryptamine by tryptophan decarboxylase (*TDC*), a rate-limiting enzyme in CPT biosynthesis (Asano et al., 2013). Species-specific diversification is evident: *O. pumila* possesses two functional *TDC* genes (Yamazaki et al., 2003a; You et al., 2021), while *C. acuminata* and *O. pumila* harbor multiple *TDC*copies (López-Meyer & Nessler, 1997; Y. Yamazaki et al., 2003).

10.5.2 Formation of secologanin via the MEP pathway

Parallel to tryptamine biosynthesis, secologanin is generated through the plastidial MEP pathway. This pathway differs from the cytosolic mevalonate (MVA) pathway (Rohmer & Rohmer, 1999), though metabolic crosstalk between them has been reported (Hemmerlin et al., 2003). Tracer studies in *O. pumila* hairy roots have demonstrated that the MEP pathway, rather than the MVA pathway, provides the secologanin moiety of CPT (Yamazaki et al., 2004). However, the situation in *C. acuminata* remains less certain. In the MEP pathway, (E)-4-hydroxy-3-methylbut-2-enyl diphosphate (HMBPP) is reduced by HDR to yield DMAPP. Together with IPP, DMAPP is condensed by geranyl diphosphate synthase (*GPPS*) to form geranyl diphosphate (GPP) (Sirikantaramas et al., 2013). Geraniol synthase (*GES*) then produces geraniol, which is hydroxylated at the C-10 position by the cytochrome P450 enzyme geraniol-10-hydroxylase (*G10H*), forming 10-hydroxygeraniol (Shi et al., 2020). Subsequent oxidation by 10-hydroxygeraniol oxidoreductase (*10HGO*) yields 10-oxogeranial, shown in *C. acuminata* to be dependent on NADP+/NADPH turnover (Awadasseid et al., 2020). The iridoid skeleton forms when iridoid synthase (*IS*) cyclizes 10-oxogeranial to nepetalactol/iridodial (Sadre et al., 2016). Progression through the iridoid pathway then involves 7-deoxyloganic acid hydroxylase (*7DLH*) producing loganic acid, and further oxidative cleavage reactions mediated by cytochrome P450s such as secologanin synthase (*SLS/SLAS*). Notably, *C. roseus* primarily produces secologanin through CYP72A1 (Irmler et al., 2000), whereas *C. acuminata* favors carboxylic acid derivatives such as secologanic acid via CaCYP72A565 and CaCYP72A610, which also display additional bifunctional activity (Yang et al., 2019). This species-specific divergence highlights how different plants employ alternative strategies to furnish the secologanin scaffold required for CPT biosynthesis (Kang et al., 2021).

10.5.3 Convergence at strictosidine: the gateway to alkaloid formation

The point of convergence between tryptamine and secologanin is mediated by strictosidine synthase (*STR*), which catalyzes their condensation into strictosidine, a central intermediate in MIA biosynthesis (Treimer & Zenk, 1979). *STR* has been characterized in several CPT-producing taxa: *O. pumila STR* encodes a 351-amino acid protein (Yamazaki et al., 2003b), while *N. nimmoniana* produces a closely related 352-amino acid enzyme (Singh et al., 2020). Expression analyses indicate that STR mRNA and proteins are most abundant in hairy roots, roots, and stems of *O. pumila*, correlating with CPT accumulation, whereas young leaves accumulate CPT without detectable *STR* activity, suggesting interorgan transport (Yamazaki et al., 2003a). Structural studies confirm conservation of the catalytic mechanism of *STR* across plant species (Ma et al., 2006). Metabolic profiling has refined our understanding of intermediate variation. In *O. pumila*, both strictosidine and strictosidinic acid are detectable (Yamazaki et al., 2003b; Yamazaki et al., 2013), but isotope feeding experiments demonstrated that only strictosidine participates directly in CPT biosynthesis, marking a divergence from *C. acuminata* (M. Yang et al., 2021).

10.5.4 Downstream transformations toward camptothecin

While strictosidine is established as the gateway intermediate, the downstream path to camptothecin remains incompletely resolved. Cyclization of strictosidine generates strictosamide , but the enzymatic sequence beyond this step remains unclear. Isolated intermediates such as 3(S)-pumiloside and 3(S)-deoxypumiloside from *Ophiorrhiza* species are proposed to bridge strictosidine and CPT (Asano et al., 2013). Moreover, there is growing evidence that cytochrome P450 monooxygenases and O-methyltransferases tailor downstream intermediates to produce derivatives including 10-hydroxycamptothecin and 10-methoxycamptothecin (Fan et al., 2022). A particularly notable advance was the identification of cytochrome P450 enzymes in *C. acuminata* capable of regioselective hydroxylations at the C-10 and C-11 positions of CPT. These enzymes yield hydroxycamptothecin, a scarce but clinically important derivative, and have been successfully exploited to generate therapeutic camptothecin analogs (Nguyen et al., 2021).

10.6 Metabolic engineering strategies

The advancement of camptothecin (CPT) production through metabolic engineering represents a critical frontier in meeting the growing demand for this potent anticancer compound. Current research focuses on primary approaches for enhancing CPT biosynthesis, utilizing multiomics data construction as a fundamental tool for identifying candidate biosynthetic genes and transcriptional regulators (Sirikantaramas, 2007). These strategies include both gene overexpression and downregulation techniques, coupled with the application of various elicitors and exploration of endophytic resources to provide additional pathways for CPT production enhancement (X. Fan et al., 2022).

The implementation of stable genetic transformation technology has proven essential for functional gene studies and germplasm improvement in medicinal plants, particularly for CPT-producing species (Hao et al., 2019). A comprehensive understanding of alkaloid biosynthetic pathway genes, transcription factors, and their regulatory mechanisms is crucial for enhancing the production of these valuable compounds. Modern biotechnological approaches, including plant tissue culture, CRISPR/Cas9 gene editing, miRNA-mediated manipulation, and nanoparticle-based enhancement strategies, are being extensively employed to enhance and stabilize secondary metabolite production for large-scale industrial applications (Bhushan et al., 2023).

10.6.1 Gene overexpression approaches

10.6.1.1 Pathway enzyme overexpression

The direct manipulation of biosynthetic enzyme-encoding genes is the most straightforward strategy for enhancing CPT production. In CPT-producing plants, overexpression of key biosynthetic genes has demonstrated significant improvements in alkaloid accumulation. Notable success has been achieved through the cross-species gene transfer, like transferring *CrSTR* and *CrG10H* from *C. roseus* to *C. acuminata* hairy roots (Cui et al., 2015) and *N. nimmoniana* STR (*NfSTR*) in transgenic *O. rugosa* (Singh et al., 2020). Co-overexpression approaches like simultaneous insertion of *OpG10H* and *OpSLS* genes in *O. pumila* hairy roots have proven particularly effective (Shi et al., 2020). Some reports of metabolic engineering involving the modulation of genes regulating CPT biosynthesis are given in Table 10.4. This synergistic effect highlights the importance of coordinated pathway enhancement rather than single-gene approaches. Beyond the conventional CPT biosynthesis genes, other overexpression targets have been discovered by recent genome-scale metabolic modeling research. The tricarboxylic acid cycle and amino acid biosynthesis enzymes are among the predicted targets, indicating that metabolic flux redirection through primary metabolism can improve the generation of secondary metabolites (Murali et al., 2023).

10.6.1.2 Engineering transcription factors

Transcription factors (TFs) are essential to the plant's regulatory machinery. They perform diverse roles in the overall plant lifecycle, including growth and development, stress response, and primary and secondary metabolism. Transcription factors represent powerful regulatory molecules capable of simultaneously modulating multiple biosynthetic genes, making them attractive targets for metabolic engineering. They tend to bind with cis-regulatory elements such as enhancers or DNA promoter regions lying adjacent to the genes, thereby affecting the biosynthesis of bioactive metabolites across various medicinal plants (Bhushan et al., 2023; X. Fan et al., 2022).

The transcription factors like *WRKY*, *ERFs*, *MYB*, *ORCA3*, *NAC1*, *CBR*, and *LMF* have been reported to play a regulatory role in monoterpene indole alkaloid (MIA) and CPT biosynthesis (Table 10.5). The *WRKY* transcription factor family has emerged as a critical regulatory network in camptothecin biosynthesis. *OpWRKY2* functions as a direct positive regulator, through direct enhancement of *OpTDC* expression (Hao et al., 2021a). This

TABLE 10.4 Genetic interventions influencing camptothecin (CPT) biosynthesis in plants. (OE- overexpression and DR- downregulation).

S. no.	Gene	OE/DR	Change in CPT	Plant species	References
1.	*Op10HGO2*	Overexpressed	Significant increase in CPT content in hairy roots	*Ophiorrhiza pumila*	Zhou et al. (2025)
2.	*Op10HGO2*	Knocked-out (CRISPR)	Marked decrease in CPT content in hairy roots	*Ophiorrhiza pumila*	Zhou et al. (2025)
3.	*OpLAMT1*	Overexpressed	1.52-fold increase in CPT in hairy roots	*Ophiorrhiza pumila*	Hao et al. (2023)
4.	*OpLAMT1*	Knocked-out (CRISPR)	CPT production nearly abolished	*Ophiorrhiza pumila*	Hao et al. (2023)
5.	*STR*	Overexpressed	~5-fold increase in CPT	*Nothapodytes nimmoniana*	Murali et al. (2023)
6.	*OpTDC2*			*Ophiorrhiza pumila*	You et al. (2021)
7.	*OpG10H* and *OpSLS*	Cooverexpressed	1.8-fold more CPT	*Ophiorrhiza pumila*	Shi et al. (2020)
8.	SLS *(NnCYP72A1)*	Overexpressed	3-fold more CPT	*Nothapodytes nimmoniana*	Rather et al. (2020)
9.	*NfSTR*	Overexpressed	2-fold increase in CPT	*Ophiorrhiza rugosa*	Singh et al. (2020)
10.	*CaG10H, Ca10HGO, CaTDC,* and *CaSTR*	VIGS	Decreased CPT production	*Camptotheca acuminata*	Jin et al. (2016)
11.	*OpTDC*	Downregulated (RNAi)	Decreased CPT production	*Ophiorrhiza pumila*	Asano et al. (2013)
12.	*OpSLS*	Downregulated (RNAi)	Decreased CPT production	*Ophiorrhiza pumila*	Asano et al. (2013)

TABLE 10.5 The transcription factors investigated influencing camptothecin (CPT) biosynthesis in plants (OE-overexpression and DR-downregulation).

S. no.	Transcription factor	OE/DR	Change in CPT	Mode of action	Plant species	References
1.	*OpCBR1*	Overexpressed	~2.2- to 2.9-fold reduction in CPT	inhibits the expression of *OpIO*	*Ophiorrhiza pumila*	Zhou et al. (2025)
2.	*OpNAC1*	Overexpressed	0.61-fold (39% decrease) in CPT in hairy roots	Decreased CPT, loganin, and biosynthetic gene expression (*OpLAMT1*, *OpSTR1*, *OpSLS1*)	*Ophiorrhiza pumila*	Hao et al. (2023)
3.	*OpNAC1*	Knockout	1.51-fold increase in CPT in hairy roots	Increased CPT and related gene expression	*Ophiorrhiza pumila*	Hao et al. (2023)
4.	*OpHD-ZIP7* and *OpHD-ZIP20*	dual-luciferase reporter assays		trans-activate the promoters of *OpIO* and *OpTDC*	*Ophiorrhiza pumila*	Wang et al. (2023)
5.	*OpWRKY6*	Overexpressed	Decreased CPT accumulation	downregulates *OpGES*, *OpTDC*, *Op10HGO* and *Op7DLH*	*Ophiorrhiza pumila*	Wang et al. (2022)
6.	*OpWRKY6*	Knockout	Increased CPT accumulation		*Ophiorrhiza pumila*	Wang et al. (2022)
7.	*OpWRKY2*	Overexpressed	1.7–3.4-fold increase	Enhances CPT by regulating *OpTDC*	*Ophiorrhiza pumila*	Hao et al. (2021b)
8.	*OpWRKY2*	Downregulated	slight decrease in CPT		*Ophiorrhiza pumila*	Hao et al. (2021b)
9.	*OpWRKY3*	Overexpressed	Increased CPT accumulation	promoting the transcription of *OpCPR*	*Ophiorrhiza pumila*	Wang et al. (2019)
10.	*CaLMF*	Overexpressed	Decreased CPT accumulation	downregulate the expression of CPT biosynthesis genes	*Camptotheca acuminata*	Chang et al. (2018)
11.	*OpMYB1*	Overexpressed	2- to 2.5-fold decrease in CPT	contains an EAR-like repression motif, characteristic of transcriptional repressors.	*Ophiorrhiza pumila*	Rohani et al. (2016)
12.	*OpERF2*	Overexpressed	Increased CPT levels	Positively regulates iridoid biosynthesis	*Ophiorrhiza pumila*	Udomsom et al. (2016)
13.	*CrORCA3*	Overexpressed	1.5-fold more CPT		*Camptotheca acuminata*	Ni et al. (2011)

transcription factor directly binds to W-box cis-elements in the *OpTDC* promoter, demonstrating the direct regulatory mechanism underlying its positive effects. Conversely, *OpWRKY6* acts as a negative regulator of CPT biosynthesis, directly binding and inhibiting multiple biosynthesis genes in the shikimate and iridoid pathways. This transcription factor negatively regulates genes, including *OpGES*, *Op10HGO*, *Op7DLH*, and *OpTDC* by binding to W-box elements in their promoter regions. Understanding such negative regulatory mechanisms provides opportunities for targeted gene silencing to enhance production. This diversity in regulatory functions within a single transcription factor family highlights the importance of comprehensive functional characterization.

10.6.2 Gene downregulation strategies

The suppression of negative regulators of the metabolic pathway represents an effective approach to enhance camptothecin biosynthesis. This strategy focuses on identifying and eliminating repressor molecules that directly inhibit the expression of key biosynthetic genes. Virus-induced gene silencing (VIGS) offers transient but highly effective gene knockdown by exploiting plant antiviral defense pathways (Zulfiqar et al., 2023). CRISPR/Cas9 technology enables precise gene editing through targeted DNA modifications, allowing both permanent gene knockout and controlled gene expression modulation (Afridi et al., 2025). Another approach is RNA interference (RNAi). RNAi usually regulates various biological processes by interfering with mRNA translation. MicroRNAs (miRNAs) and small interfering RNAs (siRNAs) are distinct classes of small noncoding RNAs that play key roles in RNA interference (RNAi)-mediated posttranscriptional gene regulation (Islam et al., 2025).

Some examples are given in Tables 10.4 and 10.5. This approach has been successfully applied in other secondary metabolite pathways and shows promise for CPT production enhancement.

10.7 Conclusion

The path toward environmentally friendly camptothecin manufacturing demonstrates how plant biotechnology might advance to address pressing environmental and medical issues. Relying on natural plant sources is neither sufficient nor sustainable, but innovative biotechnological approaches now offer a way forward. Hairy root cultures, improved tissue culture methods, and elicitation strategies have already shown real promise in generating reliable CPT yields. At the same time, deeper insights into the biosynthetic pathway and its regulation are enabling more targeted interventions, from enhancing key genes and transcription factors to minimizing competing pathways. Future success will depend on integrating these approaches with advanced bioreactor technology and process optimization. By uniting these different strands, plant biotechnology not only provides a sustainable supply of camptothecin but also helps safeguard endangered plant species that have long been its natural source.

Future prospects and challenges

CPT biosynthesis integrates the indole pathway (leading to tryptamine) with the iridoid pathway (leading to secologanin/secologanic acid) through strictosidine synthase. While the upstream routes are relatively well defined, the downstream pathway from strictosidine to camptothecin remains a frontier of active research. Key knowledge gaps include the characterization of late biosynthetic intermediates, the identity of downstream tailoring enzymes, and the mechanisms governing interorgan transport and accumulation of CPT. Recent advances in genomics, transcriptomics, and metabolic engineering, combined with elicitation strategies, are rapidly driving new discoveries. Together, these tools hold the potential to reconstitute camptothecin biosynthesis, paving the way for more effective biotechnological applications.

Acknowledgments

Ms. Kishmita Sharma thanks "Bhabha Atomic Research Centre" (BARC), Mumbai, for Senior Research Fellowship. The authors acknowledge Prof. Anand Ballal, Head, NABTD, for his constant encouragement and support.

References

Afridi, M. S. K., Farhan, U., Mushtaq, A., Ahmad, A., Rana, H., Naseer, S., Mushtaq, S., & Ahmad, M. (2025). CRISPR applications in enhancing medicinal plant yields. *Scholars Academic Journal of Biosciences, 13*(08), 1061–1066. https://doi.org/10.36347/sajb.2025.v13i08.002.

Akula, R., & Ravishankar, G. A. (2014). Influence of abiotic stress signals on secondary metabolites in plants. *Plant Signaling & Behavior, 6*(11), 1720–1731. https://doi.org/10.4161/psb.6.11.17613.

Al-Musawi, S., Albukhaty, S., Al-Karagoly, H., Sulaiman, G. M., Jabir, M. S., & Naderi-Manesh, H. (2020). Dextran-coated superparamagnetic nanoparticles modified with folate for targeted drug delivery of camptothecin. *Advances in Natural Sciences: Nanoscience and Nanotechnology, 11*(4). https://doi.org/10.1088/2043-6254/abc75b, https://iopscience.iop.org/article/10.1088/2043-6254/abc75b/pdf.

Alshammari, M. K., Alshehri, M. M., Alshehri, A. M., Alshlali, O. M., Mahzari, A. M., Almalki, H. H., Kulaybi, O. Y., Alghazwni, M. K., Kamal, M., & Imran, M. (2023). Camptothecin loaded nano-delivery systems in the cancer therapeutic domains: A critical examination of the literature. *Journal of Drug Delivery Science and Technology, 79*. https://doi.org/10.1016/j.jddst.2022.104034, http://www.editionsdesante.fr/category.php?id_category=48.

Amna, T., Amina, M., Sharma, P. R., Puri, S. C., Al-Youssef, H. M., Al-Taweel, A. M., & Qazi, G. N. (2012). Effect of precursors feeding and media manipulation on production of novel anticancer pro-drug camptothecin from endophytic fungus. *Brazilian Journal of Microbiology, 43*(4), 1476–1489. https://doi.org/10.1590/S1517-83822012000400032.

Asano, T., Watase, I., Sudo, H., Kitajima, M., Takayama, H., Aimi, N., Yamazaki, M., & Saito, K. (2004). Camptothecin production by in vitro cultures of *Ophiorrhiza liukiuensis* and O. kuroiwai. *Plant Biotechnology, 21*(4), 275–281. https://doi.org/10.5511/plantbiotechnology.21.275.

Awadasseid, A., Li, W., Liu, Z., Qiao, C., Pang, J., Zhang, G., & Luo, Y. (2020). Characterization of *Camptotheca acuminata* 10-hydroxygeraniol oxidoreductase and iridoid synthase and their application in biological preparation of nepetalactol in *Escherichiacoli* featuring NADP+ - NADPH cofactors recycling. *International Journal of Biological Macromolecules, 162*, 1076–1085. https://doi.org/10.1016/j.ijbiomac.2020.06.223.

Asano, T., Kobayashi, K., Kashihara, E., Sudo, H., Sasaki, R., Iijima, Y., Aoki, K., Shibata, D., Saito, K., & Yamazaki, M. (2013). Suppression of camptothecin biosynthetic genes results in metabolic modification of secondary products in hairy roots of Ophiorrhiza pumila. *Phytochemistry, 91*, 128–139. https://doi.org/10.1016/j.phytochem.2012.04.019.

Bai, S., Jia, D., Ma, X., Liang, M., Xue, P., Kang, Y., & Xu, Z. (2021). Cylindrical polymer brushes-anisotropic unimolecular micelle drug delivery system for enhancing the effectiveness of chemotherapy. *Bioactive Materials, 6*(9), 2894–2904. https://doi.org/10.1016/j.bioactmat.2021.02.011.

Banadka, A., Narasimha, S. W., Dandin, V. S., Naik, P. M., Vennapusa, A. R., Melmaiee, K., Vemanna, R. S., Al-Khayri, J. M., Thiruvengadam, M., & Nagella, P. (2024). Biotechnological approaches for the production of camptothecin. *Applied Microbiology and Biotechnology, 108*(1). https://doi.org/10.1007/s00253-024-13187-2, https://www.springer.com/journal/253.

Batool, S., Ali, Z., Alamri, A. H., Al Fatease, A., Lahiq, A. A., Asiri, A., Rehman, Au, & Din, Fu (2025). Development and statistical optimization of camptothecin loaded hyaluronic acid and zein polymeric nanoparticles towards the treatment of melanoma. *International Journal of Biological Macromolecules, 321*. https://doi.org/10.1016/j.ijbiomac.2025.146330, https://www.sciencedirect.com/science/journal/doi.

Bhalkar, B. N., Bedekar, P. A., Patil, S. M., Patil, S. A., & Govindwar, S. P. (2015). Production of camptothecine using whey by an endophytic fungus: Standardization using response surface methodology. *RSC Advances, 5*(77), 62828–62835. https://doi.org/10.1039/c5ra12212k, http://pubs.rsc.org/en/journals/journalissues.

Bhalkar, B. N., Patil, S. M., & Govindwar, S. P. (2016). Camptothecine production by mixed fermentation of two endophytic fungi from *Nothapodytes nimmoniana*. *Fungal Biology, 120*(6-7), 873–883. https://doi.org/10.1016/j.funbio.2016.04.003, http://www.elsevier.com/wps/find/journaldescription.cws_home/720691/description#description.

Bhushan, S., Sharma, D., Rakshant, Kaul, S., Dhar, M. K., & Sharma, M. (2023). Recent strategies to engineer alkaloid biosynthesis in medicinal plants. In *Medicinal plants: Their response to abiotic stress*. India: Springer Nature, 391–416. https://doi.org/10.1007/978-981-19-5611-9, https://doi.org/10.1007/978-981-19-5611-9_15.

Biswas, D., Chakraborty, A., Mukherjee, S., & Ghosh, B. (2023). Hairy root culture: A potent method for improved secondary metabolite production of *Solanaceous plants. Frontiers in Plant Science, 14*. https://doi.org/10.3389/fpls.2023.1197555.

Chang, C., Liu, Z., Wang, Y., Tang, Z., Yu, F., & Mock, H.-P. (2018). A b ZIP transcription factor, CaLMF, mediated light-regulated camptothecin biosynthesis in *Camptotheca acuminata. Tree Physiology, 39*(3). https://doi.org/10.1093/treephys/tpy106.

Cheng, Z., Cheng, Y., Chen, Q., Li, M., Wang, J., Liu, H., Li, M., Ning, Y., Yu, Z., Wang, Y., & Wang, H. (2020). Self-assembly of pentapeptides into morphology-adaptable nanomedicines for enhanced combinatorial chemo-photodynamic therapy. *Nano Today, 33*, 100878. https://doi.org/10.1016/j.nantod.2020.100878.

Chou, H. L., Fong, Y., Wei, C. K., Tsai, E. M., Chen, J. Y. F., Chang, W. T., Wu, C. Y., Huang, H. W., & Chiu, C. C. (2017). A quinone-containing compound enhances camptothecin-induced apoptosis of lung cancer through modulating endogenous ROS and ERK signaling. *Archivum Immunologiae et Therapiae Experimentalis, 65*(3), 241–252. https://doi.org/10.1007/s00005-016-0424-8, http://www.springer.com/sgw/cda/frontpage/0,118555,5-0-70-62708807-0,00.html?ref.

Clements, M. K., Jones, C. B., Cumming, M., & Daoud, S. S. (1999). Antiangiogenic potential of camptothecin and topotecan. *Cancer Chemotherapy and Pharmacology, 44*(5), 411–416. https://doi.org/10.1007/s002800050997.

Cortés, J., Kim, S. B., Chung, W. P., Im, S. A., Park, Y. H., Hegg, R., Kim, M. H., Tseng, L. M., Petry, V., Chung, C. F., Iwata, H., Hamilton, E., Curigliano, G., Xu, B., Huang, C. S., Kim, J. H., Chiu, J. W. Y., Pedrini, J. L., Lee, C., ... Hurvitz, S. A. (2022). Trastuzumab deruxtecan versus trastuzumab emtansine for breast cancer. *New England Journal of Medicine, 386*(12), 1142–1154. https://doi.org/10.1056/NEJMoa2115022, http://www.nejm.org/medical-index.

Cui, L., Ni, X., Ji, Q., Teng, X., Yang, Y., Wu, C., Zekria, D., Zhang, D., & Kai, G. (2015). Co-overexpression of geraniol-10-hydroxylase and strictosidine synthase improves anti-cancer drug camptothecin accumulation in *Ophiorrhiza pumila. Scientific Reports, 5*(1). https://doi.org/10.1038/srep08227.

Deb, D. B., & Mondal, D. C. (2015). Taxonomic revision of the genus <I>Ophiorrhiza L.</I> (Rubiaceae) in Indian subcontinent. *Nelumbo*, 1–148. https://doi.org/10.20324/nelumbo/v39/1997/74298.

Deepthi, S., & Satheeshkumar, K. (2016). Enhanced camptothecin production induced by elicitors in the cell suspension cultures of *Ophiorrhiza mungos* Linn. *Plant Cell, Tissue and Organ Culture (PCTOC, 124*(3), 483–493. https://doi.org/10.1007/s11240-015-0908-y.

Deepthi, S., & Satheeshkumar, K. (2017). Cell line selection combined with jasmonic acid elicitation enhance camptothecin production in cell suspension cultures of *Ophiorrhiza mungos* L. *Applied Microbiology and Biotechnology, 101*(2), 545–558. https://doi.org/10.1007/s00253-016-7808-x.

Degambada, K. D., Peramune Arachchilage, A. S. P. K., Salim, N., Chandrika, U. G., & Abeysekera, A. M. (2023). Cellular localisation and quantification of Camptothecin in different plant parts of *Nothapodytes nimmoniana* (J. Graham) Mabberley of Sri Lankan origin. *Phytochemical Analysis, 34*(4), 453–460. https://doi.org/10.1002/pca.3226, http://onlinelibrary.wiley.com/journal/10.1002/(ISSN)1099-1565.

Doan, T. N., Le, T. D., Ho, N. A., Ho, T. T., Do, T. T., Hoang, H., Nguyen, M. H., Bui, T. M., & Chu, H. H. (2024). Isolation, anticancer potency, and camptothecin—Producing ability of endophytic fungi isolated from Ixora chinensis. *Science Progress, 107*(2). https://doi.org/10.1177/00368504241253675, https://journals.sagepub.com/home/SCI.

Erb, M., & Kliebenstein, D. J. (2020). Plant secondary metabolites as defenses, regulators, and primary metabolites: The blurred functional trichotomy. *Plant Physiology, 184*(1), 39–52. https://doi.org/10.1104/pp.20.00433.

Fan, J., Zhang, K., Xuan, M., Gao, X., Vinokur, R., Göstl, R., Zheng, L., & Herrmann, A. (2024). High-Intensity Focused Ultrasound-Induced Disulfide Mechanophore Activation in Polymeric Nanostructures for Molecule Release. *CCS Chemistry, 6*(8), 1895–1907. https://doi.org/10.31635/ccschem.024.202403876.

Fan, X., Lin, X., Ruan, Q., Wang, J., Yang, Y., Sheng, M., Zhou, W., Kai, G., & Hao, X. (2022). Research progress on the biosynthesis and metabolic engineering of the anti-cancer drug camptothecin in Camptotheca acuminate. *Industrial Crops and Products, 186*, 115270. https://doi.org/10.1016/j.indcrop.2022.115270.

Farhoudi, L., Maryam Hosseinikhah, S., Vahdat-Lasemi, F., Sukhorukov, V. N., Kesharwani, P., & Sahebkar, A. (2024). Polymeric micelles paving the Way: Recent breakthroughs in camptothecin delivery for enhanced chemotherapy. *International Journal of Pharmaceutics, 659*, 124292. https://doi.org/10.1016/j.ijpharm.2024.124292.

Fulzele, D. P., Satdive, R. K., & Pol, B. B. (2001). Growth and production of camptothecin by cell suspension cultures of *Nothapodytes foetida*. *Planta Medica, 67*(2), 150–152. https://doi.org/10.1055/s-2001-11519.

Fulzele, D.P., Satdive, R., Kamble, S., Singh, S., & Singh, S. (2015). Improvement of anticancer drug camptothecin production by gamma irradiation on callus cultures of nothapodytes foetida. *International Journal of Pharmaceutical Research & Allied Sciences, 4*, 1.

Gao, F., Qiu, X., Baddi, S., He, S., Wang, S., Zhao, C., Dou, X., & Feng, C. (2025). Chiral nanofibers of camptothecin trigger pyroptosis for enhanced immunotherapy. *Angewandte Chemie International Edition, 64*(13). https://doi.org/10.1002/anie.202423446.

Gong, J., Zhang, W., & Balthasar, J. P. (2025). Camptothein-based anti-cancer therapies and strategies to improve their therapeutic index. *Cancers, 17*(6), 1032. https://doi.org/10.3390/cancers17061032.

Gonçalves, S., & Romano, A. (2018). *Production of plant secondary metabolites by using biotechnological tools*. IntechOpen. https://doi.org/10.5772/intechopen.76414.

Gunasekera, S. P., Badawi, M. M., Cordell, G. A., Farnsworth, N. R., & Chitnis, M. (1979). Plant anticancer agents x. isolation of camptothecin and 9-methoxycamptothecin from ervatamia heyxeaxa1. *Journal of Natural Products, 42*(5), 475–477. https://doi.org/10.1021/np50005a006.

Hameed, S., Dayyan, S., Maqsood, R., Younas, U., Hessien, M. M., Helal, M. H., El-Bahy, Z. M., & Pervaiz, M. (2025). Fabrication of camptothecin loaded casein nanoparticles for the treatment of colorectal cancer through targeted drug delivery. *Journal of Drug Delivery Science and Technology, 113*. https://doi.org/10.1016/j.jddst.2025.107314, https://www.sciencedirect.com/science/journal/17732247.

Hao, L., Zhou, Q., Piao, Y., Zhou, Z., Tang, J., & Shen, Y. (2021a). Albumin-binding prodrugs via reversible iminoboronate forming nanoparticles for cancer drug delivery. *Journal of Controlled Release, 330*, 362–371. https://doi.org/10.1016/j.jconrel.2020.12.035.

Hao, X., Gai, W., Wang, L., Zhao, J., Sun, D., Yang, F., Jiang, H., & Feng, Y. (2021b). 5-Boronopicolinic acid-functionalized polymeric nanoparticles for targeting drug delivery and enhanced tumor therapy. *Materials Science and Engineering: C, 119*, 111553. https://doi.org/10.1016/j.msec.2020.111553.

Hao, X., Wang, C., Zhou, W., Ruan, Q., Xie, C., Yang, Y., Xiao, C., Cai, Y., Wang, J., Wang, Y., Zhang, X., Maoz, I., & Kai, G. (2023). OpNAC1 transcription factor regulates the biosynthesis of the anticancer drug camptothecin by targeting loganic acid *O*-methyltransferase in *Ophiorrhiza pumila*. *Journal of Integrative Plant Biology, 65*(1), 133–149. https://doi.org/10.1111/jipb.13377.

Hao, X., Xie, C., Ruan, Q., Zhang, X., Wu, C., Han, B., Qian, J., Zhou, W., Nützmann, H.-W., & Kai, G. (2021c). The transcription factor OpWRKY2 positively regulates the biosynthesis of the anticancer drug camptothecin in Ophiorrhiza pumila. *Horticulture Research, 8*(1). https://doi.org/10.1038/s41438-020-00437-3.

Hao, X., Zhong, Y., Nï Tzmann, H.-W., Fu, X., Yan, T., Shen, Q., Chen, M., Ma, Y., Zhao, J., Osbourn, A., Li, L., & Tang, K. (2019). Light-induced artemisinin biosynthesis is regulated by the bZIP transcription factor AaHY5 in *Artemisia annua*. *Plant and Cell Physiology, 60*(8), 1747–1760. https://doi.org/10.1093/pcp/pcz084.

Hemmerlin, A., Hoeffler, J. F., Meyer, O., Tritsch, D., Kagan, I. A., Grosdemange-Billiard, C., Rohmer, M., & Bach, T. J. (2003). Cross-talk between the cytosolic mevalonate and the plastidial methylerythritol phosphate pathways in tobacco bright yellow-2 cells. *Journal of Biological Chemistry, 278*(29), 26666–26676. https://doi.org/10.1074/jbc.M302526200.

Hertzberg, R. P., Caranfa, M. J., & Hecht, S. M. (1989). On the mechanism of topoisomerase i inhibition by camptothecin: Evidence for binding to an enzyme-DNA complex. *Biochemistry, 28*(11), 4629–4638. https://doi.org/10.1021/bi00437a018.

Hsiang, Y. H., Hertzberg, R., Hecht, S., & Liu, L. F. (1985). Camptothecin induces protein-linked DNA breaks via mammalian DNA topoisomerase I. *Journal of Biological Chemistry, 260*(27), 14873–14878. https://doi.org/10.1016/s0021-9258(17)38654-4.

Huang, B., Cheng, J. K., Wu, C. Y., Chen, P. H., Tu, P. S., Fu, Y. S., & Wu, C. H. (2017). Camptothecin promotes the production of nitric oxide that triggers subsequent S-nitrosoproteome-mediated signaling cascades in endothelial cells. *Vascular Pharmacology, 90*, 27–35. https://doi.org/10.1016/j.vph.2015.07.014, http://www.elsevier.com/locate/genpharm.

Hussain, M. S., Fareed, S., Ansari, S., Rahman, M. A., Ahmad, I. Z., & Saeed, M. (2012). Current approaches toward production of secondary plant metabolites. *Journal of Pharmacy And Bioallied Sciences, 4*(1), 10. https://doi.org/10.4103/0975-7406.92725.

Improvement of Anticancer Drug Camptothecin Production by Gamma Irradiation on Callus Cultures of Nothapodytes Foetida. (2015). International Journal of Pharmaceutical Research & Allied Sciences. 4(1), 19–27.

Irmler, S., Schroder, G., St-Pierre, B., Crouch, N. P., Hotze, M., Schmidt, J., Strack, D., Matern, U., & Schroder, J. (2000). Indole alkaloid biosynthesis in *Catharanthus roseus*: new enzyme activities and identification of cytochrome P450 CYP72A1 as secologanin synthase. *The Plant Journal, 24*(6), 797–804. https://doi.org/10.1046/j.1365-313x.2000.00922.x.

Isah, T., Masood, S., & Umar, S. (2022). Biomass and camptothecin production in the calcium chloride elicited and liquid medium overlayed *Nothapodytes nimmoniana* (J. Graham) Mabberly callus cultures. *Vegetos, 35*(1), 104–114. https://doi.org/10.1007/s42535-021-00293-z.

Ishii, M., Iwahana, M., Mitsui, I., Minami, M., Imagawa, S., Tohgo, A., & Ejima, A. (2000). Growth inhibitory effect of a new camptothecin analog, DX-8951f, on various drug-resistant sublines including BCRP-mediated camptothecin derivative-resistant variants derived from the human lung cancer cell line PC-6. *Anti-Cancer Drugs, 11*(5), 353–362. https://doi.org/10.1097/00001813-200006000-00005.

Islam, M. S., Ahmed, M. R., Noman, M., Zhang, Z., Wang, J., Lu, Z., Cai, Y., Ahmed, T., Li, B., Wang, Y., Golam Sarwar, A. K. M., & Wang, J. (2025). Integrating RNA interference and nanotechnology: A transformative approach in plant protection. *Plants, 14*(6). https://doi.org/10.3390/plants14060977, http://www.mdpi.com/journal/plants.

Jin, Z., Yan, T., Chang, C., Liu, Z., Wang, Y., Tang, Z., & Yu, F. (2016). Application of virus-induced gene silencing approach in *Camptotheca acuminata*. *Plant Cell, Tissue and Organ Culture (PCTOC, 126*(3), 533–540. https://doi.org/10.1007/s11240-016-1022-5.

K.G. Jisha, A study on the production of camptothecin from ophiorrhiza mungos and nothapodytes foetida using cell and tissue culture. (2006).

Kai, G., Wu, C., Gen, L., Zhang, L., Cui, L., & Ni, X. (2015). Biosynthesis and biotechnological production of anti-cancer drug Camptothecin. *Phytochemistry Reviews, 14*(3), 525–539. https://doi.org/10.1007/s11101-015-9405-5.

Kamble, S., Gopalakrishnan, R., & Eapen, S. (2011). Production of camptothecin by hairy roots and regenerated transformed shoots of *Ophiorrhiza rugosa* var. decumbens. *Natural Product Research, 25*(18), 1762–1765. https://doi.org/10.1080/14786419.2011.559469.

Kanazawa, T., Taki, H., & Okada, H. (2020). Nose-to-brain drug delivery system with ligand/cell-penetrating peptide-modified polymeric nano-micelles for intracerebral gliomas. *European Journal of Pharmaceutics and Biopharmaceutics, 152*, 85–94. https://doi.org/10.1016/j.ejpb.2020.05.001.

Kang, M., Fu, R., Zhang, P., Lou, S., Yang, X., Chen, Y., Ma, T., Zhang, Y., Xi, Z., & Liu, J. (2021). A chromosome-level *Camptotheca acuminata* genome assembly provides insights into the evolutionary origin of camptothecin biosynthesis. *Nature Communications, 12*(1). https://doi.org/10.1038/s41467-021-23872-9.

Kim, S. H., Hur, B. K., & Byun, S. Y. (1999). Effect of sugar concentration on camptothecin production in cell suspension cultures of *Camptotheca acuminata*. *Biotechnology and Bioprocess Engineering, 4*(4), 277–280. https://doi.org/10.1007/BF02933753, http://www.springerlink.com/content/1226-8372.

Landgraf, M., Lahr, C. A., Kaur, I., Shafiee, A., Sanchez-Herrero, A., Janowicz, P. W., Ravichandran, A., Howard, C. B., Cifuentes-Rius, A., McGovern, J. A., Voelcker, N. H., & Hutmacher, D. W. (2020). Targeted camptothecin delivery via silicon nanoparticles reduces breast cancer metastasis. *Biomaterials, 240*. https://doi.org/10.1016/j.biomaterials.2020.119791, http://www.journals.elsevier.com/biomaterials/.

Lee, J. Y., Goto, E., Yoshida, H., & Hikosaka, S. (2022). Optimal harvest-time to maximize the annual camptothecin production by *Ophiorrhiza pumila* in a plant factory with artificial light. *Journal of Natural Medicines, 76*(4), 865–872. https://doi.org/10.1007/s11418-022-01634-1, http://www.springerlink.com/content/1861-0293/.

Lee, Y. H., & Huang, C. Y. (2025). Engineered perfluorochemical cancer-derived exosomes loaded with indocyanine green and camptothecin provide targeted photochemotherapy for effective cancer treatment. *International Journal of Nanomedicine, 20*, 327–342. https://doi.org/10.2147/IJN.S505458, https://www.dovepress.com/article/download/99068.

Li, F., Jiang, T., Li, Q., & Ling, X. (2017). Camptothecin (CPT) and its derivatives are known to target topoisomerase I (Top1) as their mechanism of action: Did we miss something in CPT analogue molecular targets for treating human disease such as cancer? *American Journal of Cancer Research, 7*(12), 2350–2394. http://www.ajcr.us/files/ajcr0067986.pdf.

Li, Y., Zheng, Y., Liu, T., Zhao, J., Zhou, L., Yang, X., Yao, Z., Wang, X., Fu, Y., Wang, J., Yuan, K., & He, Z. (2025). A soluble triantennary *N*-acetylgalactosamine camptothecin prodrug for hepatocellular carcinoma-targeted therapy. *Journal of Medicinal Chemistry*. https://doi.org/10.1021/acs.jmedchem.5c00466.

Li, Y., Zhou, Q., Fan, X., Wang, Y., Wang, C., Ruan, Q., Wu, Z., Shao, L., Li, L., Sheng, M., Tang, X., Chen, H., Hao, X., & Kai, G. (2025). Functional identification of 10-hydroxygeraniol oxidoreductases related to camptothecin biosynthesis in Ophiorrhiza pumila. *International Journal of Biological Macromolecules, 318*, 145253. https://doi.org/10.1016/j.ijbiomac.2025.145253.

Li, Z., Yang, J., Ren, B., Fan, Q., Huang, L., Guo, S., Zhou, R., Chen, S., Feng, J., Yan, C., Chen, X., & Shen, Z. (2024). Double-layered hollow mesoporous cuprous oxide nanoparticles for double drug squential therapy of tumors. *Advanced Materials, 36*(28). https://doi.org/10.1002/adma.202313212.

Li, Q. Y., Zu, Y. G., Shi, R. Z., & Yao, L. P. (2006). Review camptothecin: Current perspectives. *Current Medicinal Chemistry, 13*(17), 2021–2039. https://doi.org/10.2174/092986706777585004China, http://docserver.ingentaconnect.com/deliver/connect/ben/09298673/v13n17/s5.pdf?expires=1152709534&id=30451945&titleid=3863&accname=Elsevier&checksum=3687FA26CC6461236CEE4A31E31D6A7D.

Liu, Y. Q., Li, W. Q., Morris-Natschke, S. L., Qian, K., Yang, L., Zhu, G. X., Wu, X. B., Chen, A. L., Zhang, S. Y., Nan, X., & Lee, K. H. (2015). Perspectives on biologically active camptothecin derivatives. *Medicinal Research Reviews, 35*(4), 753–789. https://doi.org/10.1002/med.21342, http://onlinelibrary.wiley.com/journal/10.1002/(ISSN)1098-1128.

Liu, Z., & Adams, J. (1996). Camptothecin yield and distribution within *Camptotheca acuminata* trees cultivated in Louisiana. *Canadian Journal of Botany, 74*(3), 360–365. https://doi.org/10.1139/b96-045.

López-Meyer, M., & Nessler, C. L. (1997). Tryptophan decarboxylase is encoded by two autonomously regulated genes in *Camptotheca acuminata* which are differentially expressed during development and stress. *Plant Journal, 11*(6), 1167–1175. https://doi.org/10.1046/j.1365-313X.1997.11061167.x.

Lorence, A., & Nessler, C. L. (2004). Camptothecin, over four decades of surprising findings. *Phytochemistry, 65*(20), 2735–2749. https://doi.org/10.1016/j.phytochem.2004.09.001, http://www.journals.elsevier.com/phytochemistry/.

Lu, H., Gorman, E., & McKnight, T. D. (2005). Molecular characterization of two anthranilate synthase alpha subunit genes in *Camptotheca acuminata*. *Planta, 221*(3), 352–360. https://doi.org/10.1007/s00425-004-1450-z.

Lu, H., & McKnight, T. D. (1999). Tissue-specific expression of the β-subunit of tryptophan synthase in camptotheca acuminata, an indole alkaloid-producing plant1. *Plant Physiology, 120*(1), 43–52. https://doi.org/10.1104/pp.120.1.43.

Lu, Y., Wang, H., Wang, W., Qian, Z., Li, L., Wang, J., Zhou, G., & Kai, G. (2009). Molecular characterization and expression analysis of a new cDNA encoding strictosidine synthase from *Ophiorrhiza japonica*. *Molecular Biology Reports, 36*(7), 1845–1852. https://doi.org/10.1007/s11033-008-9389-y.

Lu, L., Zhao, X., Fu, T., Li, K., He, Y., Luo, Z., Dai, L., Zeng, R., & Cai, K. (2020). An iRGD-conjugated prodrug micelle with blood-brain-barrier penetrability for anti-glioma therapy. *Biomaterials, 230*, 119666. https://doi.org/10.1016/j.biomaterials.2019.119666.

Lu, Y., Jia, D., Ma, X., Liang, M., Hou, S., Qiu, W., Gao, Y., Xue, P., Kang, Y., & Xu, Z. (2021). Reduction-Responsive Chemo-Capsule-Based Prodrug Nanogel for Synergistic Treatment of Tumor Chemotherapy. *ACS Applied Materials & Interfaces, 13*(7), 8940–8951. https://doi.org/10.1021/acsami.0c21710.

Ma, X., Panjikar, S., Koepke, J., Loris, E., & Stöckigt, J. (2006). The structure of *Rauvolfia serpentina* strictosidine synthase is a novel six-bladed β-propeller fold in plant proteins. *The Plant Cell, 18*(4), 907–920. https://doi.org/10.1105/tpc.105.038018, http://www.plantcell.org/.

Mai, R., Deng, B., Zhao, H., Li, L., Fang, Y., Li, S., Deng, X., & Chen, J. (2022). Design, synthesis, and bioevaluation of novel enzyme-triggerable cell penetrating peptide-based dendrimers for targeted delivery of camptothecin and cancer therapy. *Journal of Medicinal Chemistry, 65*(7), 5850–5865. https://doi.org/10.1021/acs.jmedchem.2c00287.

Malik SS, L. J (2014). *Distribution of Camptothecin through the Plant Kingdom. International Journal of Current Research*, 6(05), 6497–6507.

Miller, J. C., Hollatz, A. J., & Schuler, M. A. (2021). P450 variations bifurcate the early terpene indole alkaloid pathway in Catharanthus roseus and Camptotheca acuminata. *Phytochemistry, 183*, 112626. https://doi.org/10.1016/j.phytochem.2020.112626.

Mingzhang, A., Jing, W., Yue, S., Wentao, G., & Longjiang, Y. (2011). Camptothecin distribution and content in*Nothapodytes nimmoniana. Natural Product Communications, 6*(2). https://doi.org/10.1177/1934578x1100600210.

Mohinudeen, I. A. H. K., Pandey, S., Kanniyappan, H., Muthuvijayan, V., & Srivastava, S. (2021). Screening and selection of camptothecin producing endophytes from*Nothapodytes nimmoniana. Scientific Reports, 11*(1). https://doi.org/10.1038/s41598-021-90778-3, http://www.nature.com/srep/index.html.

Murali, S., Ibrahim, M., Rajendran, H., Shagun, S., Masakapalli, S. K., Raman, K., & Srivastava, S. (2023). Genome-scale metabolic model led engineering of *Nothapodytes nimmoniana* plant cells for high camptothecin production. *Frontiers in Plant Science, 14*. https://doi.org/10.3389/fpls.2023.1207218, https://www.frontiersin.org/journals/plant-science.

Nagesha, B. V. (2018). Elicitors act as a signal transducer in the enhancement of camptothecin production from in vitro cultures of *Ophiorrhiza mungos* L. *Annals of Phytomedicine, 7*(2), 46–54.

Namdeo, A. G. (2007). Plant cell elicitation for production of secondary metabolites: A review. *Pharmacognosy Reviews, 1*.

Nandi, A., Halder, R., Patra, D., Hussain, A., Roy, S., Das Sarma, J., & Shunmugam, R. (2025). Unique polyester-based biodegradable multiarm nano-carrier for cancer cell specific mitochondrial delivery of chemotherapeutics. *Biomacromolecules, 26*(8), 5182–5194. https://doi.org/10.1021/acs.biomac.5c00601.

Nguyen, T. A. M., Nguyen, T. D., Leung, Y. Y., McConnachie, M., Sannikov, O., Xia, Z., & Dang, T. T. T. (2021). Discovering and harnessing oxidative enzymes for chemoenzymatic synthesis and diversification of anticancer camptothecin analogues. *Communications Chemistry, 4*(1). https://doi.org/10.1038/s42004-021-00602-2 Available from: nature.com/commschem/.

Ni, X., Wen, S., Wang, X., Wang, W., Xu, H., & Kai, G. (2011). Enhancement of camptothecin production in *Camptotheca acuminata* hairy roots by overexpressing ORCA3 gene. *Journal of Applied Pharmaceutical Science, 1*(8), 85–88. http://www.japsonline.com/admin/php/uploads/217_pdf.pdf China.

Ozyigit, I. I., Dogan, I., Hocaoglu-Ozyigit, A., Yalcin, B., Erdogan, A., Yalcin, I. E., Cabi, E., & Kaya, Y. (2023). Production of secondary metabolites using tissue culture-based biotechnological applications. *Frontiers in Plant Science, 14*. https://doi.org/10.3389/fpls.2023.1132555, https://www.frontiersin.org/journals/plant-science.

Padhi, S., & Behera, A. (2020). *Nanotechnology based targeting strategies for the delivery of camptothecin*. Springer Science and Business Media LLC, 243–272. https://doi.org/10.1007/978-3-030-41842-7_7.

Pai, S. R., Pawar, N. V., Nimbalkar, M. S., Kshirsagar, P. R., Kolar, F. K., & Dixit, G. B. (2013). Seasonal variation in content of camptothecin from the bark of *Nothapodytes nimmoniana* (Grah.) Mabb., using HPLC analysis. *Pharmacognosy Research, 5*(3), 219. https://doi.org/10.4103/0974-8490.112434.

Pan, X., Chen, M., Liu, Y., Wang, Q., Zeng, L., Li, L., & Liao, Z. (2008). A new isopentenyl diphosphate isomerase gene from*Camptotheca acuminata*: Cloning, characterization and functional expression in*Escherichia coli. DNA Sequence, 19*(2), 98–105. https://doi.org/10.1080/10425170701446509.

Pan, X.-w., Xu, H.-h., liu, X., Gao, X., & lu, Y.-t. (2004). Improvement of growth and camptothecin yield by altering nitrogen source supply in cell suspension cultures of Camptotheca acuminata. *Biotechnology Letters, 26*(22), 1745–1748. https://doi.org/10.1007/s10529-004-4580-2.

Pandey, D. K., Konjengbam, M., Ghorai, M., Dwivedi, P., Roy, D., Kant, N., Gangaprasad, A., & Dey, A. (2022). Biotechnology for micropropagation and camptothecin production in *Ophiorrhiza* sp. *Applied Microbiology and Biotechnology, 106*(11), 3851–3877. https://doi.org/10.1007/s00253-022-11941-y, https://www.springer.com/journal/253.

Pantazis, P., Han, Z., Chatterjee, D., & Wyche, J. (1999). Water-insoluble camptothecin analogues as potential antiviral drugs. *Journal of Biomedical Science, 6*(1), 1–7. https://doi.org/10.1007/bf02256417.

Paul, S., De, A., Kirar, S., Thakur, N. S., Pujari, A. K., Gogde, K., & Bhaumik, J. (2024). Co-administration of chemo-phototherapeutic loaded lignin nanoarchitecture for the treatment of skin cancer cell lines and bacterial infections. *ACS Applied Nano Materials, 7*(9), 10984–10997. https://doi.org/10.1021/acsanm.4c01936, https://pubs.acs.org/journal/aanmf6.

Pu, X., Gao, H. C., Wang, M. J., Zhang, J. H., Shan, J. H., Chen, M. H., Zhang, L., Wang, H. G., Wen, A. X., Luo, Y. G., & Huang, Q. M. (2022). Integrative analysis of elicitor-induced camptothecin biosynthesis in camptotheca acuminata plantlets through a combined omics approach. *Frontiers in Plant Science, 13*. https://doi.org/10.3389/fpls.2022.851077, https://www.frontiersin.org/journals/plant-science.

Pu, X., Qu, X., Chen, F., Bao, J., Zhang, G., & Luo, Y. (2013). Camptothecin-producing endophytic fungus Trichoderma atroviride LY357: isolation, identification, and fermentation conditions optimization for camptothecin production. *Applied Microbiology and Biotechnology, 97*(21), 9365–9375. https://doi.org/10.1007/s00253-013-5163-8.

Rani, J. (2011). Study on the production of camptothecin and its derivatives from selected Ophiorrhiza species and its in vitro culture. *Amala Cancer Research Centre*.

Rapisarda, A., Uranchimeg, B., Scudiero, D. A., Selby, M., Sausville, E. A., Shoemaker, R. H., & Melillo, G. (2002). Identification of small molecule inhibitors of hypoxia-inducible factor 1 transcriptional activation pathway. *Cancer Research, 62*(15), 4316–4324.

Rather, G. A., Sharma, A., Misra, P., Kumar, A., Kaul, V., & Lattoo, S. K. (2020). Molecular characterization and overexpression analyses of secologanin synthase to understand the regulation of camptothecin biosynthesis in *Nothapodytes nimmoniana* (Graham.) Mabb. *Protoplasma, 257*(2), 391–405. https://doi.org/10.1007/s00709-019-01440-9, http://www.springer.co.at/springer.py?Page=40&Key=456&cat=9&id_journal=4.

Rehman, S., Shawl, A. S., Kour, A., Sultan, P., Ahmad, K., Khajuria, R., & Qazi, G. N. (2009). Comparative studies and identification of camptothecin produced by an endophyte at shake flask and bioreactor. *Natural Product Research, 23*(11), 1050–1057. https://doi.org/10.1080/14786410902750944.

Reshi, Z. A., Ahmad, W., Lukatkin, A. S., & Javed, S. B. (2023). From nature to lab: A review of secondary metabolite biosynthetic pathways, environmental influences, and in vitro approaches. *Metabolites, 13*(8). https://doi.org/10.3390/metabo13080895, http://www.mdpi.com/journal/metabolites.

Rohani, E. R., Chiba, M., Kawaharada, M., Asano, T., Oshima, Y., Mitsuda, N., Ohme-Takagi, M., Fukushima, A., Rai, A., Saito, K., & Yamazaki, M. (2016). An MYB transcription factor regulating specialized metabolisms in *Ophiorrhiza pumila*. *Plant Biotechnology, 33*(1), 1–9. https://doi.org/10.5511/plantbiotechnology.15.1117a, https://www.jstage.jst.go.jp/article/plantbiotechnology/33/1/33_15.1117a/_pdf.

Rohmer, M., & Rohmer, M. (1999). The discovery of a mevalonate-independent pathway for isoprenoid biosynthesis in bacteria, algae and higher plants† *Natural Product Reports, 16*(5), 565–574. https://doi.org/10.1039/a709175c.

Roja, G., & Heble, M. R. (1994). The quinoline alkaloids camptothecin and 9-methoxycamptothecin from tissue cultures and mature trees of *Nothapodytes foetida*. *Phytochemistry, 36*(1), 65–66. https://doi.org/10.1016/s0031-9422(00)97013-4.

Sadre, R., Magallanes-Lundback, M., Pradhan, S., Salim, V., Mesberg, A., Jones, A. D., & Dellapenna, D. (2016). Metabolite diversity in alkaloid biosynthesis: A multilane (diastereomer) highway for camptothecin synthesis in *Camptotheca acuminata*. *The Plant Cell, 28*(8), 1926–1944. https://doi.org/10.1105/tpc.16.00193, http://www.plantcell.org/content/28/8/1926.full.pdf.

Saito, K., Sudo, H., Yamazaki, M., Koseki-Nakamura, M., Kitajima, M., Takayama, H., & Aimi, N. (2001). Feasible production of camptothecin by hairy root culture of *Ophiorrhiza pumila*. *Plant Cell Reports, 20*(3), 267–271. https://doi.org/10.1007/s002990100320.

Sasidharan, H., Chembrakuniyil, M., & Krishnan, S. (2023). Enhancement of camptothecin biosynthesis in the green root cultures of *Pyrenacantha volubilis* Hook. under varied concentrations of ammonium nitrate and sucrose along with exogenous elicitors. *Industrial Crops and Products, 192*, 116000. https://doi.org/10.1016/j.indcrop.2022.116000.

Schöffski, P., Wang, C. C., Schöffski, M. P., & Wozniak, A. (2024). Current role of topoisomerase i inhibitors for the treatment of mesenchymal malignancies and their potential future use as payload of sarcoma-specific antibody-drug conjugates. *S. Karger AG, Belgium Oncology Research and Treatment, 47*(1-2), 18–41. https://doi.org/10.1159/000535491, http://www.karger.com/Journal/Home/224106.

Sharma, K., Satdive, R. K., & Singh, S. (2024). Exploring the influence of organ differentiation on biosynthesis and accumulation of camptothecin in *Ophiorrhiza rugosa* var. decumbens. *3 Biotech, 14*(6). https://doi.org/10.1007/s13205-024-03999-4, https://www.springer.com/journal/13205.

Shi, M., Gong, H., Cui, L., Wang, Q., Wang, C., Wang, Y., & Kai, G. (2020). Targeted metabolic engineering of committed steps improves anti-cancer drug camptothecin production in *Ophiorrhiza pumila* hairy roots. *Industrial Crops and Products, 148*, 112277. https://doi.org/10.1016/j.indcrop.2020.112277.

Singh, S., Kamble, S. N., Satdive, R. K., & Fulzele, D. P. (2020). Heterologous overexpression of *Nothapodytes foetida* strictosidine synthase enhances levels of anti-cancer compound camptothecin in *Ophiorrhiza rugosa*. *Plant Cell, Tissue and Organ Culture, 141*(1), 67–76. https://doi.org/10.1007/s11240-020-01767-9, http://www.springer.com/journal/11240.

Sirikantaramas, S., Yamazaki, M., & Saito, K. (2013). Camptothecin: Biosynthesis, biotechnological production and resistance mechanism(s). *Advances in Botanical Research, 68*, 139–161. https://doi.org/10.1016/B978-0-12-408061-4.00005-5, http://www.elsevier.com/wps/find/bookdescription.cws_home/675801/description#description.

Sirikantaramas, S., Asano, T., Sudo, H., Yamazaki, M., & Saito, K. (2007). Camptothecin: Therapeutic potential and biotechnology. *Current Pharmaceutical Biotechnology, 8*(4), 196–202. https://doi.org/10.2174/138920107781387447.

Song, S. H., & Byun, S. Y. (1998). Elicitation of camptothecin production in cell cultures of *Camptotheca acuminata*. *Biotechnology and Bioprocess Engineering, 3*(2), 91–95. https://doi.org/10.1007/BF02932509, http://www.springerlink.com/content/1226-8372.

Sudo, H., Yamakawa, T., Yamazaki, M., Aimi, N., & Saito, K. (2002). Bioreactor production of camptothecin by hairy root cultures of Ophiorrhiza pumila. *Biotechnology Letters, 24*(5), 359–363. https://doi.org/10.1023/a:1014568904957.

Sukumar, K., Bharathi, M., Hirad, A. H., Alarfaj, A. A., Hussein-Al-Ali, S. H., & Surya, P. (2025). Development of chitosan-coated graphene oxide and iron oxide nanocomposites for targeted delivery of camptothecin to liver cancer cells. *Chemistry & Biodiversity, 22*(2). https://doi.org/10.1002/cbdv.202401817.

Sun, R., Luo, Q., Li, X., Huang, X., Teng, L., Shen, Z., & Zhu, W. (2021). Supramolecular PEGylation of camptothecin for cancer therapy. *Materials Today Nano, 14*, 100115. https://doi.org/10.1016/j.mtnano.2021.100115.

Svejstrup, J. Q., Christiansen, K., Gromova, I. I., Andersen, A. H., & Westergaard, O. (1991). New technique for uncoupling the cleavage and religation reactions of eukaryotic topoisomerase I. *Journal of Molecular Biology, 222*(3), 669–678. https://doi.org/10.1016/0022-2836(91)90503-x.

Teoh, E. S. (2016). *Secondary metabolites of plants*. Springer Science and Business Media LLC, 59–73. https://doi.org/10.1007/978-3-319-24274-3_5.

Terrier, B., Courtois, D., Hénault, N., Cuvier, A., Bastin, M., Aknin, A., Dubreuil, J., & Pétiard, V. (2007). Two new disposable bioreactors for plant cell culture: The wave and undertow bioreactor and the slug bubble bioreactor. *Biotechnology and Bioengineering, 96*(5), 914–923. https://doi.org/10.1002/bit.21187.

Thengane, S. R., Kulkarni, D. K., Shrikhande, V. A., Joshi, S. P., Sonawane, K. B., & Krishnamurthy, K. V. (2003). Influence of medium composition on callus induction and camptothecin(s) accumulation in *Nothapodytes foetida*. *Plant Cell, Tissue and Organ Culture, 72*(3), 247–251. https://doi.org/10.1023/a:1022392929406.

Titova, M., Popova, E., & Nosov, A. (2024). Bioreactor systems for plant cell cultivation at the institute of plant physiology of the russian academy of sciences: 50 years of technology evolution from laboratory to industrial implications. *Plants, 13*(3), 430. https://doi.org/10.3390/plants13030430.

Treimer, J. F., & Zenk, M. H. (1979). Purification and properties of strictosidine synthase, the key enzyme in indole alkaloid formation. *European Journal of Biochemistry, 101*(1), 225–233. https://doi.org/10.1111/j.1432-1033.1979.tb04235.x.

Tsao, Y. P., D'Arpa, P., & Liu, L. F. (1992). The involvement of active DNA synthesis in camptothecin-induced G2 arrest: altered regulation of p34cdc2/cyclin B. *Cancer Res, 52*(7), 1823–1829.

Uday Bhanu, M., & Kondapi, A. K. (2010). Neurotoxic activity of a topoisomerase-I inhibitor, camptothecin, in cultured cerebellar granule neurons. *Neurotoxicology, 31*(6), 730–737. https://doi.org/10.1016/j.neuro.2010.06.008.

Udomsom, N., Rai, A., Suzuki, H., Okuyama, J., Imai, R., Mori, T., Nakabayashi, R., Saito, K., & Yamazaki, M. (2016). Function of AP2/ERF transcription factors involved in the regulation of specialized metabolism in ophiorrhiza pumila revealed by transcriptomics and metabolomics. *Frontiers in Plant Science, 7*, 2016. https://doi.org/10.3389/fpls.2016.01861.

Van Hengel, A.J., Buitelaar, R.M., & Wichers, H.J. (1994). Camptotheca acuminata Decne: in vitro culture and the production of camptothecin. *In Medicinal and aromatic plants VII*, 98–112.

Venugopalan, A., Potunuru, U. R., Dixit, M., & Srivastava, S. (2016). Effect of fermentation parameters, elicitors and precursors on camptothecin production from the endophyte*Fusarium solani*. *Bioresource Technology, 206*, 104–111. https://doi.org/10.1016/j.biortech.2016.01.079, http://www.elsevier.com/locate/biortech.

Venugopalan, A., & Srivastava, S. (2015). Enhanced camptothecin production by ethanol addition in the suspension culture of the endophyte, *Fusarium solani*. *Bioresource Technology, 188*, 251–257. https://doi.org/10.1016/j.biortech.2014.12.106.

Wall, M. E., & Wani, M. C. (1996). Discovery to clinic a. *Annals of the New York Academy of Sciences, 803*(1), 1–12. https://doi.org/10.1111/j.1749-6632.1996.tb26371.x.

Wall, M. E., Wani, M. C., Cook, C. E., Palmer, K. H., McPhail, A. T., & Sim, G. A. (1966). Plant antitumor agents. I. The isolation and structure of camptothecin, a novel alkaloidal leukemia and tumor inhibitor from *Camptotheca acuminata* 1,2. *Journal of the American Chemical Society, 88*(16), 3888–3890. https://doi.org/10.1021/ja00968a057.

Wang, C., Hao, X., Wang, Y., Maoz, I., Zhou, W., Zhou, Z., & Kai, G. (2022). Identification of WRKY transcription factors involved in regulating the biosynthesis of the anti-cancer drug camptothecin in *Ophiorrhiza pumila*. *Horticulture Research, 9*. https://doi.org/10.1093/hr/uhac099.

Wang, C., Wu, C., Wang, Y., Xie, C., Shi, M., Nile, S., Zhou, Z., & Kai, G. (2019). Transcription factor OpWRKY3 is involved in the development and biosynthesis of camptothecin and its precursors in ophiorrhiza pumila hairy roots. *International Journal of Molecular Sciences, 20*(16), 3996. https://doi.org/10.3390/ijms20163996.

Wang, J., Li, Y., Yang, Y., Xiao, C., Ruan, Q., Li, P., Zhou, Q., Sheng, M., Hao, X., & Kai, G. (2023). Comprehensive analysis of OpHD-ZIP transcription factors related to the regulation of camptothecin biosynthesis in Ophiorrhiza pumila. *International Journal of Biological Macromolecules, 242*, 124910. https://doi.org/10.1016/j.ijbiomac.2023.124910.

Wang, J., Muhammad, N., Li, T., Wang, H., Liu, Y., Liu, B., & Zhan, H. (2020). Hyaluronic acid-coated camptothecin nanocrystals for targeted drug delivery to enhance anticancer efficacy. *Molecular Pharmaceutics, 17*(7), 2411–2425. https://doi.org/10.1021/acs.molpharmaceut.0c00161.

Yamazaki, M., Mochida, K., Asano, T., Nakabayashi, R., Chiba, M., Udomson, N., Yamazaki, Y., Goodenowe, D. B., Sankawa, U., Yoshida, T., Toyoda, A., Totoki, Y., Sakaki, Y., Góngora-Castillo, E., Buell, C. R., Sakurai, T., & Saito, K. (2013). Coupling deep transcriptome analysis with untargeted metabolic profiling in *Ophiorrhiza Pumila* to further the understanding of the biosynthesis of the anti-cancer alkaloid camptothecin and anthraquinones. *Plant and Cell Physiology, 54*(5), 686–696. https://doi.org/10.1093/pcp/pct040.

Yamazaki, Y., Kitajima, M., Arita, M., Takayama, H., Sudo, H., Yamazaki, M., Aimi, N., & Saito, K. (2004). Biosynthesis of camptothecin. In Silico and in Vivo tracer study from [1-13C]Glucose. *Plant Physiology, 134*(1), 161–170. https://doi.org/10.1104/pp.103.029389.

Yamazaki, Y., Sudo, H., Yamazaki, M., Aimi, N., & Saito, K. (2003a). Camptothecin biosynthetic genes in hairy roots of ophiorrhiza pumila: Cloning, characterization and differential expression in tissues and by stress compounds. *Plant and Cell Physiology, 44*(4), 395–403. https://doi.org/10.1093/pcp/pcg051.

Yamazaki, Y., Urano, A., Sudo, H., Kitajima, M., Takayama, H., Yamazaki, M., Aimi, N., & Saito, K. (2003b). Metabolite profiling of alkaloids and strictosidine synthase activity in camptothecin producing plants. *Phytochemistry, 62*(3), 461–470. https://doi.org/10.1016/s0031-9422(02)00543-5.

Yang, M., Wang, Q., Liu, Y., Hao, X., Wang, C., Liang, Y., Chen, J., Xiao, Y., & Kai, G. (2021). Divergent camptothecin biosynthetic pathway in Ophiorrhiza pumila. *BMC Biology, 19*(1). https://doi.org/10.1186/s12915-021-01051-y.

Yang, Y., Li, W., Pang, J., Jiang, L., Qu, X., Pu, X., Zhang, G., & Luo, Y. (2019). Bifunctional cytochrome P450 enzymes involved in camptothecin biosynthesis. *ACS Chemical Biology, 14*(6), 1091–1096. https://doi.org/10.1021/acschembio.8b01124.

Ya-ut, P., Chareonsap, P., & Sukrong, S. (2011). Micropropagation and hairy root culture of *Ophiorrhiza alata* craib for camptothecin production. *Biotechnology Letters, 33*(12), 2519–2526. https://doi.org/10.1007/s10529-011-0717-2.

You, D., Feng, Y., Wang, C., Sun, C., Wang, Y., Zhao, D., & Kai, G. (2021). Cloning, characterization, and enzymatic identification of a new tryptophan decarboxylase from *Ophiorrhiza pumila*. *Biotechnology and Applied Biochemistry, 68*(2), 381–389. https://doi.org/10.1002/bab.1935.

Zhang, T., Wang, Y., Wu, S., Tian, E., Yang, C., Zhou, Z., Yan, X., & Wang, P. (2023). Chemoproteomics reveals the epoxidase enzyme for the biosynthesis of camptothecin precursor strictosamide epoxide. bioRxiv. Available from: https://www.biorxiv.org. doi: https://doi.org/10.1101/2023.07.26.550496.

Zhou, Q., Li, Y., Peng, Y., Wang, J., Fan, X., Li, L., Sheng, M., Wang, Y., Hao, X., & Kai, G. (2025). The ERF transcription factor OpCBR1 balances root growth and biosynthesis of the anticancer drug camptothecin in *Ophiorrhiza pumila*. *Industrial Crops and Products, 232*, 121259. https://doi.org/10.1016/j.indcrop.2025.121259.

Zulfiqar, S., Farooq, M. A., Zhao, T., Wang, P. P., Tabusam, J., Wang, Y., Xuan, S., Zhao, J., Chen, X., Shen, S., & Gu, A. (2023). Virus-induced gene silencing (VIGS): A powerful tool for crop improvement and its advancement towards epigenetics. *International Journal of Molecular Sciences, 24*(6), 5608. https://doi.org/10.3390/ijms24065608.

CHAPTER

11

Rhizobium rhizogenes-mediated genetic transformation for augmented production of high-valued secondary metabolites in medicinal plants

Sayani Sanyamat[1], *Hiranmoy Bhattacharya*[3], *Somroop Chakravarti*[1], *Para Ganguly*[1], *Saikat Dewanjee*[3], *and Moumita Gangopadhyay*[1,2]

[1]Department of Biotechnology, School of Life Sciences and Biotechnology, Adamas University, Kolkata, West Bengal, India [2]Agrotechnology and Rural Development Centre, Adamas University, Kolkata, West Bengal, India [3]Advanced Pharmacognosy Research Laboratory, Department of Pharmaceutical Technology, Jadavpur University, Kolkata, West Bengal, India

11.1 Introduction

Since the beginning of human civilization, plants remain inseparable sources of healthcare for all religions and cultures due to the presence of a huge array of structurally diverse bioactive compounds, called secondary metabolites (Dasgupta, 2023). From ancient times to recent times, natural products have always represented a keystone in the pharmaceutical, nutraceutical, and cosmetic industries. It is postulated that about 80 % of the entire world's population depends on a traditional system of medicine for meeting basic health needs (Habibi et al., 2017). Moreover, 40 % of allopathic medicines including anticancer drugs are derived from medicinal plants (WHO, 2019).

Plants' underground root parts are considered as a storehouse of many therapeutically important bioactive substances (Mathew et al., 2002), which are used as potential ingredients of many life-saving drugs (Habibi et al., 2017). The global herbal medicine market, valued at more than USD 200 billion in 2022 and projected to grow at a compound annual growth rate of 6–7%, illustrates the rapid expansion in demand for herbal metabolites (Zamani et al., 2025). Increasing demand forplant-based medicine gains importance due to the consumer's

Medicinal Plant Biotechnology for Ameliorated Secondary Metabolite Production
DOI: https://doi.org/10.1016/B978-0-443-43944-5.00015-3

growing concerns regarding the adverse effects associated with synthetic pharmaceuticals (Ebadi et al., 2017; Mokhtarikhah et al., 2022). In this context, the application of biotechnological techniques using plant cell culture provides an alternative strategy to supply continuous high-quality source materials to industry without hampering the natural gene pool (Gangopadhyay et al., 2011). Among various types of cell culture, *Rhizobium rhizogenes*, previously known *Agrobacterium rhizogenes*, mediated transformation, which results in the development of hairy root culture seems to be a prospective need-based technology for the synthesis of root-specific plant secondary metabolites with therapeutic importance (Ionkova, 2007; Shanks & Morgan, 1999). *Rhizobium rhizogenes*, a gram-negative soil-grown bacterium, is the causal organism of induction of hairy roots in dicotyledonous plants (Ying et al., 2023). Among various types of plant cell culture, hairy root culture gained priority due to high proliferating characteristics and maintenance of genetic and biochemical integrity over other cultures (Merkel et al., 1997; Rhodes et al., 1990), which may be influential for the future of herbal industry inventions.

The advancement of plant metabolic engineering with integrative approach of bioprocess technology bridges the demand–supply gap, which not only makes possible the transfer of technology from lab to market butalso enables ecological sustainability by conserving highly valued endangered germplasm of therapeutic plants. Artificial intelligence (AI)-based digital agriculture using machine learning approach and sensor-enabled monitoring strategies is transforming hairy root production systems into more robust, precise, and intelligent processes capable of meeting worldwide demand for important therapeutic root-derived metabolites.

This chapter delves into *Rhizobium rhizogenes*–mediated genetic transformation as an important, realistically implemented technology for root-derived therapeutic metabolites with commercial interest. Augmentation of metabolites of interest modulating physico-chemical conditions for applying external stimulus are briefly discussed. Biosynthetic pathway reprogramming using metabolic engineering reveals innovative possibilities for enhancing synthesis of products of interest. For pilot-scale production, bioreactor-assisted scale-up and product recovery are discussed in detail. Finally, application of digital technologies by AI-driven and sensor-based monitoring establishes hairy root culture as a sustainable and economically viable platform for the production of therapeutic root-derived plant secondary metabolites.

11.2 Market demand for root-derived plant secondary metabolites used in herbal industry

The most abundant sources of physiologically active substances are plant-based natural products and their semisynthetic derivatives (John, 2009). Terpenoids, polyphenols, and alkaloids are examples of plant secondary metabolites that affect vegetable color, shield plants from herbivores and microbes, and function as signal molecules in stressful situations (Fraenkel, 1959; Al-Khayri et al., 2023). Plant secondary metabolites are used in a variety of industries, including agricultural, cosmetic, nutrition, and pharmaceutical industries. On a commercial scale, root cultures are primarily used to produce significant amounts of secondary metabolites (Choi et al., 2000; Paek et al., 2009; Baque, Elgirban, et al., 2012; Baque, Moh, et al., 2012). Many medicinally significant plants' roots are the source of a variety of bioactive compounds, including proteins, metabolites, agrochemicals, flavors, perfumes, and colors (Bais et al., 2001; Fulzele et al., 2002). The formation and expansion of the hairy root culture of plant species to enhance secondary metabolite production have been documented in a few studies (Gangopadhyay et al., 2011; Pillai et al., 2015). Hence, roots of plants can be potentially exploited as rich sources of secondary metabolites, and biotechnological interventions for root development may give a viable way for enhancing secondary metabolite production and use in pharmaceutical, nutraceutical, and cosmetic industries.

11.2.1 Significance in pharmaceutical, nutraceutical, and cosmetic industries

Since ancient times, plant secondary metabolites and their derivatives have been used as therapeutic agents to treat a wide range of maladies and diseases (Newman and Cragg, 2007). Numerous plant-based natural chemicals have been used as the primary raw ingredients for a number of medications. Additionally, plant secondary metabolites have been employed as pharmacological probes, prototypes, and drug precursors (Wilson & Danishefsky, 2006). Specifically, a sizable portion of medications produced globally are derived from plants, and the pharmaceutical industry today uses a number of bioactive chemicals that are secondary metabolites of plants (Lahlou, 2007). Morphine, the first natural product to be isolated from the opium poppy (*Papaver somniferum*) in 1806, is one particular example of a plant secondary metabolite that has been utilized as a pharmacological precursor (Elshafie et al., 2023). An antimalarial medication called Artemisinin, which was extracted from *Artemisia annua*, contains sesquiterpene lactone and is used to treat malignant cerebral malaria brought on by *Plasmodium falciparum* (Phillipson, 2001). The highly oxygenated tetracyclic diterpenoid paclitaxel, which was extracted from *Taxus brevifolia*, functions as an antimitotic by stabilizing polymerized microtubules. The same material has also been utilized as a successful medication to treat breast and ovarian malignancies (Atanasov et al., 2015). These are only a few notable mentions among numerous secondary metabolite compounds derived from plants that are currently used in the pharmaceutical industry. Table 11.1 enlists a few plant secondary metabolites with established pharmacological properties.

The advantages of a large range of fruits and vegetables, which offer various bioactive secondary metabolites such as phytochemicals, and vitamins, vital amino acids, minerals, and fibers have recently attracted a lot of attention. However, a number of ecological and physical conditions, including light, temperature, and relative humidity, might negatively affect those chemicals' stability (Banerjee et al., 2017). Histidine, isoleucine, leucine, lysine, methionine, phenylalanine, threonine, tryptophan, and valine are a few examples of essential amino acids that are crucial for nutritionand a number of body functions, including protein synthesis, tissue repair, and nutrient absorption (Young, 1994). Certain EAAs can prevent muscle loss, improve mood, sleep, and sports performance, and speed up recovery after surgery (Young, 1994). A healthy, balanced diet that includes EAAs may help most people meet their daily needs. Furthermore, eating a diet high in fiber has been suggested as a nutritional strategyto help prevent diabetes, obesity, and hypertension (Banerjee et al.,

TABLE 11.1 List of a few secondary metabolites with profound pharmacological activity.

Classes of secondary metabolites	Names of the compounds	Sources of the phytochemicals	Pharmacological actions	References
Alkaloids	Vincristine	*Catharanthus roseus*	Anticancer effect	Škubník et al. (2021)
	Reserpine	*Rauwolfia sp.*	Antihypertensive effect	Shamon and Perez (2016)
	Caffein	*Coffea arabica*	Stimulation of central nervous system, hair growth in alopecia	Fiani et al. (2021); Ruse et al. (2025)
	Atropine	*Atropa belladonna, Datura stramonium*	Anticholinergic, prevents pediatric myopia progression	Shih et al. (2016)
Carotenoids	Crocin	*Crocus sativus*	Inhibits platelet aggregation, reduces oxidative stress and apoptosis	Thushara et al. (2013)
	Lutein	*Targetes erecta*	Antidiabetic effect, beneficial effect against cataract	Seddon (2007); Dembinska-Kiec et al. (2008)
	Lycopene	*Solanum lycopersicum*	Anticancer effect, Antidiabetic effect	Gloria et al. (2014); Dembinska-Kiec et al. (2008)
Flavonoids	Phloretin	*Malus sp.*	Counteracts cardiovascular complications related to diabetes	Shen et al. (2012)
	Quercetin	*Malus sp., Alium cepa, Vitis vinifera*	Induces apoptosis, inhibits cell proliferation, angiogenesis in cancers, and blocks androgenic receptors in prostate cancer, anti-inflammatory effects, lowers cardiovascular risk	Hickson et al. (2019); Zahedi et al. (2013); Pratheeshkumar et al. (2012); Xing et al. (2001); Wang et al. (2013); Liu et al. (2014)
	Catechin	*Camellia sinensis, Senegalia catechu*	Reduces hypertension through modifying arachnoid acid metabolism, prevents rheumatoid arthritis	Elbarbry et al. (2022); Lee et al. (2021)
	Naringenin	*Citrus sp., Trigonella foenum*	Mitigates neurotoxicity through *reducing oxidative stress, mitigates* corneal neovascularization	Giradkar et al. (2024); Ma et al. (2022)

Phenolic acids	Chlorogenic acid	*Coffea arabica*, *Camellia sinensis*	Chemopreventive activity against breast cancer, wound healing effect	Olayiwola and Gollahon (2024); Chahardoli et al. (2023)
	Ferulic acid	*Ferula sp.*	Antidiabetic effect	Prabhakar et al. (2013)
Polyphenols	Curcumin	*Curcuma longa*	Anticancer efficacy *in vitro*, restores cognitive ability, decreases risk for ulcerative colitis	Manikandan et al. (2012); Rinwa and Kumar (2012); Sung and Park (2013)
	Resveratrol	*Veratrum grandiflorum*, *Vitis vinifera*	Antimicrobial activity, inhibits ferroptosis and decelerates heart failure, antioxidant effect	Abedini et al. (2021); Zhang, Gong, et al. (2023); Zhang, Qian et al. (2023); Zhang, Wiese et al. (2023)

2017). One of the primary roles of plant metabolites is to affect the physical attributes of food, including color, flavor, consistency, and aroma (Fang et al., 2019). They also provide a number of health advantages; for instance, they include antibacterial, antiinflammatory, anti-carcinogenic, and antioxidant qualities. Numerous plant metabolites can also aid in nutrition and lower human cholesterol levels (Paine et al., 2005). Golden rice, for instance, has a much higher β-carotene (provitamin A) content, which helps prevent vitamin A insufficiency (Fang et al., 2019). Numerous other phytochemicals are also thought to be healthy nutrients. These are effective at lowering the risk of developing clinical disease and include glucosinolates, flavonoids, phytosterols, phenolic acids, carotenoids, and polyunsaturated fatty acids (Bradbury et al., 2014; Wang et al., 2014). A number of plant secondary metabolites-containing nutraceutical formulations have been developed in recent times, and the market demand for nutraceutical formulations is steadily rising.

Some plant secondary metabolites are used in cosmetic preparations primarily because of their minimal toxicity to mammals (Aburjai and Natsheh, 2003). Several plant secondary metabolites such as phenolic acids, flavonoids, alkaloids, and carotenoids are incorporated in cosmetic formulations for skin and hair for their dermatological benefits and contributions to altering appearances. They are utilized for skin care, for their properties of scavenging free radicals, treating dryness, eczema, acne, and other skin protection effects. They are also utilized as hair colorants and growth promoters. Essential oils, which can be used in some cosmetic products to provide a pleasing scent, gloss, and other conditioning qualities, are among the most significant plant secondary metabolites. For instance, menthol is frequently used as a flavoring ingredient in toothpaste and mouthwashes because of its antibacterial activity. It also adds a wonderful fresh taste and keeps plaque from building up, which can cause gingivitis and foul breath (Vranić et al., 2004). For generations, people have utilized natural ingredients and traditional therapies to treat a wide range of dermatological conditions. Secondary metabolites-containing extracts are also incorporated in herbal cosmetic formulations, harnessing from such ethnopharmacological knowledge. Thus, plant secondary metabolites are currently being widely used in cosmetic and herbal cosmetic industries.

11.2.2 Market size and demand for key metabolites

According to recent market analyses, the global market for plant extracts and plant-derived ingredients is worth tens of billions of dollars. A number of companies estimate that the base market for plant extracts will be between USD 35 and USD 44 billion in 2024–2025, with multiyear growth into the 2030 s with predictions of double-digit CAGRs (https://www.precedenceresearch.com/plant-extracts-market; https://www.skyquestt.com/report/plant-extracts-market). This growth will be driven by trends in natural ingredients, nutraceuticals, and cosmetics. The industry is even bigger when viewed in a broader context that encompasses the trade in all medicinal and aromatic plants. According to current evaluations and trade analyses, the worldwide medicinal plant market is valued at around USD 200 billion in 2023 (Zamani et al., 2025). Some key high-value secondary metabolites include paclitaxel and artemisinin. Paclitaxel, an anticancer plant secondary metabolite, is one of the most used drugs in the treatment of different types of cancers. The chemotherapeutic injection containing paclitaxel already holds a multibillion-dollar market worldwide, with a rapidly growing bulk-active pharmaceutical ingredient market (https://finance.yahoo.com/news/bulk-paclitaxel-strategic-industry-report-090700181.html). With increasing global prevalence of cancer and novel delivery strategies for the delivery of anticancer drugs, the market size of paclitaxel is expected to rise even higher. Artemisinin and its derivatives are key antimalarial drugs used worldwide, with a market size of up to a few hundred million of dollars (https://www.grandviewresearch.com/industry-analysis/artemisinin-combination-therapy-act-market). Food components, cosmetics, and nutraceuticals account for the majority of the bulk market for numerous phenolics, flavonoid-rich extracts, essential oils, and saponins. Together, these product categories account for a sizable portion of the market for plant extracts (standardized polyphenol extracts and aromatic oils are particularly significant in cosmetics and functional foods). The need for these classes is implicitly captured in market studies that forecast the expansion of the plant extracts sector.

Seasonal and geoclimatic variability, the low biomass yields of certain compounds due to their restricted natural abundance, and ecological and regulatory limitations on wild harvesting are the main factors limiting the supply of secondary metabolites. Given this, a more environmentally friendly method of creating these useful molecules is provided by plant biotechnology, namely the establishment of plant biofactories, implementation of plant tissue culture techniques, and the use of metabolic engineering techniques (Mipeshwaree Devi et al., 2023). Furthermore, plant biofactories can also be effective instruments for clarifying secondary biosynthetic pathways, which broadens the scope of opportunities provided by metabolic engineering to improve secondary metabolite output. In this regard, new enzymes implicated in the production of paclitaxel have been discovered recently (Zhang, Gong, et al., 2023; Zhang, Qian, et al., 2023; Zhang, Wiese, et al., 2023). Synthetic biology studies to recreate whole biosynthetic pathways in heterologous systems, identifying the minimal gene sets needed for biosynthesis and opening the door for regulated, large-scale production, further encourage these developments (Palazon and Alcalde, 2025).

11.3 Plant transformation as a tool for genetic engineering

Plant transformation is an integral part of plant biotechnology for transgenic or genetically modified plant development by introducing foreign genes and expression of the desirable traits (Keshavareddy et al., 2018). This method is used as the key tool for understanding gene regulation, expression patterns, and biosynthetic pathway analysis. This technique is the fundamental basis of functional genomics, metabolic engineering, and molecular farming.

11.3.1 Plant transformation: its implication and types

Plant transformation plays a cornerstone technique for developing transgenic plants with augmented secondary metabolites (Jha et al., 2023). In general, vector-mediated and direct gene transfer approaches are two major types of plant transformation used to develop elite lines with improved metabolite profiles. Following are the key techniques used for plant transformation.

11.3.1.1 Vector-mediated transformation via Ti or Ri plasmids

Agrobacterium-mediated transformation is the most efficient tool for introducing foreign DNA into dicotyledonous plants. Exploring T-DNA as a natural vector of Ti or Ri plasmids of *Agrobacterium tumefaciens* and *Rhizobium rhizogenes*, this technique ensures stable integration following the desired expression (Gelvin, 2003; Goralogia et al., 2025).

Moreover, this technique empowers the development of elite lines with enhanced metabolite profiles. Especially, using *R. rhizogenes* for establishment of hairy root cultures, which shows high quality of genetic and biochemical integrity and offers a steadfast system for pilot-scale metabolite production from medicinal plants, where root is ofeconomic interest (Gutierrez-Valdes et al., 2020; Biswas et al., 2023).

11.3.1.2 Direct gene transfer methods

Direct gene transfer methods are applicable generally in monocotyledonous plants where plants are recalcitrant to *Agrobacterium*-mediated methods (Shrestha, 2018). These physical or direct gene delivery systems can use the biolistic or particle bombardment method or electroporation to deliver foreign DNA into various crop plants including medicinal plants (Jinturkar et al., 2011).

11.3.1.2.1 Biolistic gene transfer/micro particle bombardment/gene gun

This is the most useful and flexible method of gene transfer where gene gun is used to deliver foreign DNA coated with tungsten or gold forced with high velocity to invade within plant tissues (Alsaggar & Liu, 2015).

11.3.1.2.2 Electroporation

Electric field-mediated direct DNA transfer method or electroporation is considered as a direct gene transfer method, where anelectric impulse is applied to isolated protoplast for uptake of foreign DNA (Ozyigit, 2020).

11.3.1.2.3 Microinjection

It is a highly precise direct DNA transfer method where DNA is directly injected into the nucleus of a single cell using microinjection. This is the most target-based DNA delivery system where foreign gene is directly transferred at single-cell level; however, the success rate is not very high (Mathias, 1987).

11.3.1.3 CRISPR/Cas and genome editing tools

This advanced genome editing tool is now considered the most efficient and cutting-edge method for the insertion, deletion, or regulation of specific genes, therefore easier forcontrolling the metabolic pathways (Mitra et al., 2023). This technique ensures sustainable gene delivery system with high specificity and efficiency, which overcomes the problems associated with conventional transformation methods (Yuan et al., 2024).

11.3.1.3.1 Hairy root transformation coupled with CRISPR/Cas genome editing

CRISPR/Cas-mediated transformation through *Agrobacterium rhizogenes* is a useful method for achieving hairy root cultures and genome editing from the same system

(Kiryushkin et al., 2022). The approach employs *A. rhizogenes* that naturally contains the Ri (root-inducing) plasmid with a T-DNA that can be integrated into plant cells to form hairy roots. With a CRISPR/Cas construct, the bacterium is a good "delivery vehicle," transferring both the Ri T-DNA and the genome editing system components such as Cas nuclease, guide RNA, and a selectable marker into plant tissues directly. The dual utility of this delivery platform makes (Zhao et al., 2025; Li, Liu, Xu, et al., 2023; Li, Liu, He, et al., 2023) *A. rhizogenes* is a very useful tool to produce the fastest-growing genetically stable transgenic hairy roots while providing the potential to tag genome-editing into them and creating a flexible beginning for functional genomics and metabolic engineering.

11.3.2 Hairy root culture: a living bio factory for the production of high-valued root metabolites

Rhizobium rhizogenes, previously known *Agrobacterium rhizogenes*, serves as a potential living bio factory for the production of root-synthesized therapeutic metabolites (Srivastava & Srivastava, 2007). This tool provides a sustainable year-round production system for industrial exploitation without sacrificing natural resources (Roy, 2021a,b). Due to increasing market demand for high-quality, adulteration-free source materials for herbal industry, academic researchers, biotechnology businesses, and pharmaceutical corporations are interested in hairy root cultures (HRCs) for their technological and commercial benefits. SCOPUS offers 607 articles (2012–2018), whereas PubMed lists 767 (2012–2019) with the keywords "hairy roots," "hairy root," "transformed root cultures," or "transformed root culture" (Gutierrez-Valdes et al., 2020).

11.3.2.1 *History of hairy root culture*

Steward and colleagues described hairy roots in 1900 (Stewart, 1900). In 1930, Ricker and others provided the first report of the causal organism of hairy roots, named it *Phytomonas rhizogenes*, later renamed *Agrobacterium rhizogenes* (Riker, 1930). The first experimental transformation of higher plants with *A. rhizogenes* was made by Ackermann (1977). Since its discovery, research has shown that more than 450 species of higher plants, from a number of genera and families, can be infected by *A. rhizogenes* (Hamill & Lidgett, 1997). *A. rhizogenes* is a Gram-negative, motile, rod-shaped, non-spore-forming microorganism, closely related to rhizobium, and capable of inducing nitrogen-fixing nodules in legume plants.

11.3.2.2 *Systemic position of* **Rhizobium rhizogenes**

The genus Agrobacterium became defined when *A. rhizogenes* was first described as a causative organism of "hairy root disease" in dicotyledonous plants, a pathogenic symptom that serves as the current basis for taxonomic classification. For several decades, organisms, namely *A. tumefaciens, A. rhizogenes, A. rubi, and A. vitis,* were maintained as separate genera from Rhizobium, which was defined as a genus of nitrogen-fixing organisms that formed symbioses with legumes. However, molecular phylogenetic studies, particularly through comparative 16 S rRNA gene sequence analysis, have shown that Agrobacterium, Allorhizobium, and Rhizobium species belong to a single monophyletic group with very similar phylogenetic relationships and equal phenotypical traits. This meant that the

Agrobacterium genus was artificial, as its differentiation relied largely on pathogenic characteristics determined by plasmids (i.e., Ti and Ri plasmids), which are mobile and not consistent for stable taxonomy. To address this, Young et al. (2001) made a significant revision of the family Rhizobiaceae, where all species of Agrobacterium were placed in the genus *Rhizobium*. This resulted in the reclassification of *Agrobacterium rhizogenes* as *Rhizobium rhizogenes*, among others (*Rhizobium radiobacter*, *Rhizobium rubi*, and *Rhizobium vitis*) (Table 11.2). This reclassification is according to genetic evidence and natural evolutionary relationships rather than based solely on the symptoms of disease (Kersters, 1984; Riker, 1930; Willems & Collins, 1993; Young et al., 2001).

11.3.2.3 Ri plasmid and its functional genes

Agrobacterium rhizogenes Ri (root-inducing) plasmid is a big extrachromosomal DNA structure that is accountable for causing the hairy root disease in dicotyledonous plants (Moore et al., 1979). Equivalent in function to the Ti plasmid of *Agrobacterium tumefaciens*, the Ri plasmid harbors one or two T-DNA (Transferred DNA) regions that are mobilized and stably integrated into the nuclear genome of the plant during infection, leading to striking changes in plant developmental and metabolic pathways that together constitute the hairy root phenotype (Morgan et al., 1987; Veena & Taylor, 2007). Depending on their organization, Ri plasmids exist in two general forms. In mannopine-type strains, the plasmid has one T-DNA segment containing the rol (root locus) genes (Fig. 11.1) (Ozyigit et al., 2013), opine biosynthesis genes, and other uncharacterized open reading frames (ORFs) (Veena & Taylor, 2007). Agropine-type strains have a split T-DNA (Fig. 11.2) (Ozyigit et al., 2013), which contains two autonomous regions, TL (left T-DNA) and TR (right T-DNA), each of about 15–20 kb (Hooykaas & Hooykaas, 2021). Both the TL and TR T-DNAs are capable of being transferred and integrated singly into the plant genome; however, the transfer of the TL-DNA is necessary for hairy root induction (Cardarelli et al., 1987), whereas TR-DNA adds secondary traits like supplementary metabolic alterations and growth responses (Mallol et al., 2001). The TR-DNA, found in agropine-type Ri plasmids only, contains genes

TABLE 11.2 A comparison-based taxonomical information highlighting the genus transition of the causative agent of hairy root.

Rank	*Agrobacterium rhizogenes* (old classification)	*Rhizobium rhizogenes* (revised classification)
Domain	Bacteria	Bacteria
Phylum	Proteobacteria	Proteobacteria
Class	Alphaproteobacteria	Alphaproteobacteria
Order	Rhizobiales (earlier: *Rhizobiaceae group*)	Hyphomicrobiales
Family	Rhizobiaceae	Rhizobiaceae
Genus	*Agrobacterium*	*Rhizobium*
Species	*rhizogenes*	*rhizogenes*

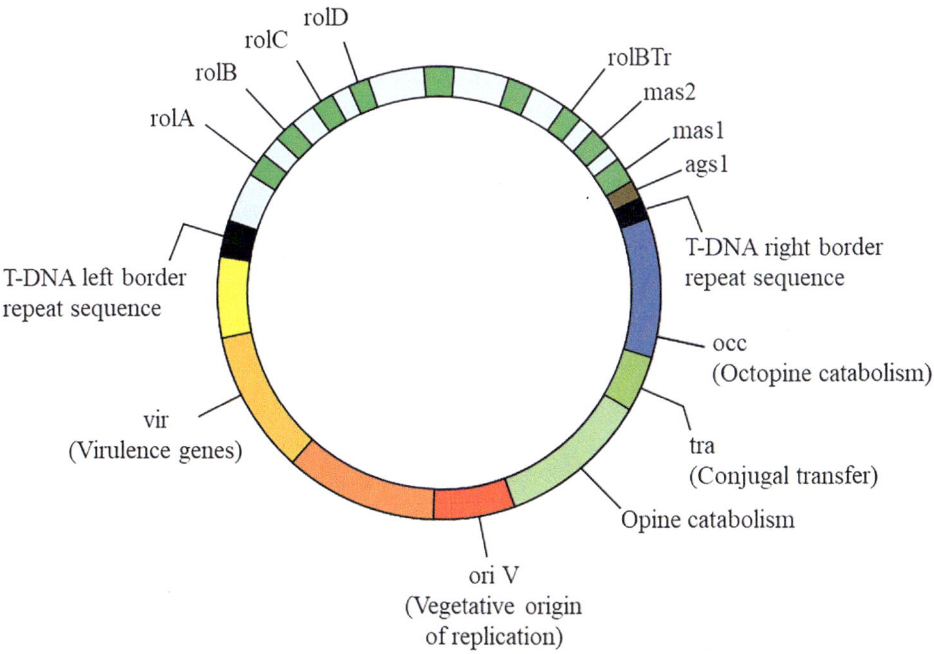

FIGURE 11.1 Mannopine type Ri plasmid (single T-DNA).

such as aux1, aux2, rolB (TR), mas1, mas2, and ags that control auxin and opine biosynthesis (Hansen et al., 1991; Nemoto et al., 2009). While the aux genes (aux1/iaaM and aux2/iaaH) increase auxin levels, they have an ancillary but not an obligate role in inducing hairy roots (Terakura et al., 2007). In mannopine- and cucumopine-type strains, the T-DNA is devoid of aux genes but has just rol genes, which are adequate to induce hairy roots (Failla et al., 1990). Comparative analysis has indicated that cucumopine- and mannopine-type T-DNAs are homologous to the agropine-type TL-DNA, but often not containing the rolD gene (Chandra, 2012; Veena & Taylor, 2007). The rol genes are the most thoroughly characterized functional elements of Ri T-DNA and are directly responsible for root initiation and modified morphogenesis (Shvets et al., 2023). Early investigations into gene function had depended extensively on transposon mutagenesis and loss-of-function approaches, which demonstrated that interfering with rolA, rolB, or rolC significantly decreased or abolished hairy root induction, making them critical determinants of the phenotype. Subsequently, molecular description confirmed these genes to correspond to distinct ORFs: rolA (ORF10), rolB (ORF11), rolC (ORF12), and rolD (ORF15) (Shvets et al., 2023; Veena & Taylor, 2007).

11.3.2.4 Functional roles of Ri plasmid rol genes

The rol genes of *Agrobacterium rhizogenes* T-DNA are involved in plant transformation and morphogenesis (Table 11.3). rolA (300 bp) codes for a 100-amino-acid regulatory protein that influences the expression of other plant genes (Mmereke et al., 2022). Its promoter contains

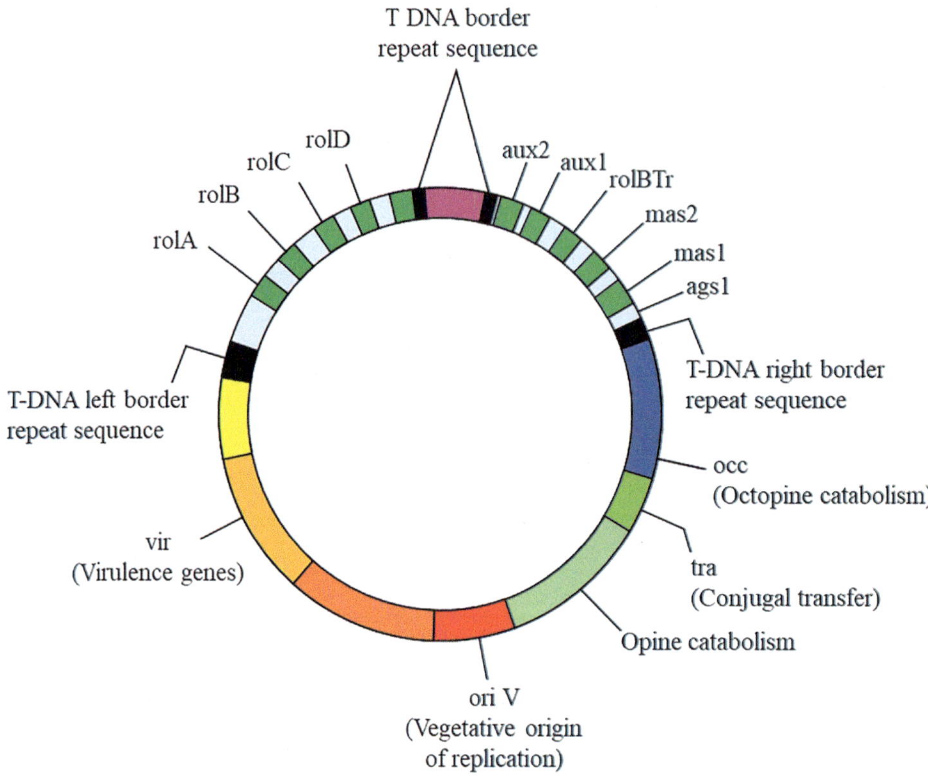

FIGURE 11.2 Agropine type Ri plasmid (split T-DNA).

TABLE 11.3 Functional characteristics of *Agrobacterium rhizogenes* rol genes.

Gene	Size (bp / aa)	Protein function/role	References
rolA	300 bp / 100 aa	Regulatory protein influencing expression of plant genes	Mmereke et al. (2022); Pandolfini et al. (2000); Shvets et al. (2023)
rolB	777 bp / 259 aa	Auxin-related responses; translocates to nucleus	Baumann et al. (1999); Moriuchi et al. (2004); Rangslang et al. (2018); Capone et al. (1994)
rolC	540 bp / 180 aa	Alters cytokinin function; affects plant morphogenesis	Mmereke et al. (2022); Matveeva and Валерьевна (2021); Otten (2018)
rolD	1032 bp / 344 aa	Homologous to ornithine cyclodeaminase; linked with proline biosynthesis (enzymatic role unclear)	Dilshad et al. (2015); Pavlova et al. (2013); Shvets et al. (2023); Altamura (2004)

three domains (domains A–C) regulating tissue-specific expression, with Domain A inducing expression in leaves, while its deletion diverts expression to roots and stems. This gene has a peculiar intron with a bacterial promoter, enabling differential regulation in bacteria and plants (Pandolfini et al., 2000; Shvets et al., 2023). rolB (777 bp) encodes a 259-amino-acid protein associated with auxin-related response; it can translocate to the nucleus, and its promoter contains five domains (domains A to E) that mediate tissue-specific and hormone-responsive expression in roots. Domain A regulates expression in root cells and shares the binding region of the transcription factor RBF1; its loss can suppress expression in protoderm and root cap cells. Domain B (–341…–306 bp) is essential for auxin-inducible expression and harbors a conserved cis-element (ACTTTA); in its absence, rolB is not expressed in meristematic cells, and the gene is not activated by the transcription factor NtBBF1. Domain C supports expression within internal meristems and represses it within protoderm, whereas Domain D guarantees expression in the main cortex (Capone et al., 1994). Lastly, Domain E, together with Domain B, represses rolB expression within meristem cells. All these domains work together with plant transcription factors and plant hormone signals to regulate rolB expression accurately across various tissues of the root as well as in different developmental stages (Baumann et al., 1999; Moriuchi et al., 2004; Rangslang et al., 2018). rolC (540 bp) codes for an 180-amino-acid protein that potentially affects cytokinin function (Mmereke et al., 2022); it is cytoplasmically localized and is able to enter the nucleus. Its promoter includes sucrose-regulated and stress-related factors that regulate gene expression (Matveeva & Валерьевна, 2021; Otten, 2018). Lastly, rolD (1032 bp) codes for a 344-amino-acid protein homologous to ornithine cyclodeaminase, perhaps playing a role in proline biosynthesis, although its enzymatic role is not established (Altamura, 2004). Its auxin-controlled promoter guarantees expression in plant cells but not bacterial cells. Altogether, these rol genes control plant growth, hormone sensitivity, and tissue-developmental pattern in transgenic plants (Dilshad et al., 2015; Pavlova et al., 2013; Shvets et al., 2023).

11.4 Mechanism of T-DNA integration into the plant genome

The infection mechanism of *Agrobacterium rhizogenes*, the pathogen causing hairy root disease, is essentially similar to that of *Agrobacterium tumefaciens* but varies in the transferred DNA content. Rather than a Ti (tumor-inducing) plasmid, *A. rhizogenes* contains an Ri plasmid (root-inducing plasmid) whose T-DNA contains the rol (root locus) genes that cause hairy root initiation (Fig. 11.3) (Yordanova & Georgiev, 2017).

11.4.1 Plant infection and wound recognition

The infection process is initiated when plant stems or roots are wounded and release phenolic compounds (e.g., acetosyringone) and sugars. These phenolics are detected by the bacterial VirA/VirG two-component regulatory system encoded on the Ri plasmid. Phenolics are detected by membrane-bound histidine kinase VirA and result in its autophosphorylation, which in turn phosphorylates the response regulator VirG, which

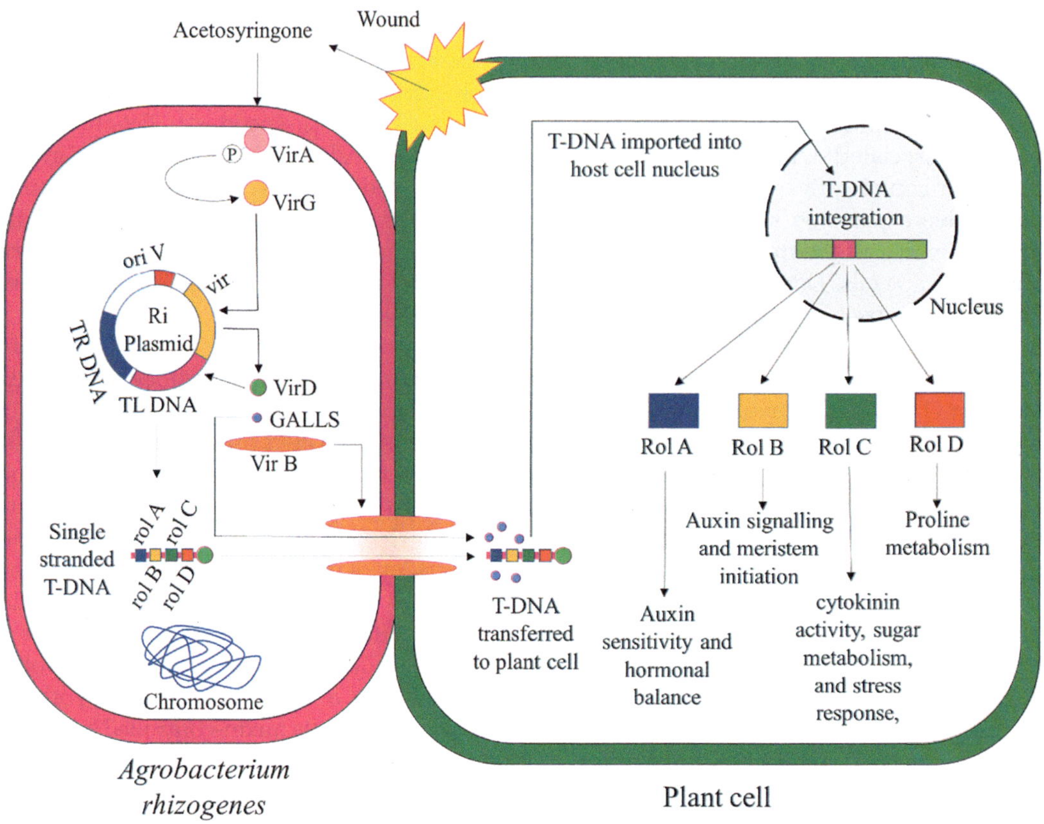

FIGURE 11.3 Mechanism of T-DNA integration into plant genome.

then triggers the transcription of other virulence (vir) genes involved in T-DNA processing and transfer (Bahramnejad et al., 2019; Massah & Rabiei, 2023).

11.4.2 Attachment to plant cells

Attachment is the initial important step, facilitated by the bacterial chromosomal virulence genes (chv genes). The chvA and chvB genes produce proteins that are involved in cyclic β-1,2-glucan synthesis, which is deposited in the periplasm of the bacterium (the area between the plasma membrane and cell wall). Tight adhesion of *Agrobacterium* to the plant cell wall is enabled by these glucans. Moreover, some plant-derived proteins are involved in bacterial adhesion (Chandra, 2012; Gelvin, 2010).

11.4.3 T-DNA processing on the Ri plasmid

The Ri plasmid (Root-inducing plasmid) has a single or double T-DNA region (strain type dependent). Every T-DNA is bordered by 24-bp border repeat sequences. The VirD1/VirD2 endonuclease complex identifies and nicks the right border sequence. This results in a single-stranded T-DNA molecule (T-strand). VirD2 covalently binds to the 5′ end of the T-strand, directing it through later processes. Contrary to *A. tumefaciens* Ti plasmids, Ri plasmids do not have overdrive sequences (which stimulate T-DNA nicking in Ti plasmids). Rather, Ri plasmids have T-DNA transfer stimulator sequences (TSS), which are a series of short repeats that stimulate transfer efficiency in a strain-specific fashion (mannopine, agropine, cucumopine, etc.)(Hansen et al., 1992; Moriguchi et al., 2001; Zhu & Welander, 2000).

11.4.4 Cytoplasmic protection and nuclear import

After entering the plant cell cytoplasm, the T-strand needs to be shielded from nucleases and transported into the nucleus. *A. tumefaciens* accomplishes this function by VirE2, an ssDNA-binding protein, and its chaperone VirE1. *A. rhizogenes* Ri plasmids, however, have no virE1 and virE2 genes. Instead, *A. rhizogenes* possesses the GALLS gene, which encodes a protein with the ability to replace VirE2 activity. GALLS has ATP-binding and helicase motifs (like TraA strand transferases), Nuclear Localization Signals (NLSs), and C-terminal T4SS secretion signals. GALLS binds to and shields the T-strand, facilitates its nuclear targeting, and also potentially helps in unwinding or strand transfer functions (Hodges et al., 2004, 2006; Veena & Taylor, 2007).

11.4.5 Expression of rol genes and hairy root formation

Once T-DNA is inserted into the plant genome, the rol genes (rolA, rolB, rolC, and rolD) found in the Ri T-DNA are transcribed and expressed by the transcriptional machinery of the host plant. These genes drastically change the hormonal balance and metabolic processes of the plant and consequently reprogram the normal growth patterns. rolA is a regulatory gene that controls plant hormone physiology and increases auxin sensitivity, which in turn influences growth habit. rolB is the primary induction determinant of hairy roots and plays a role in auxin signal perception and meristem initiation, and it induces secondary metabolism. rolC influences cytokinin activity, sugar metabolism, and stress response, resulting in altered shoot growth and flowering at an early stage. rolD, which occurs only in agropine types, is associated with proline metabolism but will not trigger root development by itself (Altamura, 2004; Matveeva & Валерьевна, 2021; Otten, 2018; Baumann et al., 1999; Moriuchi et al., 2004; Rangslang et al., 2018; Mmereke et al., 2022). The coordinated expression of these rol genes disrupts the usual auxin–cytokinin balance and induces vigorous, highly branched root growth at the point of infection, with the resulting typical hairy root habit.

11.4.6 Stable and hormone-independent hairy root induction

Hairy root cultures are induced by infecting plant explants with *Agrobacterium rhizogenes*, which copies the T-DNA region of its root-inducing (Ri) plasmid into the host genome. Expression of rol (root locus) genes from integrated T-DNA reprograms the plant cells, which results in the development of fast-growing, highly branched roots. In contrast to callus or adventitious roots, hairy roots grow abundantly without the need for exogenous application of plant growth regulators like auxins or cytokinins and represent a hormone-independent development. Established roots then have a stable phenotype, genetic stability with repeated subcultures, and often increased biosynthetic capacity for secondary metabolites. These distinctive aspects render hairy root cultures a strong and sustainable system for long-term *in vitro* experiments, functional genomics, and bulk production of pharmaceutically and industrially relevant metabolites (Guillon et al., 2006; Pistelli et al., 2010; Sonkar et al., 2023).

11.5 Stepwise protocol for establishment of hairy root culture

The protocol for hairy root transformation was followed as described by Gangopadhyay et al. (2008, 2010), with necessary modifications. In this approach, *Agrobacterium rhizogenes* carrying Ri plasmids is used to induce stable root cultures from sterile plant explants.

11.5.1 Strain selection

The establishment of hairy root cultures begins with the selection of *Agrobacterium rhizogenes* strains with Ri plasmids. Some commonly used strains are ATCC 15834, A4, or LBA 9402. The bacteria are stored on suitable sterile media, *Agrobacterium* Broth (AB), depending on the strain. Strain selection is important, since the efficiency of transformation and root induction could differ in different strains (Gangopadhyay et al., 2008; Hooykaas et al., 1977).

11.5.2 Explant selection

For plant material, *in vitro* plantlets are the explant source, which can consist of leaves, stems, hypocotyls, or cotyledons. For maintaining aseptic environments, explants are surface sterilized by using routine plant tissue culture sterilization procedures prior to inoculation (Sathyanarayana & Varghese, 2007).

11.5.3 Infection of explants with *Rhizobium rhizogenes*

The infection process involves the preparation of an inoculum of *A. rhizogenes* cultured to mid-logarithmic phase (OD600 ~0.5). This culture is delivered to the explants by wounding, which may be done by needle pricking, incision with scalpel, or injection of the bacterial suspension directly. After infection, the explants are cocultivated on hormone-free Murashige and Skoog (MS) medium, pH 5.6, and kept in darkness at 25°C ± 2°C for 2–3 days to facilitate the transfer of T-DNA (Murashige & Skoog, 1962). Subsequent to this time, explants are placed on new MS medium containing bacteriostatic antibiotics like cefotaxime or carbenicillin to remove excess bacteria. Hairy roots tend to appear from the wound sites in a matter of days or weeks, which confirms successful transformation.

11.5.4 Establishment and maintenance of hairy root cultures

When hairy roots are formed, they are removed from the explants and placed on hormone-free MS solid medium (pH 5.6). Roots are kept on antibiotic medium during early subcultures to avoid all bacterial contamination (Gantait & Mukherjee, 2021; Panda et al., 2017). Healthy and growing root tips are chosen and subcultured to liquid MS medium (pH 5.6, containing 3% sucrose) in Erlenmeyer flasks. Cultures are held on an orbital shaker at about 70 rpm, in darkness, and at 25 ± 2°C. The growth performance is measured through the assessment of fresh weight or dry biomass, especially during the exponential phase of culture.

11.5.5 Characterization and selection

11.5.5.1 Morphological characteristics

The transformed root lines are first screened for their outward characteristics, such as rapid growth, increased branching, and distinctive morphological characteristics differentiating the transformed roots from the nontransformed roots.

11.5.5.2 Molecular confirmation

Polymerase chain reaction (PCR) amplification of rol genes (e.g., rolA, rolB, and rolC) contained within the transferred T-DNA region of *Rhizobium rhizogenes* is used to confirm genetic transformation. Southern blotting is then conducted to establish stable integration of T-DNA into the plant genome (Gangopadhyay et al., 2010). Confirmation of the absence of bacterial contamination is performed by either PCR amplification of the 16 S rRNA gene or through traditional microbial culture methods (Panda et al., 2017).

11.5.5.3 Analytical confirmation

In order to validate the capacity of hairy root clones to produce secondary metabolites, it is necessary to conduct metabolic profiling using analytical methods. The most common analytical platforms for qualitative and quantitative determination of metabolites include high-performance liquid chromatography (HPLC) (Ho et al., 2018).

11.5.6 Stabilization and scale-up of hairy root cultures

Lastly, the hairy root clones of choice are multiplied and grown in hormone-free MS medium (solid or liquid, pH 5.6) in sterile conditions. Subculturing at regular intervals is essential to add nutrients and ensure growth. The cultures are kept in continuous darkness at 25 ± 2 °C to maintain genetic stability, promote prolific growth, and avoid callus formation or unguided differentiation. At this point, stable hairy root cultures are formed and can be kept for long durations. For bulk metabolite production, the most efficient root clones can then be scaled up in bioreactor cultures, under which controlled conditions promote biomass production and secondary metabolite yield (Roy, 2021a,b).

11.6 Metabolic engineering of hairy root cultures for enhanced secondary metabolite production

Plant metabolic engineering (ME) that includes hairy root (HR) culture systems is the intentional alteration of biosynthetic pathways to redirect flux toward desired metabolites. The primary aims of ME are overproduction of target compounds, elimination of unwanted by-products, and even creation of novel compounds not synthesized naturally in the host plant. In HR cultures, ME strategies have been used to improve the production of pharmaceutically and industrially valuable secondary metabolites by either enhancing precursor flux, elicitor treatment, radioisotope tracers, bypassing rate-limiting steps, minimizing competitive pathway activity, overexpressing regulatory genes or transcription factors, omics, mutant selection approaches, and/or inhibiting catabolism of the product (Table 11.4). These can be done through single-step engineering (overexpression of a single biosynthetic gene) or multistep engineering (coexpression of multiple genes in a pathway) and by manipulation of master regulators that manage whole biosynthetic networks (Table 11.5). Further, the use of heterologous enzymes possessing favorable characteristics such as resistance to feedback inhibition or greater substrate affinity, additionally increases productivity. Further, screening and selection of high-yielding lines, along with optimization of growth and production media, are still essential for continued metabolite accumulation. Although these methods are mainly aimed at improving metabolites, ME is also an effective means of enhancing agronomic quality in crop plants such as disease and insect resistance, weed suppression through allelopathic chemicals, enhanced ornamental and fruit quality (flavor, aroma, color), and even different scents for better pollination. Within HR cultures, such strategies not only increase yields of valuable secondary metabolites such as alkaloids, terpenoids, and flavonoids, but also allow stable trait regeneration of transgenic plants or transient expression systems for use in products such as edible vaccines. ME incorporated into HR culture systems is therefore an adaptable and scale-up-compatible platform for both basic research and commercial synthesis of high-value metabolites. There are many benefits to using hairy roots for metabolic engineering: they are genetically and metabolically stable over the long-term, grow quickly with and without hormone treatment, can be used with precursor feeding and elicitation, and are scalable into bioreactors for industrial applications (Aharoni et al., 2006; Buckland et al., 2000; Capell & Christou, 2004; Chandra & Chandra, 2011; Mehrotra et al., 2010).

11.7 Process optimization for augmentation of targeted product

Optimizing suitable conditions for quantitative and qualitative augmentation of the targeted product plays a crucial factor from industrial point of view concerning economic viability. It includes few importance parameters like pH, media composition, light, and elicitor type or concentration. Mathematical tools are often used to monitor quantitative growth kinetics, ensuring efficient production of high-valued root metabolites.

TABLE 11.4 Strategies in hairy root cultures for secondary metabolite pathway elucidation.

Approach	Description	Examples in hairy root cultures	References
Precursor feeding	Feeding potential precursors (amino acids, lipids, carbohydrates, or intermediates) to track metabolite accumulation, identify pathway intermediates, and rate-limiting steps.	*Catharanthus roseus* hairy roots used for monoterpenoid indole alkaloid (MIA) pathway with precursors like geraniol, loganin, tryptamine.	Herbert (1989); Morgan and Shanks (2000); Sharma et al. (2013)
Elicitor treatment	Using elicitors (biotic/abiotic agents like MeJa, yeast extract, Ag^+) to mimic stress and activate defense-related metabolic pathways.	MeJa treatment in *C. roseus* (MIA pathway); Ag^+ elicited lithospermic acid B (LAB) in *Salvia miltiorrhiza*.	Zhao et al. (2005); Xiao et al. (2009); Sharma et al. (2013)
Radioisotope tracers	Radioactive or stable isotope-labeled precursors traced to follow metabolic flux and identify intermediates/ products.	*Ophiorrhiza pumila* hairy roots used for camptothecin biosynthesis via MEP + shikimate pathways; bondomer flux analysis in *C. roseus*.	Yamazaki et al. (2004); Sriram et al. (2007); Sharma et al. (2013)
Metabolic inhibitors	Blocking specific enzymes with inhibitors to identify their role in SMP.	DXP pathway inhibition in *C. roseus* proved terpenoid precursor role in MIAs; oxygenase inhibitors used in tabersonine metabolism.	Hong et al. (2003); Sharma et al. (2013)
Mutant selection (activation tagging)	T-DNA insertion with enhancer elements activates nearby genes, creating dominant mutants with altered metabolite production.	*Panax ginseng* hairy roots with activation-tagging produced high ginsenoside mutants.	Seki et al. (2005); Sharma et al. (2013)
Gene overexpression	Overexpressing biosynthetic or regulatory genes to validate function and discover controlled genes.	*Vitis vinifera* hairy roots overexpressing MYB TFs altered anthocyanin and flavonoid biosynthesis.	Cutanda-Perez et al. (2009); Terrier et al. (2009); Sharma et al. (2013)
RNA interference (RNAi)	Gene silencing to study loss-of-function and confirm role in SMP.	*Nicotiana tabacum* hairy roots silencing NADPH-dependent reductase (NgA662) and ornithine decarboxylase in nicotine pathway.	Kajikawa et al. (2009); DeBoer et al. (2011); Sharma et al. (2013)
Omics approaches	Transcriptomics, proteomics, metabolomics integrated for large-scale SMP elucidation.	Transcriptomics in *C. roseus* (indole alkaloids), *P. ginseng* (ginsenosides), *S. miltiorrhiza* (tanshinones); metabolomics for pathway bottlenecks.	Ohlrogge and Benning (2000); Schena et al. (1995); Yang et al. (2012); Kim et al. (2003); Wang et al. (2010); Sharma et al. (2013)

TABLE 11.5 Case studies of metabolic engineering approaches for secondary metabolite production in hairy root cultures.

Approach	Gene/Enzyme targeted	Plant/HR system	Metabolite(s) affected	References
Single-step engineering	Chalcone isomerase (CHI)	*Saussurea involucrata*	Apigenin, total flavonoids	Li et al. (2006); Chandra and Chandra (2011)
	Putrescine N-methyltransferase (PMT)	*Atropa belladonna*	Tropane alkaloids	Rothe et al. (2003); Sato et al. (2001); Chandra and Chandra (2011)
	PMT	*Duboisia* hybrid	N-methylputrescine	Moyano et al. (2002); Chandra and Chandra (2011)
	PMT (CaMV 35 S)	*Datura metel*, *Hyoscyamus muticus*	Hyoscyamine, Scopolamine	Moyano et al. (2003); Chandra and Chandra (2011)
	Farnesyl diphosphate synthase (FPS)	*Centella asiatica* (PgFPS from *Panax ginseng*)	Triterpenes (asiaticoside, madecassoside), sterols	Kim et al. (2010); Chandra and Chandra (2011)
	Cytochrome P450 menthofuran synthase (MFS)	Peppermint (*Mentha piperita*)	Menthofuran	Mahmoud and Croteau (2001); Chandra and Chandra (2011)
Multigene engineering	PMT + Hyoscyamine 6β-hydroxylase (H6H)	*Hyoscyamus niger*	Scopolamine	Zhang et al. (2004); Chandra and Chandra (2011)
Novel metabolite production (mutasynthesis)	RNA silencing of tryptophan decarboxylase (TDC) + addition of unnatural tryptamine analogs	*Catharanthus roseus*	Unnatural alkaloids	Runguphan et al. (2009); Chandra and Chandra (2011)
Regulation by transcription factors	Overexpression of PAP1 (MYB TF)	*Arabidopsis thaliana*	Anthocyanins	Dare et al. (2008); Chandra and Chandra (2011)
Carbon flux engineering	ubiC (chorismate pyruvate-lyase, bacterial gene)	*Lithospermum erythrorhizon*	Shikonin precursors (4-hydroxybenzoate, 4HB)	Sommer et al. (1999); Chandra and Chandra (2011)

11.7.1 Mathematical models for biomass accumulation and metabolite yield prediction – growth kinetics study

Although numerous mathematical models exist to account for the growth of suspended plant single-cell cultures, there are only a handful of global macroscopic descriptions for the growth of hairy root plant tissue cultures, owing to their complexity. Models for hairy roots are usually confined to specific areas and are known as microscopic models. The modeling endeavors are driven by the goal of achieving a scientific understanding of processes and enhancing the conditions for culturing as well as the makeup of the nutrient medium. However, up to this point, cultures have been explored only through experiments, albeit in a highly systematic manner, leading to a vast number of experiments (Lenk and Bley, 2015).

A few microscopic mathematical models have been developed to analyze and optimize different parts of the hairy root culture of *C. roseus* to obtain the optimum quantities of alkaloids. A metabolic flux analysis (MFA) has been applied to *C. roseus* hairy root culture, and the resulting flux quantification has shed light on the "bondomers" in place of conventional isotopomers, which is a more efficient computational tool (Sriram et al., 2006; Sriram et al., 2007). Another attempt at metabolic modeling of *C. roseus* included a primary metabolic network of 31 metabolic fluxes and 20 pathways, which was efficient in predicting the growth rate of hairy root cultures (Leduc et al., 2006). Metabolites coming from similar origins were grouped together in another biogenetic organization approach for quantification of secondary metabolites of the hairy root culture of *C. roseus* (Morgan and Shanks, 2002). Cloutier et al. (2007) proposed a dynamic flux model for the plant primary metabolism that encompasses nutrient uptake (phosphate, sugars, and nitrogen sources), glycolysis and pentose phosphate pathways, TCA cycle, amino acid biosynthesis, respiratory chain, and cell building block biosynthesis. In their work, Lan et al. (2013) presented a method for engineering the salidroside biosynthetic pathway in hairy root cultures of *Rhodiola crenulata*, focusing on the metabolic characterization of tyrosine decarboxylase, a key player in salidroside production. The discovery was made that only the RcTYDC gene being overexpressed led to a rise in the production of salidroside, the end product. The objective of all the MFA approaches presented was to enhance the production of target metabolites overall. They demonstrated that MFA is an effective method for characterizing hairy root growth and biosynthesis, and that even minor alterations to metabolic pathways can result in markedly increased production yields.

Yan et al. (2011) presented a model for tanshinone synthesis and phase distribution in *S. miltiorrhiza* hairy root cultures, which includes a parameter sensitivity analysis demonstrating that the transport rate of tanshinones from the hairy roots to the medium significantly influences their accumulation in the resin. Jolicoeur et al. (2002) examined the particular symbiosis between *Glomus intraradices*, an arbuscular mycorrhizal fungus, and hairy roots of *Daucus carota*. They created a structured growth model that demonstrated how the fungus competes with the hairy roots for soluble root sugars. Experiments comparing cultures on culture dishes and in Erlenmeyer flasks confirmed the findings of model simulations. Chavarría-Krauser & Schurr (2004) devised a unique cellular growth model for root tip elongation, and the outcomes of its simulation were compared with measurements from the roots of *Arabidopsis thaliana*, *Nicotiana tabacum*, and *Pisum sativum*. A population balance model was proposed by Han et al. (2004), which offers a deeper insight into the growth of biomass in hairy root cultures. Six different hairy root types were used to calibrate the model, and model simulations were able to reproduce experimental results (Han et al., 2004). An unstructured model for growth and oxygen transfer in *Azadirachta indica* hairy roots was derived by Palavalli et al. (2012). With this model, simulations can be performed that factor in the effect of oxygen transfer limitations. The model can also predict when such a limitation will arise at the center of the growing hairy root matrix, which represents a significant bottleneck during high-density cultivations (Palavalli et al., 2012). This indicator can be utilized to enhance the aeration of the medium or to conclude the culture cycle.

Although specific facets of growth and biosynthesis in hairy root cultures are significant research areas, there is also a need for the creation of macroscopic models that can account for the development of overall hairy root morphology and its relationship with the culture environment. Inclusion of artificial intelligence (AI) has provided an avenue for macroscopic modeling. Several approaches at different aspects of hairy root culture have been incorporated, and different artificial neural networks (ANNs) have been employed in place of mathematical modeling to develop macroscopic models of analysis through machine learning. Automated image analysis of hairy root cultures is also employed to obtain data for machine learning algorithms.

11.7.2 pH and media composition: key drivers of metabolite yield

Optimization of culture conditions including pH and nutrient levels is necessary to maximize biomass as well as secondary metabolite yield in hairy root cultures (Table 11.6). Medium pH may affect enzyme function, membrane permeability, nutrient uptake, and even secretion of metabolites; different studies indicate that slightly acidic pH (5.0–6.5) is favorable for the production of metabolites, whereas pH far from this range (either acidic or basic) restricts growth as well as reduces production of secondary metabolites. For example, hairy roots of *Picrorhiza kurroa* produced the highest biomass and Picroliv (picroside-glycosides) at pH 6.0, while pH 3.0 and 8.0 had much lower production of these compounds (Verma et al., 2015). Media components such as macro- and micronutrients, carbon source (e.g., C:N), and presence of certain ions (e.g., Ca^{2+}, NH_4^+/NO_3^-) can alter precursor availability, osmotic balance, stress signaling, and change metabolic flux toward secondary metabolism pathways. For example, when the levels of NH_4NO_3, $Ca(NO_3)_2$, and sucrose in *Stephania tetrandra* hairy roots were improved through the use of response surface methodology, there was ~1.47-fold improvement in rooting dry biomass and ~1.37-fold improvement in tetrandrine production of the hairy roots compared to media without the optimizations (Kuo et al., 2025). In addition, in *Physalis peruviana* the pH of 5.5 with B5 medium and 4% sucrose gave the best compromise between biomass, phenolics, and caffeoylputrescine (Zhong et al., 2024). Therefore, pH and medium composition generally need to be cooptimized to identify a combination that produces good growth and high metabolite production (Kuo et al., 2025).

11.7.3 Optimization of hairy root cultures through light quality and intensity

Light is among the most efficient factors for the improvement of hairy root culture systems because light controls both root physiology and secondary metabolite biosynthesis

TABLE 11.6 List of examples of hairy root culture where pH & media composition plays key role for regulating plant secondary metabolism.

Plant species	Hairy root medium	Initial pH	Secondary metabolite production	Key findings	Reference
Picrorhiza kurroa	Half-strength B5 (Gamborg's B5 medium)+ 4% sucrose	6.0	Two specific glycosides	Medium pH 6.0 gave highest biomass and glycoside yield compared to pH 3–8	Verma et al. (2015)
Aconitum acutangulus	Basal medium	6.5	Moderate alkaloid production	Optimum biomass at pH 6.5; maximum alkaloid yield observed at pH 4.5	Liu et al. (2013)
Isatis tinctoria	MS/2 medium	5.8	Total alkaloids	First-time hairy root culture system (ITHRC) established using *Agrobacterium rhizogenes*; Box–Behnken design optimized culture conditions (temperature 24.7 °C, inoculum 0.75%, sucrose 3.14%) for maximum biomass and AK production; outperformed 2-year-old field-grown roots	Gai et al. (2015)
Plumbago indica	MS medium + 3% sucrose	5.6	Plumbagin	Low pH (4.6) and low sucrose favored biomass in darkness; pH 5.6 + 3% sucrose under continuous light enhanced plumbagin accumulation; direct shoot regeneration possible from hairy roots	Gangopadhyay et al. (2008)

TABLE 11.7 Influence of light on secondary metabolite production in hairy root cultures.

Type	Agents	Products	Plant species	References
Light and UV	Light of different spectral ranges (fluorescent, metal halide, blue, red, and blue + red light)	Ginseng saponin	*Panax ginseng*	Yu et al. (2005); Wang and Wu (2013)
Light	Light 385–790 nm	Artemisinin	*Artemisia annua*	Wang et al. (2001); Wang and Wu (2013)
Light	Continuous light (cool-white fluorescent lamps, 40 μM m^{-2} s^{-1}) and continuous dark	Diterpenoids (ferruginol, salvipisone, aethiopinone, 1-oxoaethiopinone); two ursene-type triterpenoids	*Salvia sclarea*	Kuźma et al. (2006); Wang and Wu (2013)
Light intensity	0–3,000 Lux	Artemisinin	*Artemisia annua*	Liu et al. (2002); Wang and Wu (2013)

(Table 11.7) (Halder et al., 2019). Not only does light supply energy, but in plant systems, it also acts as a developmental signal and controls the activity of critical enzymes and transcription factors participating in secondary metabolic processes (Chandra & Chandra, 2011). In hairy root cultures, the intensity (wavelength), quality, and duration of light exposure have been found to influence significantly biomass accumulation as well as metabolite patterns. For example, experiments in *Panax ginseng* hairy roots illustrated that a two-phase process, short-term incubation in darkness to induce biomass growth followed by illumination with fluorescent light, led to increased accumulation of ginsenosides, despite lower overall root biomass under continuous illumination. This means that light can steer carbon flux toward specialized metabolism instead of growth. Likewise, for *Artemisia annua* hairy roots, red light (660 nm) was the optimal spectral range, increasing root biomass and artemisinin content over white light. Light's differential effect is caused by its capability to drive enzymes and signaling cascades to direct precursors toward desired metabolites (Wang et al., 2001; Yu et al., 2005).

In addition, light intensity and photoperiod are also crucial. Constant light (24 h) in *A. annua* hairy roots enhanced biomass considerably but lowered levels of artemisinin in comparison with an appropriate 16 h light/8 h dark cycle and indicates that unbroken illumination may favor primary rather than secondary metabolite biosynthesis. In other *Salvia sclarea* hairy root studies, constant fluorescent light was found to increase diterpenoid content and modify triterpenoid composition slightly, and this indicates how light influences individual biosynthetic branches (Kuźma et al., 2006; Liu et al., 2002). UV light, especially UV-B (280–315 nm), has been found to be a potent factor for secondary metabolites in hairy root cultures. In *Catharanthus roseus* hairy roots, brief UV-B treatments immensely increased terpenoid indole alkaloid (TIA) biosynthesis by inducing transcripts of biosynthetic genes like strictosidine synthase (STR) and tryptophan decarboxylase (TDC). The response was time-sensitive, reflecting maintained elevations in some alkaloids such as lochnericine and declines in others, reflecting a shift in metabolic flux under UV stress (Binder et al., 2009).

In total, the results confirm light as an efficient optimization tool in hairy root culture technology. Through precise control of spectral quality, intensity, and photoperiod, researchers can effectively increase the quantity and quality of valuable metabolites like flavonoids, alkaloids, terpenoids, and anthocyanins.

11.7.4 Optimization through elicitor application in hairy root cultures

Elicitors are signal compounds that trigger plant defense pathways and induce the production of secondary metabolites in *in vitro* cultures like hairy root cultures. Instead of playing a direct role as metabolic effectors, they act as initiators of intricate cascades of signals. Elicitors are generally classified into two groups: endogenous and exogenous. Endogenous elicitors are produced within the host plant, frequently originating from fragments generated in plant–pathogen interactions or derived from signaling compounds involved in abiotic stresses. Common examples include oligogalacturonides from pectic polysaccharides, systemin, salicylic acid (SA), jasmonic acid (JA), and methyl jasmonate (MeJA). Exogenous elicitors are derived from external sources, that is, mostly microorganisms, insects, or other aggressors, and consist of substances such as microbial

TABLE 11.8 Types of biotic and abiotic elicitors for optimization of hairy root cultures.

Category	Elicitor type	Examples	Optimization	References
Biotic	Live bacterial cells	*Rhizobium* spp., *Bacillus cereus*	Early inoculation can suppress biomass; effects linked to bacterial polysaccharide–protein complexes	Cadiz et al. (2000); Wu et al. (2007); Wang and Wu (2013)
	Fungal and yeast Extracts	*Fusarium conglutinans*, *Phytophthora cinnamomi*, Yeast Extract (YE), *Penicillium chysogenum*	Crude extracts effective but composition varies → reproducibility issues	Mukundan and Hjortso (1990a); Bais et al. (2002); Liu et al. (1999); Chen et al. (2001); Wang and Wu (2013)
	Polysaccharides / oligosaccharides	Oligosaccharides from *C. gloeosporioides*, Oligogalacturonides from plant cell walls, crude polysaccharides from *B. cereus*, chitosan	Dose and molecular weight critical; crude fractions enhance both metabolites and root growth	Wang et al. (2006); Zhang et al. (2010); Zhao et al. (2010); Wang and Wu (2013)
	Lipid / protein-based	Cerebroside C (*Fusarium* sp.), glycoproteins from *Scenedesmus obliquus*	Potential eco-friendly elicitors with dual role in growth and metabolite synthesis	Wang et al. (2009); Wang and Wu (2013)
Abiotic	Heavy metal ions	Ag^{+}, Cd^{2+}, Cu^{2+}, $VOSO_4$, Se, Ni	High toxicity and growth inhibition at excess levels; optimization of dose essential	Khalili et al. (2020); Zheng and Wu (2004); Gangopadhyay et al. (2011); Furmanowa et al. (2000); Jeong and Park (2006); Wang and Wu (2013)
	Osmotic stress	NaCl, KCl, $CdCl_2$, sorbitol	Controlled osmotic stress enhances metabolites; excessive levels inhibit biomass	Jeong and Park (2006); Thakore et al. (2012); Shi et al. (2007); Wang and Wu (2013)

TABLE 11.9 Elicitation strategies as metabolic engineering tools for enhanced secondary metabolite production in hairy root cultures.

Elicitor agent	Elicitor (abiotic/biotic agent)	Plant species	Target metabolite(s)	References
Biotic elicitor	*Enterobacter sakazaki*	*Ammi majus*	Scopoletin	Staniszewska et al. (2003); Chandra and Chandra (2011)
	Lactobacillus helveticus	*Beta vulgaris*	Betalaines	Chandra and Chandra (2011); Thimmaraju et al. (2003)
	Rhizoctonia bataticola	*Solanum tuberosum*	Sesquiterpenes (rishitin, lubimin)	Komaraiah et al. (2003); Chandra and Chandra (2011)
	Phytophthora parasitica filtrate	*Cichorium intybus*	Esculin, Esculetin	Bais et al. (2000); Chandra and Chandra (2011)
	Pythium aphanidermatum	*Catharanthus roseus*	N-acetyl tryptamine	Eilert et al. (1986); Chandra and Chandra (2011)
	Trichoderma viridae	*C. roseus*	Ajmalicine	Namdeo et al. (2002); Chandra and Chandra (2011)
	Fusarium conglutinans	*Tagetes patula*	Thiophenes	Mukundan and Hjortso (1990) b; Chandra and Chandra (2011)
	Yeast extract	*Brugmansia candida*	Scopolamine, Hyoscyamine	Pitta–Alvarez et al. (2000); Chandra and Chandra (2011)
	Green alga (*Haematococcus pluvialis*) extract	*Beta vulgaris*	Betalaines	Rao et al. (2001); Chandra and Chandra (2011)
	Fungal cerebroside	*Artemisia annua*	Artemisinin	Wang et al. (2009); Chandra and Chandra (2011)
	Polysaccharides	*Salvia miltiorrhiza*	Tanshinone	Zhao et al. (2010)
	Oligogalacturonides	*Artemisia annua*	Artemisinin	Zhang et al. (2010)
	Heptasaccharide and octasaccharide	*Panax ginseng*	Ginseng saponin	Zhou et al. (2003)
	Cerebroside	*Artemisia annua*	Artemisinin	Wang et al. (2009)
	Glycoproteins	*Tanacetum parthenium*	Cis-C13-spiroketal enol ether epoxide	Wang and Wu (2013)

(*Continued*)

TABLE 11.9 (Continued)

Elicitor agent	Elicitor (abiotic/biotic agent)	Plant species	Target metabolite(s)	References
Abiotic elicitor	Chitosan, methyl jasmonate (MJ)	*Panax ginseng*	Ginsenosides	Palazón, Mallol, et al. (2003); Chandra and Chandra (2011)
	Tween 80	*Beta vulgaris*	Betalaines	Thimmaraju et al. (2003); Chandra and Chandra (2011)
	$CdCl_2$	*Brugmansia candida*	Hyoscyamine	Pitta–Alvarez et al. (2000); Chandra and Chandra (2011)
	MJ	*Centella asiatica*	Asiaticoside	Kim et al. (2007); Chandra and Chandra (2011)
	Ag^+	*Silybum marianum*	Silymarin (silybin, isosilybin, taxifolin, silycristin, silydianin)	Khalili et al. (2020); Chandra and Chandra (2011)
	Ag^+ ($AgNO_3$)	*Brugmansia candida*	Scopolamine Hyoscyamine	Pitta–Alvarez et al. (2000)
	Nitrate	*Atropa belladonna*	Tropane alkaloids	Chandra and Chandra (2011)
	Cu^{2+}, SA, MJ	*Angelica gigas*	Decursin, decursinol angelate	Rhee et al. (2010); Chandra and Chandra (2011)
	$CuCl_2$	*Plumbago indica*	Plumbagin	Gangopadhyay et al. (2011)
	Chitosan, MJ	*Artemisia annua*	Artemisinin	Putalun et al. (2007); Chandra and Chandra (2011)
	$VOSO_4$	*Panax ginseng*	Total ginsenoside	Palazón, Cusidó, et al. (2003)
	Selenium (selenite) Nickel ($NiSO_4$)	*Panax ginseng*	Ginseng saponin	Jeong and Park (2006)
	NaCl	*Panax ginseng*	Ginseng saponin	Jeong and Park (2006)
	Mannitol	*Catharanthus roseus*	Ajmalcine	Thakore et al. (2012)
	Sorbitol	*Salvia miltiorrhiza*	Tanshinones	Shi et al. (2007)

polysaccharides, glycoproteins, and peptides. For experimental and practical reasons, elicitors are generally categorized as biotic (derived from biological systems like plants, bacteria, or fungi) or abiotic (including chemical and physical factors, for example, UV radiation, heavy metals, osmotic stress) (Table 11.8). Both types have been extensively exploited in hairy root cultures as highly effective methods to increase the production of pharmaceutically and industrially significant secondary metabolites (Table 11.9) (Halder et al., 2019; Satdive et al., 2007; Wang & Wu, 2013).

11.7.4.1 Reaction kinetics of elicitor

For industrially scalable production of any system, basic foundation knowledge of reaction kinetics is essential to achieve quantitatively or qualitatively superior products. Integration of basic plant tissue culture with process engineering provides process escalation for commercial production with economic benefit.

The elicitation process can be modeled by modified first-order or logistic kinetic equations:

11.7.4.1.1 First-order kinetics

$$\frac{dP}{dt} = k_e\left(P_{\max} - P\right)$$

P=metabolite concentration at time t

P_{max}=is the maximum attainable concentration

k_e=is the elicitation rate constant

11.7.4.1.2 Logistic growth model with elicitation factor

$$P(t) = \frac{P_{\max}}{1 + e^{-k_e(t-t_0)}}$$

t_0=induction time point

k_e=reflects elicitor efficiency

11.7.4.1.3 Dual-kinetic model (growth-production coupling)

$$\frac{dP}{dt} = \alpha\mu X + \beta X$$

X=biomass concentration

μ=specific growth rate

α=growth-associated product formation

β=non-growth-associated metabolite production induced by elicitation

11.7.4.2 Molecular mechanism of elicitors

When an elicitor (which may be biotic, e.g., microbial cell wall components, or abiotic, e.g., nanoparticles, salts, or chemicals) docks onto particular elicitor-binding sites or receptors on hairy root cell plasma membrane, it triggers a cascade of intracellular signal transduction processes (Fig. 11.4) (Koutroumpi, 2021). One of the first events is the activation of nicotinamide adenine dinucleotide phosphate (NADPH) oxidase, which catalyzes the formation of reactive oxygen species (ROS) like superoxide radical and hydrogen peroxide, giving rise to so-called oxidative burst. Concurrently, the elicitor–receptor interaction initiates ion flows through the membrane: entry of calcium ions (Ca^{2+}) and exit of potassium ions (K^+), chloride ions (Cl^-), and protons (H^+). This results in cytoplasmic acidification and Ca^{2+} burst, functioning as a second messenger. The raised intracellular calcium level further stimulates guanine nucleotide-binding proteins (G-proteins) and subsequent enzymes like adenylate cyclase (AC) and other phospholipases like phospholipase A (PLA), phospholipase C (PLC), and phospholipase D (PLD). These enzymes generate crucial second messengers like cyclic adenosine monophosphate (cAMP), inositol triphosphate (IP_3), and diacylglycerol (DAG), which further amplify the signal. These events culminate in mitogen-activated protein kinases (MAPKs) getting activated, which get phosphorylated and then control several downstream pathways. Both MAPK activation and ROS signaling initiate the biosynthesis of plant hormones, that is, salicylic acid (SA), jasmonic acid (JA), and methyl jasmonate (MeJA) that help regulate gene transcription that results in secondary metabolite biosynthesis. Genes involved in secondary metabolism are expressed, ultimately providing the plant with enzymes that help synthesize secondary

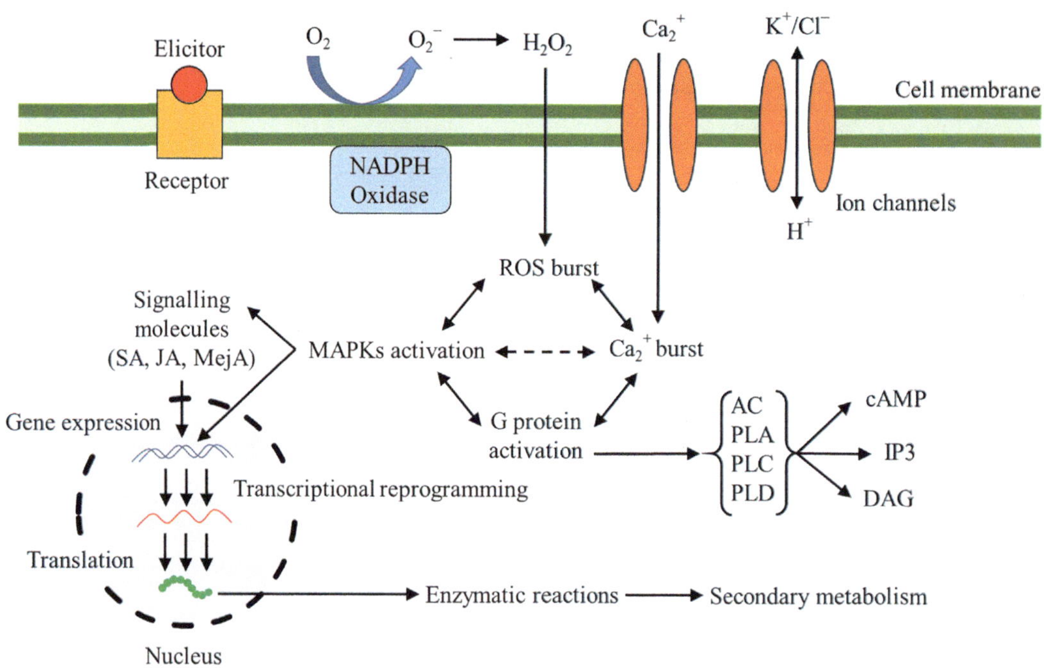

FIGURE 11.4 Molecular mechanism of elicitors.

TABLE 11.10 Nanoparticles used as elicitors in hairy root cultures for secondary metabolite production.

Nanoparticle (NPs)	Plant system (hairy root/ callus/cell suspension)	Secondary metabolites induced	Key findings / Notes	References
AgNPs	*Cucumis anguria* (hairy root)	Phenolics, flavonoids (flavonols)	Higher flavonol content (3136.20 μg/g) than control	Chung, Rajakumar, et al. (2018); Khan et al. (2021)
AgNPs	*Brassica rapa* (hairy root)	Phenolics, flavonoids (9 types: quercetin, kaempferol, rutin, etc.)	100 mg/L gave max TPC & TFC; 150 mg/L reduced content	Chung, Rekha, et al. (2018); Khan et al. (2021)
Fe_3O_4NPs	*Dracocephalum kotschyi* (hairy root)	Flavonoids, rosmarinic acid	Induced secondary metabolite gene expression	Nourozi et al. (2019)a; Khan et al. (2021)
Fe_3O_4NPs	*Cichorium intybus* (hairy root)	Flavonoids	Superior TFC: 4.65 mg g^{-1} DW	Khan et al. (2021)
SiO_2NPs	*D. kotschyi* (hairy root)	Flavonoids (cirsimaritin, xanthomicrol, isokaempferide)	Significant induction of anticancer flavonoids	Nourozi et al. (2019)b; Khan et al. (2021)
SiO_2NPs	*Hyoscyamus reticulatus* & *H. pusillus* (hairy root)	Secondary metabolites	Increased flavonoid production	Hedayati et al. (2020); Khan et al. (2021)

metabolites. After transcription and translation, secondary metabolite synthesis in hairy root cultures shifts the metabolic flux to produce bioactive molecules such as alkaloids, flavonoids, phenolic acids, terpenoids, and glycosides. As a result, perception and signaling produced by elicitors contribute to a broadly regulated biosynthetic response for secondary metabolites, enhancing the defense potential in plant cells and providing hairy root culture as a superb method for utilizing plant resources for metabolites (Koutroumpi, 2021; Liu et al., 2016; Seybold et al., 2014; Takeda et al., 2002; Trewavas & Malhó, 1998; Wojtaszek, 1997).

11.7.4.3 Nano-elicitation in hairy root cultures for secondary metabolite production

Among the strategies of elicitation, nanoparticles (NPs) have emerged as promising new elicitors because of their distinctive physicochemical characteristics. The term nano-elicitation describes the capability of nanoparticles to promote or induce secondary metabolism in plants by regulating physiological and biochemical pathways (Table 11.10) (Javed et al., 2022). When added to plant culture systems, nanoparticles can serve both as a source of micronutrients and as an eliciting agent, inducing defense-related signaling cascades. The mechanism of action of NPs in elicitation is usually attributed to the production of reactive oxygen species (ROS), which in turn trigger enzymatic and nonenzymatic antioxidant defense mechanisms (Murthy et al., 2024). Through this activation, the biosynthesis of phenolics, alkaloids, flavonoids, and other antioxidant and pharmacologically active metabolites is enhanced (Tariverdizadeh et al., 2021). NP signaling leads to ROS (reactive oxygen species) bursts, amplifies antioxidant enzymes such as catalase (CAT), superoxide dismutase (SOD), and ascorbate peroxidase (APX), and modulates secondary metabolic pathway genes of phenylalanine ammonia lyase (PAL), chalcone isomerase (CHI), and flavonoid synthase (FLS). This both maximizes the yields of metabolites and enhances antioxidant, antimicrobial, and anticancer activity of plant-derived products. New NP applications like "ligand fishing," which harness the powers of NP action, allow you to specifically pick bioactive metabolites directly from living cells without harming plant viability. Thus, NP-mediated elicitation is a new green and highly effective tool for increasing metabolite production in hairy root biotechnology (Chung, Rekha, et al., 2018; Lala, 2021). The nano-elicitation mechanism in hairy root cultures is very similar to the general elicitation mechanism. Nanoparticles (or NP-elicitor conjugates) are identified by the receptors in the plasma membrane such as the receptor-like kinases (RLKs), inducing early cellular events like ion fluxes (Ca^{2+} influx and K^+/Cl^- efflux), cytoplasmic acidification, and the activation of protein kinases and MAP kinases (MAPKs). These events stimulate phospholipase signaling cascades (PLC/IP_3–DAG/PKC), resulting in the release of intracellular Ca^{2+} and activation of calcium sensors (calmodulins and CDPKs). Increased activity of NADPH oxidase and ROS generation are induced by elevated Ca^{2+} levels, activating the defense pathways through secondary messengers derived from calcium and ROS (Marslin et al., 2017). These signaling cascades may also intersect with phytohormones (salicylic acid, jasmonic acid, and ethylene) and promote the enhancement of transcription of biosynthetic genes and contribute to the accumulation of secondary metabolites (phenolics, alkaloids, and terpenoids) (Garcia-Brugger et al., 2006; Khan et al., 2021; Marslin et al., 2017; Ramirez-Estrada et al., 2016; Zhao et al., 2005).

11.8 Pilot production of hairy root cultures using bioreactor-driven scale-up

Bioreactors have an important function in massive plant secondary metabolite production using hairy root cultures (HRC) (Mishra & Ranjan, 2008). They are specially engineered to overcome limitations like unstable productivity, sensitivity of cells, low oxygen requirement, and slow rates of growth in plant systems (Supriya et al., 2020). Various sizes and types of bioreactors are utilized to create optimal growth and metabolite production conditions. Hairy root cultures are especially useful because of their genetic constancy, high biomass production rate, and increased metabolite yield, which make them a viable and effective system for industrial purposes (Mehrotra et al., 2010). Though traditional *in vitro* culture systems such as Petri plates or flasks are appropriate for lab-scale studies, they are unable to cope with industrial needs due to space and nutrient limitations, oxygen transfer, and scalability. Bioreactors are able to circumvent these limitations through aeration, mixing, pH control, and temperature control, all of which greatly promote biomass formation and production of metabolites. This renders them especially suitable for filling the demand–supply gap of plant-derived pharmaceuticals, nutraceuticals, and other high-value products. Furthermore, hairy roots established in bioreactors ensure economic consistency by economizing on production via persistent culture, automation, and the potential for year-round production irrespective of climatic or seasonal changes. Therefore, going for bioreactors in hairy root cultivation guarantees sustainable, reproducible, and cost-saving production at an industrial level.

11.8.1 Key parameters governing hairy root growth in bioreactors

Plant cell and hairy root cultures utilize bioreactors that are typically operated under similar conditions as microbial bioreactors (Fig. 11.5) (Eibl & Eibl, 2008), but it is clear that they need to be designed more thoroughly to reflect the distinct features of plant tissues. Plant cells have a larger volume, are slow-growing, and they either aggregate in suspension or develop structured tissues, such as hairy roots, when cultured (SU, 2008). These features require the use of low-shear bioreactors that do not introduce any form of physical stress, as well as optimum mixing and aeration rates to ensure oxygen and nutrient distribution to the plant tissues. Aerobic conditions are necessary, but the method of agitation and aeration must be qualitatively thought through to maximize oxygen transfer while minimizing shear stress (Ziv, 2005). One of the biggest challenges of such systems is the fact that it is hard to directly sample biomass at set intervals because the heterogeneous and aggregated state of roots makes precise measurements problematic. An indirect approach like the measurement of the conductivity of the culture medium is therefore used, because changes in conductivity can indicate nutrient uptake and release of metabolites, and thus are a good surrogate of biomass growth (Hahlbrock et al., 1974; Suresh et al., 2001).

Successful hairy root cultivation in bioreactors relies on various physiological and engineering parameters that have direct effects on secondary metabolite formation and biomass accumulation. These parameters not only affect the growth kinetics but also the whole process scalability and efficiency. A comparative survey of the most relevant influencing parameters is shown in Table 11.11, emphasizing their respective effects on hairy root cultures.

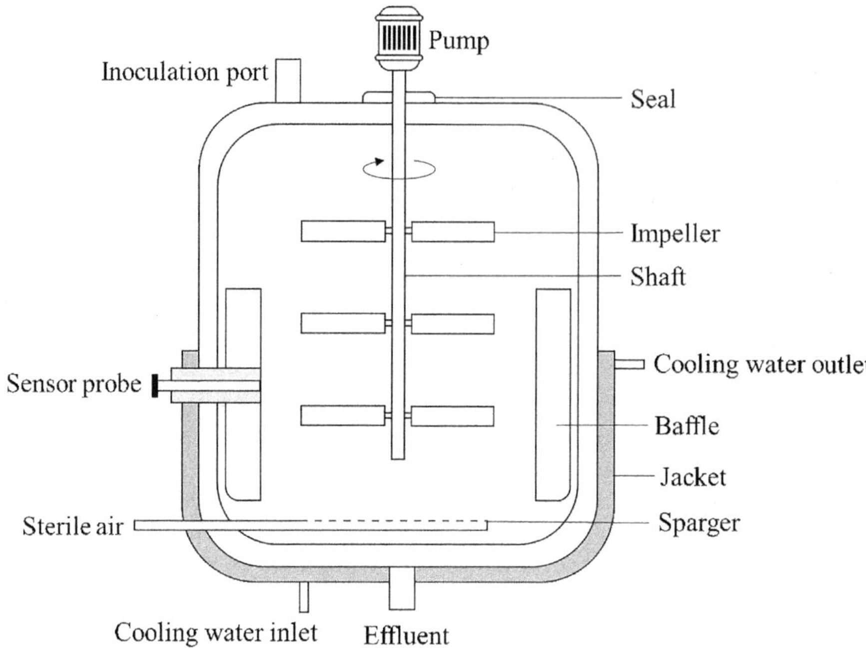

FIGURE 11.5 Schematic representation of bioreactor.

TABLE 11.11 Key factors influencing plant cell and hairy root cultivation in bioreactors.

Factor	Remarks	Reference
Oxygen transfer	Aeration strategy must be optimized	Yu and Doran (1994)
Shear stress	Use low-shear designs (airlift, bubble column)	Choi et al. (2008)
Nutrient availability	Media optimization required	Sivakumar et al. (2005)
pH control	Buffered media recommended	Neelwarne and Thimmaraju (2009)
Temperature	Generally, 23°C–28°C optimal	Buitelaar et al. (1991)
Mixing and homogeneity	Must balance with shear protection	Stiles and Liu (2013)
Foaming	Antifoam agents or design adjustments	Krasniqi et al. (2024)
Metabolite induction	Strategy depends on target product	Wang and Wu (2013)

11.8.2 Types of bioreactors used as production factory for plant-derived therapeutics from hairy root culture

There are several types of bioreactors applied for plant cell and hairy root cultivation, each having specific advantages based on the culture system and product. Stirred-tank,

TABLE 11.12 Bioreactors used for augmented secondary metabolite production from *Rhizobium rhizogenes*-induced hairy root cultures of medicinal plants.

Sl. no	Medicinal plants	Bioreactor used	Secondary metabolites	References
1.	*Beta vulgaris*	Bubble column reactor	Betalaine	Suresh et al. (2004); Supriya et al. (2020)
		Airlift bioreactor	Betacyanin	Georgiev et al. (2012), Supriya et al. (2020)
2.	*Calystegia sepium*	Stirred-tank bioreactor	Tropane alkaloids	Jung and Tepfer (1987)
3.	*Ophiorrhiza pumila*	Glass bioreactor	Camptothecin	Supriya et al. (2020); Sudo et al. (2002)
4.	*Astragalus membranaceus*	Airlift bioreactor	Astragaloside IV and polysaccharide	Wu et al. (2011); Supriya et al. (2020)
5.	*Pueraria phaseoloides*	Airlift bioreactor	Puerarin	Mirmazloum et al. (2024); Shi and Kintzios (2003); Supriya et al. (2020)
6.	*Panax ginseng*	Balloon-type bubble bioreactor	Ginsenosides	Baque, Elgirban, et al. (2012); Baque, Moh, et al. (2012); Supriya et al. (2020)
7.	*Atropa belladonna*	Stirred-tank bioreactor	Curcumin	Supriya et al. (2020); Singh et al. (2021)
8.	*Artemisia annua*	Stirred tank bioreactor	Artemisinin	Al-Khayri et al. (2022); Supriya et al. (2020)
9.	*Catharanthus roseus*	Bubble column bioreactor	Ajmalicine	Thakore et al. (2017); Verma et al. (2012); Supriya et al. (2020)
10.	*Azadirachta indica*	Stirred-tank bioreactor	Azadirachtin	Srivastava and Srivastava (2012); Supriya et al. (2020)
11.	*Astragalus membranaceus*	Airlift bioreactor	Astragalosides	Du et al. (2003)
12.	*Hypericum perforatum*	Airlift bioreactor	Hypericin	Montazeri et al. (2018)
13.	*Harpagophytum procumbens*	Bubble column bioreactor	Iridoid glycosides	Ludwig-Müller et al. (2008)
14.	*Azadirachta indica*	Nutrient mist reactors	Azadirachtin	Valdiani et al. (2019)

(Continued)

TABLE 11.12 (Continued)

Sl. no	Medicinal plants	Bioreactor used	Secondary metabolites	References
15.	*Stizolobium hassjoo*	Nutrient mist reactors	L-DOPA	Sung and Huang (2006); Huang et al. (2004)
16.	*Tagetes patula*	Nutrient mist reactors	Thiophene	Suresh et al. (2005)
17.	*Plumbago indica*	Bubble column bioreactor	Plumbagin	Gangopadhyay et al. (2011)

bubble column, and airlift bioreactors are most commonly utilized among them (Table 11.12). *Rhizobium rhizogenes*-induced hairy roots of medicinal plants are especially ideal for such systems because of their rapid growth, genetic stability, and capability to produce valuable secondary metabolites like alkaloids, flavonoids, terpenoids, and phenolics.

11.8.2.1 Stirred-tank reactors

Stirred-tank reactors (STRs) are traditionally employed in biotechnology as mechanically agitated liquid-phase bioreactors (Fig. 11.6) (Mishra & Ranjan, 2008). The design consists of an impeller system with an associated aeration device, sparging compressed air into the medium to enable efficient oxygen supply and blending (Nickel et al., 2024). Bubbles are dispersed in the impeller zone into the continuous liquid phase, providing aerobic growing conditions (Martín et al., 2008). For hairy root cultures (HRCs), though, the classical STR design is problematic because shear stress from impeller rotation can induce wound response and undesirable callus formation (Srikantan & Srivastava, 2018). To overcome this, STR designs have been modified. These involve the use of flat-blade turbines to minimize turbulence and the immobilization of the roots on support material such as steel or nylon meshes. These adjustments work to reduce shear damage, root morphology stabilization, and increase metabolite production with the benefit of retaining the advantages of STRs in mass production (Mishra & Ranjan, 2008).

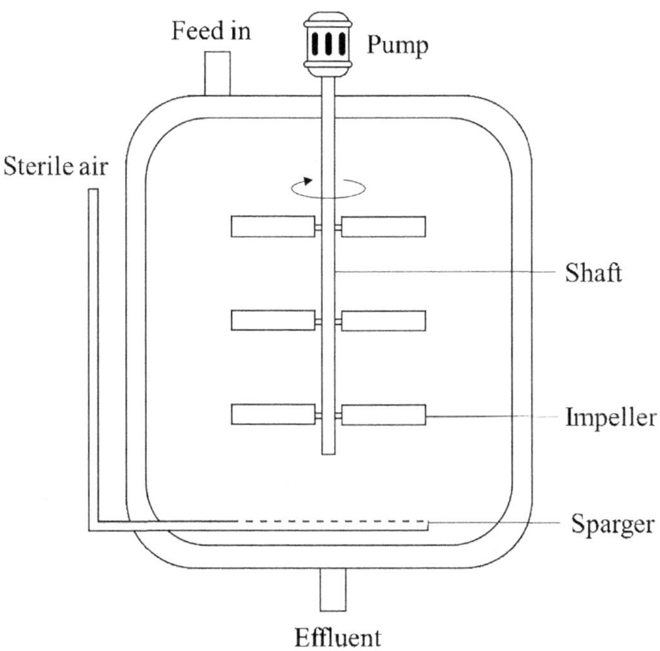

FIGURE 11.6 Schematic representation of stirred-tank reactors (STRs).

11.8.2.2 *Airlift bioreactors*

Airlift bioreactors are engineered variants of bubble column bioreactors (Doran, 1995) with a draught tube that improves mixing and oxygen transfer efficiency by providing circulation via density gradients between downcomer and riser (de Jesus et al., 2017). The air sparging decreases the liquid density within the riser, which creates an upward movement, while the heavy liquid in the downcomer moves downwards, leading to effective mixing with low-shear stress (Fig. 11.7) (Han, 2017). Depending on the design, airlift bioreactors may be of either internal loop type, with the draught tube inside the vessel, or external loop type, with the downcomer physically distant from the main column (Gumery et al., 2009). For hairy root cultures, airlift bioreactors have certain advantages over stirred-tank reactors (STRs) since they cause minimal mechanical damage while allowing for stable growth and metabolite accumulation and also reduce maintenance time and costs (Valdiani et al., 2019).

11.8.2.3 *Bubble column bioreactors*

Bubble column bioreactors are agitated pneumatic systems commonly utilized for the growth of hairy root cultures owing to their easy design and effectiveness (Stiles & Liu, 2013). They structurally comprise a cylindrical column vertically placed where roots are submerged in the nutrient-sufficient liquid medium (Fig. 11.8) (Srikantan & Srivastava, 2018). Aeration and mixing are obtained by sparging compressed gas or air through a diffuser at the bottom of the reactor, forming a stream of continuous bubbles rising (Moresi, 2025). Rising flow of bubbles encourages circulation of the medium, maintaining uniform

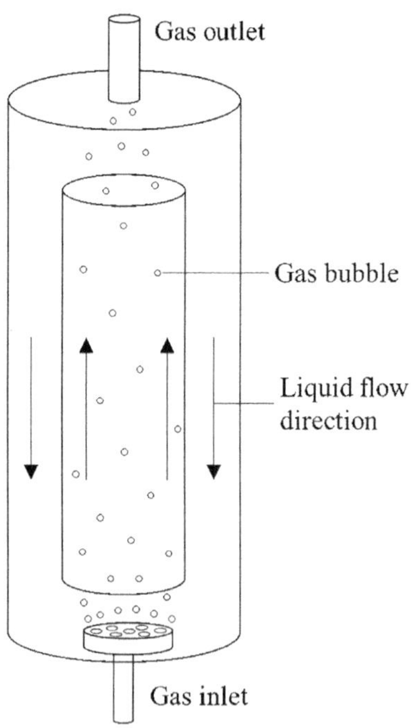

FIGURE 11.7 Schematic representation of airlift bioreactors.

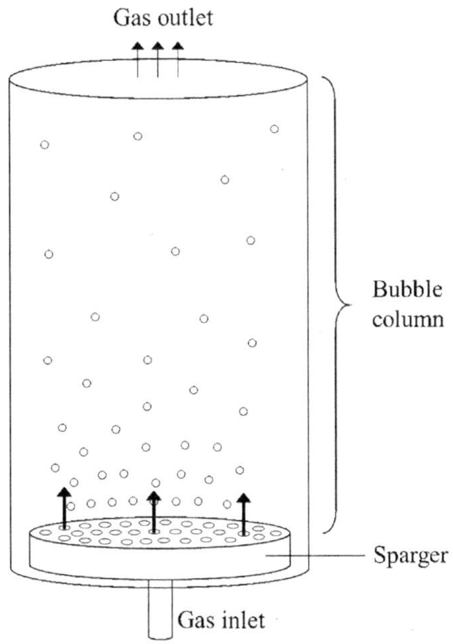

FIGURE 11.8 Schematic representation of bubble column bioreactors.

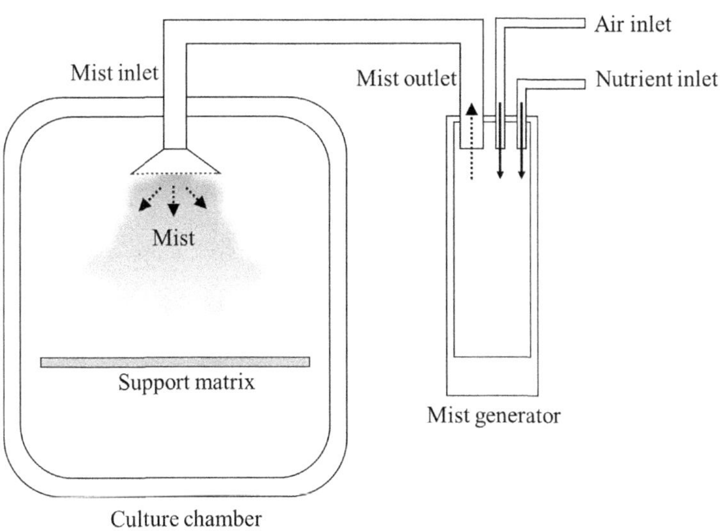

FIGURE 11.9 Schematic representation of nutrient mist reactors (NMRs).

mixing, improved mass transfer of gas and liquid, and effective oxygen delivery to the roots without mechanical agitation (Patra & Srivastava, 2016). This configuration is a major benefit for hairy root cultures since it minimizes shear damage typical in stirred-tank reactors, thus maintaining the morphology and metabolic activity of the root.

11.8.2.4 Nutrient mist reactors

Mist bioreactors offer a compelling alternative to more conventional submerged airlift systems for hairy root culture (Towler et al., 2008; Valdiani et al., 2019). In nutrient mist reactors (NMRs), the culture is anchored to a mesh support and remains suspended in the gas phase, while the nutrient medium is delivered as mist via ultrasonic transducers producing microdroplets (Fig. 11.9) (0.5–30.0 mm) (Steingroewer et al., 2013; Valdiani et al., 2019). Because the culture is suspended in continuous gas phase, structures (such as horizontal mesh supports) or hollow structures (e.g., stainless-steel meshes) are required to anchor the roots (Patra & Srivastava, 2016). NMRs also have some advantages for hairy root cultures when compared to more traditional systems, with relevant advantages including ease of operation, low-shear stress, efficient and rapid nutrient supply, an effective paradigm to remove toxic metabolites, and ease of scalability. In addition, because the culture is in continuous gas phase, oxygen transfer is improved, and the disadvantages of gas transfer in traditional bioreactor designs can be overcome (Sharma & Shahzad, 2013).

11.8.3 Product recovery and purification strategies for secondary metabolites from hairy root bioreactors

The recovery and purification of secondary metabolites from hairy root bioreactors is an important process that defines the efficiency, economy, and utilization of the production process (Fig. 11.10) (Das et al., 2025). Since hairy root cultures can accumulate

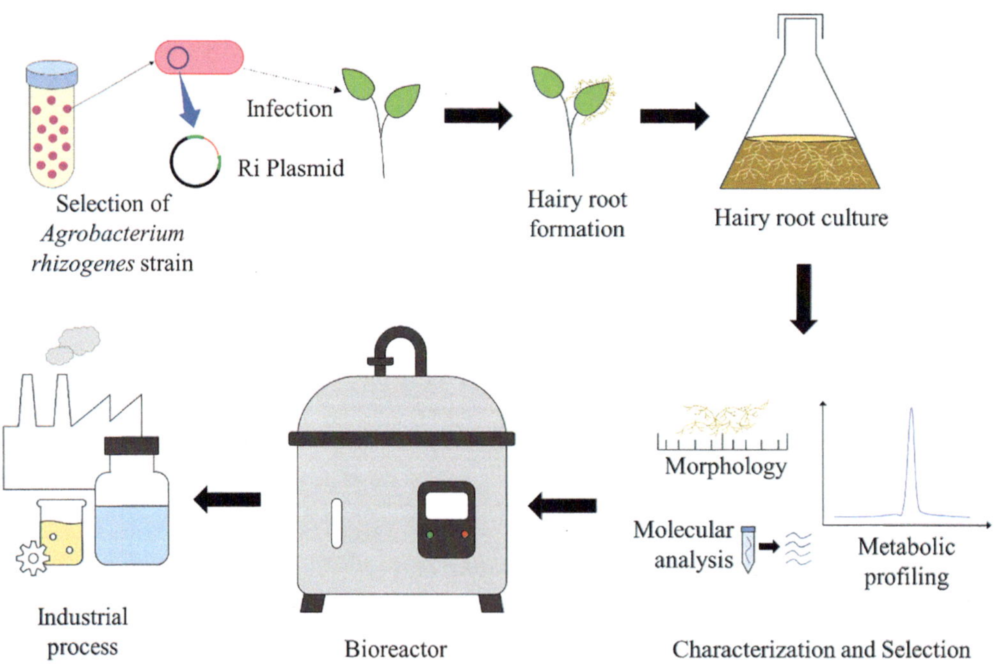

FIGURE 11.10 Subsequent steps from hairy root induction to final production.

metabolites in two locations (intracellular, within root biomass, and extracellular, secreted into the culture medium), the recovery strategy elected depends primarily on where the target compound is located and its chemical properties (Halder et al., 2019; Srivastava & Srivastava, 2007). The recovery process begins with primary recovery, whereby biomass is removed from the culture medium. The root biomass, cellular debris, and other particulates can be removed by several techniques (i.e., filtration and centrifugation) that produce a cleared broth suitable for further processing (Parkinson et al., 1990;Baek et al., 2022). For metabolites contained in the biomass, mechanical leaching of the roots followed by extraction in solvent is typically required (Gheisary et al., 2025). If sensitive, labile compounds are targeted, aqueous two-phase extraction (ATPS), which entirely avoids organic solvents, can represent a more amenable route by encouraging partitioning of metabolites in two immiscible aqueous phases that risk degradation (Fierascu et al., 2020). The method of extracting bioactive compounds has been improved thanks to newer "green" extraction methods that can provide more sustainable alternatives when compared to traditional methods. New methods include microwave extraction (ME), which utilizes microwave energy to increase the rate of cell breakage and extraction of the solvent (Afsharzadeh et al., 2025); pressurized liquid extraction (PLE), subcritical water extraction (SWE), and supercritical fluid extraction (SFE), which utilize heat and pressure to enhance solubility and diffusion and extraction under mass transfer; and ultrasound-assisted extraction (UAE), which helps further develop solvent dispersion through cavitation caused by ultrasonic waves (Fierascu et al., 2020; Ghasemy-Piranloo et al., 2022). These methods could likewise potentially reduce extraction times compared to classical methods, utilize less solvent, and provide larger yields. A thorough method of preconcentration and removal of low-level impurities can include membrane filtration and ultrafiltration. Both of these methods consist of passive membrane filtration and are based on size separation. That is, a semi-permeable membrane that would allow the passage of solvents and low-level impurities respects not permitting the larger molecules to pass. This not only facilitates concentration of the sought metabolites but also allows effective elimination of undesired low-molecular-weight compounds (Shahnawaz et al., 2021;Cai et al., 2012). Finally, the purity and increased yields of the recovered products are assured by advanced purification processes that aim to remove coextracted impurities such as proteins, polysaccharides, pigments, and phenolic compounds that may interfere with their bioactivity or stability as a final product. Accurate quantitation, purity assessments, and structural identification of the metabolites (as well as standardization for pharmaceuticals and nutraceuticals) rely upon rigorous analytical techniques such as HPLC, LC-MS, and NMR spectroscopy (Sharma et al., 2025; Tisserant et al., 2016).

11.8.4 Engineering challenges in bioreactor design for scalable hairy root cultivation

The design of bioreactors for hairy root cultures is confronted with some serious challenges. Scaling-up from shake flasks to the industrial scale is challenging due to hairy roots' intricate morphology and nonhomogeneous growth. Their fibrous and entangled network hinders nutrient and oxygen diffusion, which tends to create nonhomogeneous conditions and senescent tissues. Oxygen limitation is a key growth-limiting factor, since even small reductions in dissolved oxygen content will depress growth and synthesis of secondary metabolites. Hairy roots are also very responsive to environmental factors like temperature, shear, and medium composition, where root density, thickness, and length can have dramatic impacts on metabolite production. Intensive mixing enhances mass transfer but induces hydrodynamic shear stress, injuring delicate roots and frequently causing callus formation, hence necessitating the use of designs that avoid direct impeller contact. Increased bioreactor volume results in pressure from the weight of the biomass itself, subjecting roots to mechanical stress, and careful container and volume of medium consideration is necessary. Long-term aseptic maintenance is also critical for certain hairy root species. Besides, real-time tracking of metabolite and biomass formation is technologically complicated, and indirect methods such as conductivity changes tend to be employed. Support matrices are sometimes needed to stabilize root growth, and light conditions need to be adjusted such that some of the secondary metabolites form in light while others form in darkness. In general, the most important challenge is to design a bioreactor system that has the efficient mass transfer of gases and nutrients but minimizes shear stress, reconciling several conflicting requirements for successful large-scale hairy root culture (Goswami et al., 2018; S. A. Khan et al., 2018; Paek et al., 2014; Patra & Srivastava, 2016; Srivastava & Srivastava, 2007; Stiles & Liu, 2013).

11.8.5 Reaction kinetics for large-scale production of hairy root cultures in bioreactors

For scaling up the medicinal hairy roots, understanding and monitoring reaction kinetics of production in bioreactor seems to be the foundation of realistic industrial implementation. For smooth running of the system and getting genetically and biochemically high-yielding hairy roots, a few important parameters like biomass development, media utilization, and product development relationship need to be considered. Increasing root mass quantity follows Monod or logistic-type kinetics, emphasizing specific growth rate, which correlates with concentration of the media (Berzin et al., 1997). How much media is used by the hairy roots cultivated within bioreactors is associated stoichiometrically to the product (yield) outcome. In general, plant secondary metabolite production follows the Luedeking–Piret model (Shibasaki et al., 1993), which considers both growth-associated and non-growth-associated terms, even including the effect of any external physical or chemical stimulators. Apart from these, two important factors, oxygen (gas–liquid mass transfer and cellular oxygen uptake) and shear sensitivity (high agitation and excess aeration damage root tips and hamper production) are very important parameters for quantitative enhancement of the hairy roots with augmented secondary metabolites.

The following reaction kinetics are used to monitor bioreactor-assisted large scale of hairy roots for plant secondary metabolite production-

11.8.5.1 Biomass (hairy-root) growth

$$r_X = \mu(S)X,\ \mu(S) = \mu_{\max}\frac{S}{K_S + S}\left(1 - \frac{X}{X_{\max}}\right)$$

r_X=Biomass growth rate (g/L.h)
$\mu(S)$=Specific growth rate as a function of substrate concentration (1/h)
μ_{max}=Maximum specific growth rate (1/h)
S=Substrate concentration (g/L)
K_S=Half-saturation (Monod) constant (g/L)
X=Biomass concentration (g/L)
X_{max}=Maximum biomass concentration (g/L)

11.8.5.2 Substrate consumption

$$r_S = -\frac{1}{Y_{X/S}}r_X - m_s X$$

r_S=Substrate consumption rate (g/L.h)
$Y_{X/S}$=Biomass yield coefficient on substrate (g biomass/g substrate)
m_s=Maintenance coefficient (g substrate/g biomass.h)

11.8.5.3 Product (secondary metabolite) formation: Luedeking–Piret with elicitor term

$$r_P = \alpha r_X + \beta X + \gamma\frac{E}{K_E + E}X$$

(Luedeking–Piret with elicitor term)
r_P=Product formation rate (g/L.h)
α=Growth-associated product formation constant (g product/g biomass formed)
β=Non-growth-associated product formation constant (g product/g biomass.h)
γ=Elicitor-induced product formation constant (g product/g biomass.h)
E=Elicitor concentration (mg/L or g/L depending on system)
K_E=Elicitor saturation constant (mg/L or g/L)

11.8.5.4 Oxygen (gas–liquid transfer and uptake)

$$\frac{dC_{L,O_2}}{dt} = k_L a(C^*_{O_2} - C_{L,O_2}) - q_{O_2}X$$

$C_{L,O2}$=Dissolved oxygen concentration in liquid phase (mg/L or mmol/L)

C^*_{O2}=Saturated dissolved oxygen concentration at equilibrium with gas phase (mg/L or mmol/L)

k_La=Volumetric mass transfer coefficient (1/h)

q_{O2}=Specific oxygen uptake rate (mg O2/g biomass.h)

11.8.5.5 Decay/Shear damage (optional lumped term)

$$r_d = k_d X,\ k_d = k_{d0} + k_{shear} N^p$$

r_d=Decay (death) rate due to shear or stress (g/L.h)

k_d=Decay constant (1/h)

k_{d0}=Basal decay rate constant (1/h)

k_{shear}=Shear-induced death constant

N=Agitation speed (rpm)

p=Shear exponent (dimensionless, indicates sensitivity to agitation).

Finally, it can be said that for creating suitable bioreaction operation maintenance of low-shear sensitivity, high oxygen concentration and targeted nutrient delivery are required throughout production operation. Integrating the approach of process development with automation and AI-enabled monitoring systems may provide herbal industries an alternative strategy to minimize agricultural dependency for source material collection.

11.9 Precise monitoring, intelligent prediction, and adaptive decision-making for modular system optimization

Computers can now carry out a variety of cognitive activities, including planning, searching, representation, learning from experience, voice and pattern recognition, heuristic categorization, and genetic algorithms, which are all specifically connected to learning and problem solving (Sarker, 2022). In order to explain and forecast many unpredictable biological bases, AI is divided into a variety of methods, including machine learning, ANNs, genetic algorithms, data mining, support vector machines (SVM), Random Forest, and natural language processing. The aforementioned learning algorithms are used as a reliable and effective computational approach. ANNs are the most widely utilized and prevalent form of AI; the primary goal of which is to mimic the human brain by creating a sophisticated black-box model that is derived from intricate mathematical algorithms that functions entirely in terms of input and output provided to the system without any prior understanding of its internal workings (Prakash et al., 2010; Alzubaidi et al., 2021). A "three-layered network that consists of an input layer, a hidden or interactive layer, and an output layer" is the fundamental design of an ANN. The algorithm's complexity is defined by a number of hidden layers. Neurons in the human brain are connected by axons, which are activated electrochemically when they receive messages from other neurons through synaptic connections, and external inputs are taken up by dendrites, which then produce electrical impulses that move through the neural network and transmit data between cells. Similarly, the entire learning process of ANNs is dependent on the "weight" (Wn) of the edges that connect each ANN, just like axons do, and it is made up of multiple nodes, which mimics the neurons in the human brain (Dayhoff & DeLeo, 2001; Dias et al., 2004; Ding et al., 2013; Cabaneros et al., 2019).

Biosensors are defined as devices that detect chemical compounds, typically by means of electrical, thermal, or optical signals, using particular biochemical reactions mediated by immune systems, tissues, organelles, or entire cells (Mehrotra, 2016). Biosensors, in association with machine learning models, can be used rigorously in optimization of the plant tissue culture process and production of secondary metabolites.

11.9.1 Different machine learning models

Numerous ANNs, including generalized regression neural networks (GRNN), probabilistic neural networks (PNN), radial basis functions (RBF), and multilayer perceptrons (MLP), have vast applications in the field of plant biology (Gardner & Dorling, 1998; Ali et al., 2012; Abiodun et al., 2019; Zarbakhsh & Shahsavar, 2022). AI operates on user-provided input, after which the training algorithm adjusts the network parameters and generates output data that is as accurate as feasible. Both supervised and unsupervised learning can be used to train an ANN, and a delta rule is frequently used. In supervised learning, each neuron must be given a precise weight in order to train the network (Dayhoff & DeLeo, 2001; Zarbakhsh & Shahsavar, 2022; Brownlee, 2023). The most popular machine learning models, either alone or in conjunction with ANNs, are Random Forest and SVM. Regression analysis, clustering, and object classification are commonly accomplished using SVM and Random Forest approaches (Guyon et al., 2002; Burgos-Artizzu et al., 2011; Maxwell et al., 2018; Tufail et al., 2023). Additionally, "Decision Trees" are widely used to handle nonlinear biological data. Following the supervised technique to teach various decision tree architectures to categorize the data with higher efficiency using relatively basic rules of classification, decision trees replicate the typical tree structures made up of a leaf node, inner nodes, and root nodes (Fu et al., 2023; Akkad et al., 2023; An et al., 2023).

11.9.2 Hybrid machine learning models and their use in plant biology

As the machine learning research community grows, scientists are working to create hybrid AI models that combine neural networks and symbolic AI techniques to get beyond the limitations of deep learning and ANNs. Two hybrid models that are highly applicable to plant biological processes are Neuro-fuzzy logic and Genetic algorithms. To more effectively optimize biological data while overcoming all the constraints, ANN, Neuro-fuzzy, and Genetic algorithms are typically employed in conjunction with one another (Corchado and Aiken, 2003; Elbaz et al., 2019; Jamshidi et al., 2020). ANNs and fuzzy logic are combined to create neuro-fuzzy logic (Du and Wolfe, 1995). Neuro-fuzzy systems for function approximation, adaptive spline modeling of data, and adaptive neuro-fuzzy inference systems are a few of the neuro-fuzzy approaches that have been created and put into practice. In the study of plant biological processes, the application of optimization approaches is crucial because the optimization algorithm seeks to determine the optimal input parameter to provide the lowest and largest output, which benefits the researchers by saving money and time. There are two categories of optimization techniques: traditional algorithms and evolutionary algorithms (Hesami and Jones, 2020; Arab et al., 2018). The most dependable optimization method is a genetic algorithm since it generates the finest solutions in each generation and uses them to improve subsequent solutions, improving the population as a whole generation after generation (Hassanat et al., 2019; Katoch et al., 2021).

11.9.3 Use of AI and machine learning in plant tissue culture

The use of AI in plant tissue culture techniques is multidirectional, and it can be used to predict and optimize different outcomes at most of the steps of the operation, that is, from sterilization to optimal culture conditions and output of biomass. Preventing contamination is essential for good micropropagation in tissue culture. Both internal and exogenous sources of contamination are present in cultured biological substances, and even a minor error in the explants sterilization process results in the waste of labor, chemicals, and time. To remove the contamination, the explant must be submerged in varying concentrations of sterilizing agents and treated with a variety of disinfectants. By forecasting the immersion duration and concentration of different sterilizers, AI demonstrated its effectiveness in lowering the contamination percentage in micropropagation of plant species *Bellevalia sarmatica*, *Echinacea purpurea*, *Nigella damascena*, *Chrysanthemum*, and *Cannabis* without compromising their explant viability (Ivashchuk et al., 2018; Hesami, Naderi, Tohidfar, 2019; Aasim et al., 2022). The fundamental medium for plant micropropagation is Murashige and Skoog's media, although each plant has unique growth requirements for nutrients. Numerous factors, including the kind and concentration of nutrients, are involved in the creation of media, making it challenging to plan and analyze using conventional statistical techniques. By adjusting eight different input variables, including NO_3^-, Mg^{2+}, NH^{4+}, K^+, Ca^{2+}, PO_4^{3-}, SO_4^{2-}, and Cl^-, ANN and genetic algorithms have been successfully trained with a variety of media formulations with different nutrient variables. The media were optimized for proliferation rate, shoot tip necrosis, shoot length, vitrification, and chlorosis, as well as for the number of microshoots for each explant, shoot length, and overall productivity (Hesami, Naderi, Tohidfar, Yoosefzadeh-Najafabadi, 2019). For the micropropagation of pistachio rootstocks, researchers were able to determine changes in the quality and formation of the shoots, the pace of proliferation, the length of the shoots, and the amount of necrosis at the tips of the shoots by varying the concentrations of macronutrients and micronutrients, as well as the quantities of the 18 cations and anions involved (NH_4^+, K^+, NO_3^-, Ca^{2+}, Mg^{2+}, $S_2O_4^{2-}$, Cl^-, Zn^{2+}, Cu^{2+}, BO_3^{3-}, Mn^{2+}, Fe^{2+}, Co^{2+}, MoO_4^{2-}, I^-, Na^+ and $EDTA^-$). Finally, they optimized the media plant growth parameters by applying hybrid algorithms like Neuro-fuzzy logic with ANN (Nezami-Alanagh et al., 2018). Similarly, techniques including Genetic algorithms, Neuro-fuzzy logic, and ANN have been effectively used to optimize the media plant development parameters for the micropropagation of *Actinidia arguta* and *Pyro dwarf* pear rootstocks (Jamshidi et al., 2016; Hameg et al., 2020). As hairy root culture is an efficient technique to enhance the production of secondary metabolites, models of MLP have been created for different plant species containing important secondary metabolites for human use, which are summarized in Table 11.13, in an effort to model and forecast the mean of fresh root weight, total biomass accumulation, and secondary metabolite production.

TABLE 11.13 Optimization of hairy root cultures using different machine learning model.

Sl no.	Machine learning model	Plant species	Data provided	Data obtained	References
1.	Multilayer Perceptrons	*Glycyrrhiza glabra*	Density, size, and fresh weight of innoculum, volume and pH of the medium, month of inoculation, temperature of the culture room, and total biomass after culture	Prediction for average weight of fresh roots	Mehrotra et al. (2008)
		Evolvulus alsinoides	Concentration of carbon quantum dots as elicitors, culture conditions	Prediction for amount of triterpenoid accumulation in hairy roots	Awere et al. (2024)
2.	Multilayer Perceptrons-Recurrent Neural Networks	*Glycyrrhiza glabra*	Density of the inoculums, pH, and sucrose content of the growth medium	Fresh weight of hairy roots	Prakash et al. (2010)
3.	Artificial Neural Network-Hidden Markov Model	*Rauwolfia serpentina*	The concentration of sucrose and nitrate in the growth medium, the volume of the solution per culture vessel, and the pH of the liquid growth medium	Fresh weight of hairy roots	Mehrotra et al. (2013)
4.	Artificial Neural Network-Response Surface Methodology & Kriging	*Datura stramonium*	The concentration of elicitor (salicylic acid) and the exposure time of the cultured hairy roots to the elicitor	Prediction for amount of produced hyoscyamine	Amdoun et al. (2019)
5.	Generalized Regression Neural Network- Back Propagation Neural Network	*Vinca minor*	Amounts of precursor molecules: shikimate, tryptophan, tryptamine, loganin, and secologanin; cyclooxygenase inhibitors, hydroxylase, and peroxidase elicitors	Total accumulated biomass of hairy roots, concentration of alkaloid produced, amount of vincamine produced	Verma et al. (2016)

(Continued)

TABLE 11.13 (Continued)

Sl no.	Machine learning model	Plant species	Data provided	Data obtained	References
6.	Back Propagation Neural Network (Feed forward, cascade forward, Layer recurrent network)	*Artemisia annua*	Packing fraction, media volume, culture duration in days, sucrose content, and starting inoculum on a dry weight basis	Resulting biomass of hairy roots on a dry weight basis	Osama et al. (2013)
7.	Random Forest, Support Vector Machine, Linear Discriminant Analysis, Artificial Neural Networks	*Centella asiatica*	Morphological parameters, expressions of rolA, rolB, rolC, and aux1 genes, hormonal profiles	Prediction for centelloside-production in hairy roots of different hormonal and genetic profile, identification of high-yielding hairy root lines	Alcalde et al. (2022)
8.	Heterogeneous ensemble machine learning	*Centella asiatica*	Environmental parameters including CO_2 concentration, intensity and duration of light exposure, strain of the hairy root culture	Prediction for amount of asiaticoside production in hairy roots	Sriprateep et al. (2024)

Despite the significant level of biological variability in plant material and the tissue culture method itself, AI technologies have transformed plant tissue culture by integrating efficient and fast computation techniques. When paired with optimization techniques, AI systems can simulate the growth of plants. As previously stated, this section validates the value of AI approaches and optimization algorithms in a variety of scenarios, ranging from the prediction and optimization of sterilization conditions to the formulation of ideal culture media and evaluation of ideal culture conditions. Selecting the right solution from the wide range of plant tissue culture choices can save a significant amount of money and time. A useful framework for evaluating data from *in vitro* cultures, combining the data gathered, and developing a thorough comprehension of *in vitro* biological systems is provided by AI-based approaches (Holzinger et al., 2023; Ji et al., 2023). The effective mass multiplication of plants in bioreactors is made possible by applying the best chemical and biological parameters at the flask level, which leads to significant labor, financial, and time savings. A number of variables, including the kind of plant tissue, the culture technique, and exposure to environmental stress, might affect the buildup of secondary metabolites. The practical application of plant tissue culture has benefited considerably from the rapid advancement of multi-domain technologies. The goal of developments in plant tissue culture technology is to increase secondary metabolite production and plant biomass productivity. Furthermore, the application of gene editing techniques in the *in vitro* propagation of genetically modified plants could be automated and streamlined by AI methodologies (Mitra and Murthy, 2022).

Notwithstanding the potential, complex definitions and computational procedures have limited the usage of AI in this field. To guarantee reliable findings, it is imperative to address some issues, such as processing power limitations, inexperienced applications, and the potential for inaccurate input collection. Since AI's level of intelligence or efficacy is entirely dependent on the data it receives, results could be erroneous or biased if the program is given faulty or untrustworthy data. The ethical and privacy concerns are another drawback of AI, especially with regard to genome editing (Kaushik et al., 2025).

11.9.4 Biosensors for real-time analysis of plant secondary metabolism under tissue culture systems

Unlike traditional laboratory methods, which are not only complicated but also costly and time-consuming, biosensors have been created since the 1960s to act as ready detection tools. Since clinical biosensors flooded the global market decades ago, the biosensor industry has grown steadily. Today, biosensors powered by contemporary technologies have been successfully tested as efficient diagnostic tools in a variety of sectors, particularly in scientific research, healthcare, environmental monitoring, food safety, and agriculture (Gangopadhyay et al., 2024). Biosensors are analytical tools that convert biological reactions into measurable signals by combining a biological component with a physicochemical detector. The target analyte, the intended use, and the sensor's sensitivity and specificity requirements all influence the particular parts of a biosensor (Gangopadhyay et al., 2024). The major components of the biosensors include a biochemical component to be identified, biomimetic materials that provide an identifiable change through interaction with the biochemical component, a transducer that creates a measurable signal that can be measured through optical, thermal, mechanical or electrical measurements, a signal processing unit that processes the measurable signal created by the transducer, and an output display that displays the final analyzed result (Gangopadhyay et al., 2024). There are several types of biosensors including electrochemical biosensors, optical biosensors, piezoelectric biosensors, and immunosensors that are currently in use (Gangopadhyay et al., 2024).

In the field of plant tissue culture, biosensors may be used in monitoring and optimization of different steps and processes. Plant hormone synthesis, transport, and distribution can be controlled by visualizing the distribution and content of plant hormones in plant tissues, which can enhance the quality and yield of the cultured plant tissues. The distribution of phytohormones can be visualized and quantified by the use of biosensors and associated technologies such as luciferase, transcriptional reporters, degradation sensors, and fluorescent sensors (Isoda et al., 2021). The distribution and content of plant hormones in the subcellular space can be investigated using fluorescence biosensors, which are especially good at tracking the temporal and spatial distribution of tiny signal molecules (Wells et al., 2013). The expression of the reporter gene allows one to see the distribution of hormones in living tissues (Wells et al., 2013). Rapid changes in the distribution and concentration of plant hormones in living cells can be detected using genetically encoded biosensors (Isoda et al., 2021). A binding bag created especially for auxin was disclosed by a researcher (Herud-Sikimić et al., 2021). When auxin attaches itself to the bag, the conformation of auxin-binding portion changes to pair with the fluorescent protein. This causes the generation of a fluorescence resonance energy transfer signal, which allows the auxin level to be reflected (Herud-Sikimić et al., 2021).

Many physiochemical changes in plant tissues, such as the release of specific biomolecules or the manifestation of disease symptoms, are common responses of plants to biotic and abiotic stresses. These changes can be identified using a variety of techniques to assess the presence and occasionally severity of the stress and its effects on the plants (Gangopadhyay et al., 2024). The stress response often triggers secondary metabolic pathways and produces secondary metabolites in plant tissues. Through the detection of pathogenic antigens or stress marker molecules generated by diseased plants, biosensors have been designed to measure these changes in plants directly and evaluate the severity of the stress (Gangopadhyay et al., 2024). A biosensor has been developed to detect photosynthetic dysfunction by delayed fluorescence and reactive oxygen species (ROS) levels, which are important markers of abiotic stress in plant tissues (Zhang et al., 2009). Li et al. (2021) have compiled a number of techniques for detecting abiotic stressors in plants, such as salt, temperature, and heavy metal contamination, by measuring phytohormone levels using several biosensor types (Li et al., 2021). Several biosensors have been developed for detection of pathogen levels in plant tissues as well (Gangopadhyay et al., 2024).

Hairy root cultures using genetically modified *Rhizobium rhizogenes* are usually developed to enhance the production of several secondary metabolites of plants, which are widely used in pharmaceutical, nutraceutical, and cosmetic industries, as mentioned in previous sections. There are several biosensors that have been developed to detect the amounts of flavonoids, isoprenoids, stilbenoids, and alkaloids produced by genetically modified microbes in controlled environments, which can be used in quantification of produced secondary metabolites in hairy root cultures as well (Zhang, Gong, et al., 2023; Zhang, Qian, et al., 2023; Zhang, Wiese, et al., 2023).

11.10 Ethical concerns in the viable use of hairy roots for industrial applications

Irrespective of the fact that hairy root culture is the most suitable technique for production of root-synthesized therapeutics, itneeds the foundation of ethical regulations and consumer acceptance. Ethical considerations in the field of genetic modification of plant–microbe systems raise a few concerns including ecological imbalance caused by genetically modified bacteria species, biosafety, intellectual property rights on genetically engineered microbes, and potential socio-economic impact. In this regard, there are few biosafety guidelines, like the Cartagena Protocol (The Cartagena Protocol on Biosafety, 2025), provide a clear indication about safety regulations and management of genetically modified plants. For protecting natural germplasm from overexploitation to meet ever-increasing market demand, the Convention on Biological Diversity (Prip, 2018) and the Nagoya Protocol (UN Convention on Biological Diversity, 2011) play a crucial role for sustaining biodiversity. Patenting of genetically transformed lines or metabolites needs special attention from Intellectual Property Rights (IPR) perspective. For successful transfer of technology from laboratory innovation to consumer-accepted market exploitation, involvement of scientists, industrialists, and policymakers are utmost essential to ensure end-user acceptance of this cutting-edge technology. The implementation of this technology also aligns with United Nations (UN) Sustainable Development Goals, particularly SDG-3 (Good Health and Well-Being), SDG-12 (Responsible Consumption and Production), and SDG-15 (Life on Land), ensuring development of a green technology without hampering ecological stability, human safety, or cultural integrity.

11.11 Conclusions and future prospects

The rising global demand for root-derived secondary metabolites, including alkaloids, terpenoids, flavonoids, and phenolics, highlights their importance in pharmaceuticals, nutraceuticals, cosmetics, and herbal industries. To address the scarcity of biologically active, commercially valuable, and medicinally significant plant secondary metabolites, metabolic engineering and biotechnological methods are being utilized as an alternative production system. Hairy root culture technology has evolved into an innovative biotechnological resource for researchers. Even though issues such as chromosomal alterations and variable yields of target biochemicals arise, the cultures are more manageable and genetically stable than cell suspension cultures, and exhibit rapid proliferation while offering significant opportunities for pharmaceutical investigation. It also ensures the protection of medicinal plants due to their demand for biosynthesizing invaluable medicinal products, serving as an effective means for producing secondary metabolites. Researchers have explored the natural symbiotic association of *Rhizobium rhizogenes* with plant roots to develop hairy root cultures, and genetic modifications in the bacteria have also been explored to modify biosynthetic pathways responsible for secondary metabolite production. This biotechnological approach improves the yield, quality, and consistency of plant-derived secondary metabolites. Establishment of the hairy root culture follows a defined protocol of explant preparation, bacterial inoculation, cocultivation, and stabilization.

Advances in metabolic engineering enhance biosynthetic pathways, while optimization of pH, media composition, light conditions, and nano-elicitor applications significantly improve yields. Growth kinetics modeling further aids predictive biomass and metabolite accumulation. Bioreactor-driven scale-up, supported by efficient product recovery and purification techniques, enables industrial feasibility, while AI, machine learning, and sensor monitoring ensure adaptive, real-time optimization. Looking ahead, synthetic biology, CRISPR editing, and nanotechnology promise greater specificity and scalability. Future research is expected to focus on intricate genetic modifications aimed to maximize the production of secondary metabolites in hairy roots by using cutting-edge technologies such as omics technologies for elucidation of biosynthetic pathways and integration of synthetic biology for controlled expression of desired genes. Use of nanotechnology and bioreactors may enhance the commercial viability of this approach even further. Ethical concerns surrounding biodiversity, IPR, and benefit-sharing must be resolved through a robust regulatory framework and transparent public engagement, ensuring that hairy root culture technology develops responsibly and sustainably. Finally, genetic modification of *Rhizobium rhizogenes* to enhance the production of secondary metabolites in hairy root culture of plants holds a huge promise in finding solutions to the growing need of secondary metabolites in pharmaceutical, nutraceutical, and cosmetic industries. However, the successful application of this technology will depend on maintaining a balance amongscientific advancements, ensuring ethical responsibilities, and sustainable application of the technology.

References

Aasim, M., Katırcı, R., Akgur, O., Yildirim, B., Mustafa, Z., Nadeem, M. A., Baloch, F. S., Karakoy, T., & Yılmaz, G. (2022). Machine learning (ML) algorithms and artificial neural network for optimizing in vitro germination and growth indices of industrial hemp (*Cannabis sativa* L.). *Industrial Crops and Products, 181*, 114801. https://doi.org/10.1016/j.indcrop.2022.114801.

Abedini, E., Khodadadi, E., Zeinalzadeh, E., Moaddab, S. R., Asgharzadeh, M., Mehramouz, B., Dao, S., & Samadi Kafil, H. (2021). A comprehensive study on the antimicrobial properties of resveratrol as an alternative therapy. *Evidence-Based Complementary and Alternative Medicine, 2021*, 8866311. https://doi.org/10.1155/2021/8866311.

Abiodun, O. I., Jantan, A., Omolara, A. E., Dada, K. V., Umar, A. M., Linus, O. U., Arshad, H., Kazaure, A. A., Gana, U., & Kiru, M. U. (2019). Comprehensive review of artificial neural network applications to pattern recognition. *IEEE Access, 7*, 158820–158846. https://doi.org/10.1109/ACCESS.2019.2945545.

Aburjai, T., & Natsheh, F. M. (2003). Plants used in cosmetics. *Phytotherapy Research, 17*(9), 987–1000. https://doi.org/10.1002/ptr.1363.

Ackermann, C. (1977). Pflanzen aus agrobacterium rhizogenes-tumoren an nicotiana tabacum. *Plant Science Letters, 8*(1), 23–30. https://doi.org/10.1016/0304-4211(77)90167-5.

Afsharzadeh, N., Paltram, R., Jungwirth, A., Tabrizi, L., Nazeri, V., Kalantari, H., Halbwirth, H., Samiei, L., Sheehan, H., & Shokrpour, M. (2025). Advancing *Glycyrrhiza glabra* L. Cultivation and hairy root transformation and elicitation for future metabolite overexpression. *Horticulturae, 11*(1), 62. https://doi.org/10.3390/horticulturae11010062.

Aharoni, A., Jongsma, M. A., Kim, T.-Y., Ri, M.-B., Giri, A. P., Verstappen, F. W. A., Schwab, W., & Bouwmeester, H. J. (2006). Metabolic engineering of terpenoid biosynthesis in plants. *Phytochemistry Reviews, 5*(1), 49–58. https://doi.org/10.1007/s11101-005-3747-3.

Akkad, K., Mehboob, H., Alyamani, R., & Tarlochan, F. (2023). A machine-learning-based approach for predicting mechanical performance of semi-porous hip stems. *Journal of Functional Biomaterials, 14*(3), 156. https://doi.org/10.3390/jfb14030156.

Alcalde, M. A., Müller, M., Munné-Bosch, S., Landín, M., Gallego, P. P., Bonfill, M., Palazon, J., & Hidalgo-Martinez, D. (2022). Using machine learning to link the influence of transferred *Agrobacterium rhizogenes* genes to the hormone profile and morphological traits in *Centella asiatica* hairy roots. *Frontiers in Plant Science, 13*, 1001023. https://doi.org/10.3389/fpls.2022.1001023.

Ali, J., Khan, R., Ahmad, N., & Maqsood, I. (2012). Random forests and decision trees. *International Journal of Computer Science Issues, 9*(5), 272.

Al-Khayri, J. M., Rashmi, R., Toppo, V., Chole, P. B., Banadka, A., Sudheer, W. N., Nagella, P., Shehata, W. F., Al-Mssallem, M. Q., Alessa, F. M., Almaghasla, M. I., & Rezk, A. A. (2023). Plant secondary metabolites: The weapons for biotic stress management. *Metabolites, 13*(6), 716. https://doi.org/10.3390/metabo13060716.

Al-Khayri, J. M., Sudheer, W. N., Lakshmaiah, V. V., Mukherjee, E., Nizam, A., Thiruvengadam, M., Nagella, P., Alessa, F. M., Al-Mssallem, M. Q., Rezk, A. A., Shehata, W. F., & Attimarad, M. (2022). Biotechnological approaches for production of artemisinin, an anti-malarial drug from *Artemisia annua* L. *Molecules, 27*(9), 3040. https://doi.org/10.3390/molecules27093040.

Alsaggar, M., & Liu, D. (2015). Chapter one—physical methods for gene transfer. In L. Huang, D. Liu, & E. Wagner (Vol. Eds.), *Advances in genetics: 89*, (pp. 1–24). Academic Press https://doi.org/10.1016/bs.adgen.2014.10.001.

Altamura, M. M. (2004). Agrobacterium rhizogenes rolB and rolD genes: Regulation and involvement in plant development. *Plant Cell, Tissue and Organ Culture, 77*(1), 89–101. https://doi.org/10.1023/B:TICU.0000016609.22655.33.

Alzubaidi, L., Zhang, J., Humaidi, A. J., Al-Dujaili, A., Duan, Y., Al-Shamma, O., Santamaría, J., Fadhel, M. A., Al-Amidie, M., & Farhan, L. (2021). Review of deep learning: Concepts, CNN architectures, challenges, applications, future directions. *Journal of Big Data, 8*(1), 53. https://doi.org/10.1186/s40537-021-00444-8.

Amdoun, R., Benyoussef, E. H., Benamghar, A., & Khelifi, L. (2019). Prediction of hyoscyamine content in *Datura stramonium* L. hairy roots using different modeling approaches: Response surface methodology (RSM), artificial neural network (ANN) and kriging. *Biochemical Engineering Journal, 144*, 8–17. https://doi.org/10.1016/j.bej.2019.01.002.

An, Q., Rahman, S., Zhou, J., & Kang, J. J. (2023). A comprehensive review on machine learning in healthcare industry: Classification, restrictions, opportunities and challenges. *Sensors, 23*(9), 4178. https://doi.org/10.3390/s23094178.

Arab, M. M., Yadollahi, A., Eftekhari, M., Ahmadi, H., Akbari, M., & Khorami, S. S. (2018). Modeling and optimizing a new culture medium for in vitro rooting of G×N15 *Prunus* rootstock using artificial neural network-genetic algorithm. *Scientific Reports, 8*(1), 9977. https://doi.org/10.1038/s41598-018-27858-4.

Atanasov, A. G., Waltenberger, B., Pferschy-Wenzig, E. M., Linder, T., Wawrosch, C., Uhrin, P., Temml, V., Wang, L., Schwaiger, S., Heiss, E. H., Rollinger, J. M., Schuster, D., Breuss, J. M., Bochkov, V., Mihovilovic, M. D., Kopp, B., Bauer, R., Dirsch, V. M., & Stuppner, H. (2015). Discovery and resupply of pharmacologically active plant-derived natural products: A review. *Biotechnology Advances, 33*(8), 1582–1614. https://doi.org/10.1016/j.biotechadv.2015.08.001.

Awere, C. O., Sneha, A., Rakkammal, K., Muthui, M. M., Kumari, R. A., Govindan, S., Batur Çolak, A., Bayrak, M., Muthuramalingam, P., Anadebe, V. C., Archana, P., Sekar, C., & Ramesh, M. (2024). Carbon dot unravels accumulation of triterpenoid in *Evolvulus alsinoides* hairy roots culture by stimulating growth, redox reactions and ANN machine learning model prediction of metabolic stress response. *Plant Physiology and Biochemistry, 216*, 109142. https://doi.org/10.1016/j.plaphy.2024.109142.

Baek, S., Han, J.-E., Ho, T.-T., & Park, S.-Y. (2022). Development of hairy root cultures for biomass and triterpenoid production in *Centella asiatica*. *Plants, 11*(2), 148. https://doi.org/10.3390/plants11020148.

Bahramnejad, B., Naji, M., Bose, R., & Jha, S. (2019). A critical review on use of *Agrobacterium* rhizogenes and their associated binary vectors for plant transformation. *Biotechnology Advances, 37*(7), 107405. https://doi.org/10.1016/j.biotechadv.2019.06.004.

Bais, H. P., Govindaswamy, S., & Ravishankar, G. A. (2000). Enhancement of growth and coumarin production in hairy root cultures of witloof chicory (*Cichorium intybus* L.cv.Lucknow local) under the influence of fungal elicitors. *Journal of Bioscience and Bioengineering, 90*(6), 648–653. https://doi.org/10.1016/S1389-1723(00)90011-2.

Bais, H. P., Loyola-Vargas, V. M., Flores, H. E., & Vivanco, J. M. (2001). Root-specific metabolism: The biology and biochemistry of underground organs. *In Vitro Cellular & Developmental Biology-Plant, 37*(6), 730–741.

Bais, H. P., Walker, T. S., Schweizer, H. P., & Vivanco, J. M. (2002). Root specific elicitation and antimicrobial activity of rosmarinic acid in hairy root cultures of *Ocimum basilicum*. *Plant Physiology and Biochemistry, 40*(11), 983–995. https://doi.org/10.1016/S0981-9428(02)01460-2.

Banerjee, J., Singh, R., Vijayaraghavan, R., MacFarlane, D., Patti, A. F., & Arora, A. (2017). Bioactives from fruit processing wastes: Green approaches to valuable chemicals. *Food Chemistry, 225*, 10–22. https://doi.org/10.1016/j.foodchem.2016.12.093.

Baque, M. A., Elgirban, A., Lee, E. J., & Paek, K. Y. (2012). Sucrose regulated enhanced induction of anthraquinone, phenolics, flavonoids biosynthesis and activities of antioxidant enzymes in adventitious root suspension cultures of *Morinda citrifolia* (L.). *Acta Physiologiae Plantarum, 34*(2), 405–415.

Baque, Md. A., Moh, S.-H., Lee, E.-J., Zhong, J.-J., & Paek, K.-Y. (2012). Production of biomass and useful compounds from adventitious roots of high-value added medicinal plants using bioreactor. *Biotechnology Advances, 30*(6), 1255–1267. https://doi.org/10.1016/j.biotechadv.2011.11.004.

Baumann, K., De Paolis, A., Costantino, P., & Gualberti, G. (1999). The DNA binding site of the Dof Protein NtBBF1 Is essential for tissue-specific and auxin-regulated expression of the rolB oncogene in plants. *The Plant Cell, 11*(3), 323–333. https://doi.org/10.1105/tpc.11.3.323.

Berzin, I., Mills, D., & Merchuk, J. C. (1997). Morphological structured model for hairy root cultures. In A. Altman, & Y. Waisel (Eds.). *Biology of root formation and development* (pp. 327–334). Springer US https://doi.org/10.1007/978-1-4615-5403-5_71.

Binder, B. Y. K., Peebles, C. A. M., Shanks, J. V., & San, K. Y. (2009). The effects of UV-B stress on the production of terpenoid indole alkaloids in Catharanthus roseus hairy roots. *Biotechnol Prog, 25*(3), 861–865.

Biswas, D., Chakraborty, A., Mukherjee, S., & Ghosh, B. (2023). Hairy root culture: A potent method for improved secondary metabolite production of Solanaceous plants. *Frontiers in Plant Science, 14*. https://doi.org/10.3389/fpls.2023.1197555.

Bradbury, K. E., Appleby, P. N., & Key, T. J. (2014). Fruit, vegetable, and fiber intake in relation to cancer risk: Findings from the European prospective investigation into cancer and nutrition (EPIC). *The American Journal of Clinical Nutrition, 100*(Suppl 1), 394S–398S. https://doi.org/10.3945/ajcn.113.071357.

Brownlee, J. (2023). Supervised and unsupervised machine learning algorithms. *Machine Learning Mastery*. https://machinelearningmastery.com/supervised-and-unsupervisedmachine-learning-algorithms/ Accessed November 5, 2023.

Buckland, B. C., Robinson, D. K., & Chartrain, M. (2000). Biocatalysis for pharmaceuticals—Status and prospects for a key technology. *Metabolic Engineering, 2*(1), 42–48. https://doi.org/10.1006/mben.1999.0138.

Buitelaar, R. M., Langenhoff, A. A. M., Heidstra, R., & Tramper, J. (1991). Growth and thiophene production by hairy root cultures of *Tagetes patula* in various two-liquid-phase bioreactors. *Enzyme and Microbial Technology, 13*(6), 487–494. https://doi.org/10.1016/0141-0229(91)90007-W.

Burgos-Artizzu, X. P., Ribeiro, A., Guijarro, M., & Pajares, G. (2011). Real-time image processing for crop/weed discrimination in maize fields. *Computers and Electronics in Agriculture, 75*(2), 337–346. https://doi.org/10.1016/j.compag.2010.12.011.

Cabaneros, S. M., Calautit, J. K., & Hughes, B. R. (2019). A review of artificial neural network models for ambient air pollution prediction. *Environmental Modelling & Software, 119*, 285–304. https://doi.org/10.1016/j.envsoft.2019.06.014.

Cadiz, N. M., Vivanco, J. M., & Flores, H. E. (2000). Coculture of *Pachyrhizus erosus* (L.) hairy roots with *Rhizobium* spp. *In Vitro Cellular & Developmental Biology - Plant, 36*(4), 238–242. https://doi.org/10.1007/s11627-000-0044-0.

Cai, Z., Kastell, A., Knorr, D., & Smetanska, I. (2012). Exudation: An expanding technique for continuous production and release of secondary metabolites from plant cell suspension and hairy root cultures. *Plant Cell Reports, 31*(3), 461–477. https://doi.org/10.1007/s00299-011-1165-0.

Capell, T., & Christou, P. (2004). Progress in plant metabolic engineering. *Current Opinion in Biotechnology, 15*(2), 148–154. https://doi.org/10.1016/j.copbio.2004.01.009.

Capone, I., Frugis, G., Costantino, P., & Cardarelli, M. (1994). Expression in different populations of cells of the root meristem is controlled by different domains of the rolB promoter. *Plant Molecular Biology, 25*(4), 681–691. https://doi.org/10.1007/BF00029606.

Cardarelli, M., Mariotti, D., Pomponi, M., Spanò, L., Capone, I., & Costantino, P. (1987). *Agrobacterium* rhizogenes T-DNA genes capable of inducing hairy root phenotype. *Molecular and General Genetics MGG, 209*(3), 475–480. https://doi.org/10.1007/BF00331152.

Chahardoli, F., Pourmoslemi, S., Soleimani Asl, S., Tamri, P., & Haddadi, R. (2023). Preparation of polyvinyl alcohol hydrogel containing chlorogenic acid microspheres and its evaluation for use in skin wound healing. *Journal of Biomaterials Applications, 37*(9), 1667–1675. https://doi.org/10.1177/08853282221150845.

Chandra, S. (2012). Natural plant genetic engineer *Agrobacterium rhizogenes*: Role of T-DNA in plant secondary metabolism. *Biotechnology Letters, 34*(3), 407–415. https://doi.org/10.1007/s10529-011-0785-3.

Chandra, S., & Chandra, R. (2011). Engineering secondary metabolite production in hairy roots. *Phytochemistry Reviews, 10*(3), 371–395. https://doi.org/10.1007/s11101-011-9210-8.

Chavarría-Krauser, A., & Schurr, U. (2004). A cellular growth model for root tips. *Journal of Theoretical Biology, 230*(1), 21–32. https://doi.org/10.1016/j.jtbi.2004.04.007.

Chen, H., Chena, F., Chiu, F. C. K., & Lo, C. M. Y. (2001). The effect of yeast elicitor on the growth and secondary metabolism of hairy root cultures of Salvia miltiorrhiza. *Enzyme and Microbial Technology, 28*(1), 100–105. https://doi.org/10.1016/S0141-0229(00)00284-2.

Choi, S. M., Son, S. H., Yun, S. R., Kwon, O. W., Seon, J. H., & Paek, K. Y. (2000). Pilot-scale culture of adventitious roots of ginseng in a bioreactor system. *Plant Cell, Tissue and Organ Culture, 62*(3), 187–193.

Choi, Y.-E., Kim, Y.-S., & Paek, K.-Y. (2008). Types and designs of bioreactors for hairy root culture. In In. S. D. Gupta, & Y. Ibaraki (Eds.). *Plant tissue culture engineering* (pp. 161–172). Springer Netherlands https://doi.org/10.1007/978-1-4020-3694-1_9.

Chung, I. M., Rajakumar, G., & Thiruvengadam, M. (2018). Effect of silver nanoparticles on phenolic compounds production and biological activities in hairy root cultures of cucumis anguria. *Acta Biologica Hungarica, 69*(1), 97–109. https://doi.org/10.1556/018.68.2018.1.8.

Chung, I. M., Rekha, K., Rajakumar, G., & Thiruvengadam, M. (2018). Influence of silver nanoparticles on the enhancement and transcriptional changes of glucosinolates and phenolic compounds in genetically transformed root cultures of *Brassica rapa* ssp. Rapa. *Bioprocess and Biosystems Engineering, 41*(11), 1665–1677. https://doi.org/10.1007/s00449-018-1991-3.

Cloutier, M., Perrier, M., & Jolicoeur, M. (2007). Dynamic flux cartography of hairy roots primary metabolism. *Phytochemistry, 68*(16–18), 2393–2404. https://doi.org/10.1016/j.phytochem.2007.04.028.

Corchado, J. M., & Aiken, J. (2003). Hybrid artificial intelligence methods in oceanographic forecast models. *IEEE Transactions on Systems, Man, and Cybernetics, Part C (Applications and Reviews, 32*(4), 307–313. https://doi.org/10.1109/TSMCC.2002.806072.

Cutanda-Perez, M.-C., Ageorges, A., Gomez, C., Vialet, S., Terrier, N., Romieu, C., & Torregrosa, L. (2009). Ectopic expression of VlmybA1 in grapevine activates a narrow set of genes involved in anthocyanin synthesis and transport. *Plant Molecular Biology, 69*(6), 633–648. https://doi.org/10.1007/s11103-008-9446-x.

Dare, A. P., Schaffer, R. J., Lin-Wang, K., Allan, A. C., & Hellens, R. P. (2008). Identification of a cis-regulatory element by transient analysis of co-ordinately regulated genes. *Plant Methods, 4*(1), 17. https://doi.org/10.1186/1746-4811-4-17.

Das, S., Manna, D., Mondal, T., & Mondal, P. (2025). Advanced systems and bioreactors for large-scale secondary metabolite production in medicinal plants. In *Biotechnology, multiple omics, and precision breeding in medicinal plants*. CRC Press.

Dasgupta, S. C. (2023). Bioactive compounds from medicinal plants and its therapeutic uses in the traditional healthcare system. In S. Jha, & M. Halder (Eds.). *Medicinal plants: Biodiversity, biotechnology and conservation* (pp. 525–537). Springer Nature https://doi.org/10.1007/978-981-19-9936-9_19.

Dayhoff, J. E., & DeLeo, J. M. (2001). Artificial neural networks: Opening the black box. *Cancer, 91*(8 Suppl), 1615–1635. https://doi.org/10.1002/1097-0142(20010415)91:8+<1615::AID-CNCR1175>3.0.CO;2-L.

DeBoer, K. D., Dalton, H. L., Edward, F. J., & Hamill, J. D. (2011). RNAi-mediated down-regulation of ornithine decarboxylase (ODC) leads to reduced nicotine and increased anatabine levels in transgenic *Nicotiana tabacum* L. *Phytochemistry, 72*(4), 344–355. https://doi.org/10.1016/j.phytochem.2010.12.012.

Dembinska-Kiec, A., Mykkänen, O., Kiec-Wilk, B., & Mykkänen, H. (2008). Antioxidant phytochemicals against type 2 diabetes. *British Journal of Nutrition, 99*(Suppl 1), ES109–ES117. https://doi.org/10.1017/S000711450896579X.

Dias, F. M., Antunes, A., & Mota, A. M. (2004). Artificial neural networks: A review of commercial hardware. *Engineering Applications of Artificial Intelligence, 17*(8), 945–952. https://doi.org/10.1016/j.engappai.2004.08.011.

Dilshad, E., Cusido, R. M., Estrada, K. R., Bonfill, M., & Mirza, B. (2015). Genetic transformation of *Artemisia carvifolia* buch with rol genes enhances artemisinin accumulation. *PLOS ONE, 10*(10), e0140266. https://doi.org/10.1371/journal.pone.0140266.

Ding, S., Li, H., Su, C., Yu, J., & Jin, F. (2013). Evolutionary artificial neural networks: A review. *Artificial Intelligence Review, 39*(3), 251–260. https://doi.org/10.1007/s10462-011-9270-6.

Doran, P. M. (1995). *Bioprocess engineering principles*. Elsevier.

Du, M., Wu, X. J., Ding, J., Hu, Z. B., White, K. N., & Branford-White, C. J. (2003). Astragaloside IV and polysaccharide production by hairy roots of *Astragalus membranaceus* in bioreactors. *Biotechnology Letters, 25*(21), 1853–1856. https://doi.org/10.1023/A:1026233728375.

Du, T. C., & Wolfe, P. M. (1995). The amalgamation of neural networks and fuzzy logic systems—A survey. *Computers & Industrial Engineering, 29*(1–4), 193–197. https://doi.org/10.1016/0360-8352(95)00070-H.

Ebadi, M.-T., Sefidkon, F., Azizi, M., & Ahmadi, N. (2017). Packaging methods and storage duration affect essential oil content and composition of lemon verbena (*Lippia citriodora* Kunth. *Food Science & Nutrition, 5*(3), 588–595. https://doi.org/10.1002/fsn3.434.

Eibl, R., & Eibl, D. (2008). Design of bioreactors suitable for plant cell and tissue cultures. *Phytochemistry Reviews, 7*(3), 593–598. https://doi.org/10.1007/s11101-007-9083-z.

Eilert, U., Constabel, F., & Kurz, W. G. W. (1986). Elicitor-stimulation of monoterpene indole alkaloid formation in suspension cultures of *Catharanthus roseus*. *Journal of Plant Physiology, 126*(1), 11–22. https://doi.org/10.1016/S0176-1617(86)80211-5.

Elbarbry, F., Jones, G., & Ung, A. (2022). Catechin reduces blood pressure in spontaneously hypertensive rats through modulation of arachidonic acid metabolism. *Molecules, 27*(23), 8432. https://doi.org/10.3390/molecules27238432.

Elbaz, K., Shen, S. L., Zhou, A., Yuan, D. J., & Xu, Y. S. (2019). Optimization of EPB shield performance with adaptive neuro-fuzzy inference system and genetic algorithm. *Applied Sciences, 9*(4), 780. https://doi.org/10.3390/app9040780.

Elshafie, H. S., Camele, I., & Mohamed, A. A. (2023). A comprehensive review on the biological, agricultural and pharmaceutical properties of secondary metabolites based-plant origin. *International Journal of Molecular Sciences, 24*(4), 3266. https://doi.org/10.3390/ijms24043266.

Failla, M. C., Maimone, F., De Paolis, A., Costantino, P., & Cardarelli, M. (1990). The non-conserved region of cucumopine-type *Agrobacterium rhizogenes* T-DNA is responsible for hairy root induction. *Plant Molecular Biology, 15*(5), 747–753. https://doi.org/10.1007/BF00016124.

Fang, C., Luo, J., & Wang, S. (2019). The diversity of nutritional metabolites: Origin, dissection, and application in crop breeding. *Frontiers in Plant Science, 10*, 1028. https://doi.org/10.3389/fpls.2019.01028.

Fiani, B., Zhu, L., Musch, B. L., Briceno, S., Andel, R., Sadeq, N., & Ansari, A. Z. (2021). The neurophysiology of caffeine as a central nervous system stimulant and the resultant effects on cognitive function. *Cureus, 13*(5), e15032. https://doi.org/10.7759/cureus.15032.

Fierascu, R. C., Fierascu, I., Ortan, A., Georgiev, M. I., & Sieniawska, E. (2020). Innovative approaches for recovery of phytoconstituents from medicinal/aromatic plants and biotechnological production. *Molecules, 25*(2), 309. https://doi.org/10.3390/molecules25020309.

Fraenkel, G. S. (1959). The raison d'ětre of secondary plant substances; these odd chemicals arose as a means of protecting plants from insects and now guide insects to food. *Science, 129*(3361), 1466–1470. https://doi.org/10.1126/science.129.3361.1466.

Fu, M., Zhang, C., Hu, C., Wu, T., Dong, J., & Zhu, L. (2023). Achieving verifiable decision tree prediction on hybrid blockchains. *Entropy, 25*(7), 1058. https://doi.org/10.3390/e25071058.

Fulzele, D. P., Satdive, R. K., & Pol, B. B. (2002). Untransformed root cultures of Nothapodytes foetida and production of camptothecin. *Plant Cell, Tissue and Organ Culture, 69*(3), 285–288.

Furmanowa, M., Olędzka, H., Sykłowska-Baranek, K., Józefowicz, J., & Gieracka, S. (2000). Increased taxane accumulation in callus cultures of Taxus cuspidata and Taxus × media by some elicitors and precursors. *Biotechnology Letters, 22*(18), 1449–1452. https://doi.org/10.1023/A:1005611114628.

Gai, Q.-Y., Jiao, J., Luo, M., Wang, W., Ma, W., Zu, Y.-G., & Fu, Y.-J. (2015). Establishment of high-productive *Isatis tinctoria* L. hairy root cultures: A promising approach for efficient production of bioactive alkaloids. *Biochemical Engineering Journal, 95*, 37–47. https://doi.org/10.1016/j.bej.2014.12.006.

Gangopadhyay, M., Chakravarti, S., Sanyamat, S., & Dewanjee, S. (2024). Biosensors in agriculture: Pioneering the future of precision farming and sustainability. In R. Mathew, & J. Ajayan (Eds.). *Biosensors: Developments, challenges and perspectives* (pp. 259–301). Springer https://doi.org/10.1007/978-981-97-3048-3_13.

Gangopadhyay, M., Dewanjee, S., & Bhattacharya, S. (2011). Enhanced plumbagin production in elicited *Plumbago indica* hairy root cultures. *Journal of Bioscience and Bioengineering, 111*(6), 706–710. https://doi.org/10.1016/j.jbiosc.2011.02.003.

Gangopadhyay, M., Dewanjee, S., Bhattacharyya, S., & Bhattacharya, S. (2010). Effect of different strains of *Agrobacterium rhizogenes* and Nature of explants on plumbago indica hairy root culture with special emphasis on root biomass and plumbagin production. *Natural Product Communications, 5*(12), 1934578X1000501215. https://doi.org/10.1177/1934578X1000501215.

Gangopadhyay, M., Sircar, D., Mitra, A., & Bhattacharya, S. (2008). Hairy root culture of *Plumbago indica* as a potential source for plumbagin. *Biologia Plantarum, 52*(3), 533–537. https://doi.org/10.1007/s10535-008-0104-6.

Gantait, S., & Mukherjee, E. (2021). Hairy root culture technology: Applications, constraints and prospect. *Applied Microbiology and Biotechnology, 105*(1), 35–53. https://doi.org/10.1007/s00253-020-11017-9.

Garcia-Brugger, A., Lamotte, O., Vandelle, E., Bourque, S., Lecourieux, D., Poinssot, B., Wendehenne, D., & Pugin, A. (2006). Early signaling events induced by elicitors of plant defenses. *Molecular Plant-Microbe Interactions® 19*(7), 711–724. https://doi.org/10.1094/MPMI-19-0711.

Gardner, M. W., & Dorling, S. R. (1998). Artificial neural networks (the multilayer perceptron)—A review of applications in the atmospheric sciences. *Atmospheric Environment, 32*(14–15), 2627–2636. https://doi.org/10.1016/S1352-2310(97)00447-0.

Gelvin, S. B. (2003). Agrobacterium-mediated plant transformation: The biology behind the "gene-jockeying" tool. *Microbiology and Molecular Biology Reviews, 67*(1), 16–37. https://doi.org/10.1128/mmbr.67.1.16-37.2003.

Gelvin, S. B. (2010). Plant proteins involved in *Agrobacterium*-mediated genetic transformation. *Annual Review of Phytopathology, 48*(2010), 45–68. https://doi.org/10.1146/annurev-phyto-080508-081852.

Georgiev, V. G., Bley, T., & Pavlov, A. I. (2012). Bioreactors for the cultivation of red beet hairy roots. In B. Neelwarne (Ed.). *Red beet biotechnology: Food and pharmaceutical applications* (pp. 251–281). Springer US https://doi.org/10.1007/978-1-4614-3458-0_11.

Ghasemy-Piranloo, F., Kavousi, F., & Kazemi-Abharian, M. (2022). Comparison for the production of essential oil by conventional, novel and biotechnology methods. *Journal of Essential Oil Research, 34*(5), 455–478. https://doi.org/10.1080/10412905.2022.2120557.

Gheisary, B., Fattahi, M., & Alipour, H. (2025). Enhancing extraction of shikonin and phenolic antioxidants from *Echium italicum* L. using ultrasound and response surface methodology: Optimizing temperature, time, and liquid–solid ratio. *Biomass Conversion and Biorefinery, 15*(10), 15619–15630. https://doi.org/10.1007/s13399-024-06211-y.

Giradkar, V., Mhaske, A., & Shukla, R. (2024). Naringenin nanocrystals mitigate rotenone neurotoxicity in SH-SY5Y cell line by modulating mitophagy and oxidative stress. *AAPS PharmSciTech, 25*(7), 227. https://doi.org/10.1208/s12249-024-02936-1.

Gloria, N. F., Soares, N., Brand, C., Oliveira, F. L., Borojevic, R., & Teodoro, A. J. (2014). Lycopene and beta-carotene induce cell-cycle arrest and apoptosis in human breast cancer cell lines. *Anticancer Research, 34*(3), 1377–1386.

Goralogia, G. S., Willig, C., & Strauss, S. H. (2025). Engineering *Agrobacterium* for improved plant transformation. *The Plant Journal, 121*(5), e70015. https://doi.org/10.1111/tpj.70015.

Goswami, M., Akhtar, S., & Osama, K. (2018). Strategies for monitoring and modeling the growth of hairy root cultures: An in silico perspective. In V. Srivastava, S. Mehrotra, & S. Mishra (Eds.). *Hairy roots: An effective tool of plant biotechnology* (pp. 311–327). Springer https://doi.org/10.1007/978-981-13-2562-5_14.

Guillon, S., Trémouillaux-Guiller, J., Pati, P. K., Rideau, M., & Gantet, P. (2006). Hairy root research: Recent scenario and exciting prospects. *Current Opinion in Plant Biology, 9*(3), 341–346. https://doi.org/10.1016/j.pbi.2006.03.008.

Gumery, F., Ein-Mozaffari, F., & Dahman, Y. (2009). Characteristics of local flow dynamics and macro-mixing in airlift column reactors for reliable design and scale-up. *International Journal of Chemical Reactor Engineering, 7*(1). https://doi.org/10.2202/1542-6580.2081.

Gutierrez-Valdes, N., Häkkinen, S. T., Lemasson, C., Guillet, M., Oksman-Caldentey, K. M., Ritala, A., & Cardon, F. (2020). Hairy root cultures—A versatile tool with multiple applications. *Frontiers in Plant Science, 11*. https://doi.org/10.3389/fpls.2020.00033.

Guyon, I., Weston, J., Barnhill, S., & Vapnik, V. (2002). Gene selection for cancer classification using support vector machines. *Machine Learning, 46*(1), 389–422. https://doi.org/10.1023/A:1012487302797.

Habibi, P., De Sa, M. F. G., Makhzoum, A., Malik, S., da Silva, A. L. L., Hefferon, K., & Soccol, C. R. (2017). Bioengineering hairy roots: Phytoremediation, secondary metabolism, molecular pharming, plant-plant interactions and biofuels. In E. Lichtfouse (Ed.). *Sustainable agriculture reviews* (pp. 213–251). Springer International Publishing https://doi.org/10.1007/978-3-319-48006-0_7.

Hahlbrock, K., Ebel, J., Oaks, A., Auden, J., & Liersch, M. (1974). Determination of specific growth stages of plant cell suspension cultures by monitoring conductivity changes in the medium. *Planta, 118*(1), 75–84. https://doi.org/10.1007/BF00390504.

Halder, M., Sarkar, S., & Jha, S. (2019). Elicitation: A biotechnological tool for enhanced production of secondary metabolites in hairy root cultures. *Engineering in Life Sciences, 19*(12), 880–895. https://doi.org/10.1002/elsc.201900058.

Hameg, R., Arteta, T. A., Landin, M., Gallego, P. P., & Barreal, M. E. (2020). Modeling and optimizing culture medium mineral composition for in vitro propagation of *Actinidia arguta*. *Frontiers in Plant Science, 11*, 554905. https://doi.org/10.3389/fpls.2020.554905.

Hamill, J. D., & Lidgett, A. J. (1997). Hairy root cultures—Opportunities and key protocols for studies in metabolic engineering. In *Hairy roots*. CRC Press.

Han, B., Linden, J. C., Gujarathi, N. P., & Wickramasinghe, S. R. (2004). Population balance approach to modeling hairy root growth. *Biotechnology Progress, 20*(3), 872–879. https://doi.org/10.1021/bp0342304.

Han, M. (2017). *Hydrodynamics and mass transfer in airlift bioreactors: Experimental and numerical simulation analysis*. Lappeenranta University of Technology. https://lutpub.lut.fi/handle/10024/147646.

Hansen, G., Larribe, M., Vaubert, D., Tempé, J., Biermann, B. J., Montoya, A. L., Chilton, M. D., & Brevet, J. (1991). *Agrobacterium rhizogenes* pRi8196 T-DNA: Mapping and DNA sequence of functions involved in mannopine synthesis and hairy root differentiation. *Proceedings of the National Academy of Sciences, 88*(17), 7763–7767. https://doi.org/10.1073/pnas.88.17.7763.

Hansen, G., Tempé, J., & Brevet, J. (1992). A T-DNA transfer stimulator sequence in the vicinity of the right border of pRi8196. *Plant Molecular Biology, 20*(1), 113–122. https://doi.org/10.1007/BF00029154.

Hassanat, A., Almohammadi, K., Alkafaween, E. A., Abunawas, E., Hammouri, A., & Prasath, V. S. (2019). Choosing mutation and crossover ratios for genetic algorithms—A review with a new dynamic approach. *Information, 10*(12), 390. https://doi.org/10.3390/info10120390.

Hedayati, A., Hosseini, B., Palazon, J., & Maleki, R. (2020). Improved tropane alkaloid production and changes in gene expression in hairy root cultures of two Hyoscyamus species elicited by silicon dioxide nanoparticles. *Plant Physiology and Biochemistry, 155*, 416–428. https://doi.org/10.1016/j.plaphy.2020.07.029.

Herbert, R. B. (1989). *The biosynthesis of secondary metabolites*. Springer Science & Business Media.

Herud-Sikimić, O., Stiel, A. C., Kolb, M., Shanmugaratnam, S., Berendzen, K. W., Feldhaus, C., Höcker, B., & Jürgens, G. (2021). A biosensor for the direct visualization of auxin. *Nature, 592*(7856), 768–772. https://doi.org/10.1038/s41586-021-03425-2.

Hesami, M., & Jones, A. M. P. (2020). Application of artificial intelligence models and optimization algorithms in plant cell and tissue culture. *Applied Microbiology and Biotechnology, 104*(22), 9449–9485. https://doi.org/10.1007/s00253-020-10888-2.

Hesami, M., Naderi, R., & Tohidfar, M. (2019). Modeling and optimizing in vitro sterilization of chrysanthemum via multilayer perceptron–non-dominated sorting genetic algorithm-II (MLP-NSGAII). *Frontiers in Plant Science, 10*, 282. https://doi.org/10.3389/fpls.2019.00282.

Hesami, M., Naderi, R., Tohidfar, M., & Yoosefzadeh-Najafabadi, M. (2019). Application of adaptive neuro-fuzzy inference system–non-dominated sorting genetic algorithm-II (ANFIS-NSGAII) for modeling and optimizing somatic embryogenesis of chrysanthemum. *Frontiers in Plant Science, 10*, 869. https://doi.org/10.3389/fpls.2019.00869.

Hickson, L. J., Langhi Prata, L. G. P., Bobart, S. A., Evans, T. K., Giorgadze, N., Hashmi, S. K., Herrmann, S. M., Jensen, M. D., Jia, Q., Jordan, K. L., Kellogg, T. A., Khosla, S., Koerber, D. M., Lagnado, A. B., Lawson, D. K., LeBrasseur, N. K., Lerman, L. O., McDonald, K. M., McKenzie, T. J., Passos, J. F., Pignolo, R. J., Pirtskhalava, T., Saadiq, I. M., Schaefer, K. K., Textor, S. C., Victorelli, S. G., Volkman, T. L., Xue, A., Wentworth, M. A., Wissler Gerdes, E. O., Zhu, Y., Tchkonia, T., & Kirkland, J. L. (2019). Senolytics decrease senescent cells in humans:

Preliminary report from a clinical trial of dasatinib plus quercetin in individuals with diabetic kidney disease. *EBioMedicine, 47*, 446–456. https://doi.org/10.1016/j.ebiom.2019.08.069.

Ho, T. T., Lee, J. D., Ahn, M. S., Kim, S. W., & Park, S. Y. (2018). Enhanced production of phenolic compounds in hairy root cultures of Polygonum multiflorum and its metabolite discrimination using HPLC and FT-IR methods. *Applied Microbiology and Biotechnology, 102*(22), 9563–9575.

Hodges, L. D., Cuperus, J., & Ream, W. (2004). *Agrobacterium rhizogenes* GALLS protein substitutes for *Agrobacterium tumefaciens* single-stranded DNA-binding protein VirE2. *Journal of Bacteriology, 186*(10), 3065–3077. https://doi.org/10.1128/jb.186.10.3065-3077.2004.

Hodges, L. D., Vergunst, A. C., Neal-McKinney, J., den Dulk-Ras, A., Moyer, D. M., Hooykaas, P. J. J., & Ream, W. (2006). *Agrobacterium rhizogenes* GALLS protein contains domains for ATP binding, nuclear localization, and type IV secretion. *Journal of Bacteriology, 188*(23), 8222–8230. https://doi.org/10.1128/jb.00747-06.

Holzinger, A., Keiblinger, K., Holub, P., Zatloukal, K., & Müller, H. (2023). AI for life: Trends in artificial intelligence for biotechnology. *New Biotechnology, 74*, 16–24. https://doi.org/10.1016/j.nbt.2023.02.001.

Hong, S.-B., Hughes, E. H., Shanks, J. V., San, K.-Y., & Gibson, S. I. (2003). Role of the non-mevalonate pathway in indole alkaloid production by *Catharanthus roseus* hairy roots. *Biotechnology Progress, 19*(3), 1105–1108. https://doi.org/10.1021/bp034031k.

Hooykaas, M. J. G., & Hooykaas, P. J. J. (2021). The genome sequence of hairy root *Rhizobium rhizogenes* strain LBA9402: Bioinformatics analysis suggests the presence of a new opine system in the agropine Ri plasmid. *MicrobiologyOpen, 10*(2), e1180. https://doi.org/10.1002/mbo3.1180.

Hooykaas, P. J. J., Klapwijk, P. M., Nuti, M. P., Schilperoort, R. A., & Rörsch, A. (1977). Transfer of the *Agrobacterium tumefaciens* TI plasmid to avirulent agrobacteria and to *Rhizobium* ex planta. *Microbiology, 98*(2), 477–484. https://doi.org/10.1099/00221287-98-2-477.

Huang, S.-Y., Hung, C.-H., & Chou, S.-N. (2004). Innovative strategies for operation of mist trickling reactors for enhanced hairy root proliferation and secondary metabolite productivity. *Enzyme and Microbial Technology, 35*(1), 22–32. https://doi.org/10.1016/j.enzmictec.2004.03.011.

Ionkova, I. (2007). Biotechnological approaches for the production of lignans. *Pharmacognosy Reviews, 1*(1).

Isoda, R., Yoshinari, A., Ishikawa, Y., Sadoine, M., Simon, R., Frommer, W. B., & Nakamura, M. (2021). Sensors for the quantification, localization and analysis of the dynamics of plant hormones. *The Plant Journal, 105*(2), 542–557. https://doi.org/10.1111/tpj.15096.

Ivashchuk, O. A., Fedorova, V. I., Shcherbinina, N. V., Maslova, E. V., & Shamraeva, E. O. (2018). Microclonal propagation of plant process modeling and optimization of its parameters based on neural network. *Drug Invention Today, 10*(3), 1–6.

Jamshidi, S., Yadollahi, A., Ahmadi, H., Arab, M. M., & Eftekhari, M. (2016). Predicting in vitro culture medium macro-nutrients composition for pear rootstocks using regression analysis and neural network models. *Frontiers in Plant Science, 7*, 274. https://doi.org/10.3389/fpls.2016.00274.

Jamshidi, S., Yadollahi, A., Arab, M. M., Soltani, M., Eftekhari, M., & Shiri, J. (2020). High throughput mathematical modeling and multi-objective evolutionary algorithms for plant tissue culture media formulation: Case study of pear rootstocks. *PLoS ONE, 15*(12), e0243940. https://doi.org/10.1371/journal.pone.0243940.

Javed, R., Yucesan, B., Zia, M., & Gurel, E. (2022). Nanoelicitation: A promising and emerging technology for triggering the sustainable in vitro production of secondary metabolites in medicinal plants. In J.-T. Chen (Ed.). *Plant and nanoparticles* (pp. 265–280). Springer Nature https://doi.org/10.1007/978-981-19-2503-0_10.

Jeong, G. T. A., & Park, D. H. (2006). Enhanced secondary metabolite biosynthesis by elicitation in transformed plant root system. *Applied Biochemistry and Biotechnology, 130*(1), 436–446. https://doi.org/10.1385/ABAB:130:1:436.

de Jesus, S. S., Moreira Neto, J., & Maciel Filho, R. (2017). Hydrodynamics and mass transfer in bubble column, conventional airlift, stirred airlift and stirred tank bioreactors, using viscous fluid: A comparative study. *Biochemical Engineering Journal, 118*, 70–81. https://doi.org/10.1016/j.bej.2016.11.019.

Jha, S., Yin, H., Li, C., & Wang, G.-L. (2023). Editorial: Transgenesis and secondary metabolites: Recent developments and technological challenges. *Frontiers in Plant Science, 14*. https://doi.org/10.3389/fpls.2023.1276922.

Ji, B., Xuan, L., Zhang, Y., Mu, W., Paek, K. Y., Park, S. Y., Wang, J., & Gao, W. (2023). Application of data modeling, instrument engineering and nanomaterials in selected medicinal plant tissue culture. *Plants, 12*(7), 1505. https://doi.org/10.3390/plants12071505.

Jinturkar, K. A., Rathi, M. N., & Misra, A. (2011). Gene delivery using physical methods. In A. Misra (Vol. Ed.), *Challenges in delivery of therapeutic genomics and proteomics: 3*, (pp. 83–126). Elsevier https://doi.org/10.1016/B978-0-12-384964-9.00003-7.

John, J. E. (2009). Natural products-based drug discovery: Some bottlenecks and considerations. *Current Science, 96*(6), 753–754.

Jolicoeur, M., Bouchard-Marchand, E., Bécard, G., & Perrier, M. (2002). Regulation of mycorrhizal symbiosis: Development of a structured nutritional dual model. *Ecological Modelling, 158*(1–2), 121–142. https://doi.org/10.1016/S0304-3800(02)00168-0.

Jung, G., & Tepfer, D. (1987). Use of genetic transformation by the Ri T-DNA of *Agrobacterium rhizogenes* to stimulate biomass and tropane alkaloid production in *Atropa belladonna* and *Calystegia sepium* roots grown in vitro. *Plant Science, 50*(2), 145–151. https://doi.org/10.1016/0168-9452(87)90151-8.

Kajikawa, M., Hirai, N., & Hashimoto, T. (2009). A PIP-family protein is required for biosynthesis of tobacco alkaloids. *Plant Molecular Biology, 69*(3), 287–298. https://doi.org/10.1007/s11103-008-9424-3.

Katoch, S., Chauhan, S. S., & Kumar, V. (2021). A review on genetic algorithm: Past, present, and future. *Multimedia Tools and Applications, 80*(5), 8091–8126. https://doi.org/10.1007/s11042-020-10139-6.

Kaushik, P., Rani, M., Khurana, N., Pandey, P., Payal, & Kapoor, S. (2025). Revolutionizing plant tissue culture: Harnessing artificial intelligence for precision propagation and optimization. *The Natural Products Journal, 15*(3). https://doi.org/10.2174/0122103155302871240527094915 E040624230642.

Kersters, K. (1984). Genus III. Agrobacterium Conn 1942. *Bergey's Manual of Systematic Bacteriology, 1*, 244–254.

Keshavareddy, Kumar, A. R. V., & Ramu, V. S. (2018). Methods of plant transformation- a review. *International Journal of Current Microbiology and Applied Sciences, 7*(7), 2656–2668. https://doi.org/10.20546/ijcmas.2018.707.312.

Khalili, M., Hasanloo, T., & Kazemi, T. S. K. (2020). Ag{+} Enhanced silymarin production in hairy root cultures of "*Silybum Marianum*" (L.) Gaertn. *Plant Omics, 3*(4), 109–114. https://doi.org/10.3316/informit.282215964290718.

Khan, A. K., Kousar, S., Tungmunnithum, D., Hano, C., Abbasi, B. H., & Anjum, S. (2021). Nano-elicitation as an effective and emerging strategy for in vitro production of industrially important flavonoids. *Applied Sciences, 11*(4), 1694. https://doi.org/10.3390/app11041694.

Khan, S. A., Siddiqui, M. H., & Osama, K. (2018). Bioreactors for hairy roots culture: A review. *Current Biotechnology, 7*(6), 417–427. https://doi.org/10.2174/2211550108666190114143824.

Kim, O. T., Bang, K. H., Shin, Y. S., Lee, M. J., Jung, S. J., Hyun, D. Y., Kim, Y. C., Seong, N. S., Cha, S. W., & Hwang, B. (2007). Enhanced production of asiaticoside from hairy root cultures of *Centella asiatica* (L.) Urban elicited by methyl jasmonate. *Plant Cell Reports, 26*(11), 1941–1949. https://doi.org/10.1007/s00299-007-0400-1.

Kim, O. T., Kim, S. H., Ohyama, K., Muranaka, T., Choi, Y. E., Lee, H. Y., Kim, M. Y., & Hwang, B. (2010). Upregulation of phytosterol and triterpene biosynthesis in *Centella asiatica* hairy roots overexpressed ginseng farnesyl diphosphate synthase. *Plant Cell Reports, 29*(4), 403–411. https://doi.org/10.1007/s00299-010-0831-y.

Kim, S. I., Kim, J. Y., Kim, E. A., Kwon, K. H., Kim, K. W., Cho, K., Lee, J. H., Nam, M. H., Yang, D. C., Yoo, J. S., & Park, Y. M. (2003). Proteome analysis of hairy root from Panax ginseng C. A. Meyer using peptide fingerprinting, internal sequencing and expressed sequence tag data. *PROTEOMICS, 3*(12), 2379–2392. https://doi.org/10.1002/pmic.200300619.

Kiryushkin, A. S., Ilina, E. L., Guseva, E. D., Pawlowski, K., & Demchenko, K. N. (2022). Hairy CRISPR: Genome editing in plants using hairy root transformation. *Plants, 11*(1), 51. https://doi.org/10.3390/plants11010051.

Komaraiah, P., Reddy, G. V., Reddy, P. S., Raghavendra, A. S., Ramakrishna, S. V., & Reddanna, P. (2003). Enhanced production of antimicrobial sesquiterpenes and lipoxygenase metabolites in elicitor-treated hairy root cultures of *Solanum tuberosum*. *Biotechnology Letters, 25*(8), 593–597. https://doi.org/10.1023/A:1023038804556.

Koutroumpi, S. K. (2021). Elicitation, A mechanistic approach to change the metabolic pathway of plants to produce pharmacological important compounds in in-vitro cell cultures. *Global Journal of Engineering Sciences, 8*(1), 1–7.

Krasniqi, D. L., Kocaoğlu, B., İşleyen, K., Uras, D., Yılmaz, E., Ünek, C., Nikerel, E., & Özdemir, B. S. (2024). Scaling up plant molecular farming via bioprocessing of plant cell suspension culture. In C. Kole, A. Chaurasia, K. L. Hefferon, & J. Panigrahi (Eds.). *Applications of plant molecular farming* (pp. 163–189). Springer Nature https://doi.org/10.1007/978-981-97-0176-6_7.

Kuo, C. H., Liu, H.-C., Nargotra, P., Chan, H.-S., Shih, H. D., & Liu, Y.-C. (2025). Media composition effects on hairy root biomass and tetrandrine production in *Stephania tetrandra*. *Molecules, 30*(8), 1859. https://doi.org/10.3390/molecules30081859.

Kuźma, Ł., Skrzypek, Z., & Wysokińska, H. (2006). Diterpenoids and triterpenoids in hairy roots of *Salvia sclarea*. *Plant Cell, Tissue and Organ Culture, 84*(2), 171–179. https://doi.org/10.1007/s11240-005-9018-6.

Lahlou, M. (2007). Screening of natural products for drug discovery. *Expert Opinion on Drug Discovery, 2*(5), 697–705. https://doi.org/10.1517/17460441.2.5.697.

Lala, S. (2021). Nanoparticles as elicitors and harvesters of economically important secondary metabolites in higher plants: A review. *IET Nanobiotechnology, 15*(1), 28–57. https://doi.org/10.1049/nbt2.12005.

Lan, X., Chang, K., Zeng, L., Liu, X., Qiu, F., Zheng, W., Quan, H., Liao, Z., Chen, M., Huang, W., Liu, W., & Wang, Q. (2013). Engineering salidroside biosynthetic pathway in hairy root cultures of *Rhodiola crenulata* based on metabolic characterization of tyrosine decarboxylase. *PLoS ONE, 8*(10), e75459. https://doi.org/10.1371/journal.pone.0075459.

Leduc, M., Tikhomiroff, C., Cloutier, M., Perrier, M., & Jolicoeur, M. (2006). Development of a kinetic metabolic model: Application to *Catharanthus roseus* hairy root. *Bioprocess and Biosystems Engineering, 28*(5), 295–313. https://doi.org/10.1007/s00449-005-0034-z.

Lee, F., Bae, K. H., Ng, S., Yamashita, A., & Kurisawa, M. (2021). Hyaluronic acid–green tea catechin conjugates as a potential therapeutic agent for rheumatoid arthritis. *RSC Advances, 11*(24), 14285–14294. https://doi.org/10.1039/d1ra01491a.

Lenk, F., & Bley, T. (2015). Ramified challenges: Monitoring and modeling of hairy root growth in bioprocesses—A review. *Advances in Biochemical Engineering/Biotechnology, 149*, 253–273. https://doi.org/10.1007/10_2015_305.

Li, F. X., Jin, Z. P., Zhao, D. X., Cheng, L. Q., Fu, C. X., & Ma, F. (2006). Overexpression of the *Saussurea medusa* chalcone isomerase gene in S. involucrata hairy root cultures enhances their biosynthesis of apigenin. *Phytochemistry, 67*(6), 553–560. https://doi.org/10.1016/j.phytochem.2005.12.004.

Li, G., Liu, R., Xu, R., Varshney, R. K., Ding, H., Li, M., Yan, X., Huang, S., Li, J., Wang, D., Ji, Y., Wang, C., He, J., Luo, Y., Gao, S., Wei, P., Zong, X., & Yang, T. (2023). Development of an *Agrobacterium*-mediated CRISPR/Cas9 system in pea (*Pisum sativum* L. *The Crop Journal, 11*(1), 132–139. https://doi.org/10.1016/j.cj.2022.04.011.

Li, Q., Liu, S., He, X., Li, H., Lyu, S., & Fan, Y. (2023). *Agrobacterium rhizogenes*-mediated genetic transformation and establishment of CRISPR/Cas9 genome-editing technology in *Limonium bicolor*. *Agronomy, 13*(9), 2244. https://doi.org/10.3390/agronomy13092244.

Li, Z., Zhou, J., Dong, T., Xu, Y., & Shang, Y. (2021). Application of electrochemical methods for the detection of abiotic stress biomarkers in plants. *Biosensors and Bioelectronics, 182*, 113105. https://doi.org/10.1016/j.bios.2021.113105.

Liu, C., Guo, C., Wang, Y., & Ouyang, F. (2002). Effect of light irradiation on hairy root growth and artemisinin biosynthesis of *Artemisia annua* L. *Process Biochemistry, 38*(4), 581–585. https://doi.org/10.1016/S0032-9592(02)00165-6.

Liu, C., Xu, Y. W. X., Ouyang, F., Ye, H., & Li, G. (1999). Improvement of artemisinin accumulation in hairy root cultures of *Artemisia annua* L by fungal elicitor. *Bioprocess Engineering, 20*(2), 161–164. https://doi.org/10.1007/PL00009041.

Liu, K. C., Yen, C. Y., Wu, R. S., Yang, J. S., Lu, H. F., Lu, K. W., Lo, C., Chen, H. Y., Tang, N. Y., Wu, C. C., & Chung, J. G. (2014). The roles of endoplasmic reticulum stress and mitochondrial apoptotic signaling pathway in quercetin-mediated cell death of human prostate cancer PC-3 cells. *Environmental Toxicology, 29*(4), 428–439. https://doi.org/10.1002/tox.21769.

Liu, Q., Cui, L., Guo, Y., Ni, X., Zhang, Y., & Kai, G. (2013). Optimization of nutritive factors in culture media for growth and tropane alkaloid production from *Anisodus acutangulus* hairy roots. *Journal of Applied Pharmaceutical Science, 3*(1), 001–004. https://doi.org/10.7324/JAPS.2013.30101.

Liu, Z., Xia, J., Wang, C.-Z., Zhang, J.-Q., Ruan, C.-C., Sun, G.-Z., & Yuan, C.-S. (2016). Remarkable impact of acidic ginsenosides and organic acids on ginsenoside transformation from fresh ginseng to red ginseng. *Journal of Agricultural and Food Chemistry, 64*(26), 5389–5399. https://doi.org/10.1021/acs.jafc.6b00963.

Ludwig-Müller, J., Georgiev, M., & Bley, T. (2008). Metabolite and hormonal status of hairy root cultures of Devil's claw (*Harpagophytum procumbens*) in flasks and in a bubble column bioreactor. *Process Biochemistry, 43*(1), 15–23. https://doi.org/10.1016/j.procbio.2007.10.006.

Ma, Y., Yang, J., Zhang, Y., Zheng, C., Liang, Z., Lu, P., Song, F., Wang, Y., & Zhang, J. (2022). Development of a naringenin microemulsion as a prospective ophthalmic delivery system for the treatment of corneal neovascularization: In vitro and in vivo evaluation. *Drug Delivery, 29*(1), 111–127. https://doi.org/10.1080/10717544.2021.2021323.

Mahmoud, S. S., & Croteau, R. B. (2001). Metabolic engineering of essential oil yield and composition in mint by altering expression of deoxyxylulose phosphate reductoisomerase and menthofuran synthase. *Proceedings of the National Academy of Sciences, 98*(15), 8915–8920. https://doi.org/10.1073/pnas.141237298.

Mallol, A., Cusidó, R. M., Palazón, J., Bonfill, M., Morales, C., & Piñol, M. T. (2001). Ginsenoside production in different phenotypes of *Panax ginseng* transformed roots. *Phytochemistry, 57*(3), 365–371. https://doi.org/10.1016/S0031-9422(01)00062-0.

Manikandan, R., Beulaja, M., Arulvasu, C., Sellamuthu, S., Dinesh, D., Prabhu, D., Babu, G., Vaseeharan, B., & Prabhu, N. M. (2012). Synergistic anticancer activity of curcumin and catechin: An in vitro study using human cancer cell lines. *Microscopy Research and Technique, 75*(2), 112–116. https://doi.org/10.1002/jemt.21032.

Marslin, G., Sheeba, C. J., & Franklin, G. (2017). Nanoparticles alter secondary metabolism in plants via ROS burst. *Frontiers in Plant Science, 8*. https://doi.org/10.3389/fpls.2017.00832.

Martín, M., Montes, F. J., & Galán, M. A. (2008). Influence of impeller type on the bubble breakup process in stirred tanks. *Industrial & Engineering Chemistry Research, 47*(16), 6251–6263. https://doi.org/10.1021/ie800063v.

Massah, M., & Rabiei, M. (2023). Effect of acetosyringone, sucrose and nutrients on transgenic hairy root induction in *Chenopodium quinoa* using different *Rhizobium rhizogenesstrains*. *Research Square*. https://doi.org/10.21203/rs.3.rs-2833983/v1.

Mathew, N., Paily, K. P., Abidha, Vanamail, P., Kalyanasundaram, M., & Balaraman, K. (2002). Macrofilaricidal activity of the plant *Plumbago indica/rosea* in vitro. *Drug Development Research, 56*(1), 33–39. https://doi.org/10.1002/ddr.10056.

Mathias, R. J. (1987). Plant microinjection techniques. In J. K. Setlow (Vol. Ed.), *Genetic engineering: Principles and methods: 9*, (pp. 199–227). Springer US https://doi.org/10.1007/978-1-4684-5377-5_11.

Matveeva, T. V., & Валерьевна, M T (2021). Why do plants need agrobacterial genes? *Ecological genetics, 19*(4), 365–375. https://doi.org/10.17816/ecogen89905.

Maxwell, A. E., Warner, T. A., & Fang, F. (2018). Implementation of machine-learning classification in remote sensing: An applied review. *International Journal of Remote Sensing, 39*(9), 2784–2817. https://doi.org/10.1080/01431161.2018.1433343.

Mehrotra, P. (2016). Biosensors and their applications—A review. *Journal of Oral Biology and Craniofacial Research, 6*(2), 153–159. https://doi.org/10.1016/j.jobcr.2015.12.002.

Mehrotra, S., Prakash, O., Khan, F., & Kukreja, A. K. (2013). Efficiency of neural network-based combinatorial model predicting optimal culture conditions for maximum biomass yields in hairy root cultures. *Plant Cell Reports, 32*(2), 309–317. https://doi.org/10.1007/s00299-012-1364-3.

Mehrotra, S., Prakash, O., Mishra, B. N., & Dwevedi, B. (2008). Efficiency of neural networks for prediction of in vitro culture conditions and inoculum properties for optimum productivity. *Plant Cell, Tissue and Organ Culture, 95*(1), 29–35. https://doi.org/10.1007/s11240-008-9410-0.

Mehrotra, S., Rahman, L. U., & Kukreja, A. K. (2010). An extensive case study of hairy-root cultures for enhanced secondary-metabolite production through metabolic-pathway engineering. *Biotechnology and Applied Biochemistry, 56*(4), 161–172. https://doi.org/10.1042/BA20100171.

Merkel, A., Christen, P., & Kapetanedis, I. (1997). Production of Diosgenin by hairy root cultures of Trigonella-foenum graceum L. 16, 632–636.

Mipeshwaree Devi, A., Khedashwori Devi, K., Premi Devi, P., Lakshmipriyari Devi, M., & Das, S. (2023). Metabolic engineering of plant secondary metabolites: Prospects and its technological challenges. *Frontiers in Plant Science, 14*, 1171154. https://doi.org/10.3389/fpls.2023.1171154.

Mirmazloum, I., Slavov, A. K., & Marchev, A. S. (2024). The untapped potential of hairy root cultures and their multiple applications. *International Journal of Molecular Sciences, 25*(23), 12682. https://doi.org/10.3390/ijms252312682.

Mishra, B. N., & Ranjan, R. (2008). Growth of hairy-root cultures in various bioreactors for the production of secondary metabolites. *Biotechnology and Applied Biochemistry, 49*(1), 1–10. https://doi.org/10.1042/BA20070103.

Mitra, S., Anand, U., Ghorai, M., Kant, N., Kumar, M., Radha, Jha, N. K., Swamy, M. K., Proćków, J., de la Lastra, J. M. P., & Dey, A. (2023). Genome editing technologies, mechanisms and improved production of therapeutic phytochemicals: Opportunities and prospects. *Biotechnology and Bioengineering, 120*(1), 82–94. https://doi.org/10.1002/bit.28260.

Mitra, S., & Murthy, G. S. (2022). Bioreactor control systems in the biopharmaceutical industry: A critical perspective. *Systems Microbiology and Biomanufacturing, 2*(1), 91–112. https://doi.org/10.1007/s43393-021-00048-6.

Mmereke, K. M., Javid, A., Majumdar, S., Ghosh, I., Malik, S., Makhzoum, A., & Jha, S. (2022). Preference of *Agrobacterium rhizogenes* mediated transformation of angiosperms. In *Applications in plant biotechnology*. CRC Press.

Mokhtarikhah, G., Ebadi, M.-T., & Ayyari, M. (2022). Agro-morphological and phytochemical studies of spearmint landraces (*Mentha spicata* L.) in Iran. *Industrial Crops and Products, 176*, 114367. https://doi.org/10.1016/j.indcrop.2021.114367.

Montazeri, M., Azadbakht, M., Najafi Zarrini, H., NematZadeh, G., Pakdin Parizi, A., & Davoodi, A. (2018). Hairy root cultures of *Hypericum perforatum* L.; A promising method for the high scale production of hypericin. *Journal of Medicinal Plants, 17*(67), 55–67.

Moore, L., Warren, G., & Strobel, G. (1979). Involvement of a plasmid in the hairy root disease of plants caused by *Agrobacterium rhizogenes*. *Plasmid, 2*(4), 617–626. https://doi.org/10.1016/0147-619X(79)90059-3.

Moresi, M. (2025). Design and operation of a multifunctional pilot-scale bioreactor for enhanced aerobic fermentation. *Fermentation, 11*(2), 101. https://doi.org/10.3390/fermentation11020101.

Morgan, A. J., Cox, P. N., Turner, D. A., Peel, E., Davey, M. R., Gartland, K. M. A., & Mulligan, B. J. (1987). Transformation of tomato using an Ri plasmid vector. *Plant Science, 49*(1), 37–49. https://doi.org/10.1016/0168-9452(87)90018-5.

Morgan, J. A., & Shanks, J. V. (2000). Determination of metabolic rate-limitations by precursor feeding in *Catharanthus roseus* hairy root cultures. *Journal of Biotechnology, 79*(2), 137–145. https://doi.org/10.1016/S0168-1656(00)00221-2.

Morgan, J. A., & Shanks, J. V. (2002). Quantification of metabolic flux in plant secondary metabolism by a biogenetic organizational approach. *Metabolic Engineering, 4*(3), 257–262. https://doi.org/10.1006/mben.2002.0224.

Moriguchi, K., Maeda, Y., Satou, M., Hardayani, N. S. N., Kataoka, M., Tanaka, N., & Yoshida, K. (2001). The complete nucleotide sequence of a plant root-inducing (Ri) plasmid indicates its chimeric structure and evolutionary relationship between tumor-inducing (Ti) and symbiotic (Sym) plasmids in rhizobiaceae1. *Journal of Molecular Biology, 307*(3), 771–784. https://doi.org/10.1006/jmbi.2001.4488.

Moriuchi, H., Okamoto, C., Nishihama, R., Yamashita, I., Machida, Y., & Tanaka, N. (2004). Nuclear localization and interaction of RolB with plant 14-3-3 proteins correlates with induction of adventitious roots by the oncogene rolB. *The Plant Journal, 38*(2), 260–275. https://doi.org/10.1111/j.1365-313X.2004.02041.x.

Moyano, E., Fornalé, S., Palazón, J., Cusidó, R. M., Bagni, N., & Piñol, M. T. (2002). Alkaloid production in *Duboisia hybrid* hairy root cultures overexpressing the pmt gene. *Phytochemistry, 59*(7), 697–702. https://doi.org/10.1016/S0031-9422(02)00044-4.

Moyano, E., Jouhikainen, K., Tammela, P., Palazón, J., Cusidó, R. M., Piñol, M. T., Teeri, T. H., & Oksman-Caldentey, K. (2003). Effect of pmt gene overexpression on tropane alkaloid production in transformed root cultures of *Datura metel* and *Hyoscyamus muticus*. *Journal of Experimental Botany, 54*(381), 203–211. https://doi.org/10.1093/jxb/erg014.

Mukundan, U., & Hjortso, M. A. (1990a). Effect of fungal elicitor on thiophene production in hairy root cultures of *Tagetes patula*. *Applied Microbiology and Biotechnology, 33*(2), 145–147. https://doi.org/10.1007/BF00176515.

Mukundan, U., & Hjortso, M. A. (1990b). Thiophene content in normal and transformed root cultures of *Tagetes erecta*: A comparison with thiophene content in roots of intact plants. *Journal of Experimental Botany, 41*(11), 1497–1501. https://doi.org/10.1093/jxb/41.11.1497.

Murashige, T., & Skoog, F. (1962). A revised medium for rapid growth and bio assays with tobacco tissue cultures. | *EBSCOhost*. https://doi.org/10.1111/j.1399-3054.1962.tb08052.x.

Murthy, H. N., Joseph, K. S., Paek, K. Y., & Park, S. Y. (2024). Nanomaterials as novel elicitors of pharmacologically active plant specialized metabolites in cell and organ cultures: Current status and future outlooks. *Plant Growth Regulation, 104*(1), 5–30. https://doi.org/10.1007/s10725-023-01086-x.

Namdeo, A., Patil, S., & Fulzele, D. P. (2002). Influence of fungal elicitors on production of ajmalicine by cell cultures of *Catharanthus roseus*. *Biotechnology Progress, 18*(1), 159–162. https://doi.org/10.1021/bp0101280.

Neelwarne, B., & Thimmaraju, R. (2009). Bioreactor for cultivation of red beet hairy roots and in situ recovery of primary and secondary metabolites. *Engineering in Life Sciences, 9*(3), 227–238. https://doi.org/10.1002/elsc.200800116.

Nemoto, K., Hara, M., Goto, S., Kasai, K., Seki, H., Suzuki, M., Oka, A., Muranaka, T., & Mano, Y. (2009). The aux1 gene of the Ri plasmid is sufficient to confer auxin autotrophy in tobacco BY-2 cells. *Journal of Plant Physiology, 166*(7), 729–738. https://doi.org/10.1016/j.jplph.2008.09.006.

Newman, D. J., & Cragg, G. M. (2007). Natural products as sources of new drugs over the last 25 years. *Journal of Natural Products, 70*(3), 461–477. https://doi.org/10.1021/np068054v.

Nezami-Alanagh, E., Garoosi, G. A., Landín, M., & Gallego, P. P. (2018). Combining DOE with neurofuzzy logic for healthy mineral nutrition of pistachio rootstocks in vitro culture. *Frontiers in Plant Science, 9*, 1474. https://doi.org/10.3389/fpls.2018.01474.

Nickel, N., Fitschen, J., Haase, I., Kuschel, M., Schulz, T. W., Wucherpfennig, T., & Schlüter, M. (2024). Novel sparging strategies to enhance dissolved carbon dioxide stripping in industrial scale stirred tank reactors. *Frontiers in Chemical Engineering, 6*. https://doi.org/10.3389/fceng.2024.1470991.

Nourozi, E., Hosseini, B., Maleki, R., & Mandoulakani, B. A. (2019a). Iron oxide nanoparticles: A novel elicitor to enhance anticancer flavonoid production and gene expression in *Dracocephalum kotschyi* hairy-root cultures. *Journal of the Science of Food and Agriculture, 99*(14), 6418–6430. https://doi.org/10.1002/jsfa.9921.

Nourozi, E., Hosseini, B., Maleki, R., & Mandoulakani, B. A. (2019b). Pharmaceutical important phenolic compounds overproduction and gene expression analysis in *Dracocephalum kotschyi* hairy roots elicited by SiO2 nanoparticles. *Industrial Crops and Products, 133*, 435–446. https://doi.org/10.1016/j.indcrop.2019.03.053.

Ohlrogge, J., & Benning, C. (2000). Unraveling plant metabolism by EST analysis. *Current Opinion in Plant Biology, 3*(3), 224–228. https://doi.org/10.1016/S1369-5266(00)80069-2.

Olayiwola, Y., & Gollahon, L. (2024). Natural compounds and breast cancer: Chemo-preventive and therapeutic capabilities of chlorogenic acid and cinnamaldehyde. *Pharmaceuticals, 17*(3), 361. https://doi.org/10.3390/ph17030361.

Osama, K., Somvanshi, P., Pandey, A. K., & Mishra, B. N. (2013). Modelling of nutrient mist reactor for hairy root growth using artificial neural network. *European Journal of Scientific Research, 97*(4), 516–526.

Otten, L. (2018). The *Agrobacterium* phenotypic plasticity (Plast) Genes. In S. B. Gelvin (Ed.). *Agrobacterium biology: From basic science to biotechnology* (pp. 375–419). Springer International Publishing https://doi.org/10.1007/82_2018_93.

Ozyigit, I. I. (2020). Gene transfer to plants by electroporation: Methods and applications. *Molecular Biology Reports, 47*(4), 3195–3210. https://doi.org/10.1007/s11033-020-05343-4.

Ozyigit, I. I., Dogan, I., & Artam Tarhan, E. (2013). *Agrobacterium rhizogenes*-mediated transformation and its biotechnological applications in crops. In K. R. Hakeem, P. Ahmad, & M. Ozturk (Eds.). *Crop improvement: New approaches and modern techniques* (pp. 1–48). Springer US https://doi.org/10.1007/978-1-4614-7028-1_1.

Paek, K. Y., Murthy, H. N., Hahn, E. J., & Zhong, J. J. (2009). Large scale culture of ginseng adventitious roots for production of ginsenosides. *Advances in Biochemical Engineering/Biotechnology, 113*, 151–176. https://doi.org/10.1007/10_2008_31.

Paek, K.-Y., Murthy, H. N., & Zhong, J. J. (2014). *Production of biomass and bioactive compounds using bioreactor technology*. Springer.

Paine, J. A., Shipton, C. A., Chaggar, S., Howells, R. M., Kennedy, M. J., Vernon, G., Wright, S. Y., Hinchliffe, E., Adams, J. L., Silverstone, A. L., & Drake, R. (2005). Improving the nutritional value of Golden Rice through increased pro-vitamin A content. *Nature Biotechnology, 23*(4), 482–487. https://doi.org/10.1038/nbt1082.

Palavalli, R. R., Srivastava, S., & Srivastava, A. K. (2012). Development of a mathematical model for growth and oxygen transfer in in vitro plant hairy root cultivations. *Applied Biochemistry and Biotechnology, 167*(6), 1831–1844. https://doi.org/10.1007/s12010-011-9515-5.

Palazon, J., & Alcalde, M. A. (2025). Secondary metabolites in plants. *Plants (Basel), 14*(14), 2146. https://doi.org/10.3390/plants14142146.

Palazón, J., Cusidó, R. M., Bonfill, M., Mallol, A., Moyano, E., Morales, C., & Piñol, M. T. (2003). Elicitation of different Panax ginseng transformed root phenotypes for an improved ginsenoside production. *Plant Physiology and Biochemistry, 41*(11), 1019–1025. https://doi.org/10.1016/j.plaphy.2003.09.002.

Palazón, J., Mallol, A., Eibl, R., Lettenbauer, C., Cusidó, R. M., & Piñol, M. T. (2003). Growth and ginsenoside production in hairy root cultures of panax ginseng using a novel bioreactor. *Planta Medica, 69*, 344–349. https://doi.org/10.1055/s-2003-38873.

Panda, B. M., Mehta, U. J., & Hazra, S. (2017). Optimizing culture conditions for establishment of hairy root culture of *Semecarpus anacardium* L. *3 Biotech, 7*(1), 21. https://doi.org/10.1007/s13205-017-0608-x.

Pandolfini, T., Storlazzi, A., Calabria, E., Defez, R., & Spena, A. (2000). The spliceosomal intron of the rolA gene of *Agrobacterium rhizogenes* is a prokaryotic promoter. *Molecular Microbiology, 35*(6), 1326–1334. https://doi.org/10.1046/j.1365-2958.2000.01810.x.

Parkinson, M., Cotter, T., & Dix, P. J. (1990). Peroxidase production by cell suspension and hairy root cultures of horseradish (*Armoracia rusticana*). *Plant Science, 66*(2), 271–277. https://doi.org/10.1016/0168-9452(90)90213-8.

Patra, N., & Srivastava, A. K. (2016). Artemisinin production by plant hairy root cultures in gas- and liquid-phase bioreactors. *Plant Cell Reports, 35*(1), 143–153. https://doi.org/10.1007/s00299-015-1875-9.

Pavlova, O. A., Андреевна, П O, Matveyeva, T. V., Валерьевна, M T, Lutova, L. A., & Алексеевна, Л Л (2013). Rol-GENES of *Agrobacterium rhizogenes*. *Ecological genetics, 11*(1), 59–68. https://doi.org/10.17816/ecogen11159-68.

Phillipson, J. D. (2001). Phytochemistry and medicinal plants. *Phytochemistry, 56*(3), 237–243. https://doi.org/10.1016/S0031-9422(00)00456-8.

Pillai, D. B., Jose, B., Satheeshkumar, K., & Krishnan, P. N. (2015). Optimization of inoculum density in hairy root culture of *Plumbago rosea* L. for enhanced growth and plumbagin production towards scaling-up in bioreactor. *Indian Journal of Biotechnology, 14*(2), 264–269.

Pistelli, L., Giovannini, A., Ruffoni, B., Bertoli, A., & Pistelli, L. (2010). Hairy root cultures for secondary metabolites production. In M. T. Giardi, G. Rea, & B. Berra (Eds.). *Bio-farms for nutraceuticals: Functional food and safety control by biosensors* (pp. 167–184). Springer US https://doi.org/10.1007/978-1-4419-7347-4_13.

Pitta-Alvarez, S. I., Spollansky, T. C., & Giulietti, A. M. (2000). The influence of different biotic and abiotic elicitors on the production and profile of tropane alkaloids in hairy root cultures of *Brugmansia candida*. *Enzyme and Microbial Technology, 26*(2), 252–258. https://doi.org/10.1016/S0141-0229(99)00137-4.

Prabhakar, P. K., Prasad, R., Ali, S., & Doble, M. (2013). Synergistic interaction of ferulic acid with commercial hypoglycemic drugs in streptozotocin induced diabetic rats. *Phytomedicine, 20*(6), 488–494. https://doi.org/10.1016/j.phymed.2012.12.004.

Prakash, O., Mehrotra, S., Krishna, A., & Mishra, B. N. (2010). A neural network approach for the prediction of in vitro culture parameters for maximum biomass yields in hairy root cultures. *Journal of Theoretical Biology, 265*(4), 579–585. https://doi.org/10.1016/j.jtbi.2010.05.020.

Pratheeshkumar, P., Budhraja, A., Son, Y. O., Wang, X., Zhang, Z., Ding, S., Wang, L., Hitron, A., Lee, J. C., Xu, M., Chen, G., Luo, J., & Shi, X. (2012). Quercetin inhibits angiogenesis-mediated human prostate tumor growth by targeting VEGFR-2 regulated AKT/mTOR/P70S6K signaling pathways. *PLoS ONE, 7*(10), e47516. https://doi.org/10.1371/journal.pone.0047516.

Prip, C. (2018). The convention on biological diversity as a legal framework for safeguarding ecosystem services. *Ecosystem Services, 29*, 199–204. https://doi.org/10.1016/j.ecoser.2017.02.015.

Putalun, W., Luealon, W., De-Eknamkul, W., Tanaka, H., & Shoyama, Y. (2007). Improvement of artemisinin production by chitosan in hairy root cultures of *Artemisia annua* L. *Biotechnology Letters, 29*(7), 1143–1146. https://doi.org/10.1007/s10529-007-9368-8.

Ramirez-Estrada, K., Vidal-Limon, H., Hidalgo, D., Moyano, E., Golenioswki, M., Cusidó, R. M., & Palazon, J. (2016). Elicitation, an effective strategy for the biotechnological production of bioactive high-added value compounds in plant cell factories. *Molecules, 21*(2), 182. https://doi.org/10.3390/molecules21020182.

Rangslang, R. K., Liu, Z., Lütken, H., & Favero, B. T. (2018). *Agrobacterium* spp. genes and ORFs: Mechanisms and applications in plant science. *Ciência e Agrotecnologia, 42*, 453–463. https://doi.org/10.1590/1413-70542018425000118.

Rao, S. R., Tripathi, U., Suresh, B., & Ravishankar, G. A. (2001). Enhancement of secondary metabolite production in hairy root cultures of beta vulgaris and tagetes patula under the influence of microalgal elicitors. *Food Biotechnology, 15*(1), 35–46. https://doi.org/10.1081/FBT-100103893.

Rhee, H. S., Cho, H.-Y., Son, S. Y., Yoon, S.-Y. H., & Park, J. M. (2010). Enhanced accumulation of decursin and decursinol angelate in root cultures and intact roots of *Angelica gigas* Nakai following elicitation. *Plant Cell, Tissue and Organ Culture (PCTOC, 101*(3), 295–302. https://doi.org/10.1007/s11240-010-9688-6.

Rhodes, M. J. C., Robins, R. J., Hamill, J. D., Parr, A. J., Hilton, M. H., & Walton, N. J. (1990). Properties of transformed root cultures. In B. V. Charlwood, & M. J. C. Rhodes (Eds.). *Secondary metabolites from plant tissue culture* (pp. 201–225). Clarendon press Oxford.

Riker, A. (1930). Studies on infectious hairy root of nursery trees of apples. *Journal of Agricultural Research, 41*, 507–540.

Rinwa, P., & Kumar, A. (2012). Piperine potentiates the protective effects of curcumin against chronic unpredictable stress-induced cognitive impairment and oxidative damage in mice. *Brain Research, 1488*, 38–50. https://doi.org/10.1016/j.brainres.2012.10.002.

Rothe, G., Hachiya, A., Yamada, Y., Hashimoto, T., & Dräger, B. (2003). Alkaloids in plants and root cultures of *Atropa belladonna* overexpressing putrescine N-methyltransferase. *Journal of Experimental Botany, 54*(390), 2065–2070. https://doi.org/10.1093/jxb/erg227.

Roy, A. (2021a). Hairy root culture an alternative for bioactive compound production from medicinal plants. *Current Pharmaceutical Biotechnology, 22*(1), 136–149. https://doi.org/10.2174/1389201021666201229110625.

Roy, A. (2021b). Hairy Root Culture an Alternative for Bioactive Compound Production from Medicinal Plants. http://www.Eurekaselect.Com. Retrieved September 23, 2025, from https://www.eurekaselect.com/article/112774.

Runguphan, W., Maresh, J. J., & O'Connor, S. E. (2009). Silencing of tryptamine biosynthesis for production of nonnatural alkaloids in plant culture. *Proceedings of the National Academy of Sciences, 106*(33), 13673–13678. https://doi.org/10.1073/pnas.0903393106.

Ruse, G., Jîjie, A. R., Moacă, E. A., Pătrașcu, D., Ardelean, F., Jojic, A. A., Ardelean, S., & Tchiakpe-Antal, D. S. (2025). *Coffea arabica*: An emerging active ingredient in dermato-cosmetic applications. *Pharmaceuticals, 18*(2), 171. https://doi.org/10.3390/ph18020171.

Sarker, I. H. (2022). AI-based modeling: Techniques, applications and research issues towards automation, intelligent and smart systems. *SN Computer Science, 3*(2), 158. https://doi.org/10.1007/s42979-022-01043-x.

Satdive, R. K., Fulzele, D. P., & Eapen, S. (2007). Enhanced production of azadirachtin by hairy root cultures of *Azadirachta indica* A. Juss by elicitation and media optimization. *Journal of Biotechnology, 128*(2), 281–289. https://doi.org/10.1016/j.jbiotec.2006.10.009.

Sathyanarayana, B. N., & Varghese, D. B. (2007). *Plant tissue culture: Practices and new experimental protocols*. I. K. International Pvt Ltd.

Sato, F., Hashimoto, T., Hachiya, A., Tamura, K., Choi, K.-B., Morishige, T., Fujimoto, H., & Yamada, Y. (2001). Metabolic engineering of plant alkaloid biosynthesis. *Proceedings of the National Academy of Sciences, 98*(1), 367–372. https://doi.org/10.1073/pnas.98.1.367.

Schena, M., Shalon, D., Davis, R. W., & Brown, P. O. (1995). Quantitative monitoring of gene expression patterns with a complementary DNA microarray. *Science, 270*(5235), 467–470. https://doi.org/10.1126/science.270.5235.467.

Seddon, J. M. (2007). Multivitamin-multimineral supplements and eye disease: Age-related macular degeneration and cataract. *American Journal of Clinical Nutrition, 85*(1), 304S–307S. https://doi.org/10.1093/ajcn/85.1.304S.

Seki, H., Nishizawa, T., Tanaka, N., Niwa, Y., Yoshida, S., & Muranaka, T. (2005). Hairy root-activation tagging: A High-throughput system for activation tagging in transformed hairy roots. *Plant Molecular Biology, 59*(5), 793–807. https://doi.org/10.1007/s11103-005-1008-x.

Seybold, H., Trempel, F., Ranf, S., Scheel, D., Romeis, T., & Lee, J. (2014). Ca^{2+} signalling in plant immune response: From pattern recognition receptors to Ca^{2+} decoding mechanisms. *New Phytologist, 204*(4), 782–790. https://doi.org/10.1111/nph.13031.

Shahnawaz, Pandey, D. K., Konjengbam, M., Dwivedi, P., Kaur, P., Kumar, V., Ray, D., Ray, P., Nazir, R., Kaur, H., Parida, S., & Dey, A. (2021). Biotechnological interventions of in vitro propagation and production of valuable secondary metabolites in *Stevia rebaudiana*. *Applied Microbiology and Biotechnology, 105*(23), 8593–8614. https://doi.org/10.1007/s00253-021-11580-9.

Shamon, S. D., & Perez, M. I. (2016). Blood pressure-lowering efficacy of reserpine for primary hypertension. *Cochrane Database of Systematic Reviews*(12), CD007655. https://doi.org/10.1002/14651858.CD007655.pub3.

Shanks, J. V., & Morgan, J. (1999). Plant 'hairy root' culture. *Current Opinion in Biotechnology, 10*(2), 151–155. https://doi.org/10.1016/S0958-1669(99)80026-3.

Sharma, A. R., Gajurel, G., Abdel-Karim, S., Alam, M. A., Shields, R. C., & Medina-Bolivar, F. (2025). Production of Malheuran A, a geranylated flavonoid with antimicrobial and anti-inflammatory activities, in hairy root cultures of *Dalea purpurea*. *Plants, 14*(2), 259. https://doi.org/10.3390/plants14020259.

Sharma, P., Padh, H., & Shrivastava, N. (2013). Hairy root cultures: A suitable biological system for studying secondary metabolic pathways in plants. *Engineering in Life Sciences, 13*(1), 62–75. https://doi.org/10.1002/elsc.201200030.

Sharma, S., & Shahzad, A. (2013). Bioreactors: A rapid approach for secondary metabolite production. In M. Shahid, A. Shahzad, A. Malik, & A. Sahai (Eds.). *Recent trends in biotechnology and therapeutic applications of medicinal plants* (pp. 25–49). Springer Netherlands https://doi.org/10.1007/978-94-007-6603-7_2.

Shen, L., You, B. A., Gao, H. Q., Li, B. Y., Yu, F., & Pei, F. (2012). Effects of phlorizin on vascular complications in diabetes db/db mice. *Chinese Medical Journal, 125*(20), 3692–3696.

Shi, H. P., & Kintzios, S. (2003). Genetic transformation of *Pueraria phaseoloides* with *Agrobacterium rhizogenes* and puerarin production in hairy roots. *Plant Cell Reports, 21*(11), 1103–1107. https://doi.org/10.1007/s00299-003-0633-6.

Shi, M., Kwok, K. W., & Wu, J. Y. (2007). Enhancement of tanshinone production in *Salvia miltiorrhiza* Bunge (red or Chinese sage) hairy-root culture by hyperosmotic stress and yeast elicitor. *Biotechnology and Applied Biochemistry, 46*(4), 191–196. https://doi.org/10.1042/BA20060147.

Shibasaki, N., Yonemoto, T., & Tadaki, T. (1993). Application of five-state structured models for growth and secondary metabolite production in cultures of *Nicotiana tabacum*. *Journal of Chemical Technology & Biotechnology, 58*(2), 151–157. https://doi.org/10.1002/jctb.280580208.

Shih, K. C., Chan, T. C., Ng, A. L., Lai, J. S., Li, W. W., Cheng, A. C., & Fan, D. S. (2016). Use of atropine for prevention of childhood myopia progression in clinical practice. *Eye & Contact Lens, 42*(1), 16–23. https://doi.org/10.1097/ICL.0000000000000189.

Shrestha, M. (2018). Direct gene transfer methods in plants (Microprojectile bombardment, Electroporation; polyethylene glycol (PEG)- mediated gene transformation, Silica carbon fibres whiskers). https://miteshshrestha.wordpress.com/wp-content/uploads/2018/03/direct-gene-transfer-methods-in-plants.pdf.

Shvets, D. Yu, Berezhneva, Z. A., Musin, Kh. G., Baimukhametova, E. A., & Kuluev, B. R. (2023). Rol genes of agrobacteria: Possible biological functions. *Biology Bulletin Reviews, 13*(3), S359–S376. https://doi.org/10.1134/S2079086423090116.

Singh, S., Pandey, P., Akhtar, Md. Q., Negi, A. S., & Banerjee, S. (2021). A new synthetic biology approach for the production of curcumin and its glucoside in *Atropa belladonna* hairy roots. *Journal of Biotechnology, 328*, 23–33. https://doi.org/10.1016/j.jbiotec.2020.12.022.

Sivakumar, G., Yu, K. W., Hahn, E. J., & Paek, K. Y. (2005). Optimization of organic nutrients for ginseng hairy roots production in large-scale bioreactors. *Current Science, 89*(4), 641–649.

Škubník, J., Pavlíčková, V. S., Ruml, T., & Rimpelová, S. (2021). Vincristine in combination therapy of cancer: Emerging trends in clinics. *Biology, 10*(9), 849. https://doi.org/10.3390/biology10090849.

Sommer, S., Köhle, A., Yazaki, K., Shimomura, K., Bechthold, A., & Heide, L. (1999). Genetic engineering of shikonin biosynthesis hairy root cultures of *Lithospermum erythrorhizon* transformed with the bacterial ubiC gene. *Plant Molecular Biology, 39*(4), 683–693. https://doi.org/10.1023/A:1006185806390.

Sonkar, N., Shukla, P. K., & Misra, P. (2023). Plant hairy roots as biofactory for the production of industrial metabolites. In *Plants as bioreactors for industrial molecules*. John Wiley & Sons, Ltd, 273–297 https://doi.org/10.1002/9781119875116.ch11.

Srikantan, C., & Srivastava, S. (2018). Bioreactor design and analysis for large-scale plant cell and hairy root cultivation. In V. Srivastava, S. Mehrotra, & S. Mishra (Eds.). *Hairy roots: An effective tool of plant biotechnology* (pp. 147–182). Springer https://doi.org/10.1007/978-981-13-2562-5_7.

Sriprateep, K., Sala-Ngamand, S., Khonjun, S., Tseng, M. L., Srichok, T., Nanthasamroeng, N., Pitakaso, R., & Butploy, N. (2024). Heterogeneous ensemble machine learning to predict the asiaticoside concentration in *Centella asiatica* urban. *Intelligent Systems with Applications, 21*, 200319. https://doi.org/10.1016/j.iswa.2023.200319.

Sriram, G., Fulton, D. B., & Shanks, J. V. (2007). Flux quantification in central carbon metabolism of *Catharanthus roseus* hairy roots by 13C labeling and comprehensive bondomer balancing. *Phytochemistry, 68*(16–18), 2243–2257. https://doi.org/10.1016/j.phytochem.2007.04.009.

Sriram, G., González-Rivera, O., & Shanks, J. V. (2006). Determination of biomass composition of *Catharanthus roseus* hairy roots for metabolic flux analysis. *Biotechnology Progress, 22*(6), 1659–1663. https://doi.org/10.1021/bp060162k.

Srivastava, S., & Srivastava, A. K. (2007). Hairy root culture for mass-production of high-value secondary metabolites. *Critical Reviews in Biotechnology, 27*(1), 29–43. https://doi.org/10.1080/07388550601173918.

Srivastava, S., & Srivastava, A. K. (2012). Azadirachtin production by hairy root cultivation of *Azadirachta indica* in a modified stirred tank reactor. *Bioprocess and Biosystems Engineering, 35*(9), 1549–1553. https://doi.org/10.1007/s00449-012-0745-x.

Staniszewska, I., Królicka, A., Maliński, E., Łojkowska, E., & Szafranek, J. (2003). Elicitation of secondary metabolites in in vitro cultures of *Ammi majus* L. *Enzyme and Microbial Technology, 33*(5), 565–568. https://doi.org/10.1016/S0141-0229(03)00180-7.

Steingroewer, J., Bley, T., Georgiev, V., Ivanov, I., Lenk, F., Marchev, A., & Pavlov, A. (2013). Bioprocessing of differentiated plant in vitro systems. *Engineering in Life Sciences, 13*(1), 26–38. https://doi.org/10.1002/elsc.201100226.

Stewart, F. C. (1900). A fruit-disease survey of Western New York in. *New York Agricultural Experiment Station.*

Stiles, A. R., & Liu, C.-Z. (2013). Hairy root culture: Bioreactor design and process intensification. In P. M. Doran (Ed.). *Biotechnology of hairy root systems* (pp. 91–114). Springer https://doi.org/10.1007/10_2013_181.

SU, W. W. (2008). Bioreactor engineering for recombinant protein production using plant cell suspension culture. In S. D. Gupta, & Y. Ibaraki (Eds.). *Plant tissue culture engineering* (pp. 135–159). Springer Netherlands https://doi.org/10.1007/978-1-4020-3694-1_8.

Sudo, H., Yamakawa, T., Yamazaki, M., Aimi, N., & Saito, K. (2002). Bioreactor production of camptothecin by hairy root cultures of *Ophiorrhiza pumila*. *Biotechnology Letters, 24*(5), 359–363. https://doi.org/10.1023/A:1014568904957.

Sung, L.-S., & Huang, S.-Y. (2006). Lateral root bridging as a strategy to enhance L-DOPA production in *Stizolobium hassjoo* hairy root cultures by using a mesh hindrance mist trickling bioreactor. *Biotechnology and Bioengineering, 94*(3), 441–447. https://doi.org/10.1002/bit.20804.

Sung, M. K., & Park, M. Y. (2013). Nutritional modulators of ulcerative colitis: Clinical efficacies and mechanistic view. *World Journal of Gastroenterology, 19*(7), 994–1004. https://doi.org/10.3748/wjg.v19.i7.994.

Supriya, R., Kala, R. G., & Thulaseedharan, A. (2020). Hairy root culture: Secondary metabolite production in a biotechnological perspective. In S. T. Sukumaran, S. Sugathan, & S. Abdulhameed (Eds.). *Plant metabolites: Methods, applications and prospects* (pp. 89–110). Springer https://doi.org/10.1007/978-981-15-5136-9_5.

Suresh, B., Bais, H. P., Raghavarao, K. S. M. S., Ravishankar, G. A., & Ghildyal, N. P. (2005). Comparative evaluation of bioreactor design using *Tagetes patula* L. hairy roots as a model system. *Process Biochemistry, 40*(5), 1509–1515. https://doi.org/10.1016/j.procbio.2003.10.017.

Suresh, B., Rajasekaran, T., Rao, S. R., Raghavarao, K. S. M. S., & Ravishankar, G. A. (2001). Studies on osmolarity, conductivity and mass transfer for selection of a bioreactor for *Tagetes patula* L. hairy roots. *Process Biochemistry, 36*(10), 987–993. https://doi.org/10.1016/S0032-9592(01)00132-7.

Suresh, B., Thimmaraju, R., Bhagyalakshmi, N., & Ravishankar, G. A. (2004). Polyamine and methyl jasmonate-influenced enhancement of betalaine production in hairy root cultures of *Beta vulgaris* grown in a bubble column reactor and studies on efflux of pigments. *Process Biochemistry, 39*(12), 2091–2096. https://doi.org/10.1016/j.procbio.2003.10.009.

Takeda, J., Ito, Y., Maeda, K., & Ozeki, Y. (2002). Assignment of UVB-responsive cis-element and protoplastization-(dilution-) and elicitor-responsive ones in the promoter region of a carrot phenylalanine ammonia-lyase gene (gDcPAL1). *Photochemistry and Photobiology, 76*(2), 232–238. https://doi.org/10.1562/0031-8655(2002)076<0232:aource>2.0.co;2.

Tariverdizadeh, N., Mohebodini, M., Chamani, E., & Ebadi, A. (2021). Iron and zinc oxide nanoparticles: An efficient elicitor to enhance trigonelline alkaloid production in hairy roots of fenugreek. *Industrial Crops and Products, 162*, 113240. https://doi.org/10.1016/j.indcrop.2021.113240.

Terakura, S., Ueno, Y., Tagami, H., Kitakura, S., Machida, C., Wabiko, H., Aiba, H., Otten, L., Tsukagoshi, H., Nakamura, K., & Machida, Y. (2007). An oncoprotein from the plant pathogen *Agrobacterium* Has histone chaperone–like activity. *The Plant Cell, 19*(9), 2855–2865. https://doi.org/10.1105/tpc.106.049551.

Terrier, N., Torregrosa, L., Ageorges, A., Vialet, S., Verriès, C., Cheynier, V., & Romieu, C. (2009). Ectopic expression of VvMybPA2 promotes proanthocyanidin biosynthesis in grapevine and suggests additional targets in the pathway. *Plant Physiology, 149*(2), 1028–1041. https://doi.org/10.1104/pp.108.131862.

Thakore, D., Srivastava, A. K., & Sinha, A. (2012). Enhanced production of antihypertensive drug ajmalicine in transformed hairy root culture of catharanthus roseus by application of stress factors in statistically optimized medium. In L. D. Khemani, M. M. Srivastava, & S. Srivastava (Eds.). *Chemistry of phytopotentials: Health, energy and environmental perspectives* (pp. 39–42). Springer https://doi.org/10.1007/978-3-642-23394-4_8.

Thakore, D., Srivastava, A. K., & Sinha, A. K. (2017). Mass production of Ajmalicine by bioreactor cultivation of hairy roots of *Catharanthus roseus*. *Biochemical Engineering Journal, 119*, 84–91. https://doi.org/10.1016/j.bej.2016.12.010.

The Cartagena Protocol on Biosafety. (2025). Secretariat of the Convention on Biological Diversity. http://bch.cbd.int/protocol.

Thimmaraju, R., Bhagyalakshmi, N., Narayan, M. S., & Ravishankar, G. A. (2003). Food-grade chemical and biological agents permeabilize red beet hairy roots, Assisting the release of betalaines. *Biotechnology Progress, 19*(4), 1274–1282. https://doi.org/10.1021/bp0201399.

Thushara, R. M., Hemshekhar, M., Santhosh, M. S., Jnaneshwari, S., Nayaka, S. C., Naveen, S., Kemparaju, K., & Girish, K. S. (2013). Crocin, a dietary additive protects platelets from oxidative stress-induced apoptosis and inhibits platelet aggregation. *Molecular and Cellular Biochemistry, 373*(1–2), 73–83. https://doi.org/10.1007/s11010-012-1476-7.

Tisserant, L.-P., Hubert, J., Lequart, M., Borie, N., Maurin, N., Pilard, S., Jeandet, P., Aziz, A., Renault, J.-H., Nuzillard, J.-M., Clément, C., Boitel-Conti, M., & Courot, E. (2016). 13C NMR and LC-MS profiling of stilbenes from elicited grapevine hairy root cultures. *Journal of Natural Products, 79*(11), 2846–2855. https://doi.org/10.1021/acs.jnatprod.6b00608.

Towler, M. J., Kim, Y., Wyslouzil, B. E., Correll, M. J., & Weathers, P. J. (2008). Design, Development, and applications of mist bioreactors for micropropagation and hairy root culture. In S. D. Gupta, & Y. Ibaraki (Eds.). *Plant tissue culture engineering* (pp. 119–134). Springer Netherlands https://doi.org/10.1007/978-1-4020-3694-1_7.

Trewavas, A. J., & Malhó, R. (1998). Ca2+ signalling in plant cells: The big network!. *Current Opinion in Plant Biology, 1*(5), 428–433. https://doi.org/10.1016/S1369-5266(98)80268-9.

Tufail, S., Riggs, H., Tariq, M., & Sarwat, A. I. (2023). Advancements and challenges in machine learning: A comprehensive review of models, libraries, applications, and algorithms. *Electronics, 12*(8), 1789. https://doi.org/10.3390/electronics12081789.

UN Convention on Biological Diversity. (2011). Nagoya Protocol on access to genetic resources and the fair and equitable sharing of benefits arising from their utilization to the convention on biological diversity: Text and annex. Secretariat of the Convention on Biological Diversity.

Valdiani, A., Hansen, O. K., Nielsen, U. B., Johannsen, V. K., Shariat, M., Georgiev, M. I., Omidvar, V., Ebrahimi, M., Tavakoli Dinanai, E., & Abiri, R. (2019). Bioreactor-based advances in plant tissue and cell culture: Challenges and prospects. *Critical Reviews in Biotechnology, 39*(1), 20–34. https://doi.org/10.1080/07388551.2018.1489778.

Veena, V., & Taylor, C. G. (2007). *Agrobacterium rhizogenes*: Recent developments and promising applications. *In Vitro Cellular & Developmental Biology - Plant, 43*(5), 383–403. https://doi.org/10.1007/s11627-007-9096-8.

Verma, P., Anjum, S., Khan, S. A., Roy, S., Odstrcilik, J., & Mathur, A. K. (2016). Envisaging the regulation of alkaloid biosynthesis and associated growth kinetics in hairy roots of *Vinca minor* through the function of artificial neural network. *Applied Biochemistry and Biotechnology, 178*(6), 1154–1166. https://doi.org/10.1007/s12010-015-1935-1.

Verma, P., Mathur, A. K., & Shanker, K. (2012). Growth, alkaloid production, rol genes integration, bioreactor up-scaling and plant regeneration studies in hairy root lines of *Catharanthus roseus*. *Plant Biosystems - An International Journal Dealing with All Aspects of Plant Biology, 146*(sup1), 27–40. https://doi.org/10.1080/11263504.2011.649797.

Verma, P. C., Singh, H., Negi, A. S., Saxena, G., Rahman, L., & Banerjee, S. (2015). Yield enhancement strategies for the production of picroliv from hairy root culture of *Picrorhiza kurroa* royle ex benth. *Plant Signaling & Behavior, 10*(5), e1023976. https://doi.org/10.1080/15592324.2015.1023976.

Vranić, E., Lacević, A., Mehmedagić, A., & Uzunović, A. (2004). Formulation ingredients for toothpastes and mouthwashes. *Bosnian Journal of Basic Medical Sciences, 4*(4), 51–58. https://doi.org/10.17305/bjbms.2004.3362.

Wang, C. T., Liu, H., Gao, X.-S., & Zhang, H.-X. (2010). Overexpression of G10H and ORCA3 in the hairy roots of *Catharanthus roseus* improves catharanthine production. *Plant Cell Reports, 29*(8), 887–894. https://doi.org/10.1007/s00299-010-0874-0.

Wang, G., Song, L., Wang, H., & Xing, N. (2013). Quercetin synergizes with 2-methoxyestradiol inhibiting cell growth and inducing apoptosis in human prostate cancer cells. *Oncology Reports, 30*(1), 357–363. https://doi.org/10.3892/or.2013.2469.

Wang, J., Zheng, L.-P., & Tan, R.-X. (2006). [The preparation of an elicitor from a fungal endophyte to enhance artemisinin production in hairy root cultures of *Artemisia annua* L]. *Sheng Wu Gong Cheng Xue Bao=Chinese Journal of Biotechnology, 22*(5), 829–834.

Wang, J. W., & Wu, J. Y. (2013). Effective elicitors and process strategies for enhancement of secondary metabolite production in hairy root cultures. In P. M. Doran (Ed.). *Biotechnology of hairy root systems* (pp. 55–89). Springer https://doi.org/10.1007/10_2013_183.

Wang, J. W., Zheng, L. P., Zhang, B., & Zou, T. (2009). Stimulation of artemisinin synthesis by combined cerebroside and nitric oxide elicitation in *Artemisia annua* hairy roots. *Applied Microbiology and Biotechnology, 85*(2), 285–292. https://doi.org/10.1007/s00253-009-2090-9.

Wang, S., Moustaid-Moussa, N., Chen, L., Mo, H., Shastri, A., Su, R., Bapat, P., Kwun, I., & Shen, C. L. (2014). Novel insights of dietary polyphenols and obesity. *Journal of Nutritional Biochemistry, 25*(1), 1–18. https://doi.org/10.1016/j.jnutbio.2013.09.001.

Wang, Y., Zhang, H., Zhao, B., & Yuan, X. (2001). Improved growth of *Artemisia annua* L hairy roots and artemisinin production under red light conditions. *Biotechnology Letters, 23*(23), 1971–1973. https://doi.org/10.1023/A:1013786332363.

Wells, D. M., Laplaze, L., Bennett, M. J., & Vernoux, T. (2013). Biosensors for phytohormone quantification: Challenges, solutions, and opportunities. *Trends in Plant Science, 18*(5), 244–249. https://doi.org/10.1016/j.tplants.2012.12.005.

WHO. (2019). *WHO Global Report on Traditional and Complementary Medicine*. World Health Organization.

Willems, A., & Collins, M. D. (1993). Phylogenetic analysis of rhizobia and agrobacteria based on 16S rRNA gene sequences. *International Journal of Systematic and Evolutionary Microbiology, 43*(2), 305–313. https://doi.org/10.1099/00207713-43-2-305.

Wilson, R. M., & Danishefsky, S. J. (2006). Small molecule natural products in the discovery of therapeutic agents: The synthesis connection. *The Journal of Organic Chemistry, 71*(22), 8329–8351. https://doi.org/10.1021/jo0610053.

Wojtaszek, P. (1997). Mechanisms for the generation of reactive oxygen species in plant defence response. *Acta Physiologiae Plantarum, 19*(4), 581–589. https://doi.org/10.1007/s11738-997-0057-y.

Wu, J. Y., Ng, J., Shi, M., & Wu, S. J. (2007). Enhanced secondary metabolite (tanshinone) production of *Salvia miltiorrhiza* hairy roots in a novel root–bacteria coculture process. *Applied Microbiology and Biotechnology, 77*(3), 543–550. https://doi.org/10.1007/s00253-007-1192-5.

Wu, S. Q., Lian, M. L., Gao, R., Park, S. Y., & Piao, X. C. (2011). Bioreactor application on adventitious root culture of *Astragalus membranaceus. In Vitro Cellular & Developmental Biology - Plant, 47*(6), 719–724. https://doi.org/10.1007/s11627-011-9376-1.

Xiao, Y., Gao, S., Di, P., Chen, J., Chen, W., & Zhang, L. (2009). Lithospermic acid B is more responsive to silver ions (Ag+) than rosmarinic acid in *Salvia miltiorrhiza* hairy root cultures. *Bioscience Reports, 30*(1), 33–40. https://doi.org/10.1042/BSR20080124.

Xing, N., Chen, Y., Mitchell, S. H., & Young, C. Y. (2001). Quercetin inhibits the expression and function of the androgen receptor in LNCaP prostate cancer cells. *Carcinogenesis, 22*(3), 409–414. https://doi.org/10.1093/carcin/22.3.409.

Yamazaki, Y., Kitajima, M., Arita, M., Takayama, H., Sudo, H., Yamazaki, M., Aimi, N., & Saito, K. (2004). Biosynthesis of camptothecin in silico and in vivo tracer study from [1-13C]glucose. *Plant Physiology, 134*(1), 161–170. https://doi.org/10.1104/pp.103.029389.

Yan, Q., Wu, J., & Liu, R. (2011). Modeling of tanshinone synthesis and phase distribution under the combined effect of elicitation and in situ adsorption in *Salvia miltiorrhiza* hairy root cultures. *Biotechnology Letters, 33*(4), 813–819. https://doi.org/10.1007/s10529-010-0484-5.

Yang, D., Ma, P., Liang, X., Liang, Z., Zhang, M., Shen, S., Liu, H., & Liu, Y. (2012). Metabolic profiles and cDNA-AFLP analysis of *Salvia miltiorrhiza* and *Salvia castanea* Diel f. Tomentosa Stib. *PLOS ONE, 7*(1), e29678. https://doi.org/10.1371/journal.pone.0029678.

Ying, W., Wen, G., Xu, W., Liu, H., Ding, W., Zheng, L., He, Y., Yuan, H., Yan, D., Cui, F., Huang, J., Zheng, B., & Wang, X. (2023). *Agrobacterium rhizogenes*: Paving the road to research and breeding for woody plants. *Frontiers in Plant Science, 14*. https://doi.org/10.3389/fpls.2023.1196561.

Yordanova, Z. P., & Georgiev, M. I. (2017). Cell factories. In B. Thomas, B. G. Murray, & D. J. Murphy (Eds.). *Encyclopedia of applied plant sciences* (pp. 72–76). (Second Edition). Academic Press https://doi.org/10.1016/B978-0-12-394807-6.00158-1.

Young, J. M., Kuykendall, L. D., Martínez-Romero, E., Kerr, A., & Sawada, H. (2001). A revision of Rhizobium Frank 1889, with an emended description of the genus, and the inclusion of all species of *Agrobacterium* Conn 1942 and Allorhizobium undicola de Lajudie et al. 1998 as new combinations: Rhizobium radiobacter, R. rhizogenes, R. rubi, R. undicola and R. vitis. *International Journal of Systematic and Evolutionary Microbiology, 51*(Pt 1), 89–103. https://doi.org/10.1099/00207713-51-1-89.

Young, V. R. (1994). Adult amino acid requirements: The case for a major revision in current recommendations. *The Journal of Nutrition, 124*(8 Suppl), 1517S–1523S. https://doi.org/10.1093/jn/124.suppl_8.1517S.

Yu, K. W., Murthy, H. N., Hahn, E. J., & Paek, K. Y. (2005). Ginsenoside production by hairy root cultures of *Panax ginseng*: Influence of temperature and light quality. *Biochemical Engineering Journal, 23*(1), 53–56. https://doi.org/10.1016/j.bej.2004.07.001.

Yu, S., & Doran, P. M. (1994). Oxygen requirements and mass transfer in hairy-root culture. *Biotechnology and Bioengineering, 44*(8), 880–887. https://doi.org/10.1002/bit.260440803.

Yuan, P., Usman, M., Liu, W., Adhikari, A., Zhang, C., Njiti, V., & Xia, Y. (2024). Advancements in plant gene editing technology: From construct design to enhanced transformation efficiency. *Biotechnology Journal, 19*(12), e202400457. https://doi.org/10.1002/biot.202400457.

Zahedi, M., Ghiasvand, R., Feizi, A., Asgari, G., & Darvish, L. (2013). Does quercetin improve cardiovascular risk factors and inflammatory biomarkers in women with type 2 diabetes: A double-blind randomized controlled clinical trial. *International Journal of Preventive Medicine, 4*(7), 777–785.

Zamani, S., Fathi, M., Ebadi, M. T., & Máthé, Á. (2025). Global trade of medicinal and aromatic plants. A review. *Journal of Agricultural and Food Research, 25*, 101910. https://doi.org/10.1016/j.jafr.2025.101910.

Zarbakhsh, S., & Shahsavar, A. R. (2022). Artificial neural network-based model to predict the effect of γ-aminobutyric acid on salinity and drought responsive morphological traits in pomegranate. *Scientific Reports, 12*(1), 16662. https://doi.org/10.1038/s41598-022-21129-z.

Zhang, B., Zou, T., Lu, Y. H., & Wang, J. W. (2010). Stimulation of artemisinin biosynthesis in *Artemisia annua* hairy roots by oligogalacturonides. *African Journal of Biotechnology, 9*(23), 3437–3442.

Zhang, J., Gong, X., Gan, Q., & Yan, Y. (2023). Application of metabolite-responsive biosensors for plant natural products biosynthesis. *Biosensors, 13*(6), 633. https://doi.org/10.3390/bios13060633.

Zhang, W., Qian, S., Tang, B., Kang, P., Zhang, H., & Shi, C. (2023). Resveratrol inhibits ferroptosis and decelerates heart failure progression via Sirt1/p53 pathway activation. *Journal of Cellular and Molecular Medicine, 27*(20), 3075–3089. https://doi.org/10.1111/jcmm.17874.

Zhang, Y., Wiese, L., Fang, H., Alseekh, S., Perez de Souza, L., Scossa, F., Molloy, J., Christmann, M., & Fernie, A. R. (2023). Synthetic biology identifies the minimal gene set required for paclitaxel biosynthesis in a plant chassis. *Molecular Plant, 16*(12), 1951–1961. https://doi.org/10.1016/j.molp.2023.10.016.

Zhang, L., Ding, R., Chai, Y., Bonfill, M., Moyano, E., Oksman-Caldentey, K.-M., Xu, T., Pi, Y., Wang, Z., Zhang, H., Kai, G., Liao, Z., Sun, X., & Tang, K. (2004). Engineering tropane biosynthetic pathway in *Hyoscyamus niger* hairy root cultures. *Proceedings of the National Academy of Sciences, 101*(17), 6786–6791. https://doi.org/10.1073/pnas.0401391101.

Zhang, L., Xing, D., & Wen, F. (2009). An on-line multi-parameter analyzing optical biosensor for real-time and non-invasive monitoring of plant stress responses in vivo. *Chinese Science Bulletin, 54*(21), 4009–4016. https://doi.org/10.1007/s11434-009-0157-8.

Zhao, J., Davis, L. C., & Verpoorte, R. (2005). Elicitor signal transduction leading to production of plant secondary metabolites. *Biotechnology Advances, 23*(4), 283–333. https://doi.org/10.1016/j.biotechadv.2005.01.003.

Zhao, J.-L., Zhou, L.-G., & Wu, J.-Y. (2010). Promotion of *Salvia miltiorrhiza* hairy root growth and tanshinone production by polysaccharide–protein fractions of plant growth-promoting rhizobacterium *Bacillus cereus*. *Process Biochemistry, 45*(9), 1517–1522. https://doi.org/10.1016/j.procbio.2010.05.034.

Zhao, Y., Zhu, H., Lu, X., Anees, M., He, N., Yang, D., Chen, Z., Hong, Z., Zhang, J., & Liu, W. (2025). Streamlined *Agrobacterium rhizogenes*-mediated hairy root transformation for efficient CRISPR/Cas9-based gene editing evaluation in diverse *Citrullus cultivars*. *Horticultural Plant Journal, 11*(2), 816–826. https://doi.org/10.1016/j.hpj.2023.10.005.

Zheng, Z., & Wu, M. (2004). Cadmium treatment enhances the production of alkaloid secondary metabolites in *Catharanthus roseus*. *Plant Science, 166*(2), 507–514. https://doi.org/10.1016/j.plantsci.2003.10.022.

Zhong, Y., Wu, S., Zhang, L., Yin, Z., Xie, Y., & Chen, J. (2024). In vitro strategy to enhance the production of bioactive polyphenols and caffeoylputrescine in the hairy roots of *Physalis peruviana* L. *Scientific Reports, 14*, 27600. https://doi.org/10.1038/s41598-024-77698-8.

Zhou, L., Yang, C., Li, J., Wang, S., & Wu, J. (2003). Heptasaccharide and octasaccharide isolated from Paris polyphylla var. Yunnanensis and their plant growth-regulatory activity. *Plant Science, 165*(3), 571–575. https://doi.org/10.1016/S0168-9452(03)00216-4.

Zhu, L. H., & Welander, M. (2000). Growth characteristics of the untransformed and transformed apple rootstock M26 with rola and rolb genes under steady-state nutrient supply conditions. *Acta Horticulturae*, (521), 139–146. https://doi.org/10.17660/ActaHortic.2000.521.15.

Ziv, M. (2005). Simple bioreactors for mass propagation of plants. *Plant Cell, Tissue and Organ Culture, 81*(3), 277–285. https://doi.org/10.1007/s11240-004-6649-y.

CHAPTER

12

CRISPR/Cas9-mediated gene editing and its application for developing novel genotypes in medicinal plants

Sanjay Kumar Singh[1], *Jayoti Majumder Sarkar*[2], and *Saikat Gantait*[3]

[1]Division of Horticultural Science, Indian Council of Agricultural Research, Krishi Anusandhan Bhawan – II, New Delhi, India [2]Department of Floriculture and Landscape Architecture, Bidhan Chandra Krishi Viswavidyalaya, Mohanpur, West Bengal, India [3]Crop Research Unit (Genetics and Plant Breeding), Bidhan Chandra Krishi Viswavidyalaya, Mohanpur, West Bengal, India

12.1 Introduction

Plants possess highly dynamic metabolic systems that have evolved in response to environmental challenges, enabling them to produce a remarkable array of specialized metabolites (Sinha et al., 2019). These compounds are broadly classified into two categories: primary metabolites, such as carbohydrates, proteins, lipids, and nucleic acids, which are indispensable for plant survival and growth (Sangwan & Sangwan, 2014), and secondary (specialized) metabolites, derived from primary precursors, which provide adaptive and evolutionary advantages. Specialized metabolites are especially significant because they support plant defense mechanisms and facilitate resilience under biotic and abiotic stress (Trethewey, 2004). Plants are considered the richest source of such specialized metabolites, with nearly 100,000 compounds documented across approximately 50,000 species (Hussein & El-Anssary, 2018; Kaushik et al., 2017; Pyne et al., 2019). These compounds fall into three principal classes—nitrogen-containing compounds, terpenes, and phenolics—which include sub-groups such as alkaloids, carotenoids, flavonoids, glucosinolates, glycosides, tannins, sterols, and volatile oils (Bourgaud et al., 2001; Jadaun et al., 2017). Unlike primary metabolites, which are vital for plant growth, these secondary metabolites provide ecological fitness, enabling plants to survive fluctuating environments (Verpoorte et al., 2002; Singer et al., 2003). From a human perspective, these metabolites have immense pharmaceutical and industrial relevance. They function as anticancer, antitumor,

Medicinal Plant Biotechnology for Ameliorated Secondary Metabolite Production
DOI: https://doi.org/10.1016/B978-0-443-43944-5.00014-1

antimicrobial, and antibiotic agents, with well-known examples including arjunolic acid, bleomycin, penicillin, and taxol (Sangwan et al., 2018). Plant species such as *Artemisia* (artemisinin), *Withania* (withaferin and withanone), *Taxus* (taxol), *Centella* (asiaticoside and madecassoside), and *Cymbopogon* (essential oils) are highly valued in medicine and biotechnology (Chaurasiya et al., 2012; Sangwan & Sangwan, 2014; Sangwan et al., 2014). These therapeutic herbs have been used in traditional medicine systems for centuries, and with modern tools, they are gaining renewed significance in healthcare and allied industries.

The biotechnology revolution has accelerated the exploration of medicinal plants and their bioactive compounds. High-quality reference genomes and efficient transformation systems are now available for species like *Dendrobium officinale* (Niu et al., 2021), *Cannabis sativa* (Gao et al., 2020), *Salvia miltiorrhiza* (Xu et al., 2016), and *Papaver somniferum* (Guo et al., 2018). These advances facilitate gene discovery, functional genomics, metabolic engineering, and breeding of superior germplasm with enhanced metabolite yields (Hassan, 2012). The development of genome engineering technologies has further transformed this landscape. Earlier genome editing approaches, such as meganucleases (MN), zinc-finger nucleases (ZFNs), and transcription activator-like effector nucleases (TALENs), introduced site-specific double-strand breaks (DSBs) in DNA, repaired by either homology-directed repair (HDR) or nonhomologous end joining (NHEJ) pathways (Zhu et al., 2020). While pioneering, these tools required intricate protein engineering, were labor-intensive, and had limited efficiency (Christian et al., 2010; Wright et al., 2004). The CRISPR/Cas system has emerged as a revolutionary third-generation editing platform, offering simplicity, versatility, and cost-effectiveness. Originally discovered as an adaptive immune system in bacteria and archaea, CRISPR/Cas employs guide RNAs to recognize complementary DNA sequences and direct Cas nucleases (e.g., Cas9) to precise genomic loci, where targeted cleavage occurs (Mali et al., 2013). Unlike earlier protein–DNA interaction-based systems, CRISPR relies on RNA–DNA recognition, requiring only a 20-nucleotide guide sequence to target specific genes (Niu et al., 2021). This ease of design enables targeted gene deletions, insertions, and substitutions with minimal impact on the host genome (Lozano-Juste & Cutler, 2014).

Introduced to plant research in 2013, CRISPR/Cas9 overcame the deficiencies of ZFNs and TALENs, particularly inefficiencies in transfection, challenges in multiplex editing, and complex protein design requirements (Liu & Fan, 2014). Its wide adoption has catalyzed progress in metabolic engineering of medicinal plants, allowing researchers to identify key biosynthetic genes, reprogram metabolic pathways, and enhance production of bioactive molecules (Kiss & Piwowarski, 2018). Significant case studies illustrate its potential. *Salvia miltiorrhiza*, a traditional Chinese medicinal herb valued for treating cardiovascular and cerebrovascular diseases (Ren et al., 2019), has been successfully engineered using CRISPR/Cas9. In 2018, researchers generated hairy root mutants by knocking out the rosmarinic acid synthase (*SmRAS*) gene, a crucial enzyme in the water-soluble phenolic acid biosynthetic pathway. This targeted disruption altered levels of rosmarinic acid, demonstrating the utility of CRISPR in regulating metabolite synthesis (Ren et al., 2021). Similarly, *Dendrobium officinale*, a medicinal orchid with high commercial value, has benefited from genomic resources (Niu et al., 2021) and efficient transformation systems (Ren et al., 2021; Teixeira da Silva et al., 2016). CRISPR/Cas9 system has been applied here to edit multiple genes in the lignocellulose biosynthesis pathway, achieving mutation efficiencies ranging from 10% to 100% across different loci (Kui et al., 2017). CRISPR provides a strategy to boost metabolite production *in vitro*, ensuring a sustainable and scalable supply of valuable natural products, and the broader impact of this technology is also economic. In this chapter, it is going to be revealed how the affordability and efficiency of CRISPR editing, combined with rising global demand for plant-derived products, are expected to accelerate the growth of the medicinal plant market. Increasing investment in biotechnology and plant genomics will likely lead to the discovery of novel bioactive compounds, improved breeding strategies, and large-scale production of high-value natural products for pharmaceuticals, cosmetics, and nutraceuticals.

12.2 Gene editing techniques

Genome editing tools enable precise cutting and rejoining of DNA at defined sites, allowing targeted modifications within living cells. Traditionally, molecular biologists relied on restriction endonucleases and DNA ligases to cleave and reassemble DNA fragments. While effective for relatively small and simple genomes, such as those of bacteria and viruses, these approaches proved inadequate for the large, complex, and repetitive genomes of plants and animals. The key limitation lies in the sequence specificity of restriction enzymes: they recognize only short DNA motifs, which are sufficient for microbial genomes but occur too frequently and nonspecifically in higher organisms. Consequently, achieving predictable and targeted modifications in plants using such methods remained highly challenging.

To overcome these limitations, advanced genome editing tools were developed, including ZFNs, TALENs, and ultimately the CRISPR/Cas system. Unlike traditional restriction enzymes, these technologies allow programmable DNA recognition, greatly enhancing the precision and flexibility of editing in complex genomes. Early innovations involved the design of "artificial enzymes"—short oligonucleotides engineered to bind specific DNA sequences and equipped with chemical groups for cleavage. These synthetic molecules mimicked natural restriction enzymes but offered greater programmability, laying the groundwork for modern genome-editing systems. Subsequent developments introduced chimeric nucleases, engineered proteins composed of two functional domains: one responsible for DNA cleavage and the other for sequence-specific DNA binding. Expression of these nucleases in living cells was achieved using engineered vectors, often carrying nuclear localization signals to ensure delivery into the nucleus and access to genomic DNA. Such programmable nucleases significantly improved specificity and efficiency in genome editing.

The advent of engineered ZFNs, TALENs, and particularly CRISPR/Cas9 has transformed genetic research and paved the way for crop improvement and agricultural biotechnology (Arora & Narula, 2017; Jankele & Svoboda, 2014; Kamburova et al., 2017; Palpant & Dudzinski, 2013).

12.2.1 Zinc-finger nucleases technology

Zinc-finger nucleases (ZFNs) represent one of the earliest genome editing tools, developed using engineered nucleases. Their design is based on the functional properties of the Cys2–His2 zinc-finger (ZF) domain (Gaj et al., 2013; Kamburova et al., 2017). Each Cys2–His2 ZF protein consists of ~30 amino acids, forming a stable folded structure.

Crystallographic studies have shown that the α-helix of the ZF protein interacts with the major groove of double-stranded DNA (Kamburova et al., 2017; Pavletich & Pabo, 1991). Typically, a single ZF recognizes three base pairs, allowing arrays of 3–6 ZFs to bind 9–18 bp DNA sequences. By assembling larger arrays, researchers can expand recognition capacity to 18–24 bp, producing unique genomic target sites (Kamburova et al., 2017). A ZFN monomer generally comprises two functional domains: (1) the DNA-binding Cys2–His2 ZF array at the N-terminus and (2) a nonspecific DNA cleavage domain derived from the type II restriction enzyme FokI at the C-terminus. The FokI domain must dimerize to induce double-strand breaks, making dimerization essential for ZFN activity (Kamburova et al., 2017). Despite their versatility, ZFNs face challenges because the binding specificity of individual ZFs often depends on neighboring ZFs and surrounding DNA sequences, leading to context-dependent recognition (Kamburova et al., 2017). Nevertheless, modular design strategies and optimization of ZF arrays have enabled the targeting of diverse genomic sites. Over the years, ZF libraries have been developed to recognize most nucleotide triplets, facilitating custom assembly for precise editing.

ZFNs have been successfully applied in different plants and animals. In *Arabidopsis thaliana*, they enabled targeted gene inactivation; in *Nicotiana tabacum*, they facilitated high-frequency gene modification; and in maize (*Zea mays*), they were used for insertional inactivation, herbicide tolerance, and trait stacking (Ainley et al., 2013; Gaj et al., 2013). These applications highlight their significance as pioneering tools in genome editing, though later technologies have surpassed them in ease and efficiency.

12.2.2 Transcription activator-like effector nucleases

TALENs represent the second generation of engineered nucleases, offering a simpler and more predictable design framework compared to ZFNs. They are created by fusing transcription activator-like effector (TALE) proteins to the FokI cleavage domain (Boch, 2011; Jankele & Svoboda, 2014). The TALEs, secreted by *Xanthomonas* bacteria during plant infection, naturally function as transcriptional activators. Their DNA-binding domain consists of 33–34 amino acid repeats, where the 12th and 13th residues—known as repeat variable diresidues (RVDs)—determine nucleotide specificity (Boch & Bonas, 2010; Moscou & Bogdanove, 2009). This straightforward code allows TALE arrays to be rationally engineered to target virtually any DNA sequence. For genome editing, engineered TALE DNA-binding arrays are fused to FokI, generating TALENs. Like ZFNs, FokI requires dimerization, meaning two TALEN constructs must bind opposite strands at adjacent sites to induce double-strand breaks (Christian et al., 2010; Gaj et al., 2013). The cleavage efficiency is influenced by spacer length between the binding sites and the linker sequence between TALE and FokI (Miller et al., 2011; Mussolino et al., 2011).

The TALEN design is simplified by the clear RVD–nucleotide relationship, but technical challenges arise from repetitive sequences within TALE arrays. Software tools such as DNA Works and modular assembly systems have been developed to overcome this limitation (Cermak et al., 2011; Zhang et al., 2011). Presently, TALENs can be delivered either as plasmids or mRNAs, avoiding stable integration risks.

Like ZFNs, TALENs function by inducing targeted DNA breaks, which cells repair via nonhomologous end joining (gene knockout) or homology-directed repair (gene insertion).

Their high efficiency and programmability have made them valuable in plant biotechnology, including the creation of nutritionally enhanced and commercially valuable crops (Haun et al., 2014; Zhang et al., 2013).

12.3 CRISPR/Cas9 genome editing

The limitations of ZFNs and TALENs in terms of design complexity and assembly gave rise to the third generation of genome editing tools: CRISPR/Cas9 system. Today, CRISPR/Cas9 is widely regarded as a versatile and user-friendly platform for targeted gene modifications, including insertions, deletions, replacements, and even single-base pair edits (Dong et al., 2020; Zhang & Showalter, 2020). Originally discovered as an adaptive immune system in bacteria and archaea, CRISPR protects host DNA from invading viruses and plasmids through RNA-guided cleavage by Cas proteins (Gaj et al., 2016; Sorek et al., 2013). At its core, a CRISPR locus contains arrays of short direct repeats (21–40 bp) interspaced with variable spacer sequences (25–40 bp) that serve as "molecular memories" of past infections. When transcribed, these spacers generate CRISPR RNAs (crRNAs), which pair with trans-activating crRNAs (tracrRNAs) to guide Cas9 to complementary DNA sequences for cleavage (Arora & Narula, 2017; Jinek et al., 2012). A critical discovery was that Cas9 requires only two elements for precise targeting: (1) the presence of a protospacer-adjacent motif (PAM) near the binding site and (2) a short seed sequence within the gRNA for recognition (Cho et al., 2013; Mali et al., 2013). For practical applications, crRNA and tracrRNA were fused into a single guide RNA (gRNA), which, together with Cas9 nuclease, forms the minimal editing machinery (Jinek et al., 2012). Once the Cas9–gRNA complex binds to its target, Cas9 generates a double-strand break ~3 bp upstream of the PAM site. Repair occurs primarily *via* error-prone nonhomologous end joining (NHEJ), leading to gene disruption, or *via* homology-directed repair (HDR) when a donor template is supplied (Jiang & Doudna, 2017; Ran et al., 2013). Because of its simplicity, cost-effectiveness, and high efficiency, CRISPR/Cas9 has revolutionized plant biotechnology. Researchers now routinely employ this system to reprogram metabolic pathways, improve agronomic traits, and develop crops with enhanced resistance to biotic and abiotic stresses (Gurumurthy et al., 2016; Li et al., 2018; Makarova et al., 2018; Wang et al., 2019; Zhang & Showalter, 2020).

12.4 Merits and potential applications of CRISPR in enhancing therapeutics and medicinal compounds

The CRISPR/Cas9 system has emerged as a precise and efficient genome editing tool for medicinal plants, offering new opportunities to enrich pharmaceutically important metabolites, while reducing toxic or undesirable compounds. Gene knockouts typically occur through nonhomologous end joining (NHEJ)-induced indels, fragment deletions, or domain loss, which inactivate key biosynthetic genes. This reverse genetics strategy enables functional validation of biosynthetic genes and transcription factors (TFs) and has rapidly expanded in scope across diverse medicinal species. One well-studied example is the disruption of hyoscyamine 6β-hydroxylase (AbH6H), the enzyme that converts hyoscyamine to anisodamine and scopolamine. Using sgRNA targeting the second exon under the AtU6–26 promoter with SpCas9 driven by CaMV35S, Mitra et al. (2023) achieved a 63.6% editing efficiency, with 27.3% homozygous mutants. Edited lines accumulated 3.7- to 4.3-fold higher hyoscyamine, while anisodamine and scopolamine were undetectable, as confirmed by high-performance liquid chromatography (HPLC). This highlights how targeted mutagenesis can simplify metabolite purification and enhance commercial yields. In the orchid *D. officinale*, sgRNAs driven by the OsU3 promoter were designed against five lignocellulose-related genes (C3H, C4H, 4CL, CCR, IRX5). Editing frequencies ranged from 10% to 100%, with outcomes influenced by chromatin accessibility (Kui et al., 2017). Given *D. officinale*'s valued pharmacological properties (immunomodulatory, antitumor, and antioxidant), such edits provide insights into cell wall metabolism and potential avenues to alter bioactive compound levels. The farnesyl pyrophosphate synthase (*Dzfps*) gene was targeted to modulate diosgenin biosynthesis, a precursor for steroidal drugs. Frameshift-inducing deletions (1–10 bp) disrupted enzyme function, reducing squalene levels and confirming *Dzfps* involvement (Feng et al., 2018). Knocking out the bZIP2 transcription factor revealed its role as a negative regulator of phenolic acid synthesis. CRISPR/Cas9-mediated disruption increased rosmarinic acid and salvianolic acid B levels by 23%–53%, with PAL showing the strongest upregulation. This illustrates how TF knockouts can uncover regulators of complex secondary metabolism (Shi et al., 2021). Targeting multiple BBL family genes nearly abolished nicotine accumulation, with T3 plants retaining only 0.04 mg/g DW nicotine—a 99.7% reduction compared to wild type (Schachtsiek & Stehle, 2019). This provides a strategy for creating low-nicotine tobacco lines. The *4′OMT2* gene, linked to BIA biosynthesis, was knocked out using viral sgRNA delivery (TRV system). Edited lines showed reduced levels of morphine, thebaine, and other alkaloids, providing functional validation of BIA pathway genes (Alagoz et al., 2016). Disruption of 1-FFT reduced inulin accumulation, indirectly improving rubber biosynthesis. Editing efficiency reached 88.9% in hairy roots, with stable mutant plants regenerated in six weeks (Iaffaldano et al., 2016). Targeting THC acid synthase (THCAS) aims to yield CBD-rich, THC-free plants with higher commercial value. While transformation remains challenging, studies have demonstrated successful knockout of marker genes (*CsPDS1*), where editing efficiency depended strongly on sgRNA position (Zhang & Showalter, 2020). Across these examples, several factors emerge as critical for successful editing: promoter choice (*AtU6*, *OsU3*, *CaMV35S*), sgRNA design, chromatin accessibility, and redundancy in gene families. Functional knockouts not only clarify gene roles but also enable metabolic engineering to enhance therapeutic yields or reduce regulatory hurdles (e.g., THC in cannabis and nicotine in tobacco). Together, these studies underscore CRISPR/Cas9's transformative role in medicinal plant research, bridging functional genomics with industrial biotechnology.

12.5 Modification of biochemical pathways

The CRISPR/Cas system has emerged as a transformative technology for reprogramming plant biosynthetic pathways to enhance the production of valuable primary and secondary metabolites. Unlike conventional mutagenesis and transgenic approaches, CRISPR/Cas9 provides precise, efficient, and heritable modifications at targeted genomic loci, enabling researchers to redirect metabolic fluxes with minimal off-target effects. In plants, biosynthetic pathways often involve multiple enzymes and complex regulatory networks. By targeting structural genes, key transcription factors, or rate-limiting enzymes, CRISPR/Cas9 has been successfully employed to increase metabolite yield, modify composition, or suppress undesirable compounds. For instance, carotenoid biosynthesis has been manipulated in tomato and banana, while anthocyanin and flavonoid pathways have been engineered in morning glory and *Salvia miltiorrhiza*. Similarly, disruption of germacrene A synthase in chicory suppressed sesquiterpene lactone synthesis but increased phenolic acids and squalene. Beyond pigments and nutraceuticals, CRISPR/Cas9 has also been applied to the lignin biosynthesis pathway in poplar, achieving improved saccharification efficiency without yield penalties. These examples highlight the versatility of CRISPR/Cas as a metabolic engineering tool. By fine-tuning enzymes and regulators, the technology offers unprecedented opportunities to develop nutritionally fortified crops, enhance therapeutic phytochemicals, and optimize industrial feedstocks. As genome editing platforms evolve (e.g., CRISPR/Cas12, base editors, and prime editing), the scope for remodeling biosynthetic pathways will expand further, fostering a new era of precision metabolic engineering in plants. Some of the applications are listed hereunder (Table 12.1).

TABLE 12.1 CRISPR/Cas9 applications in plant biosynthetic pathways.

Plant species	Target gene(s)	Metabolite(s) affected	Key outcome
Tomato (*Solanum lycopersicum*)	*Psy1*, *CrtR-b2*	Carotenoids, GABA	Enhanced carotenoid and GABA production; visible pigmentation changes (Li et al., 2018; D'Ambrosio et al., 2018).
Soybean (*Glycine max*)	Isoflavone pathway genes	Isoflavones	Increased isoflavone production.
Banana (*Musa* spp.)	*LCYε*	β-carotene, α-carotene, lutein	Higher β-carotene content with reduced α-carotene and lutein (Kaur et al., 2020).
Japanese morning glory (*Ipomoea nil* (L.) Roth)	*DFR-B*, *CCD4*	Anthocyanins, carotenoids	Loss of anthocyanins (white flowers); reduced carotenoids in petals (Watanabe et al., 2017).
Chinese sage [*Salvia miltiorrhiza* (Bunge)]	*bZIP2*, *SmbHLH60*	Phenolic acids, anthocyanins	*bZIP2* knockout ↑ phenolics; overexpression ↓ phenolics. *SmbHLH60* knockout ↑ anthocyanins & phenolics (Liu & Fan, 2014; Shi et al., 2021).
Chicory (*Cichorium intybus* L.)	*CiGAS* (germacrene A synthase)	Sesquiterpene lactones, squalene	Suppressed STL biosynthesis; ↑ phenolic acids and squalene accumulation (Cankar et al., 2021).
Poplar (*Populus tremula* × *P. alba*)	*CSE1*, *CSE2* (caffeoyl shikimate esterase)	Lignin, saccharification	Weak alleles improved saccharification efficiency without yield penalties (de Vries et al., 2021).

12.6 Accomplishments across species

12.6.1 *Aloe vera*

Aloe (*Aloe vera* and *A. barbadensis* Miller) are the widely used medicinal and cosmetic plants, yet their genetic improvement through molecular tools remains limited. Ren et al. (2020) reported the complete chloroplast genome of *A. vera* (152,875 bp), revealing its quadripartite structure, GC content (37.7%), and 131 annotated genes. Phylogenetic analysis placed *A. vera* in close relation to *A. maculata* within the Asphodelaceae family, offering valuable genomic resources for future genome editing. At the same time, optimized transformation protocols have been developed, where stem explants showed higher regeneration frequency than leaves. *Agrobacterium* strain EHA105 achieved up to 80% transient *GUS* expression, outperforming C58C1, and generated 67 G418-resistant transgenic plantlets. Molecular assays confirmed successful gene integration and expression, demonstrating the feasibility of stable transformation in aloe. Earlier, Choudhri et al. (2018) advanced transcriptomic resources by sequencing root and leaf tissues, assembling over 1,13,000 and 1,41,000 unigenes, respectively. The KEGG pathway analysis identified genes involved in secondary metabolite biosynthesis, with stress-responsive transcription factors (bHLH, NAC, MYB) regulating pathways for saponins, anthraquinones, and carotenoids. Together, these studies establish a molecular foundation for deploying CRISPR/Cas genome editing in aloe, with complete chloroplast data, functional transcriptomes, and transformation protocols in place.

12.6.2 Water hyssop, *Brahmi* (*Bacopa monnieri* (L.) Pennell)

Bacopa monnieri, a key nootropic herb in Ayurveda, has long been valued for its memory-enhancing properties, though the molecular basis of its activity has only recently begun to emerge. Leung et al. (2017) used RNA-Seq on SH-SY5Y neuroblastoma cells and revealed that *B. monnieri* regulates genes involved in mRNA translation, oxidative stress response, transmembrane transport, and protein misfolding. Pathway analysis suggested protective roles in brain development and neuroprotection, offering mechanistic insights into its cognitive benefits. Expanding this understanding, Jeena et al. (2017) conducted the first *de novo* transcriptome analysis of *B. monnieri* and identified transcripts associated with triterpenoid sapogenin biosynthesis, the precursors of bacosides. Since bacosides are the principal bioactive compounds contributing to the plant's pharmacological effects, this dataset provides a foundation for metabolic engineering strategies aimed at enhancing bacoside yield for therapeutic use. Parallel biochemical studies by Muszyńska et al. (2016) highlighted the plant's metabolic diversity. *In vitro* cultures enriched with zinc hydroaspartate showed significant accumulation of L-tryptophan and serotonin, alongside phenolic compounds such as chlorogenic, neochlorogenic, and caffeic acids. The fatty acid profile was dominated by palmitic acid, while extracts demonstrated strong antiinflammatory potential by suppressing COX-2 and cPGES expression in LPS-activated A549 cells.

12.6.3 Shatavari (*Asparagus racemosus*)

CRISPR/Cas9 has emerged as a transformative tool for both plant and fungal genome editing, offering precision and efficiency far superior to traditional mutagenesis approaches.

Abdelrahman et al. (2018) highlighted its potential in crop improvement, particularly under the increasing challenges of biotic and abiotic stress. Unlike conventional breeding, which is slow and often random, CRISPR/Cas9 enables targeted gene knockouts, modifications, and regulation, thereby accelerating the development of resilient, high-yielding crops. However, regulatory frameworks and public acceptance remain significant hurdles for the widespread deployment of genome-edited plants. Ferrara et al. (2019) demonstrated the power of this approach in fungal systems by developing a novel CRISPR/Cas9 tool for *Fusarium proliferatum*, a pathogen producing toxic fumonisins. Their method employed direct delivery of Cas9 ribonucleoproteins into protoplasts, coupled with microhomology-mediated recombination using only 50-bp flanking regions. This strategy reduced off-target risks and avoided disruption of neighboring genes. Using this system, the authors successfully deleted the *FUM1* gene, a polyketide synthase essential for fumonisin biosynthesis, thereby generating mutants incapable of producing fumonisins.

12.6.4 *Artemisia annua* (L.)

Artemisinin, a sesquiterpene lactone from *Artemisia annua*, remains one of the most potent antimalarial compounds, yet its natural content is very low (0.01%–0.8% DW), making large-scale supply costly. Strategies to boost artemisinin yield have therefore attracted significant research attention. Dilshad et al. (2015) demonstrated that genetic transformation of *Artemisia carvifolia* with *rolB* and *rolC* genes enhanced artemisinin and its derivatives (artesunate, artemether, and dihydroartemisinin) by up to sevenfold compared with wild type. These increases correlated with higher expression of key biosynthetic genes, including *ADS* (amorpha-4,11-diene synthase), *CYP71AV1*, and *ALDH1*. Additionally, altered expression of *TFAR1* was linked to enhanced trichome density, confirming the importance of glandular trichomes in secondary metabolite production. Building on these advances, Lv et al. (2016) proposed CRISPR/Cas9 as a powerful approach to redirect metabolic flux toward artemisinin. By targeting genes at branching points such as β-farnesene synthase (BFS), caryophyllene synthase (CPS), and squalene synthase (SQS), competing pathways can be blocked, thereby channeling precursors toward artemisinin biosynthesis. Transgenic plants with suppressed *BFS*, *CPS*, and *SQS* exhibited elevated artemisinin levels, underscoring these genes as prime CRISPR targets.

12.6.5 *Glycirrhiza* sp. (Licorice)

Licorice (*Glycyrrhiza uralensis*, *G. glabra*, *G. inflata*) is a medicinally and commercially important legume whose roots produce glycyrrhizin, a triterpenoid saponin with pharmacological and sweetening properties. Rising global demand has threatened wild resources, highlighting the need for sustainable production methods. The draft genome sequence of *G. uralensis* (379 Mb, 34,445 predicted protein-coding genes) reported by Mochida and Tran (2017) provided essential genomic resources, including annotation of cytochrome P450 and UDP-dependent glycosyltransferase families involved in triterpenoid saponin biosynthesis. Comparative genomics also revealed conserved gene clusters related to isoflavonoid pathways, offering insights for metabolic engineering. Building on this genomic foundation, Chiyo et al. (2024) applied CRISPR/Cas9 to manipulate glycyrrhizin biosynthesis in

G. uralensis hairy root cultures. By generating CYP93E3 and CYP72A566 double knockouts, as well as quadruple knockouts including CYP716A179 and LUS1, they successfully blocked competing pathways and enhanced glycyrrhizin accumulation. Further metabolic boosting was achieved by overexpressing CYP88D6 in the double-KO lines, resulting in a threefold increase in glycyrrhizin content (~1.4 mg/g FW). These findings demonstrate the power of combining CRISPR/Cas-mediated pathway suppression with targeted overexpression for efficient metabolite enhancement

12.6.6 *Ocimum* sp.

Sweet basil (*Ocimum basilicum*), a globally important herb valued for its culinary and medicinal uses, faces major production threats from basil downy mildew caused by *Peronospora belbahrii*. Recent advances have demonstrated the potential of CRISPR/Cas9 in both functional genomics and breeding for resistance in *Ocimum*. Hasley et al. (2021) identified ObDMR6, a susceptibility (*S*) gene homologous to *Arabidopsis* DMR6, and generated 56 transgenic basil lines via *Agrobacterium*-mediated transformation. Mutations in ObDMR6 resulted in a significant reduction of pathogen biomass (up to 93%), yielding transgene-free lines with enhanced resistance. Similarly, CRISPR/Cas9 targeting of *ObDMR1*, another susceptibility gene, produced homozygous mutants in the T1 generation, with some lines being transgene-free. This work established an efficient editing system in an allotetraploid crop, demonstrating that knockout of *S* genes can deliver durable resistance. Complementary advances in CRISPR technology further support applications in *Ocimum*. Recently, Laura et al. (2023) showed that Cas9 with custom sgRNAs can streamline seamless cloning and genotyping of edited mutants, facilitating functional gene analysis. Faizal et al. (2024) highlighted CRISPR's broader utility in crop protection, including pathogen resistance, pest control, and stress resilience. Together, these studies position CRISPR/Cas9 as a transformative tool for improving *Ocimum*, enabling resistance breeding while also unlocking pathways for secondary metabolite engineering in essential oils.

Some of the other leads in CRISPR/Cas9 in interventions in different MAPs have been listed hereunder (Table 12.2).

12.7 Challenges and ethical considerations

12.7.1 Off-target effects

The CRISPR/Cas system enables precise and efficient modification of DNA sequences, but its application can be limited by off-target effects in homologous gene regions (Kleinstiver et al., 2016; Modrzejewski et al., 2020). A logistic regression model has identified five key factors influencing off-target activity: (1) the number of mismatches and the structural features of sgRNAs, (2) the position of mismatches, (3) GC content, (4) nuclease variants, and (5) delivery methods. Off-target events decrease as the number of mismatches increases, with a reduction of up to 59% observed when a single mismatch is introduced; this effect becomes even more pronounced with four or more mismatches. Importantly,

TABLE 12.2 Genome editing leads for different trait(s) made in medicinal plants.

Sl. No.	Plant species	Target trait	Salient findings	Reference (s)
1.	*Artemisia annua*	Gene silencing of cinnamate-4-hydroxylase (AaC4H), responsible for the conversion of trans-cinnamic acid (CA) to p-coumaric acid (COA)	AaC4H RNAi lines exhibited an elevation of trans-cinnamic acid content and reduction in p-coumaric acid levels, salicylic acid (SA), and artemisinin	Kumar et al. (2016)
2.	*Dendrobium officinale*	Lignocellulose biosynthesis pathway	5 target genes namely *C4H, C3H, CCR, IRX I,* and *4CL* were altered using the CRISPR/ Cas 3 system	Kui et al. (2017)
3.	*Camelina sativa*	Fatty acid chain	Fatty acid desaturase 2 (*FAD2*) genes were knocked out so that the content of oleic acid can be increased in this plant	Jiang et al. (2017)
4.	*Cannabis sativa*	Phytoene desaturase gene marker	CsPDS1-deficient seedlings	Zhang et al. (2021)
5.	*Catharanthus roseus*	RNA sequencing	Construction of a database "CathaCyc," with facilitating visualization and interpretation of metabolic pathway-related information from transcriptomic data	Strickler et al. (2012)
6.	*Dioscorea zingiberensis*	First exon	Decreased OsU3::sgRNA; squalene level 35 S::SpCas9	Feng et al. (2018)
7.	*Nicotiana tabacum* (L.)	Synthesis of glycoprotein	6 glycosyltransferase genes were suppressed by CRISPR/Cas	Gao et al. (2015)
8.	*Papaver bracteatum*	Upregulated codeinone reductase (CodR)	Elevated codeine levels by 11-fold and morphine levels with marginal increase of 0.28%	Sharafi et al. (2013)
9.	*Ophiorrhiza pumila*	Metabolomic analysis of suspension culture and hairy roots	Distinguish between potential genes regulating monoterpene indole alkaloid and anthraquinones	Higashi and Saito (2013)
10.	*Trifolium pratense* (L.)	Isoflavone synthase	CRISPR/Cas9 was utilized to inhibit the activity of isoflavone synthase, a crucial enzyme in the manufacture of isoflavones (IFS1), which leads to lower levels of biochanin A, genistein, and isoflavone mononetin	Dinkins et al. (2021)

mismatches in the seed sequence—the first eight nucleotides adjacent to the PAM site—substantially lower off-target activity. Optimizing sgRNA design is therefore critical to maximizing target specificity while minimizing unintended edits (Modrzejewski et al., 2020). However, current knowledge regarding the influence of GC content, alternative nuclease variants such as Cas9 nickases, and delivery strategies on off-target dynamics remains limited (Doench et al., 2016; Kimberland et al., 2018).

12.7.2 Efficiency of CRISPR/Cas9 tool delivery

Application of CRISPR/Cas9 tools in medicinal plants like any other plant groups has several difficulties associated with (1) delivery of editing tools, that is, getting the editing components (like CRISPR-Cas9) into the plant cells is the major obstacle, especially dependent of nature and genotype of the crop species; (2) inefficient plant regeneration system, that is, the edited plant line must be regenerated from a single cell is sometime time-consuming and labor-intensive process; (3) low editing efficiency, the success rate of the actual DNA modification can be low, with the cell often repairing the break without making the desired change; (4) limited precision, that is, achieving specific changes, like a precise base replacement is often difficult and limited some plant species; and (5) complex genomes, that is, editing polyploid plants is very difficult due to their complex genetic background, besides presence of duplicate genes. Thus the major challenge in plant genome editing lies in the efficient delivery of CRISPR/Cas9 and sgRNAs into cells. Low transformation efficiency and the difficulty of selecting suitable plasmid vectors remain critical obstacles. Two main strategies have been explored to overcome these issues. The first involves the expression of Cas9 protein directly within cells, while the second employs preassembled Cas9–sgRNA complexes prepared *in vitro*, resembling the ribonucleoprotein (RNP) system commonly used for protoplast delivery. Another promising approach is the use of nanoparticles to transport Cas9–sgRNA complexes into the nucleus (Dittmann et al., 2015; Tong et al., 2019).

12.7.3 CRISPR/Cas tool driving machinery

The efficiency of CRISPR/Cas9 editing also depends on the regulatory components driving its expression, including promoters, terminators, selectable markers, and transformation protocols. Current limitations in intrinsic regulatory mechanisms of medicinal plants hinder the biosynthesis of valuable bioactive compounds and restrict the editing efficacy of CRISPR/Cas systems (Jiang & Henry, Lynagh, et al., 2017; Tong et al., 2019; Zhou et al., 2018).

12.7.4 Linking phenotype to genotype

Although CRISPR/Cas9 offers advantages in terms of time and cost efficiency, connecting genotype modifications to observable phenotypes remains a significant challenge. To bridge this gap, researchers have developed advanced tools such as the CRISPR-MAGIC system and high-throughput biosensors to facilitate genotype–phenotype mapping (Lian et al., 2019). CRISPR/Cas9 can link a phenotype to a genotype by using it for high-throughput screening to observe how gene modifications affect an organism's traits. By using CRISPR to introduce mutations, researchers can create a population of cells with different genotypes and then observe the corresponding phenotypes to identify the causal genes. Thus techniques like single-cell sequencing can effectively be used to confirm the specific DNA changes that led to a particular phenotype.

12.7.5 Approaches to genetic transformation and regeneration

Establishing robust plant transformation and regeneration protocols is essential for improving CRISPR/Cas9 efficiency across diverse species. Many plants lack effective tissue culture or transformation systems, creating barriers to high editing efficiency. In this context, *in vitro* tissue culture and protoplast regeneration methods play a pivotal role in enhancing transformation and regeneration competence, thereby advancing CRISPR/Cas9-mediated editing (Altpeter et al., 2016; Zhang et al., 2013).

12.8 Future directions

Medicinal plants play a vital role in pharmacology, nutrition, biomedicine, and industry due to their reservoir of specialized metabolites. Engineering metabolic circuits by alleviating competitive or rate-limiting steps presents opportunities to boost metabolite yields. The CRISPR/Cas9 system has emerged as a powerful tool to achieve such modifications, offering precise editing of genes associated with biosynthetic pathways to meet commercial and pharmaceutical demands. Beyond conventional genome editing, recent advances have enabled CRISPR-based manipulation of the epigenome, transcriptional regulation, and posttranslational modifications. The development of homology-directed repair (HDR) for precision editing, chromosomal rearrangements through inversions or somatic crossovers, and single-base editing further expands the potential applications of this technology (Shrestha et al., 2018; Wolter et al., 2019). CRISPR/Cas9 also facilitates the construction of large mutant libraries, enabling gene knockouts, knockdowns (CRISPRi), and activation (CRISPRa). These approaches support metabolic engineering by introducing elite alleles, generating novel cell factories, and elucidating biosynthetic pathways. The ultimate objective is to enhance the accumulation of beneficial metabolites while minimizing harmful or undesirable compounds. Optimization of regulatory elements, such as promoters, terminators, and nuclear localization signals, is central to improving editing efficiency. For instance, Morineau et al. (2017) employed endogenous Camelina U3 and U6 promoters to induce Cas9 expression, highlighting the promise of species-specific regulatory modules. Multiplex editing further enables simultaneous manipulation of multiple genes, addressing pathway complexity. Functional studies using CRISPR/Cas have deepened the understanding of specialized metabolite pathways. In *Cannabis sativa*, for example, CRISPR has been applied to explore cannabinoid and terpenoid biosynthesis (Bonini et al., 2018), while Gracz-Bernaciak et al. (2021) proposed its use for studying latex production in *Chelidonium majus* with implications for antiviral activity. Despite challenges in HDR-based gene overexpression, CRISPR-induced mutagenesis of endogenous genes is well established. Moreover, targeting cis-regulatory elements and untranslated regions (UTRs) provides fine-tuned modulation of gene expression (Si et al., 2020). Delivery methods remain a bottleneck for plant genome editing. Common techniques include *Agrobacterium*-mediated transformation, polyethylene glycol (PEG)-mediated protoplast transfection, and particle bombardment. Of these, *Agrobacterium*-mediated transformation is the most widely used due to cost-effectiveness and efficiency, although it requires binary vectors and often results in transgene integration. PEG-mediated transformation is labor-intensive due to protoplast preparation, while particle bombardment can result in random DNA integration (Agapito-Tenfen et al., 2018; Eckerstorfer et al., 2019). Nanoparticle-based CRISPR delivery systems are emerging as alternatives to overcome these limitations. Additionally, PAM sequence requirements constrain Cas9 targeting, though newly discovered Cas proteins with varied PAM compatibilities promise to expand editing scope (Larson et al., 2013). From a regulatory standpoint, CRISPR-edited crops are increasingly considered nontransgenic if no foreign DNA remains. For example, a browning-resistant mushroom edited with CRISPR/Cas9 was granted nonregulatory status in the United States (Waltz, 2016). However, public perception remains a challenge, with surveys indicating only 11.9% of the Chinese population supporting GM foods, while a large proportion remained neutral or negative (Cui & Shoemaker, 2018). In summary, CRISPR/Cas9 represents a cost-effective and versatile platform for metabolic engineering in medicinal plants. By enabling precise pathway modifications, multiplexed gene editing, and functional genomics studies, this technology holds immense potential to enhance the production of pharmaceutically and commercially valuable metabolites.

12.9 Conclusion

In conclusion, the effective application of CRISPR/Cas9 gene editing technology in medicinal plants is currently constrained by the limited availability of comprehensive genome and transcriptome data. A deeper understanding of genetic architecture and metabolic networks is indispensable for identifying key biosynthetic genes and regulatory elements that govern the synthesis of valuable secondary metabolites. Integrating genomic and transcriptomic insights with functional analyses will enable the precise modulation of enzymatic pathways and the redirection of metabolic fluxes toward enhanced production of desired compounds. As omics-driven discoveries continue to expand, CRISPR/Cas9 emerges as a transformative tool to tailor metabolic pathways, offering possibilities for the sustainable and cost-effective synthesis of bioactive molecules with therapeutic and nutritional significance. Furthermore, as these genome editing strategies advance, it becomes crucial to align scientific innovation with robust regulatory and ethical guidelines that ensure biosafety, transparency, and societal acceptance. Thus the convergence of functional genomics, metabolic engineering, and responsible biotechnology governance holds the promise of revolutionizing phytopharmaceutical development—unlocking the full potential of medicinal plants as biofactories for high-value metabolites essential to human health and well-being.

References

Abdelrahman, M., Al-Sadi, A. M., Pour-Aboughadareh, A., Burritt, D. J., & Phan Tran, L.-S. (2018). Genome editing using CRISPR/Cas9–targeted mutagenesis: an opportunity for yield improvements of crop plants grown under environmental stresses. *Plant Physiology and Biochemistry*. https://doi.org/10.1016/j.plaphy.2018.03.012.

Agapito-Tenfen, S. Z., Okoli, A. S., Bernstein, M. J., Wikmark, O.-G., & Myhr, A. I. (2018). Revisiting risk governance of GM plants: The need to consider new and emerging gene-editing techniques. *Frontiers in Plant Science, 9*, 1874.

Ainley, W. M., Sastry-Dent, L., Welter, M. E., Murray, M. G., Zeitler, B., Amora, R., & Petolino, J. F. (2013). Trait stacking via targeted genome editing. *Plant Biotechnology Journal, 11*(9), 1126–1134. https://doi.org/10.1111/pbi.12107.

Alagoz, Y., Gurkok, T., Zhang, B., & Unver, T. (2016). Manipulating the biosynthesis of bioactive compound alkaloids for next-generation metabolic engineering in opium poppy using CRISPR-Cas9 genome editing technology. *Scientific Reports, 6*(1), 30910. https://doi.org/10.1038/srep30910.

Altpeter, F., Springer, M. N., Bartley, E. L., Blechl, E. A., Brutnell, P. T., Citovsky, V., et al. (2016). Advancing crop transformation in the era of genome editing. *The Plant Cell, 28*, 1510–1520. https://doi.org/10.1105/tpc.16.00196.

Arora, L., & Narula, A. (2017). Gene editing and crop improvement using CRISPR-Cas9 system. *Frontiers in Plant Science, 8*, 1932. https://doi.org/10.3389/fpls.2017.01932.

Boch, J. (2011). TALEs of genome targeting. *Nature Biotechnology, 29*(2), 135–136. https://doi.org/10.1038/nbt.1767.

Boch, J., & Bonas, U. (2010). *Xanthomonas* AvrBs3 family type III effectors: Discovery and function. *Annual Review of Phytopathology, 48*, 419–436. https://doi.org/10.1146/annurev-phyto-080508-081936.

Bonini, S. A., Premoli, M., Tambaro, S., Kumar, A., Maccarinelli, G., Memo, M., et al. (2018). *Cannabis sativa*: A comprehensive ethnopharmacological review of a medicinal plant with a long history. *Journal of Ethnopharmacology, 227*, 300–315. https://doi.org/10.1016/j.jep.2018.09.004.

Bourgaud, F., Gravot, A., Milesi, S., & Gontier, E. (2001). Production of plant secondary metabolites: A historical perspective. *Plant Science, 161*, 839–851. https://doi.org/10.1016/S0168-9452(01)00490-3.

Cankar, K., Paul, B., Robert, S., Suvi, T. H., Johanna, C. H., Jules, B., van der Ingrid, M. M., de Both, M., & Dirk, B. (2021). Inactivation of the germacrene A synthase genes by CRISPR/Cas9 eliminates the biosynthesis of sesquiterpene lactones in *Cichorium intybus* L. *Plant Biotechnology Journal, 19*(12), 2442–2453. https://doi.org/10.1111/pbi.13670.

Cermak, T., Doyle, E. L., Christian, M., Wang, L., Zhang, Y., Schmidt, C., Baller, J. A., Somia, N. V., Bogdanove, A. J., & Voytas, D. F. (2011). Efficient design and assembly of custom TALEN and other TAL effector-based constructs for DNA targeting. *Nucleic Acids Research, 39*(12), e82. https://doi.org/10.1093/nar/gkr218.

Chaurasiya, N. D., Sangwan, N. S., Sabir, F., Misra, L., & Sangwan, R. S. (2012). Withanolide biosynthesis recruits both mevalonate and DOXP pathways of isoprenogenesis in ashwagandha (*Withania somnifera* L. Dunal). *Plant Cell Reports, 31*, 1889–1897. https://doi.org/10.1007/s00299-012-1302-4.

Chiyo, N., Seki, H., Kanamoto, T., Ueda, H., Kojoma, M., & Muranaka, T. (2024). Glycyrrhizin production in licorice hairy roots based on metabolic redirection of triterpenoid biosynthetic pathway by genome editing. *Plant and Cell Physiology, 65*(2), 185–198. https://doi.org/10.1093/pcp/pcad161.

Choudhri, P., Rani, M., Sangwan, R. S., Kumar, R., Kumar, A., & Chhokar, V. (2018). De novo sequencing, assembly and characterisation of *Aloe vera* transcriptome and analysis of expression profiles of genes related to saponin and anthraquinone metabolism. *BMC Genomics, 19*(1), 427. https://doi.org/10.1186/s12864-018-4818-8.

Cho, S., Shin, J., & Cho, B.-K. (2018). Applications of CRISPR/Cas system to bacterial metabolic engineering. *International Journal of Molecular Sciences, 19*(4), 1089. https://doi.org/10.3390/ijms19041089.

Christian, M., Cermak, T., Erin, L. D., Schmidt, C., Zhang, F., Hummel, A., et al. (2010). Targeting DNA double-strand breaks with TAL effector nucleases. *Genetics, 186*, 757–761. https://doi.org/10.1534/genetics.110.120717.

Cong, L., Ran, F. A., Cox, D., Lin, S., Barretto, R., Habib, N., Hsu, P. D., Wu, X., Jiang, W., Marraffini, L. A., & Zhang, F. (2013). Multiplex genome engineering using CRISPR/Cas systems. *Science, 339*(6121), 819–823. https://doi.org/10.1126/science.1231143.

Cui, K., & Shoemaker, S. P. (2018). Public perception of genetically modified (GM) food: A nationwide Chinese consumer study. *npj Science of Food, 2*(1), 10. https://doi.org/10.1038/s41538-018-0018-4.

D'Ambrosio, C., Stigliani, A. L., & Giorio, G. (2018). CRISPR/Cas9 editing of carotenoid genes in tomato. *Transgenic Research, 27*(4), 367–378. https://doi.org/10.1007/s11248-018-0079-9.

de Vries, L., Brouckaert, M., Chanoca, A., Kim, H., Regner, M. R., Timokhin, V. I., Sun, Y., De Meester, B., Van Doorsselaere, J., Goeminne, G., & Chiang, V. L. (2021). CRISPR-Cas9 editing of *CAFFEOYL SHIKIMATE ESTERASE* 1 and 2 shows their importance and partial redundancy in lignification in *Populus tremula* × *P. alba*. *Plant Biotechnology Journal, 19*(11), 2221–2234. https://doi.org/10.1111/pbi.13631.

Dilshad, E., Cusido, R. M., Palazon, J., Ramirez Estrada, K., Bonfill, M., & Mirza, B. (2015). Enhanced artemisinin yield by expression of *rol* genes in *Artemisia annua*. *Malaria Journal, 14*(1), 424. https://doi.org/10.1186/s12936-015-0953-3.

Dinkins, R. D., Hancock, J., Coe, B. L., May, J. B., Goodman, J. P., Bass, W. T., Liu, J., Fan, Y., Zheng, Q., & Zhu, H. (2021). Isoflavone levels, nodulation and gene expression profiles of a CRISPR/Cas9 deletion mutant in the isoflavone synthase gene of red clover. *Plant Cell Reports, 40*(3), 517–528. https://doi.org/10.1007/s00299-020-02647-4.

Dittmann, E., Gugger, M., Sivonen, K., & Fewer, D. P. (2015). Natural product biosynthetic diversity and comparative genomics of the cyanobacteria. *Trends in Microbiology, 23*, 642–652. https://doi.org/10.1016/j.tim.2015.07.008.

Doench, J. G., Fusi, N., Sullender, M., Hegde, M., Vaimberg, E. W., Donovan, K. F., et al. (2016). Optimized sgRNA design to maximize activity and minimize off-target effects of CRISPR–Cas9. *Nature Biotechnology, 34*, 184–191. https://doi.org/10.1038/nbt.3437.

Dong, O. X., Yu, S., Jain, R., Zhang, N., Duong, P. Q., Butler, C., Li, Y., Lipzen, A., Martin, J. A., Barry, K. W., Schmutz, J., Tian, L., & Ronald, P. C. (2020). Marker-free carotenoid-enriched rice generated through targeted gene insertion using CRISPR-Cas9. *Nature Communications, 11*(1), 1178. https://doi.org/10.1038/s41467-020-14981-2.

Eckerstorfer, M. F., Dolezel, M., Heissenberger, A., Miklau, M., Reichenbecher, W., Steinbrecher, R. A., & Waßmann, F. (2019). An EU perspective on biosafety considerations for plants developed by genome editing and other new genetic modification techniques (nGMs). *Frontiers in Bioengineering and Biotechnology, 7*, 31. https://doi.org/10.3389/fbioe.2019.00031.

Faizal, A., Nugroho, S., Sembada, A. A., Theda, Y., Komariyah, T., & Esyanti, R. R. (2024). Genome editing in future crop protection: Utilizing CRISPR/Cas9 to improve crop resistance against diseases, pests, and weeds. *Discover Agriculture, 2*(1), 104. https://doi.org/10.1007/s44279-024-00124-0.

Feng, S., Song, W., Fu, R., Zhang, H., Xu, A., & Li, J. (2018). Application of the CRISPR/Cas9 system in *Dioscorea zingiberensis*. *Plant Cell, Tissue and Organ Culture, 133*(1), 133–135. https://doi.org/10.1007/s11240-017-1370-8.

Ferrara, M., Haidukowski, M., Logrieco, A. F., Leslie, J. F., & Mulè, G. (2019). A CRISPR-Cas9 system for genome editing of *Fusarium proliferatum*. *Scientific Reports, 9*(1), 19836. https://doi.org/10.1038/s41598-019-56182-8.

Gaj, T., Gersbach, C. A., & Barbas, C. F. I. I. I. (2013). ZFN, TALEN, and CRISPR/Cas-based methods for genome engineering. *Trends in Biotechnology, 31*(7), 397–405. https://doi.org/10.1016/j.tibtech.2013.04.004.

Gaj, T., Sirk, S. J., Shui, S. L., & Liu, J. (2016). Genome-editing technologies: Principles and applications. *Cold Spring Harbor Perspectives in Biology, 8*(12), a023754. https://doi.org/10.1101/cshperspect.a023754.

Gao, J., Wang, G., Ma, S., Xie, X., Wu, X., Zhang, X., Wu, Y., Zhao, P., & Xia, Q. (2015). CRISPR/Cas9-mediated targeted mutagenesis in *Nicotiana tabacum*. *Plant Molecular Biology, 87*(1), 99–110. https://doi.org/10.1007/s11103-014-0263-0.

Gao, S., Wang, B., Xie, S., Xu, X., Zhang, J., Pei, L., et al. (2020). A high-quality reference genome of wild *Cannabis sativa*. *Horticulture Research, 7*, 73. https://doi.org/10.1038/s41438-020-0295-3.

Gracz-Bernaciak, J., Mazur, O., & Nawrot, R. (2021). Functional studies of plant latex as a rich source of bioactive compounds: Focus on proteins and alkaloids. *International Journal of Molecular Sciences, 22*, 12427. https://doi.org/10.3390/ijms222212427.

Guo, L., Winzer, T., Yang, X., Li, Y., Ning, Z., He, Z., Teodor, R., Lu, Y., Bowser, T. A., Graham, I. A., & Ye, K. (2018). The opium poppy genome and morphinan production. *Science (New York, N.Y.), 362*(6412), 343–347. https://doi.org/10.1126/science.aat4096.

Gurumurthy, C. B., Grati, M., Ohtsuka, M., Schilit, S. L. P., Quadros, R. M., & Liu, X. Z. (2016). CRISPR: A versatile tool for both forward and reverse genetics research. *Human Genetics, 135*(9), 971–976. https://doi.org/10.1007/s00439-016-1687-9.

Hasley, J. A. R., Navet, N., & Tian, M. (2021). CRISPR/Cas9-mediated mutagenesis of sweet basil candidate susceptibility gene *ObDMR6* enhances downy mildew resistance. *PLoS One, 16*(6), e0253245. https://doi.org/10.1371/journal.pone.0253245.

Hassan, A. (2012). Effects of mineral nutrients on physiological and biochemical processes related to secondary metabolites production in medicinal herbs. *Medicinal and Aromatic Plant Science and Biotechnology, 6*, 105–110.

Haun, W., Coffman, A., Clasen, B. M., Demorest, Z. L., Lowy, A., Ray, E., & Zhang, F. (2014). Improved soybean oil quality by targeted mutagenesis of the fatty acid desaturase 2 gene family. *Plant Biotechnology Journal, 12*(7), 934–940. https://doi.org/10.1111/pbi.12201.

Higashi, Y., & Saito, K. (2013). Network analysis for gene discovery in plant specialized metabolism. *Plant, Cell & Environment, 36*, 1597–1606. https://doi.org/10.1111/pce.12069.

Hussein, R. A., & El-Anssary, A. A. (2018). Plants' secondary metabolites: The key drivers of the pharmacological actions of medicinal plants. In P. F. Builders (Ed.). *Herbal Medicine*. IntechOpen. https://doi.org/10.5772/intechopen.76139.

Iaffaldano, B., Zhang, Y., & Cornish, K. (2016). CRISPR/Cas9 genome editing of rubber-producing dandelion *Taraxacum kok-saghyz* using *Agrobacterium rhizogenes* without selection. *Industrial Crops and Products, 89*, 356–362. https://doi.org/10.1016/j.indcrop.2016.05.031.

Jadaun, J. S., Sangwan, N. S., Narnoliya, L. K., Tripathi, S., & Sangwan, R. S. (2016). Withania coagulans tryptophan decarboxylase gene cloning, heterologous expression, and catalytic characteristics of the recombinant enzyme. *Protoplasma, 254*(1), 181–192. https://doi.org/10.1007/s00709-015-0929-8.

Jankele, R., & Svoboda, P. (2014). TAL effectors: Tools for DNA targeting. *Briefings in Functional Genomics, 13*(5), 409–419. https://doi.org/10.1093/bfgp/elu013.

Jeena, G. S., Fatima, S., Tripathi, P., Upadhyay, S., & Shukla, R. K. (2017). Comparative transcriptome analysis of shoot and root tissue of *Bacopa monnieri* identifies potential genes related to triterpenoid saponin biosynthesis. *BMC Genomics, 18*(1), 490. https://doi.org/10.1186/s12864-017-3888-4.

Jiang, F., & Doudna, J. A. (2017). CRISPR-Cas9 structures and mechanisms. *Annual Review of Biophysics, 46*(1), 505–529. https://doi.org/10.1146/annurev-biophys-062215-010822.

Jiang, W. Z., Henry, I. M., Lynagh, P. G., Comai, L., Cahoon, E. B., & Weeks, D. P. (2017). Significant enhancement of fatty acid composition in seeds of the allohexaploid *Camelina sativa* using CRISPR/Cas9 gene editing. *Plant Biotechnology Journal, 15*, 648–657. https://doi.org/10.1111/pbi.12663.

Jinek, M., Chylinski, K., Fonfara, I., Hauer, M., Doudna, J. A., & Charpentier, E. (2012). A programmable dual-RNA-guided DNA endonuclease in adaptive bacterial immunity. *Science (New York, N.Y.), 337*(6096), 816–821. https://doi.org/10.1126/science.1225829.

Kamburova, V. S., Nikitina, E. V., Shermatov, S. E., Buriev, Z. T., Kumpatla, S. P., Emani, C., & Abdurakhmonov, I. Y. (2017). Genome editing in plants: An overview of tools and applications. *International Journal of Agronomy, 2017*, 1–15. https://doi.org/10.1155/2017/7315351.

Kaur, N., Alok, A., Shivani, Kumar, P., Kaur, N., Awasthi, P., Chaturvedi, S., Pandey, P., Pandey, A., Pandey, A. K., & Tiwari, S. (2020). CRISPR/Cas9 directed editing of lycopene epsilon-cyclase modulates metabolic flux for β-carotene biosynthesis in banana fruit. *Metabolic Engineering, 59*, 76–86. https://doi.org/10.1016/j.ymben.2020.01.008.

Kaushik, P., Gramazio, P., Vilanova, S., Raigón, M. D., Prohens, J., & Plazas, M. (2017). Phenolics content, fruit flesh colour and browning in cultivated eggplant, wild relatives and interspecific hybrids and implications for fruit quality breeding. *Food Research International, 102*, 392–401. https://doi.org/10.1016/j.foodres.2017.09.028.

Kimberland, M. L., Hou, W., Alfonso-Pecchio, A., Wilson, S., Rao, Y., Zhang, S., et al. (2018). Strategies for controlling CRISPR/Cas9 off-target effects and biological variations in mammalian genome editing experiments. *Journal of Biotechnology, 284*, 91–101. https://doi.org/10.1016/j.jbiotec.2018.08.007.

Kiss, A. K., & Piwowarski, J. P. (2018). Ellagitannins, gallotannins and their metabolites—The contribution to the anti-inflammatory effect of food products and medicinal plants. *Current Medicinal Chemistry, 25*(37), 4946–4967. https://doi.org/10.2174/0929867323666160919111559.

Kleinstiver, B. P., Pattanayak, V., Prew, M. S., Tsai, S. Q., Nguyen, N. T., Zheng, Z., et al. (2016). High-fidelity CRISPR–Cas9 nucleases with no detectable genome-wide off-target effects. *Nature, 529*, 490–495. https://doi.org/10.1038/nature16526.

Kui, L., Chen, H., Zhang, W., He, S., Xiong, Z., Zhang, Y., Yan, L., Zhong, C., He, F., Chen, J., & Zeng, P. (2017). Building a genetic manipulation toolbox for orchid biology: Identification of constitutive promoters and application of CRISPR/Cas9 in the orchid *Dendrobium officinale*. *Frontiers in Plant Science, 7*, 2036. https://doi.org/10.3389/fpls.2016.02036.

Kumar, R., Vashisth, D., Misra, A., Akhtar, M. Q., Jalil, S. U., Shanker, K., et al. (2016). RNAi down-regulation of cinnamate-4-hydroxylase increases artemisinin biosynthesis in *Artemisia annua*. *Scientific Reports, 6*, 26458. https://doi.org/10.1038/srep26458.

Larson, M. H., Gilbert, L. A., Wang, X., Lim, W. A., Weissman, J. S., & Qi, L. S. (2013). CRISPR interference (CRISPRi) for sequence-specific control of gene expression. *Nature Protocols, 8*(11), 2180–2196. https://doi.org/10.1038/nprot.2013.132.

Laura, M., Forti, C., Barberini, S., Ciorba, R., Mascarello, C., Giovannini, A., Pistelli, L., et al. (2023). Highly efficient CRISPR/Cas9-mediated gene editing in *Ocimum basilicum* 'FT Italiko' to induce resistance to *Peronospora belbahrii*. *Plants, 12*(13), 2395. https://doi.org/10.3390/plants12132395.

Leung, H. W., Foo, G., Banumurthy, G., Chai, X., Ghosh, S., Mitra-Ganguli, T., & VanDongen, A. M. (2017). The effect of *Bacopa monnieri* on gene expression levels in SH-SY5Y human neuroblastoma cells. *PLoS One, 12*(8), e0182984. https://doi.org/10.1371/journal.pone.0182984.

Li, R., Li, R., Li, X., Fu, D., Zhu, B., Tian, H., & Zhu, H. (2018). Multiplexed CRISPR/Cas9-mediated metabolic engineering of γ-aminobutyric acid levels in *Solanum lycopersicum*. *Plant Biotechnology Journal, 16*(2), 415–427. https://doi.org/10.1111/pbi.12781.

Lian, J., Carl, S., Cao, M., HamediRad, M., & Zhao, H. (2019). Multi-functional genome-wide CRISPR system for high throughput genotype–Phenotype mapping. *Nature Communications, 10*, 5794. https://doi.org/10.1038/s41467-019-13621-4.

Liu, L., & Fan, X. D. (2014). CRISPR–Cas system: A powerful tool for genome engineering. *Plant Molecular Biology, 85*(3), 209–218. https://doi.org/10.1007/s11103-014-0188-7.

Lozano-Juste, J., & Cutler, S. R. (2014). Plant genome engineering in full bloom. *Trends in Plant Science, 19*(5), 284–287. https://doi.org/10.1016/j.tplants.2014.02.014.

Lv, Z., Zhang, F., Pan, Q., Fu, X., Jiang, W., Shen, Q., Yan, T., Shi, P., Lu, X., Sun, X., & Tang, K. (2016). Branch pathway blocking in *Artemisia annua* is a useful method for obtaining high-yield artemisinin. *Plant and Cell Physiology, 57*(3), 588–602. https://doi.org/10.1093/pcp/pcw010.

Makarova, S. S., Khromov, A. V., Spechenkova, N. A., Taliansky, M. E., & Kalinina, N. O. (2018). Application of the CRISPR/Cas system for generation of pathogen-resistant plants. *Biochemistry (Moscow), 83*(12), 1552–1562. https://doi.org/10.1134/S0006297918120064.

Mali, P., Luhan, Y., Esvelt, K. M., Aach, J., Guell, M., DiCarlo, J. E., et al. (2013). RNA-guided human genome engineering via Cas9. *Science (New York, N.Y.), 339*, 823–826. https://doi.org/10.1126/science.1232033.

Miller, J. C., Tan, S., Qiao, G., Barlow, K. A., Wang, J., Xia, D. F., Meng, X., Paschon, D. E., Leung, E., Hinkley, S. J., Dulay, G. P., Hua, K. L., Ankoudinova, I., Cost, G. J., Urnov, F. D., Zhang, H. S., Holmes, M. C., Zhang, L., Gregory, P. D., & Rebar, E. J. (2011). A TALE nuclease architecture for efficient genome editing. *Nature Biotechnology, 29*(2), 143–148.

Mitra, S., Anand, U., Ghorai, M., Kant, N., Kumar, M., Jha, R., Swamy, N. K., Proćków, M. K., Lastra, J., de La, J. M. P., & Dey, A. (2023). Genome editing technologies, mechanisms and improved production of therapeutic phytochemicals: Opportunities and prospects. *Biotechnology and Bioengineering, 120*(1), 82–94. https://doi.org/10.1002/bit.28260.

Mochida, K., & Tran, L. S. P. (2017). Transcription factors in *Jatropha*. In *The jatropha genome*. Springer International Publishing, 47–60. https://doi.org/10.1007/978-3-319-49574-2_4.

Modrzejewski, D., Hartung, F., Lehnert, H., Sprink, T., Kohl, C., & Keilwagen, J. (2020). Which factors affect the occurrence of off-target effects caused by the use of CRISPR/Cas: A systematic review in plants. *Frontiers in Plant Science, 11*, 574959. https://doi.org/10.3389/fpls.2020.574959.

Morineau, C., Bellec, Y., Tellier, F., Gissot, L., Kelemen, Z., & Nogué, F. (2017). Selective gene dosage by CRISPR–Cas9 genome editing in hexaploid *Camelina sativa*. *Plant Biotechnology Journal, 15*, 729–739. https://doi.org/10.1111/pbi.12671.

Moscou, M. J., & Bogdanove, A. J. (2009). A simple cipher governs DNA recognition by TAL effectors. *Science (New York, N.Y.), 326*(5959), 1501. https://doi.org/10.1126/science.1178817.

Mussolino, C., Morbitzer, R., Lütge, F., Dannemann, N., Lahaye, T., & Cathomen, T. (2011). A novel TALE nuclease scaffold enables high genome editing activity in combination with low toxicity. *Nucleic Acids Research, 39*(21), 9283–9293. https://doi.org/10.1093/nar/gkr597.

Muszyńska, B., Łojewski, M., Sułkowska-Ziaja, K., Szewczyk, A., Gdula-Argasińska, J., & Hałaszuk, P. (2016). *In vitro* cultures of *Bacopa monnieri* and an analysis of selected groups of biologically active metabolites in their biomass. *Pharmaceutical Biology, 54*(11), 2443–2453. https://doi.org/10.3109/13880209.2016.1158843.

Niu, Z., Zhu, F., Fan, Y., Li, C., Zhang, B., Zhu, S., Hou, Z., Wang, M., Yang, J., Xue, Q., & Liu, W. (2021). The chromosome-level reference genome assembly for *Dendrobium officinale* and its utility for functional genomics research and molecular breeding study. *Acta Pharmaceutica Sinica B, 11*(7), 2080–2092. https://doi.org/10.1016/j.apsb.2021.01.019.

Palpant, N. J., & Dudzinski, D. (2013). Zinc finger nucleases: Looking toward translation. *Gene Therapy, 20*(2), 121–127. https://doi.org/10.1038/gt.2012.2.

Pavletich, N. P., & Pabo, C. O. (1991). Zinc finger–DNA recognition: Crystal structure of a Zif268–DNA complex at 2.1 Å *Science (New York, N.Y.), 252*(5007), 809–817. https://doi.org/10.1126/science.2028256.

Pyne, M. E., Narcross, L., & Martin, V. J. J. (2019). Engineering plant secondary metabolism in microbial systems. *Plant Physiology, 179*, 844–861. https://doi.org/10.1104/pp.18.01291.

Ran, F. A., Hsu, P. D., Wright, J., Agarwala, V., Scott, D. A., & Zhang, F. (2013). Genome engineering using the CRISPR-Cas9 system. *Nature Protocols, 8*(11), 2281–2308. https://doi.org/10.1038/nprot.2013.143.

Ren, J., Fu, L., Nile, S. H., Zhang, J., & Kai, G. (2019). *Salvia miltiorrhiza* in treating cardiovascular diseases: A review on its pharmacological and clinical applications. *Frontiers in Pharmacology, 10*, 753. https://doi.org/10.3389/fphar.2019.00753.

Ren, R., Gao, J., Yin, D., Li, K., Lu, C., Ahmad, S., Wei, Y., Jin, J., Zhu, G., & Yang, F. (2021). Highly efficient leaf base protoplast isolation and transient expression systems for orchids and other important monocot crops. *Frontiers in Plant Science, 12*, 626015. https://doi.org/10.3389/fpls.2021.626015.

Ren, J.-J., Wang, J., Lee, K.-K., Deng, H., Xue, H., Zhang, N., Zhao, J.-C., Cao, T., Cui, C.-L., & Zhang, X.-H. (2020). The complete chloroplast genome of *Aloe vera* from China as a Chinese herb. *Mitochondrial DNA Part B, 5*(1), 1092–1093. https://doi.org/10.1080/23802359.2020.1726229.

Sangwan, N. S., Mishra, L. N., Tripathi, S., & Kushwaha, A. K. (2014a). Omics of secondary metabolic pathways in *Withania somnifera* Dunal (ashwagandha). *Omics Applications in Crop Science*. https://doi.org/10.1201/b16352-12.

Sangwan, N. S., & Sangwan, R. S. (2014b). Secondary metabolites of traditional medicinal plants: A case study of ashwagandha (*Withania somnifera*). *Applied Plant Cell Biology*, 325–367. https://doi.org/10.1007/978-3-642-41787-0_11.

Sangwan, N. S., Jadaun, J. S., Tripathi, S., Mishra, B., Narnoliya, L. K., & Sangwan, R. S. (2018). *Omics technologies and bio-engineering*. Elsevier, 143–175. https://doi.org/10.1016/b978-0-12-815870-8.00009-7.

Schachtsiek, J., & Stehle, F. (2019). Nicotine-free, nontransgenic tobacco (*Nicotiana tabacum* L.) edited by CRISPR-Cas9. *Plant Biotechnology Journal, 17*(12), 2228–2230. https://doi.org/10.1111/pbi.13139.

Sharafi, A., Sohi, H. H., Mousavi, A., Azadi, P., Khalifani, B. H., & Razavi, K. (2013). Metabolic engineering of morphinan alkaloids by overexpression of codeinone reductase in transgenic hairy roots of *Papaver bracteatum*, the Iranian poppy. *Biotechnology Letters, 35*, 445–453. https://doi.org/10.1007/s10529-012-1080-7.

Shi, M., Du, Z., Hua, Q., & Kai, G. (2021). CRISPR/Cas9-mediated targeted mutagenesis of *bZIP2* in *Salvia miltiorrhiza* leads to promoted phenolic acid biosynthesis. *Industrial Crops and Products, 167*, 113560. https://doi.org/10.1016/j.indcrop.2021.113560.

Shrestha, A., Khan, A., & Dey, N. (2018). *Cis-trans* engineering: Advances and perspectives on customized transcriptional regulation in plants. *Molecular Plant, 11*, 886–898. https://doi.org/10.1016/j.molp.2018.05.008.

Si, X., Zhang, H., Wang, Y., Chen, K., & Gao, C. (2020). Manipulating gene translation in plants by CRISPR–Cas9-mediated genome editing of upstream open reading frames. *Nature Protocols, 15*, 338–363. https://doi.org/10.1038/s41596-019-0238-3.

Singer, A. C., Crowley, D. E., & Thompson, I. P. (2003). Secondary plant metabolites in phytoremediation and biotransformation. *Trends in Biotechnology, 21*(3), 123–130. https://doi.org/10.1016/s0167-7799(02)00041-0.

Sinha, S., Sandhu, K., Bisht, N., Naliwal, T., Saini, I., & Kaushik, P. (2019). Ascertaining the paradigm of secondary metabolism enhancement through gene-level modification in therapeutic plants. *Journal of Young Pharmacists, 11*(4), 337.

Sorek, R., Lawrence, C. M., & Wiedenheft, B. (2013). CRISPR-mediated adaptive immune systems in bacteria and archaea. *Annual Review of Biochemistry, 82*(1), 237–266. https://doi.org/10.1146/annurev-biochem-072911-172315.

Strickler, S. R., Bombarely, A., & Mueller, L. A. (2012). Designing a transcriptome next-generation sequencing project for a non-model plant species. *American Journal of Botany, 99*, 257–266. https://doi.org/10.3732/ajb.1100292.

Teixeira da Silva, J. A., Dobránszki, J., Cardoso, J. C., Chandler, S. F., & Zeng, S. (2016). Methods for genetic transformation in *Dendrobium*. *Plant Cell Reports, 35*, 483–504. https://doi.org/10.1007/s00299-015-1917-3.

Tong, Y., Weber, T., & Lee, S. Y. (2019). CRISPR/Cas-based genome engineering in natural product discovery. *Natural Product Reports, 36*, 1262–1280. https://doi.org/10.1039/c8np00089a.

Trethewey, R. N. (2004). Metabolite profiling as an aid to metabolic engineering in plants. *Current Opinion in Plant Biology, 7*, 196–201. https://doi.org/10.1016/j.pbi.2003.12.003.

Verpoorte, R., Contin, A., & Memelink, J. (2002). Biotechnology for the production of plant secondary metabolites. *Phytochemistry Reviews, 1*(1), 13–25. https://doi.org/10.1023/a:1015871916833.

Waltz, E. (2016). CRISPR-edited crops free to enter market, skip regulation. *Nature Biotechnology, 34*(6), 582–583. https://doi.org/10.1038/nbt0616-582.

Wang, C., Liu, Q., Shen, Y., Hua, Y., Wang, J., Lin, J., Wu, M., Sun, T., Cheng, Z., Mercier, R., & Wang, K. (2019). Clonal seeds from hybrid rice by simultaneous genome engineering of meiosis and fertilization genes. *Nature Biotechnology, 37*(3), 283–286. https://doi.org/10.1038/s41587-018-0003-0.

Watanabe, K., Kobayashi, A., Endo, M., Sage-Ono, K., Toki, S., & Ono, M. (2017). CRISPR/Cas9-mediated mutagenesis of the dihydroflavonol-4-reductase-B (DFR-B) locus in the Japanese morning glory Ipomoea (Pharbitis) nil. *Scientific Reports, 7*(1). https://doi.org/10.1038/s41598-017-10715-1.

Wolter, F., Schindele, P., & Puchta, H. (2019). Plant breeding at the speed of light: The power of CRISPR/Cas to generate directed genetic diversity at multiple sites. *BMC Plant Biology, 19*, 176. https://doi.org/10.1186/s12870-019-1775-1.

Wright, D. A., Townsend, A. J., Winfrey, R. J., Irwin, P. A., Rajagopal, J., Lonosky, P. M., et al. (2004). High-frequency homologous recombination in plants mediated by zinc-finger nucleases. *Plant Journal, 44*, 693–705. https://doi.org/10.1111/j.1365-313X.2005.02551.x.

Xu, H., Song, J., Luo, H., Zhang, Y., Li, Q., Zhu, Y., et al. (2016). Analysis of the genome sequence of the medicinal plant *Salvia miltiorrhiza*. *Molecular Plant, 9*, 949–952. https://doi.org/10.1016/j.molp.2016.03.010.

Zhang, F., Cong, L., Lodato, S., Kosuri, S., Church, G. M., & Arlotta, P. (2011). Efficient construction of sequence-specific TAL effectors for modulating mammalian transcription. *Nature Biotechnology, 29*(2), 149–153. https://doi.org/10.1038/nbt.1775.

Zhang, X., Xu, G., Cheng, C., Lei, L., Sun, J., Xu, Y., et al. (2021). Establishment of an *Agrobacterium*-mediated genetic transformation and CRISPR/Cas9-mediated targeted mutagenesis in hemp (*Cannabis sativa* L). *Plant Biotechnology Journal*. https://doi.org/10.1111/pbi.13611.

Zhang, Y., & Showalter, A. M. (2020). CRISPR/Cas9 genome editing technology: A valuable tool for understanding plant cell wall biosynthesis and function. *Frontiers in Plant Science, 11*, 589517. https://doi.org/10.3389/fpls.2020.589517.

Zhang, Y., Zhang, F., Li, X., Baller, J. A., Qi, Y., Starker, C. G., Bogdanove, A. J., & Voytas, D. F. (2013). Transcription activator-like effector nucleases enable efficient plant genome engineering. *Plant Physiology, 161*, 20–27. https://doi.org/10.1104/pp.112.205179.

Zhou, Z., Tan, H., Li, Q., Chen, J., Gao, S., Wang, Y., Chen, W., & Zhang, L. (2018). CRISPR/Cas9-mediated efficient targeted mutagenesis of *RAS* in *Salvia miltiorrhiza*. *Phytochemistry, 148*, 63–70. https://doi.org/10.1016/j.phytochem.2018.01.015.

Zhu, H., Li, C., & Gao, C. (2020). Applications of CRISPR–Cas in agriculture and plant biotechnology. *Nature Reviews Molecular Cell Biology, 21*(11), 661–677. https://doi.org/10.1038/s41580-020-00288-9.

Further reading

Ceasar, S. A., Rajan, V., Prykhozhij, S. V., Berman, J. N., & Ignacimuthu, S. (2016). Insert, remove or replace: A highly advanced genome editing system using CRISPR/Cas9. *Biochimica et Biophysica Acta (BBA) - Molecular Cell Research, 1863*(9), 2333–2344. https://doi.org/10.1016/j.bbamcr.2016.06.009.

Doudna, J. A., & Charpentier, E. (2014). The new frontier of genome engineering with CRISPR-Cas9. *Science (New York, N.Y.), 346*(6213), 1258096. https://doi.org/10.1126/science.1258096.

Ma, X., Kthiri, D., Gill, M. S., Pozniak, C. J., & Kagale, S. (2025). Cas9 endonuclease: A molecular tool for *in vitro* cloning and CRISPR edit detection. *Frontiers in Genome Editing, 7*, 1565297. https://doi.org/10.3389/fgeed.2025.1565297.

Mahfouz, M. M., Piatek, A., & Stewart, C. N., Jr. (2014). Genome engineering *via* TALENs and CRISPR/Cas9 systems: Challenges and perspectives. *Plant Biotechnology Journal, 12*(8), 1006–1014. https://doi.org/10.1111/pbi.12256.

Navet, N., & Tian, M. (2020). Efficient targeted mutagenesis in allotetraploid sweet basil by CRISPR/Cas9. *Plant Direct*. https://doi.org/10.1002/pld3.233.

Ren, X., Wang, M., Li, B., Jamieson, K., Zheng, L., Jones, I. R., Li, B., et al. (2021). Parallel characterization of cis-regulatory elements for multiple genes using CRISPRpath. *Science Advances, 7*(38), eabi4360. https://doi.org/10.1126/sciadv.abi4360.

CHAPTER

13

Gas chromatography–mass spectrometry-mediated plant metabolomics analysis: recent trends in *in vitro* cultures and potential pharmacological applications

Yashika Bansal, A. Mujib, Mohammad Mohsin, and Afeefa Nafees

Cellular Differentiation and Molecular Genetics Section, Department of Botany, Jamia Hamdard, New Delhi, Delhi, India

13.1 Introduction

Advancements in profiling techniques like genomics, transcriptomics, and metabolomics have facilitated research into an organism's metabolism and the molecular mechanisms that regulate its chemical composition (Fernie & Tohge, 2017). One of the most significant scientific advances in recent years has been metabolomics, which has made it possible to accurately profile the metabolites in microorganisms, plants, and animals. This approach enables quick and accurate metabolite analysis by detecting a wide variety of compounds from a single sample (Kumar et al., 2017). According to Oliver (1998), the metabolome can be defined as the quantifiable collection of low-molecular-weight compounds found in a cell or organism during a specific physiological or developmental stage. These small metabolites indicate the results of gene expression and the biochemical behavior of cells, tissues, organs, and organisms (Commisso et al., 2013). It is now possible to conduct extensive metabolite surveys that encompass both known and unknown compounds using various metabolomics platforms. For the large-scale analysis of complex mixtures of extracts, a number of integrated technologies and methodologies are employed, such as mass spectrometry (MS)-based techniques like gas chromatography–mass spectrometry (GC–MS), liquid chromatography mass–spectroscopy (LC–MS), Fourier transform ion cyclotron

Medicinal Plant Biotechnology for Ameliorated Secondary Metabolite Production
DOI: https://doi.org/10.1016/B978-0-443-43944-5.00016-5

resonance-mass spectrometry (FTICR-MS), capillary electrophoresis mass spectrometry (CE–MS), matrix-assisted laser desorption/ionization (MALDI), field asymmetric waveform ion mobility spectrometry (FAIMS) and nuclear magnetic resonance (NMR) (Patel et al., 2016). Metabolome databases, such as METLIN, NIST, GOLM can be used to identify metabolites (Johnson & Lange, 2015), which are then subject to statistical analysis using web tools and software, such as MetaboAnalyst, Cytoscape, and Statistical analysis tool, including correlation maps, principal component analysis (PCA), partial least squares (PLS), K-means clustering, boxplots, heatmaps, and reconstruction of metabolic pathways (Kumar et al., 2017).

Different constituents can be profiled at multiple levels, including "metabolite (or metabolic) fingerprinting," "metabolite profiling," and "metabolite target analysis." As per Fiehn (2001), these approaches are defined as follows:

Metabolite fingerprinting: It serves as a hypothesis-formulating method and is employed to compare the "fingerprints" or patterns of metabolites that vary within a particular biological system.

Metabolite profiling: Considered a precursor of metabolomics, metabolite profiling is the most traditional and well-established method. It analyzes a wide group of metabolites that are either associated with a certain metabolic pathway or a class of bioactive compounds.

Metabolite target analysis: It focuses on examining metabolites associated with a certain metabolic pathway. This approach aims to identify the precise metabolic alterations that could be connected to a certain hypothesis.

In the field of plant science, the study of the plant metabolome involves examining a broad variety of chemical species with various physical characteristics, such as hydrophilic carbohydrates, organic and amino acids, hydrophobic lipid-related compounds, and ionic inorganic compounds (Jorge et al., 2015). It is now known that a single plant species may contain up to 15,000 different compounds, and over 200,000 natural products have been identified so far (Wolfender et al., 2014). Plant metabolites are essential for growth, cellular regeneration, systemic resource distribution, and the adaptability of plants to an ever-evolving environment (Fang et al., 2019). In addition, the bioactive compounds found in medicinal, aromatic, and ornamental plants (MAOPs) have a variety of industrial uses, including pharmaceutics, food, flavoring, cosmetics, fragrances, and agrochemicals. These compounds, in particular secondary metabolites (SMs), are especially useful to the pharmaceutical sector because they (1) effectively alter the body's metabolic processes and treat illnesses using a multi-target spectrum, and (2) are quite popular due to their safety and minimal negative effects (Marchev et al., 2020; Saranraj et al., 2016). However, due to the constant increase in the human population and the new difficulties brought on by global warming, there are fewer natural resources available to meet the rising demand for naturally occurring bioactive chemicals. Accordingly, one of the biotechnology-based breeding techniques (BBBMs) that has the ability to develop alternate, sustainable, and economical means of producing bioactive chemicals is the plant *in vitro* culture system (Niazian and Sabbatini, 2021). Researchers around the globe have now integrated these two diverse research fields (plant biotechnology and metabolomics) to understand complex developmental processes such as *in vitro* organogenesis, embryogenesis, or dedifferentiation. Thus the present chapter discusses the developments and applications of one of the widely applied techniques, that is, GC–MS in profiling metabolites of medicinal and ornamental plants, paving the way for future research studies.

13.1.1 Gas chromatography–mass spectrometry

GC–MS is widely regarded as a dynamic analytical tool because of its superior capacity for separation, robustness, selectivity, sensitivity, and repeatability. The use of GC–MS studies for medicinal plant research has grown in recent years due to the technique's effectiveness in analyzing volatile essential oils, fatty acids, lipids, alkaloids, and nonpolar components (Al-Rubaye et al., 2017). However, GC–MS is limited to the separation and identification of volatile and low molecular weight (ca. 50–600 Da) molecules. In order to detect polar, thermolabile, nonvolatile metabolites, the application of chemical derivatisation is necessary before analysis, enhancing the detector's responsiveness, sensitivity, thermal stability, and volatility (Beale et al., 2018).

13.1.2 Gas chromatography–mass spectrometry-based plant metabolomics workflow

A flowchart of a typical GC–MS-based metabolomics experiment is shown in Fig. 13.1. Crucial points include sample preparation, chemical derivatization, GC–MS analysis, data preprocessing, data analysis, and bioinformatics.

13.1.3 Sample preparation

Sampling and extraction procedures vary based on biological sample type and target metabolite class(es). In case of plant samples, several environmental factors, including diurnal fluctuations in metabolite concentrations, temperature, light conditions, developmental stage, and harvesting time, have a qualitative and quantitative impact on metabolites (Kumar et al., 2024). It should be noted that no single sample preparation method can cover every metabolite in any kind of sample (Álvarez-Sánchez et al., 2010). This is because of the metabolome's complexity, which includes thousands of metabolites with diverse molecular weights, polarities, and chemical properties (Ghatak et al., 2018). The initial stage of any metabolomics workflow is quenching biological samples to halt metabolically active cells and tissues. Depending on the type of sample, metabolites are then extracted using appropriate solvents (water-soluble/lipophilic solvents) (Hao et al., 2025). Plant metabolomics have been studied using many solvent systems. For instance, the extraction of metabolites from leaves and callus samples of *Hibiscus sabdariffa* L. has been done by using ethyl acetate solvent (Sobrinho et al., 2022). Ralte et al. (2022) identified important phytochemicals in the methanolic extracts of different plant parts (flower, capitulum, pod, and seed) of *Parkia timoriana*.

13.2 Chemical derivatization

Another important aspect of GC–MS-based metabolomics is that it requires derivatization of polar chemicals. Various derivatising agents, such as alkylation, silylation, trimethylsilylation (TMS), methoximation, esterification, and acylation, have been

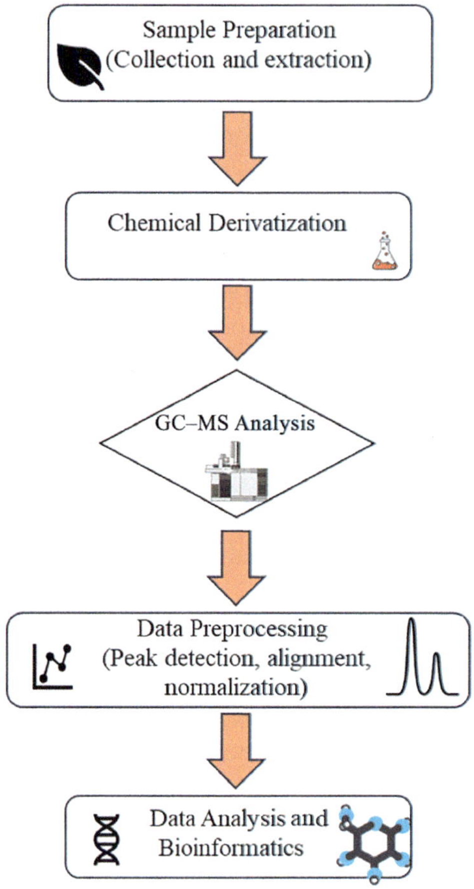

FIGURE 13.1 A typical representation of multistep workflow of gas chromatography–mass spectrometry (GC–MS) based metabolomics study.

developed so far. The frequently used derivatisation reagents in metabolomics-based research are listed in Table 13.1. Among these, TMS is widely utilized due to its relative ease of usage and thorough coverage of compound classes such as sugars, sugar alcohols, amines, amino acids, and organic acids (Patel et al., 2021). Derivatisation is frequently performed in two steps: sugars are first stabilized in the open-ring conformation by converting carbonyl groups into the appropriate oximes using hydroxylamine or alkylhydroxylamine reagents (such as O-methylhydroxylamine hydrochloride, MeOx), followed by TMS using silylating reagents like N-methyl-N-(trimethylsilyl)trifluoroacetamide (MSTFA) or alternatively N,O-bis(trimethylsilyl)trifluoroacetamide (BSTFA) (Hill & Roessner, 2013). To capture a wide range of metabolites having polarity and high boiling points on a GC–MS system, the latter step is essential for the proper derivatisation of nonvolatile molecules.

TABLE 13.1 Commonly used derivatization agents in metabolomics-based GC–MS studies (Moldoveanu & David, 2018; Orata, 2012).

Derivatization method	Reagent(s) used	Functional group derivatized									
Amines	**Amides**	**Carboxyls**	**Carbonyl**	**Phenols**	**Hydroxyls**	**Thiols**	**Carbohydrates**	**Sulfona-mides**	**Easters**		
Alkylation	Dimethylformamide			+							
	Diazomethane			+							
	Ethylchloroformate	+									
	Methyl-8 ®			+		+		+			
Silylation	BSA	+	+	+	+				+		+
	BSTFA	+	+			+				+	
	MSTFA	+	+	+	+				+		
Acylation	Fluorinated anhydride	+				+	+				
	Fluoracylimidazole	+	+			+	+				
	MBTFA	+					+	+			
Oximation	Methoxyamine HCl				+		+				

Orata, F. (2012). Derivatization reactions and reagents for gas chromatography analysis. In M.A. Mohd (Ed.), Advanced Advanced gas chromatography - Progress in agricultural, biomedical and industrial applicationsromatography-Progress in omedical and plications. InTech. 10.5772/33098.

13.2.1 Gas chromatography–mass spectrometry analysis

Fig. 13.2 illustrates GC–MS instrumentation, depicting the components of a simple GC–MS system. The analytes being examined are carried along a column coated or packed with a stationary phase using a gaseous mobile phase (generally helium), also known as an eluent. In gas chromatography, the stationary phase is typically a mixture of inert, small-diameter particles coated with a nonpolar liquid on the column's inner surface (Pramod et al., 2021). After separation, ions are produced and ionized samples are adjusted using electric and magnetic fields prior to detection in mass spectrometer (Khalifea & Ali, 2025). GC–MS uses two primary ionization techniques: chemical ionization (CI) and electron ionization (EI). But EI is still used in the majority of GC–MS metabolomics techniques, as it is regarded as a "hard" ionization technique that produces repeatable molecular fragmentation into precisely defined mass spectrum fingerprints (Zeki et al., 2020). Ionization is carried out at 70 eV for all EI instruments, and mass spectra are generally thought to be repeatable across instruments made by various vendors as well as across instruments using various mass analyzers (quadrupole, time of flight, etc.). A crucial component of a mass spectrometer is the mass analyzer that determines the mass-to-charge ratio (m/z) of molecules present as charged ions. Numerous mass analyzers with varying mass resolutions and accuracies have been connected to GC (Table 13.2). Low mass resolution quadrupole mass filters, such as GC single quadrupole (GC–Q–MS) and triple quadrupole instruments (GC–QqQ–MS), are the most widely used mass analyzers for GC–MS applications (Li et al., 2021). For instance, Abd El-Kareem et al. (2025) used a single quadruple analyzer to detect essential oils from fennel plants.

13.2.2 Data preprocessing

Biologically significant results and successful data analysis begin with efficient and credible data processing. The datasets generated by GC–MS analysis have to undergo

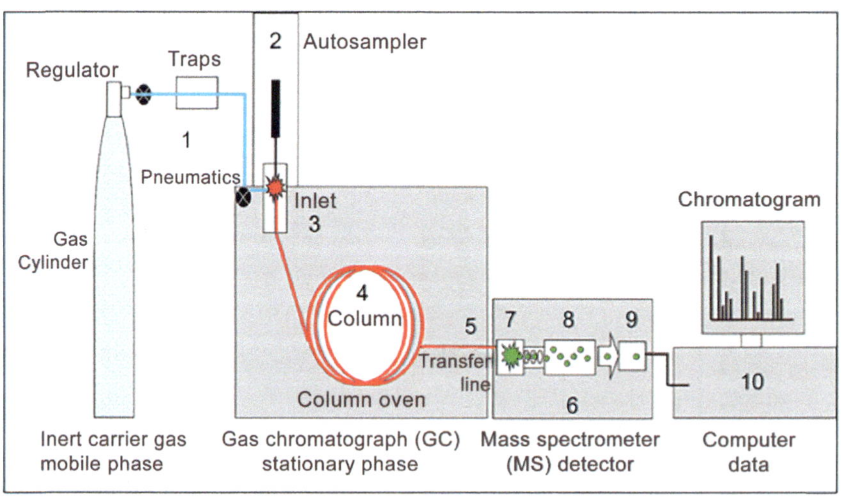

FIGURE 13.2 Schematic display of gas chromatography–mass spectrometry instrument with its specifications.

TABLE 13.2 Details of typical mass analyzers used and their configurations.

Mass analyzer	Mass resolution	Mass range (Da)	MS/MS	Operating pressure (torr)
Ion trap	~ 1000	50–4000	Yes	10^{-3}
Quadruple (Q)	~ 1000	50–6000	Yes	10^{-5}
Time of flight (TOF)	2500–40,000	20–500,000	No	10^{-6}
qTOF	2500–40,000	20–500,000	Yes	10^{-7}
FTICR	>200,000	10–10,000	Yes	10^{-9}
Orbitrap	>100,000	40–4000	Yes	10^{-10}

No permission required.

preprocessing methods such as normalization, baseline correction, smoothing and filtering, background noise reduction, mass spectral deconvolution, retention time alignment, peak detection, etc. (Feizi et al., 2021; Yan, 2025). Such data preprocessing is done by using commercial software packages including MarkerLynx (Waters), MassHunter (Agilent), MarkerView (AB Sciex), XCMS, Progenesis QI (Waters), Xcalibur (Thermo Fisher Scientific), and the freely available AMDIS (National Institute of Standards and Technology, Gaithersburg, MD, USA) software (Karaman, 2017). As the noise can significantly impact the identification and quantification of metabolites in chromatographic-based research. Thus baseline correction aims to separate metabolite signals from noise and baseline, whereas noise reduction reduces high-frequency changes in signals to enhance the signal-to-noise (S/N) ratio (Pierce et al., 2012; Tian et al., 2016). Peak picking/ detection, deconvolution, and peak alignment methods are also carried out during data preprocessing to prevent any error in metabolite identification that can have a major impact on the data analysis's outcome in GC–MS-based metabolomics analyses (Karaman et al., 2018). The process of normalization eliminates undesired variation among the samples, enabling quantitative comparison. In addition to minimizing the effect of instrument dynamic range on its efficacy, the normalization stage also prepares the dataset for statistical analysis later on (Yao et al., 2019). An example of GC–MS chromatogram obtained after preprocessing the datasets has been illustrated in the Fig. 13.3.

13.3 Data analysis and bioinformatics

The masses acquired in metabolomics investigations must be defined and interpreted using the proper databases. Numerous public databases are accessible for the annotation of metabolites (Table 13.3). Among the databases that are regularly utilized are XCMS, METLIN, Fiehn GC–MS Database, Golm Metabolome Database, MetaCyc, and NIST (Vinaixa et al., 2016). The nature of metabolomics investigations necessitates the application of the proper statistical test since hundreds or even thousands of features are examined. These statistical analyses can be carried out using programs like SIMCA (Tsugawa et al., 2011) and the web-based Metaboanalyst (Xia et al., 2009). Data produced from metabolomics

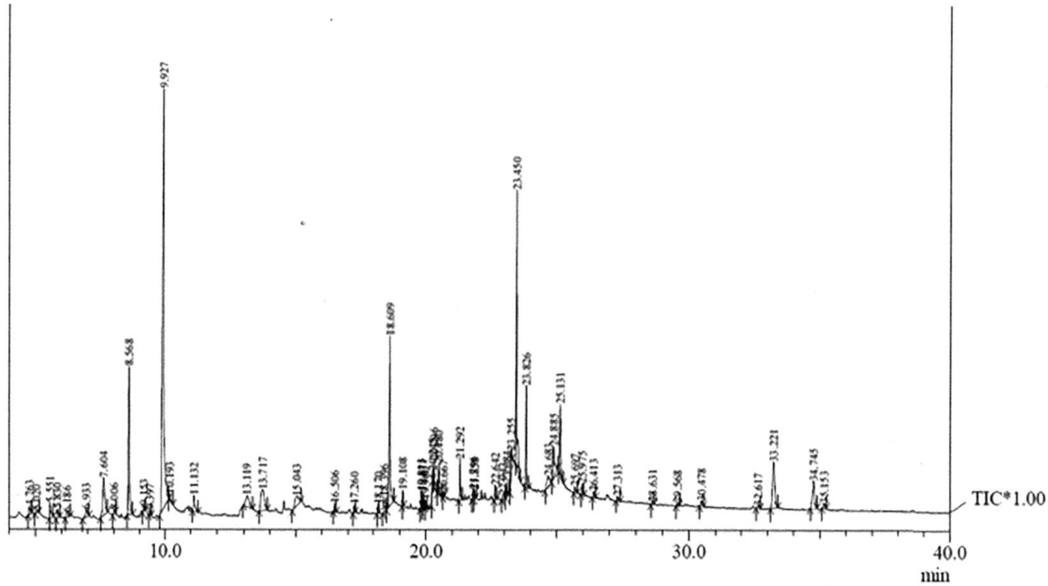

FIGURE 13.3 A gas chromatography–mass spectrometry chromatogram depicting the relative abundance of different phytocompounds in a sample (Fatima et al., 2024). X-axis represents retention time, and Y-axis represents the relative abundance of each compound identified. *Source: From Fatima, T., Mujib, A., Bansal, Y., Dewir, Y.H., & Mendler-Drienyovszki, N. (2024). Indirect organogenesis of* Calendula officinalis *L. and comparative phytochemical studies of field-grown and in vitro-regenerated tissues. Agronomy, 14(8), 1743. https://doi.org/10.3390/agronomy14081743.*

research are assessed by using correlation analysis, multivariate analysis, and univariate analysis. Both univariate and multivariate approaches measure every metabolite under study in a biological system in an effort to analyze the behavior of the system as a whole and within a system (Beale et al., 2018). Multivariate analyses generally involve principal component analysis (PCA), partial least squares-discrimination analysis (PLS-DA), and hierarchical cluster analysis (HCA). Using metabolomic methods and statistics, large datasets could then be utilized to identify the roles of such metabolites as biomarkers for the purpose of genetic and metabolic engineering (Commisso et al., 2013; Salem et al., 2020).

13.4 Application of gas chromatography–mass spectrometry technique in *in vitro* culture systems

GC–MS has been a long-standing metabolomics technology for identifying volatile and semivolatile organic chemicals in various plant tissues cultivated *in vivo*. However, in recent times, this technique has now found its immense utilization in *in vitro* plant culture systems due to its great advantages such as its high robustness, sensitivity, reliability, and reproducibility (Salem et al., 2020). Plant tissue cultures are often regarded as chemical factories that continuously generate secondary metabolites. The bioactive compounds of MAOPs can be produced *in vitro* using a variety of culture techniques, such as callus, shoot, roots, hairy roots,

TABLE 13.3 List of bioinformatics and statistical tools available for metabolite identification.

Database name	Website	Description	Data input	Major function
AMDIS	http://www.amdis.net/	It deals with modern aspects of GC–MS data interpretation, including chemometrics, liquid/gas chromatography, and mass spectrometry.	GC–MS	Data processing
BinBase	https://fiehnlab.ucdavis.edu/projects/binbase-setup	a system that combines mass spectrometric and biological metadata	GC–MS	Metabolite annotation
FiehnLib	https://fiehnlab.ucdavis.edu/projects/fiehnlib	a mass spectral and retention index library for comprehensive metabolic profiling	GC–qTOF-MS	Metabolite profiling
MetaboAnalyst	www.metaboanalyst.ca/	Comprehensive web-based tool for comprehensive metabolomics data analysis, interpretation, and integration with other omics data.	GC–MS	Statistical analysis
NIST	http://www.nist.gov/srd/nist1a.cfm	A mass spectral database/ library comprising thousands of compounds information	GC–MS	Metabolite annotation
XCMS	https://xcmsonline.scripps.edu	a software facilitating metabolite identification and pathway analysis	GC–MS, GC–MS/MS	Data processing

No permission required.

and suspension cell cultures (Niazian & Sabbatini, 2021). The variations in both quantity and quality of such metabolites can be brought about by changes in the nutrient medium's composition, plant growth regulators (single or in combination), elicitors, or stressors, etc. (Fig. 13.4) (Monfort et al., 2018). Therefore by comparing the metabolic profiles of the *in vitro* regenerated plants to those of their wild-type counterparts, GC–MS analysis can also reveal any modifications, offering valuable information about the genetic and environmental factors influencing plant growth and development (Faisal et al., 2023). By determining metabolites, it also seeks to decipher the likely impact of abiotic stressors on all dynamic biochemical processes within plant cells (Hamooh et al., 2021). Table 13.4 presents a consolidated list of *in vitro* plant tissues of various plants utilized for phytochemical profiling by GC–MS technique. There are multiple reports present for the phytochemical characterization of different *in vitro* regenerated plant parts and some of them are summarized below.

13.4.1 *Sansevieria trifasciata*

The succulent *Sansevieria trifasciata* (ST) is indigenous to the Tropical Canary Islands, subtropical regions of Brazil, India, Nigeria, Zaire, Thailand, and other countries. Due to its Crassulacean Acid Metabolism (CAM), it is included among the top 12 pollution-absorbing plants in the world by NASA. However, there are a number of limitations with natural breeding (using seeds) and/or traditional vegetative propagation (using leaf cuttings and rhizome divisions) (Kaur & Mudgal, 2021). *In vitro* rhizogenesis and subsequent shoot induction have been achieved for rapid multiplication of snake plant. According to GC–MS

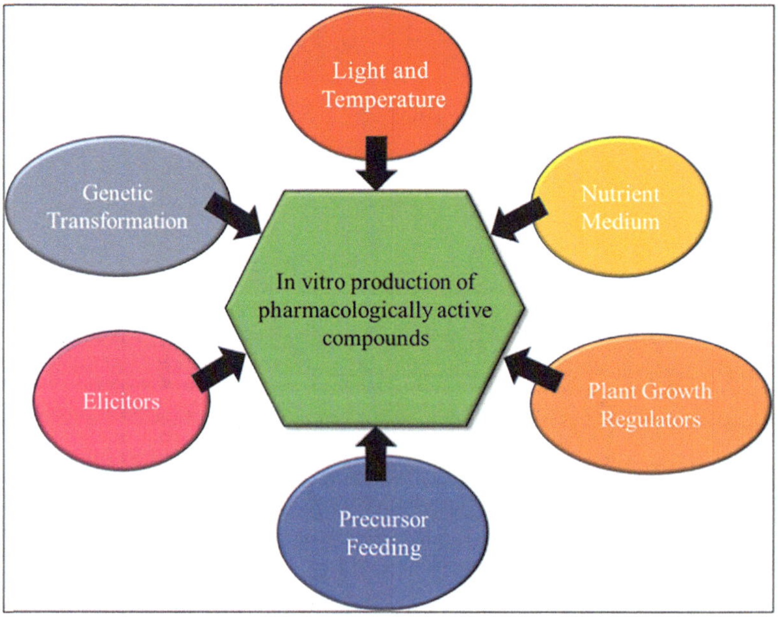

FIGURE 13.4 Factors affecting the *in vitro* production of bioactive compounds in various medicinal, ornamental, and aromatic plants.

TABLE 13.4 List of plant tissues analyzed by gas chromatography–mass spectrometry (GC–MS) technique for phytochemical profiling along with the solvent used, total compounds, and major compounds identified.

Name of the plant	*In vitro* tissue used	Solvent used	Total compounds detected	Major compounds	Reference(s)
Withania somnifera	Roots	Methanol	29	Proline, putrescine, vanillic acid	Senthil et al. (2015)
Justicia gendarussa	Leaf-derived callus	Methanol	26	Methyl linoleate, hexadecanoic acid, giberrelin A3	Wahyuni et al. (2017)
Andrographis paniculata	Callus	Methanol	24	Phytol, eicosane, sucrose	Dulara et al. (2019)
Leptadenia reticulata	Leaf-derived callus	Methanol	23	Palmitic acid methyl ester, eicosane	Godara et al. (2019)
		Ethyl acetate	19	Eicosane, ascorbic acid	
Tecoma stans	Leaf	Ethanol	50	Tecomine, palmitoleic acid, squalene, gamma-sitosterol	Hussain et al. (2019)
Clerodendrum thomsoniae	Leaf	–	27	Azelaic acid, linoleic acid, squalene, stigmasterol	Kar et al. (2019)
Echinops taeckholmiana	Roots	n-Hexane	24	Ethyl palmitate, ethyl linoleate, taraxerone, benzophenone	Hamdan et al. (2021)
Saussurea costus	Callus	Ethyl acetate	24	Selina-3,7(11)-diene, heneisane, alpha-maaliene	Khan et al. (2021)
Dianthus caryophyllus	Shootlets	Diethyl ether	40	Docosane, heptacosane, cadinol	Arafa et al. (2022)
Thymus vulgaris			30	Nonacosane, docosane, 2-undecenal	
Ocimum basilicum			20	Benzoyl isocyanate, 10-octadecenoic acid methyl ester	
Ruta chalepensis	Internodal-derived callus	Methanol	14	Dictamnine, methoxsalen, kokusaginine	Qahtan et al. (2022)
Catharanthus roseus	Anther-derived callus	Methanol	81	Guanosine, methyl palmitate, cycloartenol, ajmalicine	Bansal et al. (2023)
Urginea maritima	Callus	Methanol	25	Farnesol, ethyl iso-allocholate, oleic acid	El-Naggar et al. (2023)

(*Continued*)

TABLE 13.4 (Continued)

Name of the plant	*In vitro* tissue used	Solvent used	Total compounds detected	Major compounds	Reference(s)
Plectranthus amboinicus	Leaf	Methanol	21	5-methyl-2-(1-methylethyl)-phenol, gamma-terpenine, Trans-caryophyllene	Faisal et al. (2023)
Centratherum punctatum	Leaf	Methanol	37	Chondrillasterol, phytol, neophytadiene	Talan et al. (2023)
	Callus		57	Guanosine, tridecanoic acid, squalene	
	Roots		23	Chondrillasterol, stigmasta-5,22-dien-3-Ol, chrysantenyl 2-methuylbutanoate	
Glycyrrhiza glabra	Roots	Methanol	36	Guanosine, oleic acid, palmitic acid	Bansal, Mujib, Mamgain, et al. (2024)
	Root-derived callus		44	Methyl galactoside, oleic anhydride, glycidyl oleate	
Tylophora indica	Leaf	Methanol	50	Alpha-amyrin, lupeol, inositol, n-hexadecanoic acid	Mamgain et al. (2023)
Crinum asiaticum	Bulb-derived callus	Methanol	31	Pyranone, guanosine, methyl elaidate, 24-ethylcholesterol	Bansal et al. (2025)
Maytenus emarginata	Leaf-derived callus	Methanol	28	Beta-amyrin, squalene	Singh et al. (2025)
	Stem-derived callus		43	Stigmast-5-en-3-ol, squalene, 2-bromododecane	

No permission required.

analysis, field-raised and PTC-raised Sansevieria tissue extracts showed significant differences in the profiles of several volatile phytochemicals (Kaur et al., 2023). Various metabolites were found to be upregulated in *in vitro* raised plantlets such as citronellol, erucic acid, milbemycin b, hexadecenoic acid, and geranyl isovalerate. The authors attributed this upleveling of certain compounds to both the exogenous supply of growth regulators and modifications in the physical conditions.

13.4.2 *Gaillardia pulchella*

Gaillardia pulchella Foug. belongs to the Asteraceae family and is an annual ornamental plant known for its attractive flowers (Divyashree et al., 2021). Owing to the extensive applicability of the GC–MS technology, this plant has now attained significant prominence in the pharmaceutical sector, resulting in the identification of novel phytocompounds. Bansal, Mujib, Bansal, et al. (2024), optimized an *in vitro* protocol for indirect organogenesis from leaf explants of *G. pulchella* and compared the phytochemical profile of both *in vivo* and *in vitro* regenerated leaves and callus samples. The callus extract showed the presence of nearly 34 compounds including isoamyl acetate, palmitic acid, caprinic acid, and squalene. Comparing the *in vitro*-derived leaf tissues to the *in vivo*-grown leaves, the amount of phytocompounds produced was almost identical. These included squalene, chrysanin, neophytadiene, and 1,4-cyclohexanediol. Neophytadiene is a diterpene that is commonly used to treat rheumatism, headaches, and skin conditions. It also has antiinflammatory, antibacterial, and antipyretic properties (Gonzalez-Rivera et al., 2023). Squalene is a triterpenoid that has antiinflammatory, antineoplastic, antiatherosclerotic, and antioxidant qualities (Lou-Bonafonte et al., 2018).

13.4.3 *Pluchea lanceolata*

Pluchea lanceolata is a plant belonging to the Asteraceae family that is endangered but has significant medicinal value. Conventional methods of propagation are ineffectual since the seeds are not viable, and root sprout development is slow and yields fewer plants (Singh et al., 2021). Different *in vitro* culture techniques, namely, somatic embryogenesis, direct shoot regeneration, indirect organogenesis, etc., have been applied so far to multiply this plant rapidly (Kher & Nataraj, 2019; Mamgain, Mujib, Bansal, et al., 2023; Mamgain, Mujib, Syeed, et al., 2023). Different phytochemicals were found when the GC–MS technique was applied to plants grown both *in vitro* and *in vivo*. Several active compounds were synthesized in significant amounts (>1% peak area), including isopropyl myristate, neophytadiene, phytol acetate, n-hexadecanoic acid, squalene, vitamin E, stigmasta-5,22-dien-3-ol; alpha-amyrin; lupeol; γ-sitosterol, etc. Whereas the production of calusterone, eicosanoic acid, celidoniol, γ-tocopherol, ergost-5-en-3-ol, phytyldecanoate, and other compounds was minimal (1% peak area). Among these, alpha-amyrin and lupeol hold great medicinal importance in the pharmaceutical sector due to their anti-inflammatory and anticarcinogenic activities.

13.4.4 *Hibiscus sabdariffa*

A member of the Malvaceae Juss. family, *Hibiscus sabdariffa* L. is a naturalized medicinal species that is widely referred to as a "vinagreira" in Brazil. Sobrinho et al. (2022) established callus cultures by using hypocotyls as explants, and the bioactive compounds were then identified in the wild leaves and callus of *H. sabdariffa*. The GC–MS analysis detected 38 distinct bioactive metabolites. With a relative abundance of 26.86% and 16.68% for *H. sabdariffa* leaves and callus, respectively, protocatechuic acid is the most abundant of the compounds found. Other compounds identified in both the samples were benzoic acid, succinic acid, mallic acid, vanillic acid, etc. This ensures that through callus culture, bioactive substances can be produced consistently in a contaminated-free environment.

13.4.5 *Rubia cordifolia*

Rubia cordifolia L., a perennial flowering plant species in the Rubiaceae family, is commonly referred to as Indian Madder or Manjistha. *R. cordifolia* roots have been used in herbal formulations for centuries, including *Chadanasava* for urinogenital disorders, Jatyaadi Ghrita for chronic and septic ulcers, *Phala Sarpi Ghrita* and *Phal Kalyaan Ghrita* for amenorrhea, and *Manjisthadi Taila* (oil) for headaches, ringworm, and fungal infections (Kumari et al., 2021). *R. cordifolia* contains a variety of biologically active compounds, including anthraquinones and glycosides, naphthoquinones and glycosides, terpenes, bicyclic hexapeptides, iridoids, carboxylic acids (malic, citric, quinic, rosmarinic acids), and saccharides (xylose, ribose, fructose, glucose, sucrose, primverose) (Wen et al., 2022). Sharma et al. (2021) developed an *in vitro* propagation strategy for *R. cordifolia* cultivated in Leh, utilizing two types of explants: leaf and nodal explants, succeeded by GC–MS analysis of the methanolic extracts of *in vitro* grown leaves and roots to ascertain the phytochemical profiles. Both extracts contained 39 phytocompounds, 8 of which are common bioactive compounds with diverse biological functions. These compounds exhibit various biological activities, including antiinflammatory (octyl 2-methylpropanoate, 9-octadecenamide), antimicrobial (neophytadiene, 1, 2-benzenedicarboxylic acid, dioctyl ester; gamma-sitosterol), antioxidant (heptadecanoic acid, methyl ester; 6-octadecenoic acid, methyl ester), antitumor (squalene), and antidiabetic (stigmast-5-en-3-ol).

13.4.6 *Digitalis purpurea*

Digitalis purpurea L., a member of the Plantaginaceae family and a significant plant species in India's Nilgiris, Kashmir, and Darjeeling areas, has well-known ornamental and medicinal uses. It is now a well-known source of distinctive bioactive substances, such as cardenolides (digoxin, digitoxin, gitaloxin, gitoxin, strospeside); flavonoids (cyaniding and digicitrin); anthraquinone (digiferruginol); and phenylethanoids (cornoside and maxoside) (Amiri et al., 2023). Different micropropagation techniques (somatic embryogenesis, organogenesis, callus cultures) have been used to proliferate *D. purpurea* plants aseptically so far (Al-Oqab et al., 2023; Bansal et al., 2022; Rad et al., 2022). Using the GC–MS approach, a comparative metabolite profile of *in vivo* and *in vitro* (indirect organogenesis) regenerated leaves and leaf-derived callus of *D. purpurea* was developed (Bansal et al., 2024). More than 75 phytoconstituents from different groups, including alkaloids, phytosterols, terpenoids, steroids, phenols, sugars, and others, were found in the methanolic extracts of the samples under study. These included loliolide, stigmasterin, alpha-tocopherol, squalene, palmitic acid, linoleic acid, beta-amyrin, and campesterol, possessing immense therapeutic importance. The authors attributed a variety of factors in cultured conditions, including ecotype, genotype, explant, relative humidity, photoperiod, temperature, and exposure to PGRs, for the varied presence of metabolites in *in vitro* regenerated tissues (in contrast to the mother plant parts).

13.4.7 *Kaempferia rotunda*

The significant herb *Kaempferia rotunda* Linn. (Zingiberaceae) has both medicinal and cosmetic uses. Patients with diabetes (elevated blood sugar) can benefit greatly from the plants. This plant is used to treat tumors, malignant swellings, wounds, blood clots, inflammations, and gastropathy (Pujiyanti et al., 2025). *in vitro* propagation of *K. rotunda* utilizing axillary bud explants from unsprouted rhizomes and phytochemical analysis of micropropagated plants have been accomplished by Sahoo et al. (2023). Notably, nine phytoconstituents in all were found in CP (conventionally propagated) and IVP (in vitro propagated) plants, accounting for 96.81% and 98.00% of the total peak area, respectively, which included benzoic acid, bornyl ester, and zingiberene. The phytochemical content of *in vitro* grown *K. rotunda* was stable and identical to that of the corresponding CP plants.

13.4.8 Calendula offici*nalis*

Calendula officinalis L., also known as pot marigold, is a member of the Asteraceae family with vivid orange flowers. It is a major medicinal and ornamental herbaceous plant, cultivated widely in North America and Europe, and is endemic to the Mediterranean region (Sapkota & Kunwar, 2024). Fatima et al. (2024) analyze the phytochemical content of field-cultivated and lab-grown tissues (callus and leaf samples) using GC–MS technique. The chromatographic findings from this study showed that more than 45 significant bioactive chemicals were present in every sample that was analyzed. The callus methanolic extract displayed the presence of phytoconstituents like pyranone (7.66%), stigmasterol (4.47%), squalene (0.42%), and vitamin E (0.41%). In addition to this, *in vivo* and *in vitro* grown leaf samples revealed certain phytocompounds at elevated levels such as 2-palmitoglycerol (16.06%), neophytadiene (13.94%), phytol isomer (12.41%), 2-monopalmitin (23.59%), and guanosine (12.18%).

13.4.9 Gardenia jasmino*ides*

Cultivated for more than 2000 years in China, *Gardenia jasminoides* Ellis is a well-known medicinal plant belonging to the Rubiaceae family. Gardenia fruit extracts exhibit high antioxidant, antiinflammatory, antidepressant, neuroprotective, antiatherosclerotic, anti-thrombotic, antihypertensive, anticancer, antiangiogenic, and antidiabetic properties (Chen et al., 2023). Krasteva et al. (2022) compare the metabolite profiles of *G. jasminoides* plant leaves and *in vitro* cultures with varying degrees of differentiation (shoots, callus, and cell suspension) using GC–MS. The plant and shoots showed a total of 67 metabolites, while the callus and cell suspension cultures showed 41 and 34 metabolites, respectively. Fatty acids, hydrocarbons, and fatty alcohols were the most prevalent metabolites in the shoots. The callus culture was enriched in phytosterols and saccharides. Sugars and sugar acids

TABLE 13.5 List of therapeutically important compounds identified by gas chromatography–mass spectrometry (GC–MS) technique along with their molecular structure, formula, weight, and biological properties.

Bioactive	Chemical class	Molecular formula	Molecular weight (g/mol)	Biological activities	References
Citronellol	Monoterpenoid	$C_{10}H_{20}O$	156.26	Anticonvulsant, antihyperalgesic, antiinflammatory	Santos et al. (2019)
Neophytadiene	Diterpene	$C_{20}H_{38}$	278.5	Antiinflammatory, antimicrobial	Willie et al. (2021)
Alpha-amyrin	Triterpenoid	$C_{30}H_{50}O$	426.7	Antiinflammatory, analgesic	Nogueira et al. (2019)
Protocatechuic acid	Dihydroxybenzoic acid	$C_7H_6O_4$	154.12	Antioxidant, neuroprotective, antiviral, anticancer	Song et al. (2020)
Gamma-sitosterol	Phytosterol	$C_{29}H_{50}O$	414.7	Chemoprotective, hypocholesterolemic, Immunomodulatory	Saeidnia et al. (2014)
Loliolide	Benzofuran	$C_{11}H_{16}O_3$	196.24	Antiadipogenic, anticancer, antimicrobial	Dvorakova et al. (2024)
Zingiberene	Sesquiterpene	$C_{15}H_{24}$	204.35	Antifungal, antibacterial	Radice et al. (2022)
Guanosine	Nucleoside	$C_{10}H_{13}N_5O_5$	283.24	Antioxidant, neuroprotective	Bettio et al. (2016); Gudkov et al. (2006)
Borneol	Monoterpenoid	$C_{10}H_{18}O$	154.25	Antimicrobial, antiviral, antiinflammatory, antiulcer	Sokolova et al. (2017)
Linoleic acid	Carboxylic acid	$C_{18}H_{32}O_2$	280.4	Antibacterial, antidiabetic, anticancer	Li et al. (2023)
Ajmalicine	Monoterpenoid indole alkaloid	$C_{21}H_{24}N_2O_3$	352.4	Antihypertensive, vasodilator, *antimicrobial*	Kashyap *et al. (2020)*

No permission required.

were abundant in the cell suspension, whereas sugars, fatty acids, and organic acids dominated in the plant leaves.

13.4.10 *Echinacea purpurea*

Echinacea purpurea (L.) Moench, also known as purple coneflower, is a perennial herb from the Asteraceae family. It is native to the central and eastern United States and is cultivated in Europe (Burlou-Nagy et al., 2022). It is one of the most well-known and widely used medicinal herbs for a range of ailments, including snake bites and wound infections, as well as for its antiinflammatory, antioxidant, immune-modulatory, antiviral, and anticancer qualities (Coelho et al., 2020; Zaushintsena et al., 2019). Mokhtarzadeh (2024) micropropagated *E. purpurea* plants by using seeds and leaves as explants and also determined the volatile components of *in vitro* regenerated plantlets. Flowering samples contained major chemicals such as germacrene D (44.25%), β-caryophyllene (6.05%), borneol (5.02%), caryophyllene oxide (4.81%), and myrtenal (3.49%). Additionally, the main constituents of nonflowering samples were myrtenal (2.57%), borneol (2.09%), caryophyllene oxide (2.99%), β-caryophyllen (5.30%), and germacrene D (39%). Table 13.5 reports the details of certain pharmacological active compounds including molecular structure and weight, chemical formula, and biological properties.

13.5 Conclusion

Metabolomics has futuristic applications in a wide range of sectors of research, including agriculture, biotechnology, health, and biomedical sciences. New insights into the genetic and biochemical aspects of cellular function and metabolic network regulation can be obtained by combining the capabilities of existing analytical platforms for the analysis of complex samples with the integration of metabolomics with other "omics" approaches (Hong et al., 2016). Plant metabolomics has significantly improved recently due to the increased interest in using metabolomic technologies for a variety of biological targets. This chapter provides a detailed overview of GC–MS-based metabolomics workflow including sample preparation, chemical derivatization, chromatographic separation, data processing, and bioinformatics, as well as its applications in the plant tissue culture field for metabolite profiling. The discoveries made by combining two completely different disciplines (chromatography and mass spectrometry) in different *in vitro* culture samples have been discussed. The future studies will concentrate on the improvization of the metabolomic platform to enable the precise and efficient identification and quantification of secondary metabolites, the accurate interpretation of generated data, and the quick integration with other omics platforms. Also, the other investigations may involve the thorough examination of the molecular and biochemical mechanisms of metabolic variations in plants using both non-targeted and targeted approaches, to broaden and enhance the understanding of plant metabolism in growth and development under both normal and stressed conditions.

References

Abd El-Kareem, M. S., Rabbih, M. A., Rashad, A. M., & EL-Hefny, M. (2025). Essential oils from fennel plants as valuable chemical products: Gas chromatography–Mass spectrometry, FTIR, quantum mechanical investigation, and antifungal activity. *Biomass Conversion and Biorefinery, 15*(6), 9173–9191.

Al-Oqab, M. A., Zaid, S., Al-Ammouri, Y., & Alawdi, S. H. (2023). Effect of nutrient media enhanced with plant growth regulators on genetic stability in sub-cultures of *Digitalis purpurea* callus. *Journal of Applied Biology & Biotechnology, 11*, 244–252.

Al-Rubaye, A. F., Hameed, I. H., & Kadhim, M. J. (2017). A review: Uses of gas chromatography-mass spectrometry (GC-MS) technique for analysis of bioactive natural compounds of some plants. *International Journal of Toxicological and Pharmacological Research, 9*(1), 81–85.

Álvarez-Sánchez, B., Priego-Capote, F., & de Castro, M. L. (2010). Metabolomics analysis II. Preparation of biological samples prior to detection. *Trends in Analytical Chemistry, 29*(2), 120–127.

Amiri, F., Moghadam, A., Tahmasebi, A., & Niazi, A. (2023). Identification of key genes involved in secondary metabolite biosynthesis in *Digitalis purpurea*. *PLoS One, 18*(3), e0277293.

Arafa, N. M., Girgis, N., Ibrahem, M., Mohamed, S., & El-Bahr, M. K. (2022). Phytochemical profiling by GC-MS analysis and antimicrobial activity potential of *in vitro* derived shoot cultures of some Egyptian herbal medicinal plants. *Egyptian Journal of Chemistry, 65*(9), 155–169.

Bansal, M., Mujib, A., Bansal, Y., Dewir, Y. H., & Mendler-Drienyovszki, N. (2024). An efficient in vitro shoot Organogenesis and comparative GC-MS metabolite profiling of *Gaillardia pulchella* Foug. *Horticulturae, 10*(7), 728.

Bansal, Y., Mujib, A., Mamgain, J., Kumar, S., Dewir, Y. H., & Magyar-Tábori, K. (2024). Synthesis and accumulation of phytocompounds in field-, tissue-culture grown (Stress) root tissues and simultaneous defense response activity in *Glycyrrhiza glabra* L. *Sustainability, 16*(4), 1613.

Bansal, Y., Mujib, A., Mamgain, J., Syeed, R., Mohsin, M., Nafees, A., Dewir, Y. H., & Mendler-Drienyovszki, N. (2024). Integrated GC-MS and UPLC-ESI-QTOF-MS based untargeted metabolomics analysis of in vitro raised tissues of *Digitalis purpurea* L. *Frontiers in Plant Science, 15*, 1433634.

Bansal, Y., Mujib, A., Mamgain, J., Dewir, Y. H., & Rihan, H. Z. (2023). Phytochemical composition and detection of novel bioactives in anther callus of *Catharanthus roseus* L. *Plants, 12*(11), 2186.

Bansal, Y., Mujib, A., Mamgain, J., Mohsin, M., Nafees, A., Dewir, Y. H., & Magyar-Tabori, K. (2025). Development of a callus culture for increased yield of secondary metabolite production in spider lily (*Crinum asiaticum* L.). *HortScience: A Publication of the American Society for Horticultural Science, 60*(8), 1268–1275.

Bansal, Y., Mujib, A., Siddiqui, Z. H., Mamgain, J., Syeed, R., & Ejaz, B. (2022). Ploidy status, nuclear DNA content and start codon targeted (SCoT) genetic homogeneity assessment in *Digitalis purpurea* L., regenerated in vitro. *Genes, 13*(12), 2335.

Beale, D. J., Pinu, F. R., Kouremenos, K. A., Poojary, M. M., Narayana, V. K., Boughton, B. A., et al. (2018). Review of recent developments in GC–MS approaches to metabolomics-based research. *Metabolomics: Official Journal of the Metabolomic Society, 14*(11), 152. https://doi.org/10.1007/s11306-018-1449-2.

Bettio, L. E., Gil-Mohapel, J., & Rodrigues, A. L. (2016). Guanosine and its role in neuropathologies. *Purinergic Signalling, 12*(3), 411–426.

Burlou-Nagy, C., Bănică, F., Jurca, T., Vicaş, L. G., Marian, E., Muresan, M. E., Bacskay, I., Kiss, R., Feher, P., & Pallag, A. (2022). *Echinacea purpurea* (L.) Moench: Biological and pharmacological properties. A review. *Plants, 11*(9), 1244.

Chen, J., Tang, W., Li, C., Kuang, D., Xu, X., Gong, Y., Liu, F., & Gao, S. (2023). Multi-omics analysis reveals the molecular basis of flavonoid accumulation in fructus of Gardenia (Gardenia jasminoides Ellis). *BMC Genomics, 24*(1), 588.

Coelho, J., Barros, L., Dias, M. I., Finimundy, T. C., Amaral, J. S., Alves, M. J., Calhelha, R. C., Santos, P. F., & Ferreira, I. C. (2020). *Echinacea purpurea* (L.) Moench: Chemical characterization and bioactivity of its extracts and fractions. *Pharmaceuticals, 13*(6), 125.

Commisso, M., Strazzer, P., Toffali, K., Stocchero, M., & Guzzo, F. (2013). Untargeted metabolomics: An emerging approach to determine the composition of herbal products. *Computational and Structural Biotechnology Journal, 4*, e201301007. https://doi.org/10.5936/csbj.201301007.

Divyashree, M. S., Munikrishnappa, P. M., Seetharamu, G. K., Kumar, S. A., Ramegowda, G. K., & Basavaraj, G. (2021). Standardization of spacing and NPK levels for growth, yield and quality of gaillardia (*Gaillardia pulchella* Foug.) under eastern dry zone condition. *Environment and Ecology, 39*(1A), 281–286.

Dulara, B. K., Godara, P., & Barwer, N. (2019). In-vivo and In-vitro phytochemical GC-MS analysis of volatile constituents of *Andrographis paniculata* (Burm. f.) Nees. *The Pharma Innovation Journal, 8*, 255–261.

Dvorakova, M., Soudek, P., Pavicic, A., & Langhansova, L. (2024). The traditional utilization, biological activity and chemical composition of edible fern species. *Journal of Ethnopharmacology, 324*, 117818.

El-Naggar, H. M., Shehata, A. M., & Morsi, M. A. (2023). Micropropagation and GC–MS analysis of bioactive compounds in bulbs and callus of white squill. *In Vitro Cellular & Developmental Biology-Plant, 59*(1), 154–166.

Faisal, M., Qahtan, A. A., & Alatar, A. A. (2023). Thidiazuron induced in vitro plant regeneration, phenolic contents, antioxidant potential, GC-MS profiles and nuclear genome stability of *Plectranthus amboinicus* (Lour.) Spreng. *Horticulturae, 9*(2), 277.

Fang, C., Fernie, A. R., & Luo, J. (2019). Exploring the diversity of plant metabolism. *Trends in Plant Science 2018, 24*, 83–98. https://doi.org/10.1016/j.tplants.2018.09.006.

Fatima, T., Mujib, A., Bansal, Y., Dewir, Y. H., & Mendler-Drienyovszki, N. (2024). Indirect organogenesis of *Calendula officinalis* L. and comparative phytochemical studies of field-grown and in vitro-regenerated tissues. *Agronomy, 14*(8), 1743. https://doi.org/10.3390/agronomy14081743.

Feizi, N., Hashemi-Nasab, F. S., Golpelichi, F., Saburouh, N., & Parastar, H. (2021). Recent trends in application of chemometric methods for GC-MS and GC×GC-MS-based metabolomic studies. *TrAC Trends in Analytical Chemistry, 138*, 116239. https://doi.org/10.1016/j.trac.2021.116239.

Fernie, A. R., & Tohge, T. (2017). The genetics of plant metabolism. *Annual Review of Genetics, 51*, 287–310. https://doi.org/10.1146/annurev-genet-120116-024640.

Fiehn, O. (2001). Combining genomics, metabolome analysis, and biochemical modelling to understand metabolic networks. *Comparative and Functional Genomics, 2*, 155–168. https://doi.org/10.1002/cfg.82.

Ghatak, A., Chaturvedi, P., & Weckwerth, W. (2018). Metabolomics in plant stress physiology. In R. Varshney, M. Pandey, & A. Chitikineni (Eds.). *Plant genetics and molecular biology. Advances in biochemical engineering/biotechnology, vol 164* (pp. 187–236). Cham: Springer. https://doi.org/10.1007/10_2017_55.

Godara, P., Dulara, B. K., Barwer, N., & Chaudhary, N. S. (2019). Comparative GC-MS analysis of bioactive phytochemicals from different plant parts and callus of *Leptadenia reticulata* wight and arn. *Pharmacognosy Journal, 11*(1), 129–140. https://doi.org/10.5530/pj.2019.1.22.

Gonzalez-Rivera, M. L., Barragan-Galvez, J. C., Gasca-Martínez, D., Hidalgo-Figueroa, S., Isiordia-Espinoza, M., & Alonso-Castro, A. J. (2023). In vivo neuropharmacological effects of neophytadiene. *Molecules (Basel, Switzerland), 28*(8), 3457.

Gudkov, S. V., Shtarkman, I. N., Smirnova, V. S., Chernikov, A. V., & Bruskov, V. I. (2006). Guanosine and inosine display antioxidant activity, protect DNA in vitro from oxidative damage induced by reactive oxygen species, and serve as radioprotectors in mice. *Radiation Research, 165*(5), 538–545.

Hamdan, D. I., Fayed, M. A., & Adel, R. (2021). Echinops taeckholmiana Amin: Optimization of a tissue culture protocol, biological evaluation, and chemical profiling using GC and LC-MS. *ACS Omega, 6*(20), 13105–13115.

Hamooh, B. T., Sattar, F. A., Wellman, G., & Mousa, M. A. (2021). Metabolomic and biochemical analysis of two potato (*Solanum tuberosum* L.) cultivars exposed to in vitro osmotic and salt stresses. *Plants, 10*(1), 98.

Hao, Y., Zhang, Z., Luo, E., Yang, J., & Wang, S. (2025). Plant metabolomics: Applications and challenges in the era of multi-omics big data. *Abiotech, 6*(1), 116–132.

Hill, C. B., & Roessner, U. (2013). Metabolic profiling of plants by GC–MS. In W. Weckwerth, & G. Kahl (Eds.). *The handbook of plant metabolomics* (pp. 1–23). Wiley. https://doi.org/10.1002/9783527669882.ch1.

Hong, J., Yang, L., Zhang, D., Shi, J. (2016). Plant metabolomics: an indispensable system biology tool for plant science. International journal of molecular sciences, 17(6):767.

Hussain, S. A., Ahmad, N., Anis, M., & Alatar, A. A. (2019). Influence of meta-topolin on in vitro organogenesis in *Tecoma stans* L., assessment of genetic fidelity and phytochemical profiling of wild and regenerated plants. *Plant Cell, Tissue and Organ Culture (PCTOC), 138*(2), 339–351.

Johnson, S. R., & Lange, B. M. (2015). Open-access metabolomics databases for natural product research: Present capabilities and future potential. *Frontiers in Bioengineering and Biotechnology, 3*. https://doi.org/10.3389/fbioe.2015.00022.

Jorge, T. F., Rodrigues, J. A., Caldana, C., Schmidt, R., Van Dongen, J. T., Thomas-Oates, J., et al. (2015). Mass spectrometry-based plant metabolomics: Metabolite responses to abiotic stress. *Mass Spectrometry Reviews, 35*, 620–649. https://doi.org/10.1002/mas.21449.

Kar, P., Chakraborty, A. K., Bhattacharya, M., Mishra, T., & Sen, A. (2019). Micropropagation, genetic fidelity assessment and phytochemical studies of *Clerodendrum thomsoniae* Balf. f. with special reference to its anti-stress properties. *Research in Plant Biology, 9*, 9–15. https://doi.org/10.25081/ripb.2019.v9.3779.

Karaman, I. (2017). Preprocessing and pretreatment of metabolomics data for statistical analysis. In A. Sussulini (Ed.). *Metabolomics: From fundamentals to clinical applications. Advances in experimental medicine and biology, vol 965* (pp. 145–161). Cham: Springer. https://doi.org/10.1007/978-3-319-47656-8_6.

Karaman, I., Pinto, R. C., & Graça, G. (2018). Chapter Eight - Metabolomics data preprocessing: From raw data to features for statistical analysis. In *Comprehensive Analytical Chemistry, Volume 82*. Elsevier, 197–225. https://doi.org/10.1016/bs.coac.2018.08.003.

Kashyap, P., Kalaiselvan, V., Kumar, R., & Kumar, S. (2020). Ajmalicine and reserpine: Indole alkaloids as multi-target directed ligands towards factors implicated in Alzheimer's disease. *Molecules (Basel, Switzerland), 25*(7), 1609.

Kaur, J., Dhar, S. K., Chauhan, A., Yadav, S., Mudgal, G., Lyudmila, A., Atuchin, V., & Abdi, G. (2023). GC-MS validated phytochemical up-leveling with in vitro-raised *Sansevieria trifasciata* [Prain]: The Mother in Law's tongue gets more antibacterial. *Current Plant Biology, 35*, 100308.

Kaur, J., & Mudgal, G. (2021). An efficient and quick protocol for in vitro multiplication of snake plant, *Sansevieria trifasciata* var. Laurentii [Prain]. *Plant Cell, Tissue and Organ Culture (PCTOC), 147*(2), 405–411.

Khalifea, H., & Ali, N. (2025). Exploring the principles of GC-MS: Techniques and applications. *Physical Sciences, Life Science and Engineering, 2*(3), 10 -10.

Khan, A., Shah, A. H., & Ali, N. (2021). In-vitro propagation and phytochemical profiling of a highly medicinal and endemic plant species of the Himalayan region (*Saussurea costus). Scientific Reports, 11*(1), 23575.

Kher, M. M., & Nataraj, M. (2019). Direct somatic embryogenesis and shoot regeneration from leaves and internodes of *Pluchea lanceolata* (DC.) CB Clarke. *In Vitro Cellular & Developmental Biology-Plant, 55*(6), 720–724.

Krasteva, G., Berkov, S., Pavlov, A., & Georgiev, V. (2022). Metabolite profiling of gardenia jasminoides ellis in vitro cultures with different levels of differentiation. *Molecules (Basel, Switzerland), 27*(24), 8906.

Kumar, N., Bharat, J. S., Kumar, A., & Jaitak, V. (2024). Recent advancement in mass based plant metabolomics: Techniques, tools, and analytical approaches. *Phytochemistry Reviews, 24*, 4013–4050. https://doi.org/10.1007/s11101-024-10028-y.

Kumar, R., Bohra, A., Pandey, A. K., Pandey, M. K., & Kumar, A. (2017). Metabolomics for plant improvement: Status and prospects. *Frontiers in Plant Science, 8*. https://doi.org/10.3389/fpls.2017.01302.

Kumari, I., Kaurav, H., & Choudhary, G. (2021). Rubia cordifolia (Manjishtha): A review based upon its ayurvedic and medicinal uses. *Himalayan J Health Sci, 6*(2), 17–28.

Li, C., Chu, S., Tan, S., Yin, X., Jiang, Y., Dai, X., Gong, X., Fang, X., & Tian, D. (2021). Towards higher sensitivity of mass spectrometry: A perspective from the mass analyzers. *Frontiers in chemistry, 9*, 813359.

Li, S., Xu, L., Qing, J., Wu, X., Li, H., Chen, H., & Liu, X. (2023). Multiple biological activities and biosynthesis mechanisms of specific conjugated linoleic acid isomers and analytical methods for prospective application. *Food Chemistry, 409*, 135257.

Lou-Bonafonte, J. M., Martínez-Beamonte, R., Sanclemente, T., Surra, J. C., Herrera-Marcos, L. V., Sanchez-Marco, J., Arnal, C., & Osada, J. (2018). Current insights into the biological action of squalene. *Molecular Nutrition & Food Research, 62*(15), 1800136.

Mamgain, J., Mujib, A., Bansal, Y., Gulzar, B., Zafar, N., Syeed, R., Alsughayyir, A., & Dewir, Y. H. (2023). Elicitation induced α-amyrin synthesis in tylophora indica in vitro cultures and comparative phytochemical analyses of in vivo and micropropagated plants. *Plants, 13*(1), 122.

Mamgain, J., Mujib, A., Syeed, R., Ejaz, B., Malik, M. Q., & Bansal, Y. (2023). Genome size and gas chromatography-mass spectrometry (GC–MS) analysis of field-grown and in vitro regenerated *Pluchea lanceolata* plants. *Journal of Applied Genetics, 64*(1), 1–21.

Marchev, A. S., Yordanova, Z. P., & Georgiev, M. I. (2020). Green (cell) factories for advanced production of plant secondary metabolites. *Critical Reviews in Biotechnology, 40*, 443–458. https://doi.org/10.1080/07388551.2020.1731414.

Mokhtarzadeh, S. (2024). Phytochemical profiling of volatile components of *Echinacea purpurea* (L.) Moench multiplied by micropropagation. *Latin American Applied Research, 54*(2), 189–194.

Moldoveanu, S. C., & David, V. (2018). Derivatization methods in GC and GC/MS. In *Gas chromatography-derivatization, sample preparation, application*. IntechOpen.

Monfort, L. E., Bertolucci, S. K., Lima, A. F., de Carvalho, A. A., Mohammed, A., Blank, A. F., & Pinto, J. E. (2018). Effects of plant growth regulators, different culture media and strength MS on production of volatile fraction composition in shoot cultures of *Ocimum basilicum*. *Industrial Crops and Products, 116*, 231–239.

Niazian, M., & Sabbatini, P. (2021). Traditional in vitro strategies for sustainable production of bioactive compounds and manipulation of metabolomic profile in medicinal, aromatic and ornamental plants. *Planta, 254*. https://doi.org/10.1007/s00425-021-03771-5.

Nogueira, A. O., Oliveira, Y. I., Adjafre, B. L., de Moraes, M. E., & Aragao, G. F. (2019). Pharmacological effects of the isomeric mixture of alpha and beta amyrin from Protium heptaphyllum: A literature review. *Fundamental & Clinical Pharmacology, 33*(1), 4–12.

Oliver, S. (1998). Systematic functional analysis of the yeast genome. *Trends in Biotechnology, 16*, 373–378. https://doi.org/10.1016/s0167-7799(98)01214-1.

Orata, F. (2012). Derivatization reactions and reagents for gas chromatography analysis. In M. A. Mohd (Ed.). *Advanced gas chromatography - Progress in agricultural, biomedical and industrial applications*InTechhttps://doi.org/10.5772/33098.

Patel, M. K., Mishra, A., & Jha, B. (2016). Untargeted metabolomics of halophytes. In S.-K. Kim (Ed.). *Marine OMICS: Principles and Applications* (pp. 307–325). Boca Raton: CRC Press. https://doi.org/10.1201/9781315372303-18.

Patel, M. K., Pandey, S., Kumar, M., Haque, M. I., Pal, S., & Yadav, N. S. (2021). Plants metabolome study: Emerging tools and techniques. *Plants, 10*(11), 2409.

Pierce, K. M., Kehimkar, B., Marney, L. C., Hoggard, J. C., & Synovec, R. E. (2012). Review of chemometric analysis techniques for comprehensive two dimensional separations data. *Journal of Chromatography. A, 1255*, 3–11.

Pramod, S. K., Navnath, K. A., Pramod, S. M., Babanrao, H. S., Babanrao, S. K., & Vichardhara, P. (2021). A review on gas chromatography-mass spectrometry (GC-MS). *World J Pharm Res, 10*(3), 741–763.

Pujiyanti, A. S., Kusuma, R., Ilmi, A. N., & Hasibuan, F. R. (2025). Ethnobotanical, phytochemical, and pharmacologies aspect of *Kaempferia rotunda* L. as a local herbal plant with global potential: A review. *Biocaster: Jurnal Kajian Biologi, 5*(3), 227–239.

Qahtan, A. A., Faisal, M., Alatar, A. A., & Abdel-Salam, E. M. (2022). Callus-mediated high-frequency plant regeneration, phytochemical profiling, antioxidant activity and genetic stability in *Ruta chalepensis* L. *Plants, 11*(12), 1614.

Rad, M. M., Abdossi, V., Moradi, P., Rakhshandehroo, F., & Mehrafarin, A. (2022). Phytochemical changes of *Digitalis purpurea* L. in response to polyamines and methyl jasmonate application in callus culture. *Journal of Plant Biochemistry and Biotechnology, 31*(2), 310–319.

Radice, M., Maddela, N. R., & Scalvenzi, L. (2022). Biological activities of zingiber officinale Roscoe essential oil against *Fusarium* spp.: A minireview of a promising tool for biocontrol. *Agronomy, 12*(5), 1168.

Ralte, L., Khiangte, L., Thangjam, N. M., Kumar, A., & Singh, Y. T. (2022). GC–MS and molecular docking analyses of phytochemicals from the underutilized plant, *Parkia timoriana* revealed candidate anti-cancerous and anti-inflammatory agents. *Scientific Reports, 12*(1), 3395.

Saeidnia, S., Manayi, A., Gohari, A. R., & Abdollahi, M. (2014). The story of beta-sitosterol-a review. *European Journal of Medicinal Plants, 4*(5), 590–609.

Sahoo, S., Lenka, J., Kar, B., & Nayak, S. (2023). Clonal fidelity and phytochemical analysis of in vitro propagated *Kaempferia rotunda* Linn.—An endangered medicinal plant. *In Vitro Cellular & Developmental Biology-Plant, 59*(3), 329–339.

Salem, M. A., Perez de Souza, L., Serag, A., Fernie, A. R., Farag, M. A., Ezzat, S. M., & Alseekh, S. (2020). Metabolomics in the context of plant natural products research: From sample preparation to metabolite analysis. *Metabolites, 10*(1), 37.

Santos, P. L., Matos, J. P., Picot, L., Almeida, J. R., Quintans, J. S., & Quintans-Junior, L. J. (2019). Citronellol, a monoterpene alcohol with promising pharmacological activities-A systematic review. *Food and Chemical Toxicology, 123*, 459–469.

Sapkota, B., & Kunwar, P. (2024). A review on traditional uses, phytochemistry and pharmacological activities of *Calendula officinalis* Linn. *Natural Product Communications, 19*(6), 1934578X241259021.

Saranraj, P., Sivasakthi, S., & Deepa, M. S. (2016). Phytochemistry of pharmacologically important medicinal plants – A Review. *International Journal of Current Research in Chemistry and Pharmaceutical Sciences, 3*, 56–66. https://doi.org/10.22192/ijcrcps.2016.03.11.009.

Senthil, K., Thirugnanasambantham, P., Oh, T. J., Kim, S. H., & Choi, H. K. (2015). Free radical scavenging activity and comparative metabolic profiling of in vitro cultured and field grown *Withania somnifera* roots. *PLoS One, 10*(4), e0123360.

Sharma, A., Shukla, S., Quasim, A., Patel, M. K., Kumar, R., Chaurasia, O. P., & Saxena, S. (2021). In vitro propagation, callus culture and phytochemical profiling of Manjishtha-A invaluable medicinal species of Leh-Ladakh. *Annals of Phytomedicine, 10*(1), 230–241.

Singh, M., Yerramilli, V., Singh, J., & Nagar, L. (2025). Genetic diversity and metabolic profiling of *Maytenus emarginata* Willd. plant parts and callus using RAPD, SSR markers and GC–MS. *Vegetos, 38*(2), 552–563.

Singh, S., Mishra, R., & Agnihotri, R. (2021). In-vitro propagation of *Pluchea lanceolata* (DC) CB Clarke a potent antiarthritic medicinal herb through axillary bud. *Trends in Phytochemical Research, 5*(1), 24–30.

Sobrinho, A. C., Corpes, R. S., dos Santos, K. I., Barra, I. M., Miyagawa, H. K., & Santos, A. S. (2022). Untargeted GC-MS metabolomics applied to wild leaves and callus produced by plant tissue culture of *Hibiscus sabdariffa* L. *Arabian Journal of Chemistry, 15*(9), 104103. https://doi.org/10.1016/j.arabjc.2022.104103.

Sokolova, A. S., Yarovaya, O. I., Shtro, A. A., Borisova, M. S., Morozova, E. A., Tolstikova, T. G., Zarubaev, V. V., & Salakhutdinov, N. F. (2017). Synthesis and biological activity of heterocyclic borneol derivatives. *Chemistry of Heterocyclic Compounds, 53*(3), 371–377.

Song, J., He, Y., Luo, C., Feng, B., Ran, F., Xu, H., Ci, Z., Xu, R., Han, L., & Zhang, D. (2020). New progress in the pharmacology of protocatechuic acid: A compound ingested in daily foods and herbs frequently and heavily. *Pharmacological research, 161*, 105109.

Talan, A., Mujib, A., Ejaz, B., Bansal, Y., Dewir, Y. H., & Magyar-Tábori, K. (2023). In vitro propagation and phytochemical composition of centratherum punctatum cass—A medicinal plant. *Horticulturae, 9*(11), 1189.

Tian, T. F., Wang, S. Y., Kuo, T. C., Tan, C. E., Chen, G. Y., Kuo, C. H., Chen, C. H., Chan, C. C., Lin, O. A., & Tseng, Y. J. (2016). Web server for peak detection, baseline correction, and alignment in two-dimensional gas chromatography mass spectrometry-based metabolomics data. *Analytical Chemistry, 88*(21), 10395–10403.

Tsugawa, H., Tsujimoto, Y., Arita, M., Bamba, T., & Fukusaki, E. (2011). GC/MS based metabolomics: Development of a data mining system for metabolite identification by using soft independent modeling of class analogy (SIMCA). *BMC Bioinformatics, 12*(1), 131.

Vinaixa, M., Schymanski, E. L., Neumann, S., Navarro, M., Salek, R. M., & Yanes, O. (2016). Mass spectral databases for LC/MS-and GC/MS-based metabolomics: State of the field and future prospects. *TrAC Trends in Analytical Chemistry, 78*, 23–35.

Wahyuni, D. K., Ansori, A. N., & Vidiyanti, F. (2017). GC-MS analysis of phytocomponents in methanolic extracts of leaf-derived callus of *Justicia gendarussa* Burm. f. *Bioscience Research, 14*(3), 668–677.

Wen, M., Chen, Q., Chen, W., Yang, J., Zhou, X., Zhang, C., Wu, A., Lai, J., Chen, J., Mei, Q., & Yang, S. (2022). A comprehensive review of *Rubia cordifolia* L.: Traditional uses, phytochemistry, pharmacological activities, and clinical applications. *Frontiers in pharmacology, 13*, 965390.

Willie, P., Uyoh, E. A., & Aikpokpodion, P. O. (2021). Gas chromatography-mass spectrometry (GC-MS) assay of bio-active compounds and phytochemical analyses in three species of apocynaceae. *Pharmacognosy Journal, 13*(2), 383–392. https://doi.org/10.5530/pj.2021.13.49.

Wolfender, J.-L., Marti, G., Thomas, A., & Bertrand, S. (2014). Current approaches and challenges for the metabolite profiling of complex natural extracts. *Journal of Chromatography. A, 1382*, 136–164. https://doi.org/10.1016/j.chroma.2014.10.091.

Xia, J., Psychogios, N., Young, N., & Wishart, D. S. (2009). MetaboAnalyst: A web server for metabolomic data analysis and interpretation. *Nucleic Acids Research, 37*, W652–W660.

Yan, C. (2025). A review on spectral data preprocessing techniques for machine learning and quantitative analysis. *iScience, 28*(7), 112759. https://doi.org/10.1016/j.isci.2025.112759.

Yao, L., Sheflin, A. M., Broeckling, C. D., & Prenni, J. E. (2019). Data processing for GC-MS-and LC-MS-based untargeted metabolomics. In *High-throughput metabolomics: Methods and protocols*. New York, NY: Springer New York, 287–299.

Zaushintsena, A. V., Milentyeva, I., Babich, O., Noskova, S. Y., Kiseleva, T. F., Popova, D. G., Bakin, I. A., & Lukin, A. (2019). Quantitative and qualitative profile of biologically active substances extracted from purple echinacea (*Echinacea Purpurea* L.) growing in the Kemerovo region: Functional foods application. *Foods and Raw materials, 7*(1), 84–92.

Zeki, Ö. C., Eylem, C. C., Reçber, T., Kır, S., & Nemutlu, E. (2020). Integration of GC–MS and LC–MS for untargeted metabolomics profiling. *Journal of Pharmaceutical and Biomedical Analysis, 190*, 113509.

CHAPTER

14

Nanoparticle-mediated approaches for enhancing cryopreservation of medicinal plant germplasm: innovations and future directions

Isaac Baidoo and Paromita Sarbadhikary

Laser Research Centre, Faculty of Health Sciences, University of Johannesburg, Johannesburg, Gauteng, South Africa

14.1 Introduction

The preservation of plant genetic resources has long been recognized as a cornerstone for global biodiversity and agricultural sustainability. Plant genetic resources, also known as phylogenetic materials, refer to the hereditary material from plants that have recognized economic value or potential importance for science, culture, food, and agriculture (Mba et al., 2021). They are essential for crop development, environmental restoration, and the preservation of ecosystem functions (Milla, 2020). However, the past few decades have witnessed a concerning trend of genetic erosion, characterized by the reduction of genetic diversity and the replacement of native plant varieties with more uniform crops. This uniformity, while simplifying the agrotechnical processes, often leads to the abandonment of less profitable species and varieties.

Among the several plant groups under threat from genetic erosion, medicinal plants warrant particular attention due to their vital role in healthcare, cultural traditions, and modern drug discovery. In several developing countries, especially across Africa and Asia, over 85% of the population relies on herbal remedies as a primary healthcare source (Mekonnen et al., 2022). These plants contain a wealth of bioactive compounds that continue to serve as precursors for pharmacologically active agents used in contemporary medicine. However, the escalating demand for medicinal plant products, both locally and globally, has led to unsustainable harvesting practices, especially from wild populations. Many species are now subjected to indiscriminate collection without regard to regeneration cycles or

Medicinal Plant Biotechnology for Ameliorated Secondary Metabolite Production
DOI: https://doi.org/10.1016/B978-0-443-43944-5.00017-7

ecological balance, resulting in severe population declines (Woldearegay & Teso, 2022). In South Africa, for instance, the Natal Lily (*Clivia miniata*), a widely used medicinal plant, has experienced significant population declines due to overharvesting, habitat loss, and climate change (Williams et al., 2013). Similarly, in Ethiopia, several medicinal plant species face local extinction because of overexploitation for commercial purposes (Woldearegay & Teso, 2022). The continued loss of these plants not only erodes biodiversity but also threatens the resilience of traditional knowledge systems. There is therefore an urgent need to integrate both in situ and ex situ conservation strategies, such as cultivation, seed banking, and community-based stewardship programs, to safeguard medicinal plant diversity and ensure their long-term availability (Kiflie et al., 2024).

The consequences of genetic erosion are far-reaching. It diminishes the resilience of plant species to adapt to environmental changes, exacerbates vulnerability to emerging pests and diseases, and creates formidable challenges for future breeding programs. The importance of conserving biological diversity has been globally acknowledged, culminating in international treaties such as the Convention on Biological Diversity, ratified by 196 countries (Chand et al., 2025). These agreements emphasize the sustainable use of biodiversity and equitable sharing of benefits derived from genetic resources (U. CBD, 1982). Immiseration challenges and the accelerating impacts of climate change have added another layer of urgency. Climate change is altering the environmental conditions, frequently compelling species to relocate or risking their extinction when their required habitats disappear or become unreachable (Lambers, 2015). A growing body claims that climate change contributes to 54% of biodiversity loss, a statistic that underscores the thesis. Projections suggest that, if current trends persist, one in six species could face extinction due to habitat disruptions caused by global warming. Regions such as South America and Oceania are vulnerable, with temperature increases destabilizing ecosystems, melting polar ice, and depleting natural resources (Lambers, 2015; Prevéy et al., 2020; Ruiz-Sáenz et al., 2019).

To mitigate the risk of species extinction and genetic variability, germplasm banks have emerged as critical tools (Ruta et al., 2020). These facilities store diverse plant materials, including seeds, tissues, and embryos, ensuring the retention of genetic variation across species. In situ conservation, which involves preserving species within their natural habitats, and ex situ conservation, which includes methods like seed banks, field collections, and *in vitro* cultures, are complementary strategies used in plant genetic resource conservation (Fitzgerald et al., 2019; Vincent et al., 2019).

Cryopreservation is a dependable ex situ method used for the long-term storage of plant germplasm. It preserves biological samples at extremely low temperatures (ranging from –135°C to –196°C) in liquid nitrogen, which stops cellular activity and helps retain genetic integrity over extended periods (Acosta et al., 2020; Agrawal et al., 2019). First introduced in the 20th century, cryopreservation has evolved into an indispensable tool, enabling the preservation of both orthodox and recalcitrant seeds, as well as vegetative tissues such as shoot tips and embryos. The earliest reported success in preserving plant cells *in vitro* through cryopreservation was by Quatrano, who used dimethyl sulfoxide (DMSO) to protect callus tissue obtained from flaxseed hypocotyls from cold-related damage. At that time, it was understood that plant cells with low moisture content, such as pollen, seeds, and dormant tissues from stress-resistant species, could endure the extremely low temperatures of liquid nitrogen without serious harm. However, cells with higher water content posed major challenges, as ice crystal formation often led to cell rupture (Benson, 2008; Gang Lu & San Sun, 1992). Sakai later established that successful cryopreservation relies on the combined use of cryoprotectants and carefully controlled cooling rates, a method that has since been adapted for use with a wide variety of plant species (Sakai et al., 1990).

Cryopreservation plays a vital role in various applications. It allows for hybridization between plants that bloom at different times, supports crossbreeding between species from distant regions, helps minimize the spread of diseases through pollinators, and ensures the long-term conservation of valuable and genetically superior germplasm (Panis, 1995). Cryopreservation also helps reduce the likelihood of genetic changes, such as somaclonal variation in *in vitro* tissue cultures and gametoclonal variation in field-grown collections. This contributes to the reliable production of genetically consistent clones through vegetative propagation (Pence & Clark, 2005). Cryopreservation offers an effective method for conserving recalcitrant seeds, which are highly sensitive to cold and low moisture and have short-lived embryos. Such seeds are characteristic of several medicinal plant species propagated by seeds, including Indian snakeroot (*Rauvolfia serpentina*), Ashwagandha (*Withania somnifera*), Noni (*Morinda citrifolia*), Bael (*Aegle marmelos*), and Kalmegh (*Andrographis paniculata*), all of which present challenges for conventional seed banking due to their sensitivity to desiccation and low-temperature storage (L. Nalina et al., 2022). Preserving whole embryos, parts of embryos, or vegetative tissues like shoot tips through cryopreservation provides a reliable approach to maintaining both the genetic stability and viability of these plant species (Agrawal et al., 2014; Pence & Clark, 2005). Nanoparticles play a crucial role in cryopreservation by improving plant cell survival, preventing excessive ice crystal formation, and enhancing recovery after thawing. Their distinct characteristics, including a large surface area and adjustable thermal conductivity, facilitate regulated ice nucleation and efficient cryoprotectant distribution, thereby reducing cellular damage. These innovations are particularly beneficial for applications in biobanking, organ storage, and reproductive health (Wowk, 2010).

As part of this chapter's scope, we seek to explore the integration of nanotechnology into cryopreservation protocorms in the field by addressing challenges such as ice crystallization, oxidative stress, and suboptimal cryoprotectant delivery. Researchers have enhanced cryopreservation outcomes across various plant species by utilizing nanoparticles' unique properties, including high surface area, biocompatibility, and reactivity. Furthermore, this chapter examines the role of cryopreservation in biodiversity conservation, its applications in safeguarding agricultural resources, and the ethical considerations surrounding nanotechnology-assisted approaches.

14.2 Nanotechnology in plant science

Nanomaterials (NMs) are natural or engineered materials with dimensions typically ranging from 1 to 100 nm (Sanzari et al., 2019). Their nanoscale size and high surface-to-volume ratio endow them with exceptional chemical and physical properties compared to their bulk counterparts (Roduner, 2006). Their distinct and adaptable physicochemical characteristics have made them useful in a wide range of disciplines, such as the life sciences, electronics, and chemical engineering (Jeevanandam et al., 2018). In recent years, plant science has started to apply nanotechnology to develop effective systems that enhance seed germination, support plant growth, and provide protection against both environmental and biological stresses (Wang et al., 2016).

Metallic nanoparticles, including gold (Au) and silver (Ag), are commonly used in plant science for various purposes Table 14.1. However, their chemical production is costly and often requires the use of harmful substances (Rastogi et al., 2019; Viswanath & Kim, 2017). Green synthesis methods using plant extracts and ionizing radiation in aqueous solutions have been developed as eco-friendly alternative (Abedini et al., 2013; Mohammadlou et al., 2016). In addition, oxidized NMssuch as magnesium oxide (MgO), calcium oxide (CaO), zinc oxide (ZnO), and titanium dioxide (TiO_2) are gaining attention for their strong electrical, catalytic, and light-absorbing characteristics (Jahan et al., 2018). Polymeric NMsare receiving increasing attention due to their biocompatibility, low production costs, and ability to respond to external factors (Viswanathan et al., 2018). Core–shell nanoparticles, which combine different organic and inorganic materials, are designed for specific uses. For example, polymer-based shells can improve biocompatibility and lower toxicity levels (Wang et al., 2010). A notable advancement in this area is the development of mesoporous silica nanoparticles with customizable surfaces for targeted applications (Torney et al., 2007).

Nanogels (NGs) are an emerging type of nanomaterial gaining considerable interest in nanotechnology because of their outstanding physicochemical characteristics, stable colloidal behavior, and high capacity to encapsulate biomolecules. These nanoscale hydrogels are created from either synthetic or natural polymer chains, which are crosslinked through chemical or physical methods (Neamtu et al., 2024). Their high-water content, porous structure, and strong loading capacity make them highly suitable for various applications. Common types of nanogels include chitosan, alginate, polyvinyl alcohol, polyethylene oxide, and poly(N-isopropyl acrylamide). Hybrid nanogels, which combine either polymeric or nonpolymeric substances, are typically divided into two categories: (1) nanomaterial–nanogel hybrids that incorporate components like magnetic nanoparticles, and (2) polymer—nanogel composites, which include interpenetrating networks, copolymers, and core–shell formations (Kour et al., 2022). Core–shell nanogels are especially promising for carrying biomolecules and delivering drugs.

14.3 Mechanisms of nanoparticle action

In the past two decades, NPs have emerged as promising agents in enhancing plant cryopreservation, offering multiple benefits such as improved cryoprotectant efficiency, membrane stabilisation, oxidative stress reduction, ice nucleation inhibition, osmotic balance regulation, and gene expression modulation (Fig. 14.1) (Marcantonini et al., 2022). Cryoprotectants such as DMSO and glycerol, while effective, often exhibit cytotoxicity at high concentrations; however, nanoparticles like AuNPs, AgNPs, and SiNPs improve cryoprotectant function by reducing toxicity and enhancing permeability (Bhattacharya, 2018). Membrane stabilisation is critical during freezing and thawing to prevent lipid peroxidation and permeability loss, and nanoparticles such as chitosan-based NPs and PEG-

TABLE 14.1 Types of nanoparticles used in cryopreservation and their functions.

Nanoparticle type	Properties	Function(s) in cryopreservation	Advantage(s)	Reference(s)
Gold nanoparticles (AuNPs)	High biocompatibility, antioxidant properties	Modulates cellular uptake of cryoprotectants, enhances protection	Reduces cytotoxicity, improves postthaw viability	Kulus et al. (2024); Venzhik et al. (2022)
Silver nanoparticles (AgNPs)	Antimicrobial, antioxidant properties	Controls ice crystallization, reduces oxidative stress	Prevents microbial contamination, enhances shoot tip regeneration	Kulus et al. (2024); Singh et al. (2023)
Graphene oxide (GO)	High surface area, interacts with lipid membranes	Reinforces cell membranes, prevents ice damage	Improves postthaw recovery rates	Chang & Zhao (2021); Shayestehyekta & Moradi (2025)
Silica nanoparticles	Hydrophilic, interacts with phospholipid bilayers	Stabilizes membranes, prevents lipid peroxidation	Enhances structural integrity, improves tissue survival	Kulus et al. (2024)
Iron oxide nanoparticles	Magnetic properties, enzyme-mimicking activity	Reduces oxidative stress by scavenging reactive oxygen species (ROS)	Enhances mitochondrial function in cryopreserved cells	Yukhta et al. (2025); Tariq et al. (2020)
Hydroxyapatite nanoparticles	Biocompatible, interacts with ice surfaces	Inhibits ice nucleation, prevents intracellular ice formation	Reduces mechanical damage to plant cells	Yukhta et al. (2025)
Polymeric nanogels (e.g., PVA, PEG-based)	High-water content, encapsulation ability	Controls dehydration, maintains optimal water balance	Prevents excessive water loss, improves tissue hydration	Sun et al. (2022)

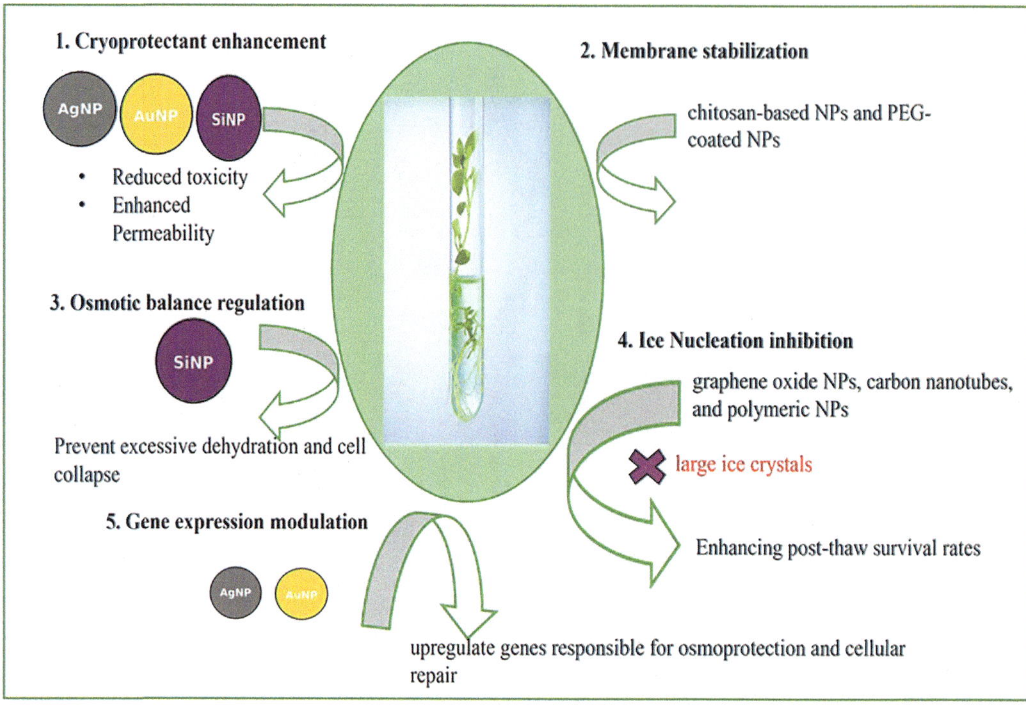

FIGURE 14.1 Mechanisms of nanoparticle-enhanced plant cryopreservation. Schematic representation of nanoparticle-assisted strategies enhancing cryopreservation efficiency in plants. Silver (AgNP), gold (AuNP), and silicon (SiNP) nanoparticles reduce toxicity and improve membrane permeability to facilitate cryoprotectant uptake. Chitosan-based and PEG-coated nanoparticles contribute to membrane stabilisation by reducing reactive oxygen species (ROS) and preserving structural integrity. SiNPs regulate osmotic balance, minimizing dehydration and preventing cell collapse. Graphene oxide, carbon nanotubes, and polymeric nanoparticles inhibit ice nucleation, preventing the formation of large intracellular ice crystals and enhancing postthaw survival. Additionally, AuNPs and AgNPs modulate gene expression by upregulating stress-responsive genes involved in osmoprotection and cellular repair. The central test tube image symbolizes the integrated role of NMs in improving cryopreservation outcomes in plant systems.

coated NPs interact with lipid bilayers to preserve cell integrity. Oxidative stress, a major challenge in cryopreservation, can be mitigated by metal-based nanoparticles like ZnO, CeO_2, and SeNPs, which act as free radical scavengers, boosting antioxidant enzyme activities such as superoxide dismutase (SOD) and catalase (CAT) (Sharma et al., 2019). Ice nucleation, a leading cause of intracellular damage, can be controlled using graphene oxide NPs, carbon nanotubes, and polymeric NPs, which prevent the formation of large ice crystals, thereby enhancing postthaw survival rates. Osmotic balance is another crucial factor, and nanoparticles such as silica NPs and quantum dots (QDs) regulate intracellular water retention, preventing excessive dehydration and cell collapse. Moreover, nanoparticles have been shown to influence gene and protein expression linked to stress tolerance and metabolic activity, with studies on AgNPs and AuNPs demonstrating upregulation of genes responsible for osmoprotection and cellular repair. The integration of nanoparticles into plant cryopreservation protocols presents a significant advancement in the field, addressing key challenges through a multi-faceted approach. Future research should focus on species-specific nanoparticle interactions, optimising formulations, and ensuring biosafety for large-scale applications. The mechanisms and specific examples of nanoparticles involved in cryopreservation are summarized in the Fig. 14.1.

14.4 Nanoparticle-assisted cryopreservation techniques

Cryopreservation techniques have been widely employed for the long-term storage of plant germplasm; however, their success depends on the ability to mitigate cellular damage caused by ice crystallization, osmotic stress, and oxidative injury (Al-Qudah et al., 2022). Traditional methods such as vitrification, encapsulation-dehydration, and slow-freezing have significantly improved the viability of preserved plant materials. However, limitations such as inadequate cryoprotectant penetration, cytotoxicity, and variability in postthaw survival rates have necessitated the exploration of innovative strategies (Kulus et al., 2025). The integration of nanotechnology into cryopreservation protocols has introduced novel solutions that enhance freezing tolerance, improve cryoprotectant efficiency, and optimize postthaw recovery. Nanoparticles (NPs) have been employed to regulate ice formation, reinforce cellular structures, and facilitate controlled dehydration, thereby overcoming some of the critical challenges associated with plant tissue cryopreservation (Mikuła et al., 2022). Table 14.2 comparatively highlights traditional and nanoparticle-assisted cryopreservation methods.

14.5 Applications in medicinal plant germplasm conservation

Cryopreservation is a critical tool for the conservation of plant genetic resources, particularly in the face of biodiversity loss, climate change, and habitat destruction (Benelli et al., 2024). The introduction of nanotechnology into cryopreservation protocols has revolutionized the field, improving plant tissue survival rates, genetic stability, and overall regeneration efficiency. The enhanced effectiveness of nanoparticle-assisted cryopreservation has led to its successful application in various areas, including the preservation of economically important crops and the conservation of rare and endangered plant species (Al-Qudah et al., 2022; Kulus et al., 2025).

14.6 Success stories in medicinal plant germplasm preservation

The increasing demand for food security and agricultural sustainability has heightened the need for effective crop germplasm conservation. Many economically significant medicinal plants, such as *Angelica glauca*, *Clivia miniata*, and *Mentha pulegium* (pennyroyal),

TABLE 14.2 Comparison of traditional and nanoparticle-assisted cryopreservation.

Cryopreservation method	Mechanism	Advantage(s)	Limitation(s)	Examples of usage	Reference(s)
Slow freezing	Gradual cooling at a controlled rate (usually 1°C/min)	Reduces ice crystal formation; widely used in plant and animal cryopreservation	Requires high concentrations of cryoprotectants, which can be toxic	Applied to seeds of *Catharanthus roseus* (Madagascar periwinkle) and *Digitalis purpurea* (foxglove)	Roque-Borda et al. (2021); Zeng et al. (2024)
Vitrification	Uses highly concentrated cryoprotectants to avoid ice formation	Prevents ice crystal damage, maintains high survival rates	High cryoprotectant concentration can lead to toxicity	Used in cryopreserving shoot tips of *Taxus baccata* (European yew) and *Withania somnifera* (Ashwagandha)	Pant et al. (2025); Roque-Borda et al. (2021)
Encapsulation-dehydration	Plant tissues encapsulated in alginate beads, then dehydrated before freezing	Mimics natural desiccation, useful for desiccation-tolerant species	Risk of excessive water loss, leading to cell damage	Applied to somatic embryos of *Panax ginseng* (ginseng) and *Rauvolfia serpentina* (Indian snakeroot)	Benelli (2021); Nalina et al. (2022); Pushpa & Didwania (2019)
Nanoparticle-assisted oxidative stress regulation	Nanoparticles regulate antioxidant enzyme activity to reduce stress-induced damage	Reduces oxidative stress, prevents lipid peroxidation	Potential accumulation of nanoparticles in plant tissues	Applied in *Nymphaea caerulea* (blue waterlily) seed cryopreservation to mitigate oxidative stress	Lee & Li (2022); Zia-ur-Rehman et al. (2023)
Nanoparticle-assisted encapsulation-vitrification	Gold nanoparticles modulate oxidative stress and enhance membrane stabilisation	Potential cytotoxicity at high concentrations	Potential cytotoxicity at high concentrations	Used in shoot tip cryopreservation of *Lamprocapnos spectabilis* (bleeding heart)	Kulus et al. (2024)
Nanoparticle-assisted cryopreservation and ex vitro performance	Nanoparticles improve the postthaw recovery and adaptation of cryopreserved plants	Enhances regeneration, reduces stress during ex-vitro acclimatization	Long-term effects on plant physiology need further study	Applied to assess postcryopreservation performance of *Lamprocapnos spectabilis* (bleeding heart)	Kulus et al. (2025)

are subject to genetic erosion due to climate variability, overharvesting, and habitat destruction (Gautam et al., 2023; Groner et al., 2022; Soto-Gómez & Pérez-Rodríguez, 2022). Cryopreservation has been instrumental in maintaining the genetic diversity of these staple crops, ensuring their availability for future breeding programs.

Nanoparticles have significantly improved the efficiency of crop germplasm cryopreservation by addressing key challenges such as ice crystallization, oxidative stress, and cryoprotectant toxicity. For example, graphene oxide (GO) and silica nanoparticles have been successfully incorporated into the cryopreservation of rice embryos, improving their postthaw survival rates and maintaining genetic fidelity (Shayestehyekta & Moradi, 2025). Similarly, iron oxide nanoparticles (SPIONs) have been used to enhance the viability of cryopreserved maize pollen, ensuring its effective use in hybridization programs (Baniasadi et al., 2023).

A notable success story in the cryopreservation of medicinal plants involves *Lamprocapnos spectabilis* (bleeding heart). Researchers have integrated silver nanoparticles (AgNPs) into the cryopreservation protocol of this species, specifically during the encapsulation-vitrification process of shoot tips. The addition of AgNPs at a concentration of 5 ppm in the alginate bead matrix resulted in enhanced shoot elongation and improved foliar growth postthaw. This approach not only mitigated oxidative stress but also maintained genetic stability, demonstrating the efficacy of nanoparticle-assisted cryopreservation in preserving medicinal plant germplasm. These findings underscore the potential of nanotechnology to enhance the conservation of valuable medicinal species through improved cryopreservation techniques (Kulus et al., 2024).

14.7 Strategies for conserving endangered medicinal plant species

The conservation of endangered plant species is a global priority, as many rare and ecologically significant plants are at risk due to deforestation, habitat fragmentation, overharvesting, and climate change. Among these are medicinal plants of high ethnopharmacological and commercial value, including *Coptis teeta* (Yunnan goldthread), *Siphonochilus aethiopicus* (African ginger), *Coscinium fenestratum* (tree turmeric), and *Angelica glauca* (Himalayan angelica), all of which are experiencing rapid population declines due to habitat degradation and unsustainable harvesting (Adebayo et al., 2021; Gautam et al., 2023; Huang & Long, 2007). Traditional conservation methods, such as seed banking and field collections, are often inadequate for these species, especially those with recalcitrant seeds or limited reproductive capacities. Cryopreservation provides a viable alternative for the long-term conservation of such vulnerable taxa by preserving genetic material at ultra-low temperatures without compromising viability or genetic integrity. This method is particularly suited for species that cannot be stored under conventional ex situ conditions, offering a stable and secure platform for maintaining biodiversity.

Nanotechnology has been particularly useful in the cryopreservation of recalcitrant seeds and somatic embryos from endangered plants. Encapsulation-dehydration methods incorporating hydrogel-based nanocomposites have been successfully applied to the preservation of *Rhododendron* species, improving dehydration control and preventing mechanical damage during freezing (Engelmann, 2011). Similarly, cerium oxide nanoparticles (CeO_2) have been used to mitigate oxidative stress in cryopreserved medicinal orchids, enhancing their regeneration potential upon thawing (Chang et al., 2025).

Nanoparticles have also played a role in improving the conservation of forest species. The application of polymeric NMs in the cryopreservation of *Castanea sativa* (sweet chestnut) has resulted in enhanced root formation and shoot development postthaw, making it a promising approach for restoring declining tree populations (Gaidamashvili et al., 2021).

14.8 Ethical and environmental implications

As nanotechnology continues to be integrated into cryopreservation protocols, it is essential to consider the ethical and environmental implications of these advancements. While nanoparticles offer significant benefits in plant germplasm conservation, concerns regarding their potential toxicity, long-term ecological impact, and regulatory considerations must be addressed to ensure responsible implementation.

14.9 Addressing nanoparticle safety and toxicity concerns

One of the primary concerns associated with the use of nanoparticles in biological systems is their potential cytotoxicity. Although many nanoparticles, such as silica and lipid-based nanocarriers, are considered biocompatible, others, such as silver and metal oxide nanoparticles, have been reported to induce oxidative stress and DNA damage at high concentrations (Janjua et al., 2023). To minimize potential risks, researchers are focusing on developing biodegradable and eco-friendly nanoparticles, such as chitosan-based and cellulose-derived NMs, which exhibit minimal cytotoxic effects while retaining their cryoprotective properties (Khedulkar et al., 2025).

Additionally, regulatory frameworks governing nanoparticle applications in agriculture and conservation are still evolving. Clear guidelines on permissible nanoparticle concentrations, degradation pathways, and risk assessment protocols are necessary to ensure the safe and ethical use of nanotechnology in germplasm conservation.

14.10 Evaluating the environmental footprint of nanotechnology in conservation

The long-term environmental impact of nanoparticle-based cryopreservation remains a topic of active research. The potential accumulation of nanoparticles in soil and water ecosystems raises concerns regarding their interactions with microbial communities, nutrient cycles, and plant health. Studies have suggested that some nanoparticles, particularly metal-based ones, may exhibit persistence in the environment, leading to unintended ecological consequences (Ameen et al., 2021; Gupta et al., 2024).

To address these concerns, researchers are exploring green nanotechnology approaches, such as the synthesis of nanoparticles using plant-derived biomolecules and biodegradable polymers. Additionally, life-cycle assessments of nanoparticle-assisted cryopreservation techniques are being conducted to evaluate their environmental sustainability and identify areas for improvement (Ulaş et al., 2025).

14.11 Future directions

The future of nanoparticle-assisted cryopreservation lies in continued innovation and interdisciplinary collaboration. Emerging trends suggest that advancements in nanomaterial design, combined with cutting-edge genomic and biotechnological tools, will further enhance the efficiency and applicability of cryopreservation techniques.

14.12 Innovations and emerging trends in nanoparticle-assisted protocols

Recent research is focused on developing multifunctional nanoparticles that combine cryoprotective, antioxidant, and antimicrobial properties into a single nanoplatform. Hybrid NMs, such as graphene-silver nanocomposites and polymer-metal oxide blends, are being explored for their ability to provide comprehensive protection to cryopreserved plant tissues (Muthukumar et al., 2021; Qaiss et al., 2021)

Additionally, the use of artificial intelligence (AI) and machine learning (ML) in nanoparticle design is expected to accelerate the development of optimized cryopreservation formulations. Predictive models can assist in identifying the most effective nanoparticle compositions and concentrations, reducing the need for extensive trial-and-error experimentation (Prasad et al., 2024).

14.13 Synergies with genomic and biotechnological tools

Integrating nanotechnology with genomic and biotechnological advancements presents new opportunities for precision cryopreservation. The use of CRISPR-based genome editing in conjunction with nanoparticle-assisted cryopreservation could allow for the preservation of genetically modified plants with enhanced stress tolerance and disease resistance (Chen et al., 2024).

Furthermore, metabolomic profiling of cryopreserved tissues is being explored to understand the biochemical responses of plant cells to nanoparticle exposure. This knowledge will aid in the refinement of cryopreservation protocols, ensuring that they are tailored to the specific metabolic and physiological needs of different plant species (Kang et al., 2025).

14.14 Conclusion

Nanotechnology has emerged as a powerful tool in plant germplasm cryopreservation, offering solutions to long-standing challenges such as ice crystallization, oxidative stress, and cryoprotectant toxicity. By enhancing vitrification efficiency, optimising encapsulation-dehydration methods, and regulating ice formation, nanoparticles have significantly improved the survival and regeneration of cryopreserved plant tissues.

While the benefits of nanoparticle-assisted cryopreservation are evident, ethical and environmental considerations must be carefully addressed to ensure responsible implementation. Future research should focus on developing biocompatible and biodegradable nanoparticles, establishing regulatory guidelines, and assessing the long-term ecological impact of nanomaterial applications.

As scientific advancements continue to refine cryopreservation protocols, the integration of nanotechnology with genomic and biotechnological tools holds immense promise for the sustainable conservation of plant biodiversity. By strengthening these innovations, researchers can enhance global efforts in safeguarding genetic resources, supporting food security, and preserving the ecological balance in an era of environmental uncertainty.

References

Abedini, A., Daud, A., Hamid, M. A. A., Othman, N. K., & Saion, E. (2013). A review on radiation-induced nucleation and growth of colloidal metallic nanoparticles. *Nanoscale Research Letters, 8*, 474. https://doi.org/10.1186/1556-276X-8-474.

Acosta, Y., Pérez, L., Linares, C., Hernández, L., Escalante, D., Pérez, A., Zevallos, B. E., Yabor, L., Martínez-Montero, M. E., Cejas, I., Fontes, D., Sershen, & Lorenzo, J. C. (2020). Effects of teramnus labialis (L.f.) Spreng seed cryopreservation on subsequent seed and seedling growth and biochemistry. *Acta Physiologiae Plantarum, 42*(1), 7. https://doi.org/10.1007/s11738-020-3012-9, https://doi.org/10.1007/s11738-020-3012-9.

Adebayo, S. A., Amoo, S. O., Mokgehle, S. N., & Aremu, A. O. (2021). Ethnomedicinal uses, biological activities, phytochemistry and conservation of African ginger *(Siphonochilus aethiopicus)*: A commercially important and endangered medicinal plant. *Journal of Ethnopharmacology, 266*. https://doi.org/10.1016/j.jep.2020.113459, www.elsevier.com/locate/jethpharm.

Agrawal, A., Sanayaima, R., Singh, R., Tandon, R., Verma, S., & Tyagi, R. K. (2014). Phenotypic and molecular studies for genetic stability assessment of cryopreserved banana meristems derived from field and in vitro explant sources. *In Vitro Cellular & Developmental Biology - Plant, 50*(3), 345–356. https://doi.org/10.1007/s11627-014-9606-4, https://doi.org/10.1007/s11627-014-9606-4.

Agrawal, A., Singh, S., Malhotra, E. V., Meena, D. P. S., & Tyagi, R. K. (2019). *In vitro conservation and cryopreservation of clonally propagated horticultural species*. Singapore: Springer Singapore, 529–578. https://doi.org/10.1007/978-981-13-3669-0_18, 10.1007/978-981-13-3669-0_18.

Al-Qudah, T., Mahmood, S. H., Abu-Zurayk, R., Shibli, R., Khalaf, A., Lambat, T. L., & Chaudhary, R. G. (2022). Nanotechnology applications in plant tissue culture and molecular genetics: A holistic approach. *Current Nanoscience, 18*(4), 442–464. https://doi.org/10.2174/1573413717666211118111333, www.eurekaselect.com/604/journal/current-nanoscience.

Ameen, F., Alsamhary, K., Alabdullatif, J. A., & ALNadhari, S. (2021). A review on metal-based nanoparticles and their toxicity to beneficial soil bacteria and fungi. *Ecotoxicology and Environmental Safety, 213*, 112027. https://doi.org/10.1016/j.ecoenv.2021.112027, https://www.sciencedirect.com/science/article/pii/S014765132100138X.

Baniasadi, F., Hajiaghalou, S., Shahverdi, A., Ghalamboran, M. R., Pirhajati, V., & Fathi, R. (2023). The beneficial effects of static magnetic field and iron oxide nanoparticles on the vitrification of mature mice oocytes. *Reproductive Sciences, 30*(7), 2122–2136. https://doi.org/10.1007/s43032-022-01144-1, https://doi.org/10.1007/s43032-022-01144-1.

Benelli, C., Tarraf, W., İzgü, T., Anichini, M., Faraloni, C., Salvatici, M. C., Jouini, N., Germanà, M. A., Danti, R., & Lambardi, M. (2024). Long-term conservation for the safeguard of *Abies nebrodensis*: An endemic and endangered species of sicily. *Plants, 13*(12). https://doi.org/10.3390/plants13121682, www.mdpi.com/journal/plants.

Benelli, C. (2021). Plant cryopreservation: A look at the present and the future. *Plants, 10*(12), 2744. https://doi.org/10.3390/plants10122744.

Benson, E. E. (2008). *Cryopreservation theory*. New York, NY: Springer New York, 15–32. https://doi.org/10.1007/978-0-387-72276-4_2, 10.1007/978-0-387-72276-4_2.

Bhattacharya, S. (2018). *Cryopretectants and their usage in cryopreservation process*. IntechOpen. 10.5772/intechopen.80477.

Chand, H. B., Kumar, A., Vatsala, Joshi, P. R., Gautam, M. K., & Singh, S. (2025). *Global efforts and framework for biodiversity conservation*. Singapore: Springer Nature Singapore, 19–34. https://doi.org/10.1007/978-981-96-7210-3_2, 10.1007/978-981-96-7210-3_2.

Chang, S.-J., Zhang, X.-W., Xu, H.-W., Zhang, S.-B., Liu, X.-W., Yi, Y.-Y., Pan, W., Li, K., & Wang, S.-J. (2025). CeO_2 nanoparticles reduce oxidative stress and delay the degeneration of intervertebral disc. *Bioinorganic Chemistry and Applications, 2025*. https://doi.org/10.1155/bca/3399767.

Chang, T., & Zhao, G. (2021). Ice inhibition for cryopreservation: Materials, strategies, and challenges. *Advanced Science, 8*, 2002425. https://doi.org/10.1002/advs.202002425.

Chen, F., Chen, L., Yan, Z., Xu, J., Feng, L., He, N., Guo, M., Zhao, J., Chen, Z., Chen, H., Yao, G., & Liu, C. (2024). Recent advances of CRISPR-based genome editing for enhancing staple crops. *Frontiers in Plant Science, 15*. https://doi.org/10.3389/fpls.2024.1478398.

Engelmann, F. (2011). Encapsulation-dehydration for cryopreservation: Past, present and future. *Acta Horticulturae, 908*, 165–171. https://doi.org/10.17660/ActaHortic.2011.908.18.

Fitzgerald, H., Palmé, A., Asdal, Å., Endresen, D., Kiviharju, E., Lund, B., Rasmussen, M., Thorbjörnsson, H., & Weibull, J. (2019). A regional approach to Nordic crop wild relative in situ conservation planning. *Plant Genetic Resources: Characterization and Utilization, 17*(2), 196–207. https://doi.org/10.1017/S147926211800059X, https://www.cambridge.org/core/product/0DF42C7069E39E0FE1E1B56B86DE07C1.

Gaidamashvili, M., Khurtsidze, E., Kutchava, T., Lambardi, M., & Benelli, C. (2021). Efficient protocol for improving the development of cryopreserved embryonic axes of chestnut (*Castanea sativa* Mill.) by encapsulation–Vitrification. *Plants, 10*(2). https://doi.org/10.3390/plants10020231.

Gang Lu, T., & San Sun, C. (1992). Cryopreservation of millet (*Setaria italica* L.). *Journal of Plant Physiology, 139*(3), 295–298. https://doi.org/10.1016/S0176-1617(11)80340-8, https://www.sciencedirect.com/science/article/pii/S0176161711803408.

Gautam, K., Raina, R., & Dikshit, N. (2023). Current knowledge on sustainability and conservation of endangered himalayan medicinal herb *Angelica glauca* Edgew.—A review. *Journal of Herbal Medicine, 42*, 100764. https://doi.org/10.1016/j.hermed.2023.100764.

Groner, V. P., Nicholas, O., Mabhaudhi, T., Slotow, R., Akçakaya, H. R., Mace, G. M., & Pearson, R. G. (2022). Climate change, land cover change, and overharvesting threaten a widely used medicinal plant in South Africa. *Ecological Applications, 32*(4). https://doi.org/10.1002/eap.2545, http://onlinelibrary.wiley.com/journal/10.1002/(ISSN)1939-5582.

Gupta, S., Kumar, D., Aziz, A., AbdelRahman, M., Mustafa, A., Scopa, A., Radice, R. P., Drosos, M., & Moursy, A. (2024). Nanoecology: Exploring engineered nanoparticles' impact on soil ecosystem health and biodiversity. *Egyptian Journal of Soil Science, 64*, 1637–1655. https://doi.org/10.21608/EJSS.2024.304704.1814.

Huang, J., & Long, C. (2007). Coptis teeta-based agroforestry system and its conservation potential: A case study from Northwest Yunnan. *AMBIO: A Journal of the Human Environment, 36*(4), 343–349. https://doi.org/10.1579/0044-7447(2007)36[343:ctasai]2.0.co;2.

Jahan, S., Alias, Y., Bakar, A. F. A., & Yusoff, I. (2018). Toxicity evaluation of ZnO and TiO_2 nanomaterials in hydroponic red bean (*Vigna angularis*) plant: Physiology, biochemistry and kinetic transport. *Journal of Environmental Sciences, 72*. https://doi.org/10.1016/j.jes.2017.12.022.

Janjua, T. I., Cao, Y., Kleitz, F., Linden, M., Yu, C., & Popat, A. (2023). Silica nanoparticles: A review of their safety and current strategies to overcome biological barriers. *Advanced Drug Delivery Reviews, 203*, 115115. https://doi.org/10.1016/j.addr.2023.115115, https://www.sciencedirect.com/science/article/pii/S0169409X23004301.

Jeevanandam, J., Barhoum, A., Chan, Y. S., Dufresne, A., & Danquah, M. (2018). Review on nanoparticles and nanostructured materials: History, sources, toxicity, and regulations. *Beilstein Journal of Nanotechnology, 9*, 1050–1074. https://doi.org/10.3762/bjnano.9.98.

Kang, P., Kim, S. J., Park, H., Han, S. J., Kim, I.-C., Lee, H., & Han Yim, J. (2025). Trends and challenges in plant cryopreservation research: A meta-analysis of cryoprotective agent development and research focus. *Plants, 14*, 447. https://doi.org/10.3390/plants14030447.

Khedulkar, A., Bobade, R., Doong, R.-an, Pandit, B., Nguyễn, M.-K., Ambare, R., Hoang, T.-D., & Kumar, K. (2025). Bio-based nanomaterials as effective, friendly solutions and their applications for protecting water, soil, and air. *Materials Today Chemistry, 46*, 102688.

Kiflie, M. A., Sharma, D. P., Haile, M. A., & Srinivasagan, R. (2024). EfficientNet ensemble learning: identifying ethiopian medicinal plant species and traditional uses by integrating modern technology with ethnobotanical wisdom. *Computers, 13*(2). https://doi.org/10.3390/computers13020038.

Kour, H., Khan, S. S., Kour, D., Singh, S., Kumari, S., Kaur, M., Khan, R. T., Yadav, A. N., Seymen, M., Kurtar, E. S., Erdinc, C., & Kumar, A. (2022). *Chapter 9 - Nanotechnologies for microbial inoculants as biofertilizers in the horticulture developments in applied microbiology and biotechnology*. Academic Press, 201–261. https://www.sciencedirect.com/science/article/pii/B9780323918619000070, 10.1016/B978-0-323-91861-9.00007-0.

Kulus, D., Tymoszuk, A., Kulpińska, A., Dębska, B., Michalska, A., Nowakowska, J., Wichrowska, D., Wojnarowicz, J., & Szałaj, U. (2025). Nanoparticles in plant cryopreservation: Effects on genetic stability, metabolic profiles, and structural integrity in bleeding heart (*Papaveraceae*) cultivars. *Nanotechnology, Science and Applications, 18*, 35–56. https://doi.org/10.2147/NSA.S485428, https://www.dovepress.com/article/download/100248.

Kulus, D., Tymoszuk, A., Kulpińska, A., Viehmannova, I., Wojnarowicz, J., Szałaj, U., & Koppala, S. (2024). Effect of nanoparticles on the ex-vitro performance of cryopreservation-derived plant material. *PLoS One, 19*(9), e0310424. https://doi.org/10.1371/journal.pone.0310424.

Lambers, J. H. R. (2015). Extinction risks from climate change. *Science (New York, N.Y.), 348*(6234), 501–502. https://doi.org/10.1126/science.aab2057, https://doi.org/10.1126/science.aab2057.

Lee, C.-H., & Li, C.-W. (2022). Cryopreservation of seeds of blue waterlily (*Nymphaea caerulea*) using glutathione adding plant vitrification solution (PVS+).

Marcantonini, G., Bartolini, D., Zatini, L., Costa, S., Passerini, M., Rende, M., Luca, G., Basta, G., Murdolo, G., Calafiore, R., & Galli, F. (2022). Natural cryoprotective and cytoprotective agents in cryopreservation: A focus on melatonin. *Molecules (Basel, Switzerland), 27*(10), 3254. http://doi.org/10.3390/molecules27103254.

Mba, C., Dulloo, M., & Nnadozie, K. (2021). Plant genetic resources for food and agriculture for sustainable development. *Annals of Science*, 3–34. https://doi.org/10.19103/AS.2020.0085.01

Mekonnen, A., Seid, A., & Tefera, A. (2022). Ethnobotanical study of traditional medicinal plants used to treat human and animal diseases in sedie Muja District, South Gondar, Ethiopia. *Evidence-Based Complementary and Alternative Medicine, 2022*, 1–22. http://doi.org/10.1155/2022/7328613.

Mikuła, A., Chmielarz, P., Hazubska-Przybył, T., Kulus, D., Maślanka, M., Pawłowska, B., Zimnoch-Guzowska, E., & Bajguz, A. (2022). Cryopreservation of plant tissues in poland: Research contributions, current status, and applications. *Acta Societatis Botanicorum Poloniae, 91*. https://doi.org/10.5586/asbp.9132.

Milla, R. (2020). Crop origins and phylo food: A database and a phylogenetic tree to stimulate comparative analyses on the origins of food crops. *Global Ecology and Biogeography, 29*, 606–614.

Mohammadlou, M., Maghsoudi, H., & Jafarizadeh, H. (2016). A review on green silver nanoparticles based on plants: Synthesis, potential applications and eco-friendly approach. *International Food Research Journal, 23*, 446–463.

C. Muthukumar, S. Krishnasamy, T. Thirugnanasambandan, S.M. K. Thiagamani, N. Jesuarockiam, S. Siengchin, Graphene and Silver Nanoparticle Based Hybrid Nanocomposites for Anti-bacterial Applications. (2021), 183–196, Muthukumar, C., Krishnasamy, S., Thirugnanasambandan, T., Thiagamani, S. M. K., Jesuarockiam, N., & Siengchin, S. (2021). Graphene and silver nanoparticle based hybrid nanocomposites for anti-bacterial applications (pp. 183–196). https://doi.org/10.1007/978-981-33-4988-9_6.

Nalina, L., Rajamani, K., Shanmugasundaram, K. A., & Boomiga, M. (2022). *Breeding and conservation of medicinal plants in India*. Springer Science and Business Media LLC, 201–236. 10.1007/978-3-030-98701-5_7.

Neamtu, I., Ghilan, A., Rusu, A. G., Nita, L. E., Chiriac, V. M., & Chiriac, A. P. (2024). Design and applications of polymer-like peptides in biomedical nanogels. *Expert Opinion on Drug Delivery, 21*(5), 713–734. https://doi.org/10.1080/17425247.2024.2364651, https://doi.org/10.1080/17425247.2024.2364651.

Panis, B. (1995). Cryopreservation of Musa germplasm. *Infomusa, 4*(1), 17–20.

Pant, M., Sohrab, S. S., & Husen, A. (2025). *Genetic improvement and conservation practices of medicinal plants genetic improvement and conservation practices of medicinal plants*. India: CRC Press, 1–272. https://www.taylorfrancis.com/books/edit/10.1201/9781003482659, 10.1201/9781003482659.

Pence, V., & Clark, J. R. (2005). Desiccation, cryopreservation and germination of seeds of the rare wetland species, *Plantago cordata* Lam. *Seed Science and Technology, 33*, 767–770. https://doi.org/10.15258/sst.2005.33.3.25.

Prasad, A., Santra, T., & Rengaswamy, J. (2024). A Study on prediction of size and morphology of ag nanoparticles using machine learning models for biomedical applications. *Metals, 14*, 539. https://doi.org/10.3390/met14050539.

Prevéy, J. S., Parker, L. E., & Harrington, C. A. (2020). Projected impacts of climate change on the range and phenology of three culturally-important shrub species. *PLoS One, 15*(5), e0232537. https://doi.org/10.1371/journal.pone.0232537.

Pushpa, P., & Didwania, N. (2019). Cryopreservation technique: A powerful tool for long-term preservation of endangered medicinal plants. *Ambient Science, 06H & 06Sp*(2 & 1). https://doi.org/10.21276/ambi.2019.06.sp1.ta02.

Qaiss, A., Bouhfid, R., & Jawaid, M. (2021). Graphene and nanoparticles hybrid nanocomposites: From preparation to applications. doi: 10.1007/978-981-33-4988-9.

Rastogi, A., Tripathi, D., Yadav, S., Chauhan, D., Zivcak, M., Ghorbanpour, M., Elsheery, N., & Brestic, M. (2019). Application of silicon nanoparticles in agriculture. *3Biotech, 9*, 90. https://doi.org/10.1007/s13205-019-1626-7.

Roduner, E. (2006). Size matters: Why nanomaterials are different. *Chemical Society Reviews, 35*, 583–592. https://doi.org/10.1039/b502142c.

Roque-Borda, C. A., Kulus, D., Vacaro de Souza, A., Kaviani, B., & Vicente, E. F. (2021). Cryopreservation of agronomic plant germplasm using vitrification-based methods: an overview of selected case studies. *International Journal of Molecular Sciences, 22*(11), 6157. https://doi.org/10.3390/ijms22116157.

Ruiz-Sáenz, D. R., Ayala-Hernández, D. D., Niino, T., Cruz-Gutiérrez, E. J., Aquino-Martínez, J. G., & López-Delgado, H. A. (2019). Salicylic acid-cryotherapy treatment for elimination of potato virus s from solanum tuberosum. *American Journal of Potato Research, 96*(3), 225–234. https://doi.org/10.1007/s12230-018-09694-4, https://doi.org/10.1007/s12230-018-09694-4.

Ruta, C., Lambardi, M., & Ozudogru, E. A. (2020). Biobanking of vegetable genetic resources by in vitro conservation and cryopreservation. *Biodiversity and Conservation, 29*(13), 3495–3532. https://doi.org/10.1007/s10531-020-02051-0, https://doi.org/10.1007/s10531-020-02051-0.

Sakai, A., Kobayashi, S., & Oiyama, I. (1990). Cryopreservation of nucellar cells of navel orange (*Citrus sinensis* Osb. var. brasiliensis Tanaka) by vitrification. *Plant Cell Reports, 9*(1), 30–33. https://doi.org/10.1007/BF00232130, https://doi.org/10.1007/BF00232130.

Sanzari, I., Leone, A., & Ambrosone, A. (2019). Nanotechnology in plant science: To make a long story short. *Frontiers in Bioengineering and Biotechnology, 7*. https://doi.org/10.3389/fbioe.2019.00120.

Sharma, S., Singh, V. K., Kumar, A., & Mallubhotla, S. (2019). Effect of nanoparticles on oxidative damage and antioxidant defense system in plants. *Molecular Plant Abiotic Stress*, 315–333. https://doi.org/10.1002/9781119463665.ch17, https://doi.org/10.1002/9781119463665.ch17.

Shayestehyekta, M., & Moradi, M. (2025). Graphene oxide and silymarin combination: A novel approach to improving post-cryopreservation quality of ram sperm. *Cryobiology, 118*, 105199. https://doi.org/10.1016/j.cryobiol.2025.105199.

Singh, Y., Kumar, U., Panigrahi, S., Balyan, P., Mehla, S., Sihag, P., Sagwal, V., Singh, K., White, J., & Dhankher, O. P. (2023). Nanoparticles as novel elicitors in plant tissue culture applications: Current status and future outlook. 108004.

Soto-Gómez, D., & Pérez-Rodríguez, P. (2022). Sustainable agriculture through perennial grains: Wheat, rice, maize, and other species. A review. *Agriculture, Ecosystems & Environment, 325*. http://doi.org/10.1016/j.agee.2021.107747 107747.

Sun, H., Cao, Y., Kim, D., & Marelli, B. (2022). Biomaterials technology for agrofood resilience (Adv. Funct. Mater. 30/2022). *Advanced Functional Materials, 32*. https://doi.org/10.1002/adfm.202270173.

Tariq, A., Ilyas, S., & Naz, S. (2020). Nanotechnology and plant tissue culture. pp. 23–35. doi: 10.1007/978-3-030-41275-3_2.

Torney, F., Trewyn, B., Lin, V., & Wang, K. (2007). Mesoporous silica nanoparticles deliver DNA and chemicals into plants. *Nature Nanotechnology, 2*, 295–300. https://doi.org/10.1038/nnano.2007.108.

U. CBD. (1982). *Convention on biological diversity (with annexes), concluded at Rio de Janeiro on 5 June 1992*. Convention on Biological Diversity. https://www.cbd.int/convention/.

Ulaş, F., Yüksel, E., Dinçer, D., Dababat, A., & İmren, M. (2025). Recent advances in plant-based green synthesis of nanoparticles: A sustainable approach for combating plant-parasitic nematodes. *Sustainability, 17*(9). https://doi.org/10.3390/su17094152.

Venzhik, Y., Deryabin, A., Popov, V., Dykman, L., & Moshkov, I. (2022). Gold nanoparticles as adaptogens increazing the freezing tolerance of wheat seedlings. *Environmental Science and Pollution Research, 29*, 1–15. https://doi.org/10.1007/s11356-022-19759-x.

Vincent, H., Amri, A., Castañeda-Álvarez, N. P., Dempewolf, H., Dulloo, E., Guarino, L., Hole, D., Mba, C., Toledo, A., & Maxted, N. (2019). Modeling of crop wild relative species identifies areas globally for in situ conservation. *Communications Biology, 2*(1), 136. https://doi.org/10.1038/s42003-019-0372-z, https://doi.org/10.1038/s42003-019-0372-z.

Viswanath, B., & Kim, S. (2017). Influence of nanotoxicity on human health and environment: The alternative strategies. *Reviews of Environmental Contamination and Toxicology, 242*, 61–104.

Viswanathan, B., Meeran, S., Shabeer, T. K., Sruthi, S., & Ali, J. (2018). Historic review on modern herbal nanogel formulation and delivery methods. *International Journal of Pharmacy and Pharmaceutical Sciences, 10*, 1–10. https://doi.org/10.22159/ijpps.2018v10i10.23071.

Wang, P., Lombi, E., & Zhao, F.-J. (2016). Nanotechnology: A new opportunity in plant sciences. *Trends in Plant Science, 21*. https://doi.org/10.1016/j.tplants.2016.04.005.

Wang, Y., Zhan, L., & Huang, C. (2010). One-pot preparation of dextran-capped gold nanoparticles at room temperature and colorimetric detection of dihydralazine sulfate in uric samples. *Analytical Methods, 2*, 1982. https://doi.org/10.1039/C0AY00470G.

Williams, V., Victor, J., & Crouch, N. (2013). Red listed medicinal plants of South Africa: Status, trends, and assessment challenges. *South African Journal of Botany, 86*, 23–35. https://doi.org/10.1016/j.sajb.2013.01.006.

Woldearegay, M., & Teso, M. (2022). Conservation strategies for medicinal plants in Ethiopia: A comprehensive review. *Ecological Insights, 5*. https://doi.org/10.59411/pm6kd039.

Wowk, B. (2010). Thermodynamic aspects of vitrification. *Special Issue: Thermodynamic Aspects of Cryobiology, 60*(1), 11–22. https://doi.org/10.1016/j.cryobiol.2009.05.007, https://www.sciencedirect.com/science/article/pii/S0011224009000674.

Yukhta, M., Bespalova, I., Koroleva, A., & Chichkov, B. (2025). Nanoparticles in low-temperature preservation of biological systems of animal origin. *Nanotechnology Reviews, 14*. https://doi.org/10.1515/ntrev-2025-0143.

Zeng, L., Sun, Z., Fu, L., Gu, Y., Li, R., He, M., & Wei, J. (2024). Cryopreservation of medicinal plant seeds: strategies for genetic diversity conservation and sustainability. *Plants, 13*(18), 2577. https://doi.org/10.3390/plants13182577.

Zia-ur-Rehman, M., Anayatullah, S., Irfan, E., Hussain, S., Rizwan, M., Sohail, M., Jafir, M., Ahmad, T., Usman, M., & Alharby, H. (2023). Nanoparticles assisted regulation of oxidative stress and antioxidant enzyme system in plants under salt stress: A review. *Chemosphere, 314*, 1–15. https://doi.org/10.1016/j.chemosphere.2022.137649.

CHAPTER

15

Spice crops with therapeutic and cosmetic uses: genetic resources, active principles, high volume macro- and micropropagation, breeding, and genomics

Sheeja T.E.[1], *Shivakumar M.S.*[2], *Sona Charles*[1], *Sharon Aravind*[1], *Muhammed Azharudheen T.P.*[1], *Aarthi S.*[1], *Vinu V.*[1], *Akshitha H.J.*[2], *Honnappa Asangi*[2], *Maneesha S.R.*[1], *Muhammed Nissar V.A.*[1], *Sankar Mukesh S.*[1], *and Divya P. Syamaladevi*[1]

[1]**ICAR–Indian Institute of Spices Research, Kozhikode, Kerala, India** [2]**ICAR–Indian Institute of Spices Research, Regional Station, Appangala, Madikeri, Karnataka, India**

15.1 Introduction

Spices are aromatic plant products derived from seeds, fruits, roots, bark, or other plant parts, valued for their flavor, color, and medicinal properties. They have played a pivotal role in human civilization, not only as essential culinary ingredients but also as key commodities in global trade. Spices are rich sources of bioactive compounds, which impart antioxidant, antimicrobial, anti-inflammatory, and nutraceutical benefits, making them integral to both traditional medicine and modern functional foods. Among the most important spices are black pepper (*Piper nigrum*), often referred to as the "King of Spices" for its pungency and wide culinary use; turmeric (*Curcuma longa*), valued for its vibrant color and bioactive curcumin content; ginger (*Zingiber officinale*), recognized for its flavor and digestive benefits; cardamom (*Elettaria cardamomum*), prized for its aromatic seeds; and

Medicinal Plant Biotechnology for Ameliorated Secondary Metabolite Production
DOI: https://doi.org/10.1016/B978-0-443-43944-5.00002-5

tree spices such as cinnamon (*Cinnamomum verum*), clove (*Syzygium aromaticum*), and nutmeg (*Myristica fragrans*), which contribute both flavor and essential oils. In this chapter, we provide a detailed discussion on the genetic resources, active principles, breeding strategies, high-volume macro- and micro-propagation techniques, and genomics research in major spices, highlighting the scientific approaches that underpin conservation, improvement, and utilization of these economically and nutritionally important crops.

15.2 Black pepper

Pepper, scientifically known as *Piper nigrum*, is a widely used spice known for its sharp, pungent flavor and aroma. Native to the Western Ghats of India, pepper is derived from the dried berries of a flowering vine belonging to the Piperaceae family. It is one of the oldest and most valuable spices in the world, often referred to as the "King of Spices."

15.2.1 Genetic resources

The conservation of genetic resources is crucial in the face of rapid gene erosion caused by biotic, abiotic, socio-political, and economic factors. Black pepper is particularly vulnerable due to devastating diseases, the spread of improved cultivars, and deforestation. Many wild Piper species, including *P. barberi, P. hapnium, P. silentvalleyensis, P. wightii*, and *P. schmidtii*, are rare and on the brink of extinction (Nirmal Babu et al., 2005). To mitigate these losses, ICAR–Indian Institute of Spices Research (ICAR-IISR) has conducted systematic surveys to collect and conserve black pepper genetic diversity.

The National Active Germplasm Site (NAGS) for black pepper has been established at ICAR-IISR, Kozhikode. Given the specific agro-climatic requirements and vegetative propagation nature of black pepper, conservation is primarily undertaken at ICAR-IISR, through clonal field repositories and *in vitro* gene banks (Nirmal Babu and Sasikumar, 2000; Ravindran & Nirmal Babu, 1994). A comprehensive crop catalog detailing black pepper genetic resources is available on the IISR website (Saji et al., 2019).

Genetic diversity in black pepper is substantial, with open-pollinated progenies displaying variability due to heterozygosity and geitonogamous pollination. Kerala and Karnataka house the largest diversity of indigenous black pepper cultivars, with over 100 identified varieties (Ravindran et al., 2000; Ravindran & Nirmal Babu, 1994). Historically, native cultivars were linked to specific regions, but human migration to the Western Ghats led to the selective propagation of higher-yielding genotypes, now dominant in pepper-growing areas. Germplasm status of major spices is summarized in Fig. 15.1.

15.2.2 Conservation strategies and characterization

Black pepper, being a vegetatively propagated crop, requires specialized conservation strategies. The ICAR–Indian Institute of Spices Research (IISR) has the world's largest black pepper germplasm collection. Conservation efforts primarily focus on ex situ conservation, where all germplasm accessions are maintained in protected shelters or polyhouses. The cuttings

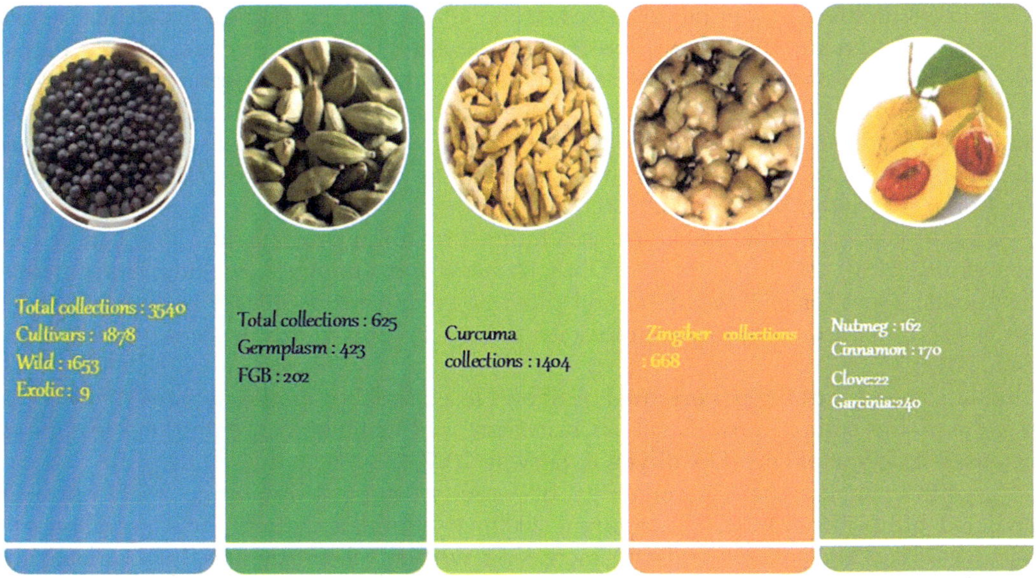

FIGURE 15.1 Current status of germplasm holdings of major spice crops conserved at ICAR-IISR, Kozhikode. Comprehensive overview of the current germplasm holdings of major spice crops conserved at ICAR–Indian Institute of Spices Research (IISR), Kozhikode, highlighting the diversity. *No permission required.*

are preserved in pots for further evaluation and characterization. Additionally, field gene banks are established where plants are grown for long-term assessment and conservation.

Every year, new accessions are collected and integrated into the conservation system to enhance genetic diversity. Evaluation of these accessions is conducted to study morphological and quality traits, including piperine content, oleoresin yield, and essential oil composition. These conservation efforts ensure the availability of diverse genetic material for breeding programs and sustainable cultivation of black pepper.

Germplasm characterization highlights variability in spike, yield, and quality traits. Sasikumar et al. (2015) documented genotypes with long spikes (up to 29.5 cm), while Shivakumar et al. (2019) found unique accessions with tri-monoecious flowers and low pungency. Shivakumar et al. (2019) identified genotypes with thin pericarp and low seed to pericarp ratio. Pannaga et al. (2021) reported high-yielding cultivars like Kurimale and Panniyur-1, and Sigandini with superior black pepper recovery. Chemical variability among cultivars affects traits such as piperine, oleoresin, protein, and phenols (Ashokkumar et al., 2021; Menon et al., 2002; Somashekar et al., 2021), underscoring the potential for selective breeding and crop improvement in black pepper.

15.2.3 Active principles

Black pepper essential oil and piperine exhibit multiple biological activities, including antioxidant, antimicrobial, anticancer, cytotoxic, insecticidal, and various health benefits (Ashokkumar et al., 2021).

Antioxidant effects: Piperine possesses strong antioxidant properties, protecting against lipid peroxidation and oxidative stress-related diseases (Vijayakumar et al., 2004). Essential oil scavenges free radicals and inhibits lipid peroxidation in tissues (Jeena et al., 2014).

Antimicrobial effects: Piperine and phenolic compounds in black pepper exhibit antibacterial activity against *Staphylococcus aureus, Bacillus subtilis, Escherichia coli,* and *Fusarium oxysporum*. Essential oil has potential as a natural preservative, inhibiting foodborne pathogens (Zhang, Ye, et al., 2017).

Anticancer effects: Piperine suppresses tumor growth in prostate and lung cancer cells (Samykutty et al., 2013). It induces DNA damage and apoptosis, making it a potential therapeutic agent for osteosarcoma, and piperine enhances the efficacy of tumor necrosis factor (TNF)-related apoptosis-cytotoxicity effects in breast cancer cells (Greenshields et al., 2015).

Insecticidal effects: Pepper essential oil shows insecticidal properties against *Amphimallon majale* and the storage pest *Tribolium castaneum*, with higher concentrations leading to increased repellent effects (Upadhyay & Jaiswal, 2007).

Piperine enhances the bioavailability of various drugs by inhibiting P-glycoprotein-mediated efflux (Vijayan & Thampuran, 2000). It stimulates gastric acid secretion and improves digestive enzyme activity, increasing nutrient absorption (Kalpana Platel and Srinivasan, 2004). Piperine is also used in traditional medicine for wound healing and pain relief. Black pepper extracts are widely used in antiaging creams, hair growth products, and skin detox formulations due to their ability to improve blood circulation and cell regeneration. Table 15.1 summarizes the major bioactive compounds and nutraceutical properties of major spice crops.

15.2.4 Breeding strategies

Varietal plant breeding is a crucial factor in agricultural progress. Genetic variability is essential for crop improvement, and the reproductive pattern plays a significant role in breeding success. Black pepper benefits from viable sexual reproduction and efficient vegetative propagation.

Breeding objectives for black pepper include high yield, improved quality (bulk density, dry recovery, essential oil, piperine, and oleoresin levels), and resistance to Phytophthora foot rot, nematodes (*Radopholus similis* and *Meloidogyne incognita*), pollu beetle (*Longitarsus nigripennis*), drought, and shade. Additionally, breeding aims to develop varieties suitable for low-input organic farming and mixed cropping (Krishnamoorthy & Parthasarathy, 2010).

15.2.5 Breeding methods

Selection is the oldest method of crop improvement, with three main approaches:

Selection from germplasm—The ICAR-Indian Institute of Spices Research (IISR) at Kozhikode, Kerala, maintains the world's largest black pepper germplasm collection. Promising selections include Panchami, Pournami (nematode-tolerant), and IISR Thevam (Phytophthora-tolerant).

TABLE 15.1 Major bioactive compounds and nutraceutical properties of major spice crops.

Spice	Major bioactive compounds	Major nutraceutical properties	References
Black pepper (*Piper nigrum L.*)	Piperine, chavicine, limonene	Enhances bioavailability of nutrients and drugs; exhibits antioxidant, anti-inflammatory, antimicrobial, and digestive stimulant properties.	Butt et al. (2013); Takooree et al. (2019)
Cardamom (*Elettaria cardamomum Maton*)	1,8-Cineole, α-terpinyl acetate, limonene	Possesses antioxidant, antihypertensive, digestive, and antimicrobial effects; supports respiratory health and lipid metabolism.	Abdullah et al. (2022); Ashokkumar et al. (2020)
Turmeric (*Curcuma longa L.*)	Curcumin, demethoxycurcumin, bisdemethoxycurcumin	Strong anti-inflammatory, antioxidant, anticancer, and hepatoprotective activities; enhances immune and neural functions.	Amalraj et al. (2017); Xu et al. (2018)
Ginger (*Zingiber officinale Roscoe*)	Gingerols, shogaols, paradols	Exhibits antiemetic, anti-inflammatory, analgesic, and digestive tonic effects; supports cardiovascular and metabolic health.	Sharifi-Rad et al. (2017); Shaukat et al. (2023)
Cinnamon (*Cinnamomum verum J. Presl*)	Cinnamaldehyde, eugenol, coumarin	Shows antidiabetic, lipid-lowering, antioxidant, and antimicrobial properties; improves glycemic control and gut health.	Shang et al. (2021); Sharifi-Rad et al. (2021)
Clove (*Syzygium aromaticum L.*)	Eugenol, β-caryophyllene,	Exhibits antioxidant, antimicrobial, analgesic, and anti-inflammatory properties; promotes oral and gastrointestinal health.	Haro-González et al. (2021); Idowu et al. (2021)
Nutmeg (*Myristica fragrans Houtt.*)	Myristicin, elemicin, sabinene	Displays antioxidant, antimicrobial, neuroprotective, and digestive properties; aids mood regulation and detoxification.	Malik et al. (2022); Sultan et al. (2023)

No permission required.

Clonal selection—Superior clones are identified from popular cultivars. Sreekara and Subhakara were selected from Karimunda, while PLD-2 were derived from Kottanadan for its high-quality oil and oleoresin content.

Selection from open-pollinated progenies—Due to residual heterozygosity, open-pollinated progenies offer variation. IISR-Shakti, selected from Perambramundi, Panniyur-2 from Balankotta, Panniyur-5 from Perumkodi, Panniyur-7 from Kalluvally, and Arka Coorg Excel, developed at CHES, Chettalli, Karnataka, are notable high-yielding selections.

TABLE 15.2 Black pepper varieties with specific quality traits.

Traits	Varieties
High piperine content	Sreekara, Panniyur-2, Panniyur-8, Panniyur-9, IISR Chandra
High oleoresin content	PLD-2, IISR Malabar Excel, Sreekara
High essential oil content	Sreekara, IISR Malabar Excel
High dry recovery	IISR Shakti and Panniyur-9

No permission required.

Hybridization Planned hybridization exploits heterosis and stabilizes hybrid vigor through vegetative propagation. The process involves parental selection, F1 progeny development, superior progeny selection, and clonal multiplication. Notable hybrids include Panniyur-1, Panniyur-3, Vijay, IISR Girimunda, and IISR Malabar Excel (Sasikumar et al., 2004). Hybrid progenies such as HP 117 × Thommonnakudi, HP 1411, and HP 780 are under multilocation trials (Shivakumar et al., 2020). Additionally, interspecific hybrids have been developed to incorporate pollu beetle resistance (Sasikumar et al., 1999).

Table 15.2 summarizes black pepper varieties developed and identified for specific quality traits such as high piperine, oleoresin, essential oil content, and dry recovery.

15.2.6 Biotechnological approaches

Marker-assisted selection (MAS) in black pepper has been advanced through molecular markers. (Kumari et al., 2019) identified 69,126 SSRs, with dinucleotides(48.6%) being the most abundant. (Joy & Soniya, 2012; Joy et al., 2013) reported SSR bias near pre-miRNA candidates. Jose et al. (2018) developed nine EST-SSR markers, validating five in 35 cultivars, yielding 18 alleles. (Das et al., 2025) identified 159,390 CNVs and 11,360 CNVRs, with 82,027 deletions and 77,363 duplications across 26 chromosomes. Anandaraj et al. (2008) identified a 360 bp RAPD marker (OPA-01) linked to Phytophthora resistance, later converted into a SCAR marker. Additionally, George et al. (2021) reported that the terpene synthase family is responsible for biosynthesizing flavor-imparting volatiles in black pepper berries. Kokkat et al. (2022) identified rotundone backbone genes viz., α-guaiene synthase and α-guaiene oxidase, in black pepper, which are responsible for aroma. CRISPR gene editing—efforts are being initiated for genome editing in Piper nigrum using CRISPR-Cas9, specifically targeting Phytophthora resistance.

15.2.7 High-volume macro- and micropropagation

Black pepper is primarily propagated vegetatively using rooted cuttings, ensuring true-to-type plants. Various propagation techniques have been developed to enhance multiplication rates and establishment success. The bamboo method (Sivaraman, 1992) enables rapid multiplication, achieving a 1:40 multiplication rate annually. The mound method involves growing runner shoots over a soil-cow dung mound, allowing all nodes to root before separating them for planting. The serpentine layering technique (Devasahayam et al., 2015) is an efficient and cost-effective method where nodes of growing vines are pressed into potting mixture-filled bags, leading to profuse rooting within three months. On average, 60 cuttings can be harvested per mother plant annually. Soil-less nursery mixtures containing partially composted coir pith and vermicompost enhance healthy planting material production. The trellis method is an alternative system for the production of runner vines, which are trained on trellises made of bamboo, metal, or synthetic materials. The vertical column method Devasahayam et al. (2015) utilizes soil-less media-filled mesh columns (2 m height, 0.3 m diameter) to produce high-quality planting materials, including single-node cuttings, laterals, and top shoots. Each column yields about 150 cuttings, 10–15 laterals, and 10 top shoots in 4–5 months, with three harvesting cycles annually. This method allows efficient use of space in polyhouses, significantly (Kadam et al., 2020) enhancing multiplication rates. Top shoots/orthotropic shoots are the best choice for early yielding.

Grafting black pepper onto resistant rootstocks like *Piper colubrinum* is an effective strategy to enhance disease resistance, particularly against Phytophthora root rot. *P. colubrinum* is known for its strong root system and resistance to soil-borne pathogens, making it a suitable rootstock for commercial cultivation (Vanaja et al., 2007).

15.2.8 Micropropagation

Various technologies for the micropropagation of black pepper using different explants have been reported (Nirmal et al., 2012). Multiple shoots can be induced in culture media (MS or SH medium) by using BA, either alone or in combination with auxins. However, endogenous contamination remains a significant challenge for the successful establishment of black pepper cultures (Bhat et al., 2024). Explants cultured in MS medium supplemented with 0.5 mg L-1 BAP produced an average of 6.5 nodes per plant and the longest shoots (10.42 cm) after 12 weeks. The regenerated shoots were then rooted in MS medium supplemented with 2.0 mg L-1 IBA, resulting in approximately 15 roots per plant and an average root length of 3.40 cm within 4 weeks (Deepak et al., 2025). Aarthi et al., 2020; Afroz et al., 2023 observed that MS medium supplemented with 2.0 mg L-1 BA and 2.0 mg L-1 IBA produced the best results for shoot induction, shoot length, and overall leaf production. Additionally, the combination of MS + 4.5 mg L-1 BAP + 1.0 mg L-1 IAA achieved the highest multiple shoot induction (69.33%) for the black pepper variety Panniyur-1 (Kadam et al., 2020).

Protocols have been standardized for the micropropagation of several endangered and medicinally important Piper species, such as *P. longum, P. chaba, P. betle, P. barberi*, and *P. colubrinum*. Nirmal Babu et al., Nirmal et al. (2012), and Umadevi et al. (2015) developed a meristem culture protocol for the rapid production of plantlets in a liquid medium containing Murashige and Skoog (MS) with 0.1 mg L-1 kinetin and 0.5 mg L-1 GA3. This was followed by direct shoot induction in ½ MS with 3 mg L-1 BA and 1 mg L-1 IAA, and shoot growth and development in ½ MS with 0.5 mg L-1 indole-3-butyric acid (IBA). Successful rooting of the meristem-derived shoots was achieved in half-strength WPM medium containing 3 mg L-1 BA and 1 mg L-1 kinetin.

Joseph et al. (1996) reported somatic embryogenesis from zygotic embryos, while Ramakrishnan Nair & Dutta Gupta (2003, 2006) documented cyclic somatic embryogenesis from maternal tissues, offering great potential for automated micro propagation. Nirmal Babu et al. (2005) reported somatic embryogenesis from mature leaf tissues, a protocol useful for transgenic experiments. Efforts to induce variability in somaclones for tolerance to Phytophthora foot rot by Shylala et al. (1996) led to the identification of tolerant somaclones through *in vitro* selection of calli, as well as somaclones using crude culture filtrate and toxic metabolites isolated from *Phytophthora capsici*.

Ramakrishnan Nair & Dutta Gupta (2007) developed an encapsulation protocol for the production of synthetic seeds from somatic embryos in black pepper. The optimal sodium alginate concentration was found to be 4% (w/v), with a calcium chloride concentration of 100 mM, producing capsules with ideal morphology and sufficient gel strength. Somatic embryos encapsulated in 4% sodium alginate could be successfully stored at culture room temperature on agar-gelled SH basal salt solution for up to 45 days. (Nair & Gupta, 2005) also studied the influence of various fruit-derived explants and genotypes on somatic embryogenesis induction in black pepper. Among the different explants cultured on plant growth regulator-free solid SH medium maintained in the dark, "abortively germinated seeds *in vitro*" produced the highest percentage response and the greatest number of somatic embryos per responding explant. Zygotic embryos, on the other hand, failed to produce any somatic embryogenic response. Of the genotypes tested, "Karimunda" was found to be highly embryogenic, while "Kutching" was completely nonembryogenic (Nair & Gupta, 2005).

Ahmad et al. (2016) found that regenerated tissues of *P. nigrum* serve as a promising source of biologically active metabolites with antimicrobial properties, with callus cultures showing potential for these activities as well. The antimicrobial activity of *in vitro* cultures (callus and shoots) and peppercorn extract was tested against *Escherichia coli, Pseudomonas aeruginosa, Salmonella typhi, Bacillus subtilis, Bacillus cereus, Staphylococcus aureus*, and *Candida albicans*. Ethanol, hexane, and chloroform extracts were evaluated using the well-diffusion method. Hexane extract from callus showed inhibition zones of 22 mm against *B. cereus* and 23 mm against *S. aureus*, while regenerated shoots and seeds showed zones of 24.3 mm and 26 mm, respectively. Ethanol extracts of shoots exhibited 25 mm activity against S. aureus and 21 mm against *B. cereus*. Peppercorn extracts showed activity against *B. cereus* (23.6 mm) and *B. subtilis* (23.5 mm). *In vitro* cultures produced phytochemicals similar to the mother plant, with growth regulators influencing tissue regeneration and secondary metabolite production.

15.2.9 Black pepper genomics

The availability of a high-quality, chromosome-scale genome assembly has provided insights into its genetic architecture, evolutionary history and biosynthetic pathways (Hu et al., 2019). The 761.2 Mb genome, organized into 26 pseudochromosomes, has revealed lineage-specific gene expansions in families related to secondary metabolism, particularly glycosyltransferases, cytochrome P450s, and acyltransferases, which are crucial for the biosynthesis of piperine, the major bioactive compound in black pepper (Hu et al., 2019). Additionally, transcriptome profiling has identified key regulatory elements governing the expression of genes involved in piperine biosynthesis, such as phenylpropanoid and polyketide pathways (Sreekumar et al., 2022). Such findings have implications for genetic improvement strategies targeting enhanced metabolite accumulation and stress resilience.

The sequencing of the chloroplast genome has further enriched phylogenetic and evolutionary studies of *Piper nigrum*, providing a 161,522 bp quadripartite structure with 131 genes, including 81 protein-coding genes, 37 tRNAs, and 4 rRNAs (Gaikwad et al., 2023). Comparative genomics within the *Piperaceae* family has identified simple sequence repeats (SSRs) that serve as valuable markers for genetic diversity assessments and breeding programs (Kumari et al., 2019). Additionally, the genome-wide identification of transcription factors such as MYB, WRKY and bHLH families has elucidated their role in stress adaptation and metabolic regulation (Qian et al., 2021). Another study by Fayad et al. (2024) identified a BAHD acyl transferase isoform on chromosome 6 of *Piper nigrum*, which exhibited the highest correlation with piperine biosynthesis, supported by expression profiling, transcription factor binding site analysis, and molecular docking.

MicroRNA (miRNA) studies have also revealed regulatory networks influencing secondary metabolite biosynthesis, with small RNA sequencing identifying miRNAs such as miR166 and miR396, which are implicated in modulating the expression of genes involved in piperine biosynthesis (Ding et al., 2021). Genomic and transcriptomic resources have greatly facilitated breeding and propagation strategies for black pepper. Furthermore, transcriptome-based studies on root and fruit development have revealed differentially expressed genes associated with growth regulation, carbohydrate metabolism, and stress tolerance (Khew et al., 2020). Such knowledge has informed breeding programs aimed at developing high-yielding and disease-resistant varieties.

15.3 Turmeric

Turmeric, scientifically known as *Curcuma longa*, is a bright yellow spice commonly used in cooking, traditional medicine, and cosmetics. Native to South Asia, particularly India, turmeric is derived from the rhizome of a plant in the Zingiberaceae family.

15.3.1 Genetic resources

Smart & Simmonds (1992) have reported about 70 curcuma species, and it is also believed that the comprehensive taxonomic revision may lead to some more species in this genus. Velayudhan et al. (1999) reported about 117 among which 40 species are reported in India, which acts as a secondary genepool for turmeric. The highest Curcuma diversity is concentrated in India and Thailand, followed by Burma, Bangladesh, Indonesia, and Vietnam.

Species such as C. aromatica (Kasturi manjal), *C. caesia* (Black turmeric), C. amada, *C. zedoaria*, *C. purpurescens*, *C. mangga*, C. heyneana, *C. xanthorrhiza*, *C. aeruginosa*, *C. phaeocaulis*, and *C. petiolata* are also cultivated in different places and regions. *C. rubrobracteata* (Mizoram, India), *C. codonantha* (Andaman Islands), and *C. mutabilis* (South India) are the new species reported in India (Škorničková et al., 2003a, 2003b) and (Škorničková et al., 2004). Detailed description (Škorničková et al., 2003b) of C. roscoeana wall is reported by Škorničková & Sabu (2005).

Cultivars are the land races cultivated by farmers, in turmeric around 50 commercial land races are available. The cultivars are classified based on the duration they are, the short duration cultivars are G.L. Puram, Kasturi CA-69, Jabedi, Dughi, Dindigram, Katigia, Udayagiri and Amalapuram. The cultivars characterize as medium duration types or Kesari types are Kesari, Kothapeta, Gorakpur, Amruthapani, and Amalapuram. The long duration types are Armoor, Duggirala, Mydukur, Tekurpet, and Avanigadda (Sasikumar et al., 1994).

(Prasath et al., 2018a) reported popular cultivars pertaining to different turmeric growing states. The cultivar Duggirala, Mydukur, Armoor local, Cuddapah, Kodur, Tekurpet, Kasturi, Chaya pasupu, Armoor, G. L. Puram, Amalapuram, Rajapur, Nandyal, Avanigaddu, Vondimitta belongs to Andhra Pradesh andamp; Telangana. Erode Local, Salem Local from Tamil Nadu. Shillong and Tall Karbi from Assam. Alleppy, Moovattupuzha, Wyanad local, Mananthody, Mannuthy local cultivars from Kerala. Rajapuri, Sangli local, Sugandham from Maharashtra.

Lakadong and Aieng from North East region. The cultivars from Odisha are Dindigam, Dubgi, Jobedi, Katingi. ICAR–Indian Institute of Spices Research (IISR), Kerala, India, is recognized as the National Active Germplasm collection center for turmeric with the prime objective of collection and conservation of genetic resources of turmeric in India, where 1404 Curcuma species are conserved in field gene bank. In addition, collections of turmeric germplasm are also maintained at various research centers under AICRPS (All India Coordinated Research Project on Spices, India). The germplasm is maintained in the field gene bank.

15.3.2 Active principles

The major bioactive compounds, are Curcuminoid, β-Caryophyllene oxide, Turmeronol B, Germacrene D, β-Caryophyllene which are source of antioxidant activity in turmeric, Curzerene, Curcumin, ar-Turmerone, Zingiberene, β-Sesquiphellandrene, Germacrone, β-Elemene, β-Turmerone, Demethoxycurcumin, Curcumenol, Eucalyptol, Bisdemethoxycurcumin, Curdione, Germacrone, β-Elemene, Tetrahydrocurcumin, β-Caryophyllene oxide; Turmeronol bioactive compounds are reported by various researchers in their clinical studies where each compounds medicinal use is explained (Chen et al., 2018; Jayaprakasha et al., 2006; Megumi et al., 2017; Radwan et al., 2014; Wang et al., 2016; Yang et al., 2022); Krup et al., (Koohpar et al., 2015; Krup et al., 2013; Wang et al., 2017; Zhang, Yang, et al., 2017). The bioactive compounds of turmeric are for the antibacterial and fungal activity, antidiabetic activity, wound healing activity, anti-obesity activity, and antiaging activity.

Aarthi et al. (2025) reported 16 volatile compounds: α-phellandrene, 1,8-cineole, α-terpinolene, α-santalene, α-bergamotene, caryophyllene, β-santalene, β-farnesene, α-humelene, α-curcumene, zingiberene, β- bisabolene, β-sesquiphellandrene, ar-turmerone, α-turmerone, and curlone. The correlation of the constituents reveals that β-bisabolene, β-farnesene, α-santalene, α- bergamotene, and β-santalene are positively correlated with each other. Zingiberene, β- sesquiphellandrene, 1,8-cineole, and α-humelene are positively correlated with each other.

The range of curcuminoid composition reported in Indian turmeric varieties was CUR 52%–63%, DMC 19%–27%, and BDMC 18%–28%, which falls in range with curcumin as a major component in the total curcuminoids (Siviero et al., 2015). Each constituent of curcuminoids has its own uses in pharmaceutical industries (Amalraj et al., 2017). (Li et al., 2014) reported that demethoxycurcumin (DMC) was quite stable than curcumin and bisdemethoxycurcumin. The selection of genotypes based on the phyto constituents can be utilized for specific industrial purposes.

15.3.3 Breeding strategies

Clonal selection, seedling selection, and induced mutation were the breeding strategies followed in turmeric for many years. With the recent report on viable seed set in turmeric, recombination breeding program is also attempted in this clonally propagated crop. The germplasm is utilized for developing different variety and breeding lines. In vegetatively propagated crops the most practiced and efficient method of breeding followed is clonal selection, through which high-yielding varieties in turmeric were developed. The selection was mainly applied to land races collected from different turmeric growing areas of the country. Earlier, the varieties were developed for the state like, high-yielding variety named Krishna was developed for general cultivation in Maharashtra through clonal selection Pujari et al., (Pujari et al., 1986). Maurya (Maurya, 1990) recommended RH-10 Rajendra Sonia for Bihar, India. Three selections, PTS-55, BSR-1, and PTS-11 were superior suitable for cultivation in high altitude areas of Visakhapatnam district, Andhra Pradesh (Naidu & Mutrhy, 2013). The varieties Rajendra Sonia, Megha Turmeric 1 and IISR Prathiba are stable and high yielders at Kozhikode, Kerala (Kandiannan et al., 2015). Considering yield and nematode resistance, two promising turmeric accessions, Acc. 48 and Acc. 79 were identified, later Acc. 48 was released as IISR Pragati (Prasath et al., 2016). Krishnamoorthy et al. (2011) have reported many improved varieties of turmeric released through clonal selection from research organizations in India.

The germplasm is the source to develop open-pollinated seedlings to induce variability, as it is being heterogenous crop. The information on pollination and seed set indicates that turmeric is a cross-pollinated crop. Where, in earlier seed set was observed and no seed set was recorded in C. longa. Nazeem et al. (1993) tried selfing, crossing, and open pollination studies conducted by using 3 cultivars of *C. longa* and 5 cultivars of *C. aromatica* resulted in no seed set by selfing, and out of the 11 cross combinations, 8 resulted in seed set, and the one open pollination produced maximum seed set. They suggested self-incompatibility as the reason for the absence of seed setting in self-pollinations. Renjith et al. (2001) did crossing studies conducted by using 2 medium-duration and 5 short-duration cultivars of turmeric resulted in seed set in 3 out of 12 cross combinations, all involving short-duration types. Thus, it appears that some compatibility mechanisms operate in the crosses involving different cultivars of turmeric, which has to be further investigated. A case of seed setting in cultivated turmeric types was reported by Lad (1993). (Nair et al., 2004) obtained seed set in two accessions of *C. longa,* which was later repeated in many germplasm collections.

Evaluation of 15 open-pollinated seedling progenies resulted in two promising lines, which were released as IISR Prabha and IISR Prathibha (Sasikumar et al., 1996). (Nair et al., 2010), revealed that the seedling progenies had significant variation from the mother variety with respect to chromosome number and other characteristics. Thus, it is evident that open-pollinated seedling progenies have the potential to generate sufficient variability for the selection of better genotypes for commercial cultivation. Mutation is another source of inducing variability in the existing germplasm. Physical mutagens were mostly successful in turmeric breeding. Effect of gamma rays in inducing mutations in turmeric was reported by Rao (1999). CO-1, BSR-1, and BSR-2 are mutant variety developed by TNAU and these were released for large-scale cultivation (Chezhiyan & Shanmugasundaram, 2000). Suroma is another mutant selection after irradiating with x-rays. The viable mutation frequency was higher (2.40%) at 2.0 kR gamma rays in the genotype, CL144 (Usha & Chezhiyan, 2007). Ravindran et al. (2007) experimented with chemical mutagens viz., Colchicine, EMS, and MNG at 20, 500, and 1000 ppm on cv. Mydukar resulted in taller and high-yielding colchiploids. Induction of mutation in combination with *in vitro* culture technique may be an effective method for crop improvement in a vegetatively propagated crop like turmeric.

The study conducted by Sharon (2012) on *in vitro* mutant of turmeric cv BSR-2 revealed that at 1.50 kR dosage of gamma rays which had excelled in performance in terms of fresh rhizome yield per plant. Also, a short-duration mutant (132 ± 5.15 days) was isolated at 1.25 kR. The best result toward development of multiple shoots and the number of leaves from irradiated explants was obtained in 5 Gy (4.15 and 10.75) gamma ray-treated explants at 60 days after inoculation, respectively (Priya et al., 2014). Heterosis breeding has not yielded any varieties as of now in turmeric, but attempt was done (Nazeem et al., 1993), who succeeded in establishing hybrid progenies in crosses between longa and aromatic type of turmeric. (Renjith et al., 2001) obtained seed set in crosses involving short-duration types. The success in developing pure line by selfing may pave the way for developing heterotic hybrids in turmeric. The seedling progeny, which had a heritable morphological marker (red color in emerging shoots), was used in developing three intervarietal hybrids. Among that, hybrid-1 showed the highest yield and was on par with pollen parent Suranjana. The varieties of turmeric with high essential oils are Sudarshana and Suvarna (7%), and high curcumin turmeric IISR Prathiba, IISR Pragati, Megha Turmeric, and Rajendra Sonia, among which IISR Prathiba and IISR Pragati are CVRC-notified varieties. Recently, IISR Surya (Acc 849) is a variety identified for powdering industry where the rhizome powder color is light yellow with unique aroma with the presence of more zingiberene (21.07%), ß-sesquiphellandrene (14.13%), 1,8-cineole (3.42%), and α humelene (6.30%) compared to the other released variety (Aarthi et al., 2025).

The difference in the profile ratio of individual curcuminoids, that is, bisdemethoxycurcumin (BDMC), demethoxycurcumin (DMC), and curcumin (CUR) is very evident in the germplasm accession studied. Most of the released varieties have the profile of CUR > DMC > BDMC (Suvarna, IISR Prathiba, SC-61, BSR-2, Varna, Duggirala Red, Megha Turmeric 1, Punjab Halid 1, and Rajapuri) and CUR>BDMC>DMC (IISR Pragati, SLP 389/1, CO2, Rajendra Sonia, and Punjab Haldi), whereas Acc 849 and Narendra Haldi 98 have a unique profile of DMC > BDMC > CUR with total curcuminoids of 1.45 % (Aarthi et al., 2020).

(Silaru et al., 2024) reported that yield and quality are affected by the production system, and yield was found to be best in the greenhouse, and quality traits performed best in field conditions. Greenhouse condition was found to be the best followed by field condition for fresh rhizome yield and essential oil content, whereas field condition was best for dry recovery, oleoresin, and curcuminoids content.

15.3.4 Macro propagation

Turmeric is a rhizomatic crop that is commonly propagated vegetatively using rhizomes as planting material. Commercially, 20–25 g rhizome pieces are used as planting units. (Hossain et al., 2005) investigated the effects of seed rhizome size on the growth and yield of turmeric and found that rhizomes with larger diameters produced more vigorous seedlings. Rhizomes weighing 30–40 g and 50 g resulted in plants with the highest plant height (140 cm), tiller number (3.5), leaf number (12–14/plant), and shoot dry weight (40 g/plant), which were significantly greater than those from lighter rhizomes. (Malhotra et al., 2021) reported that the single bud technique for seedling production in trays was more advantageous than direct planting. Traditional turmeric planting methods require a larger quantity of quality planting material, which is limited in availability and more costly. Additionally, it takes about five to six months for rhizomes to develop and for the crop to establish properly. To address these challenges, the rapid multiplication of turmeric using the single bud method is considered an effective solution (Thapa et al., 2017). The single bud technique of propagation uses only single buds with small piece of rhizomes weighing 5-7 g, which are sown in portrays one month before actual field planting, and portrays are maintained at 50% shade net house, with rose can irrigation support and transplanted to field after 30–35 days, where the seedlings will be in two- to three-leaf stage.

15.3.5 Micropropagation

The pioneering study on the micropropagation of turmeric was published by Nadgauda et al. (1978). Since then, numerous studies have focused on using rhizome buds as explants for turmeric micropropagation (Bandara et al., 2023; Das et al., 2010; Jie et al., 2019; Rahman et al., 2004; Soundar Raju et al., 2015; Tyagi et al., 2007). According to Rahman et al. (2004), the best shoot proliferation from rhizome bud explants was observed in MS medium supplemented with 2.0 mg L^{-1} BA (6-benzyladenine), where 100% of the explants produced shoots. Additionally, effective shoot regeneration was achieved when explants were treated with 2.0 mg L^{-1} BAP + 0.2 mg L^{-1} IAA (94.44%) or 2.0 mg L^{-1} BAP + 0.5 mg L^{-1} IAA (86.10%) (Sinchana et al., 2020). Evidence also suggests that a combination of 3.0 mg L^{-1} BAP + 3.0 mg L^{-1} IAA produces the highest number of shoots (4.20) and the longest shoot length (5.22 cm).

For root induction, plantlets regenerated on half MS medium with 2.00 mg L^{-1} IBA have been used by several researchers to achieve better rooting in turmeric. Rahman et al. (2004) demonstrated that 0.2 mg L^{-1} IBA resulted in 100% rooting within six weeks of culture. The highest number of shoots was observed with 5.0 mg L^{-1} IBA per shoot. Additionally, the use of 0.01 mg L^{-1} NAA also resulted in a higher number of roots in turmeric.

The induction of micro rhizomes *in vitro* has been well-documented by various researchers (Chowdhury et al., 2020; Nayak, 2000; Nirmal Babu et al., 1997; Rajan, 1997; Sarma et al., 2021; Shirgurkar et al., 2006; Sunitibala et al., 2001; Zheng et al., 2008). The most commonly used method involves culturing plantlets on MS medium containing 6-

benzyladenine (BA) (1–5 mg L^{-1}), an enhanced concentration of sucrose (50–100 g L^{-1}), and a reduced photoperiod (0–8 hours). Micro rhizomes typically form at the base of the shoots, with their weight ranging from 40 to 700 mg. The interaction of factors such as BA, sucrose concentration, and photoperiod has a significant impact on micro rhizome induction (Swarnathilaka et al., 2016). *In vitro* production of both micro rhizomes and mini rhizomes in turmeric cultivar Alleppey Supreme, along with a comparative anatomical and histochemical analysis, was reported by Archana et al. (2014). The highest response was observed in liquid MS medium with 80 g L-1 sucrose in Planton culture vessels.

The primary curcuminoids—curcumin (CUR), demethoxycurcumin (DEM), and bisdemethoxycurcumin (bis-DEM)—were analyzed in turmeric using LC-DAD-ESI-MS. Significant curcumin production (>260 μg/g fresh weight) was observed in *in vitro* micro rhizomes cultured in MS medium, supplemented with 0.1 mg L-1 kinetin, 1 mg L-1 NAA, 6% sucrose, 5% agar, and 0.1% activated charcoal. These micro rhizomes had lower levels of DEM and bis-DEM compared to curcumin. A shoot culture line was also found to produce curcuminoids in ratios similar to those in fresh rhizomes of parent plants (Pistelli et al., 2012). Phytochemical analysis of callus extracts after six subcultures showed 53.6 mg g^{-1} of total polyphenols and 36.1 mg g^{-1} of curcumin (dry weight), with higher antioxidant activity than turmeric rhizome extracts. The lateral bud callus of *C. longa* presents a promising source for synthesizing secondary metabolites with pharmaceutical potential (Gurav et al., 2020). Rostinawati et al. (2023) explored the effect of cytokinin and auxin on curcuminoid production in turmeric explants grown on MS medium. Four treatments were tested: A (control), B (9 mg L^{-1} cytokinin + 1 mg L^{-1} auxin), C (1 mg L^{-1} cytokinin + 1 mg L^{-1} auxin), and D (0.1 mg L^{-1} cytokinin + 1 mg L^{-1} auxin). After extracting with 96% ethanol, curcuminoids were analyzed via TLC-densitometry. Cytokinin and auxin significantly reduced curcumin and bisdemethoxycurcumin levels, but treatment C (1 mg L^{-1} cytokinin + 1 mg L^{-1} auxin) led to a notable increase in demethoxycurcumin content (0.78 mg L^{-1}).

DNA-based markers are critical tools for assessing the clonal fidelity of micropropagated plants, as they are unaffected by environmental factors (Linacero & Ballesteros, 2024). Rani et al. (2024) demonstrated that micropropagated turmeric shoots and their mother plant produced identical monomorphic bands with six RAPD markers, confirming genetic fidelity in the micropropagated clones. Two genome-based molecular markers, inter-simple sequence repeats (ISSR) and start codon-targeted (SCoT) markers, were used to validate the genetic stability of tissue-cultured plants from the Salem genotype. These markers showed uniformity across plants derived from all explants (shoot buds, half shoots, and shoot slices) when compared to the mother plant (Pittampalli et al., 2022).

15.3.6 Turmeric genomics

The first draft genome sequence of C. longa was assembled using 10x Genomics linked reads, revealing a genome size of approximately 1.02 gigabase pairs (Gbp) with a high percentage of repetitive sequences and 50,401 predicted protein-coding genes (Chakraborty et al., 2021). This initial sequencing laid the foundation for understanding turmerics genetic complexity. Further advancements led to a chromosome-scale genome assembly using Oxford Nanopore long reads and 10x Genomics linked reads, which improved genome contiguity and enabled the identification of key genes involved in curcumin biosynthesis and rhizome formation (Yin et al., 2022). Chloroplast genome sequencing has been instrumental in understanding the evolutionary relationships of Curcuma species. The complete chloroplast genomes of 14 Curcuma species, including C. longa, have been sequenced, revealing conserved genome organization and providing phylogenetic insights (Liang et al., 2020). This genomic data also serves as a resource for developing genetic markers and evolutionary studies. Transcriptomic studies have provided further insights into gene expression patterns in turmeric. De novo transcriptome sequencing of four elite C. longa cultivars revealed cultivar-specific sequence variations that contribute to the genetic diversity of turmeric (Sahoo et al., 2019). Such transcriptome-wide variations are crucial for understanding metabolic pathways and guiding breeding programs aimed at improving curcumin content and disease resistance. High levels of polymorphism have been detected, reflecting substantial genetic variation among turmeric genotypes. This genetic diversity is essential for the development of improved cultivars through marker-assisted selection (MAS) strategies (Verma et al., 2015).

15.4 Ginger

Ginger is a herbaceous perennial crop that belongs to Zingiberaceae family usually grown as an annual for its rhizomes. It is a valuable cash crop and widely used in food, beverage, confectionery, and medicine. Ginger is a tropical crop and can be cultivated up to an elevation of 1500 m above sea level. It requires ample moisture for normal growth, shade loving in nature, and requires cool and dry weather for rhizome development. The plant is erect with fibrous roots, aerial shoots (pseudostem) with leaves, and underground stem (rhizome), and fleshy roots are produced from rhizome. Inflorescence is a spike with bisexual flowers that emerges from the rhizome. Flowering is not common in ginger and natural seed set is not reported so far which hinders the creation of variability and crop improvement. It is clinically proven that ginger has antiemetic, antiulcer, antiplatelet, and antiinflammatory properties and antioxidant in nature.

15.4.1 Genetic resources

Ginger is originated in South-East Asia, and the major ginger growing countries are India, China, Jamaica, Taiwan, Sierra Leone, Indonesia, Brazil, and Japan. The actual origins of ginger are uncertain, and its domestication history is not definitely known. Ginger germplasm includes land races, improved varieties, polyploids, exotic collections, related species, and related taxa of *Zingiber*. Genus *Zingiber* contains 150 species, and among them, 27 are reported from India. Abundant cultivar diversity of ginger is reported in India and China, the center of origin/diversity of this crop. Ginger germplasm collection in India has been initiated by National Bureau of Plant Genetic Resources, New Delhi, in collaboration with Central Plantation Crops Research Institute, Kasaragod, and Himachal Pradesh Agricultural University, Palampur, during 1976-1977 (Thomas, 1980). At present, ICAR-Indian Institute of Spices Research, Kozhikode, is the national germplasm conservatory of ginger, and around 668 accessions are being conserved in field gene bank. Besides this, 514 ginger germplasm accessions are being conserved in various AICRP on Spices center's (Mukesh Sankar et al., 2022).

15.4.2 Active principles

Chemical analysis showed the presence of over 400 different compounds in ginger. The major constituents in ginger rhizomes are carbohydrates (50%–70%), lipids (3%–8%), proteins, minerals, fiber, terpenes, and phenolic compounds. Terpene components of ginger include zingiberene, β-bisabolene, α-farnesene, β-sesquiphellandrene, and α- curcumene, while phenolic compounds include gingerol, paradols, and shogaol. Gingerols (23–25%) and shogaol (18–25%) are found in higher quantity than others (Prasad & Tyagi, 2015). Ginger contains proteolytic enzyme (zingibain), vitamin C, vitamin B6, magnesium, iron, phosphorus, calcium, linoleic acid, and potassium (Oladosu et al., 2024). The major biochemical constituents of ginger are oleoresin, essential oil, and crude fiber (Kizhakkayil & Sasikumar, 2011). Oleoresin content of ginger varied from 3% to 11%; Crude fiber content of dried ginger ranged from 4.8% to 9%, while essential oil content of ginger varied from 0.2% to 3% (Ekundayo et al., 1988; Vernin & Parkanyl, 2005). Ginger oleoresin is extracted from dried ginger, and ginger oil is prepared by steam distillation of both fresh and dry ginger rhizomes. Shogaol and gingerol are the main active ingredients in oleoresin, which imparts its pungent taste and medicinal properties. Bisabolene, curcumene, zingiberene, farnesene, and sesquiphellandrene are the main constituents of ginger essential oil (Mao et al., 2019). Zingiberene and bisabolene are the aromatic constituents. Other phenolic constituents of ginger include camphenes, lisabolene, limoline, citrol, borneol, linaloal, cibonellol, geranial, and cineole (Oladosu et al., 2024). Fiber, volatile oil contents, and pungency level are the most important criteria in assessing suitability of ginger rhizomes for particular processing purposes. Essential oil profile of ginger varies with change in geographical region, variety, climatic conditions, and processing methods adopted.

15.4.3 Breeding strategies

Crop improvement in ginger is aimed at developing high-yielding varieties with wider adaptability, high quality, low fiber, resistance to pests and diseases. The breeding methods in ginger are confined to introduction, germplasm selection, mutation, and polyploidy breeding due to lack of seed set (Ravindran and Babu, 2005). In India, there are many local cultivars of ginger specific to different geographical locations/growing regions. These local cultivars include Assam, Burdwan, Ernad chernad, Gurubathani, Himachal Pradesh, Jorhat, Karakkal, Kuruppampadi, Manamthodi, Maran, Nadia, Narasapattam, Poona, Taffingiva, Thinladium, Thingpuri, Tura, Uttar Pradesh, Valluvanad, Wayanad local, Wayanad, and Kunnamangalam (Mohanty & Panda, 1994; Sial et al., 2016). Major exotic cultivars in ginger are Rio-de-Janeiro, China, Jamaica, etc. Improved varieties in India are developed through selection. Varieties developed from ICAR-IISR, Kozhikode, are IISR Varada, IISR Mahima, IISR Rejatha, and IISR Vajra. From High Altitude Research Station, Pottangi, four varieties *viz.*, Suprabha, Suruchi, Suravi, and Sourabh, are released. Two varieties Himagiri and Solan Giriganga are released from Dr. Y.S. Parmar University of Horticulture and Forestry, Nauni, Solan, Himachal Pradesh. From Uttarabanga Krishi Viswavidyalaya, Pundibari, West

Bengal, a ginger variety 'Mohini' with bold rhizomes was released. Three varieties of ginger *viz.*, Athira, Karthika, and Aswathy with high essential oil content, were released from Kerala Agricultural University.

15.4.4 Macropropagation

Ginger is propagated by portions of rhizomes known as seed rhizomes. Seed material should always be selected from disease-free gardens and should be stored in proper storage conditions to avoid rotting, shrivelling, and dehydration losses. Usually, the seed rhizomes are stored in well-ventilated area by covering with leaves, and the ideal temperature for storage is 22°C–25°C. Before storage, the seed rhizomes are treated with mancozeb or carbendazim 0.3% (3 g/L of water) and Quinalphos (0.05%) for 30 minutes and dried in shade for 3–4 hours. For organic cultivation, 1% Bordeaux mixture or Trichoderma can be used. The stored seed rhizomes should be examined in monthly intervals and remove the rotten ones to keep the pathogen inoculum at low level. Carefully preserved seed rhizomes are cut into small pieces of 2.5–5.0 cm length, weighing 20–25 g each with two to three good buds. The seed treatment needs to be repeated before planting. Seed rate varies from region to region and with the method of cultivation adopted. In Kerala, the seed rate of ginger varies from 1500 to 1800 kg/ha. At higher altitudes, the seed rate may vary from 2000 to 2500 kg/ha. The seed rhizome bits are placed in shallow pits of 5–10 cm depth prepared with a hand hoe and covered with well-decomposed farm yard manure and a thin layer of soil, and leveled. The application of compost/ oil cakes and a basal dose of fertilizer is to be done at the time of planting. Bolder seed rhizomes can be planted deeper, and smaller ones at shallow depths. Under irrigated conditions light irrigation after planting is beneficial. It takes 10–15 days for sprouting after planting under ideal conditions.

The main disadvantages of conventional propagation method are high seed rate, slow germination due to dormancy period, more chances for pests and diseases incidence. A rapid method of multiplication, pro-tray-based single-sprout transplanting technique, was standardized in ginger. In single-sprout transplanting technique, the seed rhizomes are cut into small pieces of rhizomes weighing 4–6 g containing a single bud and planted in 98-well pro-trays with nursery medium containing partially decomposed coir pith and vermicompost (75:25). The pro-trays need to be kept under shade net house, and need-based irrigation is given. The seedlings are ready for field planting after 30–40 days of planting. The advantage of this planting method includes reduced seed rate, disease-free planting material, 98–100 percent field establishment, high yield, less seed cost, and high cost:benefit ratio (Prasath et al., 2018b).

15.4.5 Micropropagation

Several studies have been conducted to develop efficient protocols for *in vitro* propagation of ginger. The pioneering work on the micropropagation of ginger was published by Hosoki and Sagawa (1977) and Nadgauda et al. (1980). This foundational research has been followed by numerous studies from various researchers, including (Balachandran et al., 1990; Miri, 2020; Seran, 2013; Zahid et al., 2021; Zapata et al., 2003), and (Shaaban et al., 2023), among others. Kambaska and Santilata (Kambaska & Santilata, 2009) found that MS medium containing kinetin or BAP combined with NAA produced 7.0–7.5 shoots per explant. The highest shoot number (5.33) was recorded on MS medium supplemented with 5 mg L-1 BA and 0.5 mg L^{-1} NAA (Sathyagowri & Seran, 2011), while lower concentrations of TDZ (0.5 mg L^{-1}) resulted in up to 8 shoots per explant (Hamirah et al., 2007).

In vitro propagation using inflorescence tissues such as flower buds or bracts has significant advantages over rhizome-based methods, including reduced microbial contamination and improved tissue culture success (Widyastuti et al., 2024). Floral tissues, being totipotent, can regenerate complete, virus-free plants through somatic embryogenesis and micropropagation (Babu et al., 2003). (Ramachandran & Nair, 1992) reported callus formation, root development, and rhizome-like structures in excised ginger anthers cultured on MS medium supplemented with 2,4-D and coconut milk. (Samsudeen et al., 2000) achieved successful plantlet regeneration from excised anthers containing uninucleate pollen mother cells and pollen.

Protoplasts were isolated from young *in vitro*-derived ginger leaves using an enzyme mix of 0.5% macerozyme R10, 3% hemicellulase, and 5% cellulase Onozuka R10, followed by mechanical maceration. After incubation at 15°C for 10 hours and 30°C for 6 hours, this process yielded 2.5×10^5 protoplasts per gram of leaf tissue with 55% viability. The leaf-derived protoplasts were round, chloroplast-filled, and heterogeneous in size (0.15–0.21 mm) (Babu et al., 2016). These protoplasts were cultured in MS liquid medium supplemented with 0.5 mg L^{-1} each of BAP, NAA, and 2,4-D, along with 3% sucrose and 7% mannitol. Within 2–3 days, cell walls regenerated, and cell division began within 3–4 weeks, forming microcalli in 50–70 days on MS medium with 1 mg L-1 each of NAA and BAP (Geetha et al., 2000). Somatic hybridization in ginger was achieved through polyethylene glycol (PEG)-mediated protoplast fusion. Protoplasts from three ginger cultivars, isolated from embryogenic cell suspensions, were fused with each other. The highest binary fusion rate (13.5%) was observed when fusing 'Lushan Zhangliang Jiang' and 'Chenggu Huang Jiang' (LZ + CH) using 30% PEG6000 for 15 minutes (Guan et al., 2010).

Micro rhizomes of ginger, which are small, disease-free planting materials produced through *in vitro* techniques, offer a crucial solution to challenges posed by soil-borne diseases like bacterial wilt (Aravind et al., 2023). Research has shown that optimal conditions for inducing micro rhizomes include specific concentrations of sucrose and plant growth regulators such as 6-benzylaminopurine (BAP) and zeatin, which enhance both the quantity and viability of micro rhizomes(Zahid et al., 2021). Micro rhizomes were induced at the base of *in vitro*-derived shoots by transferring them to MS medium containing varying concentrations of sucrose (30, 60, and 90 g L^{-1}) and BAP (3, 6, and 9 mg L^{-1}). Among the ginger varieties studied, IISR Varada exhibited the highest cultured plant mass (96.0 ± 4.41 g) and the greatest micro rhizome weight (32.6 ± 1.92 g), followed by Karthika (27.1 ± 1.19 g), IISR Rejatha (27.0 ± 1.79 g), and IISR Mahima (24.5 ± 1.10 g) (Aravind et al., 2024).

The production of gingerols and zingiberene was compared between callus cultures, *in vitro* regenerated plants, and *in vivo* glasshouse-grown ginger. Fresh calli, *in vitro* rhizomes, and *in vivo* rhizomes were extracted using dichloromethane and analyzed by TLC. No gingerols or zingiberene were detected in the callus cultures, whereas both *in vitro* and *in vivo* rhizomes produced these compounds. This suggests that ginger callus cultures are suitable for rapid plant cell proliferation (Dehghani et al., 2011). Phytochemical analysis revealed no significant differences in total phenolic acids, total flavonoids, or antioxidant activities between micropropagated and conventionally propagated 'Bentong' ginger (Zahid et al., 2021).

Arijanti and Suryaningsih (2019) investigated quantity and quality of callus in three ginger varieties (Gajah, Emprit, and Merah), as well as the ginger oil derived from these calluses. Gajah ginger produced the largest amount of callus, while Emprit and Merah formed more compact calluses. Emprit ginger showed the highest levels of secondary metabolites, including 1.181% gingerol, 0.118% shogaol, and 0.098% zingerone. Additionally, unique compounds such as 4-H-Pyran-4-one, 2,3-dihydro-3,5-dihydroxy-6-methyl-, phenol, and phenolic glucoside were found only in callus cultures. Methyl esters, fatty acids, and phytosterols were present in both leaf sheath and callus extracts, with callus extracts demonstrating higher antioxidant activity (Gnasekaran et al., 2023).

15.4.6 Ginger genomics

A pivotal advancement in ginger genomics is the development of a high-quality, haplotype- resolved chromosome-level genome assembly for cultivated ginger. This assembly has elucidated the complex genome structure and facilitated the identification of genes involved in key metabolic pathways, particularly those responsible for the biosynthesis of gingerol analogs. These insights are instrumental in understanding the unique characteristics of ginger and offer a platform for further biological and agronomic studies (Cheng et al., 2021). The complete chloroplast genome of Z. officinale has been sequenced, revealing a 162,621 base pair structure with a four-part circular configuration and a GC content of 36.1%. This information is crucial for molecular research and phylogenetic analyses within the Zingiberaceae family, aiding in the elucidation of evolutionary relationships and the development of genetic markers for breeding programs (Li et al., 2021). Comprehensive genome-wide analyses have identified and characterized several gene families in ginger such as bHLH transcription factors (Liu et al., 2025), MYB Transcription factors, PP2C Gene family (Zhang et al., 2024), AP2/ERF Transcription factors (Xing et al., 2021). The allele-aware genome assembly has facilitated the prediction of transcription factors potentially regulating 6-gingerol biosynthesis, the primary bioactive compound in ginger. This understanding enables targeted approaches to enhance the production of therapeutic compounds through metabolic engineering or selective breeding (Cheng et al., 2021).

15.5 Small cardamom

Small cardamom (*Elettaria cardamomum* Maton.), belonging to the Zingiberaceae family, is an important spice crop with export potential. It is popularly known as the "Queen of Spices" and is indigenous to the evergreen rainforests of the Western Ghats. Cardamom is primarily cross-pollinated, with honeybees serving as the main pollinators. The somatic chromosome number of cardamom is 2n=48. Cardamom is propagated through seeds, and there is considerable diversity in the natural population of cultivated traits from the Western Ghats.

15.5.1 Genetic resources

Efforts to systematically collect and conserve cardamom's genetic resources began in the 1960s. The plant's genetic resources are now preserved in ex situ clonal repositories and field gene banks at various centers. Germplasm collections are being maintained at National Active Germplasm Site (NAGS), ICAR–Indian Institute of Spices Research (ICAR-IISR), Regional station, Appangala, All India Coordinated Research Project on Spices (AICRPS) centers, and Indian Cardamom Research Institute (ICRI) and State Agricultural Universities (SAUs). Considerable diversity among cardamom genotypes in plant height, number of tillers, capsule size, shape, and leaf pubescence was reported (Madhusoodanan et al., 2002; Korikanthimath et al., 2006).

15.5.2 Active principles

The aroma and flavor of cardamom are primarily attributed to its essential oil, with the relative composition of the volatile oil components playing a key role in defining its distinctive characteristics. The primary bioactive compound in small cardamom is 1,8-cineole, along with various monoterpene hydrocarbons (Bhatti and JMA, 2010). The fine flavor and delicate quality of cardamom are largely dependent on its essential oil, which is typically described as camphory, sweet, and aromatic-spicy. This is due to the presence of compounds such as 1,8-cineole, α-terpineol, terpinyl acetate, limonene, sabinene, and borneol (Guenther, 1975). The dominance of 1,8-cineole, along with α-terpinyl acetate and linalyl acetate, makes cardamom oil unique (Lewis et al., 1966; Wijesekera and Jayawardena, 1973; Salzer, 1975). Variability in these quality characteristics exists across cardamom, influencing the aroma and flavor profiles of different varieties and regions.

15.5.3 Cardamom breeding strategies

Cardamom flowers are borne in panicles which emerge from the base of the tiller. Based on the panicle growth pattern cardamom varieties are categorized as Malabar with prostrate panicles, Mysore with erect panicles, and *Vazhukka* with semi erect panicles. Flowers are bisexual and zygomorphic. Prominent labellum, position of stigma above anthers, and presence of nectar glands favors insect pollination in cardamom (Madhusoodanan et al., 2002).

15.5.4 Breeding objectives

- Identify and develop cardamom genotypes with high and sustainable yield potential under diverse agro-climatic conditions.
- Evolve genotypes with synchronized and uniform flowering patterns to reduce harvesting cycles and enhance productivity.
- Varieties for bold, green capsules with an increased number of seeds per capsule, ensuring better market acceptability and profitability.
- Develop cardamom varieties resistant to major pests and diseases such as *katte* virus, rhizome rot, and thrips to minimize yield losses.

- Enhance drought tolerance and adaptability to changing environmental conditions through the selection of resilient genotypes.
- Improve the quality of cardamom by selecting genotypes with high oil content, increased oleoresin yield, and elevated α-terpenyl acetate content for superior aroma and flavor.

15.5.5 Breeding methods

Cardamom can be propagated by sexual as well as asexual methods, which benefit the crop improvement programs. Major crop improvement techniques which can be followed in cardamom are selection, heterosis breeding, mutation, and polyploidy breeding.

Selection: Selection is a widely used method for improving cardamom, with many improved and farmer-preferred varieties developed through this approach. Being a cross-pollinated crop, cardamom provides significant opportunities for enhancing productivity by selecting high-yielding genotypes (Venugopal, 1999). There are many research studies on the evaluation and selection of superior genotypes (Jacob et al., 2020; Prasath et al., 2009b; Senthil Kumar et al., 2018; Venugopal, 1999; Venugopal et al., 2006).

Heterosis breeding: In cardamom amenability for propagation through sexual as well as asexual methods is a boon for crop improvement through hybridization. Since *Elettaria* is a monotypic species, interspecific hybridization is not possible, and no seed set was observed in intergeneric hybridization, (Krishnamurthy et al., 1989). Therefore, only intraspecific crosses can be performed in cardamom. Heterosis breeding has also been reported as an effective method for developing cardamom hybrids with high yield potential, desirable quality traits, and mosaic resistance by selecting suitable parents (Prasath et al., 2009a). Cardamom variety, Appangala 2 (IC 547167) was developed through hybridization between Appangala 1 × NKE 19 (Senthil Kumar et al., 2017).

Mutation breeding: Physical mutagens such as X-rays and gamma rays, as well as chemical mutagens like ethyl methane sulfonate and maleic hydrazide, have been commonly used to induce mutations in cardamom. Large populations of selfed and open-pollinated progenies of M1 plants were repeatedly inoculated with the *Katte* virus vector, resulting in the selection of 12 plants showing tolerance to *Katte* disease (Bavappa, 1986). However, no mutants have been reported in cardamom to date.

Polyploidy breeding: Sprouting cardamom seeds were treated with a 0.5% aqueous solution of colchicine to induce polyploids (2n=4x=96). These polyploids exhibited characteristics such as gigantism, an increased number of epidermal cell layers, a thicker cuticle, and a higher wax coating on the leaves (Sudharshan, 1987, 1989).

15.5.6 Micropropagation

Micropropagation is an effective method for large-scale production of high-quality planting material for cardamom. This technique facilitates the rapid production of virus-free, high-yielding clones at a faster rate compared to traditional propagation methods. Various micropropagation techniques for cardamom have been developed using different explants *in vitro*, including those by Nadgauda and Mascarenhas (1983) using young, sprouted buds (Priyadarshan & Zachariah, 1986; Vatsya et al., 1987; Regunath and Gopalakrishnan, 1991; Nirmal Babu et al., 1997), who used rhizome bits with vegetative buds.

Malhotra et al. (2021) improved the micropropagation protocol for generating genetically uniform cardamom plants using shoot tips as explants. Optimal shoot proliferation was achieved in MS medium supplemented with 4.4 μM 6-benzylaminopurine (BAP) and 2.32 μM kinetin (Kn). Regenerated shoots rooted effectively in full-strength MS basal medium, and the genetic stability of the micropropagated plants was confirmed using 13 ISSR markers.

Plant regeneration from callus cultures of cardamom was also reported by Priyadarshan & Zachariah (1986) and Nirmal Babu et al. (1997). Morphological variability was observed among the somaclones even within the culture vessels. Some somaclones tolerant to Katte virus were identified (Peter et al., 2002). (Manohari et al., 2008) developed an efficient protocol for inducing somatic embryogenesis and plant regeneration in small cardamom using the inner core region of the rhizome. The highest frequency of embryogenic calli (68%) and plantlets (86%) was obtained from MS medium supplemented with 4.4 μM BAP and 0.5 μM NAA (Sivaranjani, 2024).

15.5.7 Cardamom genomics

Flow cytometry estimated the nuclear DNA content at 2.87 pg ($2n$=48; ~2806 Mbp), while ISSR markers revealed substantial genetic diversity, with feral accession C53 being most variable (Anjali et al., 2016). The first whole-genome sequencing of small cardamom (*Elettaria cardamomum*) assembled a 1.06 Gb genome and identified 2,50,571 SSRs (2,18,270 perfect and 32,301 compound). Wet-lab validation of 246 SSR loci led to 60 polymorphic markers, detecting an average of 14.6 alleles per locus across 60 accessions. Population structure showed high admixture due to its cross-pollination system (Gaikwad et al., 2023). Earlier, microsatellite markers developed through selective hybridization enrichment yielded 140 repeats, of which 44 primer pairs amplified expected products. Six polymorphic markers (2–7 alleles, PIC 0.14–0.38) effectively differentiated genotypes, with high transferability (71.4%) across Zingiberaceae (Cyriac et al., 2016).

Small RNA sequencing identified 150 conserved and 20 novel miRNAs, 17 of which were drought-responsive. These miRNAs regulate genes linked to stress tolerance, with miR156 and miR169 involved in mRNA cleavage and miR530b in translation inhibition (Anjali et al., 2017; Anjali et al., 2019). De novo transcriptome assembly identified 5,317 differentially expressed genes, 2,267 transcription factors, and 175,952 putative genic markers, linking marble disease to disrupted auxin and ethylene signaling. A web-genomic resource (SCMVTDb) was developed for molecular marker and transcript information related to mosaic disease (Khan et al., 2020). Genetic transformation studies introduced the *NIb* gene of Cardamom mosaic virus into calli, producing transgenic plants that were evaluated for CdMV resistance (Jebasingh et al., 2017).

15.6 Tree spices

15.6.1 Nutmeg

Nutmeg (*Myristica fragrans* Houtt.), belonging to the Myristicaceae family, originated in the Molucca Islands, Indonesia. Its seeds and arils (mace) are used as spices. It is widely used in traditional medicines. The nutmeg seed oil, known as 'nutmeg butter,' is used to prepare ointments.

15.6.2 Active ingredients

The active ingredient in nutmeg is primarily myristicin. Myristicin is a naturally occurring compound found in the essential oil of nutmeg and is responsible for much of its aroma and flavor. In small amounts, it contributes to the spice's culinary and medicinal properties, including its use as a digestive aid and mild sedative. Nutmeg has high antioxidants and antimicrobial properties. The acetone extract of nutmeg seeds showed 93.12 ± 1.48 mg gallic acid equivalents (GAE)/100 g dry weight total phenolic content, DPPH scavenging activity of 63.04 ± 1.56%, chelating activity of 64.11 ± 2.21% and 74.36 ± 1.94% inhibition of β-carotene bleaching, at 1 mg/mL extract concentration. GC-MS analysis revealed the presence of Sabinene (28.61%), β-Pinene (10.26), α-pinene (9.72), myristicin (4.30%), isoeugenol (2.72%), p-cymene (1.81%), carvacrol (1.54%), eugenol (0.89%), and β-caryophellene (0.82%) (Gupta et al., 2013).

15.6.3 Germplasm conservation

ICAR-Indian Institute of Spices Research, Kozhikode, Kerala, India, has nutmeg germplasm collection with 421 accessions.

15.6.4 Varieties

Viswashree: It is a high-yielding clonal selection released from ICAR- IISR, Kozhikode, Kerala, in the year 2002. It yields 3122 kg nuts and 480 kg of mace/ ha. It has 7.14 % nut oil and 7.13 % mace oil. Keralashree- It is the first nutmeg variety released by the farmer participatory approach. Kerala Agricultural University has released KAU Mundathanam, KAU Punnathnam, KAU Pullan. Dr. BSKKV, Dapoli, has released Konkan Swad, Konkan Suhandha, and Konkan Seemanthi. Apart from these, several farmers' varieties of nutmeg were released under PPV&FRA like Kadukkanmakkel, Edavarambel Gold, and Kochukudi.

15.6.5 Molecular markers

Sheeja et al. (2013) reported the diversity analysis of nutmeg and related genera using both RAPD and ISSR markers. Species-specific bands could be identified from all the accessions, which can be converted into SCAR markers for genotype identification and authentication.

15.6.6 Clove

Cloves (*Sysygium aromaticum*) are one of the most important spice crops used in the food and pharmaceutical industries. The mature, unopened flower buds are the economic part of the clove. This tree spice originated in Molucca Island of Indonesia and spread across the tropical India. Now, it is a commercial crop in the hilly tracts of Tamil Nadu and Kerala.

15.6.7 Active ingredients

The major bioactive compound present in cloves is eugenol, which is a phenylpropanoid derived from guaiacol. GC-MS of an ethyl acetate extract of clove buds revealed eugenol (58.88%), eugenyl acetate (23.86%), and trans-caryophyllene (14.44%) as the major components of clove oil (Yassin et al., 2020).

15.6.8 Germplasm conservation

Clove Germplasm is available at TNAU, Pechiparai, Kanyakumari, and ICAR-IISR, Calicut in India. Clove genome: A 370-Mb high-quality chromosome-scale genome assembly for clove was prepared to get insights into the eugenol biosynthesis pathway (Ouadi et al., 2022).

15.6.9 Cinnamon

Cinnamon (*Cinnamomum verum*) barks have been used as a spice from the olden days. Its origin is in Sri Lanka. Cinnamon belongs to the family Lauraceae. It has related species like *C. malabartrum* and *C. cassia*.

15.6.10 Germplasm conservation

The largest *C. zeylanicum* germplasm in the world is at the Department of Export Agriculture (DEA), Sri Lanka, with more than 600 superior accessions. ICAR- Indian Institute of Spices Research (ICAR-IISR), Calicut, Kerala, has 311 accessions, including 239 cultivated species, 58 related accessions, and 14 exotic accessions. RRL, Bhubaneswar, KAU, and TNAU are the other institutions with the Cinnamon germplasm.

15.6.11 Breeding

High variability in the Cinnamon population is the result of the natural cross-pollination due to the protogynous dichogamy, where female and male flowers are separated temporally (Kumari & Wijesinghe, 2008). Therefore, the selection of the outperforming seedlings is the best possible breeding method in this crop.

15.6.12 Varieties

YCD-1, PPI-1, Nithyasree, Navasree, Konkan Tej, Sugandhini (India), Sri gamunu (SG), and Sri wijaya (SW) (Sri Lanka).

15.6.13 Molecular markers

Kojoma et al. (2002) utilized chloroplast intergenic spacer regions (trnL 3′exon and trnF exon; trnL-trnF IGS) and the trnL intron region for the identification of *C. zeylanicum* and three other species, *C. cassia, C. burmannii*, and *Cinnamomum sieboldii*. The species-level diversity of cinnamon was studied by TrnL intron regions, intergenic spacers between trnT-trnL, trnL-trnF, trnH-psbA, and nuclear internal transcribed spacer (ITS), randomly amplified polymorphic DNA (RAPD), and sequence-related amplified polymorphic (SRAP) markers (Abeysinghe et al., 2009; Abeysinghe et al., 2014). (Swetha et al., 2014) reported adulteration detection in *C. verum* with *C. cassia* and *C. malabatrum* by barcoding loci rbcL, matK, and psbA-trnH with SNPs. RAPD technology was applied for the analysis of normal and excellent types of Cinnamomum camphora L. and genetic interrelationship of nine *Cinnamomum* species (Hui & Linshui, 2003; Joy & Maridass, 2008), respectively.

15.6.14 Genomics and gene cloning

Genes encoding cinnamomin (a type II RIP), which have three isoforms, were isolated from camphor seeds by Yang et al. (2002). A geraniol synthase gene is also isolated from Cinnamomum tenuipilum by Yang et al. (2005). Expression of CtGES was exclusively observed in the geraniol chemotype of C. tenuipilum. Furthermore, in situ hybridization analysis demonstrated that CtGES mRNA was localized in the oil cells of the leaves.

15.6.15 Garcinia

Garcinia originated in the Western Ghats of India. Its fruits are rich in anthocyanin and Hydroxy Citric Acid (HCA), an anti-obesity agent. *Garcina indica* (Kokum) and *Garcinia gummigutta/ camboge* (Malabar tamarind) are the major species of the Clusiaceae family.

15.6.16 Genetic resource management and breeding

There are a few released varieties of Kokum (*Garcinia indica* Choisy). Konkan Amrutha and Konkan Hatis were selections released from Dr. BSKKV, Dapoli. Goa Kokum-1, Goa Kokum-2, and Goa Kokum-3 were released for the state of Goa from ICAR-CCARI, Goa.

15.6.17 Micropropagation in tree spices

In vitro proliferation of nutmeg mace and the synthesis of flavor components in culture were documented by Babu (2009). Since mace contains the anti-carcinogenic compound myristicin, this technique, with further refinement, could be leveraged for myristicin production. Nutmeg micropropagation was also reported by (Mallika et al., 1997), who induced shoot initiation from explants of field-grown trees.

Somatic embryogenesis in nutmeg has been successfully induced (Iyer et al., 2000). Germination of somatic embryos derived from zygotic embryos was successfully achieved on media containing naphthalene acetic acid (NAA). Direct somatic embryo formation was observed in leaf explants from juvenile plants cultured on MS medium supplemented with kinetin, 2,4-D, and NAA (Iyer et al., 2009). Metabolite profiling of embryogenic cultures revealed the presence of various phytochemicals, including α & β-pinene, sabinene, myristicin, safrole, amphetamine derivatives, and other compounds with significant clinical potential, some of which had not been previously reported *in vivo* (Iyer, 2007). The essential oil profile of these cultures varied with age, with three-week-old cultures showing a diverse range of monoterpenes. Notably, the spent charcoal medium from these embryogenic cultures demonstrated strong antimicrobial activity against pathogens such as *Salmonella typhi* and *Staphylococcus aureus* (Iyer et al., 2009).

Research on micropropagation of cinnamon and Chinese cassia has demonstrated successful regeneration from both seedlings and mature tree explants (Huang et al., 1998; Nirmal Babu et al., 1997). In particular, *Cinnamomum camphora* has been propagated through multiple shoot induction from shoot tips and nodal segments using Woody Plant Medium (WPM) (Huang et al., 1998) and from cotyledonary nodes cultured on Murashige and Skoog (MS) medium (Azad et al., 2005; Babu et al., 2003). Additionally, *Chinese cassia* has been successfully propagated *in vitro* from nodal explants derived from seedlings using MS medium (Inomoto & Kitani, 1989).

Embryonic axes with half of the cotyledon portion were used as explants for *in vitro* establishment and were inoculated onto half-strength MS basal medium supplemented with 1.5 mg L^{-1} BAP and 0.2 mg L^{-1} IAA to initiate *in vitro* seedlings. This combination resulted in the highest culture initiation rate (90%). The addition of 1 g L^{-1} activated charcoal effectively supported the establishment of *in vitro* cultures by minimizing browning, promoting stem elongation (19.5 mm height), and encouraging leaf initiation (2.06 leaves/plantlet) after 14 days of culturing. For adventitious root elongation in *in vitro* microstem cuttings, the treatment combination of 0.1 mg L^{-1} NAA, 4.0 mg L^{-1} BAP, and 1.0 g L^{-1} activated charcoal in full-strength MS medium yielded the longest root length (6.7 cm) after 6 weeks of incubation. Coir dust was found to be the most effective potting medium for acclimatization, resulting in the highest survival rate (90%) (Subasinghe et al., 2016).

Similarly, multiple shoot induction has been achieved in *G. indica* (Kulkarni & Deodhar, 2002). The inclusion of benzylaminopurine (BAP) in Murashige and Skoog's medium resulted in optimal responses across various genotypes. Micropropagation of *Garcinia* species has been explored by Huang et al. (2000), Malik et al. (2005), and Mohan et al. (2012).

Multiple shoots were induced from nodal segments of *Syzygium aromaticum* seedlings on Murashige and Skoog's (MS) basal medium with half-strength salts and Gamborg's medium (B5), both supplemented with BAP and NAA. The optimal combination for shoot induction was 3 mg L^{-1} BAP and 0.5 mg L-1 NAA, which resulted in the formation of six to eight shoots. Both MS and B5 media produced similar results in terms of explant proliferation (Mathew & Hariharan, 1990). (Suparman, 1990) also obtained successful micropropagation of clove using seedling explants, but there are no reports on the successful regeneration of clove from mature shoot explants. Root formation in clove was stimulated by supplementing MS medium with indole-3-butyric acid (IBA) or activated charcoal.

15.6.18 Genomics of tree spices

Recent advancements in genomic research have provided significant insights into their genetic architecture, biosynthetic pathways, and stress response mechanisms, facilitating targeted breeding and conservation strategies (Das et al., 2023). The application of next-generation sequencing has enabled the assembly of draft genomes for several tree spice species. For instance, the draft genome of clove (*Syzygium aromaticum*) has been sequenced, leading to the identification of key genes involved in the biosynthesis of eugenol, the primary bioactive compound responsible for its medicinal and aromatic properties (Das et al., 2023). Similarly, the genome of cinnamon (*Cinnamomum verum*) has been explored to identify genes regulating the biosynthesis of cinnamaldehyde, the compound responsible for its characteristic flavor and antimicrobial activity (Wang et al., 2022).

Transcriptome studies have provided further insights into gene expression patterns in tree spices. In nutmeg (*Myristica fragrans*), transcriptome sequencing has identified genes involved in the biosynthesis of lignans, which exhibit antioxidant and neuroprotective properties (Rostiana et al., 2025). Additionally, differentially expressed genes under stress conditions have been studied in cinnamon and clove, providing insights into the molecular mechanisms governing their adaptability and resistance to environmental stress (Das et al., 2023).

The assessment of genetic diversity within tree spices is crucial for conservation and breeding efforts. Molecular markers, including simple sequence repeats (SSRs) and single-nucleotide polymorphisms (SNPs), have been developed for cinnamon and clove, enabling efficient evaluation of genetic variation and assisting in the selection of superior genotypes (Patel & Prajapati, 2023). These markers have also been used to track domestication patterns and identify genetically diverse populations for conservation purposes.

AI disclosure

During the preparation of this work, the authors used ChatGPT in order to paraphrase and for grammar correction. After using this tool/service, the authors reviewed and edited the content as needed and take full responsibility for the content of the publication.

References

Aarthi, Suresh, Leela, & Prasath (2020). Multi environment testing reveals genotype-environment interaction for curcuminoids in turmeric (*Curcuma longa* L.). *Industrial Crops and Products, 145*. https://doi.org/10.1016/j.indcrop.2020.112090.

Aarthi, Suresh, Premnath, A., Leela, & Prasath (2025). Genotype x environment interaction of essential oil by additive main effects and multiplicative interaction (AMMI) model and metabolomics based profiling of volatile constituents in turmeric (*Curcuma longa* L.). *Genetic Resources and Crop Evolution, 72*(1), 497–511. https://doi.org/10.1007/s10722-024-01983-8.

Abdullah, Ahmad, N., Tian, W., Zengliu, S., Zou, Y., Farooq, S., Huang, Q., & Xiao, J. (2022). Recent advances in the extraction, chemical composition, therapeutic potential, and delivery of cardamom phytochemicals. *Frontiers in Nutrition, 9*. https://doi.org/10.3389/fnut.2022.1024820.

Abeysinghe, Wijesinghe, Tachida, & Yoshda (2009). Molecular characterization of Cinnamon (*Cinnamomum verum* Presl) accessions and evaluation of genetic relatedness of Cinnamon species in Sri Lanka based on trnL intron region, intergenic spacers between trnT-trnL, trnL-trnF, trnH-psbA and nuclear ITS. *Research Journal of Agriculture and Biological Sciences, 5*(6), 1079–1088.

Abeysinghe, Samarajeewa, Li, & Wijesinghe (2014). Preliminary investigation for the identification of Sri Lankan Cinnamomum species using randomly amplified polymorphic DNA (RAPD) and sequence related amplified polymorphic (SRAP) markers. *Journal of the National Science Foundation of Sri Lanka, 42*(3), 175–182. https://doi.org/10.4038/jnsfsr.v42i3.7393.

Afroz, T., Huq, H., & Hoque, M. E. (2023). In vitro regeneration of black pepper (*Piper nigrum* L.). *Bangladesh Journal of Agriculture, 48*(2), 75–85.

Ahmad, N., Abbasi, B. H., & Fazal, H. (2016). Effect of different in vitro culture extracts of black pepper (*Piper nigrum* L.) on toxic metabolites-producing strains. *Toxicology and Industrial Health, 32*(3), 500–506. https://doi.org/10.1177/0748233713505126.

Amalraj, A., Pius, A., Gopi, S., & Gopi, S. (2017). Biological activities of curcuminoids, other biomolecules from turmeric and their derivatives – A review. *Journal of Traditional and Complementary Medicine, 7*(2), 205–233. https://doi.org/10.1016/j.jtcme.2016.05.005.

Anandaraj, M., Chandran, S., George, R. S., Bhat, A. I., & Suseela Bhai, R. (2008). Development of SCAR marker for Phytophthora resistance in black pepper (*Piper nigrum* L.). *Journal of Spices and Aromatic Crops, 17*(3), 215–222.

Anjali, Ganga, Nadiya, Shefeek, & Sabu (2016). Intraspecific variations in cardamom (*Elettaria cardamomum* Maton): Assessment of genomic diversity by flow cytometry, cytological studies and ISSR analysis. *SpringerPlus, 5*(1). https://doi.org/10.1186/s40064-016-3226-x.

Anjali, Nadiya, Thomas, J., & Sabu (2019). Identification and characterization of drought responsive microRNAs and their target genes in cardamom (*Elettaria cardamomum* Maton). *Plant Growth Regulation, 87*(2), 201–216. https://doi.org/10.1007/s10725-018-0462-9.

Anjali, N., Nadiya, F., Jinu Thomas, & Sabu, K. K. (2017). Discovery of MicroRNAs in Cardamom (*Elettaria cardamomum* Maton) under Drought Stress. *Dataset Papers in Science*. https://doi.org/10.1155/2017/9507485.

Aravind, S., Nisthar, E., Srinivasan, V., Kandiannan, K., & Babu, K. N. (2023). Micro-rhizome technology for healthy planting material of ginger and turmeric. *Indian Horticulture, 68*(4), Available from https://epubs.icar.org.in/index.php/IndHort/article/view/134509.

Aravind, S., Nisthar, Chaithanya, Sivaranjani, Kandiannan, Srinivasan, Sankar, S. M., & Babu, K. N. (2024). Induction of in vitro micro rhizomes and assessment of yield, quality, and clonal fidelity in ex vitro established plants of ginger (*Zingiber officinale* Rosc.). *Plant Cell, Tissue and Organ Culture, 157*(2). https://doi.org/10.1007/s11240-024-02751-3.

Archana, C., Geetha, S. P., & Indira, B. (2014). In vitro microrhizome and minirhizome production in turmeric (*Curcuma longa* L.) cultivar Alleppey Supreme and its comparative anatomical and histochemical analysis. *International Journal of Current Microbiology and Applied Sciences, 3*(3).

Arijanti, & Suryaningsih (2019). Biosynthesis of secondary metabolites (gingerol, shogaol, and zingerone) from callus of three ginger varieties. *Drug Invention Today, 11*(2).

Ashokkumar, K., Murugan, M., Dhanya, M. K., Pandian, A., & Warkentin, T. D. (2021). Phytochemistry and therapeutic potential of black pepper [*Piper nigrum* (L.)] essential oil and piperine: A review. *Clinical Phytoscience, 7*(52). https://doi.org/10.1186/s40816-021-00292-2.

Ashokkumar, K., Murugan, M., Dhanya, & Warkentin, T. D. (2020). Botany, traditional uses, phytochemistry and biological activities of cardamom [*Elettaria cardamomum* (L.) Maton] – A critical review. *Journal of Ethnopharmacology, 246*. https://doi.org/10.1016/j.jep.2019.112244.

Azad, Yokota, & Ishiguri (2005). *Large-scale clonal propagation of Cinnamomum camphora (L.) Nees and Eberm*. Bulletin Utsunomiya University Forestry 41.

Babu (2009). Tissue culture as a tool for invitro production of flavour, colour and nutraceuticals in spice crops. In A. Shamina, K. S. Krishnamurthy, & K. N. Shiva, (Eds.). *Flavours, nutraceuticals and food colours from horticultural crops* (pp. 108–113). Indian Institute of Spices Research.

Babu, K. N., Sajina, Minoo, John, Mini, Tushar, Rema, & Ravindran (2003). Micropropagation of camphor tree (*Cinnamomum camphora*). *Plant Cell, Tissue and Organ Culture, 74*(2), 179–183. https://doi.org/10.1023/A:1023988110064.

Babu, K. N., Samsudeen, Minoo, Ravindran, & Geetha (2016). *Gingertissue culture and biotechnology of ginger*. CRC Press, 201–230.

Balachandran, Bhat, & Chandel (1990). In vitro clonal multiplication of turmeric (*Curcuma* spp.) and ginger (*Zingiber officinale* Rosc. *Plant Cell Reports, 8*(9), 521–524. https://doi.org/10.1007/bf00820200.

Bandara, M. M. N. T., Dahanayake, N., Perera, P. C. D., & Subasinghe, S. (2023). Development of in-vitro protocol to enhance mass production of turmeric (*Curcuma longa* L.). *Tropical Agricultural Research and Extension, 26*(1).

Bavappa (1986). *Research at CPCRI. Tech. Bull.* Kasaragod, India: CPCRI.

Bhat, M., Divakaran, K. N., & Babu, K. N. (2024). Biotechnological approaches for improvement of spices. In P. N. Ravindran, K. Sivaraman, S. Devasahayam, & K. N. Babu (Eds.), *Handbook of spices in India: 75 years of research and development* (pp. xx–xx). Springer. https://doi.org/10.1007/978-981-19-3728-6_7.

Bhatti, Z. F., & JMA (2010). Evaluation of phenolic contents and antioxidant potential of methanolic extracts of green cardamom (*Elettaria cardamomum*). *Asian Journal of Chemistry, 22*(4787).

Butt, M. S., Pasha, I., Sultan, M. T., Randhawa, M. A., Saeed, F., & Ahmed, W. (2013). Black pepper and health claims: A comprehensive treatise. *Critical Reviews in Food Science and Nutrition, 53*(9), 875–886. https://doi.org/10.1080/10408398.2011.571799.

Chakraborty, A., Mahajan, S., Jaiswal, S. K., & Sharma, V. K. (2021). Genome sequencing of turmeric provides evolutionary insights into its medicinal properties. *Communications Biology, 4*(1). https://doi.org/10.1038/s42003-021-02720-y.

Chen, C., Long, L., Zhang, F., Chen, Q., Chen, C., Yu, X., Liu, Q., Bao, J., & Long, Z. (2018). Antifungal activity, main active components and mechanism of *Curcuma longa* extract against *Fusarium graminearum*. *PLoS ONE, 13*(3). https://doi.org/10.1371/journal.pone.0194284.

Cheng, S.-P., Jia, K.-H., Liu, H., Zhang, R.-G., Li, Z.-C., Zhou, S.-S., Shi, T.-L., Ma, A.-C., Yu, C.-W., Gao, C., Cao, G.-L., Zhao, W., Nie, S., Guo, J.-F., Jiao, S.-Q., Tian, X.-C., Yan, X.-M., Bao, Y.-T., Yun, Q.-Z., ... Mao, J.-F. (2021). Haplotype-resolved genome assembly and allele-specific gene expression in cultivated ginger. *Horticulture Research, 8*(1). https://doi.org/10.1038/s41438-021-00599-8.

Chezhiyan, N., & Shanmugasundaram, K. (2000). BSR.2 - a promising turmeric variety from. *Tamil Nadu*.

Chowdhury, S., Pal, K., Chakraborty, M., Chakraborty, S., Mandal, S., Pandit, G. K., Maitra, S., & Sahana, N. (2020). Conservation and in vitro propagation of an endangered wild turmeric (*Curcuma caesia* Roxb.) species from Sub-Himalayan Terai region of West Bengal. *International Journal of Current Microbiology and Applied Sciences, 9*(2), 2132–2140.

Cyriac, A., Paul, R., Anupama, S.kumar, & Sheeja, N. (2016). Babu, Parthasarathy, Isolation and characterization of genomic microsatellite markers for small cardamom (*Elettaria cardamomum* Maton) for utility in genetic diversity analysis. *Physiology and Molecular Biology of Plants, 22*(2), 219–229. https://doi.org/10.1007/s12298-016-0355-1.

Das, Kesari, & Rangan (2010). Plant regeneration in *Curcuma* species and assessment of genetic stability of regenerated plants. *Biologia Plantarum, 54*(3), 423–429. https://doi.org/10.1007/s10535-010-0077-0.

Das, P., Chandra, T., Negi, A., Jaiswal, S., Iquebal, M. A., Rai, A., & Kumar, D. (2023). A comprehensive review on genomic resources in medicinally and industrially important major spices for future breeding programs: Status, utility and challenges. *Current research in food science, 7*. https://doi.org/10.1016/j.crfs.2023.100579.

Das, P., Sheeja, Saha, B., Fayad, Chandra, T., Angadi, Shivakumar, Azharudheen, T. P. M., Jaiswal, S., Iquebal, M. A., & Kumar, D. (2025). Genome-wide identification of copy number variation in diverse black pepper accessions. *Planta, 261*(4). https://doi.org/10.1007/s00425-025-04658-5.

Deepak, D. A., Puneeth, G. M., Gowthami, R., Shankar, M., Chander, S., Malhotra, E. V., Archak, S., & Agrawal, A. (2025). Trans situ conservation of *Piper nigrum* L. in India—A review. *Genetic Resources and Crop Evolution, 72*(2), 1477–1494. https://doi.org/10.1007/s10722-024-02058-4.

Dehghani, Mostajeran, & Asghari (2011). In vitro and in vivo production of gingerols and zingiberene in ginger (*Zingiber officinale* Roscoe). *J. Spices Aromatic Crops, 1*(1).

Devasahayam, S., John Zacharaiah, T., Jayashree, E., Kandiannan, K., Prasath, D., Santhosh, J. Eapen, Sasikumar, B., Srinivasan, V., & Suseela Bhai, R. (2015). *Black pepper – Extension pamphlet*. ICAR-Indian Institute of Spices Research.).

Ding, Y., Mao, Y., Cen, Y., Hu, L., Su, Y., Ma, X., Long, L., Hu, H., Hao, C., & Luo, J. (2021). Small RNA sequencing reveals various microRNAs involved in piperine biosynthesis in black pepper (*Piper nigrum* L.). *BMC Genomics, 22*(1). https://doi.org/10.1186/s12864-021-08154-4.

Ekundayo, O., Laakso, I., & Hiltunen, R. (1988). Composition of ginger (*Zingiber officinale* roscoe) volatile oils from Nigeria. *Flavour and Fragrance Journal, 3*(2), 85–90. https://doi.org/10.1002/ffj.2730030207.

Fayad, Charles, S., Shelvy, Sheeja, Sangeetha, Angadi, Tandon, G., Iquebal, M. A., Jaiswal, S., & Kumar, D. (2024). Whole genome based identification of BAHD acyltransferase gene involved in piperine biosynthetic pathway in black pepper. *Journal of Biomolecular Structure and Dynamics*. https://doi.org/10.1080/07391102.2024.2313164.

Gaikwad, A. B., Kaila, T., Maurya, A., Kumari, R., Rangan, P., Wankhede, D. P., & Bhat (2023). The chloroplast genome of black pepper (*Piper nigrum* L.) and its comparative analysis with related Piper species. *Frontiers in Plant Science, 13*. https://doi.org/10.3389/fpls.2022.1095781.

George, J. K., Shelvy, S., Fayad, A. M., Umadevi, P., Angadi, Iquebal, M. A., Jaiswal, S., Rai, A., & Kumar, D. (2021). De novo transcriptome sequencing assisted identification of terpene synthases from black pepper (*Piper nigrum*) berry. *Physiology and Molecular Biology of Plants, 27*(5), 1153–1161. https://doi.org/10.1007/s12298-021-00986-4.

Gnasekaran, P., Rahman, Z. A., Chew, B. L., Uddain, J., Solayappan, M., Chear, N. J. Y., Appalasamy, S., Mariappan, V., Wahyuni, D. K., & Subramaniam, S. (2023). Picloram enhanced the callus induction, growth kinetics, antioxidant potentials, and secondary metabolites production of *Zingiber officinale* var. rubrum callus cultures. *Plant Cell, Tissue and Organ Culture, 155*(3), 843–859. https://doi.org/10.1007/s11240-023-02603-6.

Greenshields, A. L., Doucette, C. D., Sutton, K. M., Madera, L., Annan, H., Yaffe, P. B., Knickle, A. F., Dong, Z., & Hoskin, D. W. (2015). Piperine inhibits the growth and motility of triple-negative breast cancer cells. *Cancer Letters, 357*(1), 129–140. https://doi.org/10.1016/j.canlet.2014.11.017.

Geetha, S. P., Babu, K. N., Rema, J. et al. (2000). Isolation of protoplasts from cardamom (Elettaria cardamomum Maton.) and ginger (Zingiber officinale Rosc.). *Journal of Spices and Aromatic Crops* 9 (1):23–30.

Guan, Q., Guo, Y., Wei, Y., Meng, F., & Zhang, Z. (2010). Regeneration of somatic hybrids of ginger via chemical protoplast fusion. *Plant Cell, Tissue and Organ Culture, 102*(3), 279–284. https://doi.org/10.1007/s11240-010-9730-8.

Guenther (1975). *Cardamom-essential oil.* New York: Robert E. Krieger Publishing Vol.5.

Gupta, A. D., Bansal, V. K., Babu, V., & Maithil, N. (2013). Chemistry, antioxidant and antimicrobial potential of nutmeg (*Myristica fragrans* Houtt). *Journal of Genetic Engineering and Biotechnology, 11*(1), 25–31. https://doi.org/10.1016/j.jgeb.2012.12.001.

Gurav, S. S., Gurav, N. S., Patil, A. T., & Duragkar, N. J. (2020). Effect of explant source, culture media, and growth regulators on callogenesis and expression of secondary metabolites of *Curcuma longa*. *Journal of Herbs, Spices and Medicinal Plants, 26*(2), 172–190. https://doi.org/10.1080/10496475.2019.1689542.

Hamirah, M. N., Sani, H. B., Boyece, P. C., & Sim, S. L. (2007). Micro propagation of red ginger (*Zingiber montanum* Koenig), a medicinal plant. *Proceedings of the Asia Pacific conference on plant tissue and agribiotechnology* June 7-21, 2007, Kuala Lumpur, Malaysia.

Haro-González, J. N., Castillo-Herrera, G. A., Martínez-Velázquez, M., & Espinosa-Andrews, H. (2021). Clove essential oil (*Syzygium aromaticum* l. myrtaceae): Extraction, chemical composition, food applications, and essential bioactivity for human health. *Molecules, 26*(21). https://doi.org/10.3390/molecules26216387.

Hosoki, T., & Sagawa, Y. (1977). Clonal propagation of ginger (*Zingiber officinale* Rosc.) through tissue culture. *Hort Science, 12.*

Hossain, M. A., Ishimine, Y., Akamine, H., & Motomura, K. (2005). Effects of seed rhizome size on growth and yield of turmeric (*Curcuma longa* L. *Plant production science, 8*(1), 86–94. https://doi.org/10.1626/pps.8.86.

Hu, L., Xu, Z., Wang, M., Fan, R., Yuan, D., Wu, B., Wu, H., Qin, X., Yan, L., Tan, L., Sim, S., Li, W., Saski, C. A., Daniell, H., Wendel, J. F., Lindsey, K., Zhang, X., Hao, C., & Jin, S. (2019). The chromosome-scale reference genome of black pepper provides insight into piperine biosynthesis. *Nature Communications, 10*(1). https://doi.org/10.1038/s41467-019-12607-6.

Huang, Huang, & Murashige (1998). A micropropagation protocol for *Cinnamomum camphora. In Vitro Cellular and Developmental Biology - Plant, 34*(2), 141–146. https://doi.org/10.1007/BF02822779.

Huang, Huang, & Wang (2000). Developing an improved in vitro propagation system for slow-growing species using *Garcinia mangostana* L. (Mangosteen). *Vitro Cellular & Developmental Biology - Plant, 36*(6).

Hui, & Linshui (2003). Applying RAPD technology to analysis of normal and excellent types of *Cinnamomum camphora. Chinese Journal of Applied and Environmental Biology, 3.*

Idowu, S., Adekoya, A. E., Igiehon, O. O., & Idowu, A. T. (2021). Clove (*Syzygium aromaticum*) spices: a review on their bioactivities, current use, and potential application in dairy products. *Journal of Food Measurement and Characterization, 15*(4), 3419–3435. https://doi.org/10.1007/s11694-021-00915-9.

Inomoto, & Kitani (1989). In vitro propagation of *Cinnamomum cassia. Plant Tissue Culture Letters, 6.*

Iyer (2007). In vitro propagation of Nutmeg, Myristica Fragrans Houtt. *Protocols for Micropropagation of Woody Trees and Fruits,* 335–344. https://doi.org/10.1007/978-1-4020-6352-7_31.

Iyer, Jayaraman, & Gopinath (2000). Direct somatic embryogenesis in zygotic embryos of nutmeg (*Myristica fragrans* Houtt.). *Tropical Agriculture, 7*(2).

Iyer, I., Jayaraman, & Ramesh (2009). In vitro responses and production of phytochemicals of potential medicinal value in nutmeg, *Myristica fragrans* Houtt. *Indian Journal of Science and Technology, 2*(4), 65–70. https://doi.org/10.17485/ijst/2009/v2i4/29434.

Jacob, T. K., Senthil Kumar, C. M., Devasahayam, S., D'Silva, S., Senthil Kumar, R., Biju, C. N., Praveena, R., & Ankegowda, S. K. J. (2020). Plant morphological traits associated with field resistance to cardamom thrips (Sciothrips cardamomi) in cardamom (*Elettaria cardamomum*). *Annals of Applied Biology, 177*(1), 143–151. https://doi.org/10.1111/aab.12592.

Jayaprakasha, Jaganmohan Rao, & Sakariah (2006). Antioxidant activities of curcumin, demethoxycurcumin and bisdemethoxycurcumin. *Food Chemistry, 98*(4), 720–724. https://doi.org/10.1016/j.foodchem.2005.06.037.

Jebasingh, T., Backiyarani, S., Manohari, C., Archana Somanath, & Usha, R. (2017). Transformation of cardamom with the RNA dependent RNA polymerase gene of cardamom mosaic virus. *Journal of Applied Biotechnology & Bioengineering, 3*(3), 326–329. https://doi.org/10.15406/jabb.2017.03.00066.

Jeena, K., Liju, V. B., Umadevi, & Kuttan, R. (2014). Antioxidant, anti-inflammatory and antinociceptive properties of black pepper essential oil (*Piper nigrum* Linn). *Journal of Essential Oil-Bearing Plants, 17*(1), 1–12. https://doi.org/10.1080/0972060X.2013.831562.

Jie, E. Y., Ahn, M. S., Lee, J., Cheon, Y. I., Kim, C. Y., & Kim, S. W. (2019). Establishment of a high-frequency plant regeneration system from rhizome-derived embryogenic cell-suspension cultures of *Curcuma longa* L. *Plant Biotechnology Reports, 13*(2), 123–129. https://doi.org/10.1007/s11816-019-00519-2.

Jose, S., Sujatha, R., & Deeshma, K. (2018). Novel EST-SSR marker development and validation in black pepper cultivars and varieties. *Journal of Tropical Agriculture, 55*(2), 175–179.

Joseph, B., Joseph, D., & Philip (1996). Plant regeneration from somatic embryos in black pepper. *Plant Cell, Tissue and Organ Culture, 47*(1), 87–90. https://doi.org/10.1007/BF02318970.

Joy, & Maridass (2008). Inter species relationship of *Cinnamomum* species using RAPD marker analysis. *Ethnobotanical Leaf, 12*.

Joy, N., & Soniya, E. V. (2012). Identification of an miRNA candidate reflects the possible significance of transcribed microsatellites in the hairpin precursors of black pepper. *Functional and Integrative Genomics. 12*(2), 387–395. https://doi.org/10.1007/s10142-012-0267-2.

Joy, N., Asha, S., Mallika, V., & Soniya, E. V. (2013). De novo transcriptome sequencing reveals a considerable bias in the incidence of simple sequence repeats towards the downstream of 'Pre-miRNAs' of Black Pepper. *PLoS ONE, 8*(3). https://doi.org/10.1371/journal.pone.0056694.

Kadam, S. S., Rasam, D. V., Joshi, K. H., & Jadhav, A. D. (2020). Micropropagation of black pepper, cv. Panniyur-1: Standardization of sterilization protocol and media composition. *Global Journal of Bioscience and Biotechnology, 9*(2), 45–49.

Kalpana Platel, K. P., & Srinivasan, K. (2004). Digestive stimulant action of spices: A myth or reality? *Indian Journal of Medical Research, 119*, 167–179.

Kambaska, K. B., & Santilata, S. (2009). Effect of plant growth regulator on micropropagtion of ginger (*Zingiber officinale* Rosc.) cv- suprava and suruchi. *Journal of Agricultural Technology, 5*(2), 271–280.

Kandiannan, K., Anandaraj, M., Prasath, D., Zachariah, T., J., Krishnamurthy, K. S., & Srinivasan, V. (2015). Evaluation of short and tall true turmeric (*Curcuma longa*) varieties for growth, yield and stability. *Indian J. Agric. Sci. 85*(4), 718–720.

Khan, A., George, J., Jasrotia, R. S., Aravind, S., Angadi, Iquebal, M. A., Manju, Jaiswal, S., Umadevi, Rai, A., & Kumar, D. (2020). Plant virus interaction mechanism and associated pathways in mosaic disease of small cardamom (*Elettaria cardamomum* Maton) by RNA-Seq approach. *Genomics, 112*(2), 2041–2051. https://doi.org/10.1016/j.ygeno.2019.11.017.

Khew, C. Y., Harikrishna, J. A., Wee, W. Y., Lau, E. T., & Hwang, S. S. (2020). Transcriptional sequencing and gene expression analysis of various genes in fruit development of three different black pepper (*Piper nigrum* L.) varieties. *International Journal of Genomics, 2020*. https://doi.org/10.1155/2020/1540915.

Kizhakkayil, J., & Sasikumar (2011). Diversity, characterization and utilization of ginger: A review. *Plant Genetic Resources: Characterisation and Utilisation, 9*(3), 464–477. https://doi.org/10.1017/S1479262111000670.

Kojoma, M., Kurihara, K., Yamada, K., Sekita, S., Satake, M., & Iida, O. (2002). Genetic identification of cinnamon (*Cinnamomum* spp.) based on the trnL-trnF chloroplast DNA. *Planta Medica, 68*(1), 94–96. https://doi.org/10.1055/s-2002-20051.

Kokkat, J. G., Shelvy, S., Fayad, A. M., Shabeer, A., Palaniyandi Umadevi, T. P., Kale, R., Angadi, U. B., Iquebal, M. A., Jaiswal, S., Rai, A., & Kumar, D. (2022). In silico assisted identification of peppery aroma compound 'rotundone' backbone genes from black pepper. *Journal of Biomolecular Structure and Dynamics, 40*(14), 6398–6404. https://doi.org/10.1080/07391102.2021.1883113.

Koohpar, Z. K., Entezari, M., Movafagh, A., & Hashemi, M. (2015). Anticancer activity of curcumin on human breast adenocarcinoma: Role of Mcl-1 gene. *International Journal of Cancer Management, 8*(3). https://doi.org/10.17795/ijcp2331.

Korikanthimath, V. S., Venugopal, M. N., Sudarshan, M. R., & Cardamom (2006). In P. N. Ravindran, K. Nirmal Babu, & K. N. Shiva, (Eds.). *Advances in spices research.* Jodhpur: Agrobios.

Krishnamoorthy, & Parthasarathy (2010). Improvement of black pepper. *CAB Reviews: Perspectives in Agriculture, Veterinary Science, Nutrition and Natural Resources, 5.* https://doi.org/10.1079/PAVSNNR20105003.

Krishnamoorthy, B., Prasath, D., & Utpala, P. (2011). Genetic resources. In H. P. Singh, V. A. Parthasarathy, V. Srinivasan, & K. N. Shiva (Eds.). *Turmeric: Production and utilization.* Westville Pub. House.

Krishnamurthy, Khan, & Avadhani (1989). *Three decades of cardamom research at R.R.S., Mudigere. Tech. Bull.* Bangalore, India.: University of Agricultural Sciences.

Krup, V., Prakash, L. H., & Harini, A. (2013). Pharmacological activities of turmeric (*Curcuma longa* Linn): a review. *J Homeop Ayurv Med, 2* 2167–1206.

Kulkarni, & Deodhar (2002). In vitro regeneration and hydroxycitric acid production in tissue culture of *Garcinia indica* Choisy. *Indian Journal of Biotechnology, 1.*

Kumari, & Wijesinghe (2008). Study on floral behavior of selected ten cultivars of true cinnamon *Cinnamomum verum* (Presl.), Syn. *C. zeylanicum* (Blume). *Proceedings of the National Symposium.*

Kumari, R., Wankhede, D. P., Bajpai, A., Maurya, A., Prasad, K., Gautam, D., Rangan, P., Latha, J. K., John, S., Bhat, K. V., & Gaikwad, A. B. (2019). Genome wide identification and characterization of microsatellite markers in black pepper (*Piper nigrum*): A valuable resource for boosting genomics applications. *PLoS ONE, 14*(12). https://doi.org/10.1371/journal.pone.0226002.

Lad, S. K. (1993). A case of seed-setting in cultivated turmeric types (*Curcuma longa* Linn.). *Journal of Soils and Crops, 3*, 78–79.

Lewis, Nambudiri, & Philip (1966). Composition of cardamom oils. *Perfumery and Essential Oil Research, 57*(623).

Li, H.-X., Zhang, H.-L., Zhang, N., Wang, N., Yang, Y., & Zhang, Z.-Z. (2014). Isolation of three curcuminoids for stability and simultaneous determination of only using one single standard substance in turmeric colour principles by HPLC with ternary gradient system. *LWT, 57*(1), 446–451. https://doi.org/10.1016/j.lwt.2013.11.020.

Li, H.-L., Wu, L., Dong, Z., Jiang, Y., Jiang, S., Xing, H., Li, Q., Liu, G., Tian, S., Wu, Z., Wu, B., Li, Z., Zhao, P., Zhang, Y., Tang, J., Xu, J., Huang, K., Liu, X., Zhang, W., … Liu, Y. (2021). Correction: Haplotype-resolved genome of diploid ginger (*Zingiber officinale*) and its unique gingerol biosynthetic pathway. *Horticulture Research, 8*(1), 189. https://doi.org/10.1038/s41438-021-00700-1.

Liang, H., Zhang, Y., Deng, J., Gao, G., Ding, C., Zhang, L., & Yang, R. (2020). The complete chloroplast genome sequences of 14 curcuma species: Insights into genome evolution and phylogenetic relationships within zingiberales. *Frontiers in Genetics, 11.* https://doi.org/10.3389/fgene.2020.00802.

Linacero, R., & Ballesteros, I. (2024). Genetic basis of somaclonal variation. *Somaclonal Variation: Basic and Practical Aspects,* 1–238. https://doi.org/10.1007/978-3-031-51626-9_1.

Liu, D., Zhang, P., Zhou, T., Wu, Y., Yuan, M., Zhang, X., & Liu, Y. (2025). Genome-wide characterization and expression analysis of the bHLH gene family in response to abiotic stresses in *Zingiber officinale* Roscoe. *BMC Genomics, 26*(1). https://doi.org/10.1186/s12864-025-11284-8.

Madhusoodanan, K. J., Pradip Kumar, K., & Ravindran, P. N. (2002). Botany, crop improvement and biotechnology of cardamom. In P. N. Ravindran, & K. J. Madhusoodanan (Eds.). *Cardamom: The genus Elettaria* (pp. 11–68). London: Taylor and Francis.

Malhotra, E. V., Kamalapriya, S., Bansal, & Meena, A. (2021). Agrawal, Improved protocol for micropropagation of genetically uniform plants of commercially important cardamom (*Elettaria cardamomum* Maton). *In Vitro Cellular and Developmental Biology - Plant, 57*(3), 409–417. https://doi.org/10.1007/s11627-020-10131-7.

Malik, Chaudhury, & Kalia, R. K. (2005). Rapid in vitro multiplication and conservation of *Garcinia indica*: A tropical medicinal tree species. *Scientia Horticulturae, 106*(4), 539–553. https://doi.org/10.1016/j.scienta.2005.05.002.

Malik, T., Sharma, R., Panesar, P. S., Gehlot, R., Tokusoglu, O., Dhull, S. B., Vural, H., & Singh, A. (2022). Nutmeg nutraceutical constituents: In vitro and in vivo pharmacological potential. *Journal of Food Processing and Preservation, 46*(6). https://doi.org/10.1111/jfpp.15848.

Mallika, Rekha, & Marymol (1997). In vitro shoot initiation from explants of field-grown trees of nutmeg (Myristica fragrans Houtt.). In S. Edison, K. V. Ramana, & B. Sasikumar, (Eds.). *Biotechnology of spices, medicinal, and aromatic crops.* Calicut: Indian Society for Spices.

Manohari, Backiyarani, Jebasingh, Somanath, A., & Usha (2008). Efficient plant regeneration in small cardamom (*Elettaria cardamomum* Maton.) through somatic embryogenesis. *Indian Journal of Biotechnology, 7*, 407–409.

Mao, Q.-Q., Xu, X.-Y., Cao, S.-Y., Gan, R.-Y., Corke, H., Beta, T., & Li, H.-B. (2019). Bioactive compounds and bioactivities of ginger (*Zingiber officinale* roscoe). *Foods, 8*(6). https://doi.org/10.3390/foods8060185.

Mathew, M. K., & Hariharan, M. (1990). In Vitro multiple shoot regeneration in *Syzygium aromaticum. Annals of Botany, 65*(3), 277–279. https://doi.org/10.1093/oxfordjournals.aob.a087934.

Maurya, K. R. (1990). RH 10, A promising variety of turmeric to boost farmers economy. *Indian Cocoa Arecanut and Spices Journal, 13*, 100–101.

Megumi, C., Muroyama, K., Sasako, H., & Tsuge, N. (2017). Preventive activity of ar-turmerone and bisacurone isolated from turmeric extract against ethanol-induced hepatocyte injury. *Food Science and Technology Research, 23*(2), 275–281. https://doi.org/10.3136/fstr.23.275.

Menon, A. N., Padmakumari, & Jayalekshmy (2002). Essential oil composition of four major cultivars of black pepper (*Piper nigrum* L.). *Journal of Essential Oil Research, 14*(2), 84–86. https://doi.org/10.1080/10412905.2002.9699778.

Miri, S. M. (2020). Micropropagation, callus induction and regeneration of ginger (*Zingiber officinale* Rosc. *Open Agriculture, 5*(1), 75–84. https://doi.org/10.1515/opag-2020-0008.

Mohan, Parthasarathy, & Babu (2012). In vitro and in vivo adventitious bud differentiation from mature seeds of three Garcinia spp. *Indian Journal of Natural Products and Resources, 3*(1).

Mohanty, D. C., & Panda, B. S. (1994). Genetic resources of ginger. In advances in horticulture. In K. L. Chaddha, & Rethinam (Eds.). *Vol. 9: Plantation and spices crops. Part-1*. New Delhi: Malhotra Publishing House.

Mukesh Sankar, S., Krishnamurthy, K. S., Alagupalamuthirsolai, M., Aravind, S., George, J., & Thankamani, C. K. (2022). *Annual Report: ICAR-all india coordinated research project on spices*. Kozhikode, Kerala, India: ICAR-IISR, 168.

Nadgauda, & Mascarenhas (1983). Clonal multiplication of cardamom (*Elettaria cardamom* Maton) by tissue culture. *Journal of Plantation Crops, 11*(1), 60–64.

Nadgauda, R. S., Mascarenhas, R. M., Hendre, R. R., & Jagannathan, V. (1978). Rapid multiplication of turmeric (*Curcuma longa* Linn.) plants by tissue culture. *Indian Journal of Experimental Biology, 16*.

Nadgauda, R. S., Kulkarni, D. B., & Mascarenhas, A. F. (1980). Development of plantlets from tissue cultures of ginger. *In Proceedings, annual symposium on plantation crops*.

Naidu, M. M., & Mutrhy, G. N. (2013). Performance of different turmeric selections for high altitude areas of Andhra Pradesh. *India. Agric. Sci. Digest. 33*(3), 183–187.

Nair, R. R., Dhamayanthi, K. P. M., Babu, B., Padmini, K., & Saji, K. V. (2004). *Annual report 2003–2004*. In J. Rema (Ed.), *Annual report 2003–2004 of the Indian Institute of Spices Research, Calicut* (pp. 38–41). Indian Institute of Spices Research.

Nair, R. R., & Gupta, S. D. (2005). Effect of explants and genotypes on primary somatic embryogenesis in black pepper (*Piper nigrum* L.). *Cytologia, 70*(2), 195–202. https://doi.org/10.1508/cytologia.70.195.

Nair, R. R., Shiva, K. N., Anchu, S., & Zachariah, T. J. (2010). Characterization of open-pollinated seedling progenies of turmeric (*Curcuma longa* L.) based on chromosome number, plant morphology, rhizome yield and rhizome quality. *Cytologia, 75*(4), 443–449. https://doi.org/10.1508/cytologia.75.443.

Nayak, S. (2000). In vitro multiplication and microrhizome induction in Curcuma aromatica Salisb. *Plant Growth Regulation, 32*.

Nazeem, P. A., Menon, R., & Valsala, P. A. (1993). Blossom biological and hybridization studies in turmeric, *Curcuma* spp. *Indian Cocoa Arecanut and Spices J, 16*, 106–109.

Nirmal, K., Yamuna, G., Praveen, K., et al. (2012). Cryopreservation of spices genetic resources. *Current Frontiers in Cryobiology InTech*. https://doi.org/10.5772/35401.

Nirmal Babu, K., Ravindran, P. N., & Peter, K. V. (1997). *Protocols for micro propagation of spices and aromatic crops*. Calicut, Kerala, India.: Indian Institute of Spices Research.

Nirmal Babu, K., George, J. K., Anandaraj, Venugopal, M. N., Nair, R. R., et al. (2005). *Improvement of selected spices through Biotechnology tools – Black pepper, Cardamom, Ginger, Vanilla*, 111.

Oladosu, Y., Rafii, M. Y., Arolu, F., Murugesu, S., Chukwu, S., Salisu, M. A., Fagbohun, I. K., Muftaudeen, T. K., & Kadar, A. I. (2024). Genetic diversity and utilization of ginger (*Zingiber officinale*) for varietal improvement: A review. *AIMS Agriculture and Food, 9*(1).

Ouadi, S., Sierro, N., Goepfert, S., Bovet, L., Glauser, G., Vallat, A., Peitsch, M. C., Kessler, F., & Ivanov, N. V. (2022). The clove (*Syzygium aromaticum*) genome provides insights into the eugenol biosynthesis pathway. *Communications biology, 5*(1). https://doi.org/10.1038/s42003-022-03618-z.

Pannaga, T. S., Narayanpur, V. B., Hiremath, J. S., Hegde, L., Gandolkar, K., Rathod, V., & Chandrakala, R. (2021). Evaluation of black pepper (*Piper nigrum* L.) cultivars for yield and quality parameters under the hill zone of Karnataka. *Journal of Pharmacognosy and Phytochemistry, 10*(1), 1497–1500.

Patel, M., & Prajapati, B. G. (2023). Cinnamon oil: An insight into pharmacological and pharmaceutical potential. *Pharmacological Aspects of Essential Oils: Current and Future Trends.* 162–174. https://doi.org/10.1201/9781003389774-12.

Peter, Ravindran, Babu, N., Sasikumar, Minoo, Geetha, & Rajalakshmi (2002). *Establishing in vitro conservatory of spices germplasm.* Calicut, Kerala, India: ICAR project report. Indian Institute of Spices research.

Pistelli, L., Bertoli, A., Gelli, F., Bedini, L., Ruffoni, B., & Pistelli, L. (2012). Production of Curcuminoids in different in vitro organs of *Curcuma longa. Natural Product Communications, 7*(8), 1037–1042. https://doi.org/10.1177/1934578X1200700819.

Pittampalli, B., Jogam, P., Thampu, R. K., Abbagani, S., & Peddaboina, V. (2022). High-frequency plant regeneration and genetic homogeneity assessment of regenerants by molecular markers in turmeric (*Curcuma longa* L.). *In Vitro Cellular and Developmental Biology - Plant, 58*(1), 169–180. https://doi.org/10.1007/s11627-021-10226-9.

Prasad, S., & Tyagi, A. K. (2015). Ginger and its constituents: Role in prevention and treatment of gastrointestinal cancer. *Gastroenterology Research and Practice, 2015.* https://doi.org/10.1155/2015/142979.

Prasath, Venugopal, & Senthilkumar (2009a). Hybrid performance for yield and yield components in cardamom (*Elettaria cardamom* Maton). *Euphytica, 168*, 49–60.

Prasath, Venugopal, & Senthil Kumar (2009b). Evaluation of high yielding cardamom selection in Karnataka for yield and quality. *Indian Journal of Agricultural Research, 3*, 207–209.

Prasath, D., Eapen, S. J., & Sasikumar, B. (2016). Performance of turmeric (*Curcuma longa*) genotypes for yield and root-knot nematode resistance. *Indian J. Agril. Sci. 86*(9), 89–92.

Prasath, Kandiannan, Leela, Aarthi, Sasikumar, & Babu, K. N. (2018a). Turmeric: Botany and production practices. *Horticultural Reviews, 46*, 99–184. https://doi.org/10.1002/9781119521082.ch3.

Prasath, D., Kandiannan, K., Srinivasan, V., Leela, N. K., & Anandaraj, M. (2018). Comparison of conventional and transplant production systems on yield and quality of ginger (Zingiber officinale). *The Indian Journal of Agricultural Sciences*, 88(4), 615–620. https://doi.org/10.56093/ijas.v88i4.79135.

Priya, I. N., Devappa, V., Kulkarni, M. S., & Prabhulinga, G. (2014). Effect of gamma radiation on growth of turmeric (*Curcuma longa* L.) cv. Salem explants. *Karnataka Journal of Agricultural Sciences, 27*(2), 152–155.

Priyadarshan, P. M., & Zachariah, P. K. (1986). Studies on in vitro culture of cardamom (Elettaria cardamomum Maton: Zingiberacea) - progress and limitations. In abstract VI International congress on plant tissue and cell culture, Minneapolis, Minn. (eds. Somers et al.).

Pujari, P. P., Patil, R. B., & Sakpal, R. T. (1986). Krishna, a high yielding variety of turmeric. *Indian Cocoa Arecanut Spices J. 9.*

Qian, Y., Zhang, T., Yu, Y., Gou, L., Yang, J., Xu, J., & Pi, E. (2021). Regulatory mechanisms of bHLH transcription factors in plant adaptive responses to various abiotic stresses. *Frontiers in Plant Science, 12.* https://doi.org/10.3389/fpls.2021.677611.

Radwan, M. M., Tabanca, N., Wedge, D. E., Tarawneh, A. H., & Cutler, S. J. (2014). Antifungal compounds from turmeric and nutmeg with activity against plant pathogens. *Fitoterapia, 99*(1), 341–346. https://doi.org/10.1016/j.fitote.2014.08.021.

Rahman, M. M., Amin, M. N., Jahan, H. S., & Ahmed, R. (2004). In vitro regeneration of plantlets of *Curcuma longa* L., a valuable spice plant in Bangladesh. *Asian Journal of Plant Sciences, 3.*

Rajan, V. R. (1997). Micropropagation of turmeric (*Curcuma longa* L.) by in vitro microrhizomes. In *Biotechnology of spices, medicinal and aromatic plants.* Calicut: Indian Society for Spices.

Ramachandran, K., & Nair, C. P. N. (1992). In vitro formation of roots and rhizomes from anther explants of ginger (*Zingiber officinale* Rosc.). *Journal of Spices and Aromatic ccrops, 1*, 72–74.

Ramakrishnan Nair, & Dutta Gupta (2003). Somatic embryogenesis and plant regeneration in black pepper (*Piper nigrum* L.): I. Direct somatic embryogenesis from tissues of germinating seeds and ontogeny of somatic embryos. *Journal of Horticultural Science and Biotechnology, 78*(3), 416–421. https://doi.org/10.1080/14620316.2003.11511641.

Ramakrishnan Nair, & Dutta Gupta (2006). High-frequency plant regeneration through cyclic secondary somatic embryogenesis in black pepper (*Piper nigrum* L.). *Plant cell Reports, 24*(12), 699–707. https://doi.org/10.1007/s00299-005-0016-2.

Ramakrishnan Nair, & Dutta Gupta (2007). In vitro plant regeneration from encapsulated somatic embryos of black pepper (*Piper nigrum* L.). *Journal of Plant Sciences, 2*(3), 283–292. https://doi.org/10.3923/jps.2007.283.292.

Rani, M., Miah, M. A. H., Hasan, M. T., Rashid, M. H. O., Yasmin, S., & Haque, M. S. (2024). Clonal propagation of turmeric (*Curcuma longa*) and confirmation of genetic fidelity of the micropropagated shoots by RAPD markers. *Plant Tissue Culture and Biotechnology, 34*(1).

Rao, D. V. R. (1999). Effect of gamma irradiation on growth, yield and quality of turmeric. *Advances in Horticulture and Forestry, 6*, 107–110.

Ravindran, P. N., & Nirmal Babu, K. (1994). Chemotaxonomy of South Indian Piper. *Journal of Spices and Aromatic Crops, 3*, 6–13.

Ravindran, P. N., Nirmal Babu, K., Sasikumar, B., & Krishnamoorthy, K. S. (2000). Botany and crop improvement of black pepper. In P. N. Ravindran (Ed.). *Black Pepper, Piper nigrum* (pp. 23–142). Amsterdam, The Netherlands: Harwood Academic Publishers.

Ravindran, P. N., Nirmal Babu, K., & Shiva, K. N. (2007). Botany and crop improvement of turmeric. In: Ravindran, P.N., K. Nirmal Babu, and K. Sivaraman (eds.). Turmeric-The genus Curcuma. CRC Press, Boca Raton, 15-70.

Ravindran, P. N., & Babu, K. N. (Eds.). (2005). Ginger: The Genus Zingiber (1st ed.). CRC Press. https://doi.org/10.1201/9781420023367.

Regunath, & Gopalakrishnan (1991). Successful exploitation of in vitro culture techniques for rapid clonal multiplication and crop improvement in cardamom. In: *In proc. of Third Kerala Science Congress* March 1991.

Renjith, D., Nazeem, P. A., & Nybe, E. V. (2001). Response of turmeric (*Curcuma domestica* Val.) to in vivo and in vitro pollination. *Journal of Spices Aromatic Crop, 10*(2), 135–139.

Rostiana, O., Heryanto, R., Arlianti, T., Ruhnayat, A., Priyatno, T. P., & Hariri, M. R. (2025). De novo assembly, gene annotation and marker development using illumina paired-end transcriptome sequences in nutmeg (Myristica fragrans Houtt.). *3 Biotech, 15*(6). https://doi.org/10.1007/s13205-025-04321-6.

Rostinawati, T., Zahra, Z., Mubarok, S., & Suminar, E. (2023). Effect of cytokinin and auxin to curcuminoid content of *Curcuma domestica* explants in vitro. *International Journal of Agriculture and Biology, 30*(2), 139–144. https://doi.org/10.17957/IJAB/15.2068.

Sahoo, A., Kar, B., Sahoo, S., Jena, S., Kuanar, A., Parameswaran, S., Patnaik, J., & Nayak, S. (2019). De Novo transcriptome sequencing explored cultivar specific sequence variation and differential expression of pigment synthesis genes in turmeric (*Curcuma longa* L.). *Industrial Crops and Products, 134*, 388–402. https://doi.org/10.1016/j.indcrop.2019.04.021.

Saji, K. V., Sasikumar, B., Rema, J., Aravind, S., & Nirmal Babu, K. (2019). *Spices genetic resources: Diversity, distribution and conservation*. Springer Science and Business Media LLC, 283–320. 10.1007/978-981-13-3669-0_9.

Salzer (1975). Analytical evaluation of seasoning extracts (oleoresins) and essential oils from seasonings I. *International Flavours Food Additive, 6*(3).

Samsudeen, Babu, N., Divakaran, M., & Ravindran (2000). Plant regeneration from anther derived callus cultures of ginger (*Zingiber officinale* Rosc.). *Journal of Horticultural Science and Biotechnology, 75*(4), 447–450. https://doi.org/10.1080/14620316.2000.11511266.

Samykutty, A., Shetty, A. V., Dakshinamoorthy, G., Bartik, M. M., Johnson, G. L., Webb, B., Zheng, G., Chen, A., Kalyanasundaram, R., & Munirathinam, G. (2013). Piperine, a bioactive component of pepper spice exerts therapeutic effects on androgen dependent and androgen independent prostate cancer cells. *PLoS ONE, 8*(6). https://doi.org/10.1371/journal.pone.0065889.

Sarma, I., Deka, & Sarma (2021). A protocol for rapid clonal propagation and microrhizome production of *Curcuma caesia* Roxb. (*Zingiberaceae*): A critically endangerd medicinal plant of North East India. *Indian Journal of Agricultural Research, 55*(1), 13–22. https://doi.org/10.18805/IJARE.A-5338.

Sasikumar, B., Ravindran, P. N., George, J. K., & Peter, K. V. (1996). Ginger and turmeric breeding in Kerala. In: In P. I. Kuriachan (Ed.). *Proc. Sem. Crop Breeding in Kerala* (pp. 65–72). University of Kerala.

Sasikumar, Chempakam, George, J. K., Remashree, Devasahayam, Dhamayanthi, Ravindran, & Peter (1999). Characterization of two interspecific hybrids of Piper. *Journal of Horticultural Science and Biotechnology, 74*(1), 125–131. https://doi.org/10.1080/14620316.1999.11511084.

Sasikumar, George, J. K., Saji, Gowda, A., & Zachariah (2015). Two unique black pepper accessions with very long spikes from the centre of origin. *Plant Genetic Resources: Characterisation and Utilisation, 13*(2), 183–185. https://doi.org/10.1017/S147926211400080X.

Sasikumar, Ravindran, P. N., & George, J. K. (1994). Breeding ginger and turmeric. *Indian Cocoa Arecanut and Spices J, 18*, 10–12.

Sasikumar, B., Haridas, P., George, J. K., et al. (2004). 'IISR Thevam', 'IISR Malabar Excel', 'IISR Girimunda' – three new black pepper clones. *Journal of Spices and Aromatic Crops*, 13, 1–5.

Sathyagowri, S., & Seran, T. H. (2011). In vitro plant regeneration of ginger (*Zingiber officinale* Rosc.) with emphasis on initial culture establishment. *International Journal of Medicinal and Aromatic Plants, 1*(3), 195–202.

Senthil Kumar, Prasath, Aravind, S., Leela, Venugopal, Ankegowda, Biju, & Anandaraj (2017). Evaluation of elite cardamom (*Elettaria cardamomum*) genotypes for yield, quality and resistance to cardamom mosaic virus. *Journal of Plantation Crops, 45*(2), 129–134. https://doi.org/10.19071/jpc.2017.v45.i2.3307.

Senthil Kumar, Aravind, S., Biju, & Ankegowda (2018). Characterization of genotypes of small cardamom (*Elettaria cardamomum* Maton) for yield parameters and disease resistance. *Journal of Plantation Crops, 46*(2), 143–146. https://doi.org/10.25081/jpc.2018.v46.i2.3729.

Seran, T. H. (2013). In vitro propagation of ginger (*Zingiber officinale* Rosc.) through direct organogenesis: A review. *Pakistan Journal of Biological Sciences, 16*(24), 1826–1835. https://doi.org/10.3923/pjbs.2013.1826.1835.

Shaaban, A., Elnfishy, N., Aween, Z., Abdelah, E., & Abughni, E. (2023). In vitro micropropagation of ginger plant (*Zingiber officinale*). *Scientific Journal for Faculty of Science-Sirte University, 3*(2), 154–161.

Shang, C., Lin, H., Fang, X., Wang, Y., Jiang, Z., Qu, Y., Xiang, M., Shen, Z., Xin, L., Lu, Y., Gao, J., & Cui, X. (2021). Beneficial effects of cinnamon and its extracts in the management of cardiovascular diseases and diabetes. *Food and Function, 12*(24), 12194–12220. https://doi.org/10.1039/D1FO01935J.

Sharifi-Rad, J., Dey, A., Koirala, N., Shaheen, S., El Omari, N., Salehi, B., Goloshvili, T., Cirone Silva, N. C., Bouyahya, A., Vitalini, S., Varoni, E. M., Martorell, M., Abdolshahi, A., Docea, A. O., Iriti, M., Calina, D., Les, F., López, V., & Caruntu, C. (2021). Cinnamomum species: Bridging phytochemistry knowledge, pharmacological properties and toxicological safety for health benefits. *Frontiers in Pharmacology, 12*. https://doi.org/10.3389/fphar.2021.600139.

Sharifi-Rad, M., Varoni, E. M., Salehi, B., Sharifi-Rad, J., Matthews, K. R., Ayatollahi, S. A., Kobarfard, F., Ibrahim, S. A., Mnayer, D., Zakaria, Z. A., Sharifi-Rad, M., Yousaf, Z., Iriti, M., Basile, A., & Rigano, D. (2017). Plants of the genus zingiber as a source of bioactive phytochemicals: From tradition to pharmacy. *Molecules, 22*(12). https://doi.org/10.3390/molecules22122145.

Sharon A. (2012). In vitro mutagenesis in turmeric (*Curcuma longa* L.) cv BSR-2. PhD thesis submitted to Tamil Nadu Agricultural University.

Shaukat, M. N., Nazir, A., & Fallico, B. (2023). Ginger bioactives: A comprehensive review of health benefits and potential food applications. *Antioxidants, 12*(11). https://doi.org/10.3390/antiox12112015.

Sheeja, Sabeesh, Shabna, Shalini, & Krishnamoorthy (2013). 173. TE, C, OV, RS, B (.) Genetic diversity analysis of Myristica and related genera using RAPD and ISSR markers. *Journal of Spices and Aromatic Crops, 22*(1).

Shirgurkar, M. V., Naik, V. B., Von Arnold, S., Nadgauda, R. S., & Clapham, D. (2006). An efficient protocol for genetic transformation and shoot regeneration of turmeric (*Curcuma longa* L.) via particle bombardment. *Plant cell Reports, 25*(2), 112–116. https://doi.org/10.1007/s00299-005-0033-1.

Shivakumar, Saji, & Sasikumar (2020). Genetic variability and correlation for yield and yield attributes in promising black pepper genotypes. *Electronic Journal of Plant Breeding, 11*(1), 65–69. https://doi.org/10.37992/2020.1101.011.

Shivakumar, M. S., Saji, K., Ankegowda, S. J., & Sasi kumar, B. (2019). Source of valuable genes for augmenting breeding in black pepper (*Piper nigrum* L.). *International journal of chemical studies, 7*(6), 2930–2932.

Shylala, M. R., Nair, S. G., Nazism, P. A., Millikan, V. K., & Mathew, M. K. (1996). In vitro screening of black pepper for tolerance to *Phytophthora capsici*. *Journal of Plantation Crops, 24*, 171–178.

Sial, P., Behera, S. K., & Tarai, R. K. (2016). Ginger germplasm collection, characterization, evaluation and conservation in Eastern Ghat High Land Zone of Odisha. *International Journal of Agricultural Science, 8*(17).

Silaru, R., Duraisamy, P., Madduri, Y. K., Sounderarajan, A., Veeraraghavan, S., & Subraya, K. K. (2024). Deciphering the genotypic superiority of turmeric (*Curcuma longa* L.) for yield and quality traits under three contrasting production systems. *Journal of Applied Research on Medicinal and Aromatic Plants, 43*. https://doi.org/10.1016/j.jarmap.2024.100592.

Sinchana, N. S., Kattimani, K. N., Prabhuling, G., Sudesh, K., & Jagadeesha, N. (2020). Standardization of tissue culture protocol for turmeric (*Curcuma longa* L.) cv. Salem. *International Journal of Chemical Studies, 8*(1).

Sivaraman, K. (1992). Large-scale production of quality planting material of black pepper. In Y. R. Sarma, S. Devasahayam, & M. Anandaraj (Eds.), *Black pepper and cardamom* (pp. 68–72). Indian Society for Spices.

Sivaranjani, R. (2024). Genomics of flavor constituents of cardamom. In K. K. Sabu & C. Kole (Eds.), *The cardamom genome* (Compendium of plant genomes, pp. 77–87). Springer. https://doi.org/10.1007/978-3-031-71829-8_6.

Siviero, A., Gallo, E., Maggini, V., Gori, L., Mugelli, A., Firenzuoli, F., & Vannacci, A. (2015). Curcumin, a golden spice with a low bioavailability. *Journal of Herbal Medicine, 5*(2), 57–70. https://doi.org/10.1016/j.hermed.2015.03.001.

Škorničková, J., & Sabu, M. (2005). *Curcuma roscoeana* Wall. (Zingiberaceae) in India. *Garden Bulletin Singapore, 57*, 187–198.

Škorničková, J., Sabu, M., & Prasanthkumar, M. G. (2003a). *Curcuma codoantha* (Zingiberaceae) – A new species from Andaman Islands, India. *Garden Bulletin Singapore, 55*, 218–228.

Škorničková, J., Sabu, M., & Prasanthkumar, M. G. (2003b). A new species of *Curcuma* L. (Zingiberaceae) from Mizoram, India. *Garden Bulletin Singapore, 55*, 89–95.

Škorničková, J., Sabu, M., & Prasanthkumar, M. G. (2004). *Curcuma mutabilis* (Zingiberaceae) – A new species from South India. *Garden Bulletin Singapore, 56*.

Smart, J., & Simmonds, N. W. (1992). *Evolution of crop plants*. Edinburgh: Longman, 333.

Somashekar, S. M., Subraya, K. K., Vijayan, S. K., & Bhaskaran Pillai, S. (2021). Pericarp as a new berry trait to define dry recovery and quality in black pepper (*Piper nigrum* L. *Scientia Horticulturae, 281*. https://doi.org/10.1016/j.scienta.2021.109923.

Soundar Raju, C., Aslam, A., & Shajahan, A. (2015). High-efficiency direct somatic embryogenesis and plant regeneration from leaf base explants of turmeric (*Curcuma longa* L. *Plant Cell, Tissue and Organ Culture, 122*(1), 79–87. https://doi.org/10.1007/s11240-015-0751-1.

Sreekumar, S., Divya, K., Joy, N., & Soniya (2022). De novo transcriptome profiling unveils the regulation of phenylpropanoid biosynthesis in unripe *Piper nigrum* berries. *BMC Plant Biology, 22*(1). https://doi.org/10.1186/s12870-022-03878-1.

Subasinghe, S., Hettiarachchi, C. S., & Iddagoda, N. (2016). In-vitro propagation of cinnamon (*Cinnamomum verum* Presl.) using embryos and in vitro axillary buds. *Journal of Advanced Agricultural Technologies, 3*(3), 164–169.

Sudharshan (1987). Colchicine – induced tetraploids in cardamom (*Elettaria cardamomum* Maton). *Current Science, 56*(1).

Sudharshan (1989). Induced polyploids in cardamom. *Journal of Plant Crops, 16*.

Sultan, M. T., Saeed, F., Raza, H., Ilyas, A., Sadiq, F., Musarrat, A., Afzaal, M., Hussain, M., Raza, M. A., & Al JBawi, E. (2023). Nutritional and therapeutic potential of nutmeg (*Myristica fragrans*): A concurrent review. *Cogent Food and Agriculture, 9*(2). https://doi.org/10.1080/23311932.2023.2279701.

Sunitibala, H., Damayanti, M., & Sharma, G. (2001). In vitro propagation and rhizome formation in *Curcuma longa* Linn. *Cytobios, 105*.

Suparman, B. (1990). Studies on tissue culture of clove tree plant. *Indonesian Journal of Crop Sciences, 5*(2).

Swarnathilaka, D., Kottearachchi, N., & Weerakkody, W. J. S. K. (2016). Factors affecting induction of microrhizomes in ginger (*Zingiber officinale* Rosc.), cultivar local from Sri Lanka. *British Biotechnology Journal, 12*(2).

Swetha, Parvathy, Sheeja, & Sasikumar (2014). DNA barcoding for discriminating the economically important cinnamomum verum from its adulterants. *Food Biotechnology, 28*(3), 183–194. https://doi.org/10.1080/08905436.2014.931239.

Takooree, H., Aumeeruddy, M. Z., Rengasamy, K. R. R., Venugopala, K. N., Jeewon, R., Zengin, G., & Mahomoodally, M. F. (2019). A systematic review on black pepper (*Piper nigrum* L.): from folk uses to pharmacological applications. *Critical Reviews in Food Science and Nutrition, 59*. https://doi.org/10.1080/10408398.2019.1565489.

Thapa, A., Datta, S., Dey, A. N., & Baisare, P. (2017). Advance propagation techniques in important spice crops. *International Journal of Current Microbiology and Applied Sciences, 6*(9), 1979–1985.

Thomas, T. A. (1980). Genetic resources of ginger and turmeric in India. In: *Proc. Natt Sem. Ginger and Turmeric* Calicut.

Tyagi, R. K., Agrawal, A., Mahalakshmi, Hussain, Z., & Tyagi, H. (2007). Low-cost media for in vitro conservation of turmeric (*Curcuma longa* L.) and genetic stability assessment using RAPD markers. *In Vitro Cellular and Developmental Biology - Plant, 43*(1), 51–58. https://doi.org/10.1007/s11627-006-9000-y.

Umadevi, P., Saji, K. V., & Suraby, E. J. (2015). Meristem culture for rapid regeneration in black pepper (*Piper nigrum* Linn.). *Annals of Plant Sciences, 4*(3), 1029–1032.

Upadhyay, R. K., & Jaiswal, G. (2007). Evaluation of biological activities of *Piper nigrum* oil against Tribolium castaneum. *Bulletin of Insectology, 60*, 57–61.

Usha, N. D. H., & Chezhiyan, N. (2007). Influence of gamma rays induced mutagenesis on the frequency of viable mutants in turmeric. *The Asian Journal of Horticulture, 2*(1), 54–57.

Vanaja, T., Neema, V. P., Rajesh, R., & Mammootty, K. P. (2007). Graft recovery of *Piper nigrum* L. runner shoots on *Piper colubrinum* rootstocks as influenced by varieties and month of grafting. *Journal of Tropical Agriculture, 45*, 54–57.

Vatsya, Dinesh, Kundapurkar, & Bhaskaran (1987). Large scale plant formation of cardamom (*Elettaria cardamomum*) by shoot bud cultures. *Plant Physiology and Biochemistry, 14*.

Velayudhan, K. C., Muralidharan, V. K., Amalraj, V. A., Gautam, P. L., Mandal, S., & Kumar, D. (1999). National Bureau of Plant Genetic Resources, New Delhi, India Curcuma genetic resources. *Scientific Monograph No. 4*.

Venugopal (1999). Natural disease escapes as source ofresistance against cardamom mosaic virus causing katte disease of cardamom (*Elettaria cardamomum* Maton). *Journal of Spices and Aromatic Crops, 8*.

Venugopal, Prasath, & Mulge, R. (2006). IISR Avinash – a rhizome rot resistant and high yielding variety of cardamom (*Elettaria cardamomum* Maton). *Journal of Spices and Aromatic Crops, 15*(1).

Verma, S., Singh, S., Sharma, S., Tewari, Roy, Goel, & Rana (2015). Assessment of genetic diversity in indigenous turmeric (*Curcuma longa*) germplasm from India using molecular markers. *Physiology and Molecular Biology of Plants, 21*(2), 233–242. https://doi.org/10.1007/s12298-015-0286-2.

Vernin, G., & Parkanyl, C. (2005). Chemistry of ginger. In P. N. Ravindran, & K. Nirmal Babu (Eds.). *Ginger – The genus zingiber*. Boca Raton, FL: CRC Press.

Vijayakumar, Surya, & Nalini (2004). Antioxidant efficacy of black pepper (*Piper nigrum* L.) and piperine in rats with high fat diet induced oxidative stress. *Redox Report, 9*(2), 105–110. https://doi.org/10.1179/135100004225004742.

Vijayan, K. K., & Thampuran, R. A. (2000). Pharmacology, toxicology and clinical applications of black pepper. In P. N. Ravindran (Ed.), *Black pepper (Piper nigrum L.)* (pp. 455–466). Harwood Academic Publishers.

Wang, X.-D., Xu, C.-Y., Zheng, Y.-J., Wu, Y.-F., Zhang, Y.-T., Zhang, T., Xiong, Z.-Y., Yang, H.-K., Li, J., Fu, C., Qiu, F.-Y., Dai, X.-Y., Liu, X.-L., He, X.-S., Zhou, S.-S., Li, S.-X., Fu, T., Xie, H., Chen, Y.-L., ... Jiang, X.-M. (2022). Chromosome-level genome assembly and resequencing of camphor tree (*Cinnamomum camphora*) provides insight into phylogeny and diversification of terpenoid and triglyceride biosynthesis of Cinnamomum. *Horticulture Research, 9*. https://doi.org/10.1093/hr/uhac216.

Wang, Y., Wang, L., Zhu, X., Wang, D., & Li, X. (2016). Choleretic activity of turmeric and its active ingredients. *Journal of Food Science, 81*(7). https://doi.org/10.1111/1750-3841.13348.

Wang, Y., Li, J., Guo, J., Wang, Q., Zhu, S., Gao, S., Yang, C., Wei, M., Pan, X., Zhu, W., Ding, D., Gao, R., Zhang, W., Wang, J., & Zang, L. (2017). Cytotoxic and antitumor effects of curzerene from *Curcuma longa*. *Planta Medica, 83*(1-2), 23–29. https://doi.org/10.1055/s-0042-107083.

Widyastuti, Santosa, Nuringtyas, & Rohman (2024). Callus induction of red ginger (*Zingiber officinale* var. rubrum) with 2,4-D and Kinetin combination to enhance total phenolic and flavonoid content. *IOP Conference Series: Earth and Environmental Science, 1364*(1). https://doi.org/10.1088/1755-1315/1364/1/012051.

Wijesekera, & Jayawardena (1973). Recent developments in the production of spices and their essential oils in Ceylon. In *In Proceedings, conference on spices*. London: Tropical Products Institute.

Xing, H., Jiang, Y., Zou, Y., Long, X., Wu, X., Ren, Y., Li, Y., & Li, H.-L. (2021). Genome-wide investigation of the AP2/ERF gene family in ginger: Evolution and expression profiling during development and abiotic stresses. *BMC Plant Biology, 21*(1). https://doi.org/10.1186/s12870-021-03329-3.

Xu, X.-Y., Meng, X., Li, S., Gan, R.-Y., Li, Y., & Li, H.-B. (2018). Bioactivity, health benefits, and related molecular mechanisms of curcumin: Current progress, challenges, and perspectives. *Nutrients, 10*(10). https://doi.org/10.3390/nu10101553.

Yang, Zhang, & Zhang (2022). Bisacurone attenuates diabetic nephropathy by ameliorating oxidative stress, inflammation and apoptosis in rats. *Human and Experimental Toxicology, 41*. https://doi.org/10.1177/09603271221143713.

Yang, Q., Liu, R.-S., Gong, Z.-Z., & Liu, W.-Y. (2002). Studies of three genes encoding Cinnamomin (a type II RIP) isolated from the seeds of camphor tree and their expression patterns. *Gene, 284*(1-2), 215–223. https://doi.org/10.1016/S0378-1119(01)00890-3.

Yang, T., Li, J., Wang, H.-X., & Zeng, Y. (2005). A geraniol-synthase gene from *Cinnamomum tenuipilum*. *Phytochemistry, 66*(3), 285–293. https://doi.org/10.1016/j.phytochem.2004.12.004.

Yassin, M. T., Mostafa, A. A.-F., & Al-Askar, A. A. (2020). In vitro anticandidal potency of *Syzygium aromaticum* (clove) extracts against vaginal candidiasis. *BMC Complementary Medicine and Therapies, 20*(1). https://doi.org/10.1186/s12906-020-2818-8.

Yin, Y., Xie, X., Zhou, L., Yin, X., Guo, S., Zhou, X., Li, Q., Shi, X., Peng, C., & Gao, J. (2022). A chromosome-scale genome assembly of turmeric provides insights into curcumin biosynthesis and tuber formation mechanism. *Frontiers in Plant Science, 13*. https://doi.org/10.3389/fpls.2022.1003835.

Zahid, N. A., Jaafar, H. Z. E., & Hakiman, M. (2021). Micropropagation of ginger (*Zingiber officinale* roscoe) 'bentong' and evaluation of its secondary metabolites and antioxidant activities compared with the conventionally propagated plant. *Plants, 10*(4). https://doi.org/10.3390/plants10040630.

Zapata, E. V., Morales, G. S., Lauzardo, A. N. H., Bonfil, B. M., Tapia, G. T., Sánchez, A. D. J., Del Valle, M. V., & Aparicio, A. J. (2003). In vitro regeneration and acclimatization of plants of turmeric (*Curcuma longa* L.) in a hydroponic system. *Biotechnología Aplicada, 20*.

Zhang, J., Ye, K.-P., Zhang, X., Pan, D.-D., Sun, Y.-Y., & Cao, J.-X. (2017). Antibacterial activity and mechanism of action of black pepper essential oil on meat-borne *Escherichia coli*. *Frontiers in Microbiology, 7*(JAN). https://doi.org/10.3389/fmicb.2016.02094.

Zhang, L., Yang, Z., Chen, F., Su, P., Chen, D., Pan, W., Fang, Y., Dong, C., Zheng, X., & Du, Z. (2017). Composition and bioactivity assessment of essential oils of *Curcuma longa* L. collected in China. *Industrial Crops and Products, 109*, 60–73. https://doi.org/10.1016/j.indcrop.2017.08.009.

Zhang, P., Liu, D., Ma, J., Sun, C., Wang, Z., Zhu, Y., Zhang, X., & Liu, Y. (2024). Genome-wide analysis and expression pattern of the ZoPP2C gene family in *Zingiber officinale* Roscoe. *BMC Genomics, 25*(1). https://doi.org/10.1186/s12864-024-09966-w.

Zheng, Y., Liu, Y., Ma, M., & Xu, K. (2008). Increasing in vitro microrhizome production of ginger (*Zingiber officinale* Roscoe). *Acta Physiologiae Plantarum, 30*(4), 513–519. https://doi.org/10.1007/s11738-008-0149-3.

CHAPTER

16

Natural colorant- and dye-yielding plants: health benefits, genetic resources, micropropagation and *in vitro* pigment production

Rohini M.R.

ICAR–Indian Institute of Oil Palm Research, Research Centre, Palode, Trivandrum, Kerala, India

16.1 Introduction

The use of natural dyes dates back to the earliest days of textile-making. Since prehistoric times, humans have been fascinated by colors, using them not only for aesthetic purposes but also for cultural and symbolic expression. The practice of dyeing has a rich and diverse history, with many natural dyes derived from plants, minerals, and insects tracing their origins back thousands of years. This enduring relationship between humans and colors highlights the significance of natural pigments, not only as functional colorants but also as reflections of tradition, innovation, and human creativity across civilizations. In recent years, the global market for natural pigments has experienced rapid growth, driven by increasing consumer awareness of health and the discovery of new therapeutic effects associated with many natural pigments, such as carotenoids, anthocyanins, betalains, and chlorophyll. The barks, seeds, flowers, and leaves of plants have been identified as sources of a variety of colors. These dyes have been used as colorants in foods, garments, and cosmetics. For example, dyes for lipstick are still obtained from *Bixa orellana* L. and *Lithospermum erythrorhizon* Sieb & Zucc., and those for eye shadow from indigo (Siva, 2007). Besides the coloring property, many of the pigments are rich in medicinal properties such as antioxidant, antimicrobial, and antiinflammatory activities. It is interesting to note that plants are capable of synthesizing over 2000 distinct pigments, yet only around 150 have found commercial applications. In India alone, approximately

Medicinal Plant Biotechnology for Ameliorated Secondary Metabolite Production
DOI: https://doi.org/10.1016/B978-0-443-43944-5.00007-4

450 plant taxa are reported to yield natural dyes, with 50 species regarded as particularly significant. These valuable sources span various plant parts, including roots (10 species), barks (4), leaves (5), flowers (7), fruits (7), seeds (3), wood (8), and gums and resins (3), highlighting the remarkable diversity of plant-based colorants and their potential for sustainable utilization (Table 16.1).

The growing market demand for dyes, combined with the dwindling availability of dye-yielding plants, led to the development and widespread use of synthetic dyes, such as aniline and coal-tar derivatives. While these synthetic alternatives initially seemed to offer a practical solution, subsequent studies revealed that many of them pose serious health and environmental risks. The use of synthetic dyes has become a cause of concern due to the mutagenic, carcinogenic, and toxicological properties they possess, which can adversely affect the natural ecosystem (Gita et al., 2017). In addition, contact dermatitis has been widely reported in the textile industry (Shahid & Mohammad, 2013). Further, given the World Health Organization's prohibitions on the use of synthetic dyes, there is a growing

TABLE 16.1 Major natural colors derived from plants and their sources.

Color/Pigment	Plant source(s)	Plant part used	Notes/Uses
Red/Pink	*Bixa orellana*	Seeds (aril)	Produces bixin, used in food and cosmetics
	Hibiscus sabdariffa	Calyces	Rich in anthocyanins, used in beverages
	Rosa spp.	Flowers	Extracted for textile dyeing and cosmetics
	Beta vulgaris	Roots	Source of betalains, food coloring
Yellow/Orange	*Curcuma longa*	Rhizomes	Turmeric; curcumin pigment, culinary and medicinal use
	Tagetes spp.	Flowers	Lutein-rich dye for textiles and food
	Bixa orellana	Seeds	Carotenoid pigment, coloring agent
	Carthamus tinctorius	Flowers	Safflower yellow, textile and food coloring
Blue/Purple	*Clitoria ternatea*	Flowers	Anthocyanin-rich, used in beverages and textiles
	Hibiscus sabdariffa	Calyces	Anthocyanins, beverages
	Dioscorea alata	Tubers	Purple yam, food color
Green	*Spinacia oleracea*	Leaves	Chlorophyll-rich, food coloring
	Moringa oleifera	Leaves	Chlorophyll pigment, nutritional supplement
	Cymbopogon spp.	Leaves	Natural green dye in textiles
Brown/Black	*Acacia catechu*	Heartwood/Bark	Catechu extract, tanning and dyeing
	Terminalia chebula	Fruits	Used in textiles and leather dyeing
	Juglans regia	Hulls of nuts	Produces brown-black pigment for textiles
Orange-red	*Crocus sativus*	Stigmas	Saffron; carotenoids like crocin, culinary use
	Gardenia jasminoides	Fruits	Gardenia yellow pigment
Yellow-red	*Polygonum tinctorium*	Leaves	Indigo precursor for blue dye (oxidation gives indigo)

interest in natural dye-based goods, and demand for these products is also expanding. This realization has prompted a renewed interest in natural dyes, prompting extensive research into their sources, properties, and applications. Natural pigments are now being revisited not only as safe and eco-friendly alternatives but also as valuable bioactive compounds with potential benefits for human health. The combined focus on sustainability, safety, and functional value has reinforced the importance of conserving pigment-producing plants and developing modern technologies for their propagation and pigment production.

16.2 Health benefits of natural pigments

Natural pigments (Fig. 16.1) are vital secondary metabolites (Table 16.2) that perform diverse physiological functions throughout the plant life cycle and are widely recognized for their significant health-promoting properties (Fig. 16.2). Beyond their role in plant growth and development, these pigments exhibit strong antioxidant activity and confer numerous additional health benefits. Extensive studies have documented the biological and

COLOUR	PIGMENT	PLANT SOURCE(S)	PLANT PART USED	
RED/PINK	Bixin	*Bixa orellana* (Annatto)	Seeds (aril)	
	Anthocyanins	*Hibiscus sabdariffa* (Roselle)	Calyces	
	Anthocyanins	*Rosa* spp. (Rose).	Flowers	
	Betalains	*Beta vulgaris* (Beetroot)	Roots	
Yellow/ Orange	Curcumin	*Curcuma longa* (Turmeric)	Rhizomes	
	Lutein	*Tagetes* spp. (Marigold)	Flowers	
	Carotenoids	*Bixa orellana* (Annatto fruit)	Seeds	
	Yellow carthamin	*Carthamus tinctorius* (Safflower)	Flowers	
Blue / purple	Anthocyanins	*Clitoria ternatea* (Butterfly pea)	Flowers	
	Anthocyanins	*Hibiscus sabdariffa* (Roselle)	Calyces	
	Anthocyanins	*Dioscorea alata* (Greater yam)	Tubers	
GREEN	Chlorophyll	*Spinacia oleracea* (Spinach)	Leaves	
	Chlorophyll	*Cymbopogon* spp. (Lemongrass)	Leaves	

FIGURE 16.1 Natural plant pigments and their sources. *The figure has been created using ChatGPT*

TABLE 16.2 The major pigment groups and their plant sources include.

Pigment class	Major source plants	Key pigments	Natural habitat/ Region
Chlorophylls	*Spinacia oleracea, Moringa oleifera, Camellia sinensis*	Chlorophyll a, b	Ubiquitous, tropical to temperate
Carotenoids	*Daucus carota, Capsicum annuum, Tagetes erecta, Lycopersicon esculentum*	β-Carotene, Lutein, Zeaxanthin	Tropical/subtropical
Flavonoids/ Anthocyanins	*Hibiscus sabdariffa, Clitoria ternatea, Vitis vinifera, Vaccinium spp.*	Cyanidin, Delphinidin, Malvidin	Tropical and temperate
Betalains	*Beta vulgaris, Hylocereus spp., Opuntia spp.*	Betanin, Indicaxanthin	Arid and semi-arid ecosystems
Curcuminoids/ Phenolics	*Curcuma longa, Zingiber cassumunar*	Curcumin, Demethoxycurcumin	Indo-Malayan region
Tannins and related pigments	*Terminalia chebula, Camellia sinensis*	Ellagitannins, Catechins	Tropical forests, cultivated tea zones

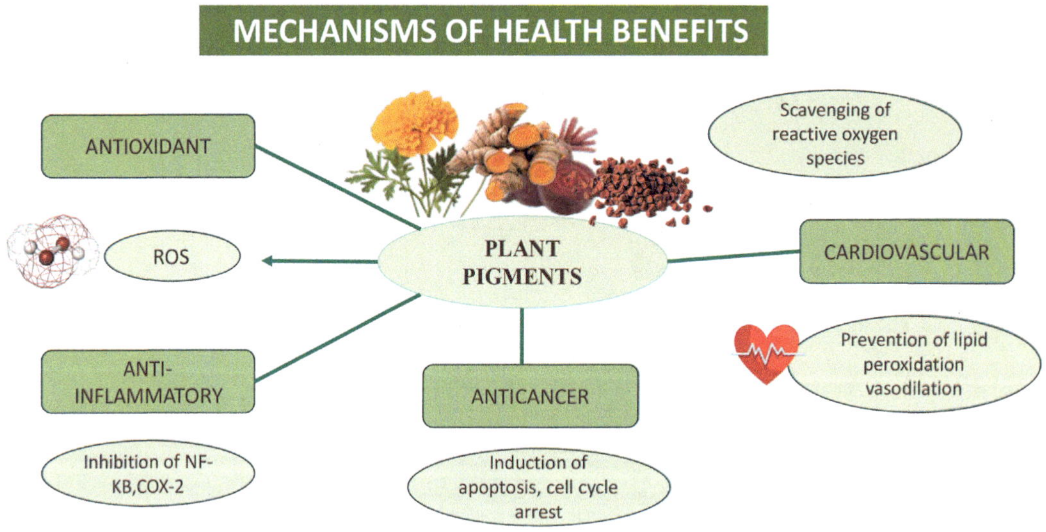

FIGURE 16.2 Mechanisms underlying the health-benefiting effects of natural plant pigments. *This figure is developed using ChatGPT*

pharmacological potential of major pigment groups such as carotenoids, flavonoids, and anthocyanidins. Reported activities include antioxidant, anticancer, antiviral, antibacterial, anti-inflammatory, antiallergic, antithrombotic, cardioprotective, hepatoprotective, neuroprotective, antimalarial, antileishmanial, antitrypanosomal, and antiamoebic effects.

Decades of systematic research have established the multifaceted contributions of plant-derived pigments to human health, leading to their increasing recognition and utilization. As a result, these bioactive compounds hold substantial promise for diverse applications across the food, pharmaceutical, cosmetic, and related industries.

16.2.1 Carotenoids

Carotenoids are natural fat-soluble pigments that impart the characteristic yellow, orange, and red hues to many fruits and vegetables. So far, more than 800 natural carotenoids have been found, with colors of red, orange, yellow, etc. (Leong et al., 2017). There are abundant and diverse carotenoids in fruits, as more than 100 kinds have been identified (Rodriguez-Concepcion et al., 2018). Carotenoids are broadly classified into two major groups based on the presence or absence of oxygen atoms in their molecular structure. Those containing oxygen are known as xanthophylls ($C_{40}H_{56}O_2$ or $C_{40}H_{56}O$), while those lacking oxygen are referred to as carotenes ($C_{40}H_{56}$) (Rodriguez-Concepcion et al. 2018). In plants, common xanthophylls include lutein, zeaxanthin, capsanthin, and β-cryptoxanthin, whereas α-carotene, β-carotene, γ-carotene, lycopene, phytoene, and phytofluene represent the principal carotenes. α-, β-, and γ-carotene serve as important biosynthetic precursors of vitamin A (DellaPenna & Pogson, 2006).

Among the major carotenoids of nutritional importance are β-carotene, lycopene, lutein, and zeaxanthin. They play an essential role in human health due to their strong antioxidant properties beneficial in the prevention of cancer and cardiovascular diseases. These compounds help neutralize free radicals, thereby reducing oxidative stress and cellular damage. β-carotene is converted into vitamin A in the body, which is vital for maintaining healthy vision, skin, and immune function. Lutein and zeaxanthin are especially important for eye health, as they protect the retina from blue light damage and lower the risk of age-related macular degeneration. Lycopene and other carotenoids have been associated with a reduced risk of several cancers, including those of the lung, prostate, and skin, by protecting DNA and reducing inflammation. In addition, carotenoids contribute to cardiovascular health by preventing arterial damage and supporting normal lipid metabolism. Major dietary sources of carotenoids include carrots, tomatoes, sweet potatoes, spinach, kale, red peppers, and egg yolks, which together provide a diverse and natural means of promoting health and preventing chronic diseases.

16.2.2 Flavonoids

Flavonoids are a diverse group of compounds belonging to the polyphenol superfamily, synthesized by plants and widely distributed in their stems, leaves, flowers, and fruits. These pigments typically exhibit yellow, pale yellow, or whitish hues, contributing to the colouration of fruits and aiding in seed dispersal by attracting animals and insects (Lu et al., 2021). Beyond their role in pigmentation, flavonoids play crucial functions in plant physiology. They are produced as a defense response against biotic and abiotic stresses and are also involved in regulating plant growth and development. To date, more than 5000 distinct flavonoid compounds have been identified in plants, reflecting their remarkable chemical diversity (Cetinkaya et al., 2017).

Their primary nutritional significance lies in their potent antioxidant and anti-inflammatory properties, which help neutralize free radicals, reduce oxidative stress, and modulate inflammatory pathways. These actions contribute to the prevention of chronic diseases such as cardiovascular disorders, diabetes, and certain types of cancer. Flavonoids also enhance vascular function, improve lipid metabolism, and protect the endothelium, thereby supporting heart health. In addition, flavonoids have been shown to exhibit anticarcinogenic, neuroprotective, and metabolic regulatory effects. They influence enzyme activity, gene expression, and cell signaling pathways involved in detoxification, apoptosis, and glucose metabolism. Flavonoids in berry fruits are mainly found in the peels and seeds, grape (*Vitis vinifera*), citrus fruit (family Rutaceae), and several tropical fruits are rich in flavonoids. Flavonoids have long been a mainstay of research in grape and wine because this class of substances not only affects the color of the wine but also the aging capacity, bitterness, and astringency (Drappier et al., 2019). Regular consumption of flavonoid-rich foods—such as fruits (berries, citrus, apples, grapes), vegetables (onions, kale, broccoli), and beverages like tea, cocoa, and red wine has been associated with improved cognitive function, reduced risk of degenerative diseases, and overall better health outcomes. Thus flavonoids represent an important class of nutraceuticals, promoting long-term wellness and serving as key dietary components in the prevention of lifestyle-related diseases.

16.2.3 Anthocyanins

Anthocyanins are natural colorants belonging to the flavonoid family of secondary metabolites. They are predominantly localized in the outer cell layers of plants, such as the epidermis and peripheral mesophyll cells. These pigments are responsible for the blue, red, purple, and occasionally green hues observed in many plant tissues. The color expression of anthocyanins is pH-dependent—they appear bright red at acidic pH (around 3), become nearly colorless at pH 5, and shift toward blue or green shades under neutral to alkaline conditions. Natural sources rich in anthocyanins include red cabbage, black currant, sweet potato, purple carrot, cherry, and red radish (Mortensen, 2006). Anthocyanins are commonly found in flowers and the fruits of many plants. Most of the red, purple, and blue-colored flowers contain anthocyanins. Red flowers include red hibiscus, red rose, red pineapple sage, red clover, and pink blossom. These red flowers are edible. Blue (cornflower, blue chicory, and blue rosemary) and purple (purple mint, purple passion flower, purple sage, common violet, and lavender) flowers are the common edible flowers. Some of these flowers have been traditionally used as folk medicine, as colorants, and as food. The colored pigments of anthocyanin from berries, blackcurrants, and other types of red to blue-colored fruits are strong antioxidants. Moreover, anthocyanin-rich black carrot, red cabbage, and purple potato are potential functional foods that have been consumed for the prevention of diseases. In humans, dietary intake of anthocyanins is associated with reduced incidences of coronary heart disease and provides protection against oxidative stress-related disorders. These compounds exhibit bactericidal, antiviral, and fungistatic activities and help prevent the oxidation of ascorbic acid, thereby protecting the body from free radicals, inhibiting oxidative enzymes, and reducing the risk of cardiovascular diseases and cancer (Delgado-Vargas & Paredes-Lopez, 2000).

Owing to their natural origin and health-promoting properties, anthocyanins are extensively used in the food and beverage industry as natural colorants and antioxidants, increasingly replacing synthetic red dyes. They are widely used in the pigmentation of fruit juices, wines, and other beverages. Among various sources, grape pomace, a by-product of wine production, represents one of the most important commercial sources of anthocyanins. Currently, anthocyanin pigments are among the few anthocyanin-derived products approved by the U.S. Food and Drug Administration (FDA) for use in foods and beverages.

16.2.4 Betalains

Betalains are a class of nitrogenous water-soluble pigments sourced mainly from *Beta vulgaris* (beetroot), *Hylocereus* spp. (Dragon fruit), *Amaranthus* spp., and *Opuntia* spp. Betalains are the main compounds responsible for the red color of flowers, fruits, and other plant tissues, commonly found in plant groups in the order of Caryophyllales. They are grouped into betacyanins, with reddish-violet hues, and betaxanthins, yellow to orange (Martinez et al., 2024). Traditionally, betalain extracts have been utilized as natural colorants in various products, including dairy items, beverages, candies, and even in some cattle-based products (Georgiev et al., 2010). However, their broader application in the food industry is limited due to their sensitivity to environmental factors such as pH and temperature. Therefore betalains are best suited for incorporation into foods that undergo minimal heat treatment and have a short shelf life. For enhanced stability, betalain-based products are often processed, packaged, and marketed in a dry form under reduced humidity, light, and oxygen conditions.

Betalains are widely used for their vibrant colors and health-promoting properties. Betalains exhibit a wide range of bioactive properties, including antiinflammatory, antioxidant, antimicrobial, and antifungal activities, in addition to being rich in essential nutrients (Kale et al., 2018). Studies have shown that beetroot utilizes betalain compounds to enhance its vital defense mechanisms and increase resistance against pathogens. These compounds also contribute to the inhibition of low-density lipoprotein (LDL) oxidation, prevention of lipid peroxidation, and reduction of chemotoxic effects. There is a considerable number of studies focusing on the cardiovascular benefits of betalain-rich species. Because betalains are major compounds in beetroot juice and cactus pear, some clinical studies have addressed the potential of betalain-rich beetroot supplements to treat cardiovascular diseases, including hypertension (Martinez et al., 2024). The commercial production of betalains, however, depends not only on the efficiency of processing techniques—such as enzymatic regulation, extraction, purification, concentration, and drying- but also on the consistent availability of highly pigmented plant sources.

16.2.5 Chlorophyll

Chlorophylls are green pigments that play a pivotal role in photosynthesis and are abundantly present in green fruits and vegetables. Two major forms, chlorophyll *a* and chlorophyll *b*, are widely utilized in the food industry as natural colorants. Being lipid-soluble compounds, chlorophylls are typically extracted using organic solvents; however, upon saponification, they can be converted into a water-soluble derivative known as chlorophyllin.

Although research on chlorophyll and its derivatives is relatively limited, existing studies highlight their therapeutic potential. These photosynthetic pigments exhibit a broad spectrum of biological activities, including antioxidant, antimutagenic, antigenotoxic, anticancer, and anti-obesogenic effects.

Dietary intake of chlorophylls primarily occurs through the consumption of green vegetables and fruits. However, their concentration and bioavailability vary significantly depending on several factors such as plant species and cultivar, maturity stage, growing conditions, harvest time, plant part utilized, storage environment, and processing methods. Furthermore, extraction and quantification techniques also influence the measured chlorophyll content. While a diet rich in green produce contributes substantially to chlorophyll intake, their metabolism, stability, and absorption efficiency ultimately determine their physiological benefits and health-promoting potential (Martins et al., 2023).

16.2.6 Curcumin

Curcumin, the principal bioactive compound of turmeric (*Curcuma longa L.*), is a bright yellow pigment belonging to the polyphenol group, widely recognized for its extensive therapeutic properties. It exhibits potent antioxidant and antiinflammatory activities by scavenging free radicals and inhibiting key inflammatory mediators such as COX-2, LOX, and NF-κB. These actions help reduce oxidative stress, cellular damage, and inflammation, thereby contributing to the prevention and management of chronic disorders like arthritis, cardiovascular diseases, and metabolic syndrome.

In addition, curcumin possesses remarkable anticancer, antimicrobial, and neuroprotective effects. It suppresses tumor growth by inducing apoptosis, inhibiting angiogenesis, and modulating signaling pathways. Its antimicrobial action extends to bacteria, fungi, and viruses, making it effective in promoting immune health and infection resistance. Curcumin also exerts neuroprotective and cognitive-enhancing effects by reducing oxidative damage and amyloid plaque formation, showing promise in managing neurodegenerative disorders such as Alzheimer's and Parkinson's diseases.

Furthermore, curcumin supports metabolic, hepatic, and skin health through its lipid-lowering, hepatoprotective, and wound-healing properties. It enhances insulin sensitivity, protects the liver from toxins, and promotes skin regeneration. Despite its broad spectrum of biological benefits, curcumin's poor bioavailability remains a challenge, which has led to the development of improved delivery systems such as nanocurcumin and liposomal formulations to enhance its therapeutic efficacy and absorption in the human body.

*Antioxidant activity, scavenging reactive oxygen species (ROS) and thereby reducing oxidative stress and cellular damage. Antiinflammatory effects are mediated by the inhibition of key pro-inflammatory enzymes and transcription factors such as COX-2 and NF-κB, which help downregulate cytokine production. Pigments also contribute to cardiovascular protection by preventing lipid peroxidation, enhancing nitric oxide bioavailability, and improving endothelial function, thus promoting vasodilation and reducing atherosclerotic risk. In addition, several pigments exhibit anticancer properties by inducing apoptosis, causing cell-cycle arrest, and reducing oxidative DNA damage. Collectively, these mechanisms illustrate how dietary intake of pigment-rich fruits and vegetables contributes to human health, emphasizing the importance of natural colorants as functional food ingredients and therapeutic agents.

16.3 Plant genetic resources of natural pigment-yielding plants

Plant genetic resources (PGR) refer to the heritable materials in plants that hold actual or potential value for food, agriculture, medicine, or industry. Among these, pigment-yielding plants constitute a significant subset, providing bioactive compounds that serve as natural colorants, antioxidants, and therapeutic agents. Natural pigment-yielding plants are widely distributed across taxonomic groups and ecosystems, particularly across tropical, subtropical, and temperate regions, reflecting a wide ecological adaptability. The Western Ghats, Eastern Himalayas, Indo-Burma region, and parts of Central India are recognized as hotspots of dye plant diversity in India. Globally, tropical forests of Southeast Asia, Africa, and South America harbor a rich gene pool of natural dye sources.

Like many other plant species, natural dye-yielding plants in the Western Ghats and other biodiversity-rich regions are increasingly threatened due to habitat loss, overharvesting, land-use changes, and climate pressures. The Western Ghats, a global biodiversity hotspot (Myers et al., 2000), hosts a rich diversity of dye-producing species such as *Bixa orellana*, *Indigofera tinctoria*, *Morinda citrifolia*, *Terminalia chebula*, and *Lawsonia inermis*. However, unsustainable extraction practices and the replacement of natural dyes by synthetic alternatives have led to genetic erosion and declining wild populations, endangering both plant diversity and associated traditional knowledge systems (Siva, 2007; Kala, 2005). Conservation of these genetic resources is essential to ensure their sustainable use and availability for future generations. Integrated conservation strategies, including in situ protection of natural habitats, ex situ maintenance in botanical gardens and gene banks, and community-based cultivation programs, are vital for conserving this valuable heritage (Heywood, 2011). Safeguarding natural dye plants not only supports biodiversity conservation but also promotes sustainable livelihoods and eco-friendly industries in regions like the Western Ghats.

The integration of biotechnology into conservation strategies has greatly enhanced the efficiency of collecting, characterizing, maintaining, exchanging, and utilizing germplasm resources of dye-yielding plants (Kameswara Rao, 2003). Analyzing genetic variation is a preliminary step in the conservation of germplasm and is important in determining which variant should be conserved and how to be conserved. Understanding genetic diversity is essential for developing cultivars with desirable traits such as high yield and stress tolerance. Comprehensive diversity data also provide insights into the evolutionary relationships and taxonomy of lesser-known dye-yielding plants. Such information supports the development of core collections of pigment-producing species with minimal redundancy (Rao and Hodgkin, 2002). Conserving diversity at the species, genus, and ecosystem levels is crucial to achieving meaningful and long-term conservation outcomes.

A range of techniques is used to assess species diversity. Morphological markers based on traits such as flower and leaf color, size, and shape are among the earliest tools employed for plant differentiation. Although time-consuming and influenced by environmental factors, they remain valuable for detecting disease symptoms and pathogenic responses (Govindaraj et al., 2015). Biochemical markers, including isoenzymes and protein profiles, constitute another important approach for evaluating diversity (Bretting and Widrlechner, 1995). For example, native varieties of *Bixa orellana*, a major pigment-yielding species, were initially classified based on fruit morphology. Later, biochemical traits such as water-soluble polysaccharide content were used to more precisely distinguish among varieties. Molecular markers have been well utilized to study genetic variations within many dye plants and are thus pivotal in framing conservation strategies. In *Curcuma longa* populations from northeastern India, molecular markers such as RAPD, ISSR, and AFLP have been extensively used to assess genetic variation, providing valuable insights into the evolutionary lineage of the genus *Curcuma* (Das et al., 2011). Similar molecular diversity studies have also been carried out in other dye - and medicinally important species, including *Carthamus tinctorius* (safflower) (Sehgal and Raina, 2005) and *Justicia adhatoda* (Kumar et al., 2014).

16.4 Micropropagation and *in vitro* pigment production of dye-yielding plants

Among the available biotechnological approaches, plant tissue culture stands out as one of the most reliable and extensively used methods for large-scale micropropagation of plants. Micropropagation plays a crucial role in the conservation and large-scale propagation of natural dye-yielding plants, many of which are threatened by overexploitation, habitat destruction, and poor natural regeneration. Several dye plants such as *Indigofera tinctoria, Bixa orellana, Lawsonia inermis, Morinda citrifolia,* and *Curcuma longa* possess high industrial value but exhibit low seed viability, seasonal growth limitations, and restricted distribution (Siva, 2007; Rout et al., 2006). Tissue culture techniques provide an efficient alternative for the rapid multiplication of elite genotypes, ensuring genetic fidelity and year-round production of planting material. Micropropagation also supports germplasm conservation through *in vitro* storage and cryopreservation, safeguarding the genetic diversity of valuable and endangered dye species. In India, successful regeneration protocols have been established for several natural dye plants such as *Bixa orellana, Indigofera tinctoria,* and *Morinda citrifolia,* highlighting their potential in sustainable utilization, conservation, and industrial applications. Thus micropropagation serves as an essential biotechnological tool for conserving biodiversity, supporting eco-friendly industries, and reducing exploitation pressure on natural populations of dye-yielding plants.

In addition to plant propagation, the method is increasingly exploited for enhanced production of commercially significant secondary metabolites. Conventional production methods for natural pigments are often constrained by the availability of specific plant sources, seasonal and climatic variations, and the generally low yield of target pigment compounds. To overcome these limitations and enhance commercial-scale production, considerable attention has shifted toward sustainable and technologically advanced approaches. In this context, plant cell, tissue, and organ culture systems have emerged as season-independent, contaminant-free, and metabolically regulated environments capable of high pigment productivity. Callus cultures, suspension cultures, hairy roots, and transformed tissues have been exploited for enhanced synthesis of anthraquinones, curcuminoids, indigo precursors, carotenoids, and betalains in several dye species. Importantly, elicitor-mediated pathways and metabolic engineering strategies have shown substantial promise for augmenting pigment biosynthesis *in vitro*, reducing dependence on wild populations.

In recent years, numerous studies have demonstrated the feasibility of cultivating plant cell and organ cultures at an industrial scale using bioreactor systems (Choi et al., 2000). Similar to microbial fermentation technology, the successful operation of plant-based bioreactors depends on maintaining optimal environmental and physiological conditions, including adequate nutrient supply, appropriate medium pH, temperature, humidity, and light. It is important to note that plant cells are generally more shear-sensitive than microbial or animal cells due to their larger size and fragile cell walls. Therefore parameters such as controlled agitation, suitable working volume, and precisely regulated gaseous composition within the microenvironment are essential to achieve high biomass and metabolite production.

Equally critical is the choice of explant source: The physiological status of the mother plant and the specific tissue used (e.g., leaves, stems, roots, or meristems) significantly influences the growth rate and metabolite accumulation in derived cell or organ cultures. Bioreactor-based cultivation has thus emerged as a promising platform for species that are difficult to propagate conventionally or display inherently slow growth, enabling the continuous production of biomass or bioactive compounds without waiting for whole plants or organs to reach maturity.

The key advantages of plant tissue culture technologies are tremendous as compared to field cultivation, as highlighted below:

Rapid multiplication of plant species that are slow-growing, difficult to propagate, or have long juvenile phases

Production systems that are not constrained by seasonal changes or geographic limitations

Consistent and reproducible synthesis of metabolites with minimal variation

Efficient, rapid, and economical extraction of phytochemicals without depleting natural plant populations

Generation of pathogen-free and genetically uniform plant material

Reduced requirement for land, manpower, and other agricultural inputs

Continuous, year-round production under controlled conditions

Advances in plant biotechnology have facilitated the maintenance and proliferation of plant cells, tissues, and organs under controlled *in vitro* conditions, including their cultivation in bioreactors for large-scale metabolite production. In recent years, the use of *in vitro* plant culture systems for the biosynthesis of high-value phytochemicals has attracted increasing scientific and industrial interest. A generalized schematic representation illustrating the production of economically important compounds, such as natural pigments, through plant cell and organ culture technologies is outlined in Fig. 16.3.

Plant

(Explant: leaf, root, stem, nodal segment, rhizome, etc.)↓Sterilization

(antifungal, antibacterial agents)↓Cell lines

(Establishment of fast-growing *in vitro* cultures)↓Quantitative analysis↓Selection of high pigment-yielding cell lines↓Establishment of suspension culture↓Understanding the growth and pigment accumulation behavior↓Sequential scale-up studies↓Downstream processing for pigment recovery

(Extraction and purification followed by characterization)

Plant

(Explant: leaf, root, stem, nodal segment, rhizome, etc.)

↓

Sterilization

(antifungal, antibacterial agents)

↓

Cell lines

(Establishment of fast-growing *in-vitro* cultures)

↓

Quantitative analysis

↓

Selection of high pigment-yielding cell lines

↓

Establishment of suspension culture

↓

Understanding the growth and pigment accumulation behaviour

↓

Sequential scale-up studies

↓

Downstream processing for pigment recovery

(Extraction and purification followed by characterization)

FIGURE 16.3 Schematic representation of *in vitro* plant culture-based pigment production.

16.4.1 Establishment of *in vitro* cultures

Plant tissue culture is fundamentally based on the principle of *totipotency*, which refers to the capacity of an individual plant cell to dedifferentiate and subsequently regenerate into a complete plant under suitable *in vitro* conditions. This unique biological property forms the basis of *in vitro* propagation techniques and also facilitates the production of valuable phytochemicals either directly from cultured tissues or indirectly via dedifferentiated callus tissues. A diverse range of *in vitro* culture systems has been developed to meet different experimental and commercial objectives. These include callus cultures, seed cultures, anther and pollen cultures, embryo and ovary cultures, protoplast cultures, as well as organ cultures such as root, shoot, and leaf cultures. The selection of an appropriate culture system largely depends on the species, the regenerative competence of the explant, and the intended application.

The initiation of an *in vitro* culture begins with the careful selection of explants, which may be obtained from aerial or subterranean, vegetative or reproductive plant organs exhibiting high morphogenic potential. Prior to culture initiation, explants must undergo rigorous surface disinfection to eliminate microbial contaminants. The sterilization process generally involves thorough washing under running tap water, followed by rinsing with deionized or distilled water containing a mild detergent such as Tween-20 to remove surface debris. This is followed by treatment with antifungal and antibacterial agents, commonly bavistin and streptomycin, for a duration of 8–10 minutes. A brief immersion in 70% ethanol is often used for rapid surface sterilization. In some cases, a short exposure to a low concentration of mercuric chloride (0.1%–0.5%) for 1–2 minutes is employed to ensure effective disinfection. The explants are then rinsed multiple times with sterile distilled water to completely remove residual sterilants.

Following sterilization, explants are aseptically inoculated into sterile culture vessels such as flasks, culture jars, or petri dishes containing nutrient media that have been autoclaved at 121°C and 15 psi for approximately 20 minutes. Depending on the experimental requirement, cultures may be established on semisolid media solidified with agar or gelrite (0.8%–1.0%) or maintained in liquid media devoid of gelling agents. Successful culture establishment is characterized by the direct formation of adventitious organs from the explant or through indirect regeneration mediated by callus formation, which consists of a mass of actively dividing, undifferentiated cells.

For the commercial-scale synthesis of pigments and other secondary metabolites, suspension cultures derived from plant cells or organized organs are particularly advantageous. These systems offer ease of manipulation, uniform growth, and suitability for scale-up in bioreactor-based production platforms.

16.4.2 Callus and cell suspension culture

Callus tissue comprises a mass of actively dividing, dedifferentiated cells capable of redifferentiating into whole plantlets under suitable conditions. However, prolonged callus culture may lead to somaclonal variations arising from mitotic irregularities and genetic instability during repeated subculturing. Cultures in which individual or small groups of cells proliferate while continuously agitated in liquid medium are termed *suspension cultures*. Their homogeneity, high proliferation rate, and reproducibility make them particularly suitable for investigating complex physiological and biochemical processes at the cellular and molecular levels (Bharathi & Bansal, 2014). Cell suspension cultures are commonly established by transferring friable callus into an appropriate liquid medium and incubating the cultures on orbital shakers. Two critical parameters for successful suspension initiation are the inoculum density and agitation speed, typically maintained between 50 and 150 rpm. Continuous shaking promotes aeration and ensures the dissociation of larger callus aggregates into smaller cell clusters or individual cells.

When grown in a suitable medium, suspension cultures exhibit rapid proliferation and therefore require regular subculturing at intervals of approximately 10–12 days. Delays in subculture often result in reduced viability and a decline in overall growth. In addition to their organogenetic potential, cell suspension cultures are widely utilized in scale-up studies using bioreactors for large-scale production of high-value phytochemicals, including pigments.

16.4.3 Organ culture

Organ culture encompasses two major categories: vegetative organ cultures, including root, shoot, and leaf cultures, and reproductive organ cultures, such as ovary, embryo, ovule, anther, and pollen cultures. The regenerative capacity of these organs depends on both the intrinsic characteristics of the plant species and the composition of the culture medium used for establishing *in vitro* growth.

In recent years, considerable emphasis has been placed on the development of adventitious root cultures and hairy root cultures (Devi et al., 2021; Shi et al., 2021). Roots of many medicinal and aromatic plant species serve as the primary site for the biosynthesis and accumulation of important secondary metabolites, including alkaloids, monoterpenes, sesquiterpenes, and naphthoquinones. Hairy roots are induced through infection of suitable explants with *Agrobacterium rhizogenes*. Once transformed, these roots can be excised and maintained indefinitely under aseptic conditions. Owing to their genetic stability, hairy roots exhibit highly consistent and enhanced phytochemical production across successive generations. Adventitious roots, on the other hand, are formed from nonradicular plant tissues through *in vitro* induction but without genetic transformation. These roots also demonstrate rapid growth, vigorous secondary metabolism, and high productivity of bioactive compounds. Grown in media supplemented with appropriate plant growth regulators, adventitious root cultures possess remarkable potential for the accumulation of secondary metabolites and, in many cases, yield levels comparable to or even greater than those obtained from field-grown plants (Devi et al., 2021). Although both root culture systems are valuable biotechnological tools, adventitious roots—being generated naturally without genetic modification—are often considered advantageous over hairy root cultures for applications where nontransgenic material is preferred.

16.4.4 Case studies

16.4.4.1 **Bixa orellana** *L. (annatto)*

Bixa orellana, commonly known as annatto or lipstick tree, is one of the most widely cultivated natural dye plants, yielding the carotenoid pigments bixin and norbixin, extensively used in food, cosmetics, and pharmaceuticals. Natural cross-pollination in *Bixa orellana* results in high levels of heterozygosity, and consequently, seed-derived progenies fail to maintain the genetic identity of elite genotypes. Although vegetative propagation through softwood and hardwood cuttings has been attempted with varying degrees of success (Joseph et al., 2011a; Joseph et al., 2012), the clonal fidelity required for large-scale cultivation cannot be reliably ensured. Therefore *in vitro* approaches, including tissue culture and micropropagation of selected genotypes, are increasingly recognized as the most reliable strategy for generating uniform planting material for agricultural production and industrial bixin extraction (Siril and Joseph, 2012). Organogenesis, both axillary and adventitious, as well as somatic embryogenesis, have been successfully induced from different explant sources, including cotyledons, hypocotyls, roots, stems, and leaves, and effective acclimatization protocols exist for regenerated plantlets.

Parimalan et al. (2009) developed an efficient micropropagation protocol using nodal shoot-tip explants in which shoot bud induction was achieved on MS medium supplemented with different combinations of IAA, BA, and triacontanol (TRIA). Later, Parimalan et al. (2010) demonstrated that supplementation of putrescine markedly improved shoot organogenesis, while the addition of silver nitrate effectively suppressed ethylene accumulation in culture vessels, thereby enhancing shoot bud induction and shoot elongation. This study highlighted the interaction between polyamines and ethylene inhibitors in regulating *in vitro* morphogenic responses, and also emphasized the role of stress modulation in achieving higher regeneration frequencies. Subsequently, Joseph et al., (2011b) optimized micropropagation from nodal explants through adjustments in cytokinin composition, reporting superior shoot multiplication on MS medium fortified with BAP in combination with kinetin, which also improved shoot vigor and reduced vitrification. Siril and Joseph (2012) further optimized *in vitro* regeneration from mature explants, reporting plants with stable pigment content. Such protocols are now integral to both conservation and commercial propagation of annatto, ensuring the consistent supply of pigment-rich planting material. Building upon this, Siril and Joseph (2012) expanded organogenic potential by employing cotyledonary and young leaf explants, which responded more efficiently to TDZ-based induction systems compared to conventional BAP-mediated pathways. Their study demonstrated that juvenile explants possess higher endogenous regenerative competence, enabling significantly greater shoot proliferation rates and reliable regeneration even from nonnodal tissues. Together, these methodological advances demonstrate the feasibility of high-frequency clonal multiplication in elite genotypes of Annatto.

Parimalan et al. (2010) evaluated the impact of abiotic elicitors on annatto pigment accumulation in *Bixa orellana* and demonstrated that methyl jasmonate, salicylic acid, and paclobutrazol markedly enhanced total pigment content and bixin proportion when applied as floral sprays. These elicitors produced substantial improvements in pigment levels compared with untreated controls, while other growth regulators, including IAA, IBA, GA_3, and TIBA, resulted in only moderate increases. Comparable enhancement was also achieved using chitin or chitosan formulations, indicating the responsiveness of the biosynthetic pathway to both hormonal and biopolymer-based elicitation. Importantly, the response to methyl jasmonate, salicylic acid, and paclobutrazol was reproducible across three successive flowering seasons, suggesting consistency and field applicability. The study concludes that abiotic elicitation represents a promising strategy for commercially scalable augmentation of annatto pigment production in *B. orellana*. In a study conducted by Mahendranath et al. (2011), normal root cultures of *B. orellana* were successfully established on MS medium supplemented with low concentrations of auxins, including NAA, IBA, and IAA. Pigment extracted from *in vitro* raised roots was qualitatively confirmed by TLC and quantified spectrophotometrically with HPLC verification. Among the tested treatments, MS medium containing 0.1 mg L^{-1} IBA supported the highest pigment accumulation (346 ± 3.8 mg per 100 g dry weight) and maximum root biomass (152 ± 2.5 mg dry weight) after 45 and 60 days. The ability of normal root cultures to synthesize annatto pigment represents a novel *in vitro* production system independent of seed arils and provides a potential year-round source of fresh pigment under controlled laboratory conditions.

16.4.4.2 *Indigofera tinctoria* L. (indigo)

Indigofera tinctoria L., commonly referred to as True Indigo, is a perennial shrubby legume of considerable economic importance due to its indigo-producing capability and diverse medicinal applications. Indigo dye has long been a significant commodity in international trade. However, indiscriminate harvesting coupled with habitat degradation has led to a decline in natural populations, raising serious conservation concerns.

To address these challenges, Jose and Nair (2018) established a reliable protocol for rapid clonal multiplication of *I. tinctoria* through high-frequency indirect somatic embryogenesis using cotyledonary explants. In their study, the highest induction of embryogenic callus was achieved on Murashige and Skoog (MS) medium supplemented with 0.5 mg L^{-1} 2,4-dichlorophenoxyacetic acid (2,4-D) in combination with 0.5 mg L^{-1} benzyladenine (BA).

In a separate investigation, Nair et al. (2016) reported an efficient *in vitro* regeneration system based on nodal explants cultured on MS medium containing 1.0 mg L^{-1} BA and 0.1 mg L^{-1} indole-3-acetic acid (IAA). Optimal root induction was observed on MS medium supplemented with 1.5 mg L^{-1} IAA. The regenerated plantlets exhibited high survival during acclimatization, with a success rate of 91.67% when transferred to sterilized sand.

16.4.4.3 Rubia cordifolia L. (Indian madder)

Rubia cordifolia produces valuable anthraquinone pigments such as alizarin and purpurin. In many species of *Rubia*, the roots serve as the primary site for pigment accumulation. Owing to their rich content of bioactive secondary metabolites, several *Rubia* species possess considerable economic and commercial importance. Among these, alizarin and purpurin are well-known anthraquinone pigments that have been traditionally used as natural dyes for coloring cotton, silk, and wool fabrics since ancient times. In addition to their textile applications, these pigments have also found extensive use in the cosmetic and food industries as natural coloring agents.

However, the natural concentration of anthraquinone pigments in *Rubia* species is generally low, limiting large-scale extraction from field-grown plants. To overcome this constraint, various *in vitro* culture systems have been developed, including cell cultures and organ cultures. Extensive research has focused on understanding the influence of different chemical and physical factors on biomass production and secondary metabolite accumulation in callus cultures, cell suspension cultures, as well as hairy and adventitious root suspension systems. Mariadoss et al. (2020) investigated the effect of gamma irradiation on *Rubia cordifolia* cell cultures grown in a bioreactor and demonstrated significant enhancement of anthraquinone accumulation under controlled conditions. Exposure to gamma rays stimulated metabolite biosynthesis without compromising biomass growth, suggesting the induction of stress-responsive pathways linked to secondary metabolism. The bioreactor system enabled scalable biomass production and pigment extraction, indicating the potential of irradiation-mediated elicitation as a practical strategy for industrial anthraquinone production in *R. cordifolia*. Veremeichik et al. (2019) reported the activation of anthraquinone biosynthesis in long-term callus cultures of *Rubia cordifolia* transformed with the rolA oncogene derived from *Agrobacterium rhizogenes*. Transformed calli exhibited elevated expression of biosynthetic genes and increased accumulation of major anthraquinone derivatives compared with nontransformed controls. The study highlighted the regulatory influence of rolA on secondary metabolism and suggested its utility as a molecular tool for enhancing pigment production in plant cell culture platforms. Shkryl et al. (2016) evaluated the impact of native and constitutively active forms of the *Arabidopsis* calcium-dependent protein kinase gene (AtCPK1) on anthraquinone production in *Rubia cordifolia* cell cultures. Genetic transformation led to markedly increased levels of anthraquinones, indicating that calcium-regulated signaling pathways play a key role in metabolic regulation in *R. cordifolia*. The results demonstrated that metabolic engineering of regulatory genes can effectively modulate secondary metabolite pathways and represents a promising strategy for enhancing pigment biosynthesis in medicinal and dye-yielding species. *In vitro* systems for *R. cordifolia* thus serve both biotechnological pigment production platforms and germplasm conservation tools, especially in the Western Ghats, where the species is under threat.

16.4.4.4 **Curcuma longa** *L. (turmeric)*

Curcuma longa are rich source of curcuminoids—notably curcumin, demethoxycurcumin, and bis-demethoxycurcumin—renowned for their antioxidant and dyeing properties. Prathanturarug et al. (2005) developed an efficient *in vitro* regeneration system for *Curcuma longa* based on enhanced multiple shoot formation from bud explants. In this approach, explants were initially cultured for one week in Murashige and Skoog (MS) liquid medium supplemented with 72.64 μM thidiazuron (TDZ) to induce shoot meristem activation. Subsequently, the explants were transferred to growth regulator-free, gelled MS medium and maintained for eight weeks, resulting in prolific shoot regeneration. Under these conditions, a maximum regeneration frequency of 11.4 ± 1.7 shoots per explant was achieved. Root induction occurred naturally without the need for exogenous auxins, and the regenerated plantlets were successfully acclimatized and established in soil. This method provides a reliable and rapid alternative for large-scale micropropagation of *C. longa*, particularly for ensuring a continuous supply of high-quality planting material for phytomedicinal applications.

Pistelli et al. (2012) induced *in vitro* plants from sprouting buds as primary explants. The major curcuminoid constituents, such as curcumin (CUR), demethoxycurcumin (DEM), and bis-demethoxycurcumin (bis-DEM) were examined in different organs by LC-DAD-ESI-MS. A significant production of curcumin (more than 260 μg g^{-1} fresh weight) was obtained from *in vitro* microrhizomes, especially grown in a Murashige and Skoog medium (MS) supplemented with kinetin (0.1 mg L^{-1}), a-naphthalene acetic acid (NAA, 1 mg L^{-1}), sucrose (6%), agar (5%), and activated charcoal (0.1%).

16.4.4.5 **Crocus sativus** *L. (saffron)*

Saffron is recognized as the costliest spice globally and is traditionally multiplied through vegetative propagation via the formation of daughter corms from the parent corm. However, this method suffers from several inherent limitations that restrict large-scale multiplication. These constraints include a very low rate of daughter corm production and a high susceptibility to fungal infections, which adversely affect both the quantity and quality of planting material. Consequently, conventional propagation practices are inadequate to meet the increasing demand for healthy saffron corms (Tahiri et al., 2022).

In this context, biotechnological interventions, particularly *in vitro* culture approaches, offer a promising alternative for rapid and disease-free propagation of saffron. Different explant types have been cultured on a range of nutrient media containing varied concentrations and combinations of plant growth regulators to identify optimal conditions for enhanced microcorm formation and the induction of stigma-like structures (SLS). Among the treatments tested, the highest morphogenic response (60%) was achieved when half-ovary explants were cultured on G-5 medium supplemented with 27 μM naphthalene acetic acid (NAA) and 44.4 μM benzyladenine (BA), followed by a 55% response on LS medium with the same growth regulator combination. The largest microcorms, with a maximum fresh weight of 1.3 g, were produced from apical bud explants cultured on LS medium fortified with 21.6 μM NAA and 22.2 μM BA. Furthermore, stigma-like structures were successfully induced from half-ovary explants through both direct organogenesis and an indirect pathway involving callus formation (Mir et al., 2010).

16.4.4.6 Morinda coreia (*Indian mulberry*)

Morinda coreia Buch.-Ham., a member of the family Rubiaceae, is a valuable medicinal plant traditionally exploited for its natural dye. In India, the plant serves as the source of morindone, a commercially important anthraquinone pigment marketed under the name "Suranji." This natural dye is extensively used for coloring cotton, silk, and wool, imparting hues ranging from red to chocolate and purple. The pigment is predominantly localized in the root bark, which is typically harvested only after the plant reaches three to four years of age. Since anthraquinone extraction requires uprooting mature plants, continuous exploitation has resulted in a rapid decline of natural populations, posing a serious threat to the species in its native habitats.

To address conservation concerns and ensure a sustainable supply of high-quality planting material, Shekhawat et al. (2015) established a robust and reproducible *in vitro* regeneration system for *M. coreia*. In this study, nodal explants were employed for shoot regeneration. Initiation of shoot buds was observed within one week of culture, and an average of 8.6 ± 0.32 shoots per explant was recorded after five weeks on Murashige and Skoog (MS) medium supplemented with 4.0 mg L^{-1} 6-benzylaminopurine (BAP). For further multiplication, regenerated shoots were subcultured onto semisolid MS medium containing 2.0 mg L^{-1} BAP in combination with 1.0 mg L^{-1} kinetin, resulting in a maximum proliferation rate of 24.5 ± 0.34 shoots per explant within five weeks. Well-developed shoots measuring 4–5 cm in length exhibited complete rooting efficiency when transferred to half-strength MS medium augmented with 1.0 mg L^{-1} indole-3-butyric acid (IBA).

16.4.4.7 *Lawsonia inermis* (henna)

Henna (*Lawsonia inermis*) is under the family Lythraceae. It is best known as a source of natural dyes. It is a native of North Africa and South-West Asia, widely cultivated as an ornamental hedge and dye plant. *L. inermis* contains naphthoquinone-based pigment lawsone, a red–orange pigment which has been used for centuries, in many cultures mainly as a dye for hair and nails. Several earlier studies attempted the micropropagation of *Lawsonia inermis* using different explants and cytokinin combinations. Early reports by Rout et al. (2001) employed apical and axillary explants on MS medium augmented with BA and kinetin and obtained approximately 4–5 shoots per explant with moderate rooting on IBA. Subsequent refinements by Kheta Ram and Shekhawat (2011) demonstrated that nodal explants cultured on MS supplemented with BA, kinetin, and a small amount of IAA, together with additives such as ammonium sulfate, considerably improved multiplication, producing nearly 18 shoots per explant along with high ex vitro rooting percentages following auxin treatment. More recently, Amit and Singh (2021) reported that the cytokinin regime of BAP in combination with kinetin enhanced shoot proliferation from nodal segments and achieved efficient rooting on half-strength MS supplemented with IBA. A major advancement was achieved by Shiji and Siril (2018), who utilized nodal explants from elite, high-lawsone genotypes and observed highly significant shoot induction on MS containing benzyladenine, while the combination of BA with 2-isopentenyladenine resulted in the production of more than forty shoots per explant with substantial elongation within a short culture period. Their protocol further demonstrated nearly complete ex vitro rooting following a brief NAA-dip, suggesting considerable practical value for large-scale propagation and field transfer. Together, these reports indicate a progressive refinement of micropropagation strategies in *L. inermis*, moving from modest multiplication rates in initial studies toward highly efficient, commercially workable protocols in recent years.

16.5 Conclusion

Natural plant pigments contribute to ecological sustainability, human health promotion, and industrial relevance. Beyond their traditional role as colorants, these bioactive compounds contribute significantly to health and wellness through their antioxidant, anti-inflammatory, and protective mechanisms. The extensive diversity of pigment-yielding plants, particularly in biodiversity hotspots such as the Western Ghats, constitutes a valuable genetic reservoir that is increasingly threatened by anthropogenic pressures and unsustainable exploitation. Conserving these plant genetic resources is therefore critical, not only for conserving biodiversity but also for sustaining and supporting eco-friendly dye, food, pharmaceutical, and cosmetic industries.

The integration of biotechnology, especially micropropagation, *in vitro* culture systems, and bioreactor-based production has emerged as a powerful strategy to overcome limitations associated with conventional cultivation, seasonal variability, and low pigment yields. These approaches enable rapid multiplication of elite genotypes, ensure genetic fidelity, reduce pressure on wild populations, and facilitate controlled, scalable production of high-value pigments. Advances in elicitation, metabolic engineering, and cell and organ culture technologies further enhance the commercial production of natural pigments while maintaining environmental safety.

References

Amit, R., & Singh, N. (2021). In vitro propagation of an economically important medicinal plant Lawsonia inermis L. through nodal segments. *Journal of Applied and Natural Science, 13*(3), 897–906. https://doi.org/10.31018/jans.v13i3.2612.

Bharathi, A. J., & Bansal, Y. K. (2014). In vitro production of flavonoids: A review. *World Journal of Pharmacy and Pharmaceutical Sciences, 3*, 508–533.

Bretting, P. K., & Widrlechner, M. P. (1995). Genetic markers and plant genetic resource management. In J. Janick (Ed.), *Plant Breeding Reviews*. Vol. *13*, (pp. 11–86). John Wiley & Sons.

Cetinkaya, H., Kulak, M., Karaman, M., Karaman, H. S., & Kocer, F. (2017). Flavonoid accumulation behavior in response to abiotic stress: Can a uniform mechanism be illustrated for all plants? In *Flavonoids—From biosynthesis to human health*. London, UK: IntechOpen.

Choi, S. M., Son, S. H., Yun, S. R., Kwon, O. W., Seon, J. H., & Paek, K. Y. (2000). Pilot-scale culture of adventitious roots of ginseng in a bioreactor system. *Plant Cell, Tissue and Organ Culture, 62*, 187–193.

Das, A., Kesari, V., Satyanarayana, V. M., Parida, A., & Reddy, M. P. (2011). Genetic diversity and phylogeography of *Curcuma longa* L. revealed by molecular markers. *Plant Systematics and Evolution, 297*, 1–16.

Delgado-Vargas, F., & Paredes-Lopez, O. (2000). Natural pigments: Carotenoids, anthocyanins, and betalains—Characteristics, biosynthesis, processing, and stability. *Critical Reviews in Food Science and Nutrition, 40*, 173–289.

DellaPenna, D., & Pogson, B. J. (2006). Vitamin synthesis in plants: Tocopherols and carotenoids. *Annual Review of Plant Biology, 57*, 711–738.

Devi, J., Kumar, R., Singh, K., Gehlot, A., Bhushan, S., & Kumar, S. (2021). In vitro adventitious roots: A non-disruptive technology for the production of phytoconstituents on an industrial scale. *Critical Reviews in Biotechnology, 41*, 564–579.

Drappier, J., Thibon, C., Rabot, A., & Geny-Denis, L. (2019). Relationship between wine composition and temperature: Impact on Bordeaux wine typicity in the context of global warming. *Critical Reviews in Food Science and Nutrition, 59*, 14–30.

Georgiev, V. G., Weber, J., Kneschke, E. M., Denev, P. N., Bley, T., & Pavlov, A. I. (2010). Antioxidant activity and phenolic content of betalain extracts from intact plants and hairy root cultures of red beetroot (*Beta vulgaris* cv. Detroit Dark Red). *Plant Foods for Human Nutrition, 65*, 105–111.

Gita, S., Hussan, A., & Choudhury, T. G. (2017). Impact of textile dyes waste on aquatic environments and its treatment. *Environment and Ecology, 35*, 2349–2353.

Govindaraj, M., Vetriventhan, M., & Srinivasan, M. (2015). Importance of genetic diversity assessment in crop plants and its recent advances. *Genetika,, 47*, 1–21.

Heywood, V. H. (2011). Ethnopharmacology, food production, nutrition and biodiversity conservation: Towards a sustainable future for indigenous peoples. *Journal of Ethnopharmacology, 137*, 1–15.

Jose, R., & Nair, M. (2018). High frequency somatic embryogenesis from cotyledonary explants of *Indigofera tinctoria* Linn. *International Journal of Advanced Research, 6*, 1054–1060.

Joseph, N., Siril, E. A., & Nair, G. M. (2011a). A comparison of conventional cloning options for annatto (*Bixa orellana* L.). *Journal of Horticultural Science and Biotechnology, 86*, 446–451.

Joseph, N., Siril, E. A., & Nair, G. M. (2011b). An efficient in vitro propagation methodology for annatto (*Bixa orellana* L.). *Physiology and Molecular Biology of Plants, 17*, 263–270.

Joseph, N., Siril, E. A., & Nair, G. M. (2012). Reproductive characterization and preliminary studies on controlled breeding of annatto (*Bixa orellana* L.). *Plant Systematics and Evolution, 298*, 239–250.

Kala, C. P. (2005). Current status of medicinal plants used by traditional Vaidyas in Uttaranchal state of India. *Ethnobotany Research and Applications, 3*, 267–278.

Kale, R., Sawate, A., Kshirsagar, R., Patil, B., & Mane, R. (2018). Evaluation of physical and chemical composition of beetroot (*Beta vulgaris* L.). *International Journal of Chemical Studies, 6*, 2977–2979.

Kameswara Rao, N. (2003). Plant genetic resources: Advancing conservation and use through biotechnology. *African Journal of Biotechnology, 3*(2), 136–145. https://doi.org/10.5897/ajb2004.000-2025.

Kumar, A., Mishra, P., & Sundaresan, V. (2014). Assessment of genetic diversity in *Justicia adhatoda* L. using RAPD and ISSR markers. *Biochemical Systematics and Ecology, 54*, 195–202.

Leong, H. Y., Show, P. L., Lim, M. H., Ooi, C. W., & Ling, T. C. (2017). Natural red pigments from plants and their health benefits: A review. *Food Reviews International, 34*, 463–482.

Lu, W., Shi, Y., Wang, R., Su, D., Tang, M., Liu, Y., & Li, Z. (2021). Antioxidant activity and health benefits of natural pigments in fruits. *International Journal of Molecular Sciences, 22*, 4945.

Mahendranath, G., Venugopalan, A., Parimalan, R., Giridhar, P., & Ravishankar, G. A. (2011). Annatto pigment production in root cultures of *Bixa orellana* L. *Plant Cell, Tissue and Organ Culture, 106*, 517–522.

Mariadoss, A., Satdive, R., Fulzele, D. P., Ramamoorthy, S., Zayed, H., Younes, S., & Rajasekaran, C. (2020). Enhanced production of anthraquinones by gamma-irradiated cell cultures of *Rubia cordifolia* in a bioreactor. *Industrial Crops and Products, 145*, 111987.

Martinez, R. M., Melo, C. P., Pinto, I. C., Mendes-Pierotti, S., Vignoli, J. A., Verri, W. A., & Casagrande, R. (2024). Betalains: A narrative review on pharmacological mechanisms supporting nutraceutical potential. *Foods, 13*, 3909.

Martins, T., Barros, A. N., Rosa, E., & Antunes, L. (2023). Enhancing health benefits through chlorophylls and chlorophyll-rich agro-food. *s. Molecules, 28*, 5344.

Mir, J. I., Ahmed, N., Wani, S. H., Rashid, R., Mir, H., & Sheikh, M. A. (2010). In vitro development of microcorms and stigma-like structures in saffron (*Crocus sativus* L. *Physiology and Molecular Biology of Plants, 16*, 369–373.

Mortensen, A. (2006). Carotenoids and other pigments as natural colorants. *Pure and Applied Chemistry, 78*, 1477–1491.

Myers, N., Mittermeier, R. A., Mittermeier, C. G., da Fonseca, G. A., & Kent, J. (2000). Biodiversity hotspots for conservation priorities. *Nature, 403*, 853–858.

Nair, D. S., Reghunath, B. R., Soni, K. B., & Swapna, A. (2016). In vitro regeneration and conservation of Indigo (Indigofera tinctoria L.) by slow growth induction. *International Journal of Environment, Agriculture and Biotechnology, 1*(4), 873–884. https://doi.org/10.22161/ijeab/1.4.35.

Parimalan, R., Giridhar, P., Gururaj, H. B., & Ravishankar, G. A. (2009). Micropropagation of *Bixa orellana* using phytohormones and triacontanol. *Biologia Plantarum, 53*, 347–350.

Parimalan, R., Giridhar, P., & Ravishankar, G. A. (2010). Enhanced shoot organogenesis in Bixa orellana L. in the presence of putrescine and silver nitrate. *Plant Cell, Tissue and Organ Culture (PCTOC), 105*(3), 285–290. https://doi.org/10.1007/s11240-010-9865-7.

Pistelli, L., Bertoli, A., Gelli, F., Bedini, L., Ruffoni, B., & Pistelli, L. (2012). Production of *Curcuminoids* in different *in vitro* organs of *Curcuma longa. Natural Product Communications, 7*(8). https://doi.org/10.1177/1934578x1200700819.

Prathanturarug, S., Soonthornchareonnon, N., Chuakul, W., Phaidee, Y., & Saralamp, P. (2005). Rapid micropropagation of Curcuma longa using bud explants pre-cultured in thidiazuron-supplemented liquid medium. *Plant Cell, Tissue and Organ Culture, 80*(3), 347–351. https://doi.org/10.1007/s11240-004-1020-x.

Rao, V. R., & Hodgkin, T. (2002). Genetic diversity and conservation and utilization of plant genetic resources. *Plant Cell, Tissue and Organ Culture, 68*(1), 1–19. https://doi.org/10.1023/a:1013359015812.

Rodriguez-Concepcion, M., Avalos, J., Bonet, M. L., Boronat, A., Gomez-Gomez, L., Hornero-Mendez, D., Limon, M. C., Meléndez-Martínez, A. J., Olmedilla-Alonso, B., Palou, A., Ribot, J., Rodrigo, M. J., Zacarias, L., & Zhu, C. (2018). A global perspective on carotenoids: Metabolism, biotechnology, and benefits for nutrition and health. *Progress in Lipid Research, 70*, 62–93. https://doi.org/10.1016/j.plipres.2018.04.004.

Rout, G. R., Mohapatra, A., & Jain, S. M. (2006). Tissue culture of ornamental pot plant: A critical review on present scenario and future prospects. *Biotechnology Advances, 24*(6), 531–560. https://doi.org/10.1016/j.biotechadv.2006.05.001.

Rout, G. R., Samantaray, S., & Das, P. (2001). In Vitro micropropagation of Lawsonia inermis (Lythraceae). *Revista de Biologia Tropical, 49*(3-4), 957–963.

Sehgal, D., & Raina, S. N. (2005). Genotyping safflower (Carthamus tinctorius) cultivars by DNA fingerprints. *Euphytica, 146*(1-2), 67–76. https://doi.org/10.1007/s10681-005-8496-2.

Shahid, M., & Mohammad, F. (2013). Recent advancements in natural dye applications: A review. *Journal of Cleaner Production, 53*, 310–331.

Shekhawat, M. S., Kannan, N., & Manokari, M. (2015). In vitro propagation of traditional medicinal and dye-yielding plant *Morinda coreia. South African Journal of Botany, 100*, 43–50.

Siril, E. A., & Joseph, Nisha (2012). Micropropagation of annatto (Bixa orellana L.) from mature tree and assessment of genetic fidelity of micropropagated plants with RAPD markers. *Physiology and Molecular Biology of Plants, 19*(1), 147–155. https://doi.org/10.1007/s12298-012-0150-6.

Shi, Min, Liao, Pan, Nile, Shivraj Hariram, Georgiev, Milen I., & Kai, Guoyin (2021). Biotechnological Exploration of Transformed Root Culture for Value-Added Products. *Trends in Biotechnology, 39*(2), 137–149. https://doi.org/10.1016/j.tibtech.2020.06.012.

Siva, R. (2007). Status of natural dyes and dye-yielding plants in India. *Current Science, 92*, 916–925.

Shiji, P. C., & Siril, E. A. (2018). An improved micropropagation and ex vitro rooting of a commercially important crop Henna (Lawsonia inermis L.). *Physiology and Molecular Biology of Plants, 24*(6), 1273–1284. https://doi.org/10.1007/s12298-018-0600-x.

Shkryl, Y. N., Veremeichik, G. N., Makhazen, D. S., Silantieva, S. A., Mishchenko, N. P., Vasileva, E. A., Fedoreyev, S. A., & Bulgakov, V. P. (2016). Increase of anthraquinone content in Rubia cordifolia cells transformed by native and constitutively active forms of the AtCPK1 gene. *Plant Cell Reports, 35*(9), 1907–1916. https://doi.org/10.1007/s00299-016-2005-z.

Tahiri, A., Mazri, M. A., Karra, Y., Ait Aabd, N., Bouharroud, R., & Mimouni, A. (2022). Propagation of saffron (*Crocus sativus* L.) through tissue culture: A review. *Journal of Horticultural Science and Biotechnology, 98*, 10–30.

Veremeichik, G. N., Bulgakov, V. P., Shkryl, Y. N., Silantieva, S. A., Makhazen, D. S., Tchernoded, G. K., & Vasileva, E. A. (2019). Activation of anthraquinone biosynthesis in long-cultured callus of *Rubia cordifolia* transformed with the rolA gene. *Journal of Biotechnology, 306*, 38–46.

Index

Note: Page numbers followed by "*f*" and "*t*" refer to figures and tables, respectively.

C

9780443439445